AF388885

Advanced Oxidation Process: Applications and Prospects

Advanced Oxidation Process: Applications and Prospects

Editors

Gassan Hodaifa
Antonio Zuorro
Joaquín R. Dominguez
Juan García Rodríguez
José A. Peres
Zacharias Frontistis
Mha Albqmi

Basel • Beijing • Wuhan • Barcelona • Belgrade • Novi Sad • Cluj • Manchester

Editors

Gassan Hodaifa
Universidad Pablo
de Olavide
Seville, Spain

Antonio Zuorro
Sapienza—University
of Rome
Rome, Italy

Joaquín R. Dominguez
Universidad de Extremadura
Badajoz, Spain

Juan García Rodríguez
Complutense University
of Madrid
Madrid, Spain

José A. Peres
University of Trás-os-Montes
e Alto Douro
Vila Real, Portugal

Zacharias Frontistis
University of
Western Macedonia
Kozani, Greece

Mha Albqmi
Jouf University
Alqurayyat, Saudi Arabia

Editorial Office
MDPI
St. Alban-Anlage 66
4052 Basel, Switzerland

This is a reprint of articles from the Topic published online in the open access journals *Catalysts* (ISSN 2073-4344), *International Journal of Environmental Research and Public Health* (ISSN 1660-4601), *Processes* (ISSN 2227-9717), *Sci* (ISSN 2413-4155), and *Water* (ISSN 2073-4441) (available at: https://www.mdpi.com/topics/advanced_oxidation).

For citation purposes, cite each article independently as indicated on the article page online and as indicated below:

Lastname, A.A.; Lastname, B.B. Article Title. *Journal Name* **Year**, *Volume Number*, Page Range.

ISBN 978-3-0365-9278-7 (Hbk)
ISBN 978-3-0365-9279-4 (PDF)
doi.org/10.3390/books978-3-0365-9279-4

Contents

About the Editors

Gassan Hodaifa

Dr. Gassan Hodaifa (h-Index of 26 in Web of Science) is a Food Engineer (University of Albaath, Syria) and Chemical Engineer (University of Granada). He received his doctoral degree from the University of Jaén. He joined the research group Bioprocesses TEP-138 (Junta de Andalucía) and works in the areas of microalgae, the treatment of wastewater, chemical oxidation, membrane technology, adsorption, pesticide removal, olive oil processing, enzyme biotechnology, and nematode biotechnology. He has participated in 27 research projects (8 as IP) and four European programs (one PRIMA Project and three in collaboration with M.A.P.A. and the INFAOLIVA and UNAPROLIVA companies), with which he has signed 16 research contracts (four as IP). He has been contracted twice as an adviser for the mega Japanese company "Asahi Kasei Corporation" and has had six projects financed by J. Andalucía (two of them as IP). He has completed five projects with the National Research Plan, one project for the Madrid Community, and the rest he conducted with companies. As a result of his research work, he has had 64 articles published. Of these articles, more than 10 articles have been published in collaboration with authors from different countries on all continents. He has published six total articles in open access journals, four articles in journals with an impact factor different from that listed in the JCR, and five other articles in outreach magazines. He has two patents for the water purification of olive mill wastewater and has written two books and had 46 book chapters published by prestigious publishers, such as the Academic Press (Elsevier), CRC Press, Taylor and Francis Group, Springer International Publishing AG, etc. He was the Editor-in-Chief of three Special Issues in J. Chemistry and Catalysts, has participated in 125 congresses, and is currently a member of 15 Editorial Boards of international journals such as Catalysts and Heliyon.

Antonio Zuorro

Prof. Antonio Zuorro (with an h-index of 36,131 scientific papers, and 5 industrial invention patents) is a Professor and Researcher at the Department of Chemical, Materials, and Environmental Engineering of Sapienza, University of Rome, in the scientific disciplinary sector of Principles of Chemical Engineering, where he leads the GreenSapiens Group and Laboratories (bit.ly/GreenSapiens). In 2017, the circular and bio-economy expert was awarded a European project under the Horizon 2020 BBI JU program, with EUR 7.2 million in funding. The individual assumed the roles of Project Manager and Scientific Coordinator, overseeing a diverse consortium of 13 partners, including public institutions and multinational private companies. He has enhanced agro-industrial residues for two decades by extracting bioactive compounds of interest in the cosmetic, nutraceutical, and pharmaceutical sectors and participated in several research projects, in many cases as the Scientific Coordinator of the research unit, experimental activities, or the project as a whole, in addition to third-sector activities with consultancy and research projects for private companies in the circular economy and green chemistry fields. His research activity has been developed in various sectors, including green processes, catalysis, chemical thermodynamics, enzyme kinetics, protein engineering, and transport phenomena in microencapsulated systems.

Currently, he is mainly focused on treating and reusing agro-industrial residues, recovering high-value bioactive compounds and degrading pollutants via advanced oxidation processes (AOP). Moreover, he worked on the production of functional products with antioxidant and/or antibacterial and anti-quorum sensing properties.

In recent years, via collaboration with important industrial partners, he has been exploring "green" nanoparticles of cosmetic and biomedical interest and optimising new-generation biorefineries operating with microalgae.

Joaquín R. Dominguez

Prof. Joaquín R. Domínguez has an undergraduate degree (1996) and a PhD (2000) in Chemistry from the University of Extremadura, Spain, both obtained with honours, and was awarded the Research Excellence Prize from his University in 2001. He completed predoctoral and postdoctoral stays at "Instituto Superior Tecnico", Lisbon (Portugal). Now, Dr. Dominguez is a Full Professor in the Chemical Engineering Department. His research interests focus on water and wastewater treatment, the removal of micropollutants and emerging contaminants from water using advanced oxidation processes, AOPs (ozone, Fenton's reagent, hydrogen peroxide, persulphate, UV and VIS radiation, and their possible combinations), and electrochemical and sonochemical treatments. Regarding his experience, Prof. Dominguez has actively participated in 24 research projects or grants within the area of environmental engineering and, more specifically, in water and wastewater treatment and removal of micropollutants from water. He has attended the biannual "International Conference on Water, Waste, and Energy Management (WWEM)" as Conference Chair since 2015. Dr Domínguez has co-authored approximately 80 papers, 9 book chapters, and 6 scientific books about his research, obtaining an h-index of 34 (WoS). He has directed or co-directed 45 advanced research works, including 20 B.Sc. and 5 Ph.D. theses. He has also presented 85 communications at National and International Scientific Congresses. Dr Dominguez is a permanent Editorial Board member of the scientific journals *Processes* (ISSN: 2227-9717) and *OpenChemistry* (ISSN: 2391-5420), and he has served as Managing Editor of seven Special Issues of internationally recognised journals, such as *Journal of Environmental Management* and *Industrial and Engineering*.

Juan García Rodríguez

Juan Garcia received a degree in Chemical Sciences (Industrial Chemistry) from the University of Extremadura, Spain, and a Ph.D. in Chemical Engineering from the same university, working on wastewater treatment by chemical and biological processes. He was an Assistant Professor at the Complutense University of Madrid, as well as a Contratado Doctor Professor and an Associate Professor. Since 2018, he has been a Full Professor in Chemical Engineering. He undertook five postdoctoral research projects (Porto University, Portugal; Bath University, United Kingdom; Concepción University, Chile; Extremadura University, Spain; and Polytechnic Institute of Bragança, Portugal) and is the author of more than 130 articles in international SCI journals, of which more than 80 are first-quartile SCI articles (Q1) of the JCR epigraph. He has an h-index of 39, with a total of 5300 citations. He has participated in over 40 research projects, led research projects since 2003, and co-supervised nine PhD theses. He has served the scientific community as Guest Editor of several international journals (JCR) and has been a member of some conference program committees. His research interests have focused on synthesising and characterising carbon materials with enhanced properties and their catalytic application in adsorption, environmental catalysis, and biomass valorisation. He is a leader–researcher of several active projects related to wastewater treatment and waste valorisation from WWTPs. Prof. Garcia has four six-year research periods and is recognised as a regular assessor for AEI (State Research Agency) and other international bodies.

José A. Peres

José Alcides Peres (h-Index of 32 in Web of Science) graduated in Chemical Engineering at the University of Coimbra, Portugal. He completed his PhD in Chemistry at the University of Trás-os-Montes and Alto Douro (UTAD), Vila Real, Portugal (2002), and obtained a Habilitation degree in 2011 at UTAD. He is an Associate Professor with habilitation at UTAD and a member of the Chemistry Research Centre of Vila Real, where he is the Scientific Coordinator of the Environmental Chemistry Research Group. He is the author and co-author of more than 270 scientific publications, including 87 papers published in SCI journals, which received more than 4100 citations, three conference books, two patents, 12 papers in national scientific periodicals, and more than 180 communications in conference abstracts and proceedings. He has been involved in 18 research and development projects and seven enterprise projects.

His main research interests focus on environmentally friendly strategies for pollution control, including the development of Advanced Oxidation Processes (AOPs), such as Fenton reagent, homogeneous and heterogeneous photo-Fenton, ozonation, ferrioxalate/solar radiation, and heterogeneous photocatalysis. He also studies combined treatments (physical, chemical, and biological), particularly for agro-industrial wastewater and water reuse. He was the Chairperson of the 5th International Conference on Green Chemistry and Sustainable Engineering (GreenChem-22), which took place in Rome (Italy) from 20 to 22 July 2022.

Zacharias Frontistis

Dr. Zacharias Frontistis is an Associate Professor specialising in Water and Wastewater Process Engineering at the Department of Chemical Engineering, University of Western Macedonia. He previously served as the Chairperson of the Department of Environmental Engineering (2018-2019) and the Department of Chemical Engineering (2019-2021) at the same university. Dr. Frontistis holds a Diploma in Environmental Engineering (2005), a master's degree (2007), and a PhD (2011), all from the Technical University of Crete. Before joining the University of Western Macedonia, he worked as a Postdoctoral Researcher at several institutions, including the Department of Civil and Environmental Engineering at the University of Cyprus and the International Water Research Centre NIREAS, the Department of Civil and Environmental Engineering at Namik Kemal University, and the Department of Chemical Engineering at the University of Patras. His research focuses on catalysis for environmental and energy applications, developing new physicochemical processes for water treatment (Advanced Oxidation Processes), combining physicochemical and biological processes for agro-industrial wastewater treatment, and simulating these processes. He has supervised and co-supervised a large number of diploma and postgraduate theses. According to Scopus, Dr. Frontistis has a significant publishing record, with over 140 articles in high-impact international journals, an h-index of 40, and approximately 4,000 citations. He has also contributed to international conference proceedings more than 60 times and authored two book chapters on advanced wastewater treatment and reuse. Dr. Frontistis has reviewed over 1,000 papers for over 50 high-impact scientific journals. He serves as an editor for the Journal of Environmental Chemical Engineering (Elsevier), sits on the editorial board of six international journals, and has been a guest editor for 11 Special Issues in his fields of interest.

Mha Albqmi

Dr. Mha Albqmi currently holds the position of Assistant Professor at Jouf University and serves as the Director of the Olive Research Centre at Jouf University, Saudi Arabia. Her educational background includes a PhD in Chemistry from Oklahoma State University, Stillwater, Oklahoma, USA, a Master of Science in Chemistry from Lamar University, Texas, USA, and a Bachelor of Science in Chemistry from Taif University, Taif, Saudi Arabia. Dr Mha Albqmi's great passion lies in the field of catalysis and studies its applications in various industrial chemical processes. Particularly, throughout her PhD program, she contributed to developing new procedures that enabled the direct conversion of organic compounds into their resulting products, eliminating the requirement for solvents and effectively reducing waste generation.

Her current research interests lie in environmental chemistry, waste materials recycling, and pollution prevention. Additionally, she actively valorises the olive waste residue on certain plants experiencing heavy-metal toxicity. She participates in studying nanotechnology applications, such as water purification and photocatalytic degradation processes. Furthermore, she is researching the direct conversion method of organic compounds using catalysts and the potential of natural products to treat metal overload diseases.

Preface

Within the last twenty years, advanced oxidation technologies have begun to be taken seriously as green technologies and are highly efficient from the point of view of final yields since they are chemical reactions that can be controlled and directed according to the end goals in each case. Thus far, they have considered emerging technologies to be of little application at the industrial level, as they are commonly considered relatively expensive compared to conventional technologies. This consideration is not entirely true since, in many cases, the costs of these technologies can be low, as they are simple technologies, and the costs of the chemical reagents used depend on each geographical area. It is also important to mention the degree of previous optimisation of these technologies in each case, which usually implies a notable cost reduction, especially when considering the circular economy of the processes in which these technologies are applied.

Advanced oxidation technologies can be used individually or incorporated into more complete processes or bioprocesses. In this sense and as a non-limiting example, they can be applied in wastewater treatment as a main technology/for pretreatment or as a final operation to adjust the final percentages required by current legislation.

This topic is characterised by being a multidisciplinary volume of the journals Catalysts, Processes, Sci, International Journal of Environmental Research and Public Health, and Water, in which the aim is to extend our knowledge of the current state-of-the-art technologies related to current and possible future applications of advanced oxidation processes.

Finally, we would like to thank the contributions of the authors participating in this volume, without whom this topic could not exist.

Gassan Hodaifa, Antonio Zuorro, Joaquín R. Dominguez, Juan García Rodríguez, José A. Peres, Zacharias Frontistis, and Mha Albqmi

Editors

water

Editorial

Advanced Oxidation Process: Applications and Prospects

Gassan Hodaifa [1,*], Antonio Zuorro [2], Joaquín R. Dominguez [3], Juan García Rodríguez [4], José A. Peres [5], Zacharias Frontistis [6] and Mha Albqmi [7]

1 Department of Molecular Biology and Biochemical Engineering, Chemical Engineering Area, Universidad Pablo de Olavide, Carretera de Utrera km 1, Building 47.1.11, ES-41013 Seville, Spain
2 Department of Chemical Engineering, Materials & Environment, Sapienza—University of Rome, Via Eudossiana 18, 00184 Rome, Italy; antonio.zuorro@uniroma1.it
3 Chemical Engineering and Chemical Physics Department, Universidad de Extremadura, Av. Elvas s/n, 06006 Badajoz, Spain; jrdoming@unex.es
4 Department of Chemical and Materials Engineering, Complutense University of Madrid, 28040 Madrid, Spain; jgarciar@ucm.es
5 Department of Chemistry, Chemistry Research Centre—Vila Real, University of Trás-os-Montes and Alto Douro (UTAD), Quinta de Prados, 5000-801 Vila Real, Portugal; jperes@utad.pt
6 Department of Chemical Engineering, University of Western Macedonia, 50100 Kozani, Greece; zfrontistis@uowm.gr
7 Chemistry Department, College of Science, Jouf University, Sakaka 2014, Saudi Arabia; maalbgmi@ju.edu.sa
* Correspondence: ghodaifa@upo.es

Citation: Hodaifa, G.; Zuorro, A.; Dominguez, J.R.; Rodríguez, J.G.; Peres, J.A.; Frontistis, Z.; Albqmi, M. Advanced Oxidation Process: Applications and Prospects. *Water* 2023, 15, 3444. https://doi.org/10.3390/w15193444

Received: 20 September 2023
Accepted: 27 September 2023
Published: 30 September 2023

The generation of waste has increased significantly over the last 50 years. This has affected the quality of the air, water, and soil, leading to climate change, the effects of which are increasingly being felt by the world's population. Faced with this scenario, society is being forced to look for solutions to maintain the health of the planet and of humans. In this sense, advanced oxidation technologies (AOTs) present solutions that are difficult to achieve through conventional technologies, given the complexity of the composition of the waste generated by industries that increasingly pollute. The scientific community is working to develop AOTs from emerging to well-established technologies, as they are a greener option and fit well into the processes that lead to a circular economy. This Topic is dedicated to understanding the current situation concerning AOTs, and what these technologies can offer us in the search for effective environmental solutions to reduce or eliminate waste generated by different industrial sectors. The papers that make up the Topic and correspond to different technologies involving advanced oxidation processes (AOPs) are as follows.

Lu et al. [1] studied the degradation effect of heat/peroxymonosulfate (PMS) on atrazine (ATZ). The results show that the heat/PMS degradation for ATZ is 96.28% when the phosphate buffer (PB) pH, temperature, PMS dosage, ATZ concentration, and reaction time are 7, 50 °C, 400 µmol/L, 2.5 µmol/L, and 60 min. Kim et al. [2] aimed to enhance the catalytic activity of a coal-based powdered activated carbon (PAC) via thermal treatment. They suggest that a simple thermal treatment can significantly change the characteristics of a PAC, to improve the removal of organic micropollutants. These changes in properties, and how they affect performance, can provide important information regarding the improvement of carbonaceous catalysts. Govindan et al. [3] examined the relationship between the intrinsic structure of a carbocatalyst and the catalytic activity of peroxomonosulfate (PMS) activation in acetaminophen degradation.

Wang et al. [4] demonstrated the potential utility of ferrate(VI)-based advanced oxidation processes for the degradation of the UV filter benzophenone-4 (BP-4). Zhu et al. [5] investigated the degradation of cetirizine (CTZ), a representative antihistamine, under UV/chlorine treatment. Liu and Sun [6] developed an efficient process for treating refractory 2,4-dichlorophenoxyacetic acid with radio frequencies.

Akter et al. [7] investigated the consequences of ozone dosage rates on the qualitative changes in organic compounds and nitrogen in anaerobic digestion effluent during the ozone process. Rafiee et al. [8] worked to reduce the chemical oxygen demand (COD) of an aeration effluent with an initial COD of 13,004 mg/L. In this study, an optimization process was conducted in order to find the quantities of H_2O_2, O_3, and UV irradiance required to reduce the COD of the effluent to the lowest possible.

Zhang et al. [9] studied doxycycline (DOX), a typical antibiotic, and its removal using potassium ferrate (Fe(VI)) and montmorillonite, and investigated the effect of the Fe(VI) dosage, reaction time, initial pH value, montmorillonite dosage, adsorption pH, time, and temperature on DOX removal.

Kastanek et al. [10] show the indispensability of the Fenton reaction in relation to environmental issues, as it represents the basis of all advanced oxidation processes around the idea of oxidative hydroxide radicals. The study aims to not only summarize the current knowledge of the Fenton process and identify its advantages, but also address the problems that remain to be solved. Zhang et al. [11] indicate that Cu and Co have shown a superior catalytic performance to those of other transition elements, and that layered double hydroxides (LDHs) have presented advantages over other heterogeneous Fenton catalysts. Manduca Artiles et al. [12] report the degradation of diazepam (DZP) in aqueous media via gamma radiation, high-frequency ultrasound, and UV radiation (artificial–solar), as well as the results when each process is intensified using oxidizing agents (H_2O_2 and Fenton reagent). Zhang et al. [13] fabricated three different types of manganese dioxide (MnO_2), with rod-like, needle-like, and mixed morphologies, via a hydrothermal method, changing the preparation conditions and adding metal ions, which were utilized as an activator of persulfate (PS) to remove aqueous dyes. In their review, do Nascimento et al. [14] comment on the importance of choosing the best synthesis method and experimental conditions to modify the structural, morphological, and electronic characteristics of semiconductors and, more specifically, tin oxide (SnO_2), as these parameters may be a determinant of a better performance in various applications, including photocatalysis. Suchanek et al. [15] have developed a new methodology for a broader assessment of the photocatalytic removal of NOx species (NO_2, HONO, and NO) from the air. The study provides important suggestions concerning the suitability of NO and NO_2 as test molecules, with NO being more suitable. Silerio-Vázquez et al. [16] analyzed a solar heterogeneous photocatalytic (HP) process for arsenite (AsIII) oxidation and coliform disinfection from a real groundwater matrix employing two reactors—a flat-plate reactor (FPR) and a compound parabolic collector (CPC)—with and without added hydrogen peroxide (H_2O_2). Arouca et al. [17] evaluated the effectiveness of white-light photolysis (WLP) via an advanced oxidation process (AOP) for removing polycyclic aromatic hydrocarbons (PAHs) from proximity firefighting protective clothing (PFPC), while maintaining the integrity of the fabric fibers. Experiments were carried out, with variations in the reaction time and the concentration of H_2O_2. With WLP (without H_2O_2), it was possible to remove more than 73% of the PAHs tested from the outer layer of PFPC in 3 days. Díez et al. [18] carried out the synthesis and characterization of novel graphene oxide coupled to TiO_2 (GO-TiO_2) in order to better understand the performance of this photocatalyst compared to the well-known TiO_2 (P25) from Degussa. Checa et al. [19] synthesized and characterized a magnetic graphene oxide titania (FeGOTi) catalyst for application in primidone removal from water, and checked for any possible enhancements. This compound was chosen because its presence is expected in urban wastewater, and its direct reaction with ozone is relatively low. Jorge et al. [20] investigated the degradation of a model agro-industrial wastewater phenolic compound (caffeic acid, CA) via a UV-A-Fenton system. According to the results, the UV-A-Fenton process at pH 3.0 achieved the highest CA degradation rate. Navidpour et al. [21] present a review of the photocatalytic processes and the mechanisms, reaction kinetics, and optical and electrical properties of semiconductors, and the unique characteristics of titanium as the most widely used photocatalyst; and compare the photocatalytic activity between different titania phases (anatase, rutile, and brookite) and between colorful and white TiO_2

nanoparticles. Haketa et al. [22] designed novel Au nano-particle catalysts immobilized on both titanium(IV)- and alkylthiol-functionalized SBA-15 type mesoporous silicates. The combination of Au nano-particle catalysts and other species that activate H_2O_2 served as an aerobic oxidation catalyst applicable to various substrates, including alkenes and alkanes.

Issa et al. [23] used the boron-doped diamond (BDD) anode, combined with a gas diffusion electrode (GDE) as a cathode, to constitute an attractive type of electrolysis system for the treatment of wastewater to remove organic pollutants. With this approach, they studied the treatment of synthetic wastewater, simulating the vacuum toilet sewage on trains via a BDD-GDE reactor, where the kinetics were presented as the abatement of the chemical oxygen demand (COD) over time.

Přibilová et al. [24] trialed the possibility of using persulfate to lower the amount of emerging contaminants released into the environment. The main disadvantage of sulfur-based AOPs is the need for activation. The authors investigated an economically and environmentally friendly solution based on hydrodynamic cavitation, which does not require heating or the additional activation of chemical substances.

Jorge et al. [25] proposed a new approach to winery wastewater treatment: (1) the application of the coagulation–flocculation–decantation (CFD) process with an organic coagulant based on almond skin extract (ASE), (2) the treatment of the organic recalcitrant matter through sulfate radical advanced oxidation processes (SR-AOPs), and (3) the evaluation of the efficiency of the combined CFD with UV-A, UV-C, and ultrasound (US) reactors.

Liu et al. [26] studied the migration/change characteristics of the centerline of the channel of the Yangtze River in Zhenjiang–Yangzhou; these characteristics are crucial for a comprehensive understanding of the river. In this study, a detailed calculation method is proposed to extract the channel centerline in Zhenjiang–Yangzhou, using old maps and remote sensing satellite maps, and to dissect it into seven parts.

Tincu et al. [27] studied mercury (Hg) exposure through a fish-based diet to evaluate the correlations between Hg blood concentrations and specific biomarkers for oxidative stress that can lead to neurological and cardiovascular diseases through the exacerbation of oxidative stress.

The scientific contributions to this Topic have shed light on the state of the art of emerging advanced oxidation technologies, which show promise for being incorporated into the industrial sector in the not-too-distant future.

We would like to express our gratitude to MDPI Editorial, through the journals *Catalysts*, *Processes*, *Sci*, *International Journal of Environmental Research and Public Health*, and *Water*, for offering us the opportunity to serve as Guest Editors and contribute to this detailed exploration of advanced oxidation technologies. In addition, we would like to thank all the authors who shared their research and the referees for their invaluable contributions.

Conflicts of Interest: The authors declare no conflict of interest.

References

1. Lu, Y.; Liu, Y.; Tang, C.; Chen, J.; Liu, G. Heat/PMS Degradation of Atrazine: Theory and Kinetic Studies. *Processes* **2022**, *10*, 941. [CrossRef]
2. Kim, D.; Kim, T.; Ko, S. Enhanced Catalytic Activity of a Coal-Based Powdered Activated Carbon by Thermal Treatment. *Water* **2022**, *14*, 3308. [CrossRef]
3. Govindan, K.; Kim, D.; Ko, S. Role of N-Doping and O-Groups in Unzipped N-Doped CNT Carbocatalyst for Peroxomonosulfate Activation: Quantitative Structure-Activity Relationship. *Catalysts* **2022**, *12*, 845. [CrossRef]
4. Wang, R.; Sun, P.; Zhai, Z.; Liu, H.; Han, R.; Liu, H.; Fang, Y. Degradation of the UV Filter Benzophenone-4 by Ferrate (VI) in Aquatic Environments. *Processes* **2022**, *10*, 1829. [CrossRef]
5. Zhu, B.; Cheng, F.; Zhong, W.; Qu, J.; Zhang, Y.; Yu, H. Mechanistic Insight into Degradation of Cetirizine under UV/Chlorine Treatment: Experimental and Quantum Chemical Studies. *Water* **2022**, *14*, 1323. [CrossRef]
6. Liu, Y.; Sun, B. Unusual Catalytic Effect of Fe^{3+} on 2,4-dichlorophenoxyacetic Acid Degradation by Radio Frequency Discharge in Aqueous Solution. *Water* **2022**, *14*, 1719. [CrossRef]
7. Akter, J.; Lee, J.; Kim, W.; Kim, I. Changes in Organics and Nitrogen during Ozonation of Anaerobic Digester Effluent. *Water* **2022**, *14*, 1425. [CrossRef]

8. Rafiee, M.; Sabeti, M.; Torabi, F.; Rahimbakhsh, A. COD Reduction of Aeration Effluent by Utilizing Optimum Quantities of $UV/H_2O_2/O_3$ in a Small-Scale Reactor. *Processes* **2022**, *10*, 2441. [CrossRef]
9. Zhang, H.; Wang, S.; Shu, J.; Wang, H. Enhanced Removal of Doxycycline by Simultaneous Potassium Ferrate(VI) and Montmorillonite: Reaction Mechanism and Synergistic Effect. *Water* **2023**, *15*, 1758. [CrossRef]
10. Kastanek, F.; Spacilova, M.; Krystynik, P.; Dlaskova, M.; Solcova, O. Fenton Reaction-Unique but Still Mysterious. *Processes* **2023**, *11*, 432. [CrossRef]
11. Zhang, R.; Liu, Y.; Jiang, X.; Meng, B. Vital Role of Synthesis Temperature in Co–Cu Layered Hydroxides and Their Fenton-like Activity for RhB Degradation. *Catalysts* **2022**, *12*, 646. [CrossRef]
12. Manduca Artiles, M.; Gómez González, S.; González Marín, M.; Gaspard, S.; Jauregui Haza, U. Degradation of Diazepam with Gamma Radiation, High Frequency Ultrasound and UV Radiation Intensified with H_2O_2 and Fenton Reagent. *Processes* **2022**, *10*, 1263. [CrossRef]
13. Zhang, X.; Gan, X.; Cao, S.; Shang, J.; Cheng, X. Efficient Removal of Rhodamine B in Wastewater via Activation of Persulfate by MnO_2 with Different Morphologies. *Water* **2023**, *15*, 735. [CrossRef]
14. do Nascimento, J.; Chantelle, L.; dos Santos, I.; Menezes de Oliveira, A.; Alves, M. The Influence of Synthesis Methods and Experimental Conditions on the Photocatalytic Properties of SnO_2: A Review. *Catalysts* **2022**, *12*, 428. [CrossRef]
15. Suchanek, J.; Vaneckova, E.; Dostal, M.; Mikyskova, E.; Brabec, L.; Zouzelka, R.; Rathousky, J. Methodology for Simultaneous Analysis of Photocatalytic deNOx Products. *Catalysts* **2022**, *12*, 661. [CrossRef]
16. Silerio-Vázquez, F.; Núñez-Núñez, C.; Proal-Nájera, J.; Alarcón-Herrera, M. Arsenite to Arsenate Oxidation and Water Disinfection via Solar Heterogeneous Photocatalysis: A Kinetic and Statistical Approach. *Water* **2022**, *14*, 2450. [CrossRef]
17. Arouca, A.; Aleixo, V.; Vieira, M.; Talhavini, M.; Weber, I. White Light-Photolysis for the Removal of Polycyclic Aromatic Hydrocarbons from Proximity Firefighting Protective Clothing. *Int. J. Environ. Res. Public Health* **2022**, *19*, 10054. [CrossRef]
18. Díez, A.; Pazos, M.; Sanromán, M.; Kolen'ko, Y. GO-TiO$_2$ as a Highly Performant Photocatalyst Maximized by Proper Parameters Selection. *Int. J. Environ. Res. Public Health* **2022**, *19*, 11874. [CrossRef]
19. Checa, M.; Montes, V.; Rivas, J.; Beltrán, F. Checking the Efficiency of a Magnetic Graphene Oxide-Titania Material for Catalytic and Photocatalytic Ozonation Reactions in Water. *Catalysts* **2022**, *12*, 1587. [CrossRef]
20. Jorge, N.; Teixeira, A.; Fernandes, J.; Oliveira, I.; Lucas, M.; Peres, J. Degradation of Agro-Industrial Wastewater Model Compound by UV-A-Fenton Process: Batch vs. Continuous Mode. *Int. J. Environ. Res. Public Health* **2023**, *20*, 1276. [CrossRef]
21. Navidpour, A.; Abbasi, S.; Li, D.; Mojiri, A.; Zhou, J. Investigation of Advanced Oxidation Process in the Presence of TiO_2 Semiconductor as Photocatalyst: Property, Principle, Kinetic Analysis, and Photocatalytic Activity. *Catalysts* **2023**, *13*, 232. [CrossRef]
22. Haketa, T.; Nozawa, T.; Nakazawa, J.; Okamura, M.; Hikichi, S. Oxidation Catalysis of Au Nano-Particles Immobilized on Titanium(IV)-and Alkylthiol-Functionalized SBA-15 Type Mesoporous Silicate Supports. *Catalysts* **2023**, *13*, 35. [CrossRef]
23. Issa, M.; Haupt, D.; Muddemann, T.; Kunz, U.; Sievers, M. The Electrochemical Reaction Kinetics during Synthetic Wastewater Treatment Using a Reactor with Boron-Doped Diamond Anode and Gas Diffusion Cathode. *Water* **2022**, *14*, 3592. [CrossRef]
24. Přibilová, P.; Odehnalová, K.; Rudolf, P.; Pochylý, F.; Zezulka, Š.; Maršálková, E.; Opatřilová, R.; Maršálek, B. Rapid AOP Method for Estrogens Removal via Persulfate Activated by Hydrodynamic Cavitation. *Water* **2022**, *14*, 3816. [CrossRef]
25. Jorge, N.; Teixeira, A.; Fernandes, L.; Afonso, S.; Oliveira, I.; Gonçalves, B.; Lucas, M.; Peres, J. Treatment of Winery Wastewater by Combined Almond Skin Coagulant and Sulfate Radicals: Assessment of HSO_5^- Activators. *Int. J. Environ. Res. Public Health* **2023**, *20*, 2486. [CrossRef] [PubMed]
26. Liu, C.; Liu, B.; Zhang, Z.; Li, C.; Wei, G.; Jiang, S. The Variations and Influences of the Channel Centerline of the Zhenjiang-Yangzhou Reach of the Yangtze River Based on Archival and Contemporary Data Sets. *Water* **2022**, *14*, 2478. [CrossRef]
27. Tincu, R.; Cobilinschi, C.; Florea, I.; Cotae, A.; Băetu, A.; Isac, S.; Ungureanu, R.; Droc, G.; Grintescu, I.; Mirea, L. Effects of Low-Level Organic Mercury Exposure on Oxidative Stress Profile. *Processes* **2022**, *10*, 2388. [CrossRef]

Article

Heat/PMS Degradation of Atrazine: Theory and Kinetic Studies

Yixin Lu [1,2,3,4], Yujie Liu [3], Chenghan Tang [3], Jiao Chen [1,2,3,4] and Guo Liu [2,*]

[1] School of Materials and Environmental Engineering, Chengdu Technological University, Chengdu 611730, China; yxlu61@163.com (Y.L.); chjicn@foxmail.com (J.C.)
[2] State Environmental Protection Key Laboratory of Synergetic Control and Joint Remediation for Soil & Water Pollution, Chengdu 610059, China
[3] Faculty of Geosciences and Environmental Engineering, Southwest Jiaotong University, Chengdu 611756, China; lyjenter@163.com (Y.L.); 15201822120@163.com (C.T.)
[4] Haitian Water Group Co., Ltd., Chengdu 610200, China
* Correspondence: liuguo@cdut.edu.cn; Tel.: +86-133-0800-0115

Abstract: The degradation effect of heat/peroxymonosulfate (PMS) on atrazine (ATZ) is studied. The results show that the heat/PMS degradation for ATZ is 96.28% at the moment that the phosphate buffer (PB) pH, temperature, PMS dosage, ATZ concentration, and reaction time are 7, 50 °C, 400 µmol/L, 2.5 µmol/L, and 60 min. A more alkaline PB is more likely to promote the breakdown of ATZ through heat/PMS, while the PB alone has a more acidic effect on the PMS than the partially alkaline solution. HO• and SO_4^-• coexisted within the heat/PMS scheme, and ATZ quantity degraded by HO• and SO_4^-• in PB with pH = 7, pH = 1.7~1. HCO_3^- makes it difficult for heat/PMS to degrade ATZ according to inorganic anion studies, while Cl^- and NO_3^- accelerate the degradation and the acceleration effect of NO_3^- is more obvious. The kinetics of ATZ degradation via heat/PMS is quasi-first-order. Ethanol (ETA) with the identical concentration inhibited ATZ degradation slightly more than HCO_3^-, and both of them reduced the degradation rates of heat/PMS to 7.06% and 11.56%. The addition of Cl^- and NO_3^- increased the maximum rate of ATZ degradation by heat/PMS by 62.94% and 189.31%.

Keywords: heat activation; PMS; atrazine; degradation mechanism; kinetics

Citation: Lu, Y.; Liu, Y.; Tang, C.; Chen, J.; Liu, G. Heat/PMS Degradation of Atrazine: Theory and Kinetic Studies. *Processes* 2022, 10, 941. https://doi.org/10.3390/pr10050941

Academic Editors: Antonio Zuorro, Gassan Hodaifa, Joaquín R. Dominguez, Juan García Rodríguez, José Alcides Peres and Zacharias Frontistis

Received: 28 March 2022
Accepted: 3 May 2022
Published: 9 May 2022

Publisher's Note: MDPI stays neutral with regard to jurisdictional claims in published maps and institutional affiliations.

1. Introduction

With the continuous growth of the world's population, the demand for crops increases, and the use of chemical fertilizers, pesticides, and herbicides has caused serious impacts on the global soil and water environment [1,2]. Herbicides remain the most effective, efficient, and economical way to control weeds, and its market continues to grow even with the plethora of generic products. With the development of herbicide-tolerant crops, use of herbicides is increasing around the world, which has resulted in severe contamination of the environment [3]. Atrazine is a commonly used chemical herbicide. This herbicide is considered moderately toxic for humans and very dangerous for the environment since significant levels persist in the environment and are highly toxic even in low concentrations [4]. ATZ is also potentially harmful to the growth and development of aquatic plants, animals, mammals, amphibians, and human cells [5,6]. Annual atrazine use in China has increased year on year since 1980 [7]. Based on this, the total amount of ATZ used nationwide could reach 10^8 kg by the end of 2018. ATZ can be excreted in wastewater on a variety of environmental substrates, for example, ground and surface water, various sediments, potable water, and soil [8]. The amount of ATZ in ground and surface water and soil can continue increasing because of its heavy use and chemical stability [9]. Inside the aqueous solution, atrazine owns a half-life of 60 to 150 days in humic acid and/or water, accordingly [10]. Due to the serious environmental hazards of ATZ, as early as 1994, the US Environmental Protection Agency issued a communique stating that only 3 µg L^{-1} ATZ is allowed in the water [11]. However, atrazine is found in 32% of U.S. water bodies, at an average quantity of 0.17 g/L [12]. Many of our rivers and reservoirs detected more than 3.9 μg/L, and in the Hygienic Standard for Drinking Water, the number is required to be limited to 2 μg/L [13–17].

Effects and endocrine disruption are the two main biological consequences of ATZ. The harmful effects of ATZ on various organisms in nature are mainly manifested in the interference of ATZ in the normal operation of the endocrine system and it causing organisms to produce toxic reactions. From the viewpoint of Sun et al. [18], ATZ with a concentration of 10 mg/L or above inhibited the germination rate of rice seeds. Fu et al. [19] found that an ATZ concentration of 0.125 mg/cm^2 inhibited the embryonic development of the red-eared turtle. The research of Benjamin and others [20] showed that ATZ can slow the vertebrae growth of zebra fish. Excessive use of ATZ (greater than 3 mmol/L) can lead to severe head defects in development zebra fish cranial [21,22]. According to the studies by Remayi and others [23], atrazine at 0.2 mg/L and 0.5 mg/L can damage the germ cell lines and spermatogenic tubules of adult frogs, and extensive connective tissue was formed around the spermatogenic tubules. Lin et al. [24] discovered that ATZ can cause ion imbalances in the heart and liver of quails, causing additional harm. Through raising the work of CYP19 enzyme in creatures' body, ATZ can disrupt the body's endocrine balance, according to the studies by Beaudoin and others. According to studies by Beaudoin and others, ATZ can disrupt the body's endocrine equilibrium through raising the action of the CYP19 proteinase among human organisms. [25]. Chao et al.'s research shows that atrazine (ATZ) poses high risks to algae and worms [26].

At present, various improved oxidation processes depending on PS/PMS, such as heat/PS [27], PBS/PMS [28], UV/PS [29], and UV/PMS [30], have been proven to have good degradation effect on ATZ in water. However, heat/PMS heat/PMS degradation of ATZ is rarely researched. Therefore, in PB, the impacts of heat/PMS upon ATZ oxidation degradation under various situations are studied. The process and dynamics of degradation are also investigated, which provides some reference value for further enriching the chemical treatment technology of pesticides containing drainage.

2. Materials and Methods

2.1. Reagents and Instruments

Reagents: All below are analytically pure, including caustic soda, sodium dihydrogen phosphate, sodium nitrite, ETA, tert-butanol (TBA), sodium chloride, baking soda, and potash nitrate. ATZ and PMS were bought on Aladdin.

Instruments: HPLC (2695–2996), electronic balance, UV lamp (Cnlight ZW5D15W-Z150), pH meter (PHSJ-3F), CNC ultrasonic cleaning machine (KH5200DB), Ultrapure Water machine by Youpu, Smart thermostat that saves energy (DC-1030), and isothermal magnetism mixer (78HW-1).

2.2. Experiment Scheme

2.2.1. Solution Preparation

A total of 100 mol/L ATZ reserve liquid was made from water through ultrapure. The water resistivity was 18.24 MΩ cm. The NaH_2PO_4 solution's concentration of $NaH_2PO_4^-$ – NaOH buffer was 0.2 mol/L. The NaOH solution's concentration was separated into two parts: 0.2 mol/L and 0.02 mol/L. Concentrations of $NaNO_2$, NaCl, $NaHCO_3$, and KNO_3 solution were configured to 0.1 mol/L, 1 mol/L, 0.5 mol/L, and 1 mol/L, respectively. PMS solution was set to 0.01 mol/L and kept in a dark place. Concentrations of tert-butyl alcohol and ethanol solution were 8 g/L and with a steady capacity of 1 L. Table 1 demonstrates the pH6, pH7, and pH8 production techniques.

Table 1. Manner of $NaH_2PO_4^-$ − NaOH buffer preparing.

pH	0.2 mol/L NaH_2PO_4 (mL)	0.2 mol/L NaOH (mL)
6	250	28.50
7	250	148.15
8	250	244.00

2.2.2. Experiments with a Heat/PMS Degradation Manner for ATZ

Heat/PMS degradation of ATZ in 1.25 mmol/L PB was investigated using a variety of temperature conditions (30, 40, 50, and 60 °C), pH value (6, 7, and 8), PMS concentration (50, 100, 200, and 400 µmol/L), and ATZ concentration (1.25, 2.5, and 5 µmol/L). Various concentrations of tertiary butyl alcohol and ethanol degrade ATZ in different ways. Below a 50 °C water bath, the process of ATZ degradation by various concentrations of TBA and ETA was studied when PB, PMS, and ATZ concentrations were 1.25 mmol/L, 100 µmol/L, and 2.5 µmol/L, accordingly. Effects of typical anions Cl^-, HCO_3^-, NO_3^- in water upon the ATZ degradation via heat/PMS were studied through adding diverse concentrations of NaCl, $NaHCO_3$ and $NaNO_3$, and 0.1 mol/L of $NaNO_2$ solution played an ending agent role.

2.3. Analysis Method

ATZ was evaluated via using a Symmetry® C18 LC column. The exact detecting methods are listed below: a 60:40 methyl alcohol to ultrapure water mobile phase ratio, a flow velocity of 0.8 mL/min, operating temperatures of 40 °C, and determinate wavelength of 225 nm.

3. Results and Discussion

3.1. The Influence of Temperature upon ATZ Degradation via Heat/PMS

In PB of pH = 7, diverse temperatures have an effect upon the ATZ degradation via heat/PMS, which is depicted in Figure 1 as ATZ and PMS concentrations are accordingly 2.5 µmol/L and 100 µmol/L. As Canlan showed in research [31], formula of the kinetics on oxidative degradation of ATZ via heat/PMS is established according to the below kinetic equation, and the first-order-kinetics are,

$$Ln(C/C_0) = -K_1 t \tag{1}$$

C: ATZ concentration at any given time, µmol/L;
C_0: The concentration of ATZ at start, µmol/L;
K_1: The pseudo first-order reaction rate constant, min^{-1}.

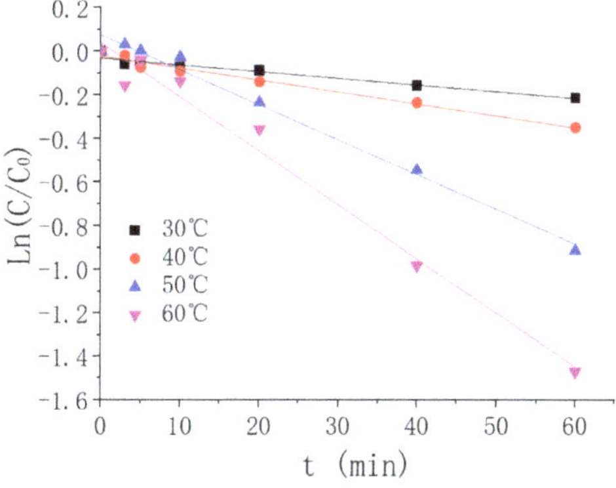

Figure 1. ATZ degradation kinetics curves under diverse temperature (C_0 = 2.5 µmol/L).

As Figure 1 shows, ATZ elimination rate is higher as the temperature increases from 40 °C to 50 °C than as the temperature ranges from 30 °C to 40 °C and 50 °C to 60 °C. The experiments of ATZ degradation via heat/PMS under various temperatures conformed to the quasi-first-order reaction kinetics. As Figure 1 depicts, reaction proceed rises 8.05 times as temperature rises from 30 °C to 60 °C. This is because temperature does have a major effect upon the outcome. As temperature rises, the level of free radicals inside the reaction rises as well, accelerating the degradation of ATZ [32,33]. Also, ATZ elimination rate at various temperatures is given. (See Appendix A, Figure A1).

3.2. The Influence of PMS Concentration upon ATZ Degradation via Heat/PMS

In PB of pH = 7, the effect of diverse PMS concentration upon ATZ degradation via heat/PMS is depicted in Figure 2 as the concentration and temperature of ATZ are 2.5 µmol/L and 50 °C, respectively. According to Figure 2, the experiments of ATZ degradation via heat/PMS under various PMS concentrations all match the pseudo-first-order reaction kinetics. As the reaction concentration increases from 0.050 mmol/L to 0.400 mmol/L, the reaction rate increases 20.19 times. This is mainly because, when other conditions remain unchanged, the higher the concentration of oxidant per unit time, the higher the concentration of free radicals generated by thermal excitation, and then the higher the oxidation efficiency. It can thus be seen that PMS concentration makes a big difference on the degradation of ATZ by heat/PMS. Also, effect of the PMS density on ATZ removal rates is given. (See Appendix A, Figure A2).

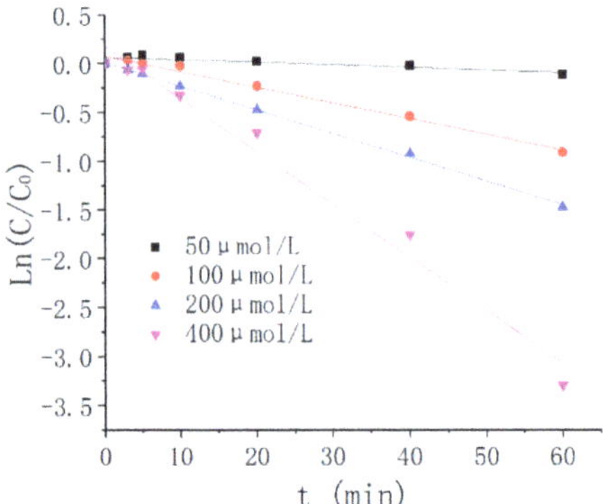

Figure 2. Effect of the PMS density on ATZ removal rates ($C_0 = 2.5$ µmol/L). The ATZ degradation kinetics curves.

3.3. The Impact of pH Value upon ATZ Degradation via Heat/PMS

As the ATZ concentration, PMS concentration, and temperature are 2.5 mol/L, 100 mol/L, and 50 °C, respectively, the influence of different pH on heat/PMS degradation of ATZ is shown in Figure 3. As Figure 3 depicts, as the reaction pH is rising, the action of ATZ degradation via heat/PMS is gradually strengthening. As the reaction pH rises from 6 to 8, ATZ elimination rate increases from 38.94% to 76.37%. The rate of ATZ elimination in an alkaline environment is greater than in an acidic environment. The following are the primary causes: heat can trigger PMS to produce $SO_4^- \bullet$ and $HO\bullet$ (as shown in Equations (1) and (2)), and $HO\bullet$ has slightly higher oxidizing ability to ATZ than to $SO_4^- \bullet$. $SO_4^- \bullet$ and $HO\bullet$ exhibit secondary reaction rates that are 3×10^9 M^{-1}s^{-1} [34] and 2.59×10^9 M^{-1}s^{-1} [35]. The $SO_4^- \bullet$ has a capacity to interact with any pH of water to generate $HO\bullet$. The reaction rate factor is 8.30 M^{-1}s^{-1} [36] (as seen in Equations (1)–(3)). However, in alkaline environments, $SO_4^- \bullet$ can interact with OH^- to generate $HO\bullet$ as well, and the reaction rate factor is 6.50×10^7 M^{-1}s^{-1} [37] (as shown in Equations (1)–(4)). In most cases, pH alterations have no effect on the production of $SO_4^- \bullet$ in heat/PMS. Thus, more $HO\bullet$ is generated in heat/PMS systems at alkaline environments, and the elimination rate of ATZ in alkaline environment is greater than that under acid environment. As is depicted in Figure 3 and Table 2, the ATZ degradations via heat/PMS experiments under diverse pH value are all in accordance with the pseudo-first-order reaction kinetics. With reaction pH rising from 6 to 8, the reaction rate increases 2.89 times, which indicates that pH value makes a big difference in ATZ degradation via heat/PMS. Also, effect of pH on ATZ removal rates is given. (See Appendix A, Figure A3).

$$HSO_5^- + Heat \rightarrow SO_4^- \bullet + HO\bullet \tag{2}$$

$$SO_4^- \bullet + H_2O \rightarrow SO_4^{2-} + HO\bullet + H^+ \tag{3}$$

$$SO_4^- \bullet + OH^- \rightarrow SO_4^{2-} + HO\bullet \tag{4}$$

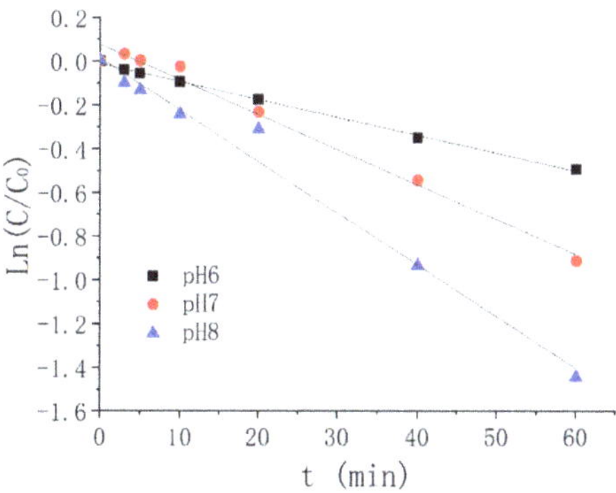

Figure 3. The kinetics of quasi-first-order reactions of heat/PMS degradation of ATZ at different pH values.

Table 2. The kinetics equations and parameters of quasi-first-order reactions of heat/PMS degradation of ATZ at various pH.

pH	Kinetic Equation	$t_{1/2}$ (min)	K_{obs} (min^{-1})	R^2
6	$Ln(C/C_0) = -0.00816t - 0.01166$	84.9	0.00185	0.99775
7	$Ln(C/C_0) = -0.01600t - 0.07571$	43.3	0.01600	0.98355
8	$Ln(C/C_0) = -0.02362t - 0.01394$	29.3	0.02362	0.98119

3.4. The Effect of ATZ Concentration upon the ATZ Degradation via Heat/PMS

Impacts of various ATZ concentrations upon the ATZ degradation via heat/PMS in PB of pH = 7 at 50 °C are depicted in Figure 4 when the PMS concentrations are 2.5 μmol/L and 10 μmol/L. As ATZ concentration rises, the influence of ATZ degradation via heat/PMS has a gradual decline, as is depicted in Figure 4a. As ATZ concentration increases from 1.25 μmol/L to 5 μmol/L, ATZ elimination rate declines from 90.77% to 56.91%. It is worth noting that the ATZ concentration tends to be balanced after 5 min when the concentration of ATZ is 5. This mainly because increasing the ATZ concentration results in the rise of collisions among ATZ with $SO_4^- \bullet$ and $HO\bullet$ in unit time, leading to rapid degradation of ATZ. After the oxidant is used up, the concentration of ATZ tends to equilibrium and does not decrease. As is shown in Figure 4b, the experiments upon the ATZ degradation via heat/PMS with different pH are all in accordance with the pseudo-first-order reaction kinetics. As ATZ concentration rises from 1.25 μmol/L to 2.5 μmol/L, the rate of reaction decreases by 2.45 times. This implies that ATZ concentration greatly affects ATZ degradation via heat/PMS.

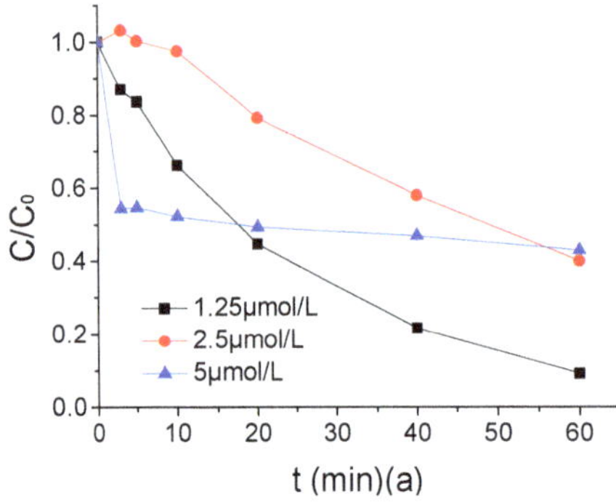
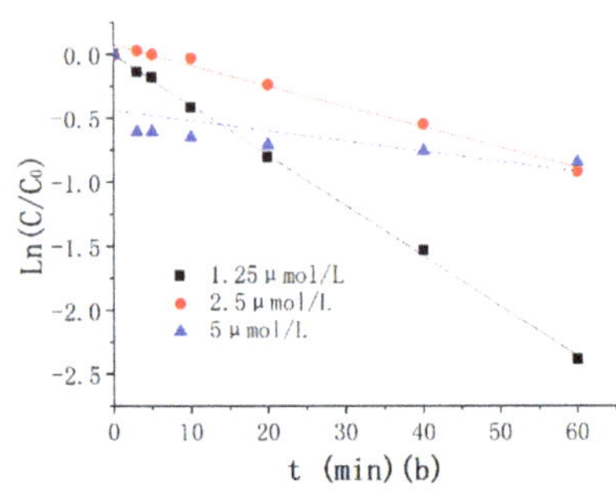

Figure 4. Effect of ATZ density (C_0) on ATZ removal rates: (**a**) experimental results; (**b**) the ATZ degradation kinetics curves.

3.5. Degradation Mechanism Study of ATZ via Heat/PMS

When the PMS and ATZ concentrations were, accordingly, 10 mol/L and 2.5 mol/L, temperature was 50 °C. The ATZ degradation process via heat/PMS was investigated using a single variable method with 1.25 mmol/L PB solution and pH values of 6, 7, and 8, and the results are shown in Figure 5.

According to Figure 5a, the degradation effect is weak and the degradation rate of ATZ by PB alone is about 3%. The ATZ would not decompose under the 50 °C water bath. Compared with adjusting the initial pH value to 7 with NaOH, using PB with the pH value of 7 has better ATZ degradation effect by heat/PMS. This is largely due to that the pH of system at the start was changed to 7 by NaOH, and the pH measured 4.5 after the reaction, which is similar to the pH gradient degradation effect demonstrated in this paper and will not be repeated here. In addition, PB can activate PMS to create $SO_4^-\bullet$ and $HO\bullet$ [28]. In accordance with studies of Dionysiou et al. [38], the tert-butanol(TBA) reaction rates with $HO\bullet$ and $SO_4^-\bullet$ were $3.8–7.6 \times 10^8$ $M^{-1}s^{-1}$ and $4–9.1 \times 10^5$ $M^{-1}s^{-1}$, while Buxton [39] and others found that ETA reaction rates with $HO\bullet$ and $SO_4^-\bullet$ were, respectively, $1.2–2.8 \times 10^9$ $M^{-1}s^{-1}$ and $1.6–7.7 \times 10^7$ $M^{-1}s^{-1}$. Therefore, when $HO\bullet$ and $SO_4^-\bullet$ cohabit, TBA can catch $HO\bullet$, while ETA can catch $HO\bullet$ and $SO_4^-\bullet$.

As depicted in Figure 5b–d, the addition of TBA and ETA have an inhibition effect upon the ATZ degradation via heat/PMS, and the inhibition effect of TBA is weaker than ETA. $HO\bullet$ and $SO_4^-\bullet$ cohabit within the heat/PMS system. Maintaining TBA and ETA with the concentration of 64 mg/L, respectively, in PB under pH 6, the ATZ degradation rate via heat/PMS is decreased to 17.44% and 5.04%, indicating that ATZ oxidative degradations by $HO\bullet$ and $SO_4^-\bullet$ are, respectively, 55.21% and 31.84%, and the oxidation ratio of the two is close to 1.7 to 1. Maintaining TBA and ETA with the concentration of 64 mg/L, respectively, in PB under pH 7, the degradation rates of ATZ by heat/PMS decrease to 42.50% and 1.66%, indicating that ATZ oxidative degradations by $HO\bullet$ and $SO_4^-\bullet$ are, respectively, 29.00% and 68.23%, and the oxidation ratio of the two is close to 1 to 2.4. Maintaining TBA and ETA with the concentration of 64 mg/L, respectively, in PB under pH 8, the ATZ degradation rates via heat/PMS decrease to 46.23% and 7.32%, indicating that ATZ oxidative degradation by $HO\bullet$ and $SO_4^-\bullet$ are, respectively, 39.47% and 50.95%, and the oxidation ratio of the two is close to 1 to 1.3. It can be seen from this that under any pH conditions, the oxidative degradation of ATZ via heat/PMS is mostly caused by free radicals, but the dominant free radical types are different under diverse pH conditions.

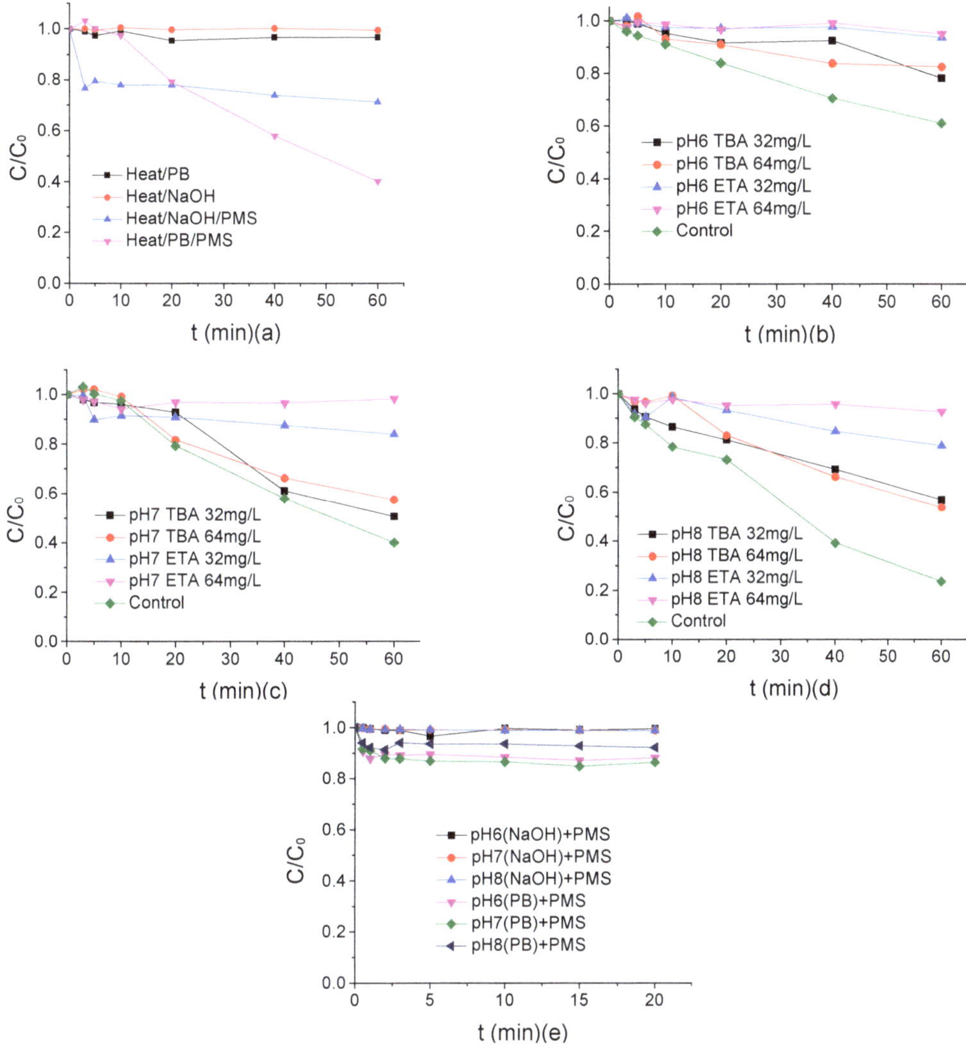

Figure 5. (**a**) Analysis of each component's oxidation impact upon the heat/PMS system. (**b**) The impact of TBA and ETA upon ATZ degradation via heat/PMS in PB of pH = 6. (**c**) The impact of TBA and ETA upon ATZ degradation via heat/PMS in PB of pH = 7. (**d**) The impact of TBA and ETA upon heat/PMS degradation ATZ in PB of pH = 8. (**e**) The impact of phosphate buffer upon the degradation of ATZ via PMS alone.

When the ATZ and PMS concentrations are, respectively, 2.5 μmol/L and 10 μmol/L, and reaction temperature is 20 °C, the degradation effect of ATZ is shown in Figure 5e in PB of 12.5 mmol/L with the pH of 6, 7, and 8. As is depicted in Figure 5e, PMS has no degradation effect on ATZ when adjusting the initial pH of the reaction system to 6, 7, and 8 by NaOH. The degradation rates of ATZ by PMS are, respectively, 11.72%, 13.5%, and 7.87% when the pHs of PB are 6, 7, and 8, respectively. The PMS oxidation mechanism is when the PMS are excited to generate strong oxidizing $SO_4^- \bullet$ to oxidize and degrade the target, indicating that the PB can excite PMS to generate $SO_4^- \bullet$. The overall effects of PMS on ATZ degradation in different PB solutions are listed as follows: degradation is best at pH 7, then followed by pH 6 and pH 8; the degradation effect is only slightly different when the PMS pH measures 6 and 7, indicating that phosphate is more likely to stimulate PMS under acidic conditions, which is consistent with the research by Gu et al. [28].

3.6. The Influence of Typical Anions Concentration in Solutions upon the ATZ Degradation via Heat/PMS

The typical anions' effects in solutions such as Cl^-, HCO_3^-, and NO_3^- upon the ATZ degradation via heat/PMS are depicted in Figure 6. The concentrations of PB, PMS, and ATZ were, respectively, 1.25 mmol/L, 10 μmol/L, and 2.5 μmol/L. PB pH value equaled 7. Reaction temperature was 50 °C.

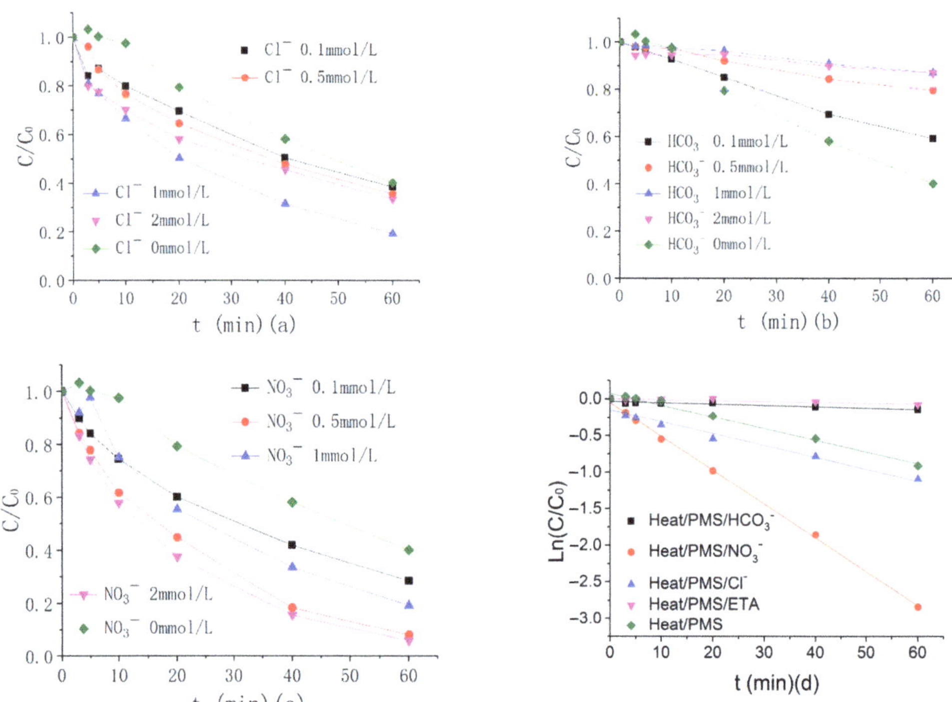

Figure 6. (**a**) The impact of Cl^- upon the ATZ degradation via heat/PMS in PB of PH = 7. (**b**) The impact of HCO_3^- upon ATZ degradation via heat/PMS in PB of pH = 7. (**c**) The impact of NO_3^- upon ATZ degradation via heat/PMS in PB of pH = 7. (**d**) The kinetics of quasi-first-order reaction of heat/PMS degradation ATZ.

On the basis of Figure 6a–c, HCO_3^- shows an inhibition influence on the ATZ degradation by heat/PMS. This is mainly because to produce $CO_3^- \bullet$, HCO_3^- could fight with ATZ for $HO\bullet$ and $SO_4^- \bullet$ within the heat/PMS system, whose reaction rate is lower than those of $HO\bullet$ and $SO_4^- \bullet$ (as shown in Equations (5)–(7)). The Cl^- and NO_3^- both have auxo-action influence upon ATZ degradation within the heat/PMS system at the identical concentration, and the promoting effect of NO_3^- is higher than Cl^-. Specific phenomenon behavior is shown below; the ATZ degradation efficiency increased from 59.86% to 80.73% and 94.15%, respectively, after adding Cl^- of 1 mmol/L and NO_3^- of 2 mmol/L to the heat/PMS system. It is worth mentioning that the Cl^- shows a promoting effect within the set concentration range, but the promoting effect first increases and then decreases. In addition, the promoting effect is strongest when the concentration is 1 mmol/L. This is largely due to that minor Cl^- could trigger PMS to produce $HO\bullet$ and $SO_4^-\bullet$ [40] to increase the $HO\bullet$ and $SO_4^-\bullet$ concentrations in the heat/PMS system and accelerate the ATZ degradation. When the Cl- increased in the reaction system, Cl^- could fight with ATZ for $HO\bullet$ and $SO_4^-\bullet$ within the heat/PMS system to produce $Cl\bullet$. Reaction rates of $HO\bullet$ and $SO_4^-\bullet$ are higher than those of Cl^-. Thus, it shows an inhibiting effect on the degradation (the main equations are shown in Equations (8) to (12)). $SO_4^-\bullet$ could react with NO_3^- to generate $NO_3\bullet$ (around 2.5 V) whose redox potential is similar to $SO_4^-\bullet$, participate in the ATZ degradation to accelerate the PMS decomposition, increase the concentration of $HO\bullet$ and $SO_4^-\bullet$ in the heat/PMS system, and accelerate the ATZ degradation (the main equations are shown in Equations (13) and (14)). The research presented by Ghauch et al. [41] also found that NO_3^- can promote the degradation of bisoprolol by heat/PS, which is similar to the experimental phenomenon. As shown in Figure 6d, comparing with HCO_3^- of the same concentration, ETA has a slightly stronger inhibitory influence upon ATZ degradation via heat/PMS; it reduced the ATZ degradation rate to 7.06% and 11.56%, respectively. The additions of Cl^- and NO_3^- increase the maximum rate of ATZ degradation by 62.94% and 189.31%, respectively. The reaction is shown in formulas (5)–(14) [36,42–49].

$$HO\bullet + HCO_3^- \rightarrow CO_3^-\bullet + H_2O, \quad K = 8.60 \times 10^6 \ M^{-1}s^{-1} \tag{5}$$

$$SO_4^-\bullet + HCO_3^- \rightarrow CO_3^-\bullet + HSO_4^-, \quad K = 2.80 \times 10^6 \ M^{-1}s^{-1} \tag{6}$$

$$CO_3^-\bullet + ATZ \rightarrow products, \quad K = 6.20 \times 10^6 \ M^{-1}s^{-1} \tag{7}$$

$$HO\bullet + Cl^- \rightarrow ClOH^-\bullet, \quad K = 4.30 \times 10^9 \ M^{-1}s^{-1} \tag{8}$$

$$ClOH^-\bullet + Cl^- \rightarrow Cl_2^-\bullet + OH^-, \quad K = 1.0 \times 10^5 \ M^{-1}s^{-1} \tag{9}$$

$$SO_4^-\bullet + Cl^- \rightarrow Cl\bullet + SO_4^{2-}, \quad K = 3.0 \times 10^9 \ M^{-1}s^{-1} \tag{10}$$

$$Cl\bullet + Cl^- \rightarrow Cl_2^-\bullet, \quad K = 8.50 \times 10^9 \ M^{-1}s^{-1} \tag{11}$$

$$Cl_2^-\bullet + ATZ \rightarrow products, \quad K = 5.0 \times 10^4 \ M^{-1}s^{-1} \tag{12}$$

$$SO_4^-\bullet + NO_3^- \rightarrow NO_3\bullet + SO_4^{2-}, \quad K = 5.0 \times 10^4 \ M^{-1}s^{-1} \tag{13}$$

$$NO_3\bullet + ATZ \rightarrow products \tag{14}$$

3.7. The AZT Degradation Products via Heat/PMS and Degradation Path Analysis

The ATZ degradation products via heat/PMS were studied and the degradation path was inferred by HPLC-ESI-MS (cationic pattern). Three samples were taken from 5, 20, and 60 min during the experiment, and first-order mass spectrometry was performed for total ion and extraction ion analysis.

It can be seen from the mass spectra, total ion flow diagram, and extract ionic flow diagram of the three samples that the charge mass ratio of the main ATZ degradation products are 128, 146, 174, 188, 198, 214, and 232 [30,31,50,51], and so on. ATZ has a relative molecular mass of 216. ATZ has a molecular mass 42 more than m/z174, and the molecular mass of isopropyl is exactly 42, so m/z174 is considered as deisopropyl ATZ (deisopropyl atrazine, DIA). ATZ has a molecular mass 28 more than m/z188 and the molecular mass of ethyl is exactly 28, so m/z188 is considered as desetyleatrazine ATZ (desetyleatrazine, DEA). M/z174 has a molecular mass 28 more than m/z146, and the molecular mass of ethyl is exactly 28; m/z188 has a molecular mass 42 more than m/z146, and the molecular mass of isopropyl is exactly 42. Therefore, m/z146 is considered as deethylation deisopropyl ATZ (deethylation deisopropyl atrazine, DEIA). m/z146 has a molecular mass 18 more than m/z128. Thus m/z128 is considered as chlorine ions of DEIA replaced by hydroxyl groups, 2-hydroxy-4,6-diaminoATZ (deethylation deisopropyl hydroxy atrazine, DEIHA). ATZ has a molecular mass 18 more than m/z198. Therefore, it is considered that in the ATZ degradation procedure, hydroxyl groups replaced the Cl atoms to generate 2-hydroxy ATZ (hydroxy atrazine, HA). ATZ has a molecular mass 16 less than m/z232 and the molecular mass of hydroxy is exactly 16. Therefore, it is considered that in the ATZ degradation procedure, a hydroxyl group replaced a hydrogen atom to generate 2-chloro-4-hydroxy-ethylene-6-isopropyl ATZ (2-Chloro-4-hydroxyethylamino-6-isopropylatrazine, CHEIA). m/z232 has a molecular mass 18 more than m/z214, so m/z128 is considered as the further dehydration and degradation product of CHEIA, 2-Chloro-4-vinylamino-6-isopropylamino ATZ (2-Chloro-4-vinylamino-6-isopropylamino atrazine, CVIA). m/198 has a molecular mass 16 less than m/z 214, so m/z214 is considered to be the product of hydroxyl substituting one of the hydrogen atoms in HA, 2,4-dihydroxy ATZ (dihydroxy atrazine, DHA).

This shows that the ATZ degradation via heat/PMS is primarily accomplished through by dealkylation and dechlorination, which matches the findings of Ji et al. [52]. Figure 7 depicts the degradation method of ATZ.

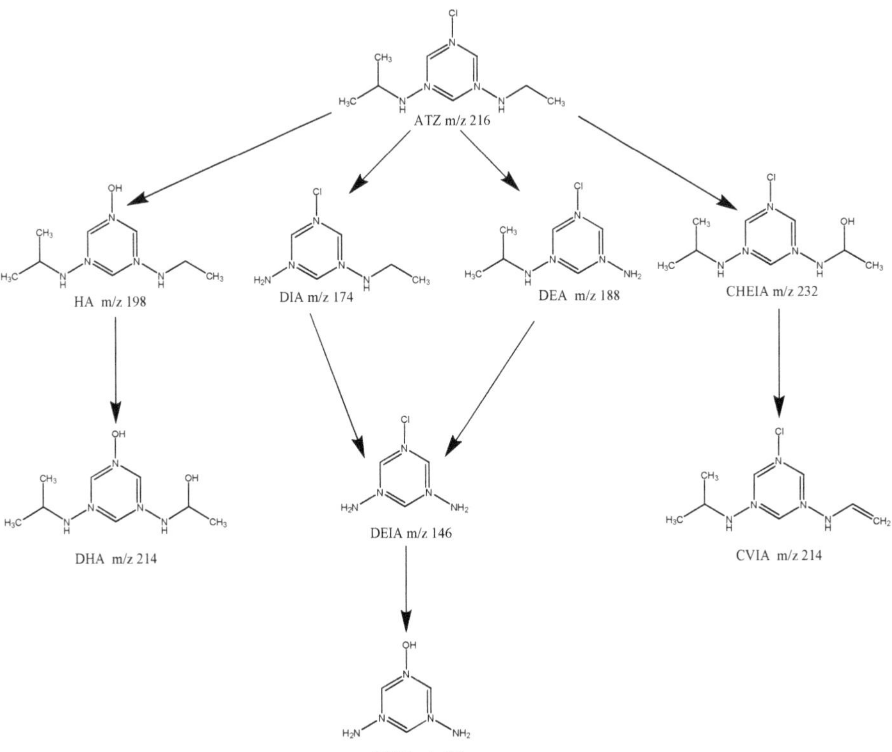

Figure 7. The possible degradation pathway of ATZ.

4. Conclusions

The ATZ degradation in PB via heat/PMS was investigated by the methods of experiment and kinetics modeling. The degradation mechanism, oxidation kinetics, and degradation products were studied. The following are the main findings of this paper. The ATZ degradation efficiency of heat/PMS is proportional to temperature and PMS concentration but inversely proportional to ATZ concentration. PB is more favorable to ATZ degradation under alkaline environments. Acidic environments are more favorable for PB to stimulate PMS. HO• and SO_4^-• cohabit within heat/PMS system. ATZ degradation via heat/PMS is mainly free radical oxidative degradation under any pH conditions, but the dominant free radical types are different under different pH conditions. HCO_3^-

has repressive impacts on the degradation of ATZ. Both Cl^- and NO_3^- show facilitating effects on the ATZ degradation via heat/PMS, and the stimulating impact of NO_3^- is more remarkable. ATZ degradation kinetics via heat/PMS corresponds with quasi-first-order reaction kinetics. Dealkylation and dechlorination are the key mechanisms through which ATZ is degraded via heat/PMS. A total of eight products of seven mass-to-charge ratio were found by the product analysis.

The degradation of atrazine by UV/PMS and US/PMS, and even the degradation of atrazine by thermal-activated persulfate, has been studied. However, there are few studies on the degradation mechanism of atrazine by heat/PMS system in PB. In this paper, the mechanism was obtained through the study of the system, and it was found that PB in alkaline condition can promote the degradation of ATZ by heat/PMS more than PB in acidic condition. The important conclusions are that PB is more likely to stimulate PMS in acidic conditions than in alkaline conditions, and that PB alone can stimulate PMS, which offers a breakthrough and contribution to previous studies and lays a foundation for subsequent studies.

Atrazine is not easy to degrade, and it can exist stably in aqueous environments for a long time, which not only affects the survival of animals and plants, but also threatens human health. Most studies do not clarify the types and hazards of ATZ degradation intermediates. Therefore, paying attention to the degradation efficiency of atrazine while taking into account the toxicity of intermediate degradation products is of great significance for optimizing ATZ degradation technology, which is also one of the future development directions.

Author Contributions: Conceptualization, Y.L. (Yixin Lu) and J.C.; data curation, G.L.; formal analysis, Y.L. (Yujie Liu); funding acquisition, Y.L. (Yixin Lu); methodology, C.T., Y.L. (Yixin Lu); supervision, G.L.; visualization, Y.L. (Yujie Liu); writing—original draft, Y.L. (Yixin Lu); writing—review and editing, J.C. All authors have read and agreed to the published version of the manuscript.

Funding: This research was funded by the Science and Technology Project of Sichuan Province, grant number: 22ZDYF2880, 22YYJC3490; Open Fund of State Environmental Protection Key Laboratory of Synergetic Control and Joint Remediation for Soil & Water Pollution, grant number: GHBK-2021-004; School Level Project of Chengdu Technological University, grant number: 2021ZR020, QM2021003, QM2021034, QM2021064, QM2021080; National Innovation Training Program for College Students, grant number: S202011116015, S202011116031.

Institutional Review Board Statement: Not applicable.

Informed Consent Statement: Not applicable.

Data Availability Statement: Data are contained within the article.

Appendix A. Single Factor Influence Diagram

As Figure A1 shows, the influence of degradation of ATZ via heat/PMS is increasing as the reaction temperature is rising. ATZ elimination rate improves from 19.22 percent to 77.01 percent as reaction temperature increases from 40 °C to 50 °C.

As Figure A2 depicts, the ATZ degradation is remarkably increasing as the PMS concentration increases. The ATZ elimination rate increases from 10.95% to 96.28% as the system's PMS concentration increases from 0.050 mmol/L to 0.400 mmol/L. Moreover, ATZ elimination rate increased most significantly as the PMS concentration improved from 0.050 mmol/L to 0.100 mmol/L.

As Figure A3 depicts, as the reaction pH rises from 6 to 8, ATZ elimination rate increases from 38.94% to 76.37%.

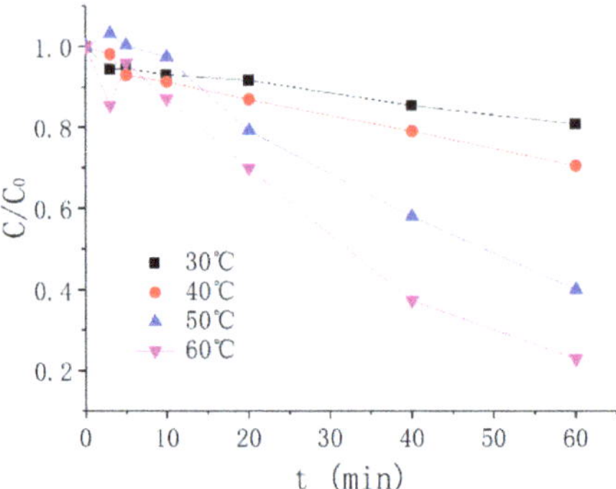

Figure A1. ATZ elimination rate at various temperatures ($C_0 = 2.5$ µmol/L).

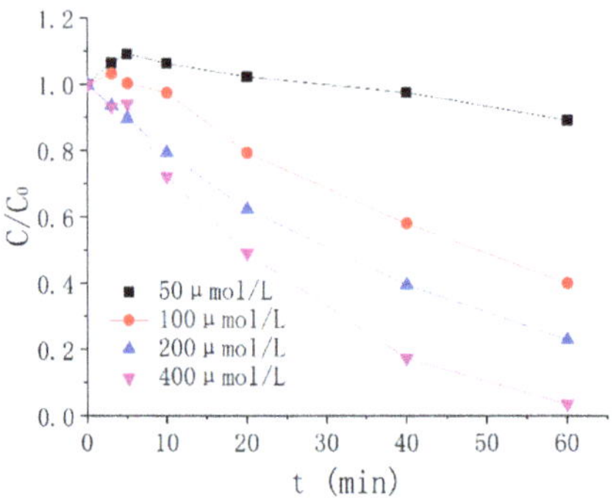

Figure A2. Effect of the PMS density on ATZ removal rates ($C_0 = 2.5$ µmol/L). Experimental findings.

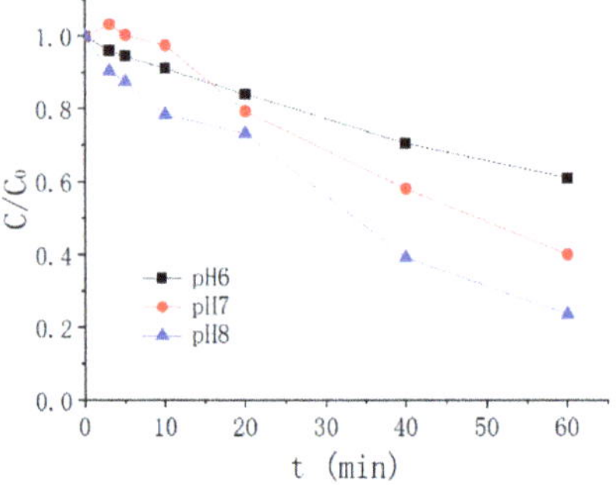

Figure A3. Effect of pH on ATZ removal rates ($C_0 = 2.5$ µmol/L). Experimental findings.

References

1. Saber, Z.; van Zelm, R.; Pirdashti, H.; Schipper, A.M.; Esmaeili, M.; Motevali, A.; Nabavi-Pelesaraei, A.; Huijbregts, M.A. Understanding farm-level differences in environmental impact and eco-efficiency: The case of rice production in Iran. *J. Sustain. Prod. Consum.* **2021**, *27*, 1021–1029. [CrossRef]
2. Tang, L.; Hayashi, K.; Inao, K.; Birkved, M.; Bruun, S.; Kohyama, K.; Shimura, M. Developing a management-oriented simulation model of pesticide emissions for use in the life cycle assessment of paddy rice cultivation. *Sci. Total Environ.* **2020**, *716*, 137034. [CrossRef]
3. Singh, B.; Singh, K. Microbial degradation of herbicides. *Crit. Rev. Microbiol.* **2016**, *42*, 245–261. [CrossRef]

4.   Esparza-Naranjo, S.B.; da Silva, G.F.; Duque-Castaño, D.C.; Araújo, W.L.; Peres, C.K.; Boroski, M.; Bonugli-Santos, R.C. Potential for the Biodegradation of Atrazine Using Leaf Litter Fungi from a Subtropical Protection Area. *Curr. Microbiol.* **2021**, *78*, 358–368. [CrossRef] [PubMed]

5.   Supraja, P.; Tripathy, S.; Krishna Vanjari, S.R.; Singh, V.; Singh, S.G. Label free, electrochemical detection of atrazine using electrospun Mn2O3 nanofibers: Towards ultrasensitive small molecule detection. *Sens. Actuators B Chem.* **2019**, *285*, 317–325. [CrossRef]

6.   Yan, P.; Jin, Y.; Xu, L.; Mo, Z.; Qian, J.; Chen, F.; Yuan, J.; Xu, H.; Li, H. Enhanced photoelectrochemical aptasensing triggered by nitrogen deficiency and cyano group simultaneously engineered 2D carbon nitride for sensitively monitoring atrazine. *Biosens. Bioelectron.* **2022**, *206*, 114144. [CrossRef] [PubMed]

7.   Huang, H.; Zhang, C.; Zhang, P.; Cao, M.; Xu, G.; Wu, H.; Zhang, J.; Li, C.; Rong, Q. Effects of biochar amendment on the sorption and degradation of atrazine in different soils. *Soil Sediment Contam. Int. J.* **2018**, *27*, 643–657. [CrossRef]

8.   Majewska, M.; Harshkova, D.; Pokora, W.; Baścik-Remisiewicz, A.; Tulodziecki, S.; Aksmann, A. Does diclofenac act like a photosynthetic herbicide on green algae? Chlamydomonas reinhardtii synchronous culture-based study with atrazine as reference. *Ecotoxicol. Environ. Saf.* **2021**, *208*, 111630. [CrossRef]

9.   Li, P.; Yao, L.Y.; Jiang, Y.J.; Wang, D.D.; Wang, T.; Wu, Y.P.; Li, B.X.; Li, X.T. Soybean isoflavones protect SH-SY5Y neurons from atrazine-induced toxicity by activating mitophagy through stimulation of the BEX2/BNIP3/NIX pathway. *Ecotoxicol. Environ. Saf.* **2021**, *227*, 112886. [CrossRef]

10. Bui, H.; Pham, V.H.; Pham, V.D.; Pham, T.B.; Nguyen, T.V. Development of nano-porous silicon photonic sensors for pesticide monitoring. *Dig. J. Nanomater. Biostruct.* **2018**, *13*, 57–65.

11. Zhang, C.-j.; Si, S.; Yang, Z. Design of molecularly imprinted TiO2/carbon aerogel electrode for the photoelectrochemical determination of atrazine. *Sens. Actuators B Chem.* **2015**, *211*, 206–212. [CrossRef]

12. Beaulieu, M.; Cabana, H.; Taranu, Z.; Huot, Y. Predicting atrazine concentrations in waterbodies across the contiguous United States: The importance of land use, hydrology, and water physicochemistry. *Limnol. Oceanogr.* **2020**, *65*, 2966–2983. [CrossRef]

13. Institute of Environmental and Health-Related Product Safety, Chinese Center for Disease Control and Prevention. *Hygienic Standards for Drinking Water*; Ministry of Health, Standardization Administration of China: Beijing, China, 2006; p. 16.

14. Ouyang, W.; Zhang, Y.; Lin, C.; Wang, A.; Tysklind, M.; Wang, B. Metabolic process spatial partition dynamics of Atrazine in an estuary-to-bay system, Jiaozhou bay. *J. Hazard Mater.* **2021**, *414*, 125530. [CrossRef] [PubMed]

15. Sun, X.; Liu, F.; Shan, R.; Fan, Y. Spatiotemporal distributions of Cu, Zn, metribuzin, atrazine, and their transformation products in the surface water of a small plain stream in eastern China. *Environ. Monit. Assess.* **2019**, *191*, 433. [CrossRef]

16. Zhang, Y.; Zhang, H.; Yang, M. Profiles and risk assessment of legacy and current use pesticides in urban rivers in Beijing, China. *Environ. Sci. Pollut. Res. Int.* **2021**, *28*, 39423–39431. [CrossRef]

17. Zheng, Q.; Wang, W.; Liu, S.; Zhang, Z.; Qu, S.; Li, F.; Wang, C.; Li, W. Modeling colloid-associated atrazine transport in sand column based on managed aquifer recharge. *Environ. Earth Sci.* **2018**, *77*, 667. [CrossRef]

18. Sun, J.T.; Pan, L.L.; Zhan, Y.; Tsang, D.C.W.; Zhu, L.Z.; Li, X.D. Atrazine contamination in agricultural soils from the Yangtze River Delta of China and associated health risks. *Environ. Geochem. Health* **2017**, *39*, 369–378. [CrossRef]

19. Fu, L.; Ni, J.; Ruan, Y.; Dong, R.; Shi, H. Effects of Atrazine on Embryonic Development and Histological Structure of Liver and Kidney in Red-eared Turtle(*Trachemys scripta elegans*). *Fish. Sci.* **2017**, *36*, 104–108.

20. Walker, B.S.; Kramer, A.G.; Lassiter, C.S. Atrazine affects craniofacial chondrogenesis and axial skeleton mineralization in zebrafish (Danio rerio). *Toxicol. Ind. Health* **2018**, *34*, 329–338. [CrossRef]

21. Hanson, M.L.; Solomon, K.R.; Van Der Kraak, G.J.; Brian, R.A. Effects of atrazine on fish, amphibians, and reptiles: Update of the analysis based on quantitative weight of evidence. *Crit. Rev. Toxicol.* **2019**, *49*, 670–709. [CrossRef]

22. Huang, W.; Wu, T.; Au, W.W.; Wu, K. Impact of environmental chemicals on craniofacial skeletal development: Insights from investigations using zebrafish embryos. *Environ. Pollut.* **2021**, *286*, 117541. [CrossRef]

23. Rimayi, C.; Odusanya, D.; Weiss, J.M.; de Boer, J.; Chimuka, L.; Mbajiorgu, F. Effects of environmentally relevant sub-chronic atrazine concentrations on African clawed frog (Xenopus laevis) survival, growth and male gonad development. *Aquat. Toxicol.* **2018**, *199*, 1–11. [CrossRef]

24. Lin, J.; Li, H.-X.; Qin, L.; Du, Z.-H.; Xia, J.; Li, J.-L. A novel mechanism underlies atrazine toxicity in quails (*Coturnix Coturnix coturnix*): Triggering ionic disorder via disruption of ATPases. *Oncotarget* **2016**, *7*, 83880. [CrossRef] [PubMed]

25. Caron-Beaudoin, E.; Denison, M.S.; Sanderson, J.T. Effects of Neonicotinoids on Promoter-Specific Expression and Activity of Aromatase (CYP19) in Human Adrenocortical Carcinoma (H295R) and Primary Umbilical Vein Endothelial (HUVEC) Cells. *Toxicol. Sci.* **2016**, *149*, 134–144. [CrossRef] [PubMed]

26. Su, C.; Cui, Y.; Liu, D.; Zhang, H.; Baninla, Y. Endocrine disrupting compounds, pharmaceuticals and personal care products in the aquatic environment of China: Which chemicals are the prioritized ones? *Sci. Total Environ.* **2020**, *720*, 137652. [CrossRef]

27. Wang, W.; Chen, M.; Wang, D.; Yan, M.; Liu, Z. Different activation methods in sulfate radical-based oxidation for organic pollutants degradation: Catalytic mechanism and toxicity assessment of degradation intermediates. *Sci. Total Environ.* **2021**, *772*, 145522. [CrossRef] [PubMed]

28. Wang, W.; Chen, M.; Wang, D.; Yan, M.; Liu, Z. On peroxymonosulfate-based treatment of saline wastewatser: When phosphate and chloride co-exist. *RSC Adv.* **2018**, *8*, 13865–13870.

29. Cui, J.Z.; Cai, S.K.; Zhang, S.R.; Wang, G.Q.; Gao, C.Z. Degradation of a non-oxidizing biocide in circulating cooling water using UV/persulfate: Kinetics, pathways, and cytotoxicity. *Chemosphere* **2022**, *289*, 133064. [CrossRef]
30. Cui, J.Z.; Cai, S.K.; Zhang, S.R.; Wang, G.Q.; Gao, C.Z. Degradation of Atrazine by UV/PMS in Phosphate Buffer. *Pol. J. Environ. Stud.* **2019**, *28*, 2735–2744.
31. Jiang, C.; Yang, Y.; Zhang, L.; Lu, D.; Lu, L.; Yang, X.; Cai, T. Degradation of Atrazine, Simazine and Ametryn in an arable soil using thermal-activated persulfate oxidation process: Optimization, kinetics, and degradation pathway. *J. Hazard Mater.* **2020**, *400*, 123201. [CrossRef]
32. Wu, S.; Li, H.; Li, X.; He, H.; Yang, C. Performances and mechanisms of efficient degradation of atrazine using peroxymonosulfate and ferrate as oxidants. *Chem. Eng. J.* **2018**, *353*, 533–541. [CrossRef]
33. Zhang, H.; Liu, X.; Lin, C.; Li, X.; Zhou, Z.; Fan, G.; Ma, J. Peroxymonosulfate activation by hydroxylamine-drinking water treatment residuals for the degradation of atrazine. *Chemosphere* **2019**, *224*, 689–697. [CrossRef] [PubMed]
34. Lutze, H.V.; Stephanie, B.; Insa, R.; Nils, K.; Rani, B.; Melanie, G.; Clemens, V.S.; Schmidt, T.C. Degradation of chlorotriazine pesticides by sulfate radicals and the influence of organic matter. *Environ. Sci. Technol.* **2015**, *49*, 1673. [CrossRef] [PubMed]
35. Khan, J.A.; He, X.; Shah, N.S.; Khan, H.M.; Hapeshi, E.; Fatta-Kassinos, D.; Dionysiou, D.D. Kinetic and mechanism investigation on the photochemical degradation of atrazine with activated H_2O_2, $S_2O_8^{2-}$ and HSO_5^-. *Chem. Eng. J.* **2014**, *252*, 393–403. [CrossRef]
36. Luo, C.; Ma, J.; Jiang, J.; Liu, Y.; Song, Y.; Yang, Y.; Guan, Y.; Wu, D. Simulation and comparative study on the oxidation kinetics of atrazine by UV/H_2O_2, UV/HSO_5^- and $UV/S_2O_8^{2-}$. *Water Res.* **2015**, *80*, 99–108. [CrossRef]
37. Hayon, E.; Treinin, A.; Wilf, J. Electronic spectra, photochemistry, and autoxidation mechanism of the sulfite-bisulfite-pyrosulfite systems. The SO_2^-, SO_3^-, SO_4^-, and SO_5^- radicals. *J. Am. Chem. Soc.* **1972**, *94*, 47–57. [CrossRef]
38. Anipsitakis, G.P.; Dionysiou, D.D. Radical generation by the interaction of transition metals with common oxidants. *Environ. Sci. Technol.* **2004**, *38*, 3705. [CrossRef]
39. Buxton, G.V.; Greenstock, C.L.; Helman, W.P.; Ross, A.B. Critical review of rate constants for reactions of hydrated electrons, hydrogen atoms and hydroxyl radicals (OH/ O^- in aqueous solution. *J. Phys. Chem. Ref. Data* **1988**, *17*, 513–886. [CrossRef]
40. Yang, S.; Wang, P.; Yang, X.; Shan, L.; Zhang, W.; Shao, X.; Niu, R. Degradation efficiencies of azo dye Acid Orange 7 by the interaction of heat, UV and anions with common oxidants: Persulfate, peroxymonosulfate and hydrogen peroxide. *J. Hazard. Mater.* **2010**, *179*, 552–558. [CrossRef]
41. Ghauch, A.; Tuqan, A.M. Oxidation of bisoprolol in heated persulfate/H_2O systems: Kinetics and products. *Chem. Eng. J.* **2012**, *183*, 162–171. [CrossRef]
42. Buxton, G.V.; Salmon, G.A.; Wood, N.D. A Pulse Radiolysis Study of the Chemistry of Oxysulphur Radicals in Aqueous Solution. In *Physico-Chemical Behaviour of Atmospheric Pollutants*; Springer: New York, NY, USA, 1990; pp. 245–250.
43. Huie, R.E.; Clifton, C.L. Temperature dependence of the rate constants for reactions of the sulfate radical, SO_4^-, with anions. *J. Phys. Chem.* **1990**, *94*, 8561–8567. [CrossRef]
44. Huang, J.; Mabury, S.A. A new method for measuring carbonate radical reactivity toward pesticides. *Environ. Toxicol. Chem. Int. J.* **2000**, *19*, 1501–1507. [CrossRef]
45. Jayson, G.G.; Parsons, B.J.; Swallow, A.J. Some simple, highly reactive, inorganic chlorine derivatives in aqueous solution. Their formation using pulses of radiation and their role in the mechanism of the Fricke dosimeter. *J. Chem. Soc. Faraday Trans. 1 Phys. Chem. Condens. Phases* **1973**, *69*, 1597–1607. [CrossRef]
46. Grigor'ev, A.E.; Makarov, I.E.; Pikaev, A.K. Formation of Cl_2^- in the bulk of solution during radiolysis of concentrated aqueous solutions of chlorides. *Khimiya Vysok. Ehnergij* **1987**, *21*, 123–126.
47. Das, T.N. Reactivity and role of SO_5^- •radical in aqueous medium chain oxidation of sulfite to sulfate and atmospheric sulfuric acid generation. *J. Phys. Chem. A* **2001**, *105*, 9142–9155. [CrossRef]
48. Yu, X.-Y.; Barker, J.R. Hydrogen peroxide photolysis in acidic aqueous solutions containing chloride ions. II. Quantum yield of HO•(Aq) radicals. *J. Phys. Chem. A* **2003**, *107*, 1325–1332. [CrossRef]
49. Gao, Y.Q.; Gao, N.Y.; Deng, Y.; Yin, D.Q.; Zhang, Y.S. Florfenicol Water by $UV/Na_2S_2O_8$ process. *Environ. Sci. Pollut. Res. Int.* **2015**, *22*, 8693–8701. [CrossRef]
50. Lu, Y.; Xu, W.; Nie, H.; Zhang, Y.; Deng, N.; Zhang, J. Mechanism and Kinetic Analysis of Degradation of Atrazine by US/PMS. *Int. J. Environ. Res. Public Health* **2019**, *16*, 1781. [CrossRef]
51. Wang, G.; Cheng, C.; Zhu, J.; Wang, L.; Gao, S.; Xia, X. Enhanced degradation of atrazine by nanoscale LaFe1-xCuxO3-delta perovskite activated peroxymonosulfate: Performance and mechanism. *Sci. Total Environ.* **2019**, *673*, 565–575. [CrossRef]
52. Ji, Y.; Dong, C.; Kong, D.; Lu, J.; Zhou, Q. Heat-activated peroulfate oxidation of atrazine: Implications for remediation of groundwater contaminated by herbicides. *Chem. Eng. J.* **2015**, *263*, 45–54. [CrossRef]

Article

Enhanced Catalytic Activity of a Coal-Based Powdered Activated Carbon by Thermal Treatment

Do-Gun Kim [1], Tae-Hoon Kim [2] and Seok-Oh Ko [2],*

[1]   Department of Environmental Engineering, Sunchon National University, Suncheon 57922, Korea
[2]   Department of Civil Engineering, Kyung Hee University, Yongin 17104, Korea
*   Correspondence: soko@khu.ac.kr

Abstract: Thermal treatment is simple and has high potential in activated carbon (AC) modification because its functional groups, structures, and pores can be significantly modified. However, the changes in characteristics of ACs, affecting catalytic activity, have not been investigated enough. Therefore, in this study, a coal-based powdered AC (PAC) was thermally treated, characterized, and subjected to the removal of an antibiotic (oxytetracycline, OTC). The PAC treated at 900 °C (PAC900) showed the best OTC removal compared to the PACs treated under lower temperatures via both adsorption and catalytic oxidation in the presence of peroxymonosulfate (PMS). The results of N_2 adsorption/desorption, Fourier transform infrared spectroscopy, X-ray photoelectron spectroscopy, Raman spectroscopy, X-ray diffraction, and Boehm titration showed increases in basicity, specific surface area, graphitic structures with higher crystallinity and more defects, and C=O in PAC900 compared to PAC. Stronger signals of DMPO-X and TEMP-$^1O_2$ were shown for PAC900+PMS compared to PAC+PMS in electron paramagnetic resonance spectroscopy. It is suggested that a simple thermal treatment can significantly change the characteristics of a PAC, which improves organic micropollutants removal. The changes in the properties, affecting the performance, would provide important information about the improvement of carbonaceous catalysts.

Keywords: activated carbon; thermal treatment; peroxymonosulfate; catalysis; antibiotic

Citation: Kim, D.-G.; Kim, T.-H.; Ko, S.-O. Enhanced Catalytic Activity of a Coal-Based Powdered Activated Carbon by Thermal Treatment. *Water* **2022**, *14*, 3308. https://doi.org/10.3390/w14203308

Academic Editors: Chengyun Zhou, Gassan Hodaifa Meri, Antonio Zuorro, Joaquín R. Dominguez, Juan García Rodríguez, José A. Peres, Zacharias Frontistis and Mha Albqmi

Received: 19 September 2022
Accepted: 17 October 2022
Published: 19 October 2022

Publisher's Note: MDPI stays neutral with regard to jurisdictional claims in published maps and institutional affiliations.

1. Introduction

A number of studies have recently reported the removal of organic micropollutants using persulfate (PS)-based advanced oxidation processes (AOPs) due to their great potential [1]. Peroxymonosulfate (PMS) and peroxydisulfate (PDS) can be activated via metallic catalysts, photocatalysts, and heat [2,3]. However, their application is limited because of the secondary contamination caused by metal leaching and high energy consumption [4,5]. In addition, the leached metals can form inactive species in the presence of an oxidant [6].

On the other hand, non-metallic carbonaceous materials have attracted considerable interest as heterogeneous catalysts because of their high chemical and thermal stability, large surface area, high electrical conductivity, and environmental friendliness [4,7]. Various carbonaceous catalysts have been studied for PS-based AOP, such as graphene oxide (GO), three-dimensional cubic mesoporous carbon, carbon nanotubes (CNTs), nanodiamonds, graphitic carbon nitride, and activated carbon (AC) [7,8]. Among them, AC has been widely used and studied as an efficient adsorbent of organic pollutants in the aqueous phase due to its large specific surface area, abundant functional groups, and well-developed pores [2,9,10]. Moreover, the specific functional groups and graphitic structure of AC can act as reactive sites for PS activation to accelerate electron transfer and/or generate reactive species, such as $SO_4^{\bullet-}$ [9,11,12].

Despite the potential of ACs as efficient heterogeneous catalysts, they are not as effective as GO and CNTs in activating PMS and PDS [8,13], and thus, they require modifications to enhance their catalytic activity. It has been demonstrated that the catalytic activity of carbonaceous materials largely depends on the degree of graphitization, the defects and crystallinity of the graphitic structures, the types and number of surface functional groups, and the surface area [2,7,8,13,14]. Therefore, properly changing those factors could improve the performance of ACs in the catalytic degradation of organic pollutants.

ACs can be modified chemically, physically, and via the introduction of metal (oxide). The basic groups and specific surface area, which are advantageous for the activation of PMS and/or PDS, can be increased by chemical modification of AC; however, the used chemicals may cause serious secondary pollution [15,16]. The immobilization of metal-based particles can enhance adsorption capacity [16–19] and catalytic activity [16,20]. However, their deactivation would be serious because of the decrease in surface area, leaching of the metals to increase the metals in treated water, and accumulation of the leached metals on ACs and the particles, as demonstrated for Fe_3O_4-loaded AC [20]. Recently, microwaves and plasma have been studied for AC modification. However, their feasibility is challenged by the intensive use of energy, and their objective is hetero-atom doping to improve adsorption capacity [21–23], not modifying the intrinsic characteristics of an AC.

On the other hand, not only the pore structure and surface functional groups but also graphitic structures can significantly be modified by simple thermal treatment. The advantages of simple thermal treatment under inert gas conditions have been demonstrated before in a limited number of studies. For example, it was reported that an AC thermally treated under N_2 showed superior performance in dyes reduction in the presence of sulphide compared to the ACs chemically treated with HNO_3 and O_3 [24]. In addition, it was reported that PDS activation by biochar was greatly improved [12], electron transfer of graphite was accelerated [25], the adsorption capacity of phenolic compounds was increased [26], and the phosphorus adsorption onto an AC was enhanced [27] via simple thermal treatment under inert gas. Those improvements are attributed to the changes in the functional groups and structures of the carbonaceous materials. Despite encouraging reports regarding the thermal modification of ACs, it is hard to find the details of the changes in the characteristics, which can be assigned to the improved catalytic activity [23,24,28,29].

Therefore, in this study, a coal-based powdered AC (PAC) was thermally treated and characterized regarding its pore structure, crystallinity, and surface functional groups. The PACs were subjected to the adsorption and catalytic degradation of an organic micropollutant, i.e., oxytetracycline (OTC). OTC was selected as a representative micropollutant. OTC is one of the most widely used antibiotics for disease prevention and growth promotion in livestock [30]. Most OTC is excreted non-metabolically, and it is frequently detected in surface water, sewage, groundwater, drinking water, seawater, and sediments worldwide [31]. In natural water environments, OTC exposure can cause eco-toxicity and the development of antibiotic resistance [30,31]. The efficient removal of OTC is of great concern because OTC is poorly biodegradable, and the ozonation and photocatalytic degradation of OTC is costly and produces toxic byproducts [32,33].

2. Results

2.1. OTC Removal by PACs Treated at Different Temperatures

OTC removal was first investigated with the PACs treated at different temperatures (Figures 1 and S1). It was shown that the pseudo second-order rate constant (k_2) of OTC removal increased with increasing treatment temperature, regardless of the presence of an oxidant, confirming the advantage of thermal treatment. The removal efficiency of OTC after 60 min by the untreated PAC, PAC treated at 500 °C (PAC500), PAC treated at 700 °C (PAC700), and PAC treated at 900 °C (PAC900) was 16.8%, 20.96%, 27.46%, and 41.24%, respectively, in the absence of an oxidant, i.e., via adsorption. It was 48.86%, 59.20%, 60.7%, and 72.01%, respectively, when PDS was introduced, while it was 68.6%, 70.2%, 70.7%, and 80.2%, respectively, in the presence of PMS.

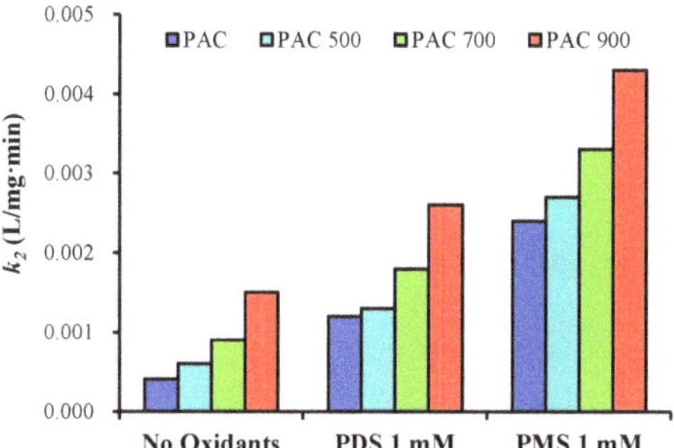

Figure 1. The pseudo second-order reaction constants (k_2) of OTC removal in the absence of an oxidant and in the presence of 1 mM PDS or PMS (PACs 0.1 g/L, OTC 25 mg/L, initial pH 6.0, 20 ± 2 °C).

It is also demonstrated that OTC removal was faster with PMS than with PDS when the same PAC was used. The k_2 of OTC removal with 1 mM PMS was 2.0, 2.1, 1.8, and 1.7 times that with 1 mM PDS when PAC, PAC500, PAC700, and PAC900 were used, respectively. It was reported that the radical generation from PMS is more difficult than from PDS because of its shorter O-O bond distance and larger bond energy [34]. Therefore, the superiority of PMS to PDS is attributable to the self-activation of PMS and the involvement of non-radical pathways [34,35]. Because PAC900 showed the best performance, the characteristics and OTC removal of PAC and PAC900 were investigated in detail, as follows.

2.2. Characterization of PAC and PAC900

The N_2 adsorption/desorption isotherm of PAC and PAC900 commonly showed Type IV isotherms with H4 hysteresis loop (Figure 2A), suggesting that the PACs are micro- and mesoporous [36,37], which was evidenced by the average pore size (d_p) (Table 1). It is suggested that micropores are dominant with a small number of mesopores for PAC and PAC900, considering the sharp increase in the adsorption amount at the P/P_0 of less than 0.1 and the faint hysteresis loop at the P/P_0 of 0.2–0.8 [38]. It was evidenced by the mesopore size distribution obtained via the Barrett–Joyner–Halenda (BJH) method (Figure 2B).

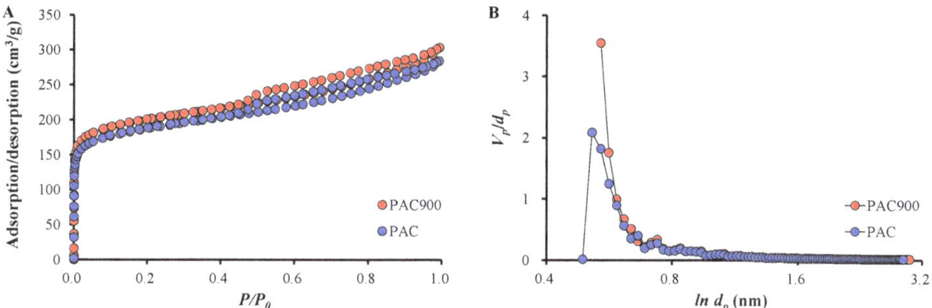

Figure 2. (**A**) N_2 adsorption–desorption isotherms and (**B**) mesopore size distribution by Barrett–Joyner–Halenda (BJH) method for PAC and PAC900.

Table 1. Specific surface areas (S_{BET}), total pore volume (V_p), and average pore size (d_p) of PAC and PAC900.

	S_{BET} (m²/g)	V_p (m³/g)	d_p (nm)
PAC	685.76	0.4375	2.552
PAC900	734.53	0.4665	2.541

Thermal treatment at 900 °C significantly increased the specific surface area (S_{BET}) and total pore volume (V_p), while the d_p decreased slightly (Table 1). However, N$_2$ adsorption–desorption isotherms and d_p did not differ significantly between PAC and PAC900, indicating that the pore structure was not notably changed by the thermal treatment.

The basicity was 872 and 1004 µmol-OH$^-$/g, while the acidity was 147 and 42 µmol-H$^+$/g for PAC and PAC900, respectively. The decrease in the acidity is attributable to a decrease in acidic functional groups, such as carboxylic, anhydrides, and lactones, which thermally decompose at 373–900 °K [39]. As a result, the fraction of basic groups, which are relatively thermally resistant, increased.

The changes in the functional groups were further investigated using Fourier transform infrared (FTIR) spectroscopy (Figure 3A). The bands were found in the FTIR spectrum of PAC at 3500–4000, 2950, 2167, 2010, 1735, 1620, 1258, and 1175 cm^{-1}, which are assigned to the stretching of O–H groups, aliphatic CH$_2$ asymmetric stretch, C≡C vibrations in alkyne groups, C=C vibrations in alkyne groups, ketonic and carboxylic C=O stretching, C=C of the aromatic ring, highly conjugated or graphitic C=C, C–O–C stretching vibrations, and phenolic C-OH, respectively [40–42]. After thermal treatment, i.e., PAC900, the intensity of the bands of OH decreased, while the bands of aliphatic CH$_2$, ketonic and carboxylic C=O, C–O–C, and phenolic C-OH disappeared. It is in agreement with the reduction in acidity. Instead, the intensity was increased for the bands of C≡C of alkyne, C=C of alkyne, and graphitic C=C, indicating an increase in graphitic structures [12,41].

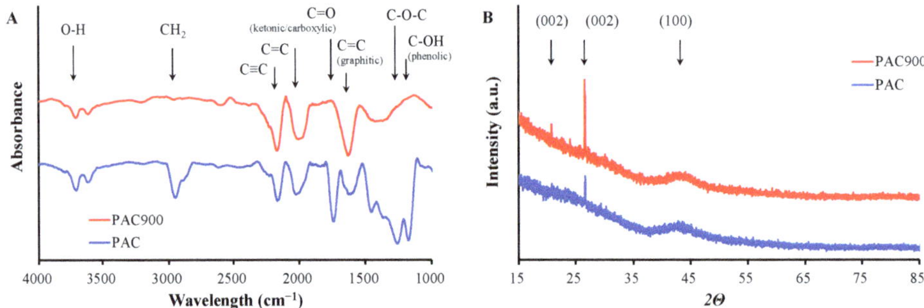

Figure 3. (A) FTIR spectra and (B) XRD patterns of PAC and PAC900.

The X-ray diffraction (XRD) patterns of both PAC and PAC900 showed three (3) common peaks (Figure 3B). The peaks at $2\theta \approx 21°$ and 26.7° represent the (002) plane of the interlayer spacing of the graphitic structures, while the broad peak at $2\theta \approx 42.6°$ is assigned to the (100) plane of the graphitic structures. The peaks of the (002) plane were significantly more intense for PAC900 than PAC, suggesting an increase in crystallinity, ordered graphite structures, and sp^2-C [43], by the thermal treatment.

The Raman spectra showed common bands at 1210, 1344, 1540–1550, 1588, 1610, 2678–2679, and 2604–2609 cm^{-1}, which were assigned to the D4, D, D3, G, D2, D+G, and 2D bands, respectively (Figure 4, Table 2). The D4, D, D3, G, and D2 bands are associated with a disordered graphitic structure and ionic impurities, disordered graphitic lattice by sp^3 hybridization including graphene layer edges, amorphous carbon, ideal graphitic structures with sp^2 C, and disordered graphitic structures in surface graphene layers, respectively [44]. The intensity of D4, D3, and D2 bands was lower for PAC900 than for PAC, indicating a decrease in disordered and amorphous graphitic structures and an increase in crystallinity by thermal treatment. In addition, the intensity of the D, G, D2, and 2D bands changed, resulting in increases in the intensity ratios of the D band to G band (I_D/I_G), of D band to D2 band (I_D/I_{D2}), and of 2D band to G band (I_{2D}/I_G). They suggest an increase in the structural defects in the graphitic structures [45], of the defects associated with charged impurities rather than the hopping defects formed by the deformation of carbon bonds [46], and of single graphene layers [44], respectively.

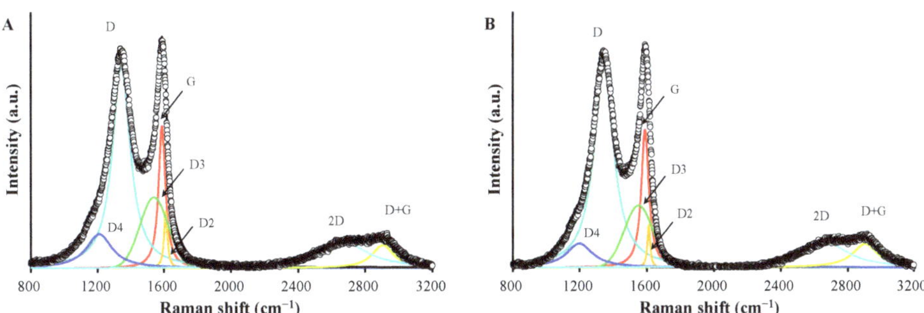

Figure 4. Raman spectra of (**A**) PAC and (**B**) PAC900.

Table 2. The results of Raman spectroscopy.

		D4	D	D3	G	D2	D+G	2D	I_D/I_G	I_D/I_{D2}	I_{2D}/I_G
PAC	Center (cm^{-1})	1210	1344	1540	1588	1610	2679	2906	3.69	19.23	0.51
	Fraction (%)	10.2	41.6	15.3	11.3	2.2	13.9	5.7			
PAC900	Center (cm^{-1})	1200	1344	1550	1588	1610	2678	2904	3.80	33.13	0.58
	Fraction (%)	7.4	46.1	13.8	12.1	1.4	12.2	7.0			

The X-ray photoelectron spectroscopy (XPS) survey spectra of PAC and PAC900 show that the C/O ratio was increased from 7.5 for PAC to 9.2 for PAC900 (Figure S2). The C1s and O1s spectra are presented in Figure 5, and the results are provided in Table 3. The C1s XPS spectrum of PAC shows peaks at 284.3, 285.0, 287.1, and 290.3 eV, which are assigned to graphitic C-C/C=C, C-O/C-N, C=O of quinone and pyrone, and O-C=O of carboxyl and ester, respectively [41]. After thermal treatment, i.e., PAC900, the fraction of the peak of graphitic C-C/C=C (284.6 eV) increased, while that of O-C=O (290.4 eV) decreased. The O1s spectra show common peaks of C=O of quinone, C-OH, and C-O at 530.3, 531.9, and 533.4–533.8 eV, respectively, for both PAC and PAC900 [47]. However, the fraction of C=O increased significantly, while that of C-OH decreased notably for PAC900, as suggested by the decrease in acidic groups and the increase in the basic groups in the results of Bohem titration.

Table 3. The results of XPS.

		C1s				O1s		
		Graphitic C-C/C=C	C-O/C-N	C=O	O-C=O	C=O	C-OH	C-O
PAC	Position (eV)	284.3	285.0	287.1	290.3	530.3	531.9	533.4
	Fraction (%)	56	24	9	11	46	49	5
PAC900	Position (eV)	284.6	285.4	287.6	290.4	530.3	531.9	533.8
	Fraction (%)	58	24	9	9	53	38	9

The scanning electron microscopy images of PAC and PAC900 are provided in Figure S3. The surface of PA900 was smoother than that of PAC, indicating more meso- and micropores in PAC900, as suggested by the results of N_2 adsorption/desorption (Table 1 and Figure 2). Meanwhile, the pH$_{pzc}$ of PAC and PAC900 was 7.46 and 9.97, respectively (Figure S4), indicating that PAC900 is more positively charged than PAC at the same pH [48]. This would affect the adsorption of OTC onto the PACs via electrostatic attraction [49].

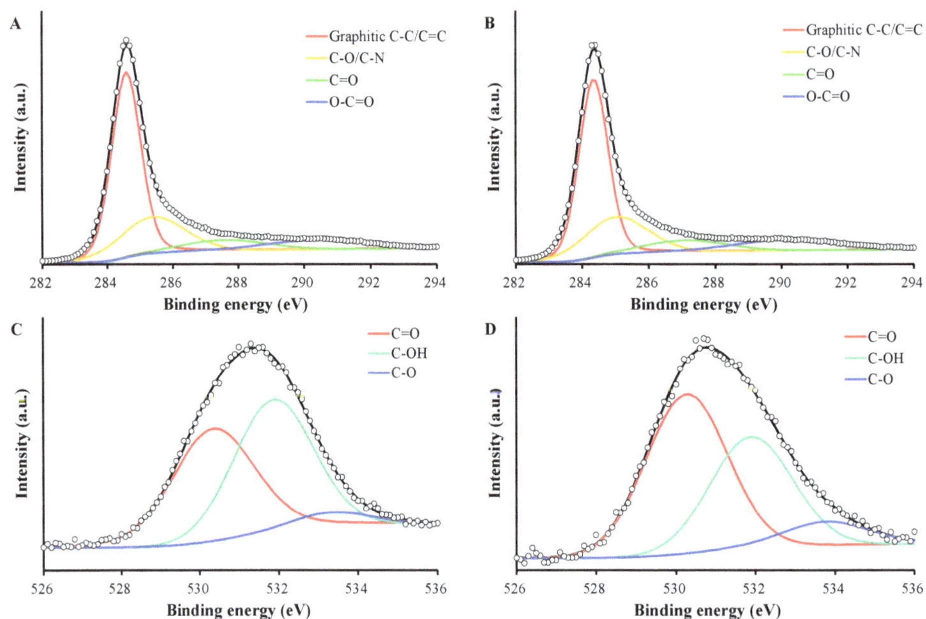

Figure 5. C1s and O1s XPS spectra of (**A**,**C**) PAC and (**B**,**D**) PAC900.

2.3. OTC Removal by PAC and PAC900 under Different Conditions

OTC removal by PAC900 was faster than by PAC under all conditions, verifying the benefits of thermal treatment (Figures 6, S5 and S6). Figure 6A shows that the OTC removal was accelerated as the dose of PAC or PAC900 was increased, regardless of the presence of PMS. It should be noted that an injection of PMS significantly improved OTC removal when the PAC900 dose was low, but the enhancement became less significant as the PAC900 dose was increased. The k_2 of PAC900+PMS was 2.8 times that of PAC (0.036 L/mg·min) when the PAC900 dose was 0.1 g/L. However, it was 1.07 times that of PAC (0.308 L/mg·min) when the PAC900 dose was 0.5 g/L. This indicates that adsorption becomes more responsible for the overall OTC removal than catalytic oxidation as the PAC900 dose was increased. However, the k_2 was increased with an increasing dose of PMS (Figure 6B), suggesting that PMS plays a significant role, i.e., catalytic oxidation, along with adsorption. This evidences that PMS was activated on the surfaces of PAC and that that was enhanced on PAC900. It was reported that surface-bound reactive species, i.e., PMS*, are formed via the electron transfer from electron donors to the PMS adsorbed on carbonaceous materials, and the process is mediated by sp^2–hybridized carbon networks [50]. This leads to the generation of $SO_4^{\bullet-}$ and/or $SO_5^{\bullet-}$, which are reactive radicals themselves and precursors of other reactive species such as $O_2^{\bullet-}$ and $^1O_2$ [12,51].

2.4. Radical Identification and the Possible Mechanisms of OTC Removal by Thermally Treated AC

The electron paramagnetic resonance (EPR) spectra using 5,5-dimethyl-1-pyrroline n-oxide (DMPO) showed signals of 1:2:1:2:1:2:1 for both PAC+PMS and PAC900+PMS, representing the nitroxide of 5,5-dimethylpyrroline-(2)-oxyl-(1) (DMPO-X) produced by DMPO oxidation [52] (Figure 7A). The DMPO-X is generated by the oxidation of DMPO by $SO_4^{\bullet-}$ and/or $SO_5^{\bullet-}$ [52,53]. On the other hand, no DMPO-OH$^{\bullet}$ signals were detected, suggesting that $SO_4^{\bullet-}$ and/or $SO_5^{\bullet-}$ is dominantly responsible for the degradation of OTC in PAC+PMS and PAC900+PMS. Figure 7B shows the signals of TEMP-$^1O_2$, i.e., 1:1:1, for both PAC+PMS and PAC900+PMS.

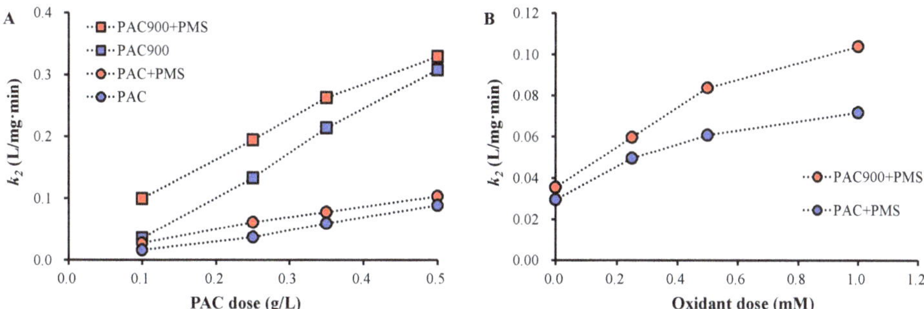

Figure 6. The pseudo second-order rate constants of OTC removal (k_2) at different dose of (**A**) PAC or PAC900 (PMS 1 mM) and (**B**) PMS (PAC or PAC900 0.1 g/L) (OTC 25 mg/L, initial pH 6.0, 20 ± 2 °C).

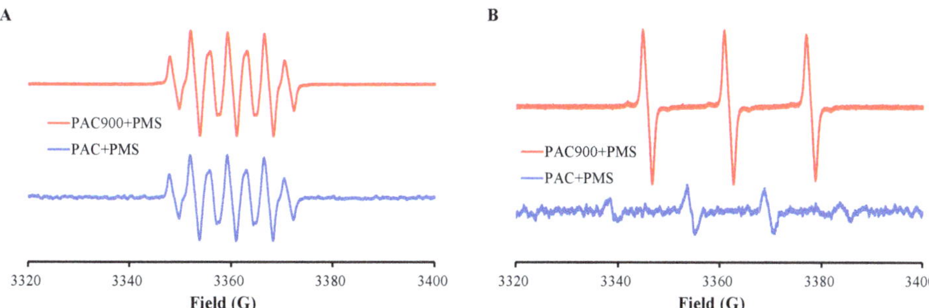

Figure 7. EPR spectra of PAC and PAC900 using (**A**) DMPO and (**B**) TMPO (PAC 0.5 g/L, PMS 1 mM, DMPO, or TMPO 10 mM).

The signals of DMPO-X and $^1O_2$ were stronger for PAC900+PMS than PAC+PMS, suggesting that the faster OTC removal in PAC800+PMS than in PAC+PMS is attributable to the enhanced generation of oxidative species, i.e., $SO_4^{\bullet-}$, $SO_5^{\bullet-}$, and $^1O_2$. In particular, the signals of $^1O_2$ were much stronger in PAC800+PMS than in PAC+PMS, indicating a significant contribution of $^1O_2$ in OTC removal in PAC800+PMS because $^1O_2$ is a strong oxidant with a standard reduction potential of 2.2 V [51]. $^1O_2$ can be generated by the recombination of $SO_5^{\bullet-}$ (Equation (1)) [54], from $SO_5^{\bullet-}$ and H_2O (Equation (2)) [55,56], and from the self-decomposition of PMS (Equation (3)) [54,55]. Therefore, the enhanced $^1O_2$ generation can be assigned to the improved PMS activation by PAC900 than by PAC.

$$SO_5^{\bullet-} + SO_5^{\bullet-} \rightarrow S_2O_8^{2-} + {}^1O_2, \tag{1}$$

$$2SO_5^{\bullet-} + H_2O \rightarrow 2HSO_4^- + 1.5\,{}^1O_2, \tag{2}$$

$$HSO_5^- + SO_5^{2-} \rightarrow HSO_4^- + SO_4^{2-} + {}^1O_2. \tag{3}$$

Enhanced PMS activation by thermal treatment is attributable to the changes in the properties of the PAC. The increase in Lewis basic sites, such as C=O, can improve PMS activation because the C=O in quinone or pyrone donates electrons to break O–O bond to generate radicals [12]. The increase in the defects in graphitic structures, as indicated by I_D/I_G (Table 3), would lead to the electron transfer to PMS due to a denser electron population [12,25]. The π electrons can form dangling bond states at the edges and defects because of the missing atoms in the C lattice so the electron transfer from the graphite structures to the PMS is accelerated [57]. Most of all, it is thought that the superior OTC removal by PAC900 is attributable to the increased graphitic structures and their crystallinity, as suggested by Table 3 and Figures 3–5. It was reported that the well-crystallized graphitic structures carry abundant π electrons, which enhance the electron transfer to PMS and reduce it to $SO_4^{\bullet-}$ and/or $SO_4^{\bullet-}$ (Equations (4) and (5)) [12,55,57,58].

$$HSO_5^- + C\text{-}\pi \rightarrow SO_4^{\bullet-} + C\text{-}\pi^+ + OH^-, \tag{4}$$

$$HSO_5^- + C\text{-}\pi^+ \rightarrow SO_5^{\bullet-} + C\text{-}\pi + H^+. \tag{5}$$

It should also be noted that the adsorption of the organic compounds with aromatic rings onto graphitic structures can be improved via enhanced π-π electron donor–acceptor (EDA) interactions because graphitic structures are effective π-electron donors [59,60]. Therefore, the adsorption of OTC onto more developed graphitic structures, i.e., PAC900, would be enhanced, as demonstrated in this study (Figures 1 and 6).

3. Materials and Methods

3.1. Materials

A coal-based PAC (NORIT® PAC 20BF) was purchased from Cabot (Alpharetta, GA, USA). The PAC of 38–72 μm was separated by sieving, washed several times using ultrapure water, and dried at 150 °C for 12 h. The PAC was subjected to thermal treatment under Ar gas at 500, 700, and 900 °C (5 °C/min) for 2 h using a tube furnace (OTF-1200X-UL, MTI Corp., Richmond, CA, USA), which is denoted as PAC 500, PAC700, and PAC900, respectively.

Potassium peroxymonosulfate ($KHSO_5 \cdot 0.5KHSO_4 \cdot 0.5K_2SO_4$, PMS), sodium peroxydisulfate ($Na_2S_2O_8$, PDS), hydrochloric acid (HCl, 37%), sodium hydroxide (NaOH, $\geq$97%), 5,5-dimethyl-1-pyrroline n-oxide (DMPO, $\geq$97%), and 2,2,6,6-tetramethyl-4-piperidinol (TEMP) were procured from Merck KGaA (Darmstadt, Germany). All chemicals were analytical grade and used as received. Ultrapure water was supplied by Aquapuri series 5 (Younglin, Anyang, Korea).

3.2. Batch Experiments

Batch experiments were performed using 250 mL round flasks at room temperature. The PACs and ultrapure water were mixed, and the pH was adjusted to 6.0 ± 0.1 using 0.5 N HCl and 0.5 N NaOH. The standard solutions of OTC (pH 6.0) and oxidant (PDS or PMS) were added to start the reaction. The concentrations of PAC, OTC, and PDS or PMS were 0.1 g/L, 25 mg/L, and 1 mM, respectively. Aliquots were taken at predetermined times and filtered through a 0.45 μm polyvinylidene fluoride (PVDF) membrane. The concentration of OTC in the filtrates was analyzed using a high-performance liquid chromatography (HPLC) system (YL9100 Plus, Youlngin, Korea). A C18 column (Eclipse Plus, 4.6 × 250 mm, 5 μm, Agilent, Santa Clara, CA, USA) was used, and the mobile phase consisted of 0.01 M oxalic acid, acetonitrile, and methanol (70:20:10, $v/v/v$). The flow rate, the column temperature, the detection wavelength, and the injection volume were 1 mL/min, 30 °C, 358 nm, and 25 μL, respectively.

3.3. Characterization

The pore structure and specific surface area of the PACs were determined using BELSORP-max (MicrotracBEL, Osaka, Japan). The total surface area and pore volume were obtained using Brunauer–Emmett–Teller (BET) method, while the mesopore size distribution was calculated using the Barrett–Joyner–Halenda (BJH) method. The acidity and basicity were measured by standardized Boehm titration [61]. FTIR spectra were recorded at 4000–400 cm^{-1} with the pellets prepared from the mixture of 0.2 mg PAC and 200 mg of KBr, using an FTIR spectrophotometer (Spectrum One System, Perkin-Elmer, Waltham, MA, USA). XRD patterns were analyzed using a DB Advance X-ray diffractometer (Bruker, Bremen, Germany), in 2θ between $3°$ and $89.14°$. Raman spectra were obtained using a Renishaw in Via Raman microscope (Renishaw Inc., West Dundee, IL, USA) with an excitation wavelength of 514 nm. XPS was carried out using a K-Alpha XPS instrument (Thermo Electron, Waltham, MA, USA) with a monochromatic Al α-Alpha radiation source, and high-resolution spectra of C 1s and O1s were obtained in 0.1 eV steps. The PAC and

PAC900 were sputtered with Pt, and the microscopic images were taken using a field emission SEM (FE-SEM; Carl Zeiss, Oberkochen, Germany).

Radicals were identified using an EPR spectrometer (JES-X320, Jeol, Akishima, Japan) at a center field, a power, and a modulation frequency of 3389.9 G, 2.5 mW, and 100 kHz, respectively. DMPO (10 mM) was used for trapping sulfate radicals ($SO_4^{\bullet-}$, $SO_5^{\bullet-}$), while TEMPO (10 mM) was used as the singlet oxygen ($^1O_2$) trapper. The pH drift method was used to determine the pH of the point of zero charge (pH_{pzc}) of the PACs [48].

4. Conclusions

A coal-based PAC was thermally treated to improve the removal of OTC in this study. It was shown that OTC removal was improved with increasing treatment temperature from 500 to 900 °C, regardless of the presence of PMS. Therefore, the PAC and PAC900 were characterized and investigated for OTC removal in detail.

N_2 adsorption/desorption isotherms showed that specific surface area and total pore volume were increased by the thermal treatment, which did not cause a notable change in pore structures after the thermal treatment at 900 °C. The results of FTIR spectra, XRD patterns, Raman spectra, Boehm titration, and XPS spectra suggest that basicity, C/O ratio, the fraction of graphitic structures, crystallinity, and defects of the graphitic structures, and the fraction of C=O were all higher for PAC900 than for PAC. Instead, carboxylic groups, phenolic groups, and disordered and amorphous graphitic structures were decreased after thermal treatment.

PAC900 showed significantly better OTC removal in the absence, i.e., adsorption, and in the presence of PMS under various doses of PACs and PMS. The EPR spectra of PAC900+PMS demonstrated stronger signals of DMPO-X and much stronger signals of TEMP-$^1O_2$ than PAC+PMS, indicating enhanced PMS activation and $^1O_2$ generation by PAC900.

The results suggest that the improved OTC removal by PAC900 is attributable to the enhanced electron transfer to PMS via increases in C=O, more crystalline graphitic structures, and defects. In addition, the improved π-π EDA interactions via the increase in the graphitic structures also contributed to the enhanced OTC adsorption onto PAC900.

The results of this study demonstrate that both the adsorption and the activation of PMS by an AC can significantly be improved by simple thermal treatment through the modification of its properties. These also would provide valuable information about the design of carbonaceous catalysts with a better performance in micropollutants removal.

Supplementary Materials: The following supporting information can be downloaded at: https://www.mdpi.com/article/10.3390/w14203308/s1, Figure S1: OTC removal using various PACs (A) in the absence of an oxidant and in the presence of (B) PDA and (C) PMS (PACs 0.1 g/L, PDS or PMS 1 mM, OTC 25 mg/L, initial pH 6.0, 20 ± 2 °C); Figure S2: XPS survey spectra of PAC and PAC900; Figure S3: SEM images of (A) PAC and (B) PAC900; Figure S4: Point of zero charge (pH_{pzc}) of PAC and PAC900; Figure S5: OTC removal at different PACs dose in (A) PAC, (B) PAC+PMS, (C) PAC900, and (D) PAC900+PMS (PMS 1 mM, OTC 25 mg/L, initial pH 6.0); Figure S6: OTC removal at different PMS dose in (A) PAC +PMS and (B) PAC900+PMS (PACs 0.1 g/L, OTC 25 mg/L, initial pH 6.0).

Author Contributions: Conceptualization and methodology, T.-H.K.; validation and formal analysis, D.-G.K. and S.-O.K.; investigation, resources, data curation, and writing—original draft preparation, T.-H.K.; writing—review and editing, D.-G.K. and S.-O.K.; visualization and supervision, S.-O.K.; project administration and funding acquisition, S.-O.K.; All authors have read and agreed to the published version of the manuscript.

Funding: This work was supported by a National Research Foundation of Korea (NRF) grant funded by the Korean government (MSIT) (No. 2022R1A2B5B02001584).

Data Availability Statement: The data presented in this study are available upon request from the corresponding author.

Conflicts of Interest: The authors declare that they have no known competing financial interests or personal relationships that could have appeared to influence the work reported in this paper.

References

1. Gogate, P.R.; Pandit, A.B. A Review of Imperative Technologies for Wastewater Treatment I: Oxidation Technologies at Ambient Conditions. *Adv. Environ. Res.* **2004**, *8*, 501–551. [CrossRef]
2. Zhao, Q.; Mao, Q.; Zhou, Y.; Wei, J.; Liu, X.; Yang, J.; Luo, L.; Zhang, J.; Chen, H.; Chen, H.; et al. Metal-Free Carbon Materials-Catalyzed Sulfate Radical-Based Advanced Oxidation Processes: A Review on Heterogeneous Catalysts and Applications. *Chemosphere* **2017**, *189*, 224–238. [CrossRef] [PubMed]
3. Guo, P.; Tang, L.; Tang, J.; Zeng, G.; Huang, B.; Dong, H.; Zhang, Y.; Zhou, Y.; Deng, Y.; Ma, L.; et al. Catalytic Reduction–Adsorption for Removal of p-Nitrophenol and Its Conversion p-Aminophenol from Water by Gold Nanoparticles Supported on Oxidized Mesoporous Carbon. *J. Colloid Interface Sci.* **2016**, *469*, 78–85. [CrossRef] [PubMed]
4. Liu, L.; Zhu, Y.-P.; Su, M.; Yuan, Z.-Y. Metal-Free Carbonaceous Materials as Promising Heterogeneous Catalysts. *ChemCatChem* **2015**, *7*, 2765–2787. [CrossRef]
5. Sun, Y.; Cho, D.-W.; Graham, N.J.D.; Hou, D.; Yip, A.C.K.; Khan, E.; Song, H.; Li, Y.; Tsang, D.C.W. Degradation of Antibiotics by Modified Vacuum-UV Based Processes: Mechanistic Consequences of H_2O_2 and $K_2S_2O_8$ in the Presence of Halide Ions. *Sci. Total Environ.* **2019**, *664*, 312–321. [CrossRef]
6. Zhang, P.; Tan, X.; Liu, S.; Liu, Y.; Zeng, G.; Ye, S.; Yin, Z.; Hu, X.; Liu, N. Catalytic Degradation of Estrogen by Persulfate Activated with Iron-Doped Graphitic Biochar: Process Variables Effects and Matrix Effects. *Chem. Eng. J.* **2019**, *378*, 122141. [CrossRef]
7. Zhu, K.; Shen, Y.; Hou, J.; Gao, J.; He, D.; Huang, J.; He, H.; Lei, L.; Chen, W. One-Step Synthesis of Nitrogen and Sulfur Co-Doped Mesoporous Graphite-like Carbon Nanosheets as a Bifunctional Material for Tetracycline Removal via Adsorption and Catalytic Degradation Processes: Performance and Mechanism. *Chem. Eng. J.* **2021**, *412*, 128521. [CrossRef]
8. Pham, V.L.; Kim, D.-G.; Ko, S.-O. Advanced Oxidative Degradation of Acetaminophen by Carbon Catalysts: Radical vs Non-Radical Pathways. *Environ. Res.* **2020**, *188*, 109767. [CrossRef]
9. Saputra, E.; Muhammad, S.; Sun, H.; Wang, S. Activated Carbons as Green and Effective Catalysts for Generation of Reactive Radicals in Degradation of Aqueous Phenol. *RSC Adv.* **2013**, *3*, 21905. [CrossRef]
10. Ghanbari, F.; Moradi, M. Application of Peroxymonosulfate and Its Activation Methods for Degradation of Environmental Organic Pollutants: Review. *Chem. Eng. J.* **2017**, *310*, 41–62. [CrossRef]
11. Yang, S.; Yang, X.; Shao, X.; Niu, R.; Wang, L. Activated Carbon Catalyzed Persulfate Oxidation of Azo Dye Acid Orange 7 at Ambient Temperature. *J. Hazard. Mater.* **2011**, *186*, 659–666. [CrossRef] [PubMed]
12. Kim, D.-G.; Ko, S.-O. Effects of Thermal Modification of a Biochar on Persulfate Activation and Mechanisms of Catalytic Degradation of a Pharmaceutical. *Chem. Eng. J.* **2020**, *399*, 125377. [CrossRef]
13. Olmez-Hanci, T.; Arslan-laton, I.; Gurmen, S.; Gafarli, I.; Khoei, S.; Safaltin, S.; Ozcelik, D.Y. Oxidative degradation of Bisphenol A by carbocatalytic activation of persulfate and peroxymonosulfate with reduced graphene oxide. *J. Hazard. Mater.* **2018**, *360*, 141–149. [CrossRef]
14. Chen, X.; Oh, W.-D.; Lim, T.-T. Graphene and CNTs-Based Carbocatalysts in Persulfates Activation: Material Design and Catalytic Mechanisms. *Chem. Eng. J.* **2018**, *354*, 941–976. [CrossRef]
15. Sultana, M.; Rownok, M.H.; Sabrin, M.; Rahaman, M.H.; NurAlam, S.M. A review on experimental chemically modified activated carbon to enhance dye and heavy metals adsorption. *Clean. Eng. Technol.* **2022**, *6*, 100382. [CrossRef]
16. Li, H.; He, X.; Wu, T.; Jin, B.; Yang, L.; Qiu, J. Synthesis, modification strategies and applications of coal-based carbon materials. *Fuel Process. Technol.* **2022**, *230*, 107203. [CrossRef]
17. Chowdhury, A.; Kumari, S.; Khan, A.A.; Chandra, M.R.; Hussain, S. Activated carbon loaded with Ni-Co-S nanoparticle for superior adsorption capacity of antibiotics and dye from wastewater: Kinetics and isotherms. *Colloids Surf. A* **2021**, *611*, 125868. [CrossRef]
18. Poudel, M.B.; Shin, M.; Kim, H.J. Interface engineering of MIL-88 derived MnFe-LDH and $MnFe_2O_3$ on three-dimensional carbon nanofibers for the efficient adsorption of Cr(VI), Pb(II), and As(III) ions. *Sep. Purif. Technol.* **2022**, *287*, 120463. [CrossRef]
19. Poudel, M.B.; Awasthi, G.P.; Kim, H.J. Novel insight into the adsorption of Cr(VI) and Pb(II) ions by MOF derived Co-Al layered double hydroxide @hematite nanorods on 3D porous carbon nanofiber network. *Chem. Eng. J.* **2021**, *417*, 129312. [CrossRef]
20. Zhao, J.; Sun, Y.; Zhang, Y.; Zhang, B.-T.; Yin, M.; Chen, L. Heterogeneous activation of persulfate by activated carbon supported iron for efficient amoxicillin degradation. *Environ. Technol. Innov.* **2021**, *21*, 101259. [CrossRef]
21. Wang, J.; Chen, Z.; Wen, H.; Cai, Z.; He, C.; Wang, Z.; Yan, W. Microwave assisted modification of activated carbons by organic acid ammoniums activation for enhanced adsorption of acid red 18. *Powder Technol.* **2018**, *323*, 230–237. [CrossRef]
22. Ji, Q.; Luo, G.; Shi, M.; Zou, R.; Fang, C.; Xu, Y.; Li, X.; Yao, H. Acceleration of the reaction of H_2S and SO_2 by non-thermal plasma to improve the mercury adsorption performance of activated carbon. *Chem. Eng. J.* **2021**, *423*, 130144. [CrossRef]
23. Chen, K.; He, Z.-J.; Liu, Z.H.; Ragauskas, A.J.; Li, B.-Z.; Yuan, Y.J. Emerging Modification Technologies of Lignin-based Activated Carbon toward Advanced Applications. *ChemSusChem* **2022**, e202201284. [CrossRef] [PubMed]
24. Pereira, L.; Pereira, R.; Pereira, M.F.R.; van der Zee, F.P.; Cervantes, F.J.; Alves, M.M. Thermal modification of activated carbon surface chemistry improves its capacity as redox mediator for azo dye reduction. *J. Hazard. Mater.* **2010**, *83*, 931–939. [CrossRef] [PubMed]
25. Nwamba, O.C.; Echeverria, E.; McIlroy, D.N.; Austin, A.; Shreeve, J.M.; Aston, D.E. Thermal Modification of Graphite for Fast Electron Transport and Increased Capacitance. *ACS Appl. Nano Mater.* **2019**, *2*, 228–240. [CrossRef]

26. Kuśmierek, K.; Świątkowski, A.; Skrzypczyńska, K.; Błażewicz, S.; Hryniewicz, J. The Effects of the Thermal Treatment of Activated Carbon on the Phenols Adsorption. *Korean J. Chem. Eng.* **2017**, *34*, 1081–1090. [CrossRef]
27. Miyazato, T.; Nuryono, N.; Kobune, M.; Rusdiarso, B.; Otomo, R.; Kamiya, Y. Phosphate Recovery from an Aqueous Solution through Adsorption-Desorption Cycle over Thermally Treated Activated Carbon. *J. Water Process Eng.* **2020**, *36*, 101302. [CrossRef]
28. Azam, K.; Shezad, N.; Shafiq, I.; Akhter, P.; Akhtar, F.; Jamil, F.; Shafique, S.; Park, Y.-K.; Hussain, M. A review on activated carbon modifications for the treatment of wastewater containing anionic dyes. *Chemosphere* **2022**, *306*, 135566. [CrossRef]
29. Macías-García, A.; Corzo, M.G.; Domínguez, M.A.; Franco, M.A.; Naharro, J.M. Study of the adsorption and electroadsorption process of Cu (II) ions within thermally and chemically modified activated carbon. *J. Hazard. Mater.* **2017**, *328*, 46–55. [CrossRef]
30. Menz, J.; Olsson, O.; Kümmerer, K. Antibiotic Residues in Livestock Manure: Does the EU Risk Assessment Sufficiently Protect against Microbial Toxicity and Selection of Resistant Bacteria in the Environment? *J. Hazard. Mater.* **2019**, *379*, 120807. [CrossRef]
31. Xu, L.; Zhang, H.; Xiong, P.; Zhu, Q.; Liao, C.; Jiang, G. Occurrence, Fate, and Risk Assessment of Typical Tetracycline Antibiotics in the Aquatic Environment: A Review. *Sci. Total Environ.* **2021**, *753*, 141975. [CrossRef] [PubMed]
32. Watkinson, A.J.; Murby, E.J.; Costanzo, S.D. Removal of Antibiotics in Conventional and Advanced Wastewater Treatment: Implications for Environmental Discharge and Wastewater Recycling. *Water Res.* **2007**, *41*, 4164–4176. [CrossRef] [PubMed]
33. Cheng, D.; Ngo, H.H.; Guo, W.; Chang, S.W.; Nguyen, D.D.; Liu, Y.; Wei, Q.; Wei, D. A Critical Review on Antibiotics and Hormones in Swine Wastewater: Water Pollution Problems and Control Approaches. *J. Hazard. Mater.* **2020**, *387*, 121682. [CrossRef]
34. Wang, J.; Wang, S. Activation of Persulfate (PS) and Peroxymonosulfate (PMS) and Application for the Degradation of Emerging Contaminants. *Chem. Eng. J.* **2018**, *334*, 1502–1517. [CrossRef]
35. Ren, W.; Nie, G.; Zhou, P.; Zhang, H.; Duan, X.; Wang, S. The Intrinsic Nature of Persulfate Activation and N-Doping in Carbocatalysis. *Environ. Sci. Technol.* **2020**, *54*, 6438–6447. [CrossRef]
36. Thommes, M.; Kaneko, K.; Neimark, A.V.; Olivier, J.P.; Rodriguez-Reinoso, F.; Rouquerol, J.; Sing, K.S.W. Physisorption of Gases, with Special Reference to the Evaluation of Surface Area and Pore Size Distribution (IUPAC Technical Report). *Pure Appl. Chem.* **2015**, *87*, 1051–1069. [CrossRef]
37. Wang, G.; Li, N.; Xing, X.; Sun, Y.; Zhang, Z.; Hao, Z. Gaseous Adsorption of Hexamethyldisiloxane on Carbons: Isotherms, Isosteric Heats and Kinetics. *Chemosphere* **2020**, *247*, 125862. [CrossRef]
38. Rezma, S.; Birot, M.; Hafiane, A.; Deleuze, H. Physically Activated Microporous Carbon from a New Biomass Source: Date Palm Petioles. *Comptes Rendus Chim.* **2017**, *20*, 881–887. [CrossRef]
39. Shafeeyan, M.S.; Daud, W.M.A.W.; Houshmand, A.; Shamiri, A. A Review on Surface Modification of Activated Carbon for Carbon Dioxide Adsorption. *J. Anal. Appl. Pyrolysis* **2010**, *89*, 143–151. [CrossRef]
40. Lazim, Z.M.; Hadibarata, T.; Puteh, M.H.; Yusop, Z. Adsorption Characteristics of Bisphenol A onto Low-Cost Modified Phyto-Waste Material in Aqueous Solution. *Water Air Soil Pollut.* **2015**, *226*, 34. [CrossRef]
41. Zhang, L.; Tu, L.; Liang, Y.; Chen, Q.; Li, Z.; Li, C.; Wang, Z.; Li, W. Coconut-Based Activated Carbon Fibers for Efficient Adsorption of Various Organic Dyes. *RSC Adv.* **2018**, *8*, 42280–42291. [CrossRef] [PubMed]
42. Mashhadimoslem, H.; Safarzadeh Khosrowshahi, M.; Jafari, M.; Ghaemi, A.; Maleki, A. Adsorption Equilibrium, Thermodynamic, and Kinetic Study of $O_2/N_2/CO_2$ on Functionalized Granular Activated Carbon. *ACS Omega* **2022**, *7*, 18409–18426. [CrossRef] [PubMed]
43. Guo, W.; Zhao, B.; Zhou, Q.; He, Y.; Wang, Z.; Radacsi, N. Fe-Doped ZnO/Reduced Graphene Oxide Nanocomposite with Synergic Enhanced Gas Sensing Performance for the Effective Detection of Formaldehyde. *ACS Omega* **2019**, *4*, 10252–10262. [CrossRef] [PubMed]
44. Jaworski, S.; Wierzbicki, M.; Sawosz, E.; Jung, A.; Gielerak, G.; Biernat, J.; Jaremek, H.; Łojkowski, W.; Woźniak, B.; Wojnarowicz, J.; et al. Graphene Oxide-Based Nanocomposites Decorated with Silver Nanoparticles as an Antibacterial Agent. *Nanoscale Res. Lett.* **2018**, *13*, 116. [CrossRef]
45. Cançado, L.G.; Jorio, A.; Ferreira, E.H.M.; Stavale, F.; Achete, C.A.; Capaz, R.B.; Moutinho, M.V.O.; Lombardo, A.; Kulmala, T.S.; Ferrari, A.C. Quantifying Defects in Graphene via Raman Spectroscopy at Different Excitation Energies. *Nano Lett.* **2011**, *11*, 3190–3196. [CrossRef]
46. Eckmann, A.; Felten, A.; Mishchenko, A.; Britnell, L.; Krupke, R.; Novoselov, K.S.; Casiraghi, C. Probing the Nature of Defects in Graphene by Raman Spectroscopy. *Nano Lett.* **2012**, *12*, 3925–3930. [CrossRef]
47. Kordek, K.; Jiang, L.; Fan, K.; Zhu, Z.; Xu, L.; Al-Mamun, M.; Dou, Y.; Chen, S.; Liu, P.; Yin, H.; et al. Two-Step Activated Carbon Cloth with Oxygen-Rich Functional Groups as a High-Performance Additive-Free Air Electrode for Flexible Zinc–Air Batteries. *Adv. Energy Mater.* **2019**, *9*, 1802936. [CrossRef]
48. Tangsir, S.; Hafshejani, L.D.; Lähde, A.; Maljanen, M.; Hooshmand, A.; Naseri, A.A.; Moazed, H.; Jokiniemi, J.; Bhatnagar, A. Water Defluoridation Using Al_2O_3 Nanoparticles Synthesized by Flame Spray Pyrolysis (FSP) Method. *Chem. Eng. J.* **2016**, *288*, 198–206. [CrossRef]
49. Scaria, J.; Anupama, K.V.; Nidheesh, P.V. Tetracyclines in the Environment: An Overview on the Occurrence, Fate, Toxicity, Detection, Removal Methods, and Sludge Management. *Sci. Total Environ.* **2021**, *771*, 145291. [CrossRef]
50. Ren, W.; Xiong, L.; Yuan, X.; Yu, Z.; Zhang, H.; Duan, X.; Wang, S. Activation of Peroxydisulfate on Carbon Nanotubes: Electron-Transfer Mechanism. *Environ. Sci. Technol.* **2019**, *53*, 14595–14603. [CrossRef]
51. Tian, L.; Chen, P.; Jiang, X.-H.; Chen, L.-S.; Tong, L.-L.; Yang, H.-Y.; Fan, J.-P.; Wu, D.-S.; Zou, J.-P.; Luo, S.-L. Mineralization of Cyanides via a Novel Electro-Fenton System Generating OH and $O_2{}^-$. *Water Res.* **2022**, *209*, 117890. [CrossRef]

52. Du, J.; Bao, J.; Liu, Y.; Kim, S.H.; Dionysiou, D.D. Facile Preparation of Porous Mn/Fe$_3$O$_4$ Cubes as Peroxymonosulfate Activating Catalyst for Effective Bisphenol A Degradation. *Chem. Eng. J.* **2019**, *376*, 119193. [CrossRef]
53. Xie, M.; Tang, J.; Kong, L.; Lu, W.; Natarajan, V.; Zhu, F.; Zhan, J. Cobalt Doped G-C3N4 Activation of Peroxymonosulfate for Monochlorophenols Degradation. *Chem. Eng. J.* **2019**, *360*, 1213–1222. [CrossRef]
54. Zhu, C.; Zhang, Y.; Fan, Z.; Liu, F.; Li, A. Carbonate-Enhanced Catalytic Activity and Stability of Co$_3$O$_4$ Nanowires for 1O$_2$-Driven Bisphenol A Degradation via Peroxymonosulfate Activation: Critical Roles of Electron and Proton Acceptors. *J. Hazard. Mater.* **2020**, *393*, 122395. [CrossRef] [PubMed]
55. Gao, Y.; Zhu, Y.; Lyu, L.; Zeng, Q.; Xing, X.; Hu, C. Electronic Structure Modulation of Graphitic Carbon Nitride by Oxygen Doping for Enhanced Catalytic Degradation of Organic Pollutants through Peroxymonosulfate Activation. *Environ. Sci. Technol.* **2018**, *52*, 14371–14380. [CrossRef]
56. Yang, Y.; Banerjee, G.; Brudvig, G.W.; Kim, J.-H.; Pignatello, J.J. Oxidation of Organic Compounds in Water by Unactivated Peroxymonosulfate. *Environ. Sci. Technol.* **2018**, *52*, 5911–5919. [CrossRef]
57. Komeily-Nia, Z.; Chen, J.-Y.; Nasri-Nasrabadi, B.; Lei, W.-W.; Yuan, B.; Zhang, J.; Qu, L.-T.; Gupta, A.; Li, J.-L. The Key Structural Features Governing the Free Radicals and Catalytic Activity of Graphite/Graphene Oxide. *Phys. Chem. Chem. Phys.* **2020**, *22*, 3112–3121. [CrossRef] [PubMed]
58. Li, J.; Li, M.; Sun, H.; Ao, Z.; Wang, S.; Liu, S. Understanding of the Oxidation Behavior of Benzyl Alcohol by Peroxymonosulfate via Carbon Nanotubes Activation. *ACS Catal.* **2020**, *10*, 3516–3525. [CrossRef]
59. Luo, H.; Liu, Y.; Lu, H.; Fang, Q.; Rong, H. Efficient Adsorption of Tetracycline from Aqueous Solutions by Modified Alginate Beads after the Removal of Cu(II) Ions. *ACS Omega* **2021**, *6*, 6240–6251. [CrossRef] [PubMed]
60. Moussavi, G.; Hossaini, Z.; Pourakbar, M. High-Rate Adsorption of Acetaminophen from the Contaminated Water onto Double-Oxidized Graphene Oxide. *Chem. Eng. J.* **2016**, *287*, 665–673. [CrossRef]
61. Li, M.; Liu, Q.; Lou, Z.; Wang, Y.; Zhang, Y.; Qian, G. Method To Characterize Acid–Base Behavior of Biochar: Site Modeling and Theoretical Simulation. *ACS Sustain. Chem. Eng.* **2014**, *2*, 2501–2509. [CrossRef]

Article

Role of N-Doping and O-Groups in Unzipped N-Doped CNT Carbocatalyst for Peroxomonosulfate Activation: Quantitative Structure–Activity Relationship

Kadarkarai Govindan [1], Do-Gun Kim [2] and Seok-Oh Ko [1,*]

[1] Environmental System Laboratory, Department of Civil Engineering, Kyung University (Global Campus), Giheung-Gu, Yongin-Si 16705, Korea; govindanmu@khu.ac.kr
[2] Department of Environmental Engineering, Sunchon National University, 255 Jungang-ro, Suncheon-Si 57922, Korea; dgkim@scnu.ac.kr
* Correspondence: soko@khu.ac.kr

Abstract: We examined the relationship between the intrinsic structure of a carbocatalyst and catalytic activity of peroxomonosulfate (PMS) activation for acetaminophen degradation. A series of nitrogen-doped carbon nanotubes with different degrees of oxidation was synthesized by the unzipping method. The linear regression analysis proposes that pyridinic N and graphitic N played a key role in the catalytic oxidation, rather than pyrrolic N and oxidized N. Pyridinic N reinforce the electron population in the graphitic framework and initiate the non-radical pathway via the formation of surface-bound radicals. Furthermore, graphitic N forms activated complexes (carbocatalyst-PMS*), facilitating the electron-transfer oxidative pathway. The correlation also affirms that -C=O was dominantly involved as a main active site, rather than -C-OH and -COOH. This study can be viewed as the first attempt to demonstrate the relationship between the fraction of N-groups and activity, and the quantity of O-groups and activity by active species (quenching studies) was established to reveal the role of N-groups and O-groups in the radical and non-radical pathways.

Keywords: N-doped CNTs; PMS activation; N-groups–activity relationship; O-groups–activity relationship

Citation: Govindan, K.; Kim, D.-G.; Ko, S.-O. Role of N-Doping and O-Groups in Unzipped N-Doped CNT Carbocatalyst for Peroxomonosulfate Activation: Quantitative Structure–Activity Relationship. *Catalysts* **2022**, *12*, 845. https://doi.org/10.3390/catal12080845

Academic Editors: Gassan Hodaifa, Antonio Zuorro, Joaquín R. Dominguez, Juan García Rodríguez, José A. Peres and Zacharias Frontistis

Received: 7 July 2022
Accepted: 26 July 2022
Published: 1 August 2022

Publisher's Note: MDPI stays neutral with regard to jurisdictional claims in published maps and institutional affiliations.

1. Introduction

In environmental remediation, metal-free carbocatalyst-based advanced oxidation processes (AOPs) with persulfates (peroxomonosulfate (PMS) and peroxodisulfate (PDS)) have received increasing attention as sustainable alternatives to metal-based AOPs [1–4]. Carbocatalysts, with a high specific surface area, unique charge carrier mobility, low mass-transfer resistance, and low-dimensional structure, have proven to be efficient in various heterogeneous catalytic processes [5–7]. Now-a-days, diverse strategies have been explored to activate PMS [8] because an asymmetric chemical structure of PMS (HSO_5^-) is believed to be effectively activated by carbocatalysts [9,10], rather than symmetric structure PDS activation [11–13].

Carbocatalyzed PMS activation and catalytic oxidation involves free radical pathways and non-free radical pathways [1,2,4]. Free radical oxidative pathways are mostly ascribed to the generation of $SO_4^{\bullet-}$, $^\bullet OH$ and $O_2^{\bullet-}$. Carbocatalysts activate PMS via the electron-conduction mechanisms, which can break the O-O bond in PMS to form free radicals. $SO_4^{\bullet-}$ is considered to be a more efficient oxidative free radical than $^\bullet OH$. $SO_4^{\bullet-}$ has higher oxidation potential (2.7–3.1 V) and longer half-life time (3.4×10^{-5} s) than $^\bullet OH$ with a low oxidation potential (1.8–2.7 V) and shorter half-life time (2×10^{-8} s) [14,15]. Furthermore, $SO_4^{\bullet-}$-mediated oxidation exhibits good catalytic efficiencies in a wide range of solutions of pH 3–9 [16]. Non-free radical pathways are predominantly attributed to the formation of singlet oxygen ($^1O_2$) [17,18], electron-transfer based oxidation [19], formation of surface-bound complexes [20], and surface-bound- $SO_4^{\bullet-}$ [21,22]. Generally, $^1O_2$ has a mild oxidation potential (0.65 V), so that the reactivity of $^1O_2$ is also relatively lower than those of $SO_4^{\bullet -}$ and $^\bullet OH$ [23]; however, $^1O_2$ is more selective to electron-rich organic compounds [24,25]. Electron-transfer mechanism occurs when the carbocatalyst surface acts as an electron-bridge (e^- acceptor) to accelerate the electron transfer from organic pollutants (e^- donor) to PMS molecules adsorbed on the catalyst surfaces. Importantly, the non-radical pathways can also sustain excellent reactivity in complex aqueous matrices, such as wide pH conditions and the presence of various co-existing anions and natural organic matters [9].

In contemporary times, carbon nanotubes (CNTs) have demonstrated to be more efficient in PMS activation and degradation of various organic compounds [19,26–30]. CNT, with a peculiar sp^2-hybridized π conjugation, along with low level defects and limited oxygen-containing functional groups, can appreciably accelerate the catalytic activity. In fact, oxygen functional groups, including hydroxyl (-OH), carbonyl (-C=O) and carboxyl (-COOH), in the graphitic network are very important for PMS activation [1,31]. In addition, the proper modification of CNT surfaces by hetero-atom doping has shown enhanced catalytic performances [16]. That is, the N-doping on CNTs can create novel catalytic active sites, good charge distribution of neighboring carbon atoms in the conjugated sp^2-carbon network, and thus induces a high chemical potential for catalytic reactions [16]. Recent investigations demonstrated that N-doping can establish electron-rich Lewis basic sites, including boundary N (pyridinic N, pyrrolic N), substitutional N (graphitic N), and oxidized N (Table S1) [32–41]. In addition, N-doping can also create oxygen functional groups (C-OH, C=O, and COOH) in the carbon framework, which also shows good catalytic performances. A few studies stated that the C=O group profoundly participated in PMS activation among the oxygen-containing functional groups [33,35].

The quantitative structure–catalytic activity relationship in NCNT/PMS has been explored to reveal the dominant species for PMS activation [34]. This study demonstrates that graphitic N in NCNT has a major role in PMS activation and facilitates non-radical oxidative pathways through NCNT–PMS* activation complexes. However, the catalytic activity of N-groups has not been explored based on the reactive oxidative species (classical quenching studies). Similarly, quantitative correlation between oxygen-containing functional groups and catalytic activity in NCNT/PMS is of fundamental importance to reveal key active oxygen groups for PMS activation. So far, the fraction of oxygen groups-activity correlation has not been reported to interpret the key active oxygen functional groups in the NCNT/PMS system.

Therefore, to explore the relationship between oxygen functional groups and catalytic activity in the NCNT/PMS system, four different NCNT materials functionalized with distinct oxygen contents were prepared through different degrees of oxidation. Furthermore, this was the first attempt to rationalize the quantitative structure–catalytic activity relationship based on the scavenging of radical and non-radical oxidative pathways by classical quenching studies. Additionally, the relationship between the intrinsic structure of N-dopants and catalytic performance was also analyzed to unveil the role of N species in NCNT/PMS catalytic activity. Radical and non-radical reactive species were determined through classical quenching study, as well as by EPR analysis. The quenching experiments were comprehensively performed with eight different quenching agents and the catalytic activity of each radical and non-radical species was rationalized.

2. Results and Discussion

2.1. Structure Characterizations

Figure S1a displays the XRD spectra of pristine CNT, NCNT, and Uz-NCNTs. The strong diffraction peak at around 25.93° is ascribed to the (002) reflection of the hexagonal graphitic structure. N-doping (NCNT) did not noticeably change the (002) diffraction; however, the main graphitic peak position gradually decreased from 25.93° to 25.87, 25.51, and 25.27°, respectively. On the other hand, the interlayer distance of 3.43 nm of CNT/NCNT slowly increased to 3.44, 3.49 and 3.52 for Uz-NCNT-2, Uz-NCNT-4 and Uz-NCNT-8, respectively. It suggests that the chemical oxidation can successively intensify the C-C bond formation and disorder the sp^2-C conjugation. Furthermore, the N-doping and chemical oxidation create defective sites (edges and vacancies) and can also establish oxygen-containing functional groups in the main graphitic network [33]. This surface modification tends to increasingly broaden the (002) diffraction peak, which substantiates the decreased graphitic degree of Uz-CNTs.

Raman spectra of the derived carbocatalysts are presented in Figure S1b. The characteristic D-band, G-band, and 2G-band of CNT were observed around 1347–1359, 1585–1612 and 2698–2732 cm^{-1}, respectively. The D-band at 1347, 1353, and 1359 cm^{-1}, and the G-band at 1501, 1612, and 1608 cm^{-1} were also noticed for Uz-NCNT-2, Uz-NCNT-4, and Uz-NCNT-8, respectively. The integrated intensity ratio of the D-band versus the G-band (I_D/I_G) is a substantial factor in determining the defect and deformation level. Table S2 shows that the I_D/I_G value of NCNT (1.03) is higher than CNT, verifying that considerable defective sites originate in the well-organized sp^2-carbon network by N-doping. The I_D/I_G value further increased for Uz-NCNT-2 (1.08) and Uz-NCNT-4 (1.13), which shows that more disorders and defective sites were established in the carbon network with the increasing oxidation levels. However, the I_D/I_G decreased to 0.96 for Uz-NCNT-8 as the degree of oxidation increased more (8 g KMnO$_4$ addition). This may be possibly due to the formation of abundant oxygenated functional groups in the carbon lattice, which could disrupt the vacancies and break the chemical bonds in the graphitic network [42]. Moreover, the intensity of the 2D-band at 2703 cm^{-1} also diminished as the amount of oxidizing agent (KMnO$_4$) increased. It implies that the chemical oxidation employed with 2, 4 and 8 g of KMnO$_4$ brings more boundaries, sp^3 domains or vacancies and disorders into the carbon network, along with edges [38].

Figure S2a shows that the absorption peak at 273 nm (CNT, and NCNT) is attributed to the π-π* excitation of the graphitic sp^2-C network. This π-π* absorption peak shifted substantially to 230 nm for Uz-CNTs. This crucial blue-shift informs us that the homogenous sp^2-C=C conjugation is broken, and sp^3-C-C domains are formed on the carbocatalyst surface by chemical oxidation. Figure S2b exhibits FT-IR spectra of the carbocatalyst. It displays limited oxygenate functional groups that exist on the CNT, as well as NCNT surfaces. A few evident bands visibly appeared with the Uz-CNT samples. In particular, the bands at around 1980 and 1891 cm^{-1} correspond to the asymmetric and symmetric stretching vibrations of the C-H bond, while the bands at around 1717, 1580, 1389, 1167 and 1038 cm^{-1} are ascribed to the -C=O, -C=N/-C=C, -C-OH, -C-N, and -C-O stretching vibrations, respectively [43].

2.2. Specific Surface Area and Chemical Status

N$_2$ sorption isotherm and BJH pore size distribution profiles are shown in Figure S3. N$_2$ adsorption/desorption isotherm results of all the samples (except Uz-NCNT-4) reveal a Type IV isotherm with a H1-type hysteresis loop when P/P_0 = 0.4–0.9. Uz-NCNT-4 exhibits a Type IV isotherm with a H3-type. Type H1 hysteresis implies that the catalyst channels have uniform pore sizes and shapes, while the type H3 hysteresis proposes that the catalyst (Uz-NCNT-4) contains a very wide distribution of pore sizes [44,45]. The evident peak between 3 and 5 nm in Figure S3f states that the synthesized carbocatalysts primarily consist of mesoporous structures (>2 nm). As shown in Table S3, the BET surface area 206.41 m^2 g^{-1} (CNT) is not substantially changed by N doping (207.06 m^2 g^{-1}, NCNT), but it is gradually reduced to 163.04 m^2 g^{-1} (Uz-NCNT-2), 140.72 m^2 g^{-1} (Uz-NCNT-4) and 4.41 m^2 g^{-1} (Uz-NCNT-8) with increasing dosages of oxidizing agent (KMnO$_4$). The decrease in BET surface area is probably due to the collapse of the carbon skeleton structure and crinkled layers during the oxidation processes. The high loss of surface area with Uz-NCNT-8 is possibly due to the intense stacking effect, resulting from strong hydrogen bonding between the oxygen functionalities on the catalyst surface. X. Duan and coworkers reported that heteroatom (N, P and S) doping into graphene greatly reduces the specific surface area due to formation of a crinkled layer and stacking effect [46,47].

The chemical compositions of the nitrogen-doped carbocatalysts were examined by XPS and are displayed in Figure S4 and Table S4. As shown in Figure S4b, the N 1s characteristic peak is noticed at around 399.9–400.8 eV, indicating that N atoms were successively introduced into the carbocatalyst. The analysis also shows the N content of 0.93, 1.07, 1.04, and 0.62% is noticed in NCNT, Uz-NCNT-2, Uz-NCNT-4 and Uz-NCNT-8, respectively (Table 1). The high resolution N 1s spectrum in a specific range (392–410 eV) was deconvoluted into four N-groups, such as pyridinic N, pyrrolic N, graphitic N or substitutional N, and oxidized N (Figure S4c–f). The deconvoluted peak center, peak area and total area of the respective sample are provided in Table 1. Each peak area was divided by the total area of the specific N 1s spectrum and calculated the fraction of N-dopants. Table 1 denotes that the pyridinic N and graphitic N contents were progressively enriched from NCNT to Uz-NCNT-2 and Uz-NCNT-4, and then drastically reduced for Uz-NCNT-8. In contrast, pyrrolic N content decreased from NCNT to Uz-NCNT-2, and Uz-NCNT-4 and increased at Uz-NCNT-8. On the other hand, oxidized N content greatly decreased from NCNT to Uz-NCNT-4 and completely vanished for Uz-NCNT-8. The analysis concludes that the amount of oxidizing agent significantly governs the N-dopants' levels.

Table 1. Atomic percentages of elements in the carbocatalysts were obtained from XPS analysis.

Carbocatalysts	N/at. %	N/at. %			
		Pyridinic N	Pyrrolic N	Graphitic N	Oxidized N
CNT	NA	0.00	0.00	0.00	0.00
NCNT	0.93	0.06	0.28	0.35	0.32
Uz-NCNT-2	1.07	0.13	0.13	0.53	0.20
Uz-NCNT-4	1.04	0.18	0.18	0.72	0.05
Uz-NCNT-8	0.62	0.15	0.67	0.18	0.00

	C at. %	O at. %	O/at. %		
			C-O	C=O	COOH
CNT	99.11	0.89	0.63	0.37	0
NCNT	96.04	3.04	0.21	0.58	0.21
Uz-NCNT-2	77.43	21.5	0.22	0.54	0.24
Uz-NCNT-4	74.39	24.58	0.08	0.59	0.33
Uz-NCNT-8	67.22	32.16	0.10	0.45	0.46

Table 1 also indicates that the oxygen atomic percentage steadily increased from 0.89 (CNT) to 3.04, 21.5, 24.58, 32.16% with NCNT, Uz-NCNT-2, Uz-NCNT-4 and Uz-NCNT-8, respectively. It confirms that extending oxidation levels establish more oxygen-containing functional groups in the carbon network. High resolution O 1s spectrum between 526 and 541 eV was deconvoluted into five major peaks, which are related to the hydroxyl group (C-O), carbonyl group (C=O), and carboxyl group (COOH) (Figure S5 and Table S5). Table 1 presents the amount of C-O group, 0.63 at. % (CNT), which steadily decreased to 0.21, 0.22, and 0.08 with NCNT, Uz-NCNT-2 and Uz-UNCT-4, respectively, although it slightly fluctuated to 0.10 at. % for Uz-NCNT-8. In contrast, the C=O group content increased from 0.37 at. % (CNT) to 0.58 for NCNT and then it fluctuated to 0.54, 0.59, and 0.45 at. % with increasing oxidizing power of 2, 4 and 8 g $KMnO_4$, respectively. The lower C=O fraction for Uz-NCNT-8 may be due to the transformation of C=O into the -COOH group at a higher degree of oxidation level. The oxidation process employed $KMnO_4$ in concentrated H_2SO_4 as an oxidizing agent; this can generate the strong oxidizer Mn_2O_7. This Mn_2O_7 oxidizer is able to oxidize the carbonyl groups in aldehyde or benzoquinone structures on carbocatalyst surfaces and converts them to -COOH [48–50]. Thus, the amount of -COOH gradually increased with the degree of oxidation level in this study. The analysis affirms that the $KMnO_4$ dosage crucially controlled the formation of oxygen-containing functional groups.

2.3. Catalytic Performance

2.3.1. Role of N-Doping on Catalytic Performances

Catalytic performance of all carbocatalysts was examined using constant pollutant concentration (C_{ACP} = 10 mg L^{-1}), and PMS dosage (C_{PMS} = 0.5 mM) with pH$_0$ = 7 ± 0.3 and is presented in Figure 1a,b. It was found that ACP degradation profiles for all catalysts were well fitted with the first-order kinetic model. The ACP degradation rate constant, k_1 (ACP) = 4.37 × 10^{-4} s^{-1}, for CNT is enhanced to 18.4 × 10^{-4} s^{-1} for NCNT, validating that the surface modification of CNT by nitrogen doping greatly increases the catalytic activity. The k_1 (ACP) of NCNT is further increased to 22.05 × 10^{-4}·s^{-1} (Uz-NCNT-2), and 28.35 × 10^{-4} s^{-1} (Uz-NCNT-4) with the increase in the degree of oxidation level, respectively. However, the k_1 (ACP) for Uz-NCNT-8 is relatively lower than k_1 for Uz-NCNT-2 and Uz-NCNT-4. This might be due to the presence of the low fraction of N (0.62 at. %) in Uz-NCNT-8, compared to Uz-NCNT-2 (1.07 at. %) and Uz-NCNT-4 (1.04 at. %). Furthermore, the graphitic N fraction is also substantially reduced for Uz-NCNT-8. The graphitic N can greatly influence the electron transfer ability among N dopants, because it can readily alter the charge distribution of adjacent carbon atoms due to the electronegativity [27]. It is also noted that the existence of abundant pyrrolic N (0.67 at. % for Uz-NCNT-8) adversely influences the catalytic activity of the carbocatalyst. Typically, a pyrrolic N atom is attached to a five membered pentagonal ring and contributes two p-electrons to the π system, increasing the electron density of the main graphitic layer [16]. However, the lone pair of electrons in the sp^2 orbital is parallel to the sp^2 orbital of adjacent C [51]. As a result, the lone pair of electrons readily contributes to the delocalization and reinforces the density of the graphitic network, rather than transporting electrons to the adsorbed organics and/or PMS. Due to the spatial arrangement of sp^2 pyrrolic N, the adsorption of organics or PDS on the pyrrolic ring is hindered.

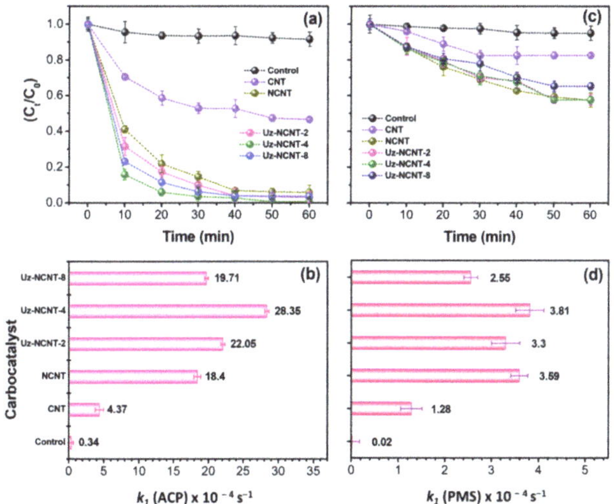

Figure 1. (**a**) ACP degradation profile versus time and (**b**) first-order kinetic rate constant, k_1 (ACP). (**c**) PMS decomposition profile versus time and (**d**) k_1 (PMS). Experimental conditions C_{ACP} = 10 mg L^{-1}, $C_{Catalyst}$ = 100 mg L^{-1}, C_{PMS} = 0.5 mM and pH$_0$ = 7 ± 0.3.

To evaluate the role of each N species on the catalytic performance, the regression analysis was applied between N-contents and k_1 (ACP). In Figure 2a, high linear correlation was obtained with pyridinic N (R^2 = 0.9219) and graphitic N (R^2 = 0.8891). Meanwhile, low correlations were obtained for pyrrolic N (R^2 = 0.1363) and oxidized N (R^2 = −0.0055). Similarly, the PMS decomposition rate constants, k_1 (PMS), of all carbocatalysts were also examined and are displayed in Figure 1c,d. It implies that the k_1 (PMS), 1.28×10^{-4} s^{-1} for CNT, is substantially enhanced with NCNT (3.59×10^{-4} s^{-1}), and slightly fluctuates for Uz-NCNT-2 (3.3×10^{-4} s^{-1}). This may be due to the crinkled graphitic layer during the chemical oxidation process. Then, k_1 (PMS) increased for Uz-NCNT-4 (3.81×10^{-4} s^{-1}) and decreased for Uz-NCNT-8 (2.55×10^{-4} s^{-1}). Furthermore, the relationship between k_1 (PMS) and N-dopants in Figure 2b exhibits a reasonable correlation ($R^2 = 0.7014$) with graphitic N, affirming that graphitic N appreciably promotes PMS activation, rather than other boundary Ns (i.e., pyrrolic N and oxidized N). Although graphitic N seems to play a main role in the catalytic system via PMS activation (0.8891 with k_1 (ACP) and 0.7014 with k_1 (PMS)), pyridinic N also slightly outperforms the other N-dopants. This may be attributed to the fact that pyridinic N has a lone pair of electrons considered as a Lewis basic site, which can effectively enhance the electron population of the main graphitic layer and improve the catalytic activity [52]. Moreover, the substitutional N (graphitic N) can also break the uniform sp^2-hybridized π-conjugations at the carbon periphery. The higher electronegative N atom (graphitic N) can induce a positive charge to adjacent carbon atoms through charge transfer. As a result, the positively charged sites can enhance the PMS adsorption and weaken the peroxide bond (O-O) in HO-SO$_4^{-}$, generating reactive oxidative species [34,36].

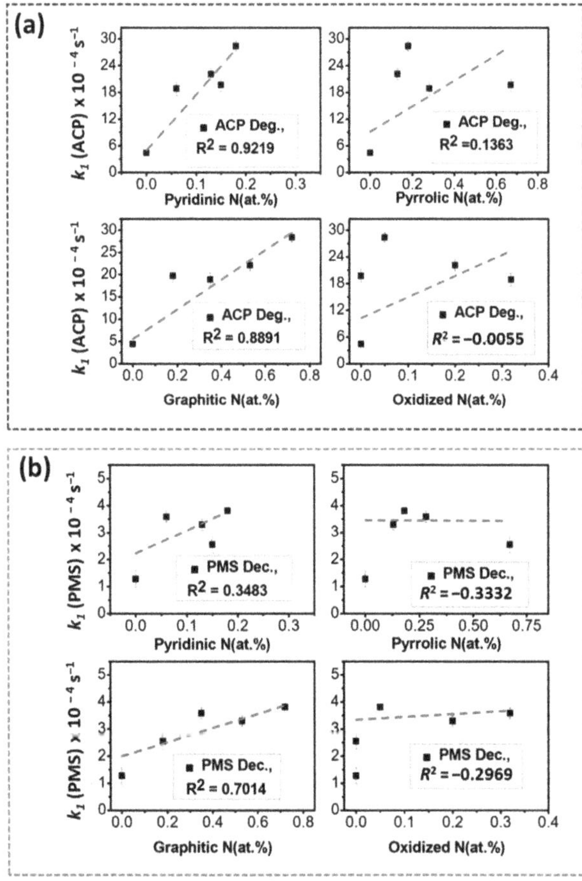

Figure 2. Role of N-content and specific N-constituent in k_1 (ACP) (**a**), and k_1 (PMS) (**b**). Experimental conditions C_{ACP} = 10 mg L^{-1}, $C_{Catalyst}$ = 100 mg L^{-1}, C_{PMS} = 0.5 mM and pH$_0$ = 7 + 0.3.

2.3.2. Role of Oxygen Functionalities on Catalytic Performances

Three oxygen-containing functional groups (C-O, C=O, and -COOH) and the amount of oxygen groups are detected from O 1s XPS spectra and are summarized in Table 1. To ascertain the effect of the O-content and three oxygen groups on catalytic performance, the k_1 (ACP)-O-content relationship was evaluated (Figure 3a,b). In Table 1, the reaction rate k_1 (ACP) gradually increased with the increasing atomic percentage of O-content from 0.89 at. % for CNT to 3.04 at. % (NCNT), 21.5 at. % (Uz-NCNT-2), and 24.58 at. % (Uz-NCNT-4). However, k_1 (ACP) decreased at 32.16 at. % of oxygen content (Uz-NCNT-8). Increasing the fraction of oxygen content could significantly enhance the electron population in the graphitic layer, which could remarkably enhance the electron transfer oxidative steps. Furthermore, the existence of a significant quantity of oxygen functional groups in the graphitic framework could also appreciably transport the electrons to PMS or organics, resulting in a higher degradation rate [53]. Nonetheless, the decrease in k_1 (ACP) for Uz-NCNT-8 may be ascribed to the existence of the many oxygen-containing functional groups, which make the electron conductivity of the graphitic layer worse by occupying the active edges and blocking the interaction between the carbocatalyst and PMS through a stereo-hindrance effect [53]. It implies that oxygen-containing functional groups crucially govern ACP degradation in CNT-based carbocatalyzed PMS systems.

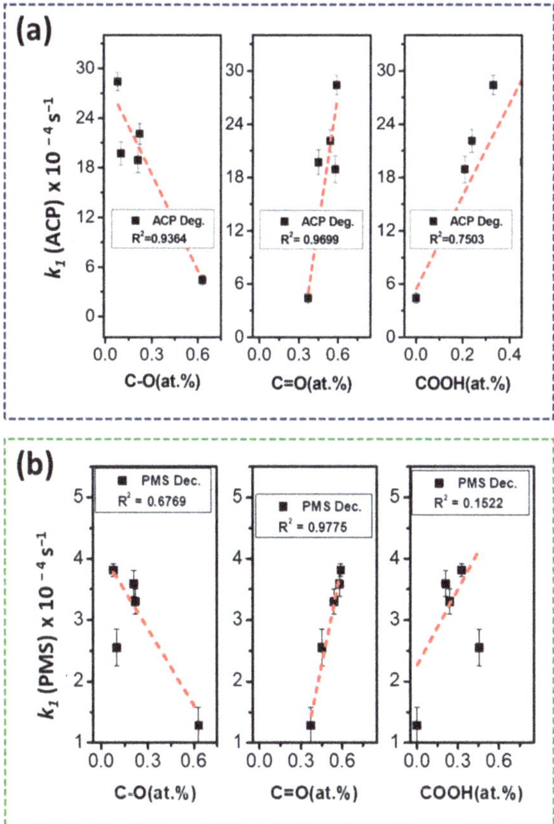

Figure 3. Role of O-contents and oxygen functionalities in k_1 (ACP) (**a**), and k_1 (PMS) (**b**). Experimental conditions C_{ACP} = 10 mg L^{-1}, $C_{Catalyst}$ = 100 mg L^{-1}, C_{PMS} = 0.5 mM and pH$_0$ = 7 ± 0.3.

A well fitted positive linear correlation (R^2 = 0.9699) with a -C=O group confirms that the carbonyl group at the edges is a dominant active site in the catalytic systems (Figure 3b). Recent investigations have also demonstrated that the electron-rich -C=O group is believed to be a paramount active site for PMS activation because it breaks the O-O bond in PMS into $SO_4^{\bullet-}$ and $^{\bullet}OH$ and then triggers the formation of $^1O_2$ [33,35]. Meanwhile, the correlation (R^2 = 0.7503) with the -COOH group suggests that carboxylic acid functionalities are also participating in the catalytic oxidation processes [54–56]. On the other hand, a negative linear fit for -C-O (R^2 = 0.9364) indicates that the excessive electron-donating hydroxyl groups in the main graphitic structure might be unfavorable for electron-transfer oxidation [53]. A good linear relationship between k_1 (PMS) and -C=O (R^2 = 0.9775) in Figure 3b proposes that the carbonyl groups at the reactive edges and defective sites can remarkably aid the PMS activation [33,35]. Furthermore, the poor correlation between -COOH and k_1 (PMS) indicates that the PMS adsorption did not significantly initiate -COOH. It may be due to the COOH group's spatially wide molecular structure and hindered broad molecular structure of the PMS. However, the meaningful R^2 between k_1 (ACP) and -COOH may be due to the electron-withdrawing carboxylic group altered the electron density in the graphitic structure and facilitates the electron transfer oxidative steps.

2.4. Identification of Main Reactive Oxidative Species in Carbocatalyst/PMS Systems

Typically, PMS activation by CNT-based carbocatalysts can generate diverse radical and non-radical oxidative components, as described in Equations (1)–(18). EPR was used to find the active species involved in the carbocatalyzed PMS activation systems. The spin-trapping agent TMP was employed to detect the $^1O_2$. In Figure 4a, a weak signal was observed in PMS without a catalyst because of the slow self-decomposition of PMS [57]. For all the carbocatalysts/PMS systems, a typical three-line TMP-$^1O_2$ adduct signal with equal intensities (1:1:1) was obtained. It substantiates the fact that the interaction between carbocatalysts and PMS can invariably generate $^1O_2$ to react with TMP and produces the stable adduct 2,2,6,6-tetramethyl-4-piperidnol-N-oxyl. Superoxide radical anion $O_2^{\bullet-}$ can also react with TMP to form equal intensities of three-line EPR signals [58]. Thus, to evaluate the contribution of $O_2^{\bullet-}$ in carbocatalyst/PMS systems, EPR analysis was carried out in the presence of an $^1O_2$ quencher (10 mM FFA). As shown in Figure 4b, the intensities of the three-line TMP-$^1O_2$ adduct signal are considerably diminished, corroborating that $O_2^{\bullet-}$ is also produced in carbocatalyzed PMS activation.

$$CNT\text{-}C=N + HSO_5^- \rightarrow CNT\text{-}C\text{-}N^+ + SO_4^{\bullet-} + OH^- \tag{1}$$

$$CNT\text{-}C=N + HSO_5^- \rightarrow CNT\text{-}C\text{-}N^+ + SO_4^{2-} + {}^{\bullet}OH \tag{2}$$

$$C=C=N\text{-}C_{surface}\text{-}C=O + HSO_5^- \rightarrow C_{surface}\text{-}C=O^+ + SO_4^{\bullet-} + OH^- \tag{3}$$

$$C=C=N\text{-}C_{surface}\text{-}C=O + HSO_5^- \rightarrow C_{surface}\text{-}C=O^+ + SO_4^{2-} + {}^{\bullet}OH \tag{4}$$

$$C_{surface}\text{-}C=O^+ + HSO_5^- \rightarrow C_{surface}\text{-}C=O^+ + SO_5^{\bullet-} + OH^- \tag{5}$$

$$C=C=N\text{-}C_{surface}\text{-}COOH + HSO_5^- \rightarrow C_{surface}\text{-}COO^+ + SO_4^{\bullet-} + 2OH^- \tag{6}$$

$$C=C=N\text{-}C_{surface}\text{-}OH + HSO_5^- \rightarrow C_{surface}\text{-}O^+ + SO_4^{\bullet-} + 2OH^- \tag{7}$$

$$C=C=N\text{-}C_{surface}\text{-}C=O + HSO_5^- \rightarrow C_{surface}\text{-}C=O=O + H^+ \tag{8}$$

$$C_{surface}\text{-}C=O\text{-}O^+ + SO_5^{2-} \rightarrow C_{surface}\text{-}C=O + {}^1O_2 \tag{9}$$

$$C=C=N\text{-}C_{surface}\text{-}C=O + HSO_5^- + 2OH^- \rightarrow C_{surface}\text{-}C\text{-}O\text{-}O^+ + H_2O + SO_4^{2-} \tag{10}$$

$$C=C=N\text{-}C_{surface}\text{-}C\text{-}O\text{-}O^+ + HSO_5^- + 2OH^- \rightarrow C_{surface}\text{-}C=O + {}^1O_2 + H_2O + SO_4^{2-} \tag{11}$$

$$SO_5^{2-} + HSO_5^- \rightarrow HSO_4\text{-}SO_4^{2-} + {}^1O_2 \tag{12}$$

$$2O_2^{\bullet-} + 2H_2O \rightarrow H_2O_2 + {}^1O_2 + OH^- \tag{13}$$

$$O_2^{\bullet-} + {}^{\bullet}OH \rightarrow {}^1O_2 + OH^- \tag{14}$$

$$HSO_5^- + e^- \rightarrow SO_4^{\bullet-} + OH^- \tag{15}$$

$$HSO_5^- + e^- \rightarrow SO_4^{2-} + {}^{\bullet}OH \tag{16}$$

$$HSO_5^- + NCNT\text{-}C \cdots \pi^+ \rightarrow SO_4^{\bullet-} + OH^- \tag{17}$$

$$HSO_5^- + NCNT\text{-}C \cdots \pi^+ \rightarrow SO_4^{2-} + {}^{\bullet}OH \tag{18}$$

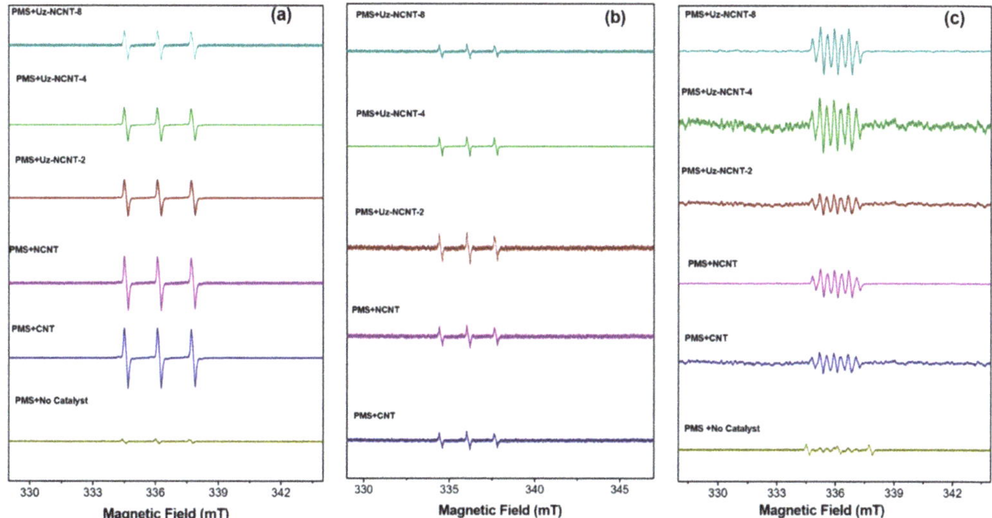

Figure 4. (a) EPR spectra of TMP-$^1O_2$ adducts under different carbocatalyst/PMS systems. (b) EPR spectra of TMP-$^1O_2$ analyzed in the presence of C_{FFA} = 10 mM, and C_{TMP} = 50 mM; (c) EPR spectra of DMPO-$SO_4^{\bullet-}$ and DMPO-$^{\bullet}OH$ adducts were analyzed using C_{DMPO} = 100 mM. $C_{Catalyst}$ = 10 mg L^{-1}, C_{ACP} = 10 mg L^{-1}; and C_{PMS} = 1 mM.

Other active free radicals $SO_4^{\bullet-}$ and $^{\bullet}OH$ were also identified by using an EPR, in which the spin-trapping agent 5,5-dimethyl-1-pyrroline N-oxide (DMPO) was employed to detect $SO_4^{\bullet-}$ and $^{\bullet}OH$. In Figure 4c, the weak characteristic peaks of DMPO-$^{\bullet}OH$ can be observed from PMS only. This can be ascribed to the hydrolysis of PMS that forms less $^{\bullet}OH$. In contrast, the prominent variation in the peak intensity is observed before and after the addition of the carbocatalyst, verifying that derived carbocatalysts (NCNT, Uz-NCNT-2, Uz-NCNT-4 and Uz-NCNT-8) are capable of generating $SO_4^{\bullet-}$ and $^{\bullet}OH$. The reaction between $^{\bullet}OH$ and DMPO can produce a DMPO-$^{\bullet}OH$ adduct signal, with a relative intensity ratio of 1:2:1:2:1:2:1 [59,60]. This may be due to the fact that DMPO can oxidize and yield 5,5, -dimethyl-2-pyrrolidon-N-oxyl (DMPOX). The presence of a DMPOX signal in Figure 4c specifies that $^{\bullet}OH$, and $SO_4^{\bullet-}$ are consistently formed in carbocatalyst/PMS systems. This may be due to the transformation of DMPO-$SO_4^{\bullet-}$ and DMPO-$^{\bullet}OH$ adducts. A fast nucleophilic substitution reaction between highly reactive DMPO-$SO_4^{\bullet-}$ and H_2O (DMPO-$SO_4^{\bullet-}$ = 95 s in water) resulted in the rapid transformation of the DMPO-$SO_4^{\bullet-}$ adduct into DMPO-$^{\bullet}OH$. Then, DMPO HO$^{\bullet}$ was oxidized into DMPOX through electron abstraction by HO$^{\bullet}$ and/or $SO_4^{\bullet-}$ [61]. EPR study recommends that $SO_4^{\bullet-}$ and $^{\bullet}OH$ are also formed through PMS activation and contribute to ACP degradation.

2.5. Classical Quenching Studies

To further identify the reactive oxidative species that participated in the catalytic systems, classical quenching experiments were also attempted in detail. The ethanol/*tert-*

BA, and p-BQ were utilized to detect $SO_4^{\bullet-}/^{\bullet}OH$ and $O_2^{\bullet-}$, respectively [62–65], and FFA and L-His were considered as efficient quenchers to identify the $^1O_2$ [66,67]. The non-free radicals, surface-bound radicals, surface-bound activated complexes, and free flowing electrons were also evaluated using specific quenchers, such as KI, phenol, and $NaClO_4$, respectively [43,68,69]. The rate constants for the reactions between the reactive oxidative species and specific quencher are briefly described in Table 2.

The effect of quenchers on the ACP degradation profiles and observed first-order rate constants are shown in Figures S6 and S7, respectively. In general, the ACP degradation rate constant is substantially inhibited by the addition of specific quenchers, suggesting that radicals ($SO_4^{\bullet-}/^{\bullet}OH$ and $O_2^{\bullet-}$), $^1O_2$, non-free radicals (surface-bound radicals, surface-bound complexes, and free flowing electrons) are intrinsically involved in the catalytic system. In particular, the high inhibition rate with the $^1O_2$ quencher (L-His, FFA) in CNT/PMS and NCNT/PMS proposes that $^1O_2$ based-oxidation process led to ACP degradation. It may be because the PMS activation by the carbocatalyst (CNT and NCNT) with a smaller number of oxygen groups was limited, resulting in the lower amount of free radical formation. Furthermore, the adsorption of organics by the carbocatalyst was ascribed to the π–π interactions and electrostatic interactions between the electron donor (ACP) and electron acceptor (catalyst). Subsequently, the organic contaminant was degraded via the electron-transfer mechanism in the surface of the graphitic layer (i.e., non-radical oxidative pathway) [42]. In contrast, the radical quenchers (ethanol, *tert*-BA and p-BQ) and non-radical quenchers ($NaClO_4$ and phenol) greatly hampered the ACP degradation rate with the oxidized NCNTs (Uz-NCNT-2, Uz-NCNT-4 and Uz-NCNT-8). The considerable number of oxygen contents in the catalyst surface can significantly enhance the electron density of the graphitic network. It can efficiently interact with the peroxo bond (-O-O-) in the PMS structure and enhance free radical generation. It is evident that free-radicals, as well as non-radical electron-transfer based oxidation, jointly facilitate ACP degradation with Uz-CNTs/PMS system.

2.5.1. Role of N-Dopants on Radical and Non-Radical Pathway

The role of N-dopants on PMS activation for ACP degradation was further rationalized based on the results in Figure S7. To investigate the relationship between N-dopants and individual oxidative species activity, the normalized inhibition ability by a specific quenching agent was calculated according to the equation given below.

$$\text{Inhibition Ability} = \frac{k_1(ACP)_C - k_1(ACP)_Q}{k_1(ACP)_C} \tag{19}$$

where $k_1(ACP)_C$ is the first-order ACP degradation rate constant for the carbocatalyst/PMS in the absence of quenchers, and $k_1(ACP)_Q$ is the ACP degradation rate constant with a specific quencher. The normalized inhibition ability with N-dopants is presented in Figures 5–7. The notable correlation of ethanol ($R^2 = 0.7359$) and *tert*-BA ($R^2 = 0.7309$) (*tert*-BA) with pyridinic N was observed in Figure 5a, while the correlation ($R^2 = -0.1122$) between p-BQ and pyridinic N-content was low. The correlation illustrates that pyridinic N substantially governs $SO_4^{\bullet-}/^{\bullet}OH$ formation and radical-based oxidation. This may be explained by the fact that the lone pair of electrons are in a sp^2 (pyridinic N) orbital perpendicular to the sp^2 orbitals of the adjacent carbon; thus, the pyridinic N can readily transport the non-bonded electrons to PMS and facilitate the formation of free radicals ($SO_4^{\bullet-}/^{\bullet}OH$) [70]. A plausible PMS activation mechanism, as described in Equations (1)–(18), suggests that no direct interaction occurs between PMS and the N-functional group to form superoxide anion radicals. Furthermore, p-BQ yields a good R^2 value (0.9256) with graphitic N, proving that graphitic N plays a crucial role in $O_2^{\bullet-}$ mediative oxidation. The increased inhibition ability by the $O_2^{\bullet-}$ quencher (p-BQ) indicates that a great number of $O_2^{\bullet-}$ contributes to efficient ACP degradation. The relationship between graphitic N and *tert*-BA seems to be that inhibition ability increased with the increasing graphitic N content, suggesting that graphitic N heavily influences the formation of $SO_4^{\bullet-}/^{\bullet}OH$. Figure 5 also implies that pyrrolic N and oxidized N did not appreciably participate in the radical oxidative pathway, as they show very low R^2 values with ethanol and *tert*-BA (Figure 5b,d). Figure 5a,c also imply that the low slope value between pyridinic N vs. p-BQ and graphitic N vs. p-BQ suggested that pyridinic N and graphitic N did not control the formation of $O_2^{\bullet-}$.

Table 2. The second-order rate constants for the reaction between reactive oxidative species and specific quenchers.

	Scavengers	Molecular Formula	ROS Species	Reaction Rate Constant ($M^{-1}\,s^{-1}$)	References
1	Ethanol	C_2H_5OH	$SO_4^{\bullet-}$ and $^{\bullet}OH$	$k_{SO4\bullet-} = 9 \times 10^8$ $k_{HO\bullet} = 1.1 \times 10^6$	[63,71]
2	*tert*-Butyl alcohol	$C_4H_{10}O$	$^{\bullet}OH$	$k_{HO\bullet} = 4.5 \times 10^8$	[63]
3	*p*-Benzoquinone	$C_6H_4O_2$	$O_2^{\bullet-}$	$k_{obs.} = 9 \times 10^8$	[64,65]
4	L-histidine	$C_6H_9N_3O_2$	$^1O_2$	$k_{obs.} = 3 \times 10^8$	[16,67]
5	Furfuryl alcohol	$C_5H_6O_2$		$k_{obs.} = 1.2 \times 10^8$	
6	Phenol	C_5H_5OH	Surface-bound radicals ($SO_4^{\bullet-}$ and $^{\bullet}OH$)	$k_{SO4\bullet-} = 8.8 \times 10^9$ $k_{HO\bullet} = 6.6 \times 10^9$	[68,69]
7	Potassium iodide	KI	Surface bound complexes	–	[72,73]
8	Sodium perchlorate	$NaClO_4$	Free electrons	–	[43]

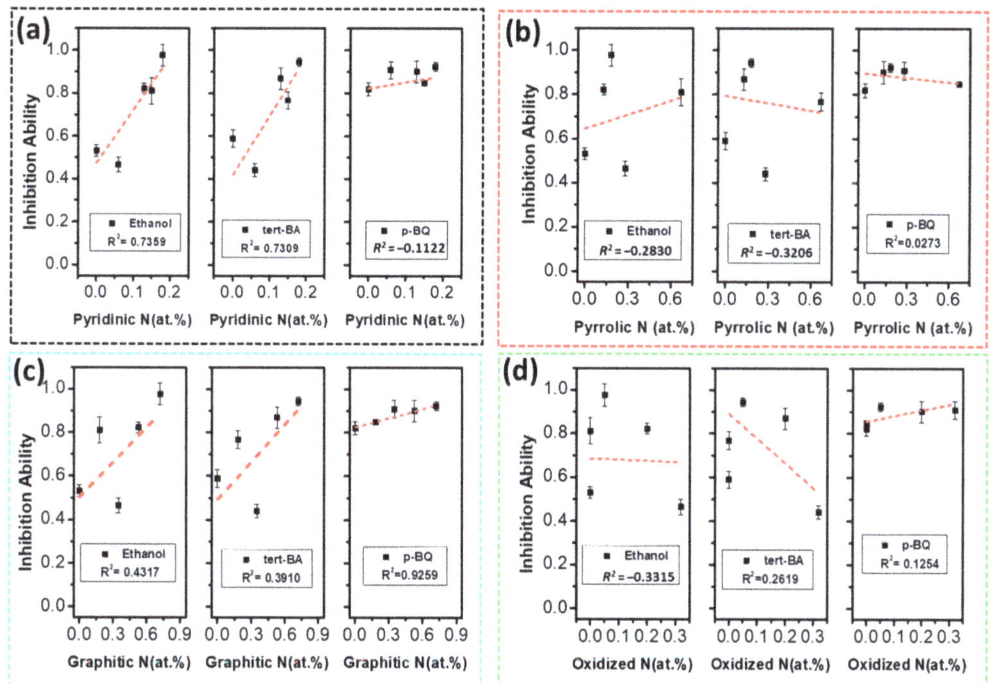

Figure 5. Role of radical ($O_2^{\bullet-}$, $^{\bullet}OH$ and $SO_4^{\bullet-}$) quenchers on ACP degradation rate (k_1(ACP)). Correlation between k_1 (ACP) and pyridinic N (**a**), k_1 (ACP) and pyrrolic N (**b**), k_1 (ACP) and graphitic N (**c**), and k_1 (ACP) and oxidized N (**d**). Quenching studies were carried out at C_{ACP} = 10 mg L^{-1}, $C_{Catalyst}$ = 100 mg L^{-1}, C_{PMS} = 0.5 mM, $C_{Ethanol}$ or $C_{tert-BA}$ = 500 mM, C_{p-BQ} = 50 mM.

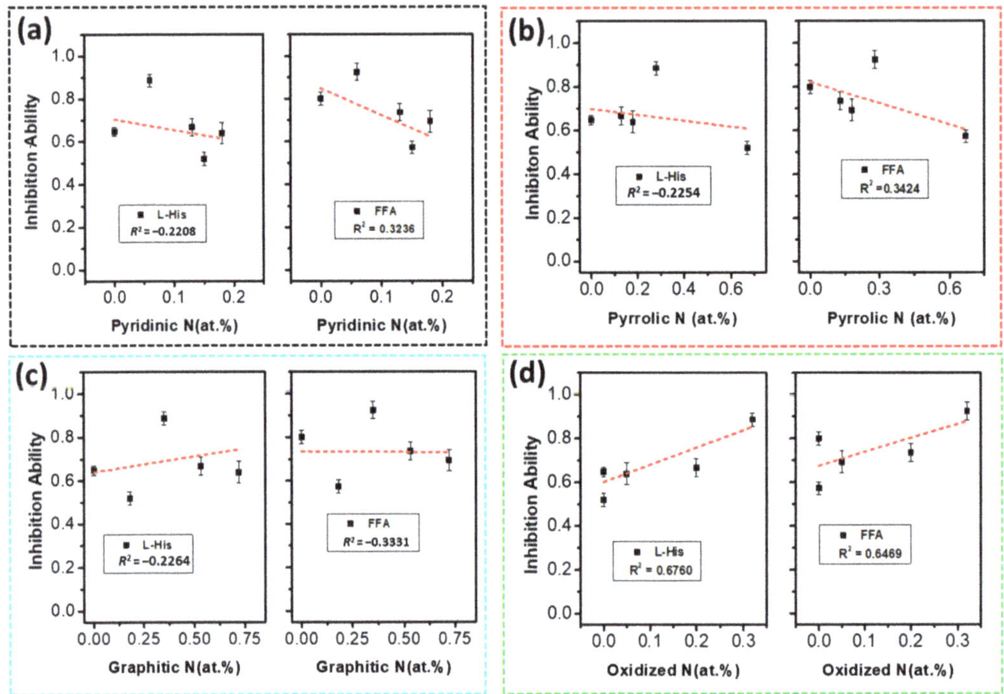

Figure 6. Role of singlet oxygen ($^1O_2$) quenchers on ACP degradation rate (k_1(ACP)). Correlation between k_1 (ACP) and pyridinic N (**a**), k_1 (ACP) and pyrrolic N (**b**), k_1 (ACP) and graphitic N (**c**), and k_1 (ACP) and oxidized N (**d**). Quenching studies were carried out at C_{ACP} = 10 mg L^{-1}, $C_{Catalyst}$ = 100 mg L^{-1}, C_{PMS} = 0.5 mM, C_{FFA} = $C_{L\text{-}His}$ = 50 mM.

Interestingly, the oxidized N exhibited meaningful correlation of R^2 = 0.6760 and R^2 = 0.6469 with L-His and FFA, respectively (Figure 6d), while other N-dopants provided a low R^2 value with $^1O_2$ quenchers. This proposes that the oxidized N has an auxiliary role in the $^1O_2$ mediative oxidative pathway. The inhibition ability of the non-free radicals in Figure 7a showed a high correlation (R^2 = 0.8929) between $NaClO_4$ and pyridinic N, emphasizing that pyridinic N accelerates catalytic oxidation by electron-transfer-based oxidation.

It also shows a decent correlation with phenol (R^2 = 0.8901), suggesting that the surface-bound radicals were also controlled via boundary pyridinic N. Moreover, graphitic N also yields notable R^2 = 0.8572 (KI), which implies that graphitic N can form activated complexes (carbocatalyst-PMS*) and initiate the electron-transfer process (Figure 7c). The N-dopant non-free radical activity correlation indicates that pyridinic N delivers more electron density to the carbon periphery and improves electron conductivity in the carbon framework. It could propagate the non-radial oxidative pathway through the formation of surface-bound radicals. Moreover, graphitic N facilitates electron-transfer based oxidation preferentially via the formation of carbocatalyst-PMS* activated complexes. W. Ren and his co-workers demonstrated that N-doped CNT profoundly enhances the PMS adsorption quantity and forms NCNT-PMS* complexes, which significantly raised the phenol oxidation efficiency through the electron-transfer regime [34].

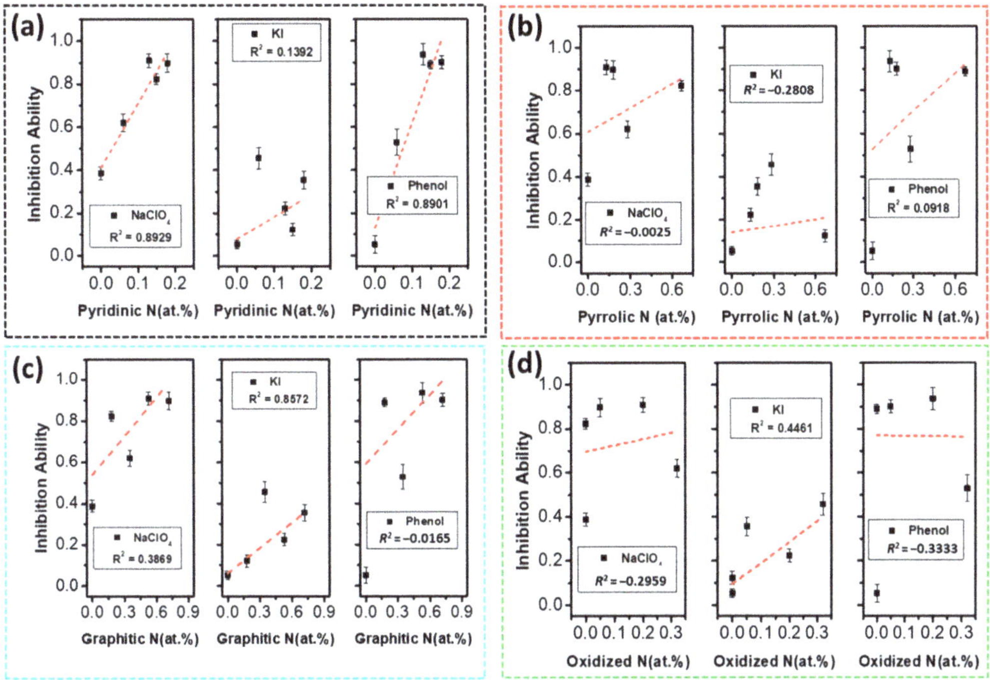

Figure 7. Role of non-radical (free electrons, surface-bound $^{\bullet}$OH and $SO_4^{\bullet -}$ radicals and surface-bound activated complexes) quenchers on ACP degradation (k_1(ACP)). Correlation between k_1 (ACP) and pyridinic N (**a**), k_1 (ACP) and pyrrolic N (**b**), k_1 (ACP) and graphitic N (**c**), and k_1 (ACP) and oxidized N (**d**). Quenching studies were carried out at C_{ACP} = 10 mg L^{-1}, $C_{Catalyst}$ = 100 mg L^{-1}, C_{PMS} = 0.5 mM, C_{FFA} = $C_{L\text{-}His}$ = 50 mM.

2.5.2. Role of O-Contents on Radical and Non-Radical Pathway

To investigate the oxygen functionalities–activity relationship, the normalized inhibition ability was correlated with O-contents. The obtained result was scrutinized alongside the individual oxygen functional groups and presented in Figure 8a–c. A good R^2 value is obtained between *p*-BQ and the carbonyl group (0.9827), while the hydroxyl group and carboxyl group yield very low R^2. This corroborates that a greater quantity of $O_2^{\bullet -}$ is produced with the increasing carbonyl content, highlighting that the carbonyl group crucially leads the $O_2^{\bullet -}$ -based oxidation. Furthermore, the inhibition ability of ethanol and *tert*-BA correlated with all the oxygen groups and showed no notable linear relationship (Figure 8b,c), implying that the $SO_4^{\bullet -}$/$^{\bullet}$OH-based oxidation pathway is not directly dependent on any specific oxygenated functional groups. The inhibition ability of L-His/FFA is also shown in Figure 9. Surprisingly, the negligible R^2 values of L-His/FFA with all the oxygen-containing functional group reveals that oxygen-containing functional groups did not directly govern the $^1O_2$ oxidation pathway. Figure 10b shows a high R^2 value with the carbonyl group (0.9472) and KI, suggesting that the -C=O group may play a major role in the surface-bound radical-based oxidation pathway. Furthermore, the hydroxyl group and carbonyl group exhibit a low R^2 value with KI, implying that -C-OH and COOH do not considerably regulate the surface-bound radicals (Figure 10a,c). On the other hand, the hydroxyl group provides a negative linear relationship with $NaClO_4$ (0.7485) and phenol (0.8932). It indicates that the presence of more electron-donating groups (-C-OH) could largely hinder the electron-transfer oxidative pathway, as well as attenuating the formation of NCNTs-PMS* activated complexes. This may be ascribed to the interplaying of oxygen functional groups and the fact that the primary carbon periphery adversely affected the surface-activated complex mediative oxidative process [16,42,53]. This study also demonstrates that formation of more hydroxyl groups on graphitic layer deliberately interrupts the non-free radical pathway (electrons and surface-bound complex). This observation also matches with Figure 3a, which shows the oxygen functional groups–catalytic activity relationship.

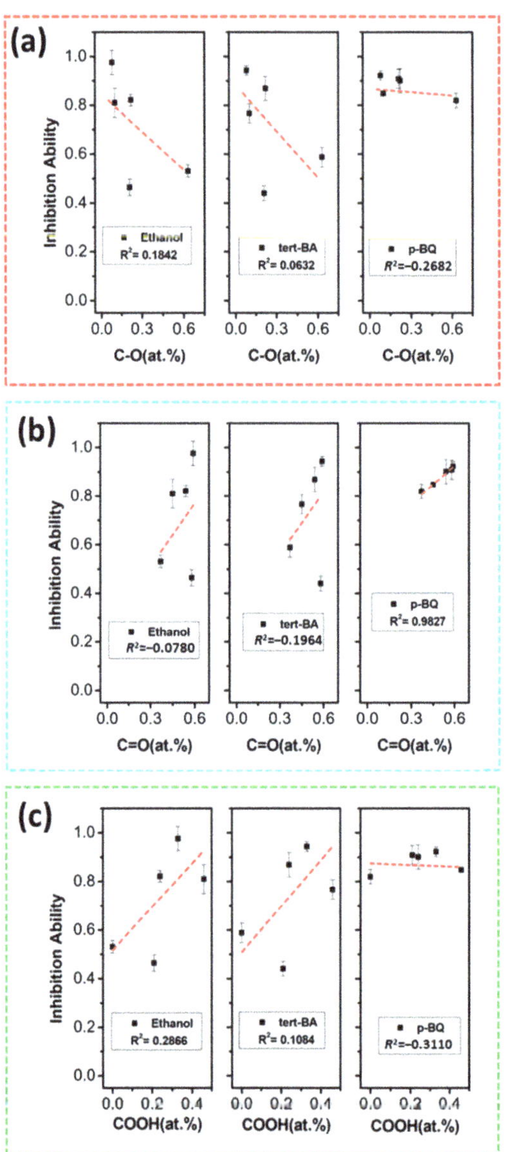

Figure 8. Role of radical ($O_2^{\bullet-}$, $^{\bullet}OH$ and $SO_4^{\bullet-}$) quenchers on ACP degradation rate (k_1). Correlation between k_1 -C-O (**a**), k_1 and -C=O (**b**) and k_1 and -COOH (**c**). Quenching studies were carried out at C_{ACP} = 10 mg L^{-1}, $C_{Catalyst}$ = 100 mg L^{-1}, C_{PMS} = 0.5 mM, $C_{Ethanol}$ or $C_{tert-BA}$ = 500 mM, C_{p-BQ} = 50 mM.

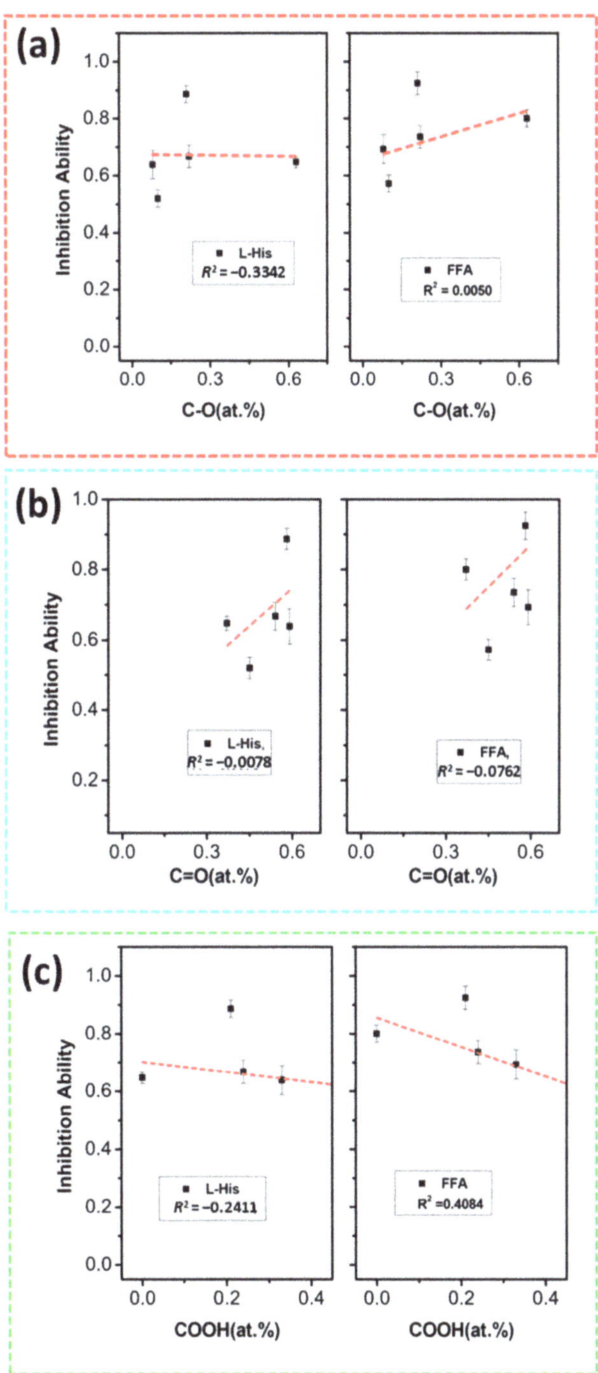

Figure 9. Role of singlet oxygen ($^1O_2$) on ACP degradation rate (k_1). Correlation between k_1 and -C-O (**a**), k_1 and -C=O (**b**) and k_1 and -COOH (**c**). Quenching studies were carried out at C_{ACP} = 10 mg L^{-1}, $C_{Catalyst}$ = 100 mg L^{-1}, C_{PMS} = 0.5 mM, C_{FFA} = $C_{L\text{-}His}$ = 50 mM.

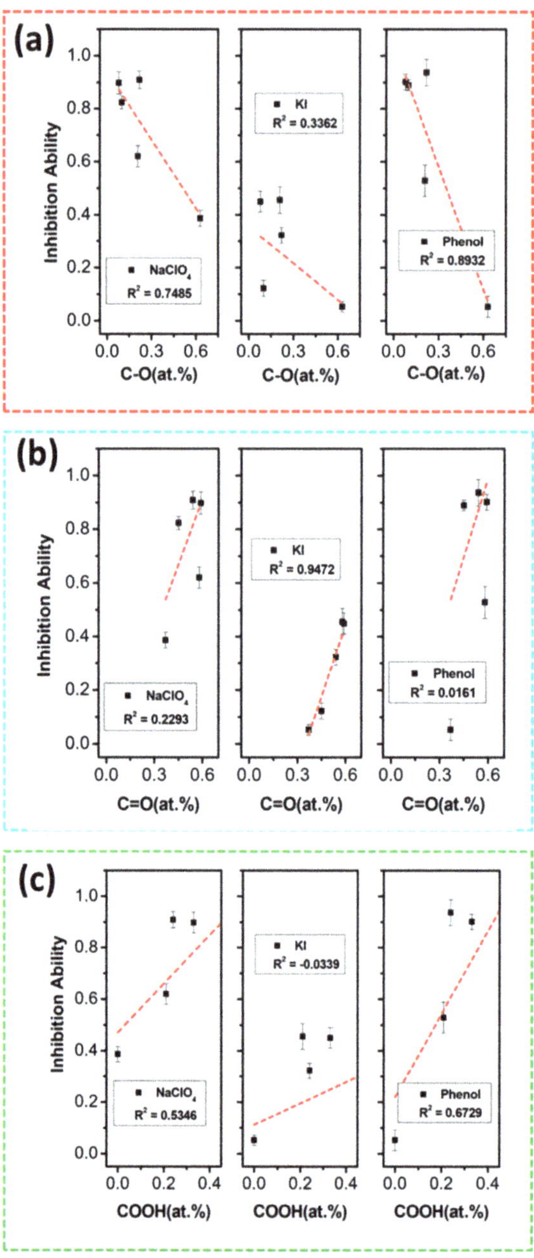

Figure 10. Role of non-radical (free electrons, surface-bound radicals ($^{\bullet}$OH and $SO_4^{\bullet-}$) and surface-bound activated complex) quenchers on ACP degradation rate (k_1). Correlation between k_1 -C-O (**a**), k_1 and -C=O (**b**) and k_1 and -COOH (**c**). Quenching studies were carried out at C_{ACP} = 10 mg L^{-1}, $C_{Catalyst}$ = 100 mg L^{-1}, C_{PMS} = 0.5 mM, C_{FFA} = $C_{L\text{-His}}$ = 50 mM.

2.6. Insights into NCNTs/PMS and Unveiling the Active Sites

The catalytic performance of nitrogen-doped CNTs is fundamentally determined by their intrinsic electronic configurations (sp^2- and sp^3- hybridization). N-groups and the oxygen-containing functional groups, and the structural defects (edges and vacancies) are believed to be key active sites for catalytic oxidation [32–37]. This present catalytic system demonstrates that high catalytic activity is noticed with Uz-NCNT-4/PMS (k_1(ACP) = 28.35 × 10^{-4} s^{-1}), rather than pristine CNT, NCNT, Uz-NCNT-2 and Uz-NCNT-4. The summary of the structure–catalytic activity regression analysis (Table S6) affirms that pyridinic N and graphitic N groups are major active sites for high catalytic oxidation. Thus, the presence of a high content of pyridinic N (0.18 at. %) and graphitic N (0.72 at. %) with Uz-NCNT-4 shows high catalytic activity. This hypothesis also matched with the structure–catalytic activity by the specific reactive species relationship, as depicted in Table 3. The assessment demonstrates that pyridinic N, as well as substitutional N (graphitic N), cooperatively lead the catalytic oxidation through radical as well as non-radical pathways.

Table 3. Role of N-dopants and oxygen functional groups on carbocatalyst/PMS activation pathway.

Carbocatalyst/PMS Activation Pathway	Reactive Oxidative Species	Governing N-Dopants	Correlation Coefficient (R^2)	Governing O-Groups	Correlation Coefficient (R^2)
Radical Species	$^{\bullet}OH/SO_4^{\bullet -}$	Pyridinic N	0.7359 (Ethanol) 0.7309 (*tert*-BA)	—	—
	$O_2^{\bullet -}$	Graphitic N	0.9259 (*p*-BQ)	-C=O	0.9827 (*p*-BQ)
Singlet Oxygen	$^1O_2$	—	—	—	—
Non-Radical Species	Free flowing electrons	Pyridinic N	0.8929 (NaClO$_4$)	-C-O	0.7485 (NaClO$_4$) (Downtrend)
	Surface-bound radicals	Graphitic N	0.8572 (KI)	-C=O	0.9472 (KI)
	Carbocatalyst-PMS* activated complex	Pyridinic N	0.8901 (Phenol)	-C-O	0.8932 (Phenol) (Downtrend)

The sp^2-hybridized N atom in a six-membered pyridinic ring is typically produced at the edges and defects of the main graphitic network and provides one p-electron to the conjugated π system. Consequently, pyridinic N can increase the density of π states of the graphitic network, which can facilitate the notable electron-transfer oxidative pathway (Figure 11a). Pyridinic N activity-inhibition ability with NaClO$_4$ correlation (R^2 = 0.8929, Table 3) supports this assumption. It means that the free flowing π electrons from sp^2-C in the graphitic carbon network can be activated through conjugating with the lone-pair electrons of pyridinic N [16]. The detectable R^2 value of pyridinic N with ethanol/*tert*-BA recommends that the terminal and edges of pyridinic N may also help to generate $SO_4^{\bullet -}/^{\bullet}OH$ through the interaction between the O-O bond of HO-O-SO$_3^-$ and pyridinic N (Figure 11b). Luo and coworkers suggested that pyridinic N with a lone pair electron can significantly improve the π conjugation [52]. Graphitic N can also be another key active site, as shown in this study. The smaller atomic radius and high electronegativity graphitic N (3.04) compared to the adjacent carbon atoms (2.55) can stretch the electrons from neighboring carbon atoms [34]. These activated carbon centers can potentially interact with HO-O-SO$_3^-$ to form a highly reactive complex (NCNT–PMS*) via the direct electron-abstraction step and generate surface-confined free radicals on the catalyst surface (Figure 11c). This hypothesis is witnessed from the correlation between graphitic N-inhibition ability by KI (R^2 = 0.8572, Table 3). Structure–oxidative species activity also substantiates that graphitic N can induce the radical pathway via $O_2^{\bullet -}$ in the catalytic system (Figure 11e).

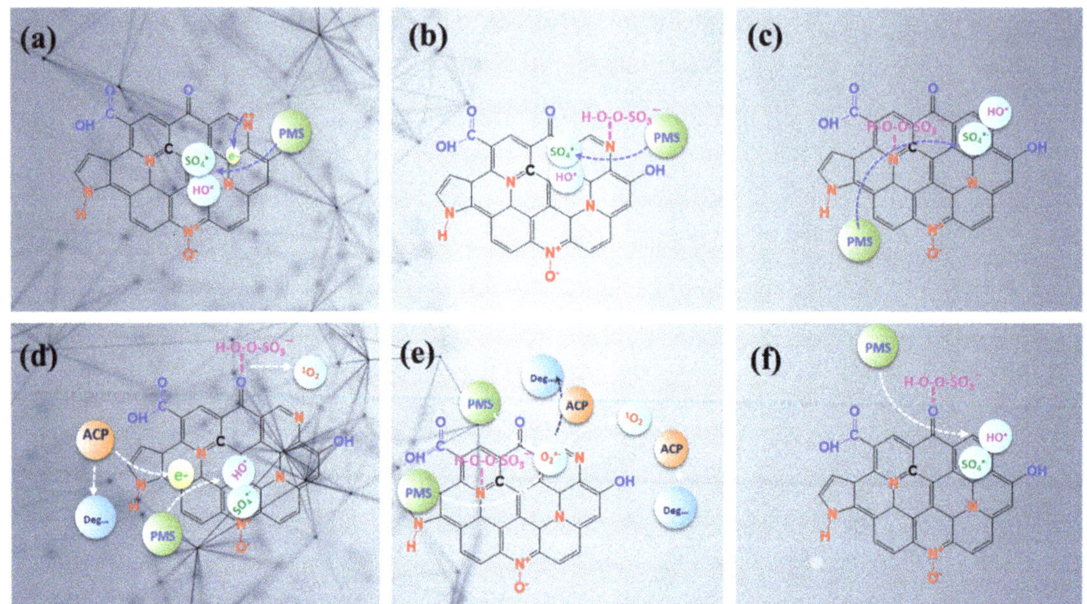

Figure 11. Proposed mechanism in PMS activation by NCNTs, non-radical pathway (**a–d**) and radical pathway (**e–f**).

Uz-NCNT-4 with the highest content of carbonyl groups (0.59 at. %), along with low contents of the hydroxyl group (0.08 at. %), shows superior catalytic performances. Tables S6 and 3 show that the carbonyl group can act as the principal active site rather than the other oxygen-containing functional groups. The electron-rich carbonyl group can act as an electron donor. Thus, the electrophilic group (peroxide bond) in PMS can be activated with the nucleophilic carbonyl group by electron-transfer and then produces reactive free radicals (Figure 11d–f). A high R^2 value (0.9427) between KI and the carbonyl group affirms that electron transfer between PMS and the carbonyl group can form surface-bound radicals. These electron-transfer processes and active surface-confined free radicals perform remarkable catalytic oxidation. J. Li and co-workers revealed that nucleophilic carbonyl groups can provoke a redox cycle to produce $SO_4^{\bullet-}/^{\bullet}OH$ [26].

Furthermore, the abundant hydroxyl group can relay a contrary role to catalytic oxidation. This may be explained by the fact that the electrophilic O-O group in PMS can be readily adsorbed with the electron-rich hydroxyl group [34]. Because the H atom in the hydroxyl group dissociates itself and binds with the PMS molecule, the electron transfer process is relatively low; thus, -C-OH in the graphitic network can convert to -C=O. This functional group conversion step may unfeasible for efficient PMS activation, as reported from the experimental observation with graphene-catalyzed PMS by X-Duan et.al [74]. The present study also substantiates that there is no good correlation between the $^1O_2$ quencher and the N-groups or O-groups, suggesting that N-dopants/oxygen-containing functional groups do not directly initiate the $^1O_2$. However, the EPR results show that $^1O_2$ is intrinsically produced in the catalytic system (Figure 3a). Typically, the formation of $^1O_2$ can include different pathways, including self-decomposition of PMS, the electron-transfer reaction from carbocatalyst to PMS and the coupling or direct oxidation of $O_2^{\bullet-}$ [9,75]. In addition, the detectable R^2 value with the carboxyl group (Figures 3b and 10c) indicates that non-bonded lone pair electrons in oxygen atoms in the carboxyl group can partially support the electron-transfer process for catalytic activity. Overall, Tables 3 and S6 clearly demonstrate that N-doping and oxygen functional groups synergistically contribute to the radical and non-radical formation and ACP degradation process.

2.7. Activity Stability Test

To determine the activity stability of the carbocatalysts, the carbocatalysts were evaluated through a reusability test over three consecutive runs and this is depicted in Figure S8. Around 53, 94, 96, 99, and 96% of ACP degradation efficiency was observed with CNT, NCNT, Uz-NCNT-2, Uz-NCNT-4 and Uz-NCNT-8, respectively. However, approximately <5% of ACP degradation efficiency was reduced at the 3rd cycle. The deactivation of the carbocatalyst may be due to the intricate influences of surface characteristics and structural changes, including the adsorption of intermediates/by-products, coverage of active sites and loss of a small quantity of catalyst during filtration/separation. This phenomenon can inhibit the charge transfer process between the carbocatalyst and PMS, resulting in suppressed ACP oxidation [46]. Furthermore, Uz-NCNT-4 demonstrated a high ACP degradation efficiency with all runs compared to the other catalyst, confirming that the chemical oxidation of NCNT with 4 g KMnO$_4$ could remarkably improve the reusability and stability.

3. Materials and Methods

3.1. Materials

PMS (available as Oxone, KHSO$_5$·0.5KHSO$_4$·0.5K$_2$SO$_4$), dicyandiamide (99.5%, C$_2$H$_4$N$_4$), ethyl alcohol (99.5%, C$_2$H$_5$OH), phosphoric acid (85%, H$_3$PO$_4$), sulfuric acid (Conc. H$_2$SO$_4$), hydrogen peroxide (30%, H$_2$O$_2$) solution and sodium hydroxide (NaOH) were purchased from Sigma Aldrich, Seoul, South Korea. In addition, HPLC grade methanol (99%, CH$_3$OH) for HPLC was purchased from Honeywell, Seoul, South Korea. Pristine multiwalled carbon nanotubes (MWCNT), with a diameter of 5–15 nm and length of ~10 μm, were obtained from Carbon Nano-Materials Technology Co., Ltd., Seoul, South Korea. All chemicals were laboratory grade and were used without further purifications.

3.2. Preparation of NCNT and Uz-NCNTs with Different Degree of Oxidation

First, 1 g of pristine CNT powder and 0.48 g dicyandiamide were dispersed into 100 mL ethanol solution. The solution mixture was constantly stirred at 80 °C until the solution was evaporated to obtain a homogeneous mixture. The mixture was kept into a tubular furnace and pyrolyzed for 3 h at 600 °C under N$_2$ atmosphere, with a gas flow rate of 20 mL min^{-1}. After cooling down to room temperature, the obtained powder was sufficiently ground and labelled as NCNT. Later, an unzipped NCNT (Uz-NCNT) carbocatalyst was synthesized based on a previously reported chemical oxidation method [39,40]. Then, 1 g of as-prepared NCNT was dispersed into 100 mL of concentrated H$_2$SO$_4$ solution for 1 h, then 20 mL of H$_3$PO$_4$ was slowly added to the above mixture and continuously stirred in an oil bath at 80 °C. Subsequently, a desirable amount of oxidizing agent KMnO$_4$ was slowly introduced, and the reaction mixture was left to react for 2 h. Later, the warm reaction mixture was carefully transferred into 500 mL of cold water containing 15 mL of H$_2$O$_2$. Then, the solution was centrifuged, washed with dilute HCl solution and again 3–5 times with distilled water. Finally, the obtained precipitates were dried by freeze-drying. Different degrees of oxidized unzipped NCNT carbocatalyst named as Uz-NCNT-2, Uz-NCNT-4 and Uz-NCNT-8 were synthesized by varying the amount of oxidizing agent KMnO$_4$ with 2, 4 and 8 g, respectively.

3.3. Characterizations of Carbocatalyst

Surface functional groups of the carbocatalyst were analyzed using Fourier-transform infrared spectroscopy (FT-IR) on a PerkinElmer instrument (Spectrum One System, Perkin-Elmer, Waltham, MA, USA). UV-visible spectra were recorded using a UV-Visible spectrometer (UV 1240, Shimadzu, Kyoto, Japan). N$_2$ adsorption/desorption was measured by BELSORP-max, BEL, Japan, Inc., Tokyo, Japan. The Brunauer-Emmett-Teller (BET) equation and Barrett-Joyner-Halenda (BJH) method was applied to determine the specific surface area and pore size distribution of the carbocatalyst, respectively. X-ray diffraction (XRD) patterns were obtained on a D8-Advanced diffractometer system from Bruker (Bruker, Karlsruhe, Germany) with $k\alpha$ radiation (λ = 1.5418 Å). The interlayer distance of the graphitic plane was determined by Bragg's equation, $n\lambda = 2d\sin\theta$, where λ is the incident wavelength (1.5406 Å), θ is the peak position, n is the order of diffraction (1), and d is the interlayer spacing or d-spacing value. Surface composition of the catalyst was examined with a K-Alpha TM X-ray photoelectron spectrometer (XPS) system (Thermo Electron, Waltham, MA, USA).

3.4. Evaluation of Catalytic Performance

Acetaminophen (ACP, $C_8H_9NO_2$, 151.16 g mol^{-1}) degradation tests were carried out to explore the catalytic performance of the NCNT and Uz-NCNTs regarding PMS activation. A total of 100 mg L^{-1} of carbocatalyst was added to 100 mL of ACP solution (10 mg L^{-1}) and stirred constantly (500 rpm) under dark conditions. At a given interval, 3 mL of solution was taken and filtered with a 0.22 µm pore size PVDF membrane filter. The concentration of ACP was analyzed using high-performance liquid chromatography (HPLC, Ultimate 3000, Dionex, Thermo Fisher Scientific, Waltham, MA, USA) with UV absorbance at 230 nm. The reverse-based separation was carried out on an Eclipse XDB-C18 analytical column (Eclipse Plus, 4.6 × 250 mm, 5 µm, Agilent, Santa Clara, CA, USA). The mobile phase was prepared with methanol and deionized water (1:2, *v:v*), and the mixed solution was adjusted by phosphoric acid until the pH reached 3. The flow rate, temperature, and injected volume were 0.8 mL min^{-1}, 25 °C, and 25 µL, respectively. All kinetic catalytic experiments were performed in triplicate and the relative standard errors were less than 5%.

The first-order kinetic equation was used and expressed as $-\ln(C_t/C_0) = k_1 t$, where k_1 represents the apparent rate constant (s^{-1}), and C_0 and C_t are the initial concentration (mg L^{-1}) and concentration at time t of ACP in solution, respectively [25]. The PMS concentration was also measured by KI, Na_2HCO_3, and a UV–Vis spectrophotometer (UV 1240, Shimadzu, Tokyo, Japan) [41]. The yellow color developed by the reaction between PS and iodide in the presence of bicarbonate exhibits an absorbance at 395 nm. PMS decomposition rate was also determined using first order kinetics. Moreover, the activity stability of the carbocatalyst was also determined by reusability tests with four consecutive steps.

3.5. Evolution of Reactive Species Quenching Study and EPR Analysis

Ethanol (C_5H_5OH), *tert*-butyl alcohol (*tert*-BA, $(CH_3)_3OH$), *para*-benzoquinone (*p*-BQ, $C_6H_4O_2$), furfuryl alcohol (FFA, $C_5H_6O_2$), L-histidine (L-His, $C_6H_9N_3O_2$), potassium iodide (KI), phenol (C_6H_5OH), and sodium perchlorate ($NaClO_4$) were employed to detect the reactive oxidative species involved in the reaction. Furthermore, electron spin resonance spectra were observed with a JES-FA200 (Joel, Tokyo, Japan, X-band) at ambient temperature under visible light irradiation (λ > 420 nm), using 5,5-dimethyl-1-pyrroline-*N*-oxide (DMPO) and 2,2,6,6-tetramethyl-4-pi-peridone (TMP) as the spin trapping agents.

4. Conclusions

Different levels of oxidized nitrogen-doped CNTs (Uz-NCNTs) were synthesized by varying the oxidizing agent $KMnO_4$. The specific surface area substantially decreased with increasing oxidation levels of N-doped CNT. On the other hand, the amount of oxygen 0.89 at. % (CNT) evidently increased to 3.04, 21.5, 24.58, and 32.16 at. % with NCNT, Uz-NCNT-2, Uz-NCNT-4 and Uz-NCNT-8, respectively. Catalytic performance of the carbocatalyst was examined via PMS activation for ACP degradation. Uz-NCNT-4 exhibited high catalytic activity for ACP degradation through PMS activation with a 6.5-fold enhancement, compared to the CNT activity.

The structure–catalytic activity relationship exposes the fact that pyridinic N and carbonyl groups play a key role in the catalytic system. EPR and classical quenching studies reveal that the radical ($SO_4^{\bullet -}$, $^{\bullet}OH$ and $O_2^{\bullet -}$), $^1O_2$ and non-radical species (surface-bound radical, electrons, and surface-bound complex) intrinsically contribute to catalytic oxidation. The mechanism of PMS activation on the carbocatalyst was illustrated based on the structure-oxidative species activity correlation. N-doping and oxygen-containing functional groups propagate both radial and non-radical mediative oxidation pathways. Existence of electron rich Pyridinic N and graphitic N configuration can boost up the greater electron density of the graphitic network and lead to significant electron transfer-based oxidation via surface-bound radicals. Moreover, the quantity of $O_2^{\bullet -}$ generation is steadily raised with the increase in graphitic N and the carbonyl group. A large amount of oxygen-containing functional groups greatly inhibits the $SO_4^{\bullet -}/^{\bullet}OH$ mediative oxidation pathway. The linear regression analysis also emphasized that the presence of more hydroxyl groups might be unfavorable for catalytic activity.

Supplementary Materials: The following supporting information can be downloaded at: https://www.mdpi.com/article/10.3390/catal12080845/s1, Figure S1: (a) XRD patterns and (b) Raman spectrum of CNT, NCNT and unzipped NCNTs; Figure S2: (a) UV-visible spectrum and (b) FT-IR spectrum of CNT, NCNT and unzipped NCNTs; Figure S3: BET N_2 adsorption/desorption profile of (a) CNT, (b) NCNT, (c) Uz-NCNT-2, (d) Uz-NCNT-4, (e) Uz-NCNT-8 and (f) BJH pore diameter.; Figure S4: (a) XPS survey spectrum, (b) N 1s core-level spectra, deconvoluted N 1s core-level spectra of (c) NCNT, (d) Uz-NCNT-2, (e) Uz-NCNT-4 and (f) Uz-NCNT-8; Figure S5: (a) C 1s spectra, and deconvoluted O 1s core-level spectra of (b) NCNT, (c) Uz-NCNT-2, (d) Uz-NCNT-4 and (e) Uz-NCNT-8; Figure S6: (a) Effect of various chemical quenchers on normalized ACP concentration at (a) CNT/PMS, (b) NCNT/PMS, (c) Uz-NCNT-2/PMS, (d) Uz-NCNT-4/PMS and (e) Uz-NCNT-8/PMS systems. Quenching studies were performed with C_{ACP} = 10 mg L^{-1}, $C_{Catalyst}$ = 100 mg L^{-1}, C_{PMS} = 0.5 mM, $C_{Ethanol}$ or $C_{tert-BA}$ = C_{p-BQ} = 500 mM or C_{FFA} = C_{L-His} = C_{NaClO4} = C_{phenol} C_{KI} = 50 mM; Figure S7: Impact of chemical scavengers on k_1 (ACP) under different catalytic systems. Quenching studies were carried out at C_{ACP} = 10 mg L^{-1}, $C_{Catalyst}$ = 100 mg L^{-1}, C_{PMS} = 0.5 mM, $C_{Ethanol}$ or $C_{tert-BA}$ = C_{p-BQ} = 500 mM or C_{FFA} = C_{L-His} = C_{NaClO4} = C_{phenol} = C_{KI} = 50 mM; Figure S8: ACP degradation efficiency by reusability test in three consecutive runs. Experimental conditions C_{ACP} = 10 mg L^{-1}, $C_{Catalyst}$ = 100 mg L^{-1}, C_{PMS} = 0.5 mM and pH_0 = 7 $\pm$ 0.3; Table S1: Summary of N-doped CNT and proposed mechanistical pathway; Table S2: Details of Raman spectra and I_D/I_G ratio; Table S3: BET surface area and BJH pore diameter of derived carbocatalysts; Table S4: Atomic percentage of N-dopants in carbocatalysts was determined from N 1s spectra; Table S5: Atomic percentage of oxygen containing functional groups in carbocatalysts was determined from O 1s spectra; Table S6: Summary of key active sites in carbocatalyst/PMS activation for ACP degradation.

Author Contributions: Conceptualization and methodology, K.G.; validation and formal analysis, D.-G.K. and S.-O.K.; investigation, resources, data curation and writing—original draft preparation, K.G.; writing—review and editing, D.-G.K. and S.-O.K.; visualization and supervision, S.-O.K.; project administration, and funding acquisition, S.-O.K. All authors have read and agreed to the published version of the manuscript.

Funding: This work was supported by a National Research Foundation of Korea (NRF) grant funded by the Korean government (MSIT) (No. 2022R1A2B5B02001584).

Data Availability Statement: The data presented in this study are available upon request from the corresponding author.

Conflicts of Interest: The authors declare that they have no known competing financial interests or personal relationships that could have appeared to influence the work reported in this paper.

References

1. Oyekunle, D.T.; Zhou, X.; Shahzad, A.; Chen, Z. Review on Carbonaceous Materials as Persulfate Activators: Structure-Performance Relationship, Mechanism and Future Perspectives on Water Treatment. *J. Mater. Chem. A* **2021**, *9*, 8012–8050. [CrossRef]
2. Gao, Y.; Wang, Q.; Ji, G.; Li, A. Degradation of Antibiotic Pollutants by Persulfate Activated with Various Carbon Materials. *Chem. Eng. J.* **2022**, *429*, 132387. [CrossRef]
3. Wu, S.; Yu, L.; Wen, G.; Xie, Z.; Lin, Y. Recent Progress of Carbon-Based Metal-Free Materials in Thermal-Driven Catalysis. *J. Energy Chem.* **2021**, *58*, 318–335. [CrossRef]
4. Huang, W.; Xiao, S.; Zhong, H.; Yan, M.; Yang, X. Activation of Persulfates by Carbonaceous Materials: A Review. *Chem. Eng. J.* **2021**, *418*, 129297. [CrossRef]
5. Chen, Y.; Wei, J.; Duyar, M.S.; Ordomsky, V.V.; Khodakov, A.Y.; Liu, J. Carbon-Based Catalysts for Fischer-Tropsch Synthesis. *Chem. Soc. Rev.* **2021**, *50*, 2337–2366. [CrossRef]
6. Hu, H.; Du, Y.; Duan, X. Functional Materials for Eco-catalysis of Small Molecules. *EcoMat* **2021**, *3*, e12121. [CrossRef]
7. Huang, X.; Wu, S.; Xiao, Z.; Kong, D.; Liang, T.; Li, X.; Luo, B.; Wang, B.; Zhi, L. Predicting the Optimal Chemical Composition of Functionalized Carbon Catalysts towards Oxidative Dehydrogenation of Ethanol to Acetaldehyde. *Nano Today* **2022**, *44*, 101508. [CrossRef]
8. Ding, Y.; Wang, X.; Fu, L.; Peng, X.; Pan, C.; Mao, Q.; Wang, C.; Yan, J. Nonradicals Induced Degradation of Organic Pollutants by Peroxydisulfate (PDS) and Peroxymonosulfate (PMS): Recent Advances and Perspective. *Sci. Total Environ.* **2021**, *765*, 142794. [CrossRef]
9. Duan, X.; Sun, H.; Shao, Z.; Wang, S. Nonradical Reactions in Environmental Remediation Processes: Uncertainty and Challenges. *Appl. Catal. B Environ.* **2018**, *224*, 973–982. [CrossRef]
10. Hu, P.; Su, H.; Chen, Z.; Yu, C.; Li, Q.; Zhou, B.; Alvarez, P.J.J.; Long, M. Selective Degradation of Organic Pollutants Using an Efficient Metal-Free Catalyst Derived from Carbonized Polypyrrole via Peroxymonosulfate Activation. *Environ. Sci. Technol.* **2017**, *51*, 11288–11296. [CrossRef]
11. Zhang, T.; Zhu, H.; Croué, J.P. Production of Sulfate Radical from Peroxymonosulfate Induced by a Magnetically Separable CuFe2O4 Spinel in Water: Efficiency, Stability, and Mechanism. *Environ. Sci. Technol.* **2013**, *47*, 2784–2791. [CrossRef]
12. Anipsitakis, G.P.; Dionysiou, D.D.; Gonzalez, M.A. Cobalt-Mediated Activation of Peroxymonosulfate and Sulfate Radical Attack on Phenolic Compounds. Implications of Chloride Ions. *Environ. Sci. Technol.* **2006**, *40*, 1000–1007. [CrossRef]
13. Anipsitakis, G.P.; Stathatos, E.; Dionysiou, D.D. Heterogeneous Activation of Oxone Using Co3O4. *J. Phys. Chem. B* **2005**, *109*, 13052–13055. [CrossRef]
14. Wang, J.; Wang, S. Activation of Persulfate (PS) and Peroxymonosulfate (PMS) and Application for the Degradation of Emerging Contaminants. *Chem. Eng. J.* **2018**, *334*, 1502–1517. [CrossRef]
15. Liu, Y.; Wang, L.; Dong, Y.; Peng, W.; Fu, Y.; Li, Q.; Fan, Q.; Wang, Y.; Wang, Z. Current Analytical Methods for the Determination of Persulfate in Aqueous Solutions: A Historical Review. *Chem. Eng. J.* **2021**, *416*, 129143. [CrossRef]
16. Chen, X.; Oh, W.D.; Lim, T.T. Graphene- and CNTs-Based Carbocatalysts in Persulfates Activation: Material Design and Catalytic Mechanisms. *Chem. Eng. J.* **2018**, *354*, 941–976. [CrossRef]
17. Cheng, X.; Guo, H.; Zhang, Y.; Wu, X.; Liu, Y. Non-Photochemical Production of Singlet Oxygen via Activation of Persulfate by Carbon Nanotubes. *Water Res.* **2017**, *113*, 80–88. [CrossRef]
18. Liang, P.; Zhang, C.; Duan, X.; Sun, H.; Liu, S.; Tade, M.O.; Wang, S. An Insight into Metal Organic Framework Derived N-Doped Graphene for the Oxidative Degradation of Persistent Contaminants: Formation Mechanism and Generation of Singlet Oxygen from Peroxymonosulfate. *Environ. Sci. Nano* **2017**, *4*, 315–324. [CrossRef]
19. Yun, E.T.; Yoo, H.Y.; Bae, H.; Kim, H.I.; Lee, J. Exploring the Role of Persulfate in the Activation Process: Radical Precursor Versus Electron Acceptor. *Environ. Sci. Technol.* **2017**, *51*, 10090–10099. [CrossRef]
20. Lee, H.; Lee, H.J.; Jeong, J.; Lee, J.; Park, N.B.; Lee, C. Activation of Persulfates by Carbon Nanotubes: Oxidation of Organic Compounds by Nonradical Mechanism. *Chem. Eng. J.* **2015**, *266*, 28–33. [CrossRef]
21. Wang, X.; Qin, Y.; Zhu, L.; Tang, H. Nitrogen-Doped Reduced Graphene Oxide as a Bifunctional Material for Removing Bisphenols: Synergistic Effect between Adsorption and Catalysis. *Environ. Sci. Technol.* **2015**, *49*, 6855–6864. [CrossRef]
22. Duan, X.; Su, C.; Zhou, L.; Sun, H.; Suvorova, A.; Odedairo, T.; Zhu, Z.; Shao, Z.; Wang, S. Surface Controlled Generation of Reactive Radicals from Persulfate by Carbocatalysis on Nanodiamonds. *Appl. Catal. B Environ.* **2016**, *194*, 7–15. [CrossRef]
23. Lee, J.; Von Gunten, U.; Kim, J.H. Persulfate-Based Advanced Oxidation: Critical Assessment of Opportunities and Roadblocks. *Environ. Sci. Technol.* **2020**, *54*, 3064–3081. [CrossRef]
24. Pham, V.L.; Kim, D.G.; Ko, S.O. Catalytic Degradation of Acetaminophen by Fe and N Co-Doped Multi-Walled Carbon Nanotubes. *Environ. Res.* **2021**, *201*, 111535. [CrossRef]
25. Kim, D.G.; Ko, S.O. Advanced Oxidative Degradation of Acetaminophen by Carbon Catalysts: Radical vs Non-Radical Pathways. *Environ. Res.* **2020**, *188*, 109767.
26. Li, J.; Li, M.; Sun, H.; Ao, Z.; Wang, S.; Liu, S. Understanding of the Oxidation Behavior of Benzyl Alcohol by Peroxymonosulfate via Carbon Nanotubes Activation. *ACS Catal.* **2020**, *10*, 3516–3525. [CrossRef]
27. Duan, X.; Sun, H.; Wang, S. Metal-Free Carbocatalysis in Advanced Oxidation Reactions. *Acc. Res.* **2018**, *51*, 678–687. [CrossRef]

28. Duan, X.; Ao, Z.; Zhou, L.; Sun, H.; Wang, G.; Wang, S. Occurrence of Radical and Nonradical Pathways from Carbocatalysts for Aqueous and Nonaqueous Catalytic Oxidation. *Appl. Catal. B Environ.* **2016**, *188*, 98–105. [CrossRef]
29. Ma, W.; Wang, N.; Fan, Y.; Tong, T.; Han, X.; Du, Y. Non-Radical-Dominated Catalytic Degradation of Bisphenol A by ZIF-67 Derived Nitrogen-Doped Carbon Nanotubes Frameworks in the Presence of Peroxymonosulfate. *Chem. Eng. J.* **2018**, *336*, 721–731. [CrossRef]
30. Chen, J.; Zhang, L.; Huang, T.; Li, W.; Wang, Y.; Wang, Z. Decolorization of Azo Dye by Peroxymonosulfate Activated by Carbon Nanotube: Radical versus Non-Radical Mechanism. *J. Hazard. Mater.* **2016**, *320*, 571–580. [CrossRef]
31. Yan, Y.; Miao, J.; Yang, Z.; Xiao, F.X.; Yang, H.B.; Liu, B.; Yang, Y. Carbon Nanotube Catalysts: Recent Advances in Synthesis, Characterization and Applications. *Chem. Soc. Rev.* **2015**, *44*, 3295–3346. [CrossRef] [PubMed]
32. Guan, C.; Jiang, J.; Luo, C.; Pang, S.; Yang, Y.; Wang, Z.; Ma, J.; Yu, J.; Zhao, X. Oxidation of Bromophenols by Carbon Nanotube Activated Peroxymonosulfate (PMS) and Formation of Brominated Products: Comparison to Peroxydisulfate (PDS). *Chem. Eng. J.* **2018**, *337*, 40–50. [CrossRef]
33. Duan, X.; Sun, H.; Wang, Y.; Kang, J.; Wang, S. N-Doping-Induced Nonradical Reaction on Single-Walled Carbon Nanotubes for Catalytic Phenol Oxidation. *ACS Catal.* **2015**, *5*, 553–559. [CrossRef]
34. Ren, W.; Nie, G.; Zhou, P.; Zhang, H.; Duan, X.; Wang, S. The Intrinsic Nature of Persulfate Activation and N-Doping in Carbocatalysis. *Environ. Sci. Technol.* **2020**, *54*, 6438–6447. [CrossRef] [PubMed]
35. Sun, H.; Kwan, C.K.; Suvorova, A.; Ang, H.M.; Tadé, M.O.; Wang, S. Catalytic Oxidation of Organic Pollutants on Pristine and Surface Nitrogen-Modified Carbon Nanotubes with Sulfate Radicals. *Appl. Catal. B Environ.* **2014**, *154–155*, 134–141. [CrossRef]
36. Duan, X.; Ao, Z.; Sun, H.; Zhou, L.; Wang, G.; Wang, S. Insights into N-Doping in Single-Walled Carbon Nanotubes for Enhanced Activation of Superoxides: A Mechanistic Study. *Chem. Commun.* **2015**, *51*, 15249–15252. [CrossRef]
37. Wang, C.; Kang, J.; Liang, P.; Zhang, H.; Sun, H.; Tadé, M.O.; Wang, S. Ferric Carbide Nanocrystals Encapsulated in Nitrogen-Doped Carbon Nanotubes as an Outstanding Environmental Catalyst. *Environ. Sci. Nano* **2017**, *4*, 170–179. [CrossRef]
38. Yang, Q.; Chen, Y.; Duan, X.; Zhou, S.; Niu, Y.; Sun, H.; Zhi, L.; Wang, S. Unzipping Carbon Nanotubes to Nanoribbons for Revealing the Mechanism of Nonradical Oxidation by Carbocatalysis. *Appl. Catal. B Environ.* **2020**, *276*, 119146. [CrossRef]
39. Xiao, B.; Li, X.; Li, X.; Wang, B.; Langford, C.; Li, R.; Sun, X. Graphene Nanoribbons Derived from the Unzipping of Carbon Nanotubes: Controlled Synthesis and Superior Lithium Storage Performance. *J. Phys. Chem. C* **2014**, *118*, 881–890. [CrossRef]
40. Liu, M.; Song, Y.; He, S.; Tjiu, W.W.; Pan, J.; Xia, Y.Y.; Liu, T. Nitrogen-Doped Graphene Nanoribbons as Efficient Metal-Free Electrocatalysts for Oxygen Reduction. *ACS Appl. Mater. Interfaces* **2014**, *6*, 4214–4222. [CrossRef]
41. Wacławek, S.; Grübel, K.; Černík, M. Simple Spectrophotometric Determination of Monopersulfate. *Spectrochim. Acta Part A Mol. Biomol. Spectrosc.* **2015**, *149*, 928–933. [CrossRef]
42. Li, H.; Zhang, J.; Gholizadeh, A.B.; Brownless, J.; Fu, Y.; Cai, W.; Han, Y.; Duan, T.; Wang, Y.; Ling, H.; et al. Photoluminescent Semiconducting Graphene Nanoribbons via Longitudinally Unzipping Single-Walled Carbon Nanotubes. *ACS Appl. Mater. Interfaces* **2021**, *13*, 52892–52900. [CrossRef]
43. Chen, X.; Oh, W.D.; Hu, Z.T.; Sun, Y.M.; Webster, R.D.; Li, S.Z.; Lim, T.T. Enhancing Sulfacetamide Degradation by Peroxymono-sulfate Activation with N-Doped Graphene Produced through Delicately-Controlled Nitrogen Functionalization via Tweaking Thermal Annealing Processes. *Appl. Catal. B Environ.* **2018**, *225*, 243–257. [CrossRef]
44. Chang, S.S.; Clair, B.; Ruelle, J.; Beauchêne, J.; Di Renzo, F.; Quignard, F.; Zhao, G.J.; Yamamoto, H.; Gril, J. Mesoporosity as a New Parameter for Understanding Tension Stress Generation in Trees. *J. Exp. Bot.* **2009**, *60*, 3023–3030. [CrossRef]
45. Groen, J.C.; Peffer, L.A.A.; Pérez-Ramírez, J. Pore Size Determination in Modified Micro- and Mesoporous Materials. Pitfalls and Limitations in Gas Adsorption Data Analysis. *Microporous Mesoporous Mater.* **2003**, *60*, 1–17. [CrossRef]
46. Duan, X.; Ao, Z.; Sun, H.; Indrawirawan, S.; Wang, Y.; Kang, J.; Liang, F.; Zhu, Z.H.; Wang, S. Nitrogen-Doped Graphene for Generation and Evolution of Reactive Radicals by Metal-Free Catalysis. *ACS Appl. Mater. Interfaces* **2015**, *7*, 4169–4178. [CrossRef]
47. Duan, X.; Indrawirawan, S.; Sun, H.; Wang, S. Effects of Nitrogen-, Boron-, and Phosphorus-Doping or Codoping on Metal-Free Graphene Catalysis. *Catal. Today* **2015**, *249*, 184–191. [CrossRef]
48. Krishnamoorthy, K.; Veerapandian, M.; Yun, K.; Kim, S.J. The Chemical and Structural Analysis of Graphene Oxide with Different Degrees of Oxidation. *Carbon* **2013**, *53*, 38–49. [CrossRef]
49. Dash, S.; Patel, S.; Mishra, B.K. Oxidation by Permanganate: Synthetic and Mechanistic Aspects. *Tetrahedron* **2009**, *65*, 707–739. [CrossRef]
50. Russ, M.T. Dimanganese Heptoxide for the Selective Oxidaion of Organic Substrates. *Angew. Chem. Int. Ed.* **1987**, *26*, 1007–1009.
51. Ma, Y.; Wu, X.; Yu, M.; Li, S.; Liu, J. Turning free-standing three dimensional graphene into electrochemically active by nitrogen doping during chemical vapor deposition process. *J. Mater. Sci. Mater. Electron.* **2020**, *31*, 3759–3768. [CrossRef]
52. Luo, Z.; Lim, S.; Tian, Z.; Shang, J.; Lai, L.; MacDonald, B.; Fu, C.; Shen, Z.; Yu, T.; Lin, J. Pyridinic N Doped Graphene: Synthesis, Electronic Structure, and Electrocatalytic Property. *J. Mater. Chem.* **2011**, *21*, 8038–8044. [CrossRef]
53. Ren, W.; Xiong, L.; Nie, G.; Zhang, H.; Duan, X.; Wang, S. Insights into the Electron-Transfer Regime of Peroxydisulfate Activation on Carbon Nanotubes: The Role of Oxygen Functional Groups. *Environ. Sci. Technol.* **2020**, *54*, 1267–1275. [CrossRef]
54. Frank, B.; Zhang, J.; Blume, R.; Schlögl, R.; Su, D.S. Heteroatoms Increase the Selectivity in Oxidative Dehydrogenation Reactions on Nanocarbons. *Angew. Chem. Int. Ed.* **2009**, *48*, 6913–6917. [CrossRef]
55. Frank, B.; Blume, R.; Rinaldi, A.; Trunschke, A.; Schlögl, R. Oxygen Insertion Catalysis by Sp2 Carbon. *Angew. Chem. Int. Ed.* **2011**, *50*, 10226–10230. [CrossRef]

56. Liu, S.; Peng, W.; Sun, H.; Wang, S. Physical and Chemical Activation of Reduced Graphene Oxide for Enhanced Adsorption and Catalytic Oxidation. *Nanoscale* **2014**, *6*, 766–771. [CrossRef]
57. Gao, Y.; Chen, Z.; Zhu, Y.; Li, T.; Hu, C. New Insights into the Generation of Singlet Oxygen in the Metal-Free Peroxymonosulfate Activation Process: Important Role of Electron-Deficient Carbon Atoms. *Environ. Sci. Technol.* **2020**, *54*, 1232–1241. [CrossRef]
58. Gao, Y.; Zhu, Y.; Chen, Z.; Zeng, Q.; Hu, C. Insights into the Difference in Metal-Free Activation of Peroxymonosulfate and Peroxydisulfate. *Chem. Eng. J.* **2020**, *394*, 123936. [CrossRef]
59. Tian, X.; Gao, P.; Nie, Y.; Yang, C.; Zhou, Z.; Li, Y.; Wang, Y. A Novel Singlet Oxygen Involved Peroxymonosulfate Activation Mechanism for Degradation of Ofloxacin and Phenol in Water. *Chem. Commun.* **2017**, *53*, 6589–6592. [CrossRef]
60. Tang, W.; Zhang, Y.; Guo, H.; Liu, Y. Heterogeneous Activation of Peroxymonosulfate for Bisphenol AF Degradation with BiOI0.5Cl0.5. *RSC Adv.* **2019**, *9*, 14060–14071. [CrossRef]
61. Kim, D.G.; Ko, S.O. Effects of thermal modification of a biochar on persulfate activation and mechanisms of catalytic degradation of a pharmaceutical. *Chem. Eng. J.* **2020**, *399*, 125377. [CrossRef]
62. Buxton, G.V.; Greenstock, C.L.; Helman, W.P.; Ross, A.B. Critical Review of Rate Constants for Reactions of Hydrated Electrons, Hydrogen Atoms and Hydroxyl Radicals ($\cdot$OH/$\cdot$O$-$ in Aqueous Solution. *J. Phys. Chem. Ref. Data* **1988**, *17*, 513–886. [CrossRef]
63. Eibenberger, H.; Steenken, S.; O'Neill, P.; Schulte-Frohlinde, D. Pulse Radiolysis and Electron Spin Resonance. *J. Phys. Chem. C* **1978**, *82*, 749–750. [CrossRef]
64. Greenstock, C.L.; Ruddock, G.W. Determination of Superoxide (O2-) Radical Anion Reaction Rates Using Pulse Radiolysis. *Int. J. Radiat. Phys. Chem.* **1976**, *8*, 367–369. [CrossRef]
65. Manring, L.E.; Kramer, M.K.; Foote, C.S. Interception of O2- by Benzoquinone in Cyanoaromatic-Sensitized Photooxygenations. *Tetrahedron Lett.* **1984**, *25*, 2523–2526. [CrossRef]
66. Wilkinson, F.; Brummer, J.G. Rate Constants for the Decay and Reactions of the Lowest Electronically Excited Singlet State of Molecular Oxygen in Solution. *J. Phys. Chem. Ref. Data* **1981**, *10*, 809–999. [CrossRef]
67. Haag, W.R.; Hoigné, J. Singlet Oxygen in Surface Waters. 3. Photochemical Formation and Steady-State Concentrations in Various Types of Waters. *Environ. Sci. Technol.* **1986**, *20*, 341–348. [CrossRef]
68. Lindsey, M.E.; Tarr, M.A. Inhibition of Hydroxyl Radical Reaction with Aromatics by Dissolved Natural Organic Matter. *Environ. Sci. Technol.* **2000**, *34*, 444–449. [CrossRef]
69. Ziajka, J.; Pasiuk-Bronikowska, W. Rate Constants for Atmoshpheric Trace Organics Scavenging SO4- in the Fe-Catalysed Autoxidation of S(IV). *Atmos. Environ.* **2005**, *39*, 1431–1438. [CrossRef]
70. Lin, Y.; Sun, X.; Su, D.S.; Centi, G.; Perathoner, S. Catalysis by Hybrid Sp2/Sp3 Nanodiamonds and Their Role in the Design of Advanced Nanocarbon Materials. *Chem. Soc. Rev.* **2018**, *47*, 8438–8473. [CrossRef]
71. Buxton, G.V.; Elliot, A.J. Rate Constant for Reaction of Hydroxyl Radicals with Bicarbonate Ions. *Int. J. Radiat. Appl. Instrum. Part C* **1986**, *27*, 241–243. [CrossRef]
72. Liang, J.; Xu, X.; Qamar Zaman, W.; Hu, X.; Zhao, L.; Qiu, H.; Cao, X. Different Mechanisms between Biochar and Activated Carbon for the Persulfate Catalytic Degradation of Sulfamethoxazole: Roles of Radicals in Solution or Solid Phase. *Chem. Eng. J.* **2019**, *375*, 121908. [CrossRef]
73. Huang, Z.; Bao, H.; Yao, Y.; Lu, W.; Chen, W. Novel Green Activation Processes and Mechanism of Peroxymonosulfate Based on Supported Cobalt Phthalocyanine Catalyst. *Appl. Catal. B Environ.* **2014**, *154–155*, 36–43. [CrossRef]
74. Duan, X.; Sun, H.; Ao, Z.; Zhou, L.; Wang, G.; Wang, S. Unveiling the Active Sites of Graphene-Catalyzed Peroxymonosulfate Activation. *Carbon* **2016**, *107*, 371–378. [CrossRef]
75. Evans, D.F.; Upton, M.W. Studies on Singlet Oxygen in Aqueous Solution. Part 3. The Decomposition of Peroxy-Acis. *J. Chem. Soc. Dalton Trans.* **1985**, *6*, 1151–1153. [CrossRef]

Article

Degradation of the UV Filter Benzophenone-4 by Ferrate (VI) in Aquatic Environments

Rouyi Wang [1], Ping Sun [2], Zhicai Zhai [2], Hui Liu [2], Ruirui Han [1], Hongxia Liu [1,*] and Yingsen Fang [2,*]

1   College of Advanced Materials Engineering, Jiaxing Nanhu University, Jiaxing 314001, China
2   College of Biological and Chemical Engineering, Jiaxing University, Jiaxing 314001, China
*   Correspondence: liuhongixa@zjxu.edu.cn (H.L.); fangyingsen@zjxu.edu.cn (Y.F.)

Abstract: This work demonstrates the potential utility of ferrate(VI)-based advanced oxidation processes for the degradation of a representative UV filter, BP-4. The operational parameters of oxidant dose and temperature were determined with kinetic experiments. In addition, the effects of water constituents including anions (Cl^-, HCO_3^-, NO_3^-, SO_4^{2-}), cations (Na^+, K^+, Ca^{2+}, Mg^{2+}, Cu^{2+}, Fe^{3+}), and humic acid (HA) were investigated. Results suggested that the removal rate of BP-4 (5 mg/L) could reach 95% in 60 min, when [Fe(VI)]:[BP-4] = 100:1, T = 25 °C and pH = 7.0, The presence of K^+, Cu^{2+} and Fe^{3+} could promote the removal of BP-4, but Cl^-, SO_4^{2-}, NO_3^-, HA and Na^+ could significantly inhibit the removal of BP-4. Furthermore, this Fe(VI) oxidation processes has good feasibility in real water samples. These results may provide useful information for the environmental elimination of benzophenone-type UV filters by Fe(VI).

Keywords: benzophenone-4; ferrate (VI); oxidation; kinetics

Citation: Wang, R.; Sun, P.; Zhai, Z.; Liu, H.; Han, R.; Liu, H.; Fang, Y. Degradation of the UV Filter Benzophenone-4 by Ferrate (VI) in Aquatic Environments. *Processes* **2022**, *10*, 1829. https://doi.org/10.3390/pr10091829

Academic Editors: Gassan Hodaifa, Antonio Zuorro, Joaquín R. Dominguez, Juan García Rodríguez, José A. Peres, Zacharias Frontistis and Mha Albqmi

Received: 15 August 2022
Accepted: 5 September 2022
Published: 10 September 2022

Publisher's Note: MDPI stays neutral with regard to jurisdictional claims in published maps and institutional affiliations.

1. Introduction

Benzophenone (BP)-type UV filters are a kind of oil soluble substance that is easy to dissolve in aromatic hydrocarbons such as benzene and toluene but difficult to dissolve in water. Therefore, it is easier for this substance to enter the organism than water-soluble substances, thus causing toxic effects (Table 1 lists the structure and solubility of some substances). BPs are discharged into the sewage treatment system in a large amount and are difficult to degrade, and there are a large number of residues in the activated sludge. At the same time, BPs can also be brought into water through water entertainment activities such as swimming and bathing [1]. BPs have been detected in various water samples [2] and other environmental media [3,4]. Furthermore, BPs can also enter the human body through the skin, diet and air, which poses a threat to human health. BP residues have been detected in human samples [5,6]. Song et al. [7] reported that BP-3 concentrations showed positive correlations between maternal serum (MS) and cord serum (CS), while the CS/MS ratios of BP-1, BP-3, BP-8 and 4-OH-BP were affected by molecular weight or $\log K_{ow}$, along with negative correlations.

Table 1. The structure and solubility of Part BPs.

NO.	Name	Structure	Solubility
1	Benzophenone		Insoluble in water (<0.1 g/100 mL at 25 °C)

Table 1. *Cont.*

NO.	Name	Structure	Solubility
2	Benzophenone-3(BP-3)		Insoluble in water (<0.1 g/100 mL at 20 °C)
3	Benzophenone-4(BP-4)		DMSO (Slightly), Methanol (Slightly)
4	4-hydroxy benzophenone		Insoluble in water

Passive diffusion may play an important role in the placental transfer of these BP-type UV filters. BP-3, BP-4, BP2 and 4-OH-BP were frequently detected in featured aquatic environments of Shanghai, and BP, BP-4 and BP-3 might adversely affect fish and other aquatic organisms [8].

Since BP substances are harmful to organisms and human, removing them from the environment is very important. Photochemical transformation is an important transformation pathway of UV sunscreen in natural water, which affects its environmental fate and ecological risk. The reaction rate and pathway of photochemical transformation are affected by pH and soluble substances in the water environment, and products with greater ecological risk can be produced under the light of BPs [9]. Vione et al. researched the light conversion process of BP-3 under the conditions related to surface water and found that BP-3 can be directly photodegradated and can also be indirectly photodegradated by hydroxyl radical (OH) and soluble organic matter DOM* [10]. The photolysis half-life of BPs varies with the environmental conditions and is possible in a few days to several months. Generally, organic pollutants can be removed by microbial degradation. Gago Ferrero et al. [11] studied the degradation and photodegradation processes of BP-1 and BP-3 by aerobic bacteria. The results showed that more than 99% of BP-1 and BP-3 could be degraded by biological treatment within 24 h, while the removal efficiency of phase reflective degradation was very low, especially for BP-3, which had little degradation effect. The results of metabolite analysis showed that BP-1 was decomposed by *T. versicolor* of BP-3, but the glycoconjugate derivative was the main metabolite. Moreover, there was no metabolite formation with a higher estrogen effect during the biodegradation of BP-1 and BP-3. Liu et al. [12] used activated sludge and digested sludge to study the biodegradation process under anaerobic and aerobic conditions. The results showed that anaerobic biodegradation was more suitable for the removal of BP-3. Beel et al. [13] explained the anaerobic biodegradation process of BP-4 in the activated sludge of the urban sewage treatment system and found nine kinds of transformation products that showed a higher toxicity of bacteria (*Vibrio fischeri*) than BP-4.

As this emerging contaminant cannot be eliminated effectively by conventional processes, it is necessary to explore efficient measures for the degradation of BPs from water. In order to remove UV filters from water, many measures have been researched, including ozonation, chlorine, persulfate and ferrate [14,15]. Amongst them, the products of ferrate (FeO_4^{2-}, Fe(VI)) reaction are ferric oxides/hydroxides, which can act as coagulants/precipitants. Therefore, as an effective green oxidant, ferrate(VI) (Fe(VI)) has attracted much attention [16,17]. Research works have proved that Fe(VI) is very promising for the degradation of organic materials, such as PPCPs, endocrine disrupting chemicals and pesticides [18–21]. However, information on the degradation of BP-4 by Fe(VI) is scarce.

In this study, we attempted to investigate the oxidation of BP-4 by Fe(VI) in aquatic environment. First, experiments were implemented in order to obtain the optimal reaction conditions of oxidant dose and temperature. Then, coexisting water components such as metal cations (Na^+, K^+, Ca^{2+}, Mg^{2+}, Cu^{2+}, Fe^{3+}), inorganic anions (Cl^-, NO_3^-, HCO_3^-, SO_4^{2-}) and humic acid (HA) were tested for their effects on BP-4 removal. Finally, the removal of BP-4 in natural waters was also evaluated.

2. Materials and Methods

2.1. Chemicals and Reagents

BP-4 (CAS no: 4065-45-6, 98% purity) and potassium ferrate (K_2FeO_4, CAS no: 39469-86-8 Fe(VI), purity > 95%) were obtained from J&K Company (Shanghai, China). Methanol and formic acid gained from Merck Company (Darmstadt, Germany) (for HPLC) were of HPLC grade. The rest of the reagents were of analytical grade or higher. Ultrapure water (18.2 MU cm) was prepared with a Milli-Q system (Millipore, Bedford, MA, USA).

2.2. Removal of BP-4 by Fe(VI)

Degradation experiments were carried out in batches with 100 mL brown glass in a rocking bed (150 r/min) at 25.0 $\pm$ 0.2 °C. The pH of solution was initially adjusted by HCl or NaOH. At the specified time interval, an 0.8 mL sample was collected and immediately filtered into a 2.0 mL vial containing 0.2 mL chromatographic grade methanol to quench the reaction (through a 0.22 μm filter). In order to study the potential effects of various environmental factors, the initial BP-4 solution was adjusted to a different pH value or pre-added with 0.5 mM ions (Fe^{3+}, Ca^{2+}, Mg^{2+}, Cu^{2+}, Na^+, K^+, SO_4^{2-}, Cl^-, NO_3^-, HCO_3^-) and 1–30 mg/L humic acid (HA). The degradation of BP-4 in two environmental water samples in Jiaxing (one surface water sample from canal, one sample of effluent of Jiaxing domestic sewage treatment plant in Jiaxing, China) was also studied. All experiments were carried out twice, and the average values are presented in the paper.

2.3. Analytical Methods

The concentrations of BP-4 were measured by an Agilent 1200 high performance liquid chromatograph (HPLC) equipped with a quaternary pump and a diode array detector (Agilent Technologies, Palo Alto, CA, USA). Chromatographic analysis was performed with a 1.0 mL/min flow rate on a Zorbax Eclipse XDB-C18 analytical column (4.6 mm $\times$ 150 mm, particle size 5 mm) (Agilent Technologies, CA, USA) at 30 °C. The injection volume was 20 μL, and the elution time was 10 min for all samples. The mobile phase was methanol (A) and water (B) (60:40).

3. Results

3.1. Removal of BP-4 from Aqueous Solution

Initially, removal of BP-4 by Fe(VI) was investigated with different molar ratios of Fe(VI) to BP-4. The results are presented in Figure 1. The concentration of BP-4 decreased rapidly within 10 min and changed slightly after 60 min. This is because the strong oxidation of Fe(VI) can effectively remove the target pollutants in a very short time when the oxidant has a high initial concentration in the reaction solution. As time goes by, the concentration of Fe(VI) decreases rapidly, and the reaction tends to stop. In addition, the removal rate of BP-4 increases remarkably with increasing Fe(VI) (except [Fe(VI)]:[BP-4]) = 200:1). For example, the removal rate BP-4 was from 30% to 95% (at 60 min) when the molar ratio was from 25: 1 to 100: 1. At a higher ratio of 200:1, the degradation efficiency will be lower. This is because Fe(VI) reacting with the target pollutant is enough in the reaction solution and will only overflow as the dosage increases. In addition, the instability of ferrate will lead to the auto-degradation of Fe(VI), and the auto-degradation rate will also increase with the increase of the concentration of Fe(VI) solution. The ratio of Fe(VI) used to degrade BP-4 decreases, meaning that the degradation efficiency decreases. Feng et al. also reported this phenomenon in their research [20].

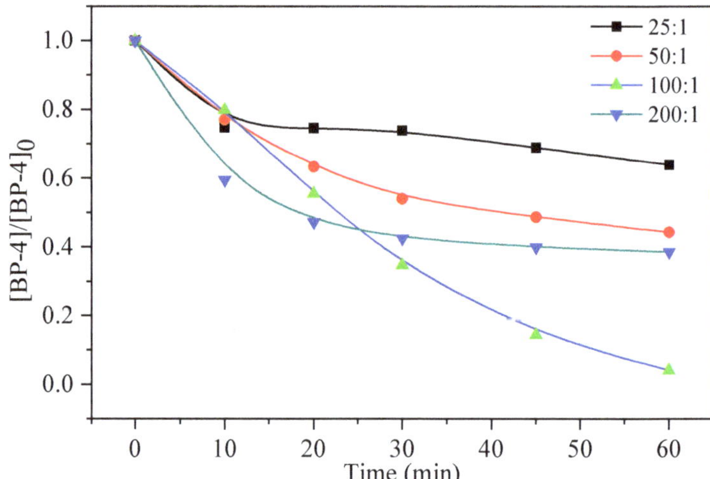

Figure 1. Effect of dose of oxidants ([Fe(VI)]) on the removal of BP-4 (experimental conditions: $[BP-4]_0$ = 5 mg/L; pH = 7.0 ± 0.2; T = 25.0 °C; reaction time = 90 min).

In the next set of experiments, the effects of pH on the degradation were determined, as shown in Figure 2A. The removal rate of BP-4 was greatly affected by the pH value of the solution. The reason is that the pH value of the solution will affect the oxidation ability and stability of potassium ferrate. In acid conditions, potassium ferrate has the highest oxidation potential (2.2 ev) in conventional water treatment agent, which is much higher than 0.72 ev in alkaline condition, meaning the degradation rate of pH = 3 (highest) is much higher than that of pH = 11 (lowest) in the first 10 min. There are different forms of ferrate solution in different pH environments—generally, the forms are $H_3FeO_4^+$, H_2FeO_4 and $HFeO_4^-$. In the acidic environment, the ferrate in the aqueous solution is mainly in the form of H_2FeO_4 and $HFeO_4^-$, at which time the ferrate is active and easy to decompose; in the alkaline condition, the ferrate is mainly in the form of stable FeO_4^{2-}. Therefore, the stability of potassium ferrate is better when the pH is higher than 7. When the pH is 3, the ferrate root is extremely unstable, and its self-decomposition is intensified, which leads to basic non-degradation after 10 min, and the final degradation rate is the lowest. Potassium ferrate with pH = 7 and pH = 9 has both high oxidation potential and good stability, meaning these should be the best pH values for degradation of BP-4. At pH = 11, its good stability enables ferrate to react with BP-4 continuously. Thus, the next experiment was studied with pH = 7.0.

The reaction temperature also significantly influenced the degradation of BP-4. As shown in Figure 2B, the BP-4 degradation rate is faster with the temperature increasing. The reason is that heating can accelerate the efficient degradation of BP-4 by Fe(VI), and the results are in accordance with the degradation of organic chemicals by other oxidants, such as potassium permanganate and persulfate [22,23]. The degradation rate constants (k) were 0.0204, 0.0290 and 0.0407 min^{-1} at 25 °C, 35 °C and 45 °C. In addition, based on the experimental data, the activation energy (Ea) of the BP-4 oxidative degradation by Fe(VI) was estimated to be 27.2 KJ/mol (R^2 = 0.9999) with the Arrhenius equation. It can be speculated that higher temperatures were of benefit to the removal of BP-4.

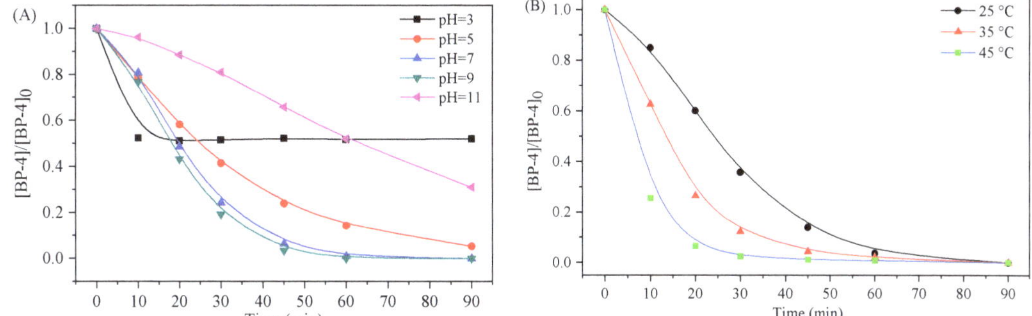

Figure 2. Effect of initial solution pH (**A**) and temperature (**B**) on BP-4 removal by Fe(VI) (experimental conditions: $[BP\text{-}4]_0 = 5$ mg/L; T = 25.0 °C for pH (initial pH 7.0 ± 0.2 for temperature); $[Fe(VI)]_0:[BP\text{-}4]_0 = 100:1$; reaction time = 90 min).

Based on the above results, experimental conditions were the following: $[BP\text{-}4] = 5$ mg/L, initial pH 7.0 ± 0.2, $[Fe(VI)]_0:[BP\text{-}4]_0 = 100:1$; T = 25 °C, reaction time = 90 min.

3.2. Effect of Inorganic Ions on Degradation

The effect of inorganic ions on the degradation of BP-4 was evaluated (as seen in Figure 3). Experimental conditions were as follows: $[BP\text{-}4] = 5$ mg/L, $[Fe(VI)]_0:[BP\text{-}4]_0 = 100:1$, T = 25 °C, [Anions] = [Cations] = 5 mM. As shown in Figure 3A, Cl^-, SO_4^{2-} and NO_3^- ions significantly inhibited the removal efficiency of BP-4 by Fe(VI), but the impact on BP-4 removal was not apparent with the addition of HCO_3^- at 5 mM. For monovalent cations such as Na^+ and K^+, they had an obvious influence on BP-4 removal (Figure 3B). Na^+ ions reduced the removal of BP-4 by Fe(VI), while K^+ ions could promote its degradation. When Mg^{2+} and Ca^{2+} were added in reaction solution, the effects were not obvious. Ca^{2+} ions inhibited the removal of BP-4 by Fe(VI), and Mg^{2+} ions presented no obvious effect on BP-4 removal. Moreover, the effect of Cu^{2+} and Fe^{3+} (transition metal ions) on the removal of BP-4 was also investigated. When these ions were present in the reaction solution, the removal rate of BP-4 increased (Figure 3B). Cu^{2+} showed more increase than Fe^{3+}.

According to Figure 3, K^+, Cu^{2+} and Fe^{3+} could promote the removal of BP-4, but Cl^-, SO_4^{2-}, NO_3^- and Na^+ could significantly inhibit the removal of BP-4. The specific reason was that K^+, Cu^{2+} and Fe^{3+} play a catalytic role in the oxidative degradation of BP-4 by Fe(VI).

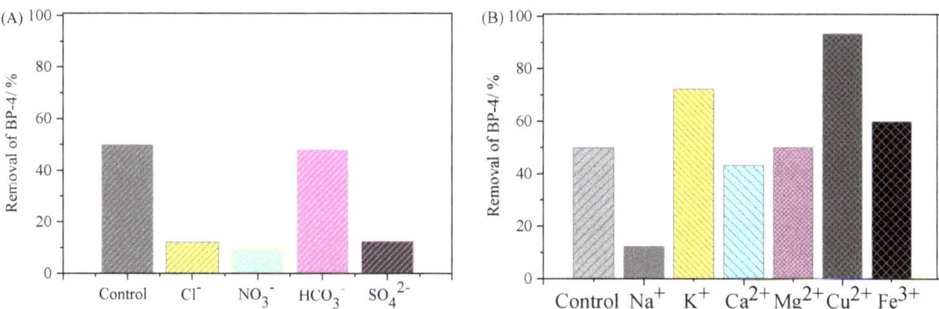

Figure 3. Effect of anions (**A**) and cations (**B**) on the removal of BP-4 after 10 min. [Anions] = [Cations] = 5 mM.

3.3. Effect of HA on Degradation

The effect of dissolved organic compounds on the removal of BP-4 by Fe(VI) was measured by adding HA with a concentration of 1–30 mg/L. The addition of HA decreased the efficiency of removing BP-4 by Fe(VI) (Figure 4). With the increase of HA concentration, the removal rate of BP-4 nonlinearly decreased. The results show that HA competes with BP-4 to react with Fe(VI). In other words, Fe(VI) reacts not only with BP-4 but also with HA.

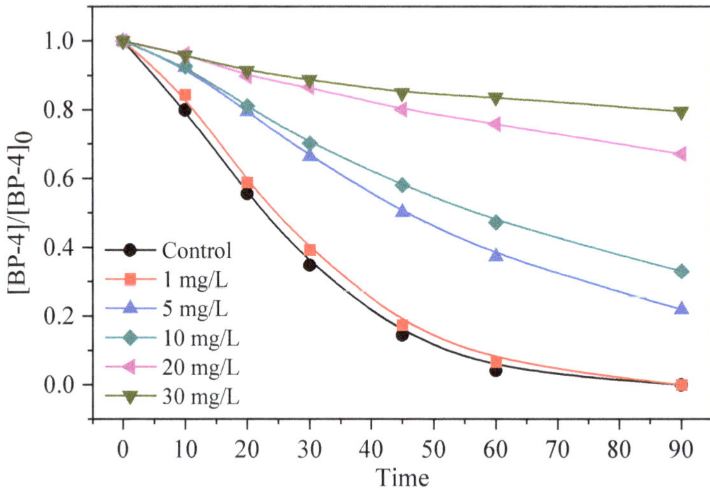

Figure 4. Effect of humic acid (HA) concentration on the removal of BP-4.

In Figure 4, when the HA concentration was lower than 1 mg/L, the removal rate of BP-4 did not show an obvious effect due to HA. The removal rate of BP-4 decreased with the increase of concentration of HA. When the concentration of HA increased to 20 mg/L, the removal rate of BP-4 changed from 95.82% to 24.09% in 60 min, and when the concentration of HA reached 30 mg/L, the removal rate of BP-4 was reduced to 16.31%. The main reason was that HA has a variety of functional groups, such as -COOH, -OH, etc. The organic matter with these functional groups could have different chemical reactions with oxidants in water, including ferrates. The HA could be oxidized by FeO_4^{2-} in the reaction solution, and that would reduce its reaction with BP-4. At the same time, the HA could not have been oxidized completely, and its products could also form complexes or be subject to adsorption with the final product $Fe(OH)_3$ colloid of potassium ferrate, thus competing with BP-4 and reducing the removal rate of BP-4.

3.4. Removal of BP-4 in Environmental Water Samples

It is necessary to assess the feasibility of this oxidative technique to completely eliminate low levels of sunscreen agents in different waters. Initially, the removal of BP-4 by Fe(VI) in different waters was evaluated at pH 7.0. The molar ratio of Fe(VI) to BP-4 was 100:1. The results of the removal of BP-4 as a function of time from various water matrices are shown in Figure 5. Fe(VI) can almost completely remove BP-4 from most water samples in 90 min, except for secondary effluent water samples. The reason may be the coexisting constituents, which could significantly inhibit the removal of BP-4.

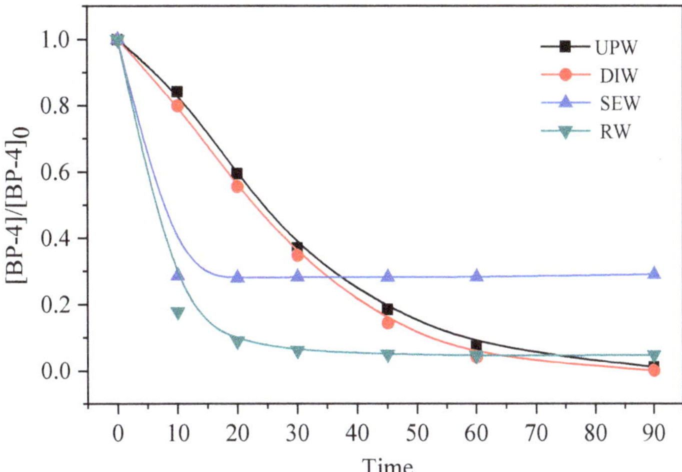

Figure 5. Removal of BP-4 from water samples by Fe(VI) at a molar ratio of 100:1 ($[Fe(VI)]_0$:$[BP-4]_0$). (abbreviations: UPW—ultrapure water; DIW—deionized water; SEW–secondary effluent of sewage treatment plant; RW—Jiaxing Canal River Water.).

3.5. Oxidation Products of BP-4 and Possible Reaction Pathways

To identify the identification of oxidation products, mass analysis experiments were performed at 5 mg/L BP-4 (initial concentration), [Fe(VI)]:[BP-4] = 100:1, T = 25 °C and pH = 7.0. A total of three products were identified in positive mode by LC-TOF-MS, and structural assignments were achieved with the product ion scan. The MS/MS spectra and the proposed fragmentation patterns of BP-4 and its reaction intermediates are illustrated in Figure 6.

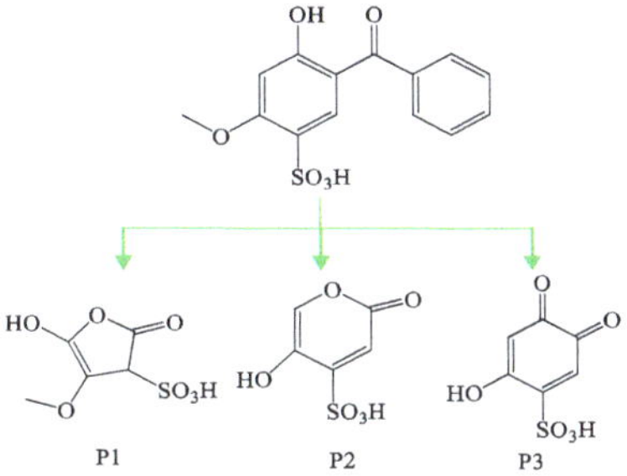

Figure 6. Degradation pathways of BP-4 by ferrate (VI) process.

Figure 6 shows that the ring of benzene can be oxidized, and the results are in accordance with references with PMS and ozone [14,24]. While the ring of benzene is not oxidized with the biological method [12], the ability of the chemical oxidizing agent is stronger than biodegradation, so more effective methods need to be developed.

4. Conclusions

The research suggests that Fe(VI) has great potential to remove BP-4 and can be used as a rapid and effective method. The presence of inorganic ions (i.e., K^+, Cu^{2+} and Fe^{3+}) in water increased the removal efficiency of BP-4 by Fe(VI). In contrast, Cl^-, NO_3^-, SO_4^{2-}, Na^+, Mg^{2+} and Ca^{2+} ions decreased the removal efficiency of BP-4 by Fe(VI). HA may affect the removal efficiency of BP-4 by Fe(VI). The removal of BP-4 by Fe(VI) varied with the water type and constituents of water matrices.

Author Contributions: H.L. (Hongxia Liu): conceptualization, validation, investigation, data curation, writing—original draft preparation, writing—review and editing, visualization, supervision, project administration, funding acquisition; Y.F.: conceptualization, validation, investigation, data curation, writing—original draft preparation, writing—review and editing, visualization, supervision, project administration, funding acquisition; R.W.: formal analysis, data curation, writing—original draft preparation, writing—review and editing; P.S.: formal analysis, investigation, writing—review and editing; Z.Z.: formal analysis, data curation, writing—review and editing; H.L. (Hui Liu): methodology, formal analysis, writing—review and editing; R.H.: formal analysis, writing—review and editing. All authors have read and agreed to the published version of the manuscript.

Funding: This work was funded by the National Natural Science Foundation of China—Project-ID 21607058; the Zhejiang Provincial Natural Science Foundation of China—Project-ID LY21B070008 and LY21B070007; the Department of Education of Zhejiang Province—Project-ID Y201840526; the Scientific Research Startup Foundation for Leading Professor from Jiaxing University—Project-ID CD70519063; the Student Research Innovation Team Funding Project of Zhejiang—Project-ID 2019R417027; the Public Welfare Research Project of Jiaxing—Project-ID 2020AY10005, 2021AY10069 and 2021AD10008; and National innovation and entrepreneurship training program for College Students—Project-ID 202113291013.

Institutional Review Board Statement: Not applicable.

Informed Consent Statement: Not applicable.

Data Availability Statement: All the data are available within the manuscript.

Conflicts of Interest: The authors declare no conflict of interest.

References

1. Sánchez-Quiles, D.; Blasco, J.; Tovar-Sánchez, A. Sunscreen Components Are a New Environmental Concern in Coastal Waters: An overview. *Sunscreens Coast. Ecosyst. Occur. Behav. Eff. Risk* **2020**, 1–14.
2. Mao, F.; He, Y.; Gin, K.Y.H. Occurrence and fate of benzophenone-type UV filters in aquatic environments: A review. *Environ. Sci. Water Res. Technol.* **2019**, *5*, 209–223. [CrossRef]
3. Wu, Y.; Venier, M.; Hites, R.A. Broad exposure of the North American environment to phenolic and amino antioxidants and to ultraviolet filters. *Environ. Sci. Technol.* **2020**, *54*, 9345–9355. [CrossRef] [PubMed]
4. Wang, L.; Asimakopoulos, A.G.; Moon, H.B.; Nakata, H.; Kannan, K. Benzotriazole, benzothiazole, and benzophenone compounds in indoor dust from the United States and East Asian countries. *Environ. Sci. Technol.* **2013**, *47*, 4752–4759. [CrossRef] [PubMed]
5. Martín-Pozo, L.; del Carmen Gómez-Regalado, M.; Cantarero-Malagón, S.; Navalón, A.; Zafra-Gómez, A. Determination of ultraviolet filters in human nails using an acid sample digestion followed by ultra-high performance liquid chromatography–mass spectrometry analysis. *Chemosphere* **2021**, *273*, 128603. [CrossRef] [PubMed]
6. Fernández, M.F.; Mustíeles, V.; Suárez, B.; Reina-Pérez, I.; Olivas-Martinez, A.; VelaSoria, F. Determination of bisphenols, parabens, and benzophenones in placenta by dispersive liquid-liquid microextraction and gas chromatographytandem mass spectrometry. *Chemosphere* **2021**, *274*, 129707. [CrossRef]
7. Song, S.; He, Y.; Huang, Y.; Huang, X.; Guo, Y.; Zhu, H.; Kannan, K.; Zhang, T. Occurrence and transfer of benzophenone-type ultraviolet filters from the pregnant women to fetuses. *Sci. Total Environ.* **2020**, *726*, 138503. [CrossRef]
8. Wu, M.H.; Xie, D.G.; Xu, G.; Sun, R.; Xia, X.Y.; Liu, W.L.; Tang, L. Benzophenone-type UV filters in surface waters: An assessment of profiles and ecological risks in Shanghai, China. *Ecotox. Environ. Safe.* **2017**, *141*, 235–241. [CrossRef]
9. Hayashi, T.; Okamoto, Y.; Ueda, K.; Kojima, N. Formation of estrogenic products from benzophenone after exposure to sunlight. *Toxicol. Lett.* **2006**, *167*, 1–7. [CrossRef]
10. Vione, D.; Caringella, R.; De Laurentiis, E.; Pazzi, M.; Minero, C. Phototransformation of the sunlight filter benzophenone-3 (2-hydroxy-4-methoxybenzophenone) under conditions relevant to surface waters. *Sci. Total Environ.* **2013**, *463–464*, 243–251. [CrossRef] [PubMed]
11. Gago-Ferrero, P.; Badia-Fabregat, M.; Olivares, A.; Piña, B.; Blánquez, P.; Vicent, T.; Caminal, G.; Díaz-Cruz, M.S.; Barceló, D. Evaluation of fungal- and photo-degradation as potential treatments for the removal of sunscreens BP3 and BP1. *Sci. Total Environ.* **2012**, *427–428*, 355–363. [CrossRef] [PubMed]
12. Liu, Y.S.; Ying, G.G.; Shareef, A.; Kookana, R.S. Biodegradation of the ultraviolet filter benzophenone-3 under different redox conditions. *Environ. Toxicol. Chem.* **2012**, *31*, 289–295. [CrossRef] [PubMed]
13. Beel, R.; Eversloh, C.L.; Ternes, T.A. Biotransformation of the UV-Filter sulisobenzone: Challenges for the identification of transformation products. *Environ. Sci. Technol.* **2013**, *47*, 6819–6828. [CrossRef] [PubMed]

14. Liu, H.; Sun, P.; Feng, M.B.; Liu, H.X.; Yang, S.G.; Wang, L.S.; Wang, Z.Y. Nitrogen and sulfur co-doped CNT-COOH as an efficient metal-free catalyst for the degradation of UV filter BP-4 based on sulfate radicals. *Appl. Catal. B Environ.* **2016**, *187*, 1–10. [CrossRef]
15. Jia, X.R.; Jin, J.; Gao, R.; Feng, T.Y.; Huang, Y.; Zhou, Q.; Li, A.M. Degradation of benzophenone-4 in a UV/chlorine disinfection process:Mechanism and toxicity evaluation. *Chemosphere* **2019**, *222*, 494–502. [CrossRef] [PubMed]
16. Sharma, V.K. Ferrate(VI) and ferrate(V) oxidation of organic compounds: Kinetics and mechanism. *Coordin. Chem. Rev.* **2013**, *257*, 495–510. [CrossRef]
17. Sun, S.; Liu, Y.; Ma, J.; Pang, S.; Huang, Z.; Gu, J.; Gao, Y.; Xue, M.; Yuan, Y.; Jiang, J. Transformation of substituted anilines by ferrate(VI): Kinetics, pathways, and effect of dissolved organic matter. *Chem. Eng. J.* **2018**, *332*, 245–252.
18. Jiang, W.; Chen, L.; Batchu, S.R.; Gardinali, P.R.; Jasa, L.; Marsalek, B.; Zboril, R.; Dionysiou, D.D.; O'Shea, K.E.; Sharma, V.K. Oxidation of Microcystin-LR by Ferrate(VI): Kinetics, Degradation Pathways, and Toxicity Assessment. *Environ. Sci. Technol.* **2014**, *48*, 12164–12172. [CrossRef] [PubMed]
19. Sharma, V.K.; Zboril, R.; Varma, R.S. Ferrates: Greener oxidants with multimodal action in water treatment technologies. *Acc. Chem. Res.* **2015**, *48*, 182–191. [CrossRef]
20. Feng, M.B.; Wang, X.H.; Chen, J.; Qu, R.J.; Sui, Y.; Cizmas, L.; Wang, Z.Y.; Sharma, V.K. Degradation of fluoroquinolone antibiotics by ferrate(VI): Effects of water constituents and oxidized products. *Water Res.* **2016**, *103*, 48–57. [CrossRef] [PubMed]
21. Liu, H.X.; Pan, X.X.; Chen, J.; Qi, Y.M.; Qu, R.J.; Wang, Z.Y. Kinetics and mechanism of the oxidative degradation of parathion by Ferrate(VI). *Chem. Eng. J.* **2019**, *365*, 142–152.
22. Pan, X.X.; Yan, L.Q.; Li, C.G.; Qu, R.J.; Wang, Z.Y. Degradation of UV-filter benzophenone-3 in aqueous solution using persulfate catalyzed by cobalt ferrite. *Chem. Eng. J.* **2017**, *326*, 1197–1209.
23. Feng, M.B.; Qu, R.J.; Zhang, X.L.; Sun, P.; Sui, Y.X.; Wang, L.S.; Wang, Z.Y. Degradation of flumequine in aqueous solution by persulfate activated with common methods and polyhydroquinone-coated magnetite/multi-walled carbon nanotubes catalysts. *Water Res.* **2015**, *85*, 1–10. [PubMed]
24. Liu, H.; Sun, P.; He, Q.; Feng, M.; Liu, H.; Yang, S.; Wang, L.; Wang, Z. Ozonation of the UV filter benzophenone-4 in aquatic environments: Intermediates and pathways. *Chemosphere* **2016**, *149*, 76–83. [PubMed]

Article

Mechanistic Insight into Degradation of Cetirizine under UV/Chlorine Treatment: Experimental and Quantum Chemical Studies

Boyi Zhu [†], Fangyuan Cheng [†], Wenjing Zhong, Jiao Qu, Ya-nan Zhang * and Hongbin Yu *

State Environmental Protection Key Laboratory of Wetland Ecology and Vegetation Restoration, School of Environment, Northeast Normal University, Changchun 130117, China; zhuby551@nenu.edu.cn (B.Z.); chengfy310@nenu.edu.cn (F.C.); zhongwj598@nenu.edu.cn (W.Z.); quj100@nenu.edu.cn (J.Q.)
* Correspondence: zhangyn912@nenu.edu.cn (Y.-n.Z.); yuhb108@nenu.edu.cn (H.Y.)
† These authors contributed equally to this work.

Abstract: UV/chlorine treatment is an efficient technology for removing organic pollutants in wastewater. Nevertheless, degradation of antihistamines in the UV/chlorine system, especially the underlying reaction mechanism, is not yet clear. In this study, the degradation of cetirizine (CTZ), a representative antihistamine, under UV/chlorine treatment was investigated. The results showed that CTZ could undergo fast degradation in the UV/chlorine system with an observed reaction rate constant (k_{obs}) of (0.19 ± 0.01) min^{-1}, which showed a first-increase and then-decrease trend with its initial concentration increased. The degradation of CTZ during the UV/chlorine treatment was attributed to direct UV irradiation (38.7%), HO$^\bullet$ (35.3%), Cl$^\bullet$ (7.3%), and ClO$^\bullet$ (17.1%). The k_{obs} of CTZ decreased with the increase in pH and the increase in concentrations of a representative dissolved organic matter, Suwannee River natural organic matter (SRNOM), due to their negative effects on the concentrations of reactive species generated in the UV/chlorine system. The detailed reaction pathways of HO$^\bullet$, ClO$^\bullet$, and Cl$^\bullet$ with CTZ were revealed using quantum chemical calculation. This study provided significant insights into the efficient degradation and the underlying mechanism for the removal of CTZ in the UV/chlorine system.

Keywords: UV/chlorine; cetirizine; reactive species; dissolved organic matter; reaction pathway

Citation: Zhu, B.; Cheng, F.; Zhong, W.; Qu, J.; Zhang, Y.-n.; Yu, H. Mechanistic Insight into Degradation of Cetirizine under UV/Chlorine Treatment: Experimental and Quantum Chemical Studies. *Water* **2022**, *14*, 1323. https://doi.org/10.3390/w14091323

Academic Editors: Gassan Hodaif, Antonio Zuorro, Joaquín R. Dominguez, Juan García Rodríguez, José A. Peres and Zacharias Frontistis

Received: 26 March 2022
Accepted: 18 April 2022
Published: 19 April 2022

Publisher's Note: MDPI stays neutral with regard to jurisdictional claims in published maps and institutional affiliations.

1. Introduction

Pharmaceuticals and personal care products (PPCPs) have attracted widespread attention because of their ubiquity in the aqueous environment and potentially high risks to the ecosystem [1–3]. Antihistamines, as one kind of PPCPs, are mainly used to alleviate human allergies [4]. In recent years, antihistamines have been frequently detected in the environment due to the extensive production and widespread use [5]. The concentrations of antihistamines in surface water have been found to be mainly at ng L^{-1} level, and high concentrations up to μg L^{-1} level were also reported [6,7]. The antihistamines in surface water are mainly discharged from wastewater treatment plants (WWTPs), of which effluent contains antihistamines with concentrations as high as mg L^{-1} level [7]. Although antihistamines have been produced and used as specific histamine H1-receptor antagonists, the toxic effects, e.g., cardiotoxicity and sublethal effects, of frequently used antihistamines to aquatic organisms have been observed [6,8,9]. Thus, it is of great significance to remove antihistamines in wastewater before their discharging into natural waters.

Nowadays, the mainly used treatment technologies in WWTPs cannot effectively remove organic pollutants [10,11], especially PPCPs [12,13]. Antihistamines have been also ineffectively or inefficiently degraded by the traditionally used technologies due to their high concentration in the effluent from WWTPs [7,14]. Advanced oxidation processes (AOPs), including ozonation, ultraviolet (UV)/H$_2$O$_2$, etc., have been proven to be effective technologies for degrading refractory PPCPs, including antihistamines [5,15,16]. The antihistamines, e.g., cimetidine (CMD), cetirizine (CTZ), and diphenhydramine (DPD), have been shown to undergo degradation in TiO_2-initiated photocatalysis [17], ozonation and peroxymonosulfate oxidation [5,18], and UV/H_2O_2 treatment [19], respectively.

Among various AOPs, UV/chlorine technology is attractive due to its efficient degradation of various pollutants [20–22]. In a UV/chlorine system, reactive species with high reactivity such as hydroxyl radicals ($HO^\bullet$) and chlorine free radicals (including $Cl^\bullet$, $ClO^\bullet$, and $Cl_2^{\bullet-}$) were generated, which dominated the degradation of organic contaminants [23]. For example, Lei et al. [24] found that two representative antihistamines, cimetidine and famotidine, could undergo fast reaction with $HO^\bullet$, $Cl^\bullet$, and $Cl_2^{\bullet-}$, and the reaction rate constants were 6.50×10^9 M^{-1} s^{-1}, 1.46×10^{10} M^{-1} s^{-1}, and 0.43×10^{10} M^{-1} s^{-1} and 1.72×10^{10} M^{-1} s^{-1}, 2.78×10^9 M^{-1} s^{-1}, and 1.65×10^9 M^{-1} s^{-1}, respectively. Therefore, it can be reasonably speculated that antihistamines can be efficiently degraded in the UV/chlorine process. However, little attention has been paid on the removal of antihistamines. Furthermore, the contribution of these reactive species to the degradation of antihistamines in the UV/chlorine system is also unclear.

The degradation of PPCPs in a UV/chlorine system was reported to be influenced by aqueous environmental factors [25]. For example, pH can affect free chlorine's dissociation and subsequently influence the generation of reactive species [26], and also change the reactivity of ionizable active pharmaceuticals with the generated reactive species during the UV/chlorine process [27]. During the UV/chlorine treatment, dissolved organic matter (DOM), which is a ubiquitous component in wastewater, was shown to inhibit the degradation of two PPCPs (naproxen and gemifibrozil) via inner filter effect and quenching of the generated reactive species [25]. The water matrix could not only influence the degradation efficiency but also affect the degradation pathways of PPCPs by changing the generation and presence of reactive species during the UV/chlorine treatment. The reactivity of the generated reactive species toward PPCPs was shown to be structure-dependent [28].

The reaction of the reactive species in a UV/chlorine system with PPCPs could lead to the formation of toxic by-products [29], and the formation of halogenated intermediates in the reactions of reactive chlorine radicals with PPCPs is a great concern in the use of the UV/chlorine technology [28]. Therefore, it is essential to investigate the detailed reaction pathways of the reactive species such as $HO^\bullet$, $Cl^\bullet$, $Cl_2^{\bullet-}$, and $ClO^\bullet$ with PPCPs. The reactions of these reactive species with organic pollutants mainly occur through single electron transfer, abstraction of hydrogen (H-abstraction), and addition pathways [30–32]. Quantum chemical calculation has been previously applied to investigate the reactions of reactive species with organic contaminants successfully [32,33], and the results demonstrated that H-abstraction and addition are the major reaction pathways [21]. Nevertheless, the reaction pathways of antihistamines with the generated reactive species in the UV/chlorine system are still unclear and worthy of urgent research.

Thus, the degradation of antihistamines in a UV/chlorine system was investigated with CTZ as a representative, which is among the most detected antihistamines with the highest concentrations in the effluent and surface water [7]. The effect of pH, initial concentration, and SRNOM (a representative of DOM) on the degradation of CTZ was revealed. Furthermore, the contribution of the reactive species on the degradation of CTZ was investigated, and the detailed reaction pathways of these reactive species with CTZ were calculated using quantum chemical calculation. The results of this study were helpful for providing alternative technology to remove antihistamines in wastewater as well as deep insight into the degradation mechanisms of emerging PPCPs in the UV/chlorine system.

2. Materials and Methods

2.1. Chemicals

CTZ (98%), sodium hypochlorite solution (available chlorine 5%), nitrobenzene (NB, 98%), and tert-butanol (TBA, chromatographical purity) were purchased from J&K Scientific Ltd. (Beijing, China). SRNOM (2R101N) was obtained from the International Humic

Substances Society. Na_2HPO_4 and NaH_2PO_4 were purchased from Tianjin Damao Chemical (Tianjin, China). Methanol (chromatographicfal purity) was purchased from TEDIA (Fairfield, CT, USA). Ultrapure water was produced using an instrument purchased from Chengdu Ultrapure Technology Co., Ltd. (Chengdu, China).

2.2. UV/Chlorine Degradation Experiments

The experiments were performed in an OCRS-PX32T rotatable photochemical reactor that was obtained from Kaifeng HXsei Science Instrument Factory (Kaifeng, China) (Figure S1). The light source used in this study was a 500 W mercury lamp with main wavelengths of 254 nm and 297 nm, of which light intensity was determined to be (1.45 ± 0.07) mW cm^{-2} and (0.57 ± 0.03) mW cm^{-2} using a UV-B dual-channel ultraviolet radiation meter (Photoelectric Instrument Factory of Beijing Normal University, Beijing, China). During the treatment, the temperature was controlled at (25 ± 1) °C using a circulating cooling water system. All experiments were conducted in triplicate.

The initial concentration of CTZ and free chlorine was 10 μM and 100 μM, respectively (Phosphate buffer solution (PBS), pH 7.0). The UV-Vis absorption spectra of NaClO and CTZ are shown in Figure S2. UV/chlorine treatment was performed in the photo-reactor for 10 min. During the irradiation, dark control experiments were conducted with quartz tubes (30 mL) covered using aluminum foil. Samples were obtained and placed in dark conditions at room temperature during the UV/chlorine treatment. The concentration of residual chlorine was determined with the method that we used in our previous study [34].

The effects of the initial concentration of CTZ (2, 5, 10, 25, and 50 μM) and free chlorine (100, 200, 500, and 700 μM) on the degradation of CTZ was investigated by changing the concentrations. SRNOM was added with concentrations of 2.0, 5.0, 10.0, and 15.0 mg L^{-1} to investigate its effect on the degradation of CTZ. NB (2 μM) was used as the quencher of $HO^\bullet$ [35], sodium bicarbonate ($NaHCO_3$, 100 mM) was used to scavenge the $HO^\bullet$, $Cl^\bullet$, and $Cl_2^{\bullet-}$ [28], and TBA (100 mM) was employed as the quencher of $HO^\bullet$, $Cl^\bullet$, and $ClO^\bullet$ [36]. During the degradation of CTZ, Cl− with high concentration of 50 mM was added to investigate the roles of $Cl^\bullet$ and $Cl_2^{\bullet-}$ as it can convert $Cl^\bullet$ to $Cl_2^{\bullet-}$ [27].

2.3. Analytical Methods

An Agilent 1260 II HPLC (Agilent Technologies Inc., Santa Clara, CA, USA) with a diode array detector was used to quantify CTZ. During the quantification, a Welch Ultimate™ AQ-C18 (250 mm × 4.6 mm, 5 m) analysis column (Welch Materials Inc., Maryland, USA) was used. The mobile phase for the detection of CTZ was methanol and 0.1% phosphoric acid solution aqueous (pH 2.3) with a ratio of 60:40, and the wavelength selected for the detection of CTZ was 200 nm.

2.4. Quantum Chemical Calculation Methods

All quantum chemical calculation was performed with Gaussian 16 software [37]. The optimization and single-point energy calculation were performed using density functional theory (DFT) with the function of B3LYP. The optimization was performed at the level of 6-31+G(d, p), and the single-point energy calculation was performed at the level of 6-311++G(3df, 2p). The integral equation form of the polarization continuum model (IEFPCM) was used to consider the solvent effect of water. The transition state (TS) was obtained and characterized by the virtual vibration frequency (only one), and was verified with intrinsic reaction coordinate (IRC) analysis. The thermodynamic energies, i.e., Gibbs free energy and enthalpy, were obtained and zero-point correction was performed while calculating these energies.

3. Results and Discussion

3.1. Degradation of CTZ and Influencing Factors during UV/Chlorine Treatment

During direct UV irradiation, obvious degradation of CTZ was observed and no obvious degradation in the dark controls was observed (Figure 1). These results indicate that CTZ was photodegradable under direct UV irradiation, and the degradation of CTZ followed pseudo first-order kinetics. The observed degradation rate constant (k_{obs}) of CTZ under direct UV irradiation was calculated to be (0.08 ± 0.01) min^{-1}. In a previous study, the photodegradation of CTZ under UV irradiation was also observed [38]. In the UV/chlorine system, the degradation rate of CTZ was faster than that under direct UV irradiation ($p < 0.05$, Figure 1). The k_{obs} of CTZ increased to (0.19 ± 0.01), (0.29 ± 0.02), (0.42 ± 0.03), and (0.50 ± 0.02) min^{-1} with concentrations of free chlorine increased to 100, 200, 500, and 700 µM, respectively. These results demonstrated the important role of free chlorine in the removal of CTZ. The degradation efficiency reached up to 84.9% within 10 min of treatment and exceeded 90% within 15 min of treatment at the free chlorine concentration of 100 µM in the UV/chlorine system. Thus, the free chlorine concentration of 100 µM was selected in the following experiments.

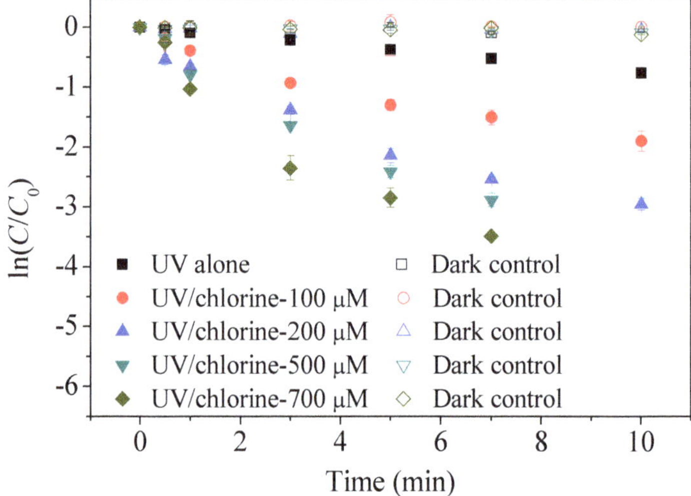

Figure 1. Degradation kinetics of CTZ in UV/chlorine system with different concentration of free chlorine (pH = 7.0, $[CTZ]_0 = 10$ µM. The error bars represent the 95% confidence interval ($n = 3$)).

In the UV/chlorine system, the initial concentration of organic pollutants was shown to affect their degradation kinetics and efficiencies [39]. Therefore, the effect of initial concentrations of CTZ on its degradation was studied, and the results are exhibited in Figure 2 and Figure S3 in the Supplementary Materials. As shown in Figure 2, the k_{obs} of CTZ showed a first-increase and then-decrease trend under UV irradiation as well as the UV/chlorine treatment. The fastest degradation rate of CTZ was observed with its initial concentration of 10 µM. With the increase in initial concentration of CTZ from 10 µM to 50 µM, the k_{obs} of CTZ under direct UV irradiation and in the UV/chlorine system decreased 36.4% and 64.4%, respectively.

The effect of initial concentration on the degradation of CTZ under direct UV irradiation was attributed to the competitive light absorption effect at the CTZ with high concentrations. The differences in k_{obs} values in the UV/chlorine system compared with those under direct UV irradiation also showed a first-increase and then decrease trend (Figure 2), indicating that the concentration-dependence in the degradation of CTZ was not only attributed to the UV-induced degradation but also the competing reactions to the generated reactive species.

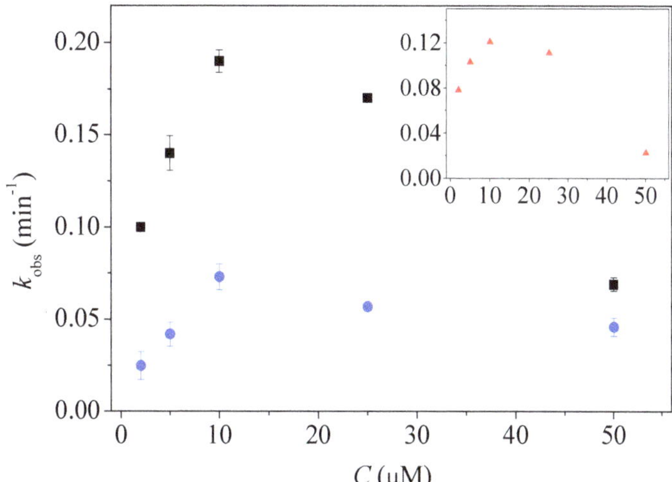

Figure 2. Observed degradation rate constant (k_{obs}) of CTZ with different initial concentration in UV (circle) and UV/chlorine (square) system (pH = 7.0, (Free chlorine)$_0$ = 100 μM. The insert figure represents the differences in k_{obs} values in UV/chlorine system compared with those under UV treatment alone. The error bars represent the 95% confidence interval (n = 3)).

The influence of pH on the degradation of CTZ was studied and the results showed that the k_{obs} of CTZ decreased with the increase in pH in the UV/chlorine system (Figure 3 and Figure S4 in the Supplementary Materials). There were two possible mechanisms for the pH-dependence of CTZ degradation: (1) the different reactivities of the generated reactive species toward CTZ of different forms, i.e., neutral form and anionic form; (2) the dissociation of HOCl/OCl$^-$ with pKa of 7.5 [40] led to different concentrations of the generated reactive species [41]. In this case, the former was not suitable as more than 99.7% CTZ (pKa = 3.5) existed with anionic form in the solutions with pH above 6.0. Thus, in the UV/chlorine system, the decrease in k_{obs} of CTZ with the increase in pH was due to the dissociation of HOCl to OCl$^-$ as the generation quantum yield of reactive species was higher and radical scavenging effect was weaker for HOCl compared with those of OCl$^-$ [42].

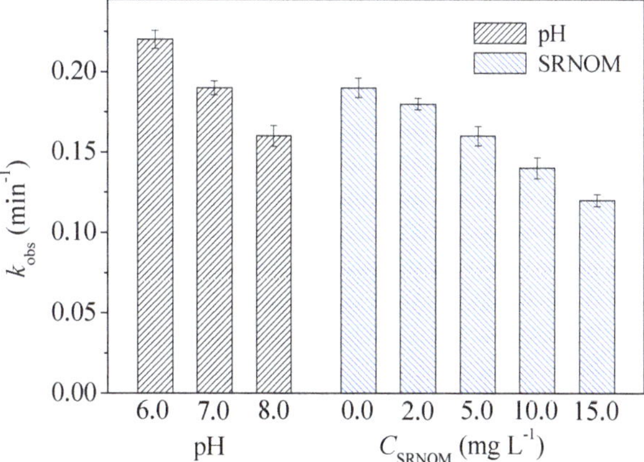

Figure 3. Observed degradation rate constant (k_{obs}) of CTZ in UV/chlorine system in different pH or in the presence of SRNOM with different initial concentrations ([CTZ]$_0$ = 10 μM, (Free chlorine)$_0$ = 100 μM. The error bars represent the 95% confidence interval (n = 3)).

The DOM in the wastewater, which was shown to significantly inhibit the removal of primidone and caffeine in the UV/chlorine system, is considered to be a great defect during the practical application of UV/chlorine technology [39]. As exhibited in Figure 3 and Figure S5 in the Supplementary Materials, in the UV/chlorine system, the presence of SRNOM obviously decreased the k_{obs} of CTZ ($p < 0.05$), and the inhibitory effect of SRNOM increased with the increase in its concentration. The k_{obs} of CTZ decreased from (0.19 ± 0.01) min^{-1} without SRNOM to (0.18 ± 0.01), (0.16 ± 0.01), (0.14 ± 0.01), and (0.12 ± 0.01) min^{-1} in the presence of SRNOM with concentrations of 2.0, 5.0, 10.0, and 15.0 mg L^{-1}, respectively, of which decreased rates were 7.3%, 17.8%, 28.4%, and 38.9%, respectively. The inhibition on the degradation of CTZ induced by SRNOM was attributed to: (1) the light screening effect of SRNOM as it can absorb light with wavelengths from 200 to 400 nm (Figure S1); (2) the competition with reactive radicals as reported in a previous study [39]. Thus, it could be concluded that the DOM in wastewater is an unfavorable factor during the degradation of antihistamines in wastewater using UV/chlorine technology.

3.2. Roles of Reactive Species in Degradation of CTZ

In the UV/chlorine system, the underlying degradation mechanisms of CTZ were revealed by performing quenching experiments. As can be seen in Figure 4, the presence of Cl$^-$ significantly decreased the k_{obs} of CTZ in the UV/chlorine system ($p < 0.05$), indicating the involvement of Cl$^\bullet$ in the degradation of CTZ as Cl$^-$ can quench Cl$^\bullet$ to generate Cl$_2^{\bullet-}$ rapidly [27]. In the presence of NB, the k_{obs} of CTZ decreased to (0.12 ± 0.01) min^{-1} from (0.19 ± 0.01) min^{-1} in the PBS ($p < 0.05$), which indicated the involvement of HO$^\bullet$ during the degradation of CTZ. The presence of HCO$_3^-$, which could quench HO$^\bullet$, Cl$^\bullet$, and Cl$_2^{\bullet-}$ [28], decreased the k_{obs} of CTZ to (0.11 ± 0.01) min^{-1} ($p < 0.05$). The contribution of Cl$^\bullet$ and Cl$_2^{\bullet-}$ (k_{obs}(NB)$-k_{obs}$(HCO$_3^-$)) was comparable with that of Cl$^\bullet$ (k_{obs}(PBS)$-k_{obs}$(Cl$^-$)) in the degradation of CTZ, which demonstrated the negligible effect of Cl$_2^{\bullet-}$ on the degradation of CTZ. The addition of TBA, which quench HO$^\bullet$, Cl$^\bullet$, and ClO$^\bullet$ [36], further decreased the k_{obs} of CTZ compared with that of HCO$_3^-$, which indicated the involvement of ClO$^\bullet$ in the degradation of CTZ.

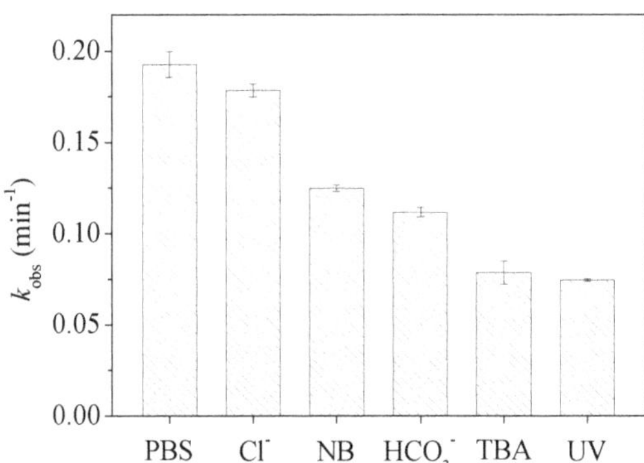

Figure 4. Observed degradation rate constant (k_{obs}) of CTZ (10 μM) in UV/chlorine system ((Free chlorine)$_0$ = 100 μM) under different conditions (PBS, pH = 7.0; (Cl$^-$) = 50 mM; (NB (nitrobenzene)) = 2 μM; (HCO$_3^-$) = 100 mM; (TBA (*tert*-butanol)) = 100 mM) and in UV system as a control (the error bars represent the 95% confidence interval ($n = 3$)).

In the UV/chlorine system, the presence of SRNOM could inhibit the degradation of CTZ (Figure 3). Thus, the involvement of the reactive species in the degradation of CTZ in SRNOM solutions was also investigated. The addition the quenchers decreased the k_{obs} of CTZ in the SRNOM solutions (Figure 5), which was similar with those in the PBS. Thus, in the presence of SRNOM, HO$^\bullet$, Cl$^\bullet$, and ClO$^\bullet$ were also involved in the degradation of CTZ.

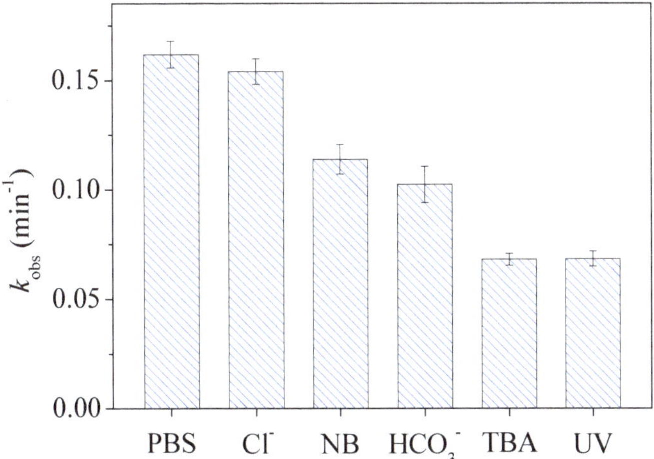

Figure 5. Observed degradation rate constant (k_{obs}) of CTZ (10 μM) in the presence of SRNOM (5 mg L$^-$) in UV/chlorine system ((Free chlorine)$_0$ = 100 μM) under different conditions (PBS, pH = 7.0; (Cl$^-$) = 50 mM; (NB (nitrobenzene)) = 2 μM; (HCO$_3^-$) = 100 mM; (TBA (*tert*-butanol)) = 100 mM) and in UV system as a control (the error bars represent the 95% confidence interval (n = 3)).

Contribution ratios of the reactive species (HO$^\bullet$, Cl$^\bullet$, and ClO$^\bullet$) and direct UV irradiation to the degradation of CTZ were calculated, and the values are shown in Table S1 in the SI. The contribution ratio of UV-induced degradation of CTZ was 38.7% in the PBS. For the reactive species, the contribution ratio of HO$^\bullet$, Cl$^\bullet$, and ClO$^\bullet$ was 35.3%, 7.3%, and 17.1%, respectively. The remaining 1.6% degradation of CTZ was attributed to other reactive species. Among these reactive species, HO$^\bullet$ played a crucial role during the degradation of CTZ. In SRNOM solutions, the contribution ratio of direct UV-induced degradation of CTZ increased to 42.2% compared with that in the PBS. The contribution ratio of HO$^\bullet$ and Cl$^\bullet$ decreased to 29.6% and 4.8%, respectively, compared with the values in the PBS. However, the contribution ratio of ClO$^\bullet$ increased to 21.2% from 17.1%. These results demonstrated the different role of SRNOM on the quenching of these reactive species.

3.3. Degradation Pathways of CTZ in UV/Chlorine System

Considering the complexity of the reactions for CTZ in the UV/chlorine system, the DFT method was used to investigate the degradation pathways of CTZ in the UV/chlorine system. In wastewater, CTZ mainly exists in anionic form as it is with low pKa of 3.5. Therefore, the reaction pathways of anionic CTZ with the important reactive species (HO$^\bullet$, Cl$^\bullet$, and ClO$^\bullet$) were calculated with DFT methods. As shown in a previous study, major reactions of these reactive species with small organic chemicals are through H-abstraction and addition pathways [21]. For anionic CTZ, there are 24 H-abstraction and 12 addition pathways (Figure 6). All these possible reaction pathways of HO$^\bullet$, Cl$^\bullet$, and ClO$^\bullet$ with CTZ were calculated, and the calculated values of Gibbs free energy change (ΔG), activation free energy ($\Delta G^\ddagger$), and enthalpy change (ΔH) for these reaction pathways are listed in Tables S2 and S3. The reaction complexes, transition states, and reaction intermediates of these reactions are shown in Figures S6–S10 in the Supplementary Materials.

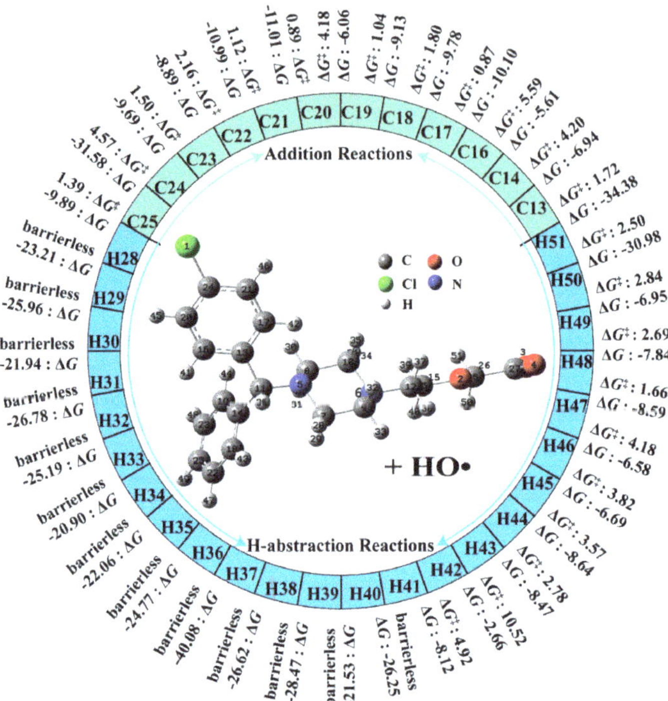

Figure 6. Addition and H-abstraction reaction pathways for reaction of CTZ with HO$^\bullet$ (ΔG: Gibbs free energy change; $\Delta G^{\ddagger}$: activation free energy; unit in kcal mol^{-1}).

For the HO$^\bullet$-initiated H-abstraction and addition reactions, the values of ΔG and ΔH were all < 0 (Figure 6 and Table S2), implying that H-abstraction from CTZ and HO$^\bullet$ addition to the unsaturated bonds of the phenyl group of CTZ were all thermodynamically spontaneous. The values of $\Delta G^{\ddagger}$ for the addition reactions were from 0.87 to 5.59 kcal mol^{-1}, indicating that these reactions were also dynamically possible. The HO$^\bullet$ additions on the C16 and C20 sites were with low $\Delta G^{\ddagger}$ values (0.87 and 0.89 kcal mol^{-1}, respectively), which indicated that the addition reactions at the two sites were more favorable. For the H-abstraction reactions, the $\Delta G^{\ddagger}$ values for the pathways from the phenyl group (H41-49) and α-carbon (α-C) are from 1.72 to 10.52 kcal mol^{-1}, and the pathways from the hydrocarbon chain (except that from the α-C) were all barrierless, which indicated that the H-abstraction reaction from the hydrocarbon chain of CTZ was more favorable. These results demonstrated the high reactivity of HO$^\bullet$ with CTZ, which is in accordance with the extremely high reaction rate constant of HO$^\bullet$ with CTZ (8.94 × 10^9 M^{-1} s^{-1}) [43]. Meanwhile, these results are also in agreement with the experimental results that HO$^\bullet$ plays an important role during the degradation of CTZ in the UV/chlorine system.

For the ClO$^\bullet$ addition pathways, only the ΔG and ΔH for the reaction at the C24 site were negative (Figure 7 and Table S3), indicating that ClO$^\bullet$ addition to the unsaturated bonds at the C24 site was thermodynamically spontaneous and exothermic. Thus, although the $\Delta G^{\ddagger}$ values of the addition reaction at the C24 site (17.21 kcal mol^{-1}) were relatively higher compared with that at other sites, ClO$^\bullet$ addition reaction could only occur spontaneously at this site. This was due to the generation of a thermodynamically more stable intermediate with an oxygen atom on C24 instead of Cl. The H-abstraction pathways on the phenyl group of CTZ were all thermodynamically nonspontaneous and endothermic as they were all with positive ΔG and ΔH (Table S3). The abstraction reactions of H39, H40, H50, and H51 at the hydrocarbon chain were thermodynamically spontaneous and exothermic due to their negative ΔG and ΔH (Figure 7 and Table S3). Among these reactions, the abstraction of H50 and H51 were more favorable due to the lower $\Delta G^{\ddagger}$ values (2.22 and 1.83 kcal mol^{-1}, respectively).

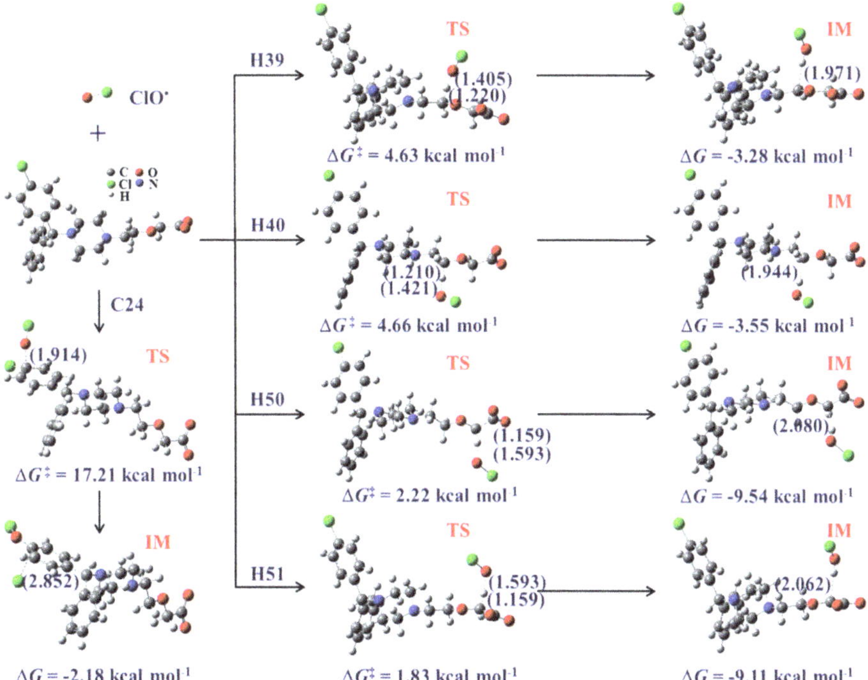

Figure 7. Detailed addition (C24 site) and H-abstraction (H39, H40, H50, and H51) pathways for the reaction of CTZ + ClO• (the values in the brackets are length of bonds marked with the dashed lines, and the unit is Å; TS and IM denote transition state and intermediate, respectively).

For the reaction of CTZ with Cl•, only H-abstraction of H36 was with negative ΔG (-3.08 kcal mol^{-1}), and of which $\Delta G^{\ddagger}$ was 9.54 kcal mol^{-1}, implying that H-abstraction of H36 was thermodynamically spontaneous and dynamically possible. The reaction complex, transition state, and intermediate of this reaction are shown in Figure S11. The results implied that the Cl• showed low reactivity toward CTZ, which is consistent with the experimental results that Cl• played a minor role (7.3%) in the degradation of CTZ in the UV/chlorine system.

For the reactions of CTZ with HO•, Cl•, and ClO•, H-abstraction is the main reaction pathway, which leads to the generation of C-centered radicals. C-centered radicals undergo subsequent reactions with dissolved O_2 or H_2O to form hydroxylated products [30,32]. Addition reaction of HO• and ClO• with CTZ could also lead to the generation of hydroxylated products and the dechlorinated product. Meanwhile, no chlorinated products were generated during the reaction of these reactive species with CTZ. Thus, the degradation of CTZ during the UV/chlorine treatment could decrease the toxicity induced by CTZ.

4. Conclusions

The degradation of an emerging organic pollutant CTZ in the UV/chlorine system was investigated and the underlying reaction mechanism was revealed by combining the methods of experiment and quantum chemical calculation. The results demonstrated that CTZ could be removed efficiently in the UV/chlorine system, of which degradation was initial-concentration dependent. In the UV/chlorine system, besides direct UV-induced degradation, the generated HO$^{\bullet}$, Cl$^{\bullet}$, and ClO$^{\bullet}$ contributed greatly to the degradation of CTZ, among which HO$^{\bullet}$ played a crucial role in the degradation of CTZ. The HO$^{\bullet}$ and ClO$^{\bullet}$ initiated the degradation of CTZ mainly through addition and H-abstraction reactions, and the Cl$^{\bullet}$ could only initiate H-abstraction reaction.

Supplementary Materials: The following supporting information can be downloaded at: https://www.mdpi.com/article/10.3390/w14091323/s1, Figure S1: Photochemical reactor system diagram; Figure S2: UV-Vis absorption spectra of NaClO, CTZ, and SRNOM; Figure S3: Degradation kinetics of CTZ with different concentrations in UV/chlorine system; Figure S4: Degradation kinetics of CTZ in UV/chlorine system with different pH values; Figure S5: Degradation kinetics of CTZ in UV/chlorine system in the presence of SRNOM with different concentration; Figure S6: Reaction complexes, transition states and intermediates of the addition reaction pathways of CTZ with HO•; Figure S7: Reaction intermediates of barrierless H-abstraction reaction pathways of CTZ with HO•; Figure S8: Reaction complexes, transition states and intermediates of the hydrogen abstraction pathways of CTZ with HO•; Figure S9: Reaction complexes, transition states and intermediates of the addition reaction pathways of CTZ with ClO•; Figure S10: Reaction complexes, transition states and intermediates of the hydrogen abstraction reaction pathways of CTZ with ClO•; Figure S11: Reaction complexes, transition states and intermediates of the hydrogen abstraction reaction of H36 of CTZ with Cl•; Table S1: Contribution ratio of different reactive species and UV degradation to the degradation of CTZ in UV/chlorine system; Table S2: Calculated Gibbs free energy change (ΔG), enthalpy change (ΔH) and activation free energy ($\Delta G^{\ddagger}$) values for possible reaction pathways of HO• with CTZ; Table S3: Calculated Gibbs free energy change (ΔG), enthalpy change (ΔH) and activation free energy ($\Delta G^{\ddagger}$) values for possible reaction pathways of ClO• with CTZ (in kcal mol^{-1}).

Author Contributions: Conceptualization, Y.-n.Z. and H.Y.; methodology, B.Z. and F.C.; software, B.Z. and F.C.; validation, B.Z., F.C. and W.Z.; formal analysis, B.Z.; investigation, B.Z., F.C. and W.Z.; resources, J.Q., Y.-n.Z. and H.Y.; data curation, B.Z. and F.C.; writing—original draft B.Z.; writing—review and editing, Y.-n.Z. and H.Y.; supervision, Y.-n.Z. and H.Y.; funding acquisition, J.Q., Y.-n.Z. and H.Y. All authors have read and agreed to the published version of the manuscript.

Funding: This research was funded by the National Natural Science Foundation of China, grant number 22176030, 42130705, and 21976027; the Fundamental Research Funds for the Central Universities, grant number 2412019FZ019 and 2412020FZ015, and the Jilin Province Science and Technology Development Projects, grant number 20210101110JC.

Acknowledgments: This work was supported by the National Natural Science Foundation of China (22176030, 42130705, and 21976027), the Fundamental Research Funds for the Central Universities (2412019FZ019 and 2412020FZ015), and the Jilin Province Science and Technology Development Projects (20210101110JC).

Conflicts of Interest: The authors declare no conflict of interest.

References

1. Cheng, Y.-X.; Chen, J.; Wu, D.; Liu, Y.-S.; Yang, Y.-Q.; He, L.-X.; Ye, P.; Zhao, J.-L.; Liu, S.-S.; Yang, B.; et al. Highly enhanced biodegradation of pharmaceutical and personal care products in a novel tidal flow constructed wetland with baffle and plants. *Water Res.* **2021**, *193*, 116870. [CrossRef] [PubMed]
2. Ren, B.; Shi, X.; Jin, X.; Wang, X.C.; Jin, P. Comprehensive evaluation of pharmaceuticals and personal care products (PPCPs) in urban sewers: Degradation, intermediate products and environmental risk. *Chem. Eng. J.* **2021**, *404*, 127024. [CrossRef]
3. Zhang, R.; Meng, T.; Huang, C.-H.; Ben, W.; Yao, H.; Liu, R.; Sun, P. PPCP degradation by chlorine–UV processes in ammoniacal water: New reaction insights, kinetic modeling, and DBP formation. *Environ. Sci. Technol.* **2018**, *52*, 7833–7841. [CrossRef]
4. Jonsson, M.; Fick, J.; Klaminder, J.; Brodin, T. Antihistamines and aquatic insects: Bioconcentration and impacts on behavior in damselfly larvae (Zygoptera). *Sci. Total Environ.* **2014**, *472*, 108–111. [CrossRef] [PubMed]
5. He, X.; O'Shea, K.E. Rapid transformation of H1-antihistamines cetirizine (CET) and diphenhydramine (DPH) by direct peroxymonosulfate (PMS) oxidation. *J. Hazard. Mater.* **2020**, *398*, 123219. [CrossRef]
6. Jonsson, M.; Andersson, M.; Fick, J.; Brodin, T.; Klaminder, J.; Piovano, S. High-speed imaging reveals how antihistamine exposure affects escape behaviours in aquatic insect prey. *Sci. Total Environ.* **2019**, *648*, 1257–1262. [CrossRef]
7. Kristofco, L.A.; Brooks, B.W. Global scanning of antihistamines in the environment: Analyses of occurrence and hazards in aquatic systems. *Sci. Total Environ.* **2017**, *592*, 477–487. [CrossRef]

8.   Bittner, L.; Teixidó, E.; Keddi, I.; Escher, B.I.; Klüvera, N. pH-Dependent uptake and sublethal effects of antihistamines in Zebrafish (*Danio rerio*) embryos. *Environ. Toxicol. Chem.* **2019**, *38*, 1012–1022. [CrossRef]

9.   Paakkari, I. Cardiotoxicity of new antihistamines and cisapride. *Toxicol. Lett.* **2002**, *127*, 279–284. [CrossRef]

10.   Zuorro, A.; Maffei, G.; Lavecchia, R. Kinetic modeling of azo dye adsorption on non-living cells of *Nannochloropsis oceanica*. *J. Environ. Chem. Eng.* **2017**, *5*, 4121–4127. [CrossRef]

11.   Montanaro, D.; Lavecchia, R.; Petrucci, E.; Zuorro, A. UV-assisted electrochemical degradation of coumarin on boron-doped diamond electrodes. *Chem. Eng. J.* **2017**, *323*, 512–519. [CrossRef]

12.   Mar-Ortiz, A.F.; Salazar-Rábago, J.J.; Sánchez-Polo, M.; Rozalen, M.; Cerino-Córdova, F.J.; Loredo-Cancino, M. Photodegradation of antihistamine chlorpheniramine using a novel iron-incorporated carbon material and solar radiation. *Environ. Sci. Water Res. Technol.* **2020**, *6*, 2607. [CrossRef]

13.   Ennouri, R.; Lavecchia, R.; Zuorro, A.; Elaoud, S.C.; Petrucci, E. Degradation of chloramphenicol in water by oxidation on a boron-doped diamond electrode under UV irradiation. *J. Water Process Eng.* **2021**, *41*, 101995. [CrossRef]

14.   Chen, W.-H.; Huang, J.-R.; Lin, C.-H.; Huang, C.-P. Catalytic degradation of chlorpheniramine over GO-Fe$_3$O$_4$ in the presence of H$_2$O$_2$ in water: The synergistic effect of adsorption. *Sci. Total Environ.* **2020**, *736*, 139468. [CrossRef]

15.   Sun, P.; Meng, T.; Wang, Z.; Zhang, R.; Yao, H.; Yang, Y.; Zhao, L. Degradation of organic micropollutants in UV/NH$_2$Cl advanced oxidation process. *Environ. Sci. Technol.* **2019**, *53*, 9024–9033. [CrossRef] [PubMed]

16.   Yang, X.; Sun, J.; Fu, W.; Shang, C.; Li, Y.; Chen, Y.; Gan, W.; Fang, J. PPCP degradation by UV/chlorine treatment and its impact on DBP formation potential in real waters. *Water Res.* **2016**, *98*, 309–318. [CrossRef] [PubMed]

17.   Park, Y.-K.; Ha, H.-H.; Yu, Y.H.; Kim, B.-J.; Bang, H.-J.; Lee, H.; Jung, S.-C. The photocatalytic destruction of cimetidine using microwave-assisted TiO$_2$ photocatalysts hybrid system. *J. Hazard. Mater.* **2020**, *391*, 122568. [CrossRef]

18.   Borowska, E.; Bourgin, M.; Hollender, J.; Kienle, C.; McArdell, C.S.; von Gunten, U.D. Oxidation of cetirizine, fexofenadine and hydrochlorothiazide during ozonation: Kinetics and formation of transformation products. *Water Res.* **2019**, *94*, 350–362. [CrossRef]

19.   Yuan, F.; Hu, C.; Hu, X.; Qu, J.; Yang, M. Degradation of selected pharmaceuticals in aqueous solution with UV and UV/H$_2$O$_2$. *Water Res.* **2009**, *43*, 1766–1774. [CrossRef]

20.   Guo, K.; Wu, Z.; Yan, S.; Yao, B.; Song, W.; Hua, Z.; Zhang, X.; Kong, X.; Li, X.; Fang, J. Comparison of the UV/chlorine and UV/H$_2$O$_2$ processes in the degradation of PPCPs in simulated drinking water and wastewater: Kinetics, radical mechanism and energy requirements. *Water Res.* **2018**, *147*, 184–194. [CrossRef]

21.   Minakata, D.; Kamath, D.; Maetzold, S. Mechanistic insight into the reactivity of chlorine-derived radicals in the aqueous-phase UV−chlorine advanced oxidation process: Quantum mechanical calculations. *Environ. Sci. Technol.* **2017**, *51*, 6918–6926. [CrossRef] [PubMed]

22.   Yin, K.; Li, T.; Zhang, T.; Zhang, Y.; Yang, C.; Luo, S. Degradation of organic filter 2-Phenylbenzidazole-5-Sulfonic acid by light-driven free chlorine process: Reactive species and mechanisms. *Chem. Eng. J.* **2022**, *430*, 132684. [CrossRef]

23.   Wu, Z.; Fang, J.; Xiang, Y.; Shang, C.; Li, X.; Meng, F.; Yang, X. Roles of reactive chlorine species in trimethoprim degradation in the UV/chlorine process: Kinetics and transformation pathways. *Water Res.* **2016**, *104*, 272–282. [CrossRef] [PubMed]

24.   Lei, Y.; Cheng, S.; Luo, N.; Yang, X.; An, T. Rate constants and mechanisms of the reactions of Cl$^\bullet$ and Cl$_2^{\bullet-}$ with trace organic contaminants. *Environ. Sci. Technol.* **2019**, *53*, 11170–11182. [CrossRef] [PubMed]

25.   Liu, H.; Hou, Z.; Li, Y.; Lei, Y.; Xu, Z.; Gu, J.; Tian, S. Modeling degradation kinetics of gemfibrozil and naproxen in the UV/chlorine system: Roles of reactive species and effects of water matrix. *Water Res.* **2021**, *202*, 117445. [CrossRef] [PubMed]

26.   Remucal, C.K.; Manley, D. Emerging investigators series: The efficacy of chlorine photolysis as an advanced oxidation process for drinking water treatment. *Environ. Sci. Water Res. Technol.* **2016**, *2*, 565–579. [CrossRef]

27.   Wang, W.L.; Wu, Q.Y.; Huang, N.; Wang, T.; Hu, H.Y. Synergistic effect between UV and chlorine (UV/chlorine) on the degradation of carbamazepine: Influence factors and radical species. *Water Res.* **2016**, *98*, 190–198. [CrossRef]

28.   Guo, K.; Wu, Z.; Shang, C.; Yao, B.; Hou, S.; Yang, X.; Song, W.; Fang, J. Radical chemistry and structural relationships of PPCP degradation by UV/chlorine treatment in simulated drinking water. *Environ. Sci. Technol.* **2017**, *51*, 10431–10439. [CrossRef]

29.   Pan, Y.; Cheng, S.; Yang, X.; Ren, J.; Fang, J.; Shang, C.; Song, W.; Lian, L.; Zhang, X. UV/chlorine treatment of carbamazepine: Transformation products and their formation kinetics. *Water Res.* **2017**, *116*, 254–265. [CrossRef]

30.   Li, M.; Mei, Q.; Han, D.; Wei, B.; An, Z.; Cao, H.; Xie, J.; He, M. The roles of HO$^\bullet$, ClO$^\bullet$ and BrO$^\bullet$ radicals in caffeine degradation: A theoretical study. *Sci. Total Environ.* **2021**, *768*, 144733. [CrossRef]

31.   Ren, X.; Sun, Y.; Fu, X.; Zhu, L.; Cui, Z. DFT comparison of the OH-initiated degradation mechanisms for five chlorophenoxy herbicides. *J. Mol. Modeling* **2013**, *19*, 2249–2263. [CrossRef]

32.   Zhang, Y.; Wang, J.; Chen, J.; Zhou, C.; Xie, Q. Phototransformation of 2,3-dibromopropyl-2,4,6-tribromophenyl ether (DPTE) in natural waters: Important roles of dissolved organic matter and chloride ion. *Environ. Sci. Technol.* **2018**, *52*, 10490–10499. [CrossRef] [PubMed]

33.   Li, C.; Xie, H.-B.; Chen, J.; Yang, X.; Zhang, Y.; Qiao, X. Predicting gaseous reaction rates of short chain chlorinated paraffins with $\bullet$OH: Overcoming the difficulty in experimental determination. *Environ. Sci. Technol.* **2014**, *48*, 13808–13816. [CrossRef] [PubMed]

34.   Zhou, Y.; Cheng, F.; He, D.; Zhang, Y.; Qu, J.; Yang, X.; Chen, J.; Peijnenburg, W.J.G.M. Effect of UV/chlorine treatment on photophysical and photochemical properties of dissolved organic matter. *Water Res.* **2021**, *192*, 116857. [CrossRef]

35. Wang, J.; Wang, K.; Guo, Y.; Ye, Z.; Guo, Z.; Lei, Y.; Yang, X.; Zhang, L.; Niu, J. Dichlorine radicals ($Cl_2^{\bullet-}$) promote the photodegradation of propranolol in estuarine and coastal waters. *J. Hazard. Mater.* **2021**, *414*, 125536. [CrossRef] [PubMed]
36. Yang, T.; Mai, J.; Wu, S.; Liu, C.; Tang, L.; Mo, Z.; Zhang, M.; Guo, L.; Liu, M.; Ma, J. UV/chlorine process for degradation of benzothiazole and benzotriazole in water: Efficiency, mechanism and toxicity evaluation. *Sci. Total Environ.* **2021**, *760*, 144304. [CrossRef]
37. Frisch, M.J.; Trucks, G.W.; Schlegel, H.B.; Scuseria, G.E.; Robb, M.A.; Cheeseman, J.R.; Scalmani, G.; Barone, V.; Petersson, G.A.; Nakatsuji, H.; et al. *Gaussian 16, B.01*; Gaussian Inc.: Wallingford, CT, USA, 2016.
38. Gadipelly, C.R.; Rathod, V.K.; Marathe, K.V. Persulfate assisted photo-catalytic abatement of cetirizine hydrochloride from aqueous waste: Biodegradability and toxicityanalysis. *J. Mol. Catal. A Chem.* **2016**, *414*, 116–121. [CrossRef]
39. Wang, Y.; dos Santos, M.M.; Ding, X.; Labanowski, J.; Gombert, B.; Snyder, S.A.; Croué, J.-P. Impact of EfOM in the elimination of PPCPs by UV/chlorine: Radical chemistry and toxicity bioassays. *Water Res.* **2021**, *204*, 117634. [CrossRef] [PubMed]
40. Dodd, M.C.; Huang, C. Aqueous chlorination of the antibacterial agent trimethoprim: Reaction kinetics and pathways. *Water Res.* **2007**, *41*, 647–655. [CrossRef]
41. Fang, J.; Fu, Y.; Shang, C. The roles of reactive species in micropollutant degradation in the UV/free chlorine system. *Environ. Sci. Technol.* **2014**, *48*, 1859–1868. [CrossRef]
42. Xiang, Y.; Fang, J.; Shang, C. Kinetics and pathways of ibuprofen degradation by the UV/chlorine advanced oxidation process. *Water Res.* **2016**, *90*, 301–308. [CrossRef] [PubMed]
43. Cui, D.; DeCaprio, A.; Tarifa, A.; O'Shea, K. Ultrasound-induced remediation of the second-generation antihistamine, Cetirizine. *J. Environ. Chem. Eng.* **2020**, *8*, 103680. [CrossRef]

Article

Unusual Catalytic Effect of Fe^{3+} on 2,4-dichlorophenoxyacetic Acid Degradation by Radio Frequency Discharge in Aqueous Solution

Yongjun Liu * and Bing Sun

College of Environmental Science & Engineering, Dalian Maritime University, Dalian 116026, China; sunb88@dlmu.edu.cn
* Correspondence: lyjglow@dlmu.edu.cn; Tel.: +86-411-847-252-75; Fax: +86-411-847-276-70

Abstract: 2,4-dichlorophenoxyacetic acid (2,4-D) is a widely used herbicide for controlling broad-leaved weeds. The development of an efficient process for treating the refractory 2,4-D wastewater is necessary. In this study, liquid-phase degradation of 2,4-D induced by radio frequency discharge (RFD) was studied. Experimental results showed that the degradation was more effective in acidic than in neutral or alkaline solutions. During the degradation, a large amount of hydrogen peroxide (H_2O_2, 1.2 mM/min, almost equal to that without 2,4-D) was simultaneously produced, and catalytic effects of both ferric (Fe^{3+}) and ferrous (Fe^{2+}) ions on the degradation were examined and compared. It was found that 2,4-D degraded more rapidly in the case of Fe^{3+} than the that of Fe^{2+}. Such a scenario is explained that Fe^{3+} was successively reduced to Fe^{2+} by the atomic hydrogen (•H) and •OH-adducts of 2,4-D resulting from RFD, which in turn catalyzed the H_2O_2 to form more •OH radicals through Fenton's reaction, indicating that Fe^{3+} not only accelerates the degradation rate but also increases the amount of •OH available for 2,4-D degradation by suppressing the back reaction between the •H and •OH. 2,4-dichlorophenol, 4,6-dichlororesorcinol, 2-hydroxy-4-chloro- and 2-chloro-4-hydroxy- phenoxyacetic acids, hydroxylated 2,4-Ds, and carboxylic acids (glycolic, formic and oxalic) were identified as the byproducts. Energy yields of RFD have been compared with those of other nonthermal plasma processes.

Keywords: aqueous solution; radio frequency discharge; 2,4-D; degradation; Fenton

Citation: Liu, Y.; Sun, B. Unusual Catalytic Effect of Fe^{3+} on 2,4-dichlorophenoxyacetic Acid Degradation by Radio Frequency Discharge in Aqueous Solution. *Water* **2022**, *14*, 1719. https://doi.org/10.3390/w14111719

Academic Editors: Gassan Hodaifa, Antonio Zuorro, Joaquín R. Dominguez, Juan García Rodríguez, José A. Peres and Zacharias Frontistis

Received: 2 May 2022
Accepted: 23 May 2022
Published: 27 May 2022

Publisher's Note: MDPI stays neutral with regard to jurisdictional claims in published maps and institutional affiliations.

1. Introduction

2,4-dichlorophenoxyacetic acid (2,4-D) is a systemic herbicide extensively used for broad-leaved weeds control in cereal crops, pastures, and orchards. 2,4-D kills dicots (but not grasses) by mimicking the growth hormone auxin, causing uncontrolled growth and eventual death of the susceptible plants [1]. In 2015, the International Agency for Research on Cancer (IARC) confirmed its 1987 classification of 2,4-D as a Group 2B carcinogen [2,3]. Due to its massive use and high solubility in water, 2,4-D can easily seep into the aquatic environment during its production and application. 2,4-D is hardly biodegradable at concentrations higher than 1.0 mg/L [4]. Acid-washed powdered activated carbon (PAC) can effectively adsorb 2,4-D from aqueous media [5]. However, as PAC adsorption is just a phase transferring process, the adsorbent becomes a hazardous waste when it is saturated with pollutants that need to be properly disposed of. In this context, some energetic methods such as TiO_2 photocatalysis [6], electrochemical oxidation [7], gamma irradiation [8], sonolysis or their combination [9] have been developed. Although showing some promising, these processes suffer respective deficiencies of either process complexity or low energy utilization, and the search for efficient and effective 2,4-D decomposition processes is necessary.

In the past decades, electrical discharges for water purification have been investigated extensively [10–12]. Pulsed corona discharge (PCD) inside water, glow discharge in contact with water and dielectric barrier discharge (DBD) close to water surface etc. were usually employed for the decomposition of harmful pollutants [12]. Degradation of 2,4-D by PCD over water surface [13,14] and by DBD in planar falling film reactor [15] have been evaluated and compared with some typical water treatment technologies. Recently, radio frequency discharge (RFD) [16,17] is receiving increasing attention as a novel plasma water treatment process due to its easy formation both in pure water and in solution with high conductivity [17], low breakdown voltage [18] and experimental simplicity. Several chemically active radicals (•OH, H•, •O, etc.) derived from H_2O molecules were in situ generated, which can initiate the degradation of organic pollutants such as methylene blue [16] and Congo Red [18]. Wang et al. investigated the RFD of metronidazole [19] and found that reducing species played a significant role during the degradation. The formation of hydrogen peroxide (H_2O_2) during RFD of pure water was reported, and some catalytic effects of ferrous sulfate on the dye discoloration were observed. However, H_2O_2 quantification with the presence of the dye was not performed. In addition, when ferrous ion (Fe^{2+}) was added to the solution, it would be rapidly oxidized by H_2O_2 into ferric ion (Fe^{3+}) [16]. However, at present, no comparison in catalytic effect between Fe^{2+} and Fe^{3+} has been made. In this study, both 2,4-D degradation and H_2O_2 formation were investigated. Besides optimizing the experimental conditions, the catalytic mechanism of iron salts, especially that of the Fe^{3+} on the degradation was fully explored.

2. Experimental

The experimental apparatus, consisting of a radio frequency power supply (RSG-300, Reshige Electronics Inc., Changzhou, China), a vacuum matching box, a quartz reactor and a coaxial electrode system, is shown in Figure 1.

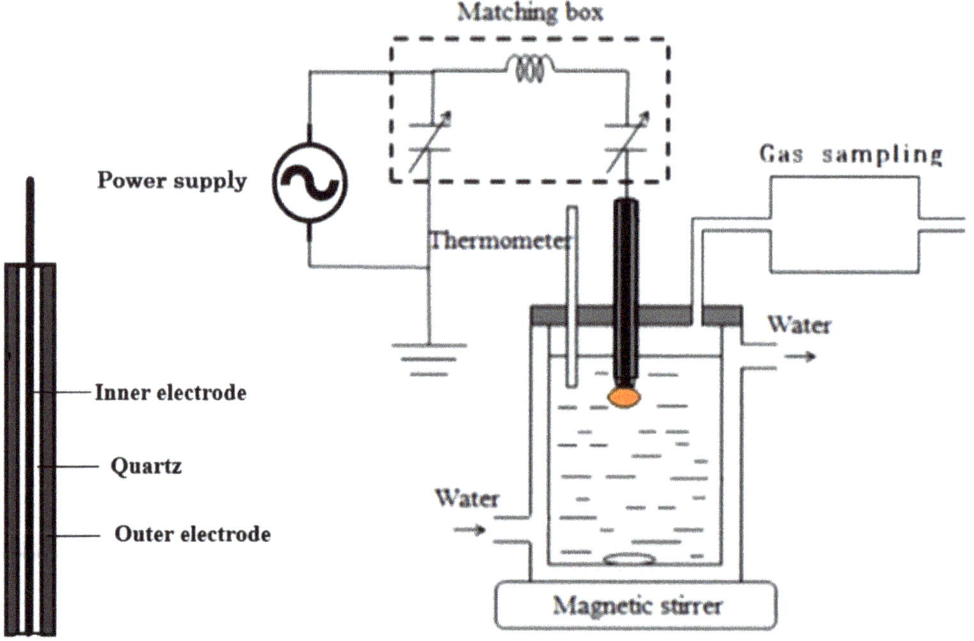

Figure 1. Experimental apparatus for 2,4-D degradation.

The reactor was a quartz cylinder with 60 mm of inner diameter and 150 mm in height. The coaxial electrode system consists of an inner electrode (pointed Pt wire, $\Phi = 1.0$ mm), a quartz tube (i. d. = 1.0 mm, o. d. = 2.0 mm), and an outer electrode (copper tube, i. d. = 2.2 mm, o. d. = 3.0 mm). Discharge is generated at the tip of the inner electrode in contact with the liquid. In such an electrode system, the discharge can be stably sustained without the shielding box needed in conventional RFD [18,19]. 2,4-D was dissolved in distilled water and a 250 mL portion was poured into the reactor for treatment. Na_2SO_4 was used to adjust the solution conductivity. The pH of the solution was adjusted with dilute H_2SO_4 or NaOH to the expected value. During the reaction, the plasma heats the water in the reaction cell [16]. The reaction vessel was coated by a water jacket, where the temperature of the solution was kept at $298 \pm 2K$ by running with cooling water.

During the discharge, the solution in the reactor was constantly agitated by a magnetic stirrer and the aliquots were periodically taken out for analysis. 2,4-D and the phenolic byproducts were analyzed by a reversed-phase HPLC (SHIMADZU LC-20A) coupled with a diode array UV detector. A C18 (4.6 × 25 mm) column was used for the separation. The eluent was composed of 40% acetonitrile and 59.9% H_2O and 0.1% H_3PO_4. The flow rate was 1.0 mL/min. The detection wavelength was usually set at 283 nm. Removal of 2,4-D ($\eta_{2,4\text{-D}}$) was calculated according to Equation (1).

$$\eta_{2,4\text{-D}}(\%) = \frac{(C_0 - C_t)}{C_0} \times 100 \tag{1}$$

where C_t (mM) was the concentration of 2,4-D at discharge time t and C_0 (mM) was the initial concentration of 2,4-D. Energy yields for 2,4-D removal ($J_{2,4\text{-D}}$) is calculated as Equation (2) [11,14].

$$J_{2,4-D} = \frac{C_0 Vol}{2Pt_{1/2}} \tag{2}$$

where *Vol* represents the solution volume (L), P the input power (W) and $t_{1/2}$, the discharge time required for 50% removal of 2,4-D (s).

The phenolic byproducts were identified by comparison of their retention times and UV spectra with those of the authentic standard samples or with the help of LC-MS [20]. Determination of organic acids and chloride ions (Cl^-) was performed by ion chromatography (IC, DIONEX ICS-1100) equipped with an Ion Pac AG-23 column. 0.01 mol/L KOH solution was used as the eluent with a flow rate of 1.0 mL/min. H_2O_2 was determined by a colorimetric method where the yellow color pertitanic acid formed by mixing the treated solution with titanium sulfate was determined by absorption at 410 nm [21]. The concentration of Fe^{2+} was determined colorimetrically where 1,10-phenanthroline was used as the color reagent [22]. It is noted that some catalase should be immediately added to the solution to decompose the residue H_2O_2 as it interferes with the determination of the Fe^{2+}. Total organic carbon (TOC) was measured by a TOC analyzer. The gaseous hydrogen evolved from the solution was determined by gas chromatography (GC-2014C, SHIMADZU). Each determination was performed three times and the average value was adopted, with a standard deviation of less than 4.5%.

3. Results and discussion

3.1. 2,4-D Degradation and Chloride Formation

2,4-D undergoes efficient degradation when subjected to RFD. Figure 2 shows the decays of 2,4-D and TOC and the formation of Cl^- during the RFD treatment.

As depicted in Figure 2, both 2,4-D and TOC declined exponentially while the yield of Cl^- increased gradually with increasing discharge time. The TOC decreased less rapidly than 2,4-D, indicating that intermediate byproducts were formed during the discharge. At the initial stage (<60 min), the yield of Cl^- is less than that expected from the decomposition of the parent compound. With 60 min of discharge treatment, 61% of 2,4-D was degraded and 0.32 mM of Cl^- was generated, meaning that only 26% of the organic chlorine of the transformed 2,4-D was mineralized into Cl^-. After 140 min of discharge, the concentration of 2,4-D was below the detection limit and 1.63 mM of Cl^- was generated, where the yield of Cl^- reached 82%. After 210 min, the Cl^- reached its stoichiometric point (2.0 mM), and after 300 min. (not shown in the figure), the concentration of TOC reached zero, indicating that 2,4-D was totally transformed into inorganic carbon and Cl^- under the action of RFD.

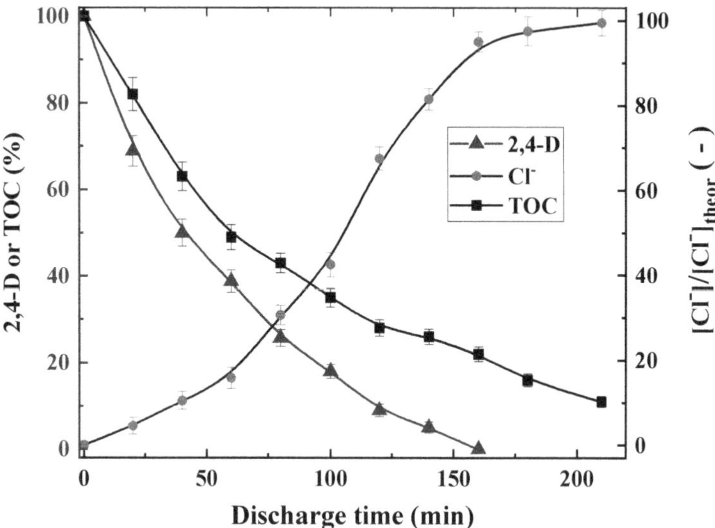

Figure 2. Decays of 2,4-D and TOC and formation of Cl^- during RFD treatment. ([2,4-D]$_0$, 1.0 mM; input power, 200 W; pH$_0$, 6.51).

3.2. Effects of Initial pH on 2,4-D Degradation

pH is an important parameter affecting the degradation process. In order to better elucidate the effects of pH on 2,4-D degradation, pH variations during the RFD in the presence and absence of 2,4-D are illustrated in Figure 3.

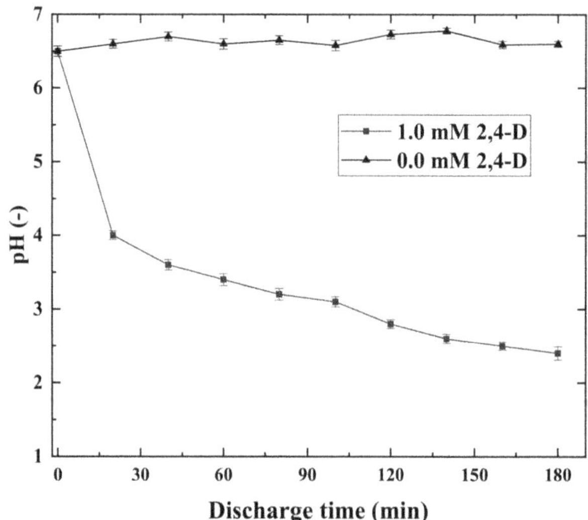

Figure 3. Variations of pH in the presence and absence of 2,4-D during RFD treatment (input power, 200 W; [2,4-D]$_0$, 1.0 mM; initial pH, 6.51).

As shown in Figure 3, without 2,4-D, the solution pH varied little during the discharge. However, in the presence of 2,4-D, the pH dropped dramatically. The decrease of pH in presence of 2,4-D can be explained by the formation of HCl and organic acids (*c.f.* 3.4). In order to better examine the effect and mechanism of pH on 2,4-D degradation, the experiments were performed both in buffered and unbuffered solutions. Here, phosphates were chosen as the buffering agents as the rate constants for the reactions of phosphates with •OH (e.g., Equations (3) and (4)) [23] are much smaller than that for the reaction of SO_4^{2-} (Equation (5)) [14].

$$H_2PO_4^- + \bullet OH \rightarrow H_2PO_4\bullet + OH^- \quad k_3 = 2.0\times10^4 \text{ M}^{-1}\cdot\text{s}^{-1} \tag{3}$$

$$HPO_4^{2-} + \bullet OH \rightarrow HPO_4^-\bullet + OH^- \quad k_4 = 1.5\times10^5 \text{ M}^{-1}\cdot\text{s}^{-1} \tag{4}$$

$$SO_4^{2-} + \bullet OH \rightarrow SO_4^-\bullet + OH^- \quad k_5 = 1.5\times10^6 \text{ M}^{-1}\cdot\text{s}^{-1} \tag{5}$$

Figure 4 shows the 2,4-D removal under different initial pH (pH_0) values with 10 min of RFD treatment in unbuffered and phosphate buffered (prepared with 2.0 mM NaH_2PO_4, no data from pH_0 3.0 to 6.0 as it has little buffering in the range) solutions.

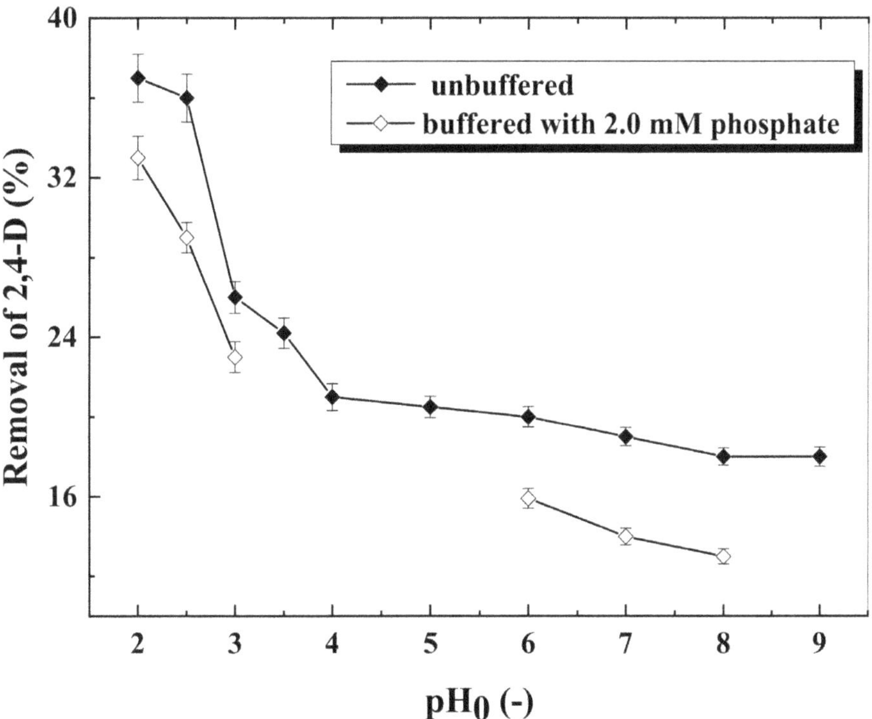

Figure 4. 2,4-D removal under different initial pH values in buffered and unbuffered solutions (input power, 200 W; [2,4-D]$_0$, 1.0 mM; discharge time, 10 min).

It can be observed from Figure 4 that the 2,4-D removal decreases with increasing pH_0, both in the cases of the buffered and the unbuffered solutions. The removal of 2.4-D declined from 37% at pH 2.0 to almost half (18%) at pH_0 9.0 and from 33% at pH 2.0 to 13% at pH_0 8.0, meaning that the decrease is more pronounced in the buffered solution. Such phenomena can be explained that the degradation of 2.4-D in RFD is mainly an •OH radical process (degradation rate decreased in the presence of ·OH scavenger, not shown in the figure), [20] it is more advantageous to the decomposition under acidic conditions as the alkoxyl group is more easily to be released from the C1 ·OH adducts in the presence of H^+ [24,25] (Scheme 1):

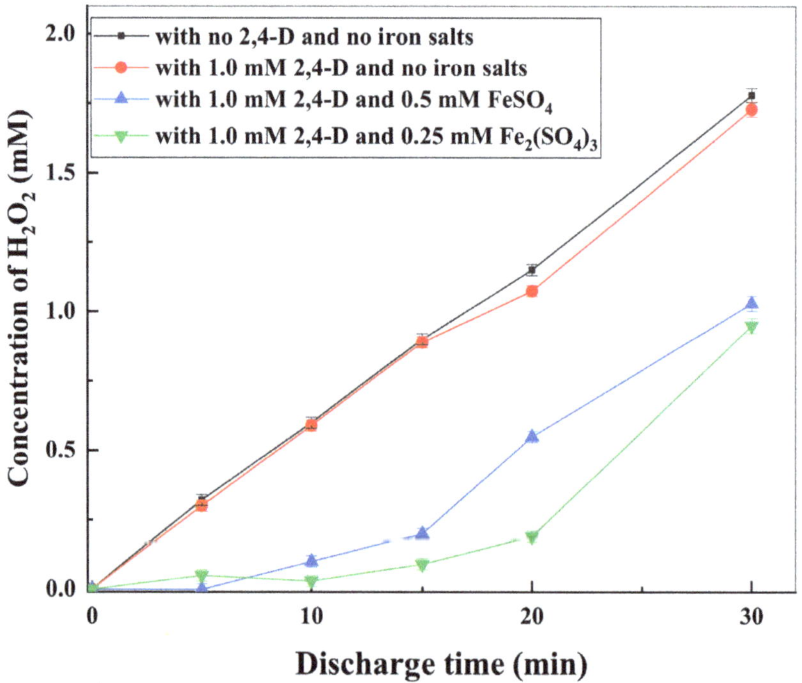

Scheme 1. H^+ catalyzed cleavage of C1 ·OH adducts.

Figure 4 shows that the difference in 2,4-D degradation between pH 4.0 and 9.0 was less noticeable in the case of unbuffered solution than in buffered ones, possibly due to the formation of acids which reduced the pH gap in the unbuffered system.

3.3. H_2O_2 Formation and Effects of Iron Salts on 2,4-D Degradation

Previous studies showed that the major molecular product formed in the liquid phase is H_2O_2. [16,17] However, the formation of H_2O_2 in the presence of organic pollutants has rarely been reported. [16] In this investigation, H_2O_2 generated both in the presence and absence of 2,4-D is presented in Figure 5.

Figure 5. H_2O_2 generation in the presence and absence of 2,4-D in RFD (input power, 200 W; $[2,4-D]_0$, 1.0 mM; pH_0, 5.6).

It can be seen in Figure 5 that less H_2O_2 was generated in 2,4-D solution than in solution without it, meaning that •OH radicals are the primary precursors of H_2O_2 and are the active species for 2,4-D degradation. However, the discrepancy is little in the early stage (<15 min), and the gap between the two concentrations of H_2O_2 was less than 10%, meaning that the concentration of •OH radicals near the plasma is so high that only a small part of them was consumed by 2,4-D and the rest dimerized to form H_2O_2.

The concentration of H_2O_2 in the presence of 2,4-D traces the curve for experiments in its absence, indicating that H_2O_2 alone cannot lead to rapid degradation of 2,4-D. It is desirable to convert H_2O_2 into reactive •OH radicals for pollutant degradation. It is desirable to add iron ions to enhance the •OH radical formation and increase the 2,4-D decomposition through Fenton's reaction (Reactions (6) and (7)) [26].

$$Fe^{2+} + H_2O_2 \rightarrow Fe^{3+} + \bullet OH + OH^- \quad k_6 = 40\text{-}80 \text{ M}^{-1} \cdot s^{-1} \quad (6)$$

$$Fe^{3+} + H_2O_2 \rightarrow Fe^{2+} + HO_2\bullet + H^+ \quad k_7 = 0.001\text{-}0.01 \text{ M}^{-1} \cdot s^{-1} \quad (7)$$

The 2,4-D degradation in the presence of 0.5 mM iron ions is shown in Figure 6.

It can be observed from Figure 6 that both Fe^{2+} and Fe^{3+} showed remarkable catalytic effects on 2,4-D degradation. With 5 min of discharge, 2,4-D removal in the presence of Fe^{3+} (38%) was a little higher than that of Fe^{2+} (36%), where it was only 20% without a catalyzer. However, the catalytic effect of Fe^{3+} was more obvious than that of Fe^{2+} with increasing discharge time. At 10 min. the 2,4-D was totally removed in Fe^{3+} while only 70% was achieved for Fe^{2+}. Such a phenomenon is abnormal in conventional Fenton reactions, as the reaction (7) is several orders of magnitude slower than the reaction (6). The reason may be that Fe^{3+} has been reduced by the ·OH-adducts formed by addition to the ring at positions unoccupied by the substituted groups (hydroxycyclohexadienyl radicals), producing the corresponding hydroxylated 2,4-Ds and Fe^{2+} [25,27] (Scheme 2):

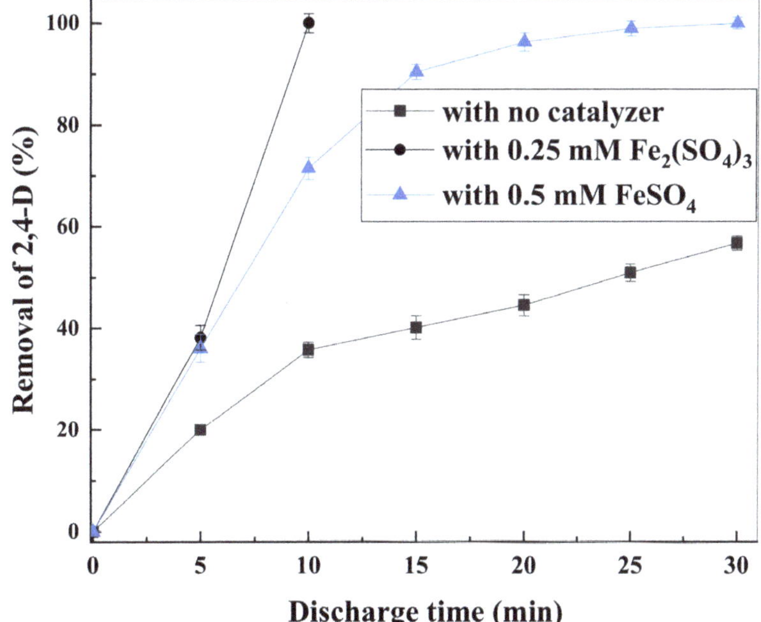

Figure 6. Effect of iron ions on 2,4-D removal in RFD treatment. (Conductivity 1.5 mS/cm, pH_0 2.0, input power 200 W).

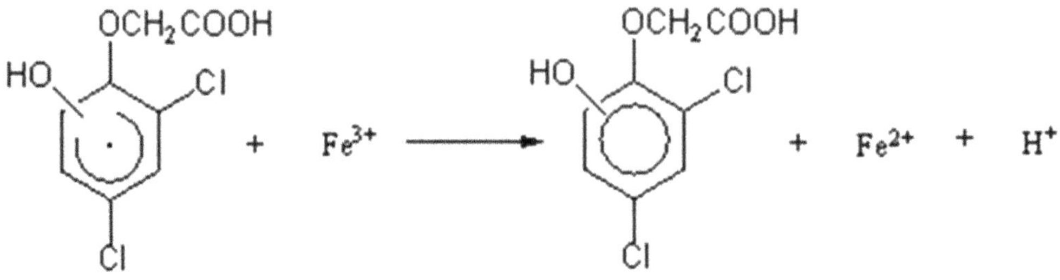

Scheme 2. Fe^{3+} reduction by ·OH-adducts formed by addition to the ring at positions unoccupied by the substituted groups (hydroxycyclohexadienyl radicals).

On the other hand, in the present pH range, much of Fe^{3+} is present in the form of $FeOH^{2+}$, which can be efficiently reduced by the ·H atoms generated by the RFD to Fe^{2+} [28]:

$$·H + FeOH^{2+} \rightarrow Fe^{2+} + H_2O \tag{8}$$

Fe^{2+} ion resulting from the reactions above reacts with H_2O_2 (reaction (6)) to convert the H_2O_2 back to ·OH radicals, ultimately leading to the decrease in H_2O_2 formation (Figure 5) and the increase in 2,4-D degradation (Figure 6) [13].

In order to prove the above assumptions, formations of Fe^{2+} and molecular hydrogen (H_2) during RFD were investigated, and the results were shown in Figure 7a and b, respectively.

It can be observed from Figure 7a that, in the case of Fe^{3+}, a lot of Fe^{2+} was formed during the RFD in the presence of 2,4-D and little Fe^{2+} was detected in the absence of 2,4-D, indicating the UV radiation from RFD is too weak to effectively induce the photoFenton activation ($FeOH^{2+} + UV \rightarrow Fe^{2+} + ·OH$) [29]. Figure 7b shows that the amount of H_2 in the gas phase decreased in the presence of Fe^{3+}. At 10 min. about 0.3 mmol H_2 was evolved from the solution in the absence of Fe^{3+}. However, only 0.19 mmol H_2 was detected when 0.5 mM Fe^{3+} was added to the solution, which means that at least 0.22 mmol Fe^{3+} was reduced to Fe^{2+} by the ·H atoms. As the reaction between ·H and ·OH is suppressed due to the scavenging effect of Fe^{3+} [30], which in turn increases the amount of ·OH radicals available for 2,4-D degradation.

$$H + ·OH \rightarrow H_2O \tag{9}$$

However, as shown in Figure 7b, the amount of H_2 formed with the case of Fe^{2+} is higher than that without it. This can be explained that the ·OH produced by RFD can rapidly oxidize the Fe^{2+} to Fe^{3+}, which suppressed the Reaction (9) and increase the amount of ·H and ultimately the amount of H_2 via dimerization [30].

$$OH + Fe^{2+} \rightarrow Fe^{3+} + OH^- \tag{10}$$

$$H + H \rightarrow H_2 \tag{11}$$

Fenton reagent ($Fe^{2+} + H_2O_2$) oxidation is an effective method for degradation of the refractory organic pollutants. However, the reaction has to be undertaken in acidic conditions (pH: 2–4). In addition, as the Fenton reaction proceeds, the Fe^{2+} rapidly oxidizew to Fe^{3+} and the reaction becomes a Fenton-like one, which leads to a substantial decrease in the reaction rate. In order to accelerate the reaction, ultraviolet light (UV) was usually employed (UV/Fenton). However, UV/Fenton was only effective in solutions with high UV transmittance. In RFD of 2,4-D, Fe^{3+} showed an extraordinary catalytic effect due to its successive reduction to the Fe^{2+}, meaning that RFD may be used to promote the Fenton-like reactions, especially in liquids with poor UV transmittance, which is of interest because most real wastewater is UV opaque. In addition, the pH of the solution can spontaneously drop to the Fenton's operation range, indicating that Fe^{3+}/RFD is a promising process for 2,4-D degradation. In order to further clarify the potential application of RFD in accelerating the Fenton-like process, concentrated 2,4-D degradation by Fe^{3+}/H_2O_2 in the presence and absence of RFD was performed, and the results were shown in Figure 8.

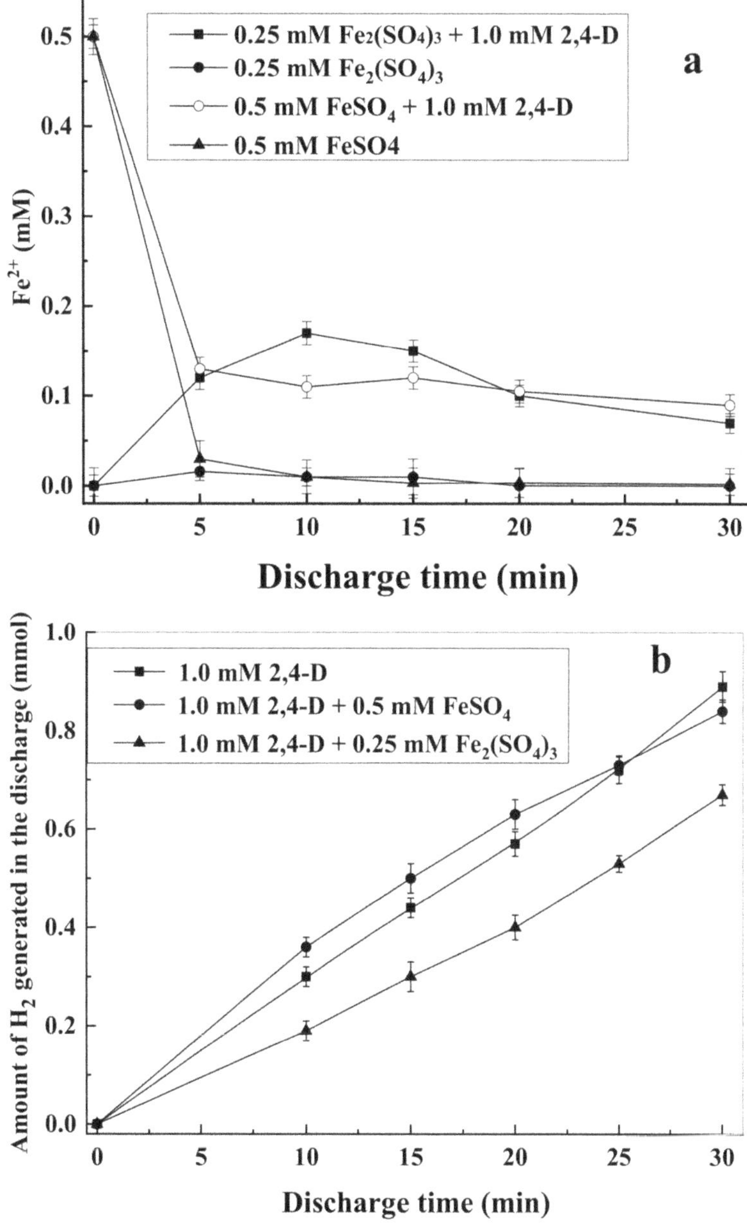

Figure 7. Formations of Fe^{2+} (**a**) and H_2 (**b**) during RFD (input power, 200 W; pH_0, 2.0).

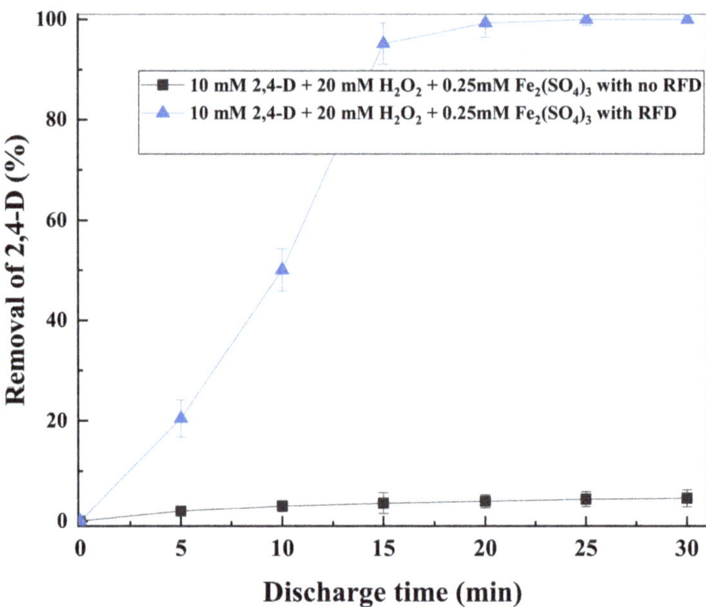

Figure 8. Accelerating effect of RFD for Fenton-like process (input power, 200 W; pH_0, 2.0).

It can be observed from Figure 8 that without RFD, little 2,4-D degradation proceeded. When RFD was applied in the Fenton-like process, the 2,4-D degradation was significantly enhanced where it could be removed completely in 25 min. of RFD, indicating that RFD was a useful tool for accelerating the Fenton-like process.

3.4. Intermediate Products Formation and Possible Decomposition Pathway

In order to gain insight into the 2,4-D decomposition mechanism, HPLC and IC were utilized to help identify the intermediate products. Figure 9 shows the HPLC chromatograms of the solution within 40 min of RFD treatment monitored at 283 nm.

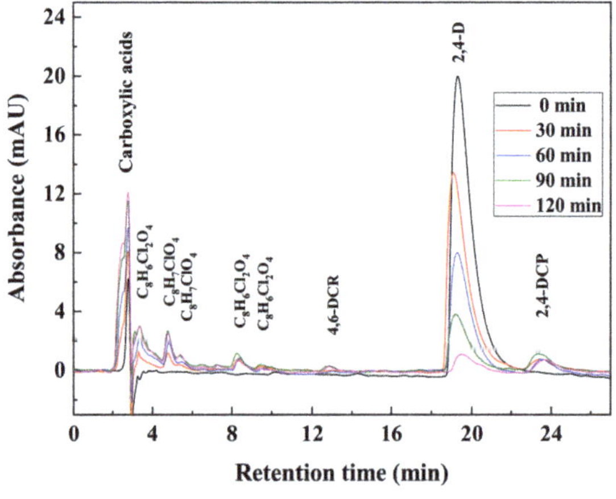

Figure 9. HPLC chromatograms during the RFD treatment (input power, 200 W; [2,4-D]$_0$, 1.0 mM; pH_0, 2.0).

As shown in Figure 9, several byproducts were observed. The peaks at retention times 12.8 min and 23.5 min are 4,6-dichlororesorcinol (4,6-DCR) and 2,4-Dichlorophenol (2,4-DCP), respectively. Three isomers of hydroxylation products of 2,4-D, i.e., 3-, 5- and 6-hydroxy-2,4-D ($C_8H_6Cl_2O_4$) and two chlorohydroxyl isomers (2-hydroxy-4-chloro- and 2-chloro-4-hydroxyphenoxyacetic acid: $C_8H_7ClO_4$) were roughly identified by the LC-MS [20]. As no standards are available for these products, their precise identification and quantification cannot be performed. As 2,4-DCP is found as a common intermediate byproduct in other methods, [6–9] the concentration variation of 2,4-DCP as a function of discharge time is given in Figure 10.

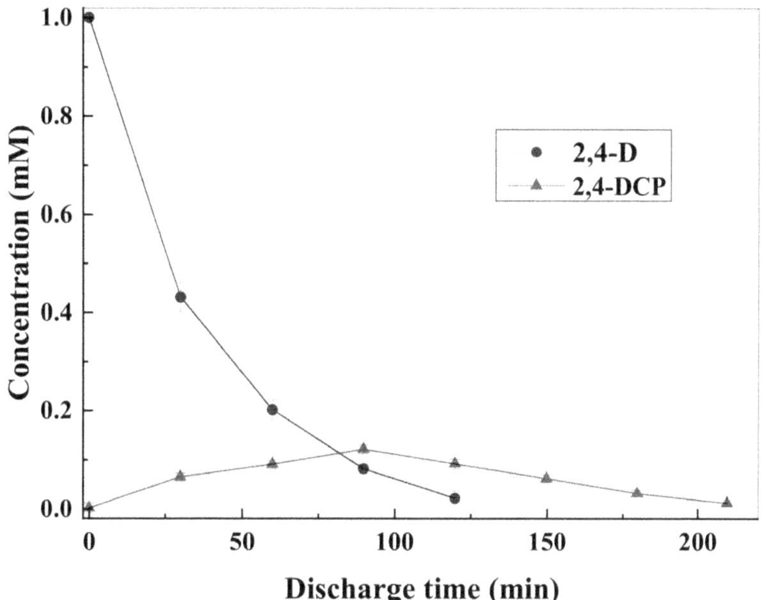

Figure 10. Appearance and decay profile of 2,4-DCP during RFD (input power, 200 W; [2,4-D]$_0$, 1.0 mM; pH$_0$, 2.0).

It can be observed from Figure 10 that the 2,4-DCP forms immediately after the discharge takes place. Its concentration increases gradually with discharge time trough to 90 min and then progressively diminishes. Such a gradual increasing and then decreasing trend is always observed in other treatment processes. The maximum concentration of 2,4-DCP formed is 0.11 mM in RFD of 1.0 mM 2,4-D. In comparison, more than 0.5 mM 2,4-DCP was formed in TiO_2 photocatalysis of 1.0 mM 2,4-D [6], 0.04 mM in γ radiolysis of 0.5 mM 2,4-D [8] and 0.02 mM in sonolysis of 0.22 mM of 2,4-D [9], respectively, indicating that the 2,4-DCP buildup/decomposition pattern in RFD is similar to those in γ radiolysis and sonolysis (the maximum concentration of 2,4-DCP is about 10% of the parent compound).

Aliphatic acids with a retention time of less than 3 min are difficult to quantify as HPLC cannot separate them well. Therefore, IC was employed to determine the organic acids, and results are shown in Figures 11 and 12, respectively.

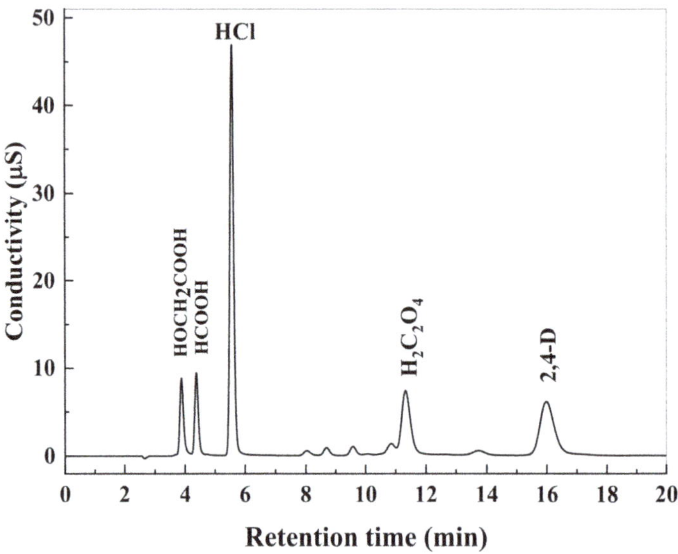

Figure 11. IC chromatogram of the RFD of 1.0 mM 2,4-D in pure water (input power, 200 W; discharge time, 40 min).

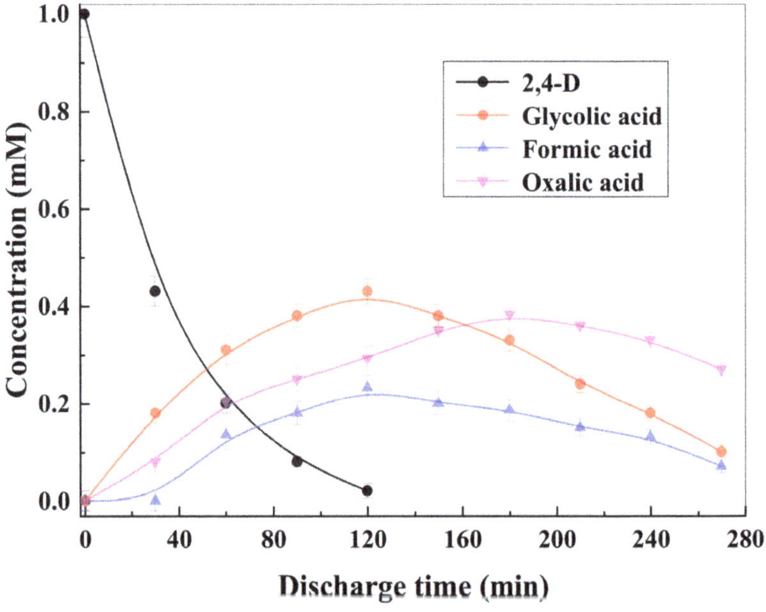

Figure 12. Carboxylic acids determined by IC during RFD (input power, 200 W; $[2,4\text{-}D]_0$, 1.0 mM; pH_0, 2.0).

As shown in Figure 11, glycolic acid, formic acid and oxalic acid were produced. The yields of formic acids and glycolic acid increased to the maximums within 120 min and then diminished gradually. On the other hand, the yield of oxalic acid increased smoothly within 180 min and then declined. At 120 min, they are present with: glycolic acid at 0.31 mM, formic acid at 0.20 mM, and oxalic acid at 0.13 mM (Figure 12).

Figure 12 shows that the amount of glycolic acid accounts for almost half of the degraded 2,4-D, indicating that its formation results from the primary reaction step. Glycolic acid was formed by cleavage of ipso-•OH-adducts on position 1 of 2,4-D [25] (Scheme 3):

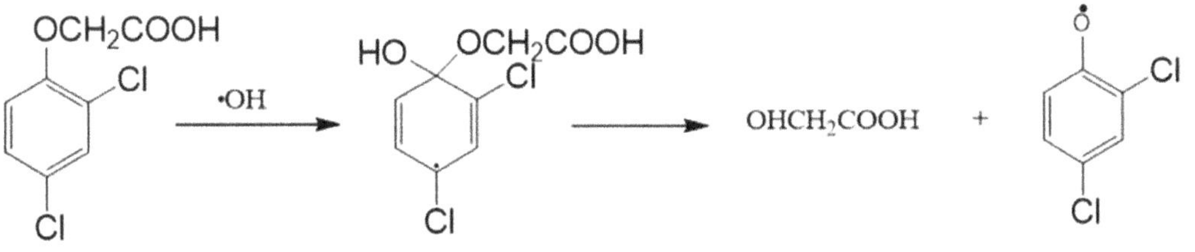

Scheme 3. Glycolic acid formation by cleavage of ipso-•OH-adducts on position 1 of 2,4-D.

Formic and oxalic acids are common ring opening byproducts of aromatic compounds in advanced oxidation processes [12]. However, it should be noted that acetic acid observed in the ionizing radiation of 2,4-D [20] was not detected in the present investigation. This may be because the retention time of acetic acid is very close to that of glycolic acid in IC.

The formation of chloride ion (Cl$^-$) was attributed to the addition of •OH on the position of 2 or 4, followed by the fast elimination of HCl [20] (Scheme 4):

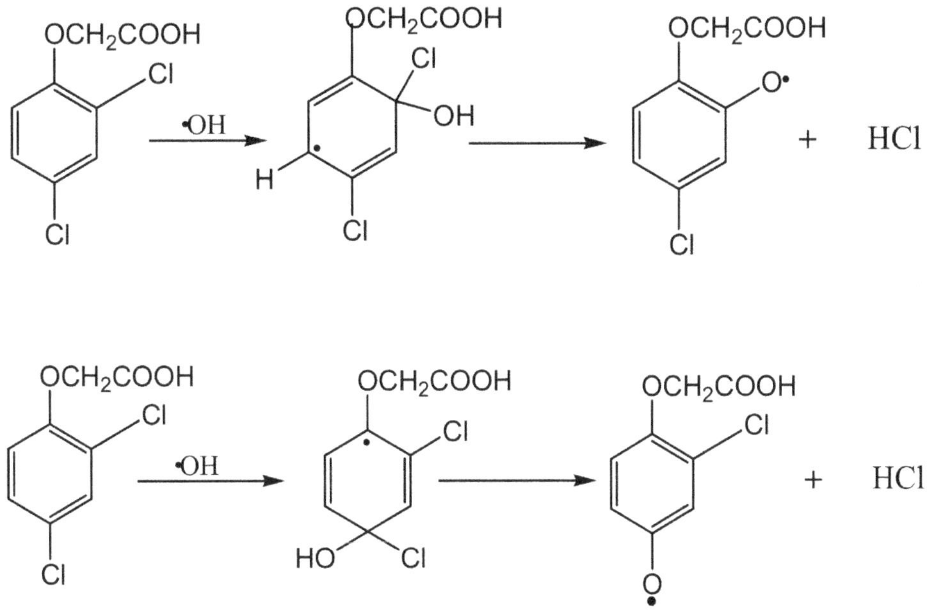

Scheme 4. Formation of chloride ion (Cl$^-$) due to the addition of •OH on the position of 2 or 4, followed by the fast elimination of HCl.

•OH radicals addition to the unsubstituted position of 2,4-D and then oxidized by phenoxyl radical to produce the hydroxylated 2,4-Ds and 2,4-dichlorophenol or chlorohydroxyphenoxyacetic acid [31] (Scheme 5):

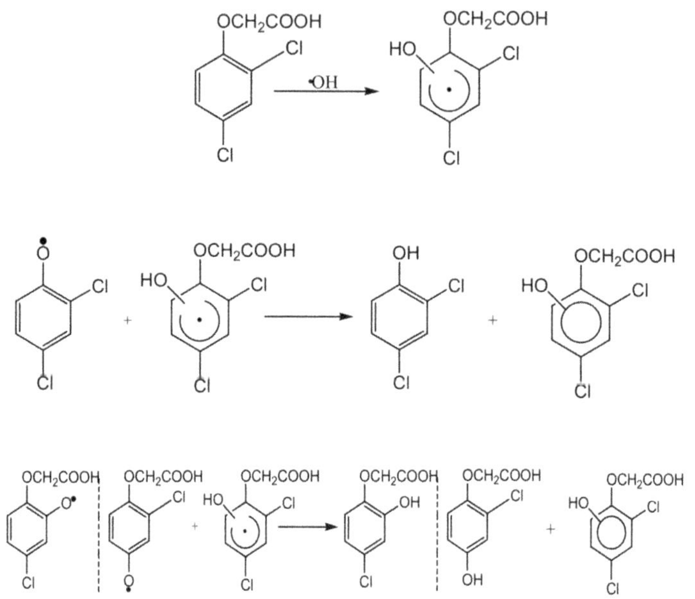

Scheme 5. Mechanism for formation of hydroxylated 2,4-Ds, 2,4-dichlorophenol and chlorohydroxyphenoxyacetic acid.

Further hydroxylating of 2,4-DCP or ydroxyl-2,4-D leads to the formation of 4,6-DCR [32] (Scheme 6):

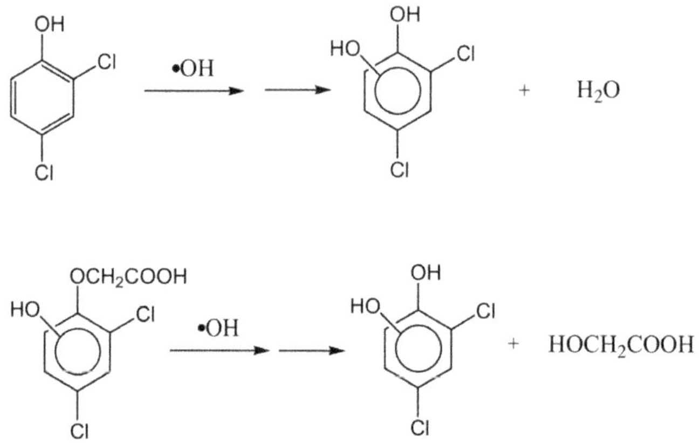

Scheme 6. Mechanism for the formation of 4,6-DCR.

It is noted that typical phenolic byproducts such as 4-chlorocatechol, 2-chlorohydroquinone and 1, 2, 4-trihydroxybenzene commonly found in other radical processes were not observed in the present investigation, possibly because of their unstable nature [7].

Based on the above observations, the probable reaction mechanism for 2,4-D decomposition by RFD is proposed in Scheme 7.

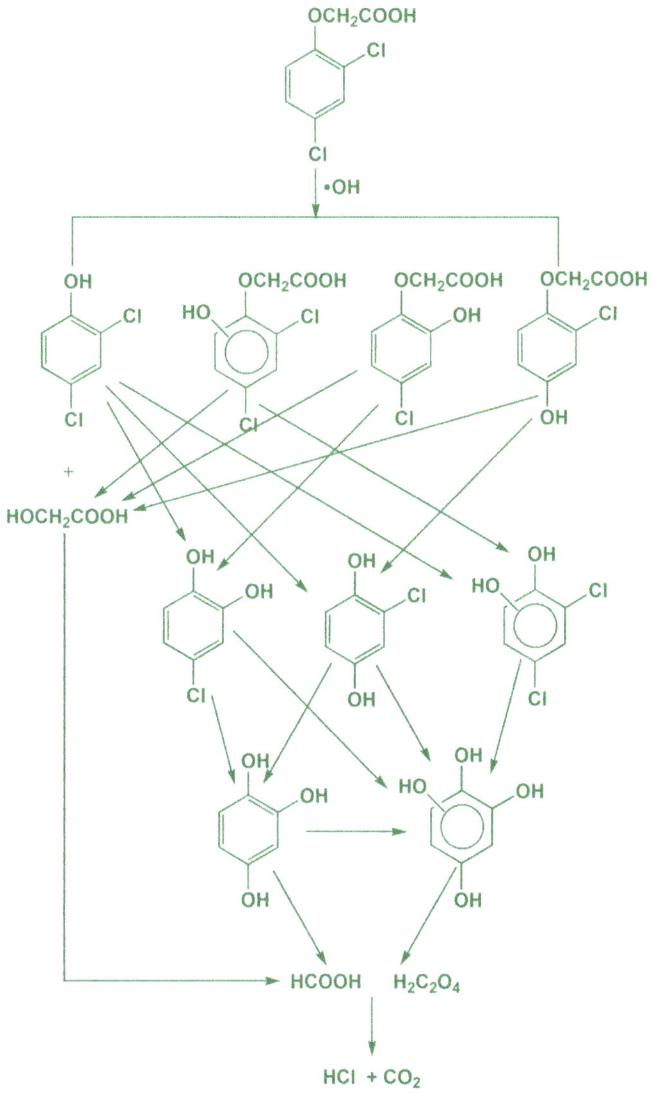

Scheme 7. Proposed decomposition pathway for 2,4-D decomposition in RFD.

Scheme 7 shows that degradation of 2,4-D in RFD proceeds as follows: initiation of the reaction, ring destruction and oxidation of the organic acids. At the beginning stage, •OH radicals resulting from the RFD reacting with 2,4-D, forming the hydroxycyclohexadienyl radicals (•OH adducts). The •OH adducts formed on ipso positions subsequently undergo HCl and glycolic elimination, forming the corresponding phenoxyl radicals. The resulting phenoxyl radicals are unstable and rapidly oxidize the •OH adducts of the unsubstituted positions to form the corresponding phenols (2,4-DCP, 2-hydroxy-4-chloro- and 2-chloro-4-hydroxy- phenoxyacetic acids) and hydroxylated 2,4-Ds. Further hydroxylating of the aromatic ring leads to the ring cleavage to form the organic acids. The organic acids are eventually oxidized by •OH radicals into CO_2 and H_2O. As indicated in Figure 6, the solution would be acidic at the end of the discharge if the 2,4-D decomposition was conducted in neutral pH. Therefore, the final product in the solution would be HCl.

3.5. Energy Yields Evaluation

Energy yields ($J_{2,4\text{-}D}$) is a significant factor in evaluating the feasibility of practical application. $J_{2,4\text{-}D}$ is calculated according to Equation (2). $J_{2,4\text{-}D}$ and reaction rate constant (k, min^{-1}) [11] of RFD, other nonthermal plasma systems and some competitive methods are summarized in Table 1.

Table 1. Energy yields for 2,4-D removal of RFD and other methods.

C_0 (mM)	Method	k	$J_{2,4\text{-}D}$	References
		min^{-1}	g/kWh	
1.0	RFD 200 W, pH_0 2.0	0.018	0.35	This work
1.0	RFD 200 W, pH_0 2.0, 0.5 mM Fe^{3+}	0.078	1.51	This work
1.0	RFD 200 W, pH_0 2.0, 0.5 mM Fe^{2+}	0.063	1.22	This work
10.0	RFD 200 W, pH_0 2.0, 20 mM H_2O_2 2,4-D, 0.5 mM Fe^{3+}	0.047	9.06	This work
0.0045	PCD over water surface, air as discharge gas (PCD/Air)		1.50	[19]
0.25	TiO_2 photocatalysis, 125 W, 2.0g/L TiO_2, pH_0 3.3		0.76	[6]
0.22	Sonolysis, 50 W, O_2 sparging		1.75	[9]
0.45	DBD/Ar-Fenton		8.83	[15]
0.45	DBD/Ar		3.85	[15]
0.45	DBD/Air, 200 W		0.25	[15]

Table 1 shows that the $J_{2,4\text{-}D}$ of RFD is comparable to that of sonolysis and is higher than that of TiO_2 photocatalysis. Among the nonthermal plasma systems, $J_{2,4\text{-}D}$ follows the order of DBD/Ar-Fenton $\approx$ RFD-Fenton like > DBD/Ar > PCD/Air $\approx$ RFD/Fe^{3+} > DBD/Air. It is shown that the $J_{2,4\text{-}D}$ of RFD/H_2O_2/Fe^{3+} is comparable to that of DBD/Ar-Fenton, the most efficient nonthermal plasma process reported. However, DBD/Ar-Fenton has to be operated in a pure noble argon atmosphere, instead of air, which would greatly increase the experimental complexity and the operation cost, indicating that RFD is not only a competitive process for 2,4-D degradation itself but also a useful tool to accelerate the Fenton-like process.

4. Conclusions

2,4-D can be efficiently degraded and dechlorinated under the action of RFD. The successive attack of •OH radicals on the benzene ring was the key step. Decreasing solution pH was favorable for the degradation. During the discharge, pH gradually decreases as HCl and organic acids are formed, and significant quantities of hydrogen peroxide are produced. Both Fe^{3+} and Fe^{2+} displayed marked catalytic effects. However, due to the reducing actions of the •OH-adducts and the hydrogen atoms, Fe^{3+} performed better than Fe^{2+}, which indicates that RFD may also be used to promote the Fenton-like reactions. Major intermediate byproducts are 2,4-dichlorophenol, 4,6-dichlororesorcinol, hydroxylated 2,4-Ds, 2-hydroxy-4-chloro- and 2-chloro-4-hydroxyphenoxyacetic acids, glycolic acid, formic acid, oxalic acid and chloride ion. RFD has comparable energy yields with those of pulsed corona discharge over a water surface and sonolysis and is higher than that of TiO_2 photocatalysis, indicating that RFD is a useful process for 2,4-D decomposition.

Author Contributions: Conceptualization, Y.L.; methodology, Y.L.; software, Y.L.; validation, Y.L.; formal analysis, Y.L.; investigation, Y.L.; resources, B.S.; data curation, Y.L.; writing—original draft preparation, Y.L.; writing—review and editing, Y.L.; visualization, Y.L.; supervision, Y.L.; project administration, Y.L.; funding acquisition, Y.L. All authors have read and agreed to the published version of the manuscript.

Funding: This research was funded by the National Natural Science Foundation of China grant number [11005014, 11675031] and The APC was funded by [11005014].

Institutional Review Board Statement: Not applicable.

Informed Consent Statement: Not applicable.

Data Availability Statement: Not applicable.

Acknowledgments: The authors are thankful for the support of the National Natural Science Foundation of China (11005014, 11675031).

Conflicts of Interest: The authors declare no conflict of interest.

References

1. Available online: https://www.invasive.org/gist/products/handbook/10.24-d.pdf (accessed on 24 April 2001).
2. Loomis, D.; Guyton, K.; Grosse, Y.; Ghissasi, F.E.; Bouvard, V.; Benbrahim-Tallaa, L.; Guha, N.; Mattock, H.; Straif, K. Carcinogenicity of lindane, DDT, and 2,4-Dichlorophenoxyacetic acid. *Lancet Oncol.* **2015**, *16*, 891–892. [CrossRef]
3. Song, Y. Insight into the mode of action of 2,4-Dichlorophenoxyacetic acid (2,4-D) as an herbicide. *J. Integr. Plant Biol.* **2014**, *56*, 106–113. [CrossRef] [PubMed]
4. Hoover, D.G.; Borgonovi, G.E.; Jones, S.H.; Alexander, M. Anomalies in mineralization of low concentrations of organic compounds in lake water and sewage. *Appl. Environ. Microbiol.* **1986**, *51*, 226–232. [CrossRef] [PubMed]
5. Aksu, Z.; Kabasakal, E. Adsorption characteristics of 2,4-Dichlorophenoxyacetic acid (2,4-D) from aqueous solution on powdered activated carbon. *J. Environ. Sci. Health B* **2005**, *40*, 545–570. [CrossRef]
6. Trillas, M.; Peral, J.; Domènech, X. Redox photodecomposition of 2,4-Dichlorophenoxyacetic acid over TiO_2. *Appl. Catal. B Environ.* **1995**, *5*, 377–387. [CrossRef]
7. Brillas, E.; Calpe, J.C.; Casado, J. Mineralization of 2,4-D by advanced electrochemical oxidation processes. *Water Res.* **2000**, *34*, 2253–2262. [CrossRef]
8. Zona, R.; Solar, S. Oxidation of 2,4-Dichlorophenoxyacetic acid by ionizing radiation: Decomposition, detoxification and mineralization. *Radiat. Phys. Chem.* **2003**, *66*, 137–143. [CrossRef]
9. Peller, J.; Wiest, O.; Kamat, P.V. Synergy of combining sonolysis and photocatalysis in the decomposition and mineralization of chlorinated aromatic compounds. *Environ. Sci. Technol.* **2003**, *37*, 1926–1932. [CrossRef]
10. Aziz, K.H.H.; Miessner, H.; Mahyar, A.; Mueller, S.; Moeller, D.; Mustafa, F.; Omer, K.M. Degradation of perfluorosurfactant in aqueous solution using non-thermal plasma generated by nano-second pulse corona discharge reactor. *Arab. J. Chem.* **2021**, *14*, 103366. [CrossRef]
11. Aziz, K.H.H.; Miessner, H.; Mahyar, A.; Mueller, S.; Kalass, D.; Moeller, D.; Omer, K.M. Removal of dichloroacetic acid from aqueous solution using non-thermal plasma generated by dielectric barrier discharge and nano-pulse corona discharge. *Sep. Purif. Technol.* **2019**, *216*, 51–57. [CrossRef]
12. Aggelopoulos, C.A. Recent advances of cold plasma technology for water and soil remediation: A critical review. *Chem. Eng. J.* **2022**, *428*, 131657. [CrossRef]
13. Bradu, C.; Magureanu, M.; Parvulescu, V.I. Degradation of the chlorophenoxyacetic herbicide 2,4-D by plasma-ozonation system. *J. Hazard. Mater.* **2017**, *336*, 52–56. [CrossRef]
14. Singh, R.K.; Philip, L.; Ramanujam, S. Removal of 2,4-dichlorophenoxyacetic acid in aqueous solution by pulsed corona discharge treatment: Effect of different water constituents, decomposition pathway and toxicity assay. *Chemosphere* **2017**, *184*, 207–214. [CrossRef]
15. Aziz, K.H.H.; Miessner, H.; Mueller, S.; Mahyar, A.; Kalass, D.; Moeller, D.; Khorshid, I.; Rashid, M.A.M. Comparative study on 2,4-dichlorophenoxyacetic acid and 2,4-dichlorophenol removal from aqueous solutions via ozonation, photocatalysis and non-thermal plasma using a planar falling film reactor. *J. Hazard. Mater.* **2018**, *343*, 107–115. [CrossRef]
16. Maehara, T.; Miyamot, I.; Kurokawa, K.; Hashimoto, Y.; Iwamae, A.; Kuramoto, M.; Yamashita, H.; Mukasa, S.; Toyota, H.; Nomura, S.; et al. Decomposition of methylene blue by RF plasma in water. *Plasma Chem. Plasma Process.* **2008**, *28*, 467–482. [CrossRef]
17. Amano, T.; Mukasa, S.; Honjoya, N.; Okumura, H.; Maehara, T. Generation of radio frequency plasma in high-conductivity NaCl solution. *Jpn. J. Appl. Phys.* **2012**, *51*, 978–982. [CrossRef]
18. Ji, L. *Research on Characteristics of Radio Frequency Plasma Discharge Underwater and Its Application of Azo Dyes Decomposition*; Soochow University: Suzhou, China, 2012. (In Chinese)
19. Wang, L.; Wang, J.; Lin, J.; Zhang, S.; Liu, Y. Decomposition of metronidazole by radio frequency discharge in an aqueous solution. *Plasma Proces. Polym.* **2018**, *15*, 1700176. [CrossRef]
20. Zona, R.; Solar, S.; Gehringer, P. Decomposition of 2,4-dichlorophenoxyacetic acid by ionizing radiation: Influence of oxygen concentration. *Water Res.* **2002**, *36*, 1369–1374. [CrossRef]
21. Gao, Q.; Liu, Y.; Sun, B. Characteristics of gas-liquid diaphragm discharge and its application on decolorization of brilliant red B in aqueous solution. *Plasma Sci. Technol.* **2017**, *19*, 115404. [CrossRef]

22. Harvey, A.E.; Smart, J.A.; Amis, E.S. Simultaneous spectrophotometric determination of iron (II) and total iron with 1,10-phenanthroline. *Anal. Chem.* **1953**, *26*, 1854–1856. [CrossRef]

23. Marotta, E.; Ceriani, E.; Schiorlin, M.; Ceretta, C.; Paradisi, C. Comparison of the rates of phenol advanced oxidation in deionized and tap water within a dielectric barrier discharge reactor. *Water Res.* **2012**, *46*, 6239–6246. [CrossRef]

24. Li, X.; Jenks, W.S. Isotope studies of photocatalysis: Dual mechanisms in the conversion of anisole to phenol. *J. Am. Chem. Soc.* **2000**, *122*, 11864–11870. [CrossRef]

25. Zona, R.; Solar, S.; Sehested, K.; Holcman, J.; Mezyk, S.P. OH-radical induced oxidation of phenoxyacetic acid and 2,4-Dichlorophenoxyacetic acid. Primary radical steps and products. *J. Phys. Chem. A* **2002**, *106*, 6743–6749. [CrossRef]

26. Pignatello, J.J.; Oliveros, E.; MacKay, A. Advanced oxidation processes for organic contaminant destruction based on the Fenton reaction and related chemistry. *Crit. Rev. Environ. Sci. Technol.* **2006**, *36*, 1–84. [CrossRef]

27. Wang, L.; Jiang, X.; Liu, Y. Degradation of bisphenol A and formation of hydrogen peroxide induced by glow discharge plasma in aqueous solutions. *J. Hazard. Mater.* **2008**, *154*, 1106–1114. [CrossRef] [PubMed]

28. Baxendale, J.H.; Dixon, R.S.; Stott, D.A. Reactivity of hydrogen atoms with Fe^{3+}, $FeOH^{2+}$ and Cu^{2+} in aqueous solutions. *Trans. Faraday Soc.* **1968**, *64*, 2398–2401. [CrossRef]

29. Assadi, A.A.; Bouzaza, A.; Soutrel, I.; Petit, P.; Medimagh, K.; Wolbert, D. A study of pollution removal in exhaust gases from animal quartering centers by combining photocatalysis with surface discharge plasma: From pilot to industrial scale. *Chem. Eng. Processing Process Intensif.* **2017**, *111*, 1–6. [CrossRef]

30. Buxton, G.V.; Greenstock, C.L.; Helman, W.P.; Ross, A.B. Critical-review of rate constants for reactions of hydrated electrons, hydrogen-atoms and hydroxyl radicals ($\bullet OH/\bullet O^{-}$) in aqueous-solution. *J. Phys. Chem. Refer. Data* **1988**, *17*, 513–886. [CrossRef]

31. Quint, R.M.; Park, H.R.; Krajnik, P.; Solar, S.; Getoff, N.; Sehested, K. γ-radiolysis and pulse radiolysis of aqueous 4-chloroanisole. *Radiat. Phys. Chem.* **1996**, *47*, 835–845. [CrossRef]

32. Albarrán, G.; Mendoza, E. Radiolytic oxidation and degradation of 2,4-dichlorophenol in aqueous solutions. *Environ. Sci. Pollut. Res.* **2019**, *26*, 17055–17065. [CrossRef]

Article

Changes in Organics and Nitrogen during Ozonation of Anaerobic Digester Effluent

Jesmin Akter, Jaiyeop Lee, Weonjae Kim and Ilho Kim *

Department of Civil and Environmental Engineering, Korea Institute of Civil Engineering and Building Technology (KICT) School, University of Science and Technology, Goyang-si 10223, Gyeonggi-do, Korea; jesmin@kict.re.kr (J.A.); pas2myth@kict.re.kr (J.L.); wjkim1@kict.re.kr (W.K.)
* Correspondence: ihkim@kict.re.kr

Abstract: The objective of this study is to investigate the consequence of ozone dosage rate on the qualitative change in organic compounds and nitrogen in anaerobic digester effluent during the ozone process. Therefore, ozonation improves the biodegradability of recalcitrant organic compounds, quickly oxidizes the unsaturated bond, and forms radicals that continue to deteriorate other organic matter. In this study, ozonation was performed in a microbubble column reactor; the use of microbubble ozone improves the status of chemical oxygen demand (COD) and changes of organic nitrogen to inorganic compounds. The ozone injection rates were 1.0, 3.2, and 6.2 mg/L/min. The samples obtained during the ozone treatments were monitored for COD_{Mn}, COD_{Cr}, TOC, $NO_2{}^{-}$-N, $NO_3{}^{-}$-N, $NH_4{}^{+}$-N, T-N, and Org-N. The ozone dose increased 1.0 to 6.2 mg/L and it increased the degradation ratio 40% and the total organic carbon 20% during 20 min of reaction time. During the ozonation, the COD_{Cr} and COD_{Mn} values were increased per unit of ozone consumption. The ozone treatment showed organic nitrogen mineralization and degradation of organic compounds with the contribution of the microbubble ozone oxidation process and is a good option for removing non-biodegradable organic compounds. The original application of the microbubble ozone process, with the degradation of organic compounds from a domestic wastewater treatment plant, was investigated.

Keywords: anaerobic digester effluent; biodegradability; nitrogen; ozone; organic compound

Citation: Akter, J.; Lee, J.; Kim, W.; Kim, I. Changes in Organics and Nitrogen during Ozonation of Anaerobic Digester Effluent. *Water* **2022**, *14*, 1425. https://doi.org/10.3390/w14091425

Academic Editor: Gassan Hodaifa

Received: 28 March 2022
Accepted: 26 April 2022
Published: 29 April 2022

Publisher's Note: MDPI stays neutral with regard to jurisdictional claims in published maps and institutional affiliations.

1. Introduction

Millions of tons of organic waste are produced by humans every day, a significant portion of which is dumped into domestic wastewater. In domestic wastewater treatment plants, the principal treatment system is mostly biological [1]. The conventional biological wastewater treatment process employs a group of various microorganisms to degrade organic wastewater compounds aerobically [2]. Effluent from anaerobic digestion tank and composting processes effluent is more specific type of wastewater, the properties primarily rely on the oxygenation and moisture of the treated substance. Recalcitrant organic compounds are compounds that resist the atmosphere and are especially conducive to the treatment of aerobic microbial wastewater [3]. Recalcitrant compounds consistently emerge from biological treatment processes, and they create a potential problem for water reuse [4–7]. Anaerobic digestion is a possible technique for processing many types of recalcitrant organic matter. In anaerobic digesters, nitrogenous substances are converted to ammonium nitrogen ($NH_4{}^{+}$-N) by a biodegradation process [8]. Therefore, the dissolved oxygen level in the anaerobic digester effluent is not sufficient, due to the anaerobic process. The high concentrations of nutrients and low oxygen availability in the anaerobic digester effluent can affect aquatic organisms that receive the natural water, reducing biodiversity. However, wastewater from anaerobic digesters requires further treatment to protect the aquatic environment [9].

Currently, ozonation is a popular technology for the neutralization of fluid process residues. Most commonly, ozone reduces the number of organic pollutants or the toxicity

---

of wastewater. In addition, organic compounds can be oxidized and become degradable by ozonation [10]. Ozone has been successfully used for the disinfection and decomposition of dissolved organic pollutants [11–16]. Ozone is a strong oxidant that reacts as molecular ozone or through the formation of secondary oxidants, such as free radicals [17]. The principle of the ozone process is based on the release of hydroxyl radicals, which accelerate the degradation of organic compounds in an aquatic environment. The main factors affecting ozonation performance were pH, the nature and concentration of oxidizable organics, ozone dose, the presence of oxidant scavengers, and the efficiency of ozone mass transfer. Ozone pretreatment, to improve biodegradation via partial oxidation, is a potential solution for recalcitrance.

Therefore, it is possible to use the ozone oxidation system for preliminary preparation of the digestion tank effluent from the composting procedure for neutralization in a professional wastewater treatment plant. Ozonation is among the most effective technologies to oxidize recalcitrant and non-degradable substances converted to a biodegradable form [1], and is also considered a good treatment option, as ozone is a strong oxidant that converts organic and inorganic pollutants into non-toxic by-products [18,19]. Ozone has been used in potable water treatment for disinfection, odor treatment, and color removal [20]. The ozonation of organic compounds involves a stepwise process, where the oxygen particles are gradually integrated into the compound [2]. Therefore, ozone treatment shows a strong positive impact on digestion wastewater, due to the high content of aerobic bacteria and non-degraded organic matter. The objective of this study is to investigate the change in recalcitrant organic carbons and nitrogen in effluent from an anaerobic digestion tank. Anaerobic digestion has been used for decades to treat leachate from municipal solid waste landfills, and leachate is well known for its high COD and ammonium concentration and low biodegradability [21]. Due to the characteristics of this kind of effluent, the improvement in organic characteristics by the ozone process is the focus of this study. This study aimed to develop a more effective and economic ozone treatment process for sewage water, with a significant component of non-biodegradable substances, to establish an ozone dosage rate for improving organic removal and conditions for the ozone operation. However, the experimental results also efficiently evaluate the ozone dosage's performance to enhance the biodegradability of effluent from the organic waste digestion tank at a laboratory scale.

2. Materials and Methods

2.1. Experimental Setup

The experimental devices shown in Figure 1 are installed at the domestic sewage treatment plant (STP), located in South Korea. In this experiment, the effluent from the anerobic digestion tank was used as the target water for treatment. The influents used in this experiment, from the anaerobic digestion tank composting process effluents, depended mainly on the oxygenation and moisture of the processed material.

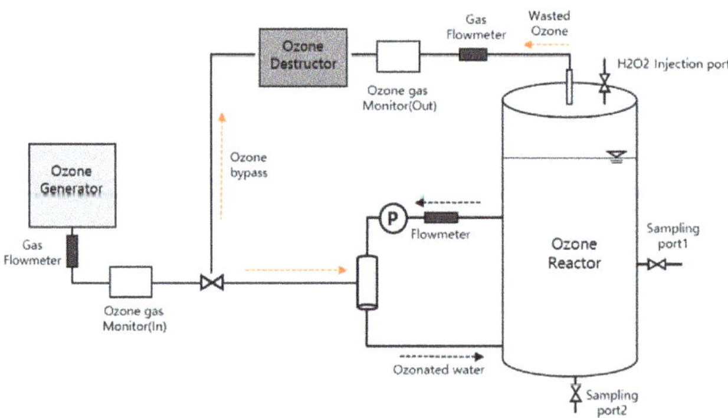

Figure 1. Schematic diagram of experimental setup.

2.2. Operational Condition

The batch test was conducted with 3 different ozone feed rate and with 1 L/min fixed ozone flow rate. The raw water volume and water flow rate were fixed at 20 L and 25 L/min, respectively, for all conditions. Operating conditions are listed in Table 1.

Table 1. Operational conditions.

Operating Conditions	Unit	Ozone Feed Rate (mg/L/min)		
		1.0	3.2	6.2
Oxygen flow rate	L/min	1.0	1.0	1.0
Oxygen tank flow rate	L/min	3.5	4.0	4.0
Pressure	(Kgf/cm^2)	0.7	0.7	0.7

2.3. Characteristics of Wastewater

The STP anerobic digestion tank effluent contained a high COD concentration and low biodegradability. This wastewater composition was consistent with the pollutant characteristic of domestic sewage. Water temperature (T) and pH were measured in the process of sampling. Detailed properties are shown in Table 2.

Table 2. Characteristics of influent.

Parameter	Concentration	Parameter	Concentration
pH	7.80	T-N (mg/L)	350–355
Temp	30–35 °C	NH_4^+-N (mg/L)	270–280
COD_{Cr} (mg/L)	180–280	NO_3^--N (mg/L)	0.007–0.009
COD_{Mn} (mg/L)	95–105	NO_2^--N (mg/L)	3.0–8.0
TOC (mg/L)	130–140	SS (mg/L)	40–45

2.4. Experimental Procedure

The ozone system consists of an ozone generator, ozone reaction column (effective volume 20 L), sampling port, residual ozone meter (electrode method, measuring range 0.5 ppm), flowmeter (20 LPM), ejector, ozone gas monitor, gas flow meter, and ozone destructor. Ozone was generated from oxygen conversion in the ozone generator and simultaneously entered the ozonation reaction column.

The ozone process was fed domestic wastewater from the anaerobic digestion tank effluent and contained a high COD concentration and had low biodegradability. This wastewater composition was consistent with the pollutant characteristics of domestic sewage. Water temperature (T) and pH were measured in the process of sampling.

2.5. Analytical Methods

The ozonation process was performed in a reactor and was fixed with a microbubble generator. Samples were collected at 2–5 min intervals after starting the ozonation using 1 L plastic bottles from sampling ports of the reactor. Temperature and pH were measured using a pH meter (S-610H) immediately after sampling. Soluble COD_{Cr}, COD_{Mn}, NO_2^--N, and NO_3^--N were measured after filtration using 47 mm microfiber filter paper. Total COD_{Cr}, COD_{Mn} and T-N were measured using the CMAC standard method by the HACH DR-5000 spectrophotometer. Dissolved ozone concentration gas was captured from the ozone gas monitor by electrode method; the measuring range was 0.5 ppm. For the analysis of TOC, the total organic high-temperature carbon combustion oxidation method (ES 04311.1b) was applied among the water pollution process test standards. The instrument used for the analysis was a TOC-V CPH (SHIMADZU), and the combustion temperature was 680 °C. The amount of sample used for the analysis was 40 mL, filtered once using a 0.45 μm syringe filter, transferred to a vial, mounted on an auto-sampler, and measured by the non-purifying organic carbon method (NPOC method) at 680 °C, and the pH was controlled by 85% H_3PO_4 (automatic injection of the instrument).

3. Results

3.1. Changes in Organic Compounds

The chemical oxygen demand potassium dichromate index (COD_{Cr}) measures the oxygen equivalent of the amount of organic matter oxidizable by potassium dichromate in a 50% sulfuric acid solution. The chemical oxygen demand potassium permanganate index (COD_{Mn}) measurement of a sample is an alternative method for measuring the oxygen requirement and is an indication of the oxidative degradation potential of wastewater [2]. The COD_{Mn} measurement requires a shorter oxidation time, simpler apparatus, and produces no hazardous chromium waste. Due to its high reproduction potential and reactivity, ozone reacts with both organic and inorganic compounds [22–24]. In addition, ozone is a strong oxidant that is capable of causing cell lysis and disinfection and increasing the content of soluble COD [17,25–27].

Figure 2a shows that when the ozone dosage was 1.0 mg/L/min, the efficiency of improving organic compounds increased slowly, and the soluble COD_{Cr} value increased from 190 to 275 mg/L. This results occurred due to dissolved ozone was present, organic matter was decomposed by the ozone, and the degradation performance of the ozone process was enhanced [28,29]. However, under the different ozone conditions, the feed rate had an almost similar efficiency in improving organic compounds with increasing reaction time. Here, mainly in the ozone system, ozone oxidized the organics directly [28]. Ozone utilization is more effective at lower ozone concentrations, but longer retention times are required for optimal efficiency [30]. When the dosage of ozone was increased from 3.2 to 6.2 mg/L/min (Figure 2b), the soluble COD_{Cr} changed only slightly after 10 to 20 min. This phenomenon happened because the O_3 solubility in water is constant; therefore, an overly excessive dose of O_3 cannot improve its oxidation performance. The COD_{Mn} determination of wastewater is the amount of oxidizing organic matter in the sample [31] and indicates the potential extent of biological oxidation. An increase in the COD_{Mn} of a sample would indicate its higher amenability to biodegradation.

Figure 2b indicates that during the ozone dosing from 1.0 to 6.2 mg/L/min, the efficiency of improving organic compounds increased from 32 to 51%. On the other hand, when the ozone dosage increased from 3.2 to 6.2 mg/L/min (Figure 2b), the soluble COD_{Mn} did not increase compared to the ozone dosage rate. The graph indicates that the 3.2 mg/L/min O_3 feed rate has the maximum increment, and increased rapidly between 10 min and 20 min. In this stage, the O_3 solubility in water is increased; therefore, a 3.2 mg/L/min dose of O_3 can improve its oxidation performance. However, in all conditions of ozone feed rate, the conversion of material into biodegradable matter enhanced after 10 min of ozone reaction time. Comparatively, in low O_3, the feed rate had better efficiency in improving organic compounds with increasing reaction time.

The total organic carbon (TOC) content of a sample is the amount of organic matter potentially available for microbial mineralization. The COD/TOC ratio is an indication of biodegradability. An increase in the proportion after ozone treatment indicates better biodegradability because of an increase in the COD and TOC proportions, which can be modified by biological mineralization [2]. From Figure 2d, it can be observed that the ozone experiments decreased the recalcitrant organic load of the anaerobic digestion tank effluent and increased the COD_{Mn}/TOC ratio. These results show that inorganic compounds were changed into organic substances; therefore, it could easily be biologically degraded. Although there is a crucial improvement of COD_{Mn}/TOC ratio which is higher than 0.4 and this is the minimum value considered apposite for the efficient implementation of a biological treatment [32].

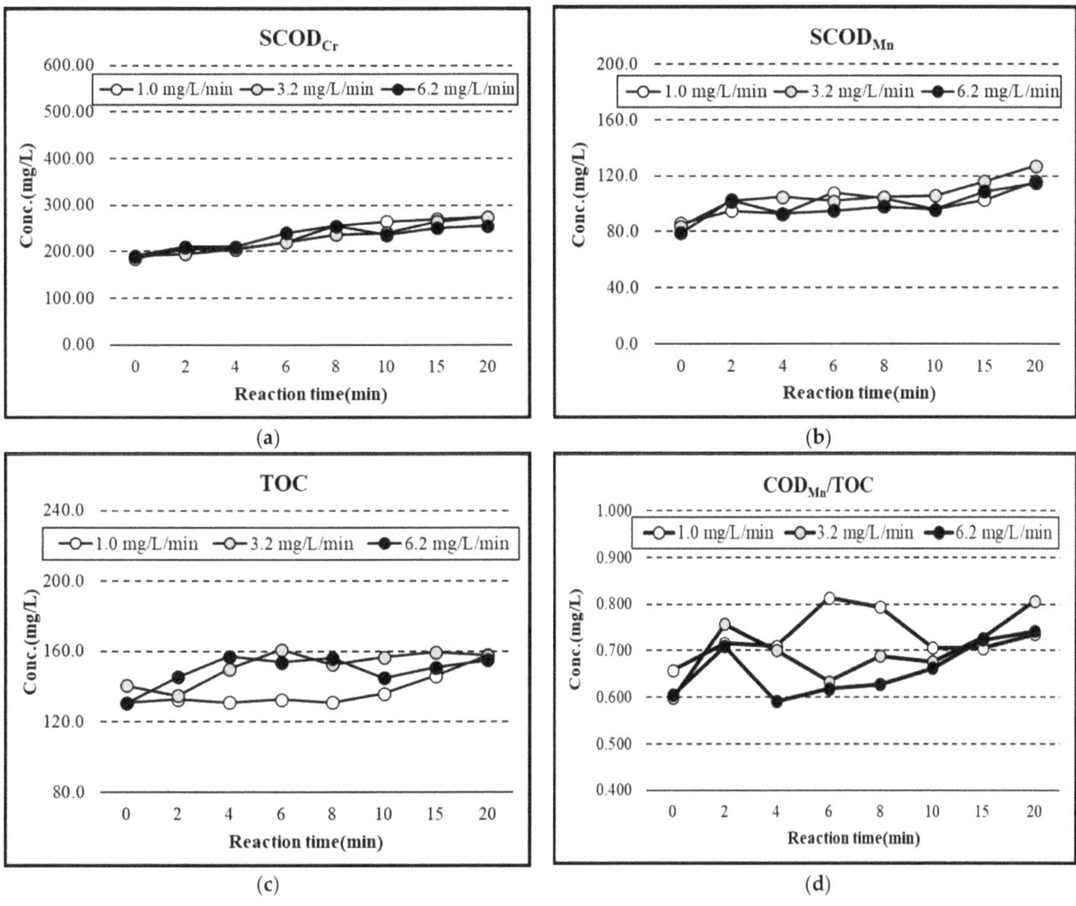

Figure 2. Effect of ozone feed rate on (**a**) $SCOD_{Cr}$, (**b**) COD_{Mn}, (**c**) TOC, and (**d**) COD_{Mn}/TOC.

3.2. Changes in Nitrogen

To demonstrate the transformation of nitrogen in the ozonation process, the anaerobic digestion tank effluent was exposed by ozone at feed rates of 1.0, 3.2, and 6.2 mg/L/min, and the total reaction time of each experiment was 20 min. Figure 3 presents NO_2^--N, NO_3^--N, NH_4^+-N, and T-N variations at three different feed rates. Due to the inhibitory characteristics of ammonium nitrogen, during 20 min of reaction, the NH_4^+-N concentration was nearly similar compared to all the feed rates of ozone oxidation. This indicated that the ozone could hardly convert the NH_4^+-N into NO_3^--N or NO_2^--N in this case. In contrast, ozone may be more involved in the oxidation of nitrite, nitrate, and organic nitrogen. Hence, the ozone had minimal effect on T-N removal and changes in NH_4^+-N. During 20 min of ozone reaction, the variation in the T-N concentration was inconspicuous, which indicated that the effect of ozone on nitrogen removal was not significant. However, the nitrate and nitrite concentrations of the anaerobic digestion tank effluent were below than 0.5 mg/L and, when ozone applied the values were increased per unit of ozone consumption. This showed that the pre-ozone treatment mainly had a significant changing effect on organic nitrogen in wastewater.

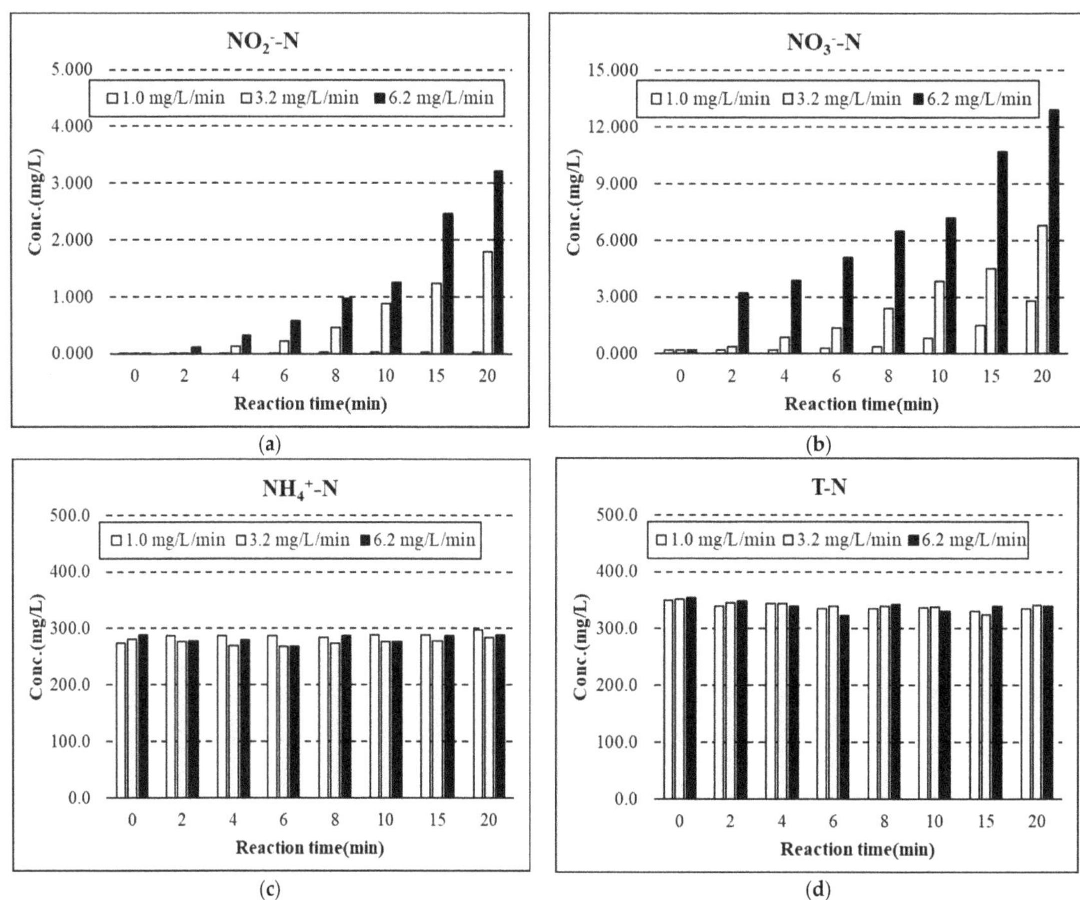

Figure 3. Nitrogen changes by ozone: (**a**) NO_2^--N, (**b**) NO_3^--N, (**c**) NH_4^+-N, and (**d**) T-N.

3.3. Effect of Ozone Dosage on Nitrogen

From Figure 3, the changes in NO_2^--N, NO_3^--N, T-N, and NH_4^+-N can be observed, and the values were monitored to investigate the effect of ozone treatment. The average changes according to the total composition of NH_4^+-N, NO_3^--N, NO_2^--N, and T-N during the ozonation treatment process are shown in Figure 4. After ozone treatment, the results present increasing NO_3^--N and NO_2^--N, slightly decreasing NH_4^+-N, and almost consistent T-N. In the process of this treatment, a small amount of ammonia nitrogen can be oxidized to nitrogen nitrate. The radicals produced by ozone treatment potentially resist ammonification from organic nitrogen, a small amount of ammonia nitrogen can be oxidized to nitrogen, and the combination of the two will reduce the concentration of ammonia nitrogen, so the amount of ammonia nitrogen declined slightly [33,34]. The ozone process could not remove nitrogen directly, which led to an almost similar concentration of T-N, and [35] also reported that nitrate formation is enhanced by the direct oxidation of ammonium by ozone.

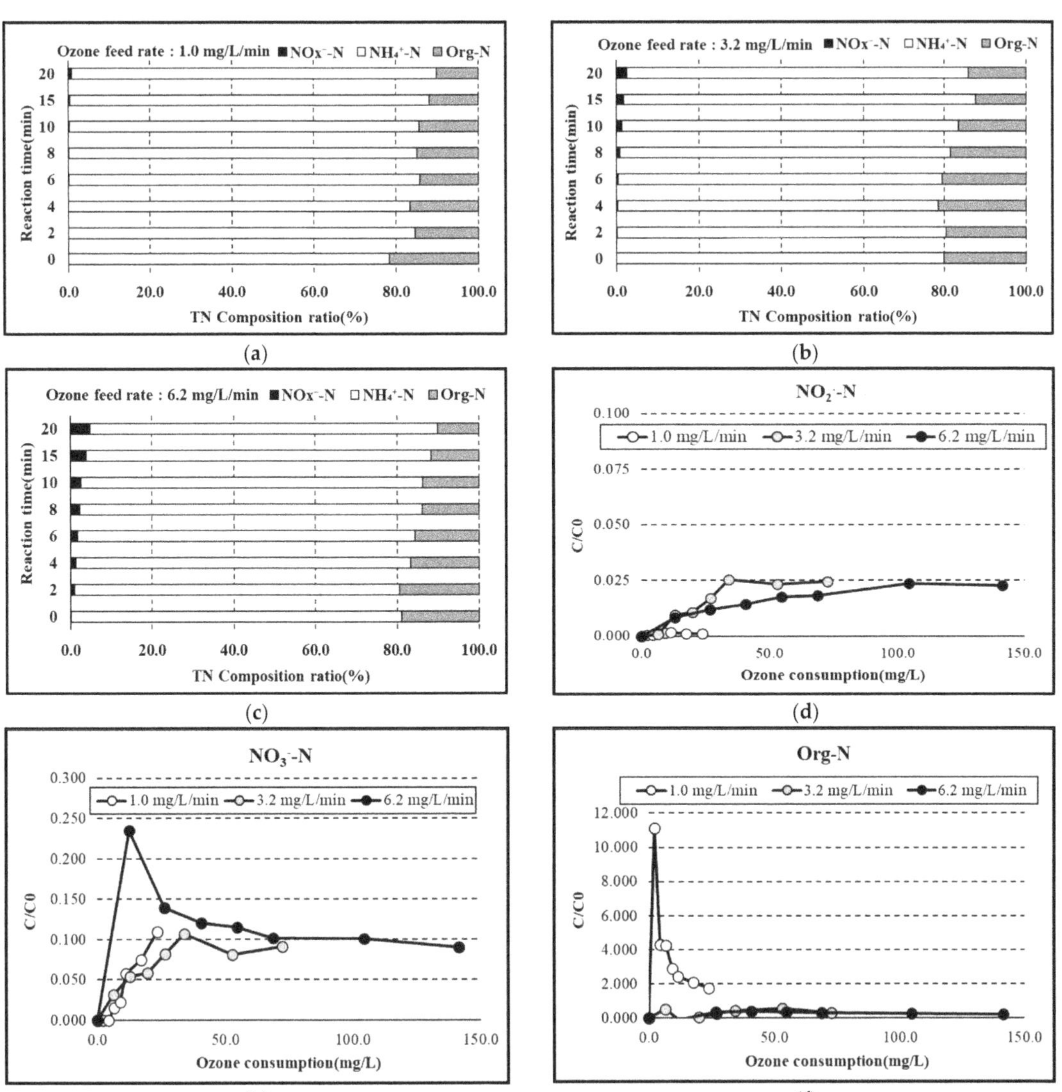

Figure 4. Effect of ozone feed rates on (**a**) NO_2^--N, (**b**) NO_3^-N, and (**c**) Org-N, and ozone consumption vs. generated (**d**) NO_2^--N, (**e**) NO_3^-N, and (**f**) Org-N.

3.4. Effect of Ozone Dosage on COD_{Cr}, COD_{Mn}, and TOC

To demonstrate the effect of ozone dosage and reaction time, the anaerobic digestion tank effluent was exposed to ozone dosage at 6.2, 3.2, and 1.0 mg/L/min, respectively, for 20 min of reaction time. Figure 5 shows that when the ozone dosage increased from 1.0 to 6.2 mg/L/min, the ozone consumption also increased from 2.1 to 141.4 mg/l. With the increase in ozone consumption, the soluble COD_{Cr} and COD_{Mn} concentrations also increased. Consequently, the effect of low ozone dosage can achieve high degradation of organic compounds. The result shows that ozonation is more effective at degrading organic pollutants in under 20 min of reaction time. In these experimental conditions, ozonation allowed direct oxidation by microbubble ozone; the pH value was around 7.8

and there was almost zero sludge production. The applied ozone dosage was determined upon TOC and COD content in the effluent. The result shows the effectivity of direct ozone reactions on TOC increment efficiency. Studying the COD and TOC variations helped to better understand the effect of ozonation. The increase in COD was insignificant when the ozone dose of 6.2 mg/L/min was introduced. An excess ozone dosage lowered the total process utilization efficiency.

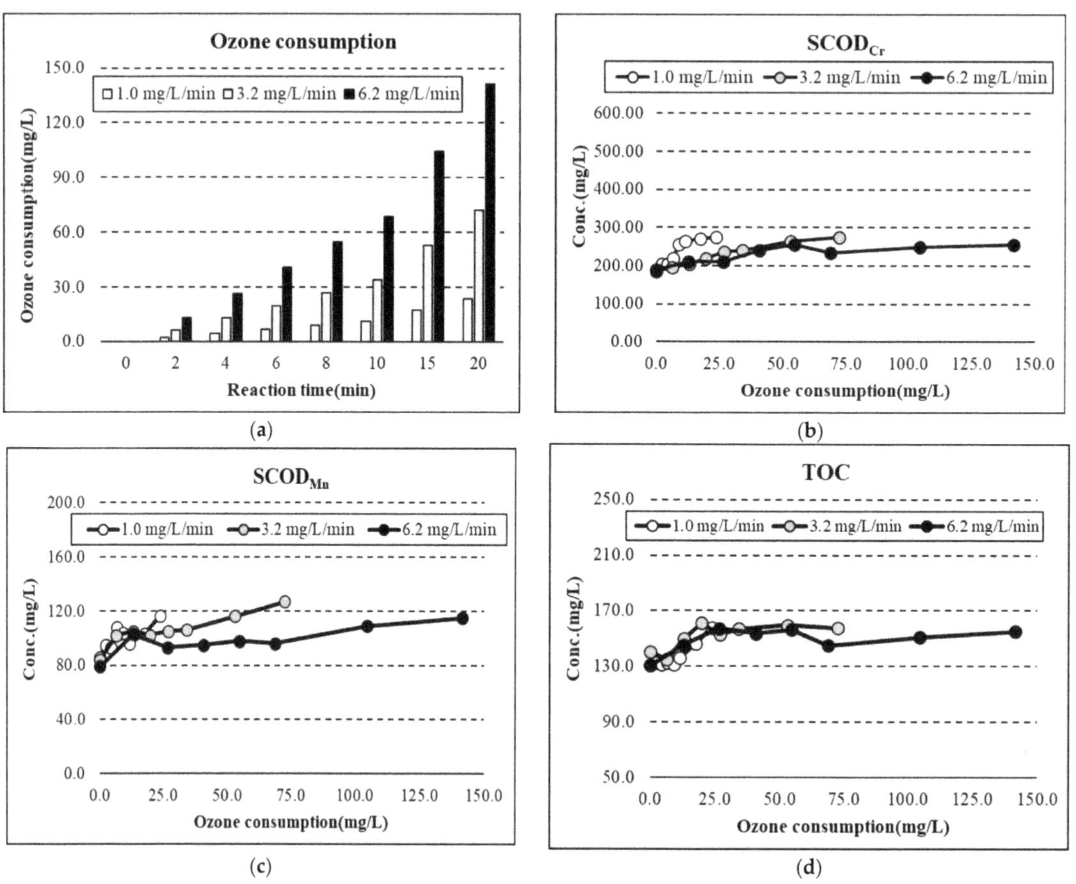

Figure 5. Effect of (**a**) ozone consumption and feed rate on (**b**) COD_{Cr}, (**c**) COD_{Mn}, and (**d**) TOC.

3.5. Ozone Dosage on Ozone Consumption and Reaction Time Effects on Biodegradability

Figure 6 indicates that when the reaction time of ozonation was 60 min, the efficiencies of improving the biodegradability COD_{Mn}/COD_{Cr} ratios were 1.08%, 0.85%, and 0.72%, respectively, but at 20 min of reaction time, the COD_{Mn}/COD_{Cr} ratios were high (0.99%, 0.92%, and 0.81%), and from this, the ratios declined to 0.83%, 0.84%, and 0.72%, respectively. With the high ozone feed rate, the efficiency of improvement was comparably high, and the high ozone concentration may have promoted the degradation of organic matter. Therefore, at a low feed rate of ozone gas, the reaction time was a maximum of 180 min and the ratio of COD_{Mn}/COD_{Cr} was increased from 0.39 to 0.77, but the maximum value of 96 appeared at 120 min of reaction time. When the initial ozone concentration was 3.4 mg/L/min, the COD_{Mn}/COD_{Cr} ratios were increasing up to 60 min, and then became stable, and, thus, difficult to decompose organic matter. In this situation, the total COD_{Cr} and COD_{Mn} values were also decreased. In opposition, at a high ozone concentration, the ratio increases from

0.79 to 1.29, which shows high biodegradability compared to the other feed rate of ozone gas. In other cases, the biodegradability parameters are estimated as a function of ozone dosage and show the optimum ozone dose for maximum biodegradability improvement [2].

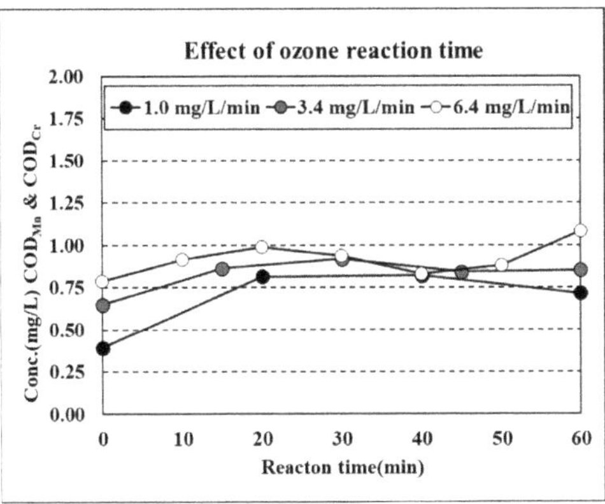

Figure 6. Ozone dosage and reaction time effects.

4. Discussion

Ozone-based oxidation processes are currently being assessed and increasingly applied for strengthened treatment of domestic wastewater effluents [36]. Initially, reduction or conversion of the parent compound was the only goal of such treatment, but micropollutants do not usually mineralize and less transformation products arise during the reaction with ozone [36,37]. Previous studies have shown that the formed transformation products have less biological activity, compared to the original compounds [38]. However, some literature showed enhanced toxicity after ozonation [39,40], and in these observations, it is unknown which transformation products from micropollutants, or by-products were formed during oxidation [36,38,41–44]. Additionally, the increased ozonation-induced toxicity is often mitigated by biological after-treatment [40].

The results of the present study confirm that ozone can effectively degrade the organic compounds from the anaerobic digestion tank and assist in biodegradability. Moreover, the results showed the significant changing effect on organic nitrogen during the ozone process. Microbubble ozone technology was chosen for ozonation because of the reliability and efficiency; the use of the microbubble type of ozone also overcomes the limitation of solubility, which leads to low utilization efficiency during wastewater treatment. During ozonation, an appropriate ozone flow rate selection for the decomposition of organic pollutants is very important for optimizing ozone utilization. An increase in ozone dose enhances the driving force for ozone to move into wastewater, which increases the production of hydroxyl groups that ultimately improve the decomposition efficiency and mineralization rate of organic matter [45,46]. In this experiment, ozonation was used to improve the characteristics of organic pollutants from non-degradable to biodegradable, by changing the biological properties during the oxidation process. The ozone feed rate and ozone reaction time affect the change of organics [2] and the ratio of organics increases over time [22]. In previous studies of wastewater treatment, various parameters have been used to measure improved biodegradability ratios (BOD/COD, BOD/TOC, and BOD/DOC) [2]. In addition, the biodegradability parameters identify the optimum ozone conditions for the removal of recalcitrance. Overall, the COD_{Cr}, COD_{Mn}, and TOC rates increased after the main experiment, to 30%, 40%, and 20%, respectively. However, the initial concentration levels of COD and TOC were higher during the main experiment, compared to the preliminary experiment, due to the different sampling times.

To demonstrate the ozone dosage effect on ozone consumption and reaction time on the improvement of biodegradability, the anaerobic digestion tank effluent was subjected to ozonation at 1.0, 3.2, and 6.2 mg/L for 20 min of reaction time. The results indicate that ozonation is more effective at degrading the non-degradable pollutants under 20 min of reaction time. However, the ozone concentration increasing because of the ozonation utilization rate decreasing is a sign of a suitable situation at which to stop ozonation and commence biodegradation. In these experimental conditions, ozonation allows direct oxidation by microbubble ozone; the pH value was around 7.8 and there was much less sludge production.

However, there are also a few limitations for the ozone treatment, such as the application of excessive ozone dosage to compensate for the low ozone solubility. This leads to the high cost of the ozonation process. Ozonation may form toxic by-products and the partial oxidation of biodegradable matter may form recalcitrant compounds [2].

5. Conclusions

The effect of ozone dose on organic compound and nitrogen changes has been investigated by using anaerobic digester effluent. Under the appropriate parameter conditions (ozone gas flow of 1.0 L/min and ozone dosage of 1.0–6.2 mg/L), the effluent soluble COD_{Cr}, COD_{Mn}, and TOC values were increasing, which indicated that the system improved the biodegradability of organic substances. The ozone dose of 6.2 mg/L increased the degradation ratio by 40% and increased the total organic carbon by 20% over 20 min of reaction time. Therefore, the COD_{Cr} and COD_{Mn} values will increase per unit of ozone consumption with the increase in reaction time. On the other hand, the new findings of this study were that the appropriate ozone feed rate confirms to change the nitrite and nitrate ions under ozonation. The ozone treatment showed organic nitrogen mineralization and degradation of organic compounds, with the contribution of the microbubble ozone oxidation process, and is a good option for removing non-biodegradable organic compounds. The application of the microbubble ozone process, with the degradation of organic compounds from a domestic wastewater treatment plant, was investigated. It is well known that ozonation can oxidize non-degradable organic compounds to degradable by-products. In addition, ozone promotes the partial oxidation of pollutants, and by partial oxidation, wastewater organic pollutants improve their characteristics. Finally, the above-mentioned results indicated that the process was more suitable for the pre-treatment of anaerobic digester effluent to treat refractory organics of domestic wastewater. Further studies should be focused on the cost effectiveness and optimization of ozone in the following treatment process. Further, it is necessary to carry out more pilot plant-scale experiments with actual domestic wastewater implementations of scale-up parameters.

Author Contributions: J.A.: sample collection, sample analysis, writing—original draft preparation, J.L.: conceptualization, investigation, methodology, resources. W.K.: conceptualization, supervision, funding acquisition. I.K.: conceptualization, investigation, methodology, resources, supervision, funding acquisition. All authors have read and agreed to the published version of the manuscript.

Funding: This research was supported by the Korea Institute of Civil Engineering and Building Technology (KICT), research project entitled 'Development of practical technology for resource and energy recycling system using organic waste biogas', project number #20220081-001.

Institutional Review Board Statement: Not applicable.

Informed Consent Statement: Not applicable.

Data Availability Statement: Not applicable.

Conflicts of Interest: The authors declare no conflict of interest.

Abbreviations

COD	Chemical oxygen demand
COD_{Cr}	Chemical oxygen demand by dichromate
COD_{Mn}	Chemical oxygen demand by permanganate
TOC	Total organic carbon
$NO_2{}^--N$	Nitrite nitrogen
$NO_3{}^--N$	Nitrate nitrogen
T-N	Total nitrogen
$NH_4{}^+-N$	Ammonia nitrogen
Org-N	Organic nitrogen
SS	Suspended solids
BOD	Biochemical oxygen demand
DOC	Dissolved organic carbon

References

1. van Leeuwen, J.; Sridhar, A.; Harrata, A.K.; Esplugas, M.; Onuki, S.; Cai, L.; Koziel, J.A. Improving the Biodegradation of Organic Pollutants with Ozonation during Biological Wastewater Treatment. *Ozone Sci. Eng.* **2009**, *31*, 63–70. [CrossRef]
2. Alvares, A.B.C.; Diaper, C.; Parsons, S.A. Partial Oxidation by Ozone to Remove Recalcitrance from Wastewaters—A Review. *Environ. Technol.* **2001**, *22*, 409–427. [CrossRef] [PubMed]
3. Fewson, C.A. Biodegradation of zenobiotic and other persistent compounds: The cause of recalcitrance. *Trends Biotechnol.* **1988**, *6*, 148–153. [CrossRef]
4. Qian, Y.; Yen, Y.; Zhang, H. Efficiency of pre-treatment methods in the activated sludge removal of refractory compounds in coke-plant wastewater. *Water Res.* **1994**, *28*, 701–707. [CrossRef]
5. Sevimli, M.F.; Aydin, A.F.; Ozturk, I.; Sarikaya, H.Z. Evaluation of the alternative treatment processes to upgrade an opium alkaloid wastewater treatment plant. *Water Sci. Technol.* **2000**, *41*, 223–230. [CrossRef]
6. Puig, A.; Ormad, P.; Roche, P.; Sarasa, J.; Gimenno, J.L.; Ovelleiro, J.L. Wastewater from the manufacture of rubber vulcanization accelerators: Characterization, downstream monitoring, and chemical treatment. *J. Chromatogr. A* **1996**, *733*, 511–522. [CrossRef]
7. Schroder, H. Characterisation and monitoring of persistent toxic organics in the aquatic environment. *Water Sci. Technol.* **1998**, *38*, 151–158. [CrossRef]
8. Kayhanian, M. Ammonium inhibition in high solids biogasification: An overview and practical solutions. *Environ. Technol.* **1999**, *20*, 355–365. [CrossRef]
9. Pincam, T.; Brix, H.; Jampeetong, A. Treatment of Anaerobic Digester Effluent Using *Acorus calamus*: Effects on Plant Growth and Tissue Composition. *Plants* **2018**, *7*, 36. [CrossRef]
10. Sontheimer, H.; Heilker, E.; Jekel, M.R.; Nolte, H.; Vollmer, F.H. The Mulheim Process. *J. Am. Water Works Assoc.* **1978**, *60*, 393–396. [CrossRef]
11. Paucar, N.E.; Kim, I.; Tanaka, H.; Sato, C. Effect of O_3 Dose on the O_3/UV Treatment Process for the Removal of Pharmaceuticals and Personal Care Products in Secondary Effluent. *ChemEngineering* **2019**, *3*, 53. [CrossRef]
12. Huber, M.M.; Korhone, S.; Ternes, T.A.; von Gunten, U. Oxidation of pharmaceuticals during water treatment with chlorine dioxide. *Water Res.* **2005**, *39*, 3607–3617. [CrossRef] [PubMed]
13. Epold, I.; Dulova, N.; Veressinina, V.; Trapido, M. Application of ozonation, UV photolysis, Fenton treatment and other related processes for degradation of ibuprofen and sulfamethoxazole in different aqueous matrices. *J. Adv. Oxid. Technol.* **2012**, *15*, 354–364. [CrossRef]
14. Klavarioti, M.; Mantzavinos, D.; Kassinos, D. Removal of residual pharmaceuticals from aqueous systems by advanced oxidation processes. *Environ. Int.* **2009**, *35*, 402–417. [CrossRef] [PubMed]
15. Ternes, T.A.; Heidenheimer, M.; McDowell, D.; Sacher, F.; Brauch, H.J.; Gulden, B.H.; Preuss, G.; Wilme, U.; Seibert, N.Z. Removal of pharmaceuticals during drinking water treatment. *Environ. Sci. Technol.* **2002**, *36*, 3855–3863. [CrossRef]
16. Zwiener, C.; Frimmel, F.H. Oxidative treatment of pharmaceuticals in water. *Water Res.* **2000**, *34*, 1881–1885. [CrossRef]
17. Derco, J.; Gotvajn, A.Z.; Cizmarova, O.; Dudas, J.; Sumegova, L.; Simovicova, K. Removal of Micropollutants by Ozone-Based Processes. *Chem. Eng. Process.* **2015**, *94*, 78–84. [CrossRef]
18. Loeb, B.L.; Thompson, C.M.; Drago, J.; Takahara, H.; Baig, S. Worldwide Ozone Capacity for Treatment of Drinking Water and Wastewater: A Review. *Ozone Sci. Eng.* **2012**, *34*, 64–77. [CrossRef]
19. Tekile, A.; Kim, I.; Lee, J.Y. Applications of Ozone Micro- and Nanobubble Technologies in Water and Wastewater Treatment: Review. *J. Korean Soc. Water Wastewater* **2017**, *31*, 481–490. [CrossRef]
20. Churchley, J. Ozone for dye waste color removal: Four years at Leed STW. *Ozone Sci. Eng.* **1998**, *20*, 111–120. [CrossRef]
21. Barraza, S.X.; Saez-Navarrete, C.; Torres-Castillo, R. Anaerobic biodegradability of leachate from MSW intermediate landfill. *Afinidad Barc.* **2019**, *75*, 585.
22. Hoigne, J.; Bader, H. Rate constants of reactions of ozone with organic and inorganic compounds in water—I. Non-dissociating organic compounds. *Water Res.* **1983**, *17*, 173–183. [CrossRef]

23. Hoigne, J.; Bader, H. Rate constants of reactions of ozone with organic and inorganic compounds in water—II. Dissociating organic compounds. *Water Res.* **1983**, *17*, 185–194. [CrossRef]

24. Hoigne, J.; Bader, H. Rate constants of reactions of ozone with organic and inorganic compounds in water—III. Inorganic compounds and radicals. *Water Res.* **1985**, *19*, 993–1004. [CrossRef]

25. Chu, L.; Wang, J.; Wang, B.; Xing, X.H.; Yan, S.; Sun, X.; Jurick, B. Changes in biomass activity and characteristics of activated sludge exposed to low ozone dose. *Chemosphere* **2009**, *77*, 269–272. [CrossRef]

26. Ahn, K.H.; Yeom, I.T.; Park, K.Y.; Maeng, S.K.; Lee, Y.; Song, K.G.; Hwang, J.H. Reduction of sludge by ozone treatment and production of carbon source for denitrification. *Water Sci. Technol.* **2002**, *46*, 121–125. [CrossRef]

27. Lee, J.W.; Cha, H.Y.; Park, K.Y.; Song, K.G.; Ahn, K.H. Operational strategies for an activated sludge process in conjunction with ozone oxidation for zero excess sludge production during winter season. *Water Res.* **2005**, *39*, 1199–1204. [CrossRef]

28. Huang, Z.; Gu, Z.; Wang, Y.; Zhang, A. Improved oxidation of refractory organics in concentrated leachate by a Fe^{2+}-enhanced O_3/H_2O_2 process. *Environ. Sci. Pollut. Res.* **2019**, *26*, 35797–35806. [CrossRef]

29. Tizaoui, C.; Bouselmi, L.; Mansouri, L.; Ghrabi, A. Landfill leachate treatment with ozone and ozone/hydrogen peroxide systems. *J. Hazard. Mater.* **2007**, *140*, 316–324. [CrossRef]

30. Shin, H.K.; Lim, J.L. Improving biodegradability of naphthalene refinery wastewater by pre-ozonation. *J. Environ. Sci. Health* **1996**, *A31*, 1009–1024.

31. Tian, J.; Hu, Y.; Zhang, J. Chemiluminescence detection of permanganate index (COD_{Mn}) by a luminol-$KMnO_4$ based reaction. *J. Environ. Sci.* **2008**, *20*, 252–256.

32. Tchobanoglous, G.; Burton, F. *Wastewater Engineering: Treatment, Disposal and Reuse*, 3rd ed.; McGraw-Hill Inc.: New York, NY, USA, 1991.

33. Chen, Y.; Xiao, Y.; Wang, G.; Shi, W.; Sun, L.; Chen, Y.; Miao, A. A pilot-scale test on the treatment of biological pretreated leachate by the synergy of ozonation-biological treatment-catalytic ozonation. *Environ. Eng. Res.* **2021**, *26*, 200349. [CrossRef]

34. Plosz, B.G.; Ried, A.; Lopez, A.; Liltved, H.; Vogelsang, C. Ozonation as a Means to Optimize Biological Nitrogen Removal from Landfill Leacate. *Ozone Sci. Eng.* **2010**, *32*, 313–322. [CrossRef]

35. Singer, P.C.; Zilli, W.B. Ozonation of ammonia in wastewater. *Water Res.* **1975**, *9*, 127–134. [CrossRef]

36. Von Gunten, U. Oxidation Processes in Water Treatment: Are We on Track? *Environ. Sci. Technol.* **2018**, *52*, 5062–5075. [CrossRef] [PubMed]

37. Von Gunten, U. Ozonation of drinking water: Part I. Oxidation kinetics and product formation. *Water Res.* **2013**, *37*, 1443–1467. [CrossRef]

38. Lee, Y.; Von Gunten, U. Advances in predicting organic contaminant abatement during ozonation of municipal wastewater effluent: Reaction kinetics, transformation products, and changes of biological effects. *Environ. Sci. Water Res. Technol.* **2016**, *2*, 421–442. [CrossRef]

39. Magdeburg, A.; Stalter, D.; Schlüsener, M.; Ternes, T.; Oehlmann, J. Evaluating the efficiency of advanced wastewater treatment: Target analysis of organic contaminants and (geno-) toxicity assessment tell a different story. *Water Res.* **2014**, *50*, 35–47. [CrossRef]

40. Stalter, D.; Magdeburg, A.; Weil, M.; Knacker, T.; Oehlmann, J. Toxication or detoxication? In vivo toxicity assessment of ozonation as advanced wastewater treatment with the rainbow trout. *Water Res.* **2010**, *44*, 439–448. [CrossRef]

41. Mestankova, H.; Escher, B.; Schirmer, K.; von Gunten, U.; Canonica, S. Evolution of algal toxicity during (photo)oxidative degradation of diuron. *Aquat. Toxicol.* **2011**, *101*, 466–473. [CrossRef]

42. Mestankova, H.; Schirmer, K.; Canonica, S.; von Gunten, U. Development of mutagenicity during degradation of N-nitrosamines by advanced oxidation processes. *Water Res.* **2014**, *66*, 399–410. [CrossRef] [PubMed]

43. Mestankova, H.; Schirmer, K.; Escher, B.I.; von Gunten, U.; Canonica, S. Removal of the antiviral agent oseltamivir and its biological activity by oxidative processes. *Environ. Pollut.* **2012**, *161*, 30–35. [CrossRef] [PubMed]

44. Ghuge, S.P.; Saroha, A.K. Catalytic ozonation for the treatment of synthetic and industrial effluents—Application of mesoporous materials: A review. *J. Environ. Manag.* **2018**, *211*, 83–102. [CrossRef] [PubMed]

45. Karrer, N.J.; Ryhiner, G.; Heinzle, E. Applicability test for combined biological-chemical treatment of wastewaters containing bio refractory compounds. *Water Res.* **1997**, *31*, 1013–2020. [CrossRef]

46. Stokinger, H.; Heinzle, E.; Kut, O.M. Removal of choloro and nitro aromatic wastewater pollutants by ozonation and biotreatment. *Environ. Sci. Technol.* **1995**, *29*, 2016–2022. [CrossRef] [PubMed]

processes

Article

COD Reduction of Aeration Effluent by Utilizing Optimum Quantities of UV/H$_2$O$_2$/O$_3$ in a Small-Scale Reactor

Mehdi Rafiee [1], Morteza Sabeti [1,2], Farshid Torabi [2,*] and Aria Rahimbakhsh [2]

[1]   Department of Chemical Engineering, Isfahan University of Technology, Isfahan 8415683111, Iran
[2]   Petroleum Systems Engineering Department, Faculty of Engineering and Applied Science,
      University of Regina, Regina, SK S4S 0A2, Canada
*   Correspondence: farshid.torabi@uregina.ca; Tel.: +1-306-585-5667

Abstract: Extensive research has been carried out to figure out safe means of disposing various industrial effluents. Industrial wastewaters from the aeration industry such as heavy metals and oily substances contain a high degree of contamination. The advanced oxidation process is one of the most effective and rapid methods of removing contaminations, which can lead to a high chemical oxygen demand (COD). The aim of the present study is to reduce the COD of an aeration effluent with the initial COD of 13,004 mg/L. About 20 sets of experimental tests were conducted to identify the contribution of H$_2$O$_2$, O$_3$, and UV to the treatment process. The influence of the quantities of additives and the dose of the UV irradiance were, too, among the subjects of the study. These factors were altered throughout the experiments and their mutual effects were measured. To design the experiments, Minitab software 16 was utilized. The experimental conditions were set at the standard values of 25 °C and 1 bar to minimize any uncertainty. Based on the results, a correlation was derived, which was capable of expressing the effects of the input parameters (AOPs parameters) on the response (the COD level). Finally, the optimization process was conducted to find the quantities of H$_2$O$_2$, O$_3$, and UV irradiance required to decrease the CODs of the effluent to their lowest possible. Based on the findings, when the doses of H$_2$O$_2$, O$_3$, and UV to the treatment process were 40 mg/L, 8 mg/L and 86 mWs/cm^2, respectively, the COD percent change was 51.5%.

Keywords: aeration contaminants; advanced oxidation processes (AOPs); COD reduction; UV; ozone; H$_2$O$_2$

Citation: Rafiee, M.; Sabeti, M.; Torabi, F.; Rahimbakhsh, A. COD Reduction of Aeration Effluent by Utilizing Optimum Quantities of UV/H$_2$O$_2$/O$_3$ in a Small-Scale Reactor. *Processes* **2022**, *10*, 2441. https://doi.org/10.3390/pr10112441

Academic Editors: Gassan Hodaifa, Antonio Zuorro, Joaquín R. Dominguez, Juan García Rodríguez, José A. Peres, Zacharias Frontistis and Mha Albqmi

Received: 16 October 2022
Accepted: 15 November 2022
Published: 18 November 2022

Publisher's Note: MDPI stays neutral with regard to jurisdictional claims in published maps and institutional affiliations.

1. Introduction

Water is used in many industrial operations [1]. Industrial wastewater streams containing harmful organic and inorganic substances are generated from various industries such as aeration, metal-working, and metallurgy. Several aeration industries exist in places that are regarded as most hazardous locations due to the discharge of large volumes of wastewater with high levels of hardness, total dissolved solids, total suspended solids, chemical oxygen demand (COD), biological oxygen demand, and pH. The sources of high COD include reactive peroxometal complexes and greasy substances. Substances with high COD levels not only are harmful for the environment, but also can accumulate in human body through the natural food chain [2]. When their concentration surpasses the determined standards, serious health problems will arise among the people [3]. Hence, it is needless to emphasize that wastewater from aeration industries must be treated prior to the disposal of the waste. Often, physical means are not efficient enough in ridding of these extra substances. To remove them, one can rely on the advanced oxidation process (AOPs) [4]. AOPs contribute to the cleaning process by generating such highly reactive radical species as OH• [5,6]. To put it in the same context, AOPs involve the production of OH• through Injecting H$_2$O$_2$, ozone, and oxidants in combination via ultraviolet (UV) irradiance. To perform the operation more proficiently, researchers have suggested combining the three methods above [7] resulting in the release of HO•, which is mainly responsible for the degradation of organic compounds.

As OH• is reactive electrophiles, it is counted as a highly effective compound in eliminating organic chemicals [8]. It none-selectively reacts rapidly with almost all electron-rich organic compounds. As a result, OH• causes the breakdown of the organic compound and diminishes the pollutants concentrations [9]. It grasps a hydrogen atom from the organic compound (R−H) in wastewaters, and then leads to the production of an organic radical (•R). Following that, the organic radical is motivated to chemically react with other compounds to be neutralized [10–12]. This procedure is concisely shown in the equation below:

$$R - H + HO\cdot \ \rightarrow H_2O + \cdot R \tag{1}$$

Due to some complex reactions, OH• and O_2^-• are generated when ozone is added to wastewater. These radical compounds cause the oxidation of organic species. By introducing a small volume of H_2O_2, the rate of OH• production could be enhanced through the following stoichiometry [13].

$$2O_3 + H_2O_2 \ \rightarrow 2HO\cdot +3O_2 \tag{2}$$

The performance of the ozone can also be boosted when the wastewater is exposed to the UV irradiance [14]. The application of UV in treatment is wide, and a comparative study among selected AOPs integrated with UV was done by some researchers [15–17]. The UV irradiance interferes in the process by generating H_2O_2 as an intermediate; following that, OH• is composed. The reactions involved are written as follows:

$$\begin{aligned} O_3 + H_2O + h\nu \ &\rightarrow \ O_2 + H_2O_2 \\ 2O_3 + H_2O_2 \ &\rightarrow 2HO\cdot +3O_2 \end{aligned} \tag{3}$$

Equation (3) shows the photolysis of ozone and the production of hydrogen peroxide, which eventually leads to the generation of the HO• radical [18].

H_2O_2 may also be introduced into the wastewater system as a single oxidant or one of the included oxidants [18]. The appropriate concentration of H_2O_2 is crucial during the degradation of pollutants. If an unexpected quantity of H_2O_2 is added to the wastewater, it can react with other contaminants to produce oxidizable materials, which is not the desired goal for the process. While easy to perform and beneficial to apply, the combination of H_2O_2 and UV is also used for decreasing the pollutant volume. Karci et al. [19] reported the polyethoxy chain of the surfactant is more susceptible to degradation in the H_2O_2 and UV treatment process. Antonopoulou [20] found the application of UV and H_2O_2 proved influential in reducing the odorous aldehyde concentrations.

Although this method of AOP is less costly and easier to carry out, it suffers some drawbacks. One of its main issues lies within its inability to absorb the UV light when large quantities of H_2O_2 are present in the wastewaters. This causes the loss of most of the light input [18]. Such an undesirable phenomenon can be hindered when the pH values of the wastewaters are decreased by altering the ratio of peroxide to ozone. AOPs have an oxidation potential of 2.33 V, approximately, and hence, compared to conventional oxidants such as $KMnO_4$, show faster oxidation reaction rates [21,22]. Ozone, with its ability to react with organic compounds including polyphenols, is known as a strong oxidizer [8]. The combination of ozone, UV irradiance, and H_2O_2 has been suggested as an effective method for the treatment of wastewater with polyphenol content in documents such as those used by manufactures working with metal materials [23]. Therefore, several researchers have tried to determine the optimized parameters in terms of the concentrations of H_2O_2, O_3, and the dose of UV when lowering the level of COD in effluents [10,19,20]. It should be mentioned that the response surface methodology (RSM) is a common way for optimization in an analytical chemistry application such as water treatment [24–26]. RSM was successfully applied in some experimental works as its responses are influenced by

AOP variables, and optimization of the levels of these variables can be obtained through the fit of a polynomial equation to the experimental data.

The idea of applying UV, H_2O_2, and O_3 to decrease the COD level seems reasonable, but is it efficient? What is the interaction of the mentioned parameters when the three AOP parameters are combined for a wastewater treatment? Do they cancel each other's effects or strengthen treatment capabilities? In the present study, 20 experimental tests were designed to observe and detect the effects of H_2O_2, O_3, and UV altogether on the treatment of aeration factory's produced effluents. The main objective was to find the most efficient process with the lowest energy consumption for eliminating organic compounds and lowering the level of COD. Therefore, the experimental conditions were designed with the help of the RSM method, which was introduced to the Minitab software 16. After conducting the tests, the optimized point at which the definitive treatment could be conducted was determined.

2. Materials and Methods

The materials, apparatus and measuring methods used in the experiments are briefly introduced as below.

2.1. The Apparatus

- COD Reactor (Box 389, Loveland, Hach Co., CO, USA)
- Spectrophotometer (Jenway 6305, UV/Vis. Spectrophotometer, UK)
- Centrifuge (Werk NT.BaujharEkin, Universal 320 R, Hettich, Germany)
- Lab oven
- Furnace (Aria-Electric, EMTC model, Iran)
- Ph meter (pH 162, Iran)
- Laboratory scales with measurement accuracy of 0.00001 g (Kyoto Co., Electronic balance, AEL-40SM, Japan)
- Ozone generator (ARDA Ozone- ONE)
- UV bulb (33 W, AQUA SAFE, Taiwan)
- RO Pump (TY-2800- 24 VDC)

2.2. The Input Materials/Chemicals

- Potassium Dichromate (Zigma), 99% (W/W)
- Concentrated sulfuric acid (Merck), 98% (W/W)
- Mercury sulfate (Zigma), 98% (W/W)
- Silver sulfate (Zigma), >99% (W/W)
- Potassium Hydrogen Phthalate (Merck), >99% (W/W)
- Hydrogen peroxide (Merck), >30% (W/W)

2.3. The Studied Wastewater

Several wastewater samples were taken from an aeration factory. The COD of the wastewater samples was measured at 13,004 mgO_2/L. Due to the high value of COD in the effluent from the aeration industry, the wastewater taken was regarded as appropriate for the purpose of the present study. They were maintained at a temperature of 4 °C to prevent any changes in the wastewater's COD.

Due to its high COD level, wastewaters are not allowed to be released into the environment without any pretreatment actions. Therefore, advanced treatment methods were employed in the study to reduce the value of the COD to a standard level of 250 mg/L [27] before disposing the wastewater.

2.4. Measurement Methods

In order to evaluate the performance of the advanced oxidation process, it is essential for the key factors of the process to be measured. To this end, the most important property, the quantity of organic substances in the wastewater, should be measured. There are various techniques for measuring the quantity of organic substances such as: (i) the chemical oxygen demand, and (ii) the total volatile solid. To determine the quantity of the wastewater's COD, a method similar to the D-5220 closed reflux, known as the colorimetric method, was employed [27].

2.5. Construction of the Base Line

Among the available treatment approaches, those combined methods, which include UV, O_3, and H_2O_2, were employed as the basis of the present work. To find the best combination, in terms of UV dose and the amount of O_3 as well as H_2O_2, several tests were conducted. The chosen ranges were picked up as per the literature [28–33].

A sample of the untreated wastewater was poured into a 2000-mL Erlenmeyer flask. This Erlenmeyer flask was the storage of the system, and the untreated water started to be pumped out through a container with a volume capacity of 1500 mL to be exposed to a high level of UV irradiance. The UV irradiance was emitted from a 33 W bulb with a length of 80 cm. The flow rate of the fluid was kept at 200 mL/min and some baffles were put on the way of the flowing water in the container to generate turbulency in the flowing fluid. The level of the water in the container was kept lower than 0.5 cm in order to increase the surface area of the exposed water to the UV irradiance. At the same time, based on the plan mentioned in Table 1, sufficient quantities of H_2O_2 and O_3 were added to the fluid. The container output was led to a tee. A branch of the tee was used to collect the treated water at the end of the process. The other branch was used to recycle back the fluid to the storage to increase the efficiency of the process. The COD of the treated product was measured after 40 min of the disinfection operation. In total, 20 tests were conducted when using various combinations of $UV/H_2O_2/O_3$. With the aim of minimizing the inaccuracies in the measurement results, and to ensure the repeatability of the tests, all of the tests were performed three times, and then the average values of the three tests were regarded as the final figures. Figures 1 and 2 represent a simple schematic of the treatment system used during the process.

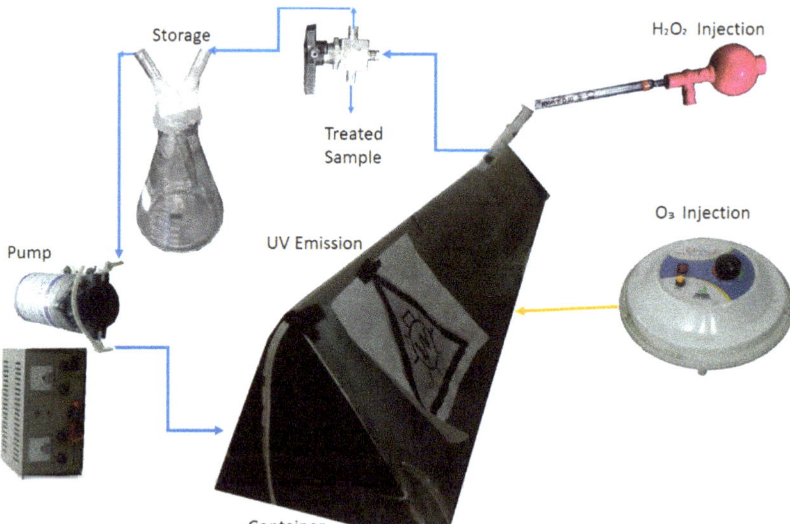

Figure 1. The disinfecting system at the experimental scale.

Table 1. The real values associated with each coded value in Minitab software 16.

The Coded Value	The Dose of UV (mWs/cm^2)	The Concentration of O_3 (mg/L)	The Concentration of H_2O_2 (mg/L)
+2	400	8	120
+1	310	6.5	95
0	220	5	70
−1	130	3.5	45
−2	40	2	20
Δζ	90	1.5	25

Experimental Conditions

In the present work, the concentration of H_2O_2 and O_3, as well as the dose of UV were studied under five levels tabulated in Table 1. To adapt the tests and designs with harmony, X domain was set between −2 and +2 in Minitab. Therefore, according to the table, each of the values in the AOP process calls a number between −2 and +2. Following the definition of the parameters' ranges, 15 tests with different values of parameters were randomly selected and conducted. Together with the test conditions, they were recommended by the Minitab software and are listed in Table 2. Accordingly, the last five tests (from case 16 to case 20) have (0, 0, 0) values. Please note that these sets of tests were designed to determine the level of errors.

Table 2. The tests plan recommended by Minitab 16.

The Test Number	The Parameters' Coded Values		
	X_1	X_2	X_3
1	−1	−1	−1
2	1	−1	−1
3	−1	1	−1
4	1	1	−1
5	−1	−1	1
6	1	−1	1
7	−1	1	1
8	1	1	1
9	−2	0	0
10	2	0	0
11	0	−2	0
12	0	2	0
13	0	0	−2
14	0	0	2
15	0	0	0
16	0	0	0
17	0	0	0
18	0	0	0
19	0	0	0
20	0	0	0

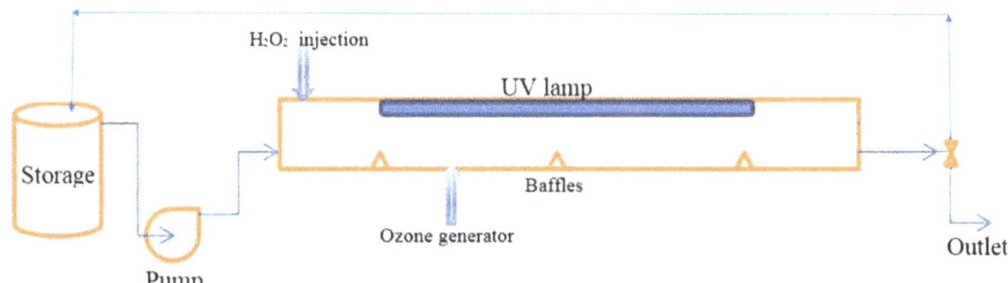

Figure 2. The schematic view of the disinfecting system.

3. Results and Discussion

Experiments were carried out by applying UV irradiance and adding H_2O_2 and O_3 compounds at the natural pH of the effluents taken from the aeration field. This enabled the identification of the power of AOPs at different conditions of removing the COD amounts. Figure 3 shows a wastewater sample before and after disinfection. The mentioned process almost removed the bad smell, dark color, COD amount, and other heavy metal effects on the water. Table 3 highlights the ability of each set of tests in removing the COD from the studied wastewater.

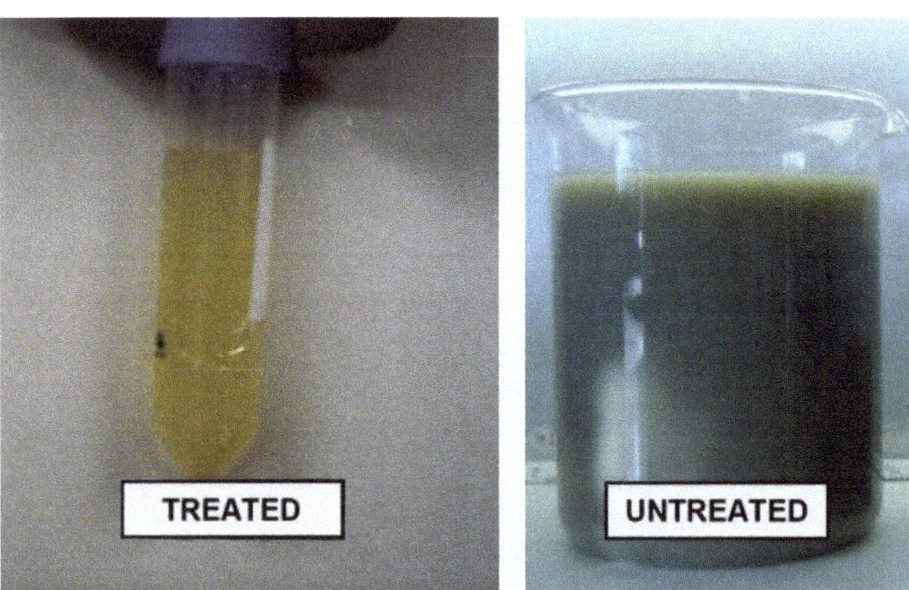

Figure 3. The appearance of the aeration wastewater before and after the treatment.

The treatment cost of each individual test was evaluated based on the electrical energy per-order (*EEO*) according to the following equation

$$EEO = \frac{W_{UV} + W_{H2O2} + W_{O3}}{V \times \log\left(\frac{COD_i}{COD_f}\right)} \tag{4}$$

where *EEO* stands for the electrical energy consumption per order (kWh/m³), W_{UV} is the electrical energy consumption of the UV lamp (kWh), W_{H2O2} shows the equivalent electrical energy of the applied H_2O_2 (kWh), W_{O3} represents the equivalent electrical energy of the applied O_3 (kWh), V presents the volume of the sample (m^3), and COD_i and COD_f are the initial and final concentration of the sample, respectively. The ozone conversion to equivalent electrical energy unit was done by assuming that the used ozone generator produced 50 mgO_3 per hour, and that the energy consumption of the equipment is 20 kWh. The electrical energy conversion for the applied peroxide is based on the cost of peroxide, 0.01 $/g, and the charge of electrical energy 0.14 $/h.

Table 3. Results of the 20 sets of tests.

| The Test Number | The Parameters' Real Values | | | COD after Treatment (mgO_2/L) | COD Percent Change (%) | Energy Consumption (kWh/m^3) |
	The Dose of UV (mWs/cm^2)	The Concentration of O_3 (mg/L)	The Concentration of H_2O_2 (mg/L)			
1	130	3.5	45	6960	46.5	23.2
2	310	3.5	45	6680	48.6	30.9
3	130	6.5	45	6470	50.2	24.8
4	310	6.5	45	6485	50.1	33.5
5	130	3.5	95	6700	48.5	33.5
6	310	3.5	95	6582	49.4	41.5
7	130	6.5	95	6435	50.5	35.5
8	310	6.5	95	6470	50.2	44.4
9	40	5	70	6985	46.3	26.9
10	400	5	70	6640	48.9	42.9
11	220	2	70	6905	46.9	31.6
12	220	8	70	6405	50.7	36.1
13	220	5	20	6908	46.9	23.9
14	220	5	70	6665	48.7	34.1
15	220	5	70	6687	48.6	34.2
16	220	5	70	6690	48.6	34.3
17	220	5	70	6692	48.5	34.3
18	220	5	70	6685	48.6	34.2
19	220	5	70	6685	48.6	34.2
20	220	5	70	6692	48.5	34.3

The COD reduction percent and the electrical energy consumption for the 20 tests done are shown in Figure 4. As seen in the figure, the best and the worst test numbers for the water treatment without paying attention to their energy consumption are tests 9 and 12, respectively. On the other hand, tests 8 and 1 are the most and the least energy-consuming processes in this work, respectively. The figure does not show a pattern to understand where the optimum point for the purpose of treatment and energy saving is. Therefore, these data were given to Minitab for more analysis.

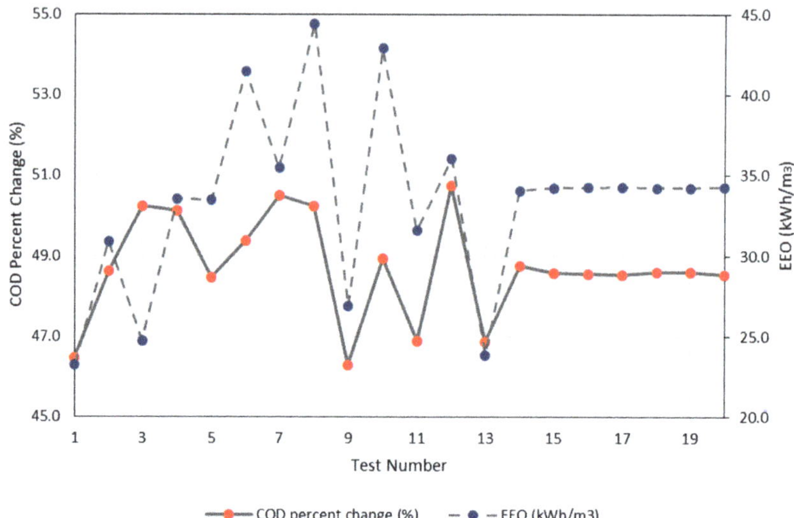

Figure 4. Comparison of COD percent change with electrical energy order over the 20 tests mentioned in Table 3.

3.1. Residual Diagrams

In order to reassure all the model assumptions for the present model have been met, the residual plots should be validated. There are essentially four residual plots in Minitab that guide the users to clarify if they follow a normal distribution or the confidence intervals and p-values can be accurate. The residual diagrams for this experiment (COD reduction) are shown in Figure 5. The first plot is the normal probability plot of the residuals, which represents the residuals versus their expected values. The plot verifies the assumption that the residuals are normally distributed if it approximately follows a straight line without any abnormality. As it can be observed from Figure 5a, data are in their normal condition and are close to the cross line.

The residuals versus fits graph indicates the fitted values on the horizontal axis and the residuals on the vertical axis. The question regarding the residual variance whether being constant or not can be addressed by residuals versus fits graph shown in Figure 5b. Considering the points on the graph, one could recognize the residual variance is reasonable and constant in the present study. Not only are the data points scattered around the horizontal zero line, but they also do not seem to follow a unique pattern.

The distribution of the residuals for all observations can be seen in a histogram plot. Theoretically, when the curvature of the histogram chart is close to bell-shaped, it represents the normal distribution of the data in the design of experiments (DOE) media. Figure 5c shows the histogram bar chart of the work. As it can be observed, although the chart does not resemble a perfect bell-shaped histogram, it is within a reasonable range to conclude the correctness of the residual data.

The residuals versus order plot displays the residuals in the order that the data were collected. The residuals versus order plot is used to confirm the assumption that the residuals are independent from one another. There are no trends or patterns in independent residuals when shown in time order. The patterns in the data may suggest that residuals close to one another are likely connected and hence, not independent. Ideally, the residuals on the plot should be distributed around the zero line randomly. Figure 5d shows data points in the order do not follow a logical pattern. On the other hand, Figure 5d confirms that the residuals are independent from one another.

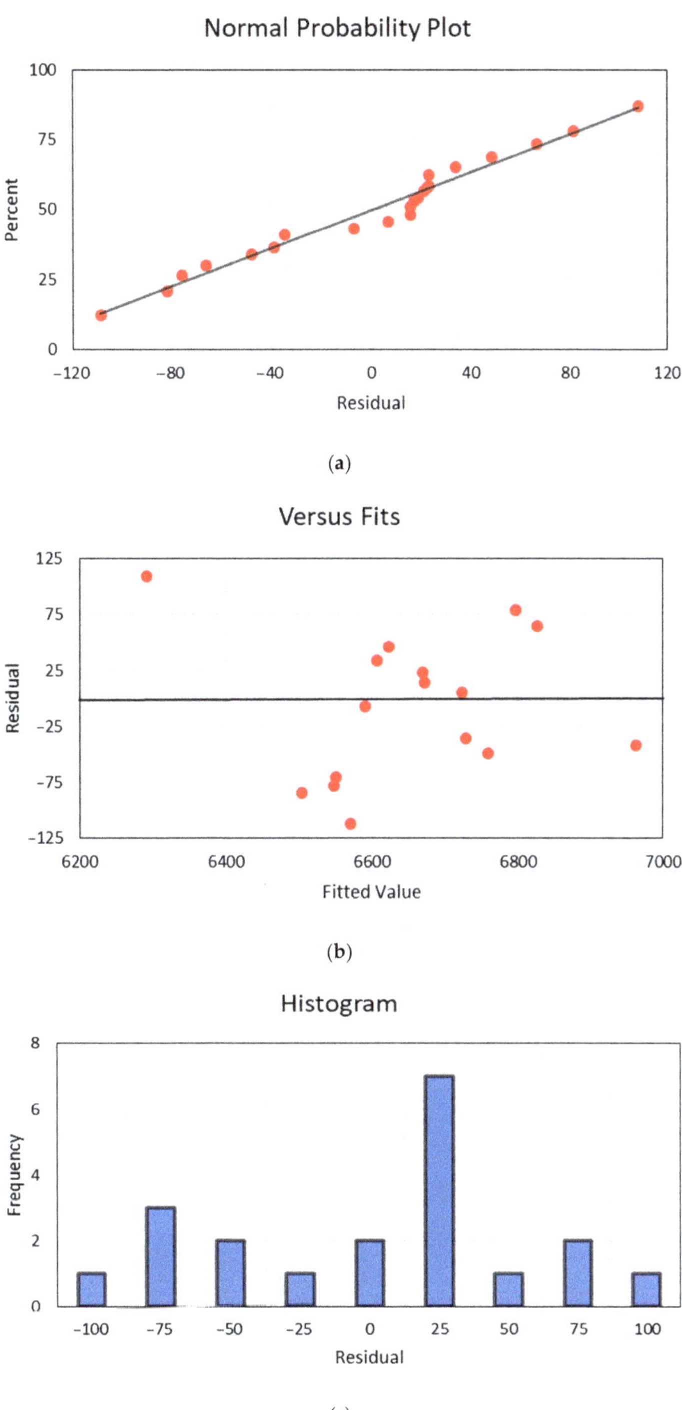

Figure 5. *Cont.*

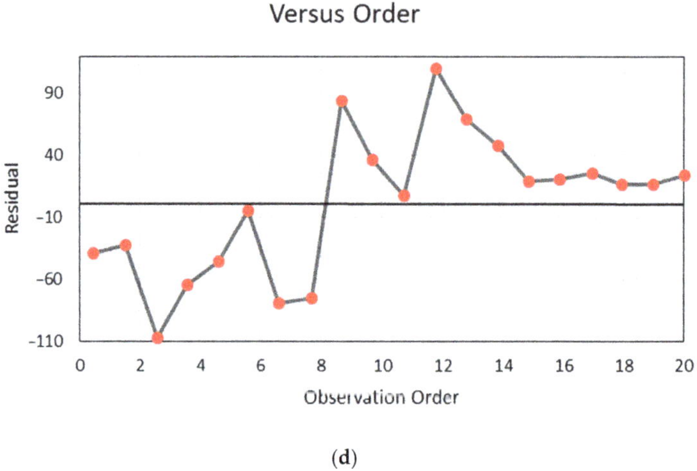

(**d**)

Figure 5. The residual diagrams related to the final COD after the water treatment process. (**a**) Normal probability plot. (**b**) Versus Fits plot. (**c**) Histogram plot (**d**) Versus order plot.

3.2. Developed Correlation

The analysis of the results for chemical oxygen demand was conducted by the Minitab software 16. Table 4 shows the analysis of variance (ANOVA) results displayed in the session window of Minitab. The p-value is a measure of how likely the sample results are, assuming the null hypothesis is true. Its value is somewhere between 0 and 1. The p-value shows the power of each parameter participating in the input-response parameters. In the present study, if the p-value of an input parameter, all of which are tabulated in Table 4, is greater than 0.001, that parameter is regarded as weak and without a significant influence on the results. Indeed, if the p-value is 0.05 or more, its corresponding parameter is regarded as insignificant, which can lead to ignoring its influence on the response's value. Based on this analogy, the following correlation, which relates the AOPs parameters to the COD value, was obtained.

$$Y = 6663 - 64.87X_1 - 128.87X_2 - 55.87X_3 + 56X_1X_2 + 38.5X_2X_3 \tag{5}$$

Table 4. Second order regression coefficients for the input factors and response.

Variables	Regression Coefficients	t-Value	p-Value
y-intercept	6663	172.812	0.000
X_3 = H_2O_2 (mg/L)	-64.87	-2.684	0.023
X_2 = O_3 (mg/L)	-128.87	-5.332	0.000
X_1 = UV (mWs/cm^2)	-55.87	-2.312	0.043
$X_3 \times X_3$	18.47	0.958	0.361
$X_2 \times X_2$	-20.91	-1.085	0.304
$X_1 \times X_1$	11.97	0.621	0.549
$X_2 \times X_3$	56	0.638	0.032
$X_1 \times X_3$	22.75	0.666	0.521
$X_2 \times X_1$	38.5	1.126	0.028
R^2 = 0.83	ADD = 0.00	R^2(ADJ) = 0.83	

Equation (5) was obtained with an adjusted R-squared value of 0.83 and a coefficient of determination of 0.72, which states that the above equation can generate fairly accurate values that match the experimental data.

3.3. Analyses of the Response's Surface Plots

It is worth noting that the synergy of independent parameters over a response can be studied through contour or 3D plots. To do so, one variable has to be maintained constant while letting the other two vary within their defined ranges in a system of three variables. The figures generated from this method show not only the effects of two independent parameters simultaneously, but also the patterns of the response.

Figure 6a shows the combined effect of UV and O_3 on the level of COD when H_2O_2 was held at 0 (the coded value). From there, one may indicate that the lowest COD is obtained through applying the maximum quantity of O_3 along with the emission of medium UV dose. Due to relative obscurity of the effluent as well as its high concentration of heavy metal, the UV irradiance does not work properly. As a result, O_3 and UV do not have an effective synergy to lower the level of COD when the effluent sample possesses low transparency and high heavy metal content.

Assuming a constant value of 0 (the coded value) for O_3, both of UV and H_2O_2 have been modified to observe their effects on the level of COD. The results are summarized in Figure 6b. While Figure 6a does not show a recognizable symmetry and pattern for the response, Figure 6b demonstrates a linear change in the value of COD (as the studied response). In other words, the higher the amounts of UV and H_2O_2 used, the lower the COD levels generated. Obviously, symmetry exists throughout the plot. Although the overall COD reduction comes from using UV, and H_2O_2 is lower than that when UV and O_3 simultaneously are employed, the COD reduction with UV and H_2O_2 is done recognizably well, which shows the strong positive synergy between UV and H_2O_2.

The advantages and disadvantages of utilizing H_2O_2 can be perceived by observing Figure 6c. In spite of the fact that the use of O_3 itself improves the level of purification and also reduces the level of COD, H_2O_2 is effective if only the optimized volume of H_2O_2 is added to the effluent. Higher volumes of H_2O_2 decrease its removal power since a competition between carbonate and bicarbonate in the mixture comes into the scene, which in turn prevents AOPs from generating more HO• radicals. Furthermore, similar to Figure 6a, Figure 6c does not indicate a predictable response pattern. The interaction of H_2O_2 and O_3 changes with varying the concentration of injected O_3, and hence, an optimization procedure needs to be conducted to figure the appropriate volume of H_2O_2 required to reduce the COD level.

3.4. Optimization Conditions

The optimized amount of AOPs can be obtained by Minitab's response optimizer. In fact, by employing Equation (5), Minitab can introduce the best AOP's combination according to which the wastewater from the aeration industry can be cleaned more effectively. The optimization outcomes and the after-treatment COD level are presented in Table 5. The optimized value for UV emission is 86 mWs/cm^2, which can be counted as a medium UV emission. The used wastewater was too dark and obscure, and therefore, UV was not effective enough to remove the pollution compared with O_3 and H_2O_2 when high chemical additives were added to the wastewater. In addition, the level of optimized O_3 should be higher than H_2O_2 level to decrease competition between bicarbonate and carbonate ions in the fluid. Thus, while 8 mg/L O_3 as the highest applied amount is the optimum point for the COD reduction, 40 mg/L H_2O_2 as a medium value is recommended by Minitab.

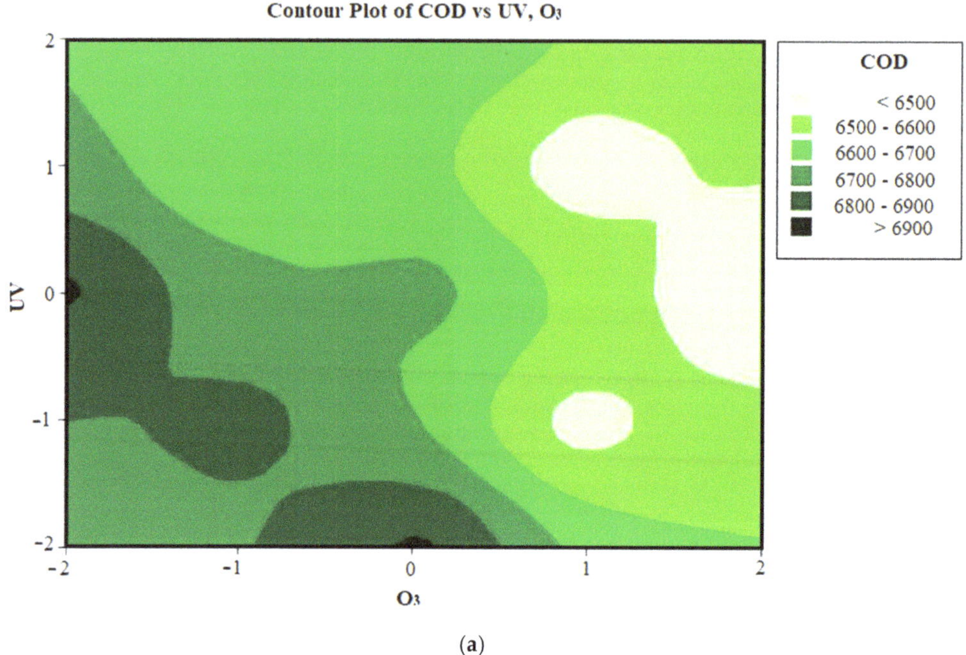

(**a**)

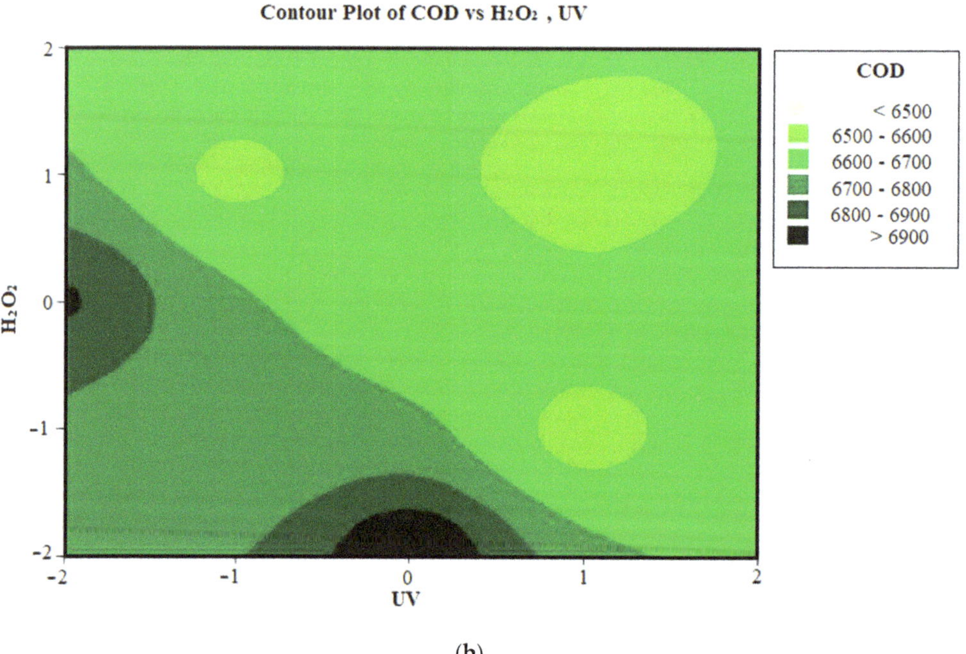

(**b**)

Figure 6. *Cont.*

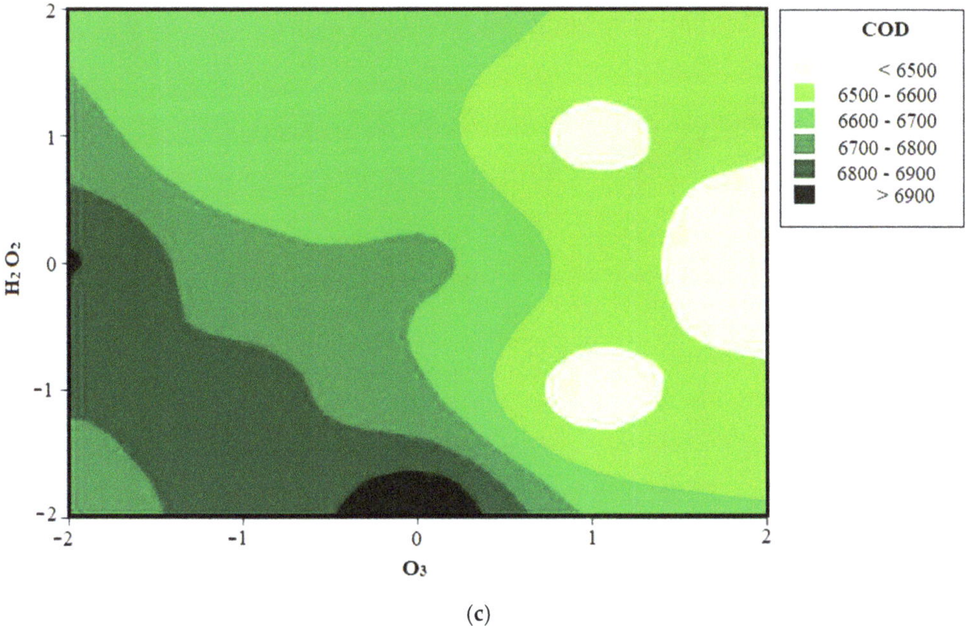

(c)

Figure 6. The contour plots showing the reduction of COD in the effluent after the AOP treatment, when (**a**) O_3 and UV change (**b**) H_2O_2 and UV change (**c**) H_2O_2 and O_3 change.

Table 5. The optimized values defined for AOP.

UV (mWs/cm^2)	O_3 (mg/L)	H_2O_2 (mg/L)	COD (mgO$_2$/L) after Treatment	COD Percent Change (%)	Energy Consumption (kWh/m^3)
86	8	40	6313	51.5	22.7

Figure 7 shows three 3D plots of the response graphs of applying O_3 and UV at three levels of H_2O_2. Based on the output depicted in Figure 7, it can be concluded that the efficiency of COD removal will be optimum when the O_3 and H_2O_2 quantities are around 2, and -1, respectively, while the UV irradiance varies between -1.5 and -2. By converting the mentioned coded numbers using Table 2 to real values, the optimum values mentioned in Table 5 are found. This verifies the optimum AOP values in Table 5.

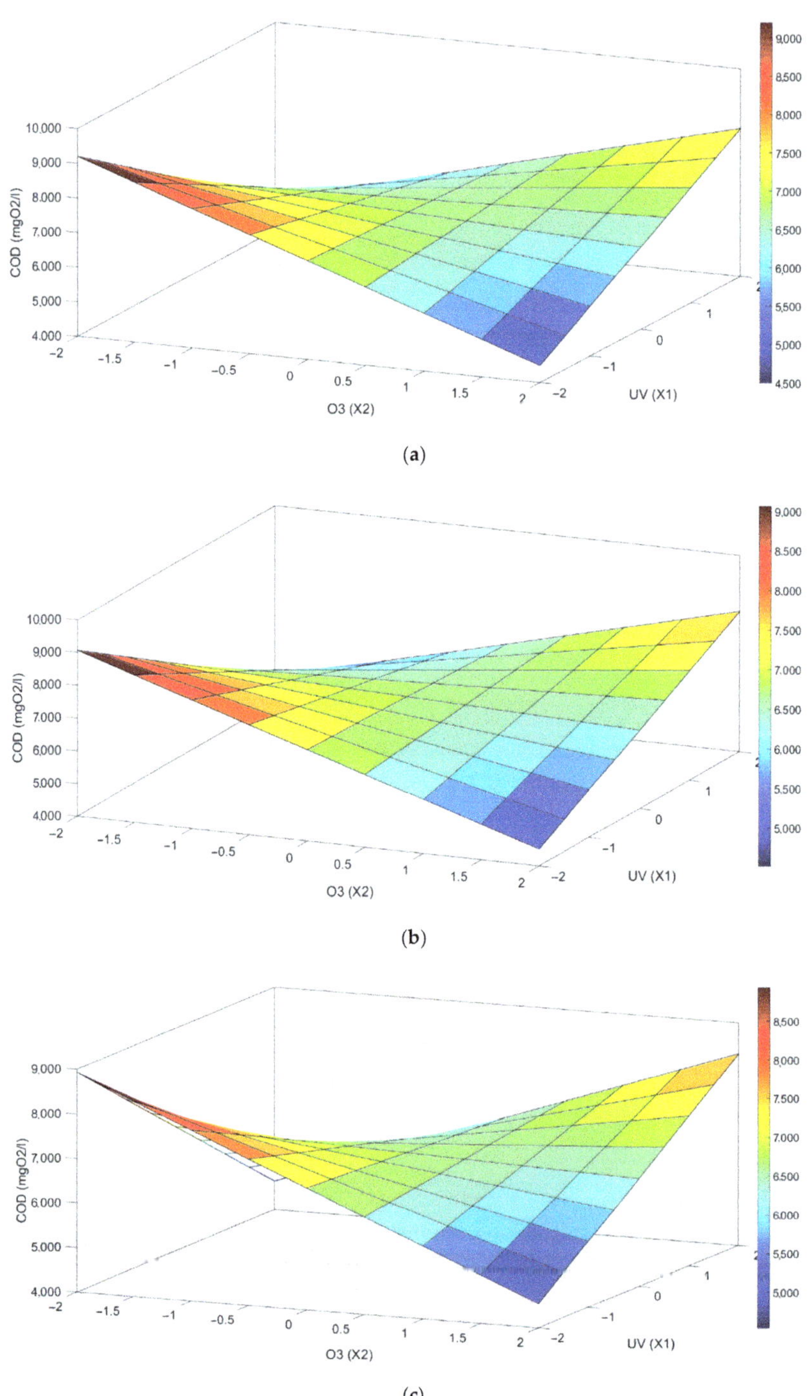

Figure 7. 3D plots of response surface graphs for the combined effect of H_2O_2, O_3, and UV on COD removal. (**a**) H_2O_2 (X_3) = -1, (**b**) H_2O_2 (X_3) = 0, and (**c**) H_2O_2 (X_3) = 1.

4. Conclusions

A total of 20 sets of experimental tests were conducted in order to study the performance of AOPs, including H_2O_2, O_3, and UV, when disinfecting the effluent released from an aeration industry. The experimental tests were planned by the Minitab software. The mutual effects of the additives over the water treatment process were examined through analyzing the contour plots. It was observed that the use of H_2O_2 and O_3 together can remarkably enhance the cleaning process. However, when the quantity of H_2O_2 exceeded what is known as the optimized amount, it unexpectedly lost the potential to reduce the level of COD in the wastewater. It was shown that at constant levels of O_3 and UV, doubling the amount of H_2O_2 from 45 to 95 mg/L only decreased the wastewater COD insignificantly from 50.1% to 50.2% while the energy consumption was increased from 33.5 to 44.4 kWh/m^3. Similar to the case of H_2O_2, the UV irradiance yielded the same results when it was employed in the treatment process together with a high level of O_3. Unlikely, H_2O_2 or UV irradiance alone linearly decreased the level of COD in the effluent when a low concentration of O_3 was used. The results showed that by increasing the dose of UV emission from 40 to 400 mWs/cm^2, the COD level was decreased from 46.3% to 48.9% when the chemical additives were low or medium. However, this positive effect could not be extended when the additive levels are high. According to their effects on the process, a polynomial correlation was derived through which the response (COD value) could be found by simply substituting the variables' quantities in the correlation. Following that, to obtain the optimized quantities of H_2O_2, O_3, and UV altogether, the optimization task was completed with the help of Minitab's response optimizer. A UV emission of 86 mWs/cm^2, O_3 dose of 8 mg/L, and H_2O_2 dose of 40 mg/L were the optimum values reported by Minitab. It was found that by using the optimized AOP values, the COD reduction reached the highest value of 51.5% with the lowest energy consumption.

Author Contributions: Conceptualization, M.R.; methodology, M.R., M.S. and F.T.; software, M.R.; validation, M.R. and M.S.; formal analysis, M.R. and M.S.; investigation, M.R.; writing—original draft preparation, A.R. and M.S.; writing—review and editing, A.R. and M.S.; supervision, F.T. All authors have read and agreed to the published version of the manuscript.

Funding: This research received no external funding.

Institutional Review Board Statement: Not applicable.

Informed Consent Statement: Not applicable.

Data Availability Statement: The data used to support the findings of this study are available from the corresponding author upon request.

Conflicts of Interest: The authors declare no conflict of interest.

References

1. Saffari, P.R.; Salarian, H.; Lohrasbi, A.; Salehi, G.; Manesh, M.H.K. Optimization of a Thermal Cracking Reactor Using Genetic Algorithm and Water Cycle Algorithm. *ACS Omega* **2022**, *7*, 12493–12508. [CrossRef]
2. Rafiee, M.; Salehi, E.; Sharifi, K.; Mohammadi, A.H.; Rohani, A.; Sabeti, M. Enhancement of biogas production efficiency using appropriate low-temperature pretreatments of municipal treatment plants' excess sludge. *Environ. Prog. Sustain. Energy* **2019**, *38*, e13072. [CrossRef]
3. Halder, J.N.; Islam, M.N. Water pollution and its impact on the human health. *J. Environ. Hum.* **2015**, *2*, 36–46. [CrossRef]
4. Ameta, S.C.; Ameta, R. *Advanced Oxidation Processes for Wastewater Treatment: Emerging Green Chemical Technology*; Academic Press: Cambridge, MA, USA, 2018.
5. Ganiyu, S.O.; Van Hullebusch, E.D.; Cretin, M.; Esposito, G.; Oturan, M.A. Coupling of membrane filtration and advanced oxidation processes for removal of pharmaceutical residues: A critical review. *Sep. Purif. Technol.* **2015**, *156*, 891–914. [CrossRef]
6. Muruganandham, M.; Swaminathan, M. Advanced oxidative decolourisation of Reactive Yellow 14 azo dye by UV/TiO2, UV/H2O2, UV/H2O2/Fe2+ processes—A comparative study. *Sep. Purif. Technol.* **2006**, *48*, 297–303. [CrossRef]
7. Gorito, A.M.; Pesqueira, J.F.; Moreira, N.F.; Ribeiro, A.R.; Pereira, M.F.R.; Nunes, O.C.; Almeida, C.M.R.; Silva, A.M. Ozone-based water treatment (O3, O3/UV, O3/H2O2) for removal of organic micropollutants, bacteria inactivation and regrowth prevention. *J. Environ. Chem. Eng.* **2021**, *9*, 105315. [CrossRef]

8. Issaka, E.; Amu-Darko, J.N.-O.; Yakubu, S.; Fapohunda, F.O.; Ali, N.; Bilal, M. Advanced catalytic ozonation for degradation of pharmaceutical pollutants—A review. *Chemosphere* **2022**, *289*, 133208. [CrossRef]
9. Mohajerani, M.; Mehrvar, M.; Ein-Mozaffari, F. An overview of the integration of advanced oxidation technologies and other processes for water and wastewater treatment. *Int. J. Eng.* **2009**, *3*, 120–146.
10. Bokare, A.D.; Choi, W. Review of iron-free Fenton-like systems for activating H2O2 in advanced oxidation processes. *J. Hazard. Mater.* **2014**, *275*, 121–135. [CrossRef]
11. Yao, H. Application of advanced oxidation processes for treatment of air from livestock buildings and industrial facilities. *Tech. Rep. Biol. Chem. Eng.* **2013**, *2*, 1–36.
12. Yao, H.; Feilberg, A. Application of Advanced Oxidation Processes for Treatment of Air from Livestock Facilities. In Proceedings of the International Conference on Agricultural Engineering, Aarhus, Denmark, 26 June 2016.
13. Andreozzi, R.; Caprio, V.; Insola, A.; Marotta, R. Advanced oxidation processes (AOP) for water purification and recovery. *Catal. Today* **1999**, *53*, 51–59. [CrossRef]
14. Cuerda-Correa, E.M.; Alexandre-Franco, M.F.; Fernández-González, C. Advanced oxidation processes for the removal of antibiotics from water. An overview. *Water* **2019**, *12*, 102. [CrossRef]
15. Buthiyappan, A.; Aziz, A.R.A.; Daud, W.M.A.W. Degradation performance and cost implication of UV-integrated advanced oxidation processes for wastewater treatments. *Rev. Chem. Eng.* **2015**, *31*, 263–302. [CrossRef]
16. Madihi-Bidgoli, S.; Asadnezhad, S.; Yaghoot-Nezhad, A.; Hassani, A. Azurobine degradation using Fe_2O_3@multi-walled carbon nanotube activated peroxymonosulfate (PMS) under UVA-LED irradiation: Performance, mechanism and environmental application. *J. Environ. Chem. Eng.* **2021**, *9*, 106660. [CrossRef]
17. Hassani, A.; Eghbali, P.; Mahdipour, F.; Wacławek, S.; Lin, K.-Y.A.; Ghanbari, F. Insights into the synergistic role of photocatalytic activation of peroxymonosulfate by UVA-LED irradiation over $CoFe_2O_4$-rGO nanocomposite towards effective Bisphenol A degradation: Performance, mineralization, and activation mechanism. *Chem. Eng. J.* **2023**, *453*, 139556. [CrossRef]
18. Kommineni, S.; Zoeckler, J.; Stocking, P.; Liang, A.; Flores, R.; Rodriguez, T.; Brown, R.; Brown, A. *Advanced Oxidation Processed*; Center for Groundwater Restoration and Protection, National Water Research Institute; Academia: San Francisco, CA, USA, 2000.
19. Karci, A.; Arslan-Alaton, I.; Bekbolet, M.; Ozhan, G.; Alpertunga, B. H2O2/UV-C and Photo-Fenton treatment of a nonylphenol polyethoxylate in synthetic freshwater: Follow-up of degradation products, acute toxicity and genotoxicity. *Chem. Eng. J.* **2014**, *241*, 43–51. [CrossRef]
20. Antonopoulou, M.; Evgenidou, E.; Lambropoulou, D.; Konstantinou, I. A review on advanced oxidation processes for the removal of taste and odor compounds from aqueous media. *Water Res.* **2014**, *53*, 215–234. [CrossRef]
21. Gogate, P.R.; Pandit, A.B. A review of imperative technologies for wastewater treatment I: Oxidation technologies at ambient conditions. *Adv. Environ. Res.* **2004**, *8*, 501–551. [CrossRef]
22. Gogate, P.R.; Pandit, A.B. A review of imperative technologies for wastewater treatment II: Hybrid methods. *Adv. Environ. Res.* **2004**, *8*, 553–597. [CrossRef]
23. Oller, I.; Malato, S.; Sánchez-Pérez, J. Combination of advanced oxidation processes and biological treatments for wastewater decontamination—A review. *Sci. Total Environ.* **2011**, *409*, 4141–4166. [CrossRef]
24. Bezerra, M.A.; Santelli, R.E.; Oliveira, E.P.; Villar, L.S.; Escaleira, L.A. Response surface methodology (RSM) as a tool for optimization in analytical chemistry. *Talanta* **2008**, *76*, 965–977. [CrossRef] [PubMed]
25. Rekhate, C.V.; Srivastava, J.K. Effectiveness of O_3/Fe^{2+}/H_2O_2 process for detoxification of heavy metals in municipal wastewater by using RSM. *Chem. Eng. Process. Process Intensif.* **2021**, *165*, 108442. [CrossRef]
26. Masouleh, S.Y.; Mozaffarian, M.; Dabir, B.; Ramezani, S.F. COD and ammonia removal from landfill leachate by UV/PMS/Fe2+ process: ANN/RSM modeling and optimization. *Process Saf. Environ. Prot.* **2022**, *159*, 716–726. [CrossRef]
27. Rice, E.W.; Baird, R.B.; Eaton, A.D.; Clesceri, L.S. *Standard Methods for the Examination of Water and Wastewater*; American Public Health Association: Washington, DC, USA, 2012; Volume 10.
28. Da Silva, G.H.R.; Daniel, L.A.; Contrera, R.C.; Bruning, H. Anaerobic effluent disinfected with ozone/hydrogen peroxide efluente anaeróbio desinfetado com ozônio/peróxido de hidrogênio. *Cienc. Y Eng. Sci. Eng. J.* **2012**, *21*, 1–7.
29. Farooq, S.; Engelbrecht, R.S.; Chian, E.S. Influence of temperature and UV light on disinfection with ozone. *Water Res.* **1977**, *11*, 737–741. [CrossRef]
30. Hassen, A.; Mahrouk, M.; Ouzari, H.; Cherif, M.; Boudabous, A.; Damelincourt, J.J. UV disinfection of treated wastewater in a large-scale pilot plant and inactivation of selected bacteria in a laboratory UV device. *Bioresour. Technol.* **2000**, *74*, 141–150. [CrossRef]
31. Hébert, N.; Gagné, F.; Cejka, P.; Bouchard, D.; Hausler, R.; Cyr, D.; Blaise, C.; Fournier, M. Effects of ozone, ultraviolet and peracetic acid disinfection of a primary-treated municipal effluent on the immune system of rainbow trout (Oncorhynchus mykiss). *Comp. Biochem. Physiol. Part C Toxicol. Pharmacol.* **2008**, *148*, 122–127. [CrossRef]
32. Macauley, J.J.; Qiang, Z.; Adams, C.D.; Surampalli, R.; Mormile, M.R. Disinfection of swine wastewater using chlorine, ultraviolet light and ozone. *Water Res.* **2006**, *40*, 2017–2026. [CrossRef]
33. Rosal, R.; Rodríguez, A.; Perdigón-Melón, J.A.; Petre, A.; García-Calvo, E. Oxidation of dissolved organic matter in the effluent of a sewage treatment plant using ozone combined with hydrogen peroxide (O3/H2O2). *Chem. Eng. J.* **2009**, *149*, 311–318. [CrossRef]

Article

Enhanced Removal of Doxycycline by Simultaneous Potassium Ferrate(VI) and Montmorillonite: Reaction Mechanism and Synergistic Effect

Hangli Zhang [1], Shujuan Wang [2], Ji Shu [3] and Hongyu Wang [3,*]

[1] College of Civil Engineering and Architecture, Zhejiang Tongji Vocational College of Science and Technology, Hangzhou 311231, China
[2] College of Civil Engineering and Architecture, Zhejiang University of Technology, Hangzhou 310014, China
[3] College of Environment, Zhejiang University of Technology, Hangzhou 310014, China
* Correspondence: hywang@zjut.edu.cn

Abstract: Doxycycline (DOX), a typical antibiotic, is harmful to aquatic ecosystems and human health. This study presents DOX removal by potassium ferrate (Fe(VI)) and montmorillonite and investigates the effect of Fe(VI) dosage, reaction time, initial pH value, montmorillonite dosage, adsorption pH, time and temperature on DOX removal. The results show that DOX removal increases when increasing the Fe(VI) dosage, with the optimal condition for DOX removal (~97%) by Fe(VI) observed under a molar ratio ([Fe(VI)]:[DOX]) of 30:1 at pH 7. The reaction of DOX with Fe(VI) obeyed second-order kinetics with a rate constant of $10.7 \pm 0.45 \ \mathrm{M^{-1} \ s^{-1}}$ at pH 7. The limited promotion (~4%) of DOX adsorption by montmorillonite was observed when the temperature increased and the pH decreased. Moreover, the synergetic effect of Fe(VI) and montmorillonite on DOX removal was obtained when comparing the various types of dosing sequences (Fe(VI) oxidation first and then adsorption; adsorption first and then Fe(VI) oxidation; simultaneous oxidation and adsorption). The best synergistic effect of DOX removal (97%) was observed under the simultaneous addition of Fe(VI) and montmorillonite, maintaining the Fe(VI) dosage (from 30:1 to 5:1). Five intermediates were detected during DOX degradation, and a plausible DOX degradation pathway was proposed.

Keywords: advanced oxidation; potassium ferrate(VI); doxycycline degradation; synergistic reaction mechanism

Citation: Zhang, H.; Wang, S.; Shu, J.; Wang, H. Enhanced Removal of Doxycycline by Simultaneous Potassium Ferrate(VI) and Montmorillonite: Reaction Mechanism and Synergistic Effect. *Water* **2023**, *15*, 1758. https://doi.org/10.3390/w15091758

Academic Editors: Gassan Hodaifa, Antonio Zuorro, Joaquín R. Dominguez, Juan García Rodríguez, José A. Peres, Zacharias Frontistis and Mha Albqmi

Received: 26 February 2023
Revised: 28 April 2023
Accepted: 1 May 2023
Published: 3 May 2023

1. Introduction

The use of drugs and personal care products (PPCPs) as emerging contaminants has attracted extensive attention from around the world [1]. The "pseudo-durability" of antibiotics caused by the large-scale use of antibiotics in human, veterinary, agricultural and aquaculture activities [2], as well as the residual and continuous testing of antibiotics in the environment, have also attracted widespread concern [3–5]. Doxycycline (DOX), as one of the most important antibiotics, has been widely used and transferred to the aquatic environment, resulting in a high concentration in pharmaceutical wastewater, ranging from 3.9 ng/L to 500.0 µg/L [6–9]. Studies have also indicated that residual DOX in water threatens aquatic ecosystems and human health in the form of endocrine infections, as well as promoting bacterial resistance [10]. Therefore, the remediation of DOX is urgent.

Ferrate(VI) (Fe(VI)), a green and environmentally friendly agent, has multiple functions, such as oxidation, flocculation and adsorption, leading to its increasing use in water treatment. Wang et al. conducted a study of the mechanism of fluoroquinolone antibiotic and sulfonamides degradation by Fe(VI), which revealed the reaction rate constant for the fluoroquinolone reaction with Fe(VI) [11,12]. Su et al. focused on the trimethoprim and carbamazepine degradation by Fe(VI) and proposed a degradation pathway [13]. Yunho et al. assessed a technology for enhancing the dual functions of Fe(VI) to oxidize micropollutants by the formation of ferric phosphates [14]. However, its redox potential decreases from +2.2 to +0.7 V when the environmental pH turns from acid to alkaline [13,15]. This results in the inefficient mineralization of target contaminations, leaving potentially toxic small molecular substances. For this reason, novel technology must be introduced to account for these limitations.

Adsorption technology plays an important role in wastewater treatment. The most important factor of adsorption between adsorbents and adsorbates is their hydrophobic and electrostatic interaction [16]. Montmorillonite, as a natural silicate mineral, has a relatively large specific surface area (SSA), with a high cation exchange capacity (CEC), which can be used as an efficient adsorption agent [17]. Therefore, the addition of montmorillonite may be an effective approach to the adsorptive removal of small molecular substances stemming from Fe(VI) degradation. Furthermore, montmorillonite can not only adsorb target pollutants in solution, but can also enhance the overall adsorption and flocculation capability of Fe(VI) by activating adsorption bridging, net flapping and sweeping and electrical neutralization with in situ Fe(III) particles from Fe(VI) decomposition.

This work's objectives were to: (i) explore the degradation performance of Fe(VI) by assessing the oxidation efficiency and mineralization of DOX; (ii) investigate the adsorption performance of DOX by montmorillonite; (iii) elucidate the mechanism of the synergistic effect of the oxidation and adsorption of DOX by Fe(VI) and montmorillonite; and (iv) identify the degradation products of DOX and propose a degradation pathway for DOX, considering the molecular structure of target pollutants and the oxidation mechanism of Fe(VI).

2. Materials and Methods

2.1. Chemicals and Materials

Doxycycline (DOX) with a purity of 98% and montmorillonite with a purity of over 95% were purchased from Aladdin (Shanghai, China). Fe(VI) was prepared by electrochemical methods with a purity of at least 90% [18]. Other chemical reagents were of analytical grade, purchased at Xiaoshan Chemical Reagent Factory (Hangzhou, China). All chemical solutions were prepared in Milli-Q water, with the initial concentration of DOX being 11.25 μM. The Fe(VI) stock solution was prepared in a borate buffer solution composed of 81% 0.05 M borax and 19% 0.2 M boric acid, with a pH of 9.2.

2.2. DOX Removal by Fe(VI)

In this study, the Fe(VI) concentration was determined by ABTS methods and detailed information is given in Text S1 [19]. Subsequently, a 1.0 mM Fe(VI) stock solution was prepared in borate buffer solution at pH 9.1 prior to the experiment. A total of 50 mL of 11.25 μM DOX was dispensed into a beaker and stirred using a magnetic stirrer (Thermo Scientific, MA, USA) at 200 rpm. The pH of the solution was adjusted using 0.1 M hydrochloric acid and 0.05 M sodium hydroxide to various levels, ranging from 4.0 to 10.0. Subsequently, various volumes of a 1 mM Fe(VI) stock solution were introduced to the reaction to maintain an initial Fe(VI) concentration range of 56.25 to 337.50 μM. At predetermined intervals, samples were collected, the reaction was quenched with 30.0 μL of 0.1 M sodium thiosulfate and filtered through a 0.22 μm membrane filter (Xingya, Shanghai, China). These samples were further used to quantify the residual DOX concentration and total organic carbon (TOC), to determine the DOX removal efficiency and mineralization. To exclude the effect of filtering, the recovery rate after filtering was measured (Figure S1). The result showed that the recovery rate was close to 100% with various initial DOX concentration injections, indicating that the filtering process did not affect the subsequent experiments.

2.3. DOX Removal by Montmorillonite Adsorption

Before the experiment, 50 mL of 11.25 μM DOX was added to a beaker. The pH of the solution was adjusted, as described in Section 2.2. Subsequently, the solution was supplemented with montmorillonite powder. The mixture was then stirred using a magnetic stirrer (IKA, Schwarzwald, Germany) at 250 rpm, at a temperature of 25 °C. After 24 h of adsorption, the supernatant was withdrawn and centrifugated at 200 rpm for 10 min. The centrifuged sample was then filtered through a 0.22 μm membrane filter (MF-Millipore, Hongkong, China), to determine the DOX concentration.

2.4. Kinetics Study

The reaction kinetics of Fe(VI) with DOX were investigated at 20 °C under pseudo-order conditions, with an excess Fe(VI) concentration of [Fe(VI)]:[DOX] = 49.8:1 to 199.2:1 at pH 7.0. The initial concentration of DOX was 2.25 μM. Samples were taken from the mixture at pre-determined time intervals, up to 1800 s, and then immediately quenched with 30.0 μL of a 0.10 M sodium thiosulfate solution. The rate expression for the reaction of Fe(VI) with DOX is represented by Equation (1):

$$r = -d[DOX]/dt = -k[ferrate(VI)]^a \cdot [DOX]^b \tag{1}$$

The reaction rate between Fe(VI) and DOX is represented by r, while k stands for the reaction rate constant. The parameters a and b in Equation (1) denote the reaction orders relating to the concentrations of Fe(VI) and DOX, respectively. It was widely reported that the oxidation of an organic contaminant by Fe(VI) follow second-order kinetics [18–20]. Therefore, Equation (1) can be reduced to Equation (2):

$$\ln\left(\frac{[DOX]_t}{[DOX]_0}\right) = -k_{app}\int_0^t [Fe(VI)]dt \tag{2}$$

where $[DOX]_t$ and $[DOX]_0$ are the concentrations of DOX at different reaction time, t and zero, respectively. All the samples were withdrawn and analyzed in triplicate.

2.5. Analytical Methods
2.5.1. The DOX Concentration

The DOX concentration was determined using high-performance liquid chromatography (HPLC) (Shimadzu, LC-20A), as referred to in the previous literature [21].Under the conditions in Table S1, the characteristic peak of DOX appeared at about 3.9 min, and the calibration curve of DOX concentration is presented in Figure S2.

2.5.2. Identification of Oxidized Products

The intermediate products of DOX were characterized by GC–MS with the conditions in Table S1 (Shimadzu GCMS-QP2010plus, Kyoto, Japan), using parameters referenced previously [22]. All samples were acidified with hydrochloric acid to pH < 1 and concentrated by using a liquid–liquid method, extracted using trichloromethane.

2.6. Characterization

The specific surface area and porosity was carried out using ASAP 2460 devices (Micromeritics, Atlanta, GA, USA) by the Brunauer–Emmett–Teller (BET) and Barret–Joyner–Halender (BJH) methods. Fourier transform infrared (FTIR) spectra were measured by Nicolet iS20 devices (400–4000 cm^{-1}, Thermo Scientific, MA, USA).

3. Results and Discussion
3.1. DOX Removal by Fe(VI) Oxidation
3.1.1. The Impact of Fe(VI) Dosage

In this experiment, various quantities of Fe(VI) were introduced to a solution containing 11.25 μM DOX at pH 7.0, with the objective of removing it within 60 min. The result in Figure 1 showed that DOX removal increased from 82.47 to 96.98% as the molar ratio ([Fe(VI)]:[DOX]) increased from 5:1 to 30:1. This demonstrated that DOX was almost fully removed by Fe(VI) at a molar ratio of 30:1. This could be attributed to the increased probability of collisions between Fe(VI) and DOX molecules when increasing the Fe(VI) dosage. In addition, the results showed that DOX removal can reach more than 70% within 5 min under all conditions. However, at lower Fe(VI) doses, DOX removal was quicker in the initial stages of the reaction, which might be attributed to its rapid self-decomposition (Equations (3) and (4)) as the Fe(VI) concentration increased [22]. This led to the ability of per unit Fe(VI) to remove DOX reduced [23].

$$2H_3FeO_4^+ \rightarrow [H_4Fe_2O_7]^{2+} + H_2O \tag{3}$$

$$[H_4Fe_2O_7]^{2+} + 2H^+ + 6H_2O \rightarrow Fe(OH)_2(H_2O)_8^{4+} + {}^3\!/_2O_2 \tag{4}$$

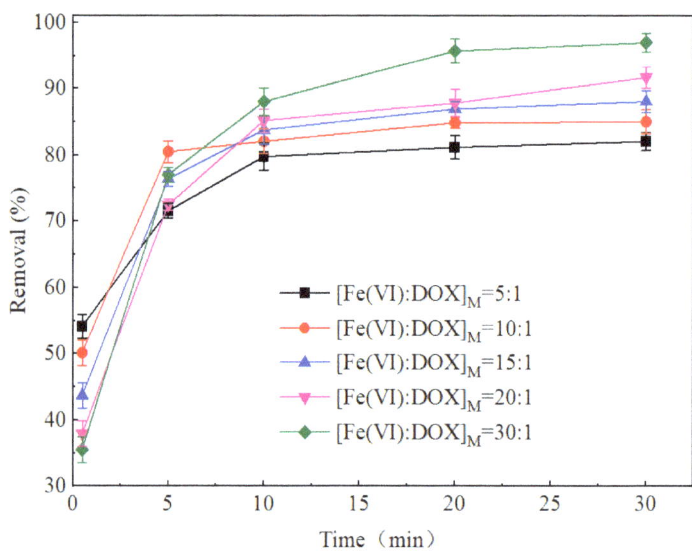

Figure 1. Effect of Fe(VI) dose on DOX degradation.

3.1.2. Impact of pH

Given that pH greatly influences the redox potential and stability of Fe(VI), the impact of pH on the removal of DOX by Fe(VI) was explored (Figure 2). In the experiment, the pH ranged from 4.0 to 10.0, the initial concentration of DOX was 11.25 µM and the molar ratio ([Fe(VI)]:[DOX]) was set at 5:1. The reaction between DOX and Fe(VI) ended within 10 min, and DOX removal reached more than 80% under all pH conditions. With the increase in pH, DOX removal showed a tendency to decrease first and then rise. At pH 7.0, DOX removal was the lowest (80.92%), and reached its maximum at pH 9.0 (97.48%). DOX removal under the other pH conditions were 91.04 (pH 4.0), 88.53 (pH 5.0), 86.61 (pH 6.0), 95.12 (pH 8.0) and 93.01% (pH 10.0). The reaction rates between DOX and Fe(VI) gradually decreased as pH increased during the initial reaction period (within the first 2 min), but DOX removal by Fe(VI) was more efficient under alkaline conditions. This may attributed to the higher redox potential (+2.2 V) of Fe(VI) [15] under acidic conditions compared to alkaline conditions (0.72 eV). Another reasonable explanation of the decrease in the reaction rate under acid conditions with the increase in pH could be that it is due to the conversion of the protonated form of Fe(VI) ($HFeO_4^-$) to the less reactive deprotonated Fe(VI) species (FeO_4^{2-}) in alkaline conditions [24]. Moreover, the potential explanation for improved DOX removal at pH 9 could be due to the pH value being close to DOX's dissociation constant of 9.15 (Table S2) [25], causing the best DOX degradation effect at pH 9.

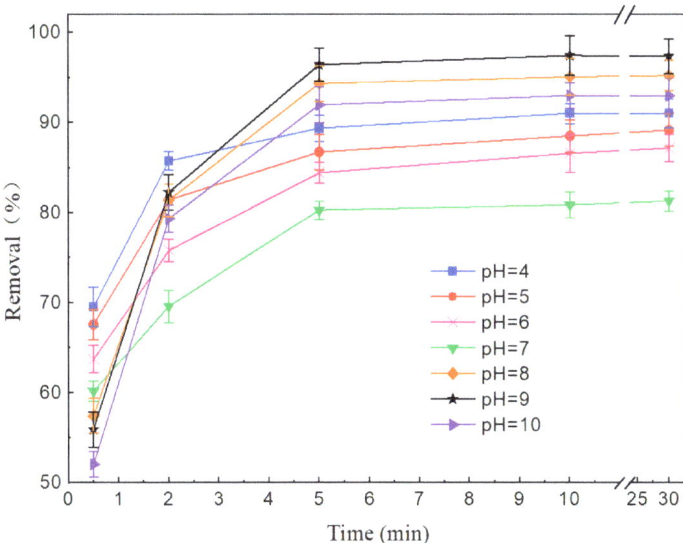

Figure 2. Effect of different pH values on DOX degradation.

3.1.3. Kinetics

Initially, the decline over time in both DOX and Fe(VI) in the reaction with excess Fe(VI) was determined at pH 7.0. The changes of Fe(VI) and DOX concentration over time with different dosage of Fe(VI) in Figure S3. The plots of $\ln([DOX]_t/[DOX]_0)$ versus $\int_0^t [Fe(VI)]$ are displayed in Figure 3. A linear relationship can be observed ($R^2 = 0.99$), indicating that the reaction of Fe(VI) with DOX followed second-order kinetics. The slope represents the second-order rate constant (k_{app}), which was 10.7 ± 0.45 M^{-1} s^{-1}.

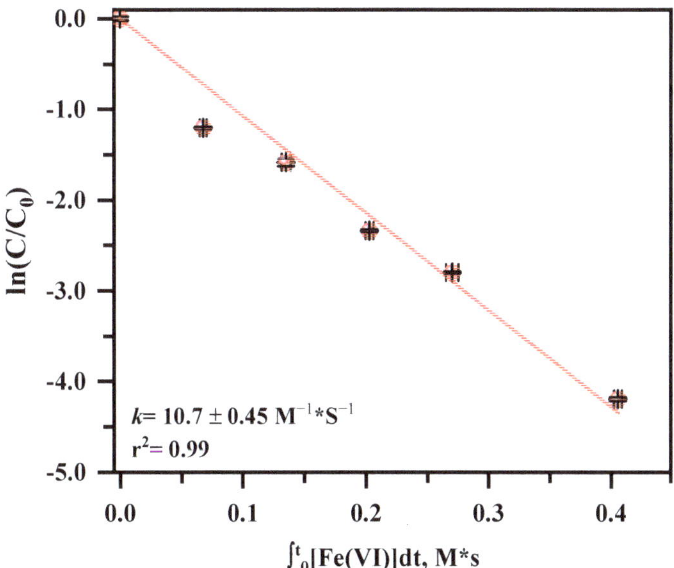

Figure 3. Kinetics analyses of DOX degradation by potassium Fe(VI).

3.1.4. Mineralization

In order to investigate the degradation degree of DOX by Fe(VI), the mineralization changes of DOX were recorded (Figure 4) [26–29]. The degradation of DOX reached a maximum within 5 min. At this time, DOX removal exceeded 77.0%, while TOC removal was only 39.2%. At 30 min, DOX removal reached 83%, while TOC removal in the solution barely increased. The results showed that, although Fe(VI) was able to remove DOX in a short time, it was limited in its ability eliminate DOX degradation products. Fe(VI) only oxidized DOX into other organic substances, and the mineralization degree was low. One study showed that that the mineralization efficiency of pristine Ag/AgCl and composites ACM-5 reached 18.97% and 46.65%, respectively, within 60 min [26]. Another study found that the transformation products of DOX with a small molecular weight were not observed even after 60 min of ozonation, suggesting that mineralization might be difficult [30]. Additionally, another study did not observe DOX degradation by peroxydisulfate oxidation [31]. The reasonable explanation for this may be that the structure of the naphthalene benzene ring in DOX is very stable and difficult to degrade, which is investigated further in Section 3.4.

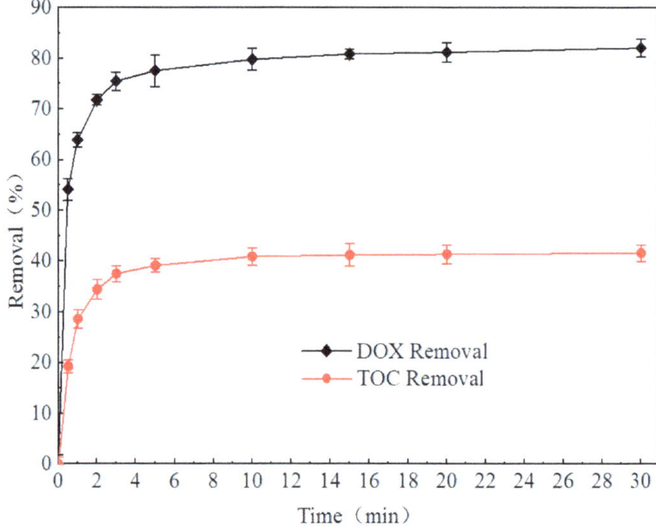

Figure 4. TOC and DOX removal at different reaction times.

3.2. Adsorption of DOX by Montmorillonite

3.2.1. Impact of Montmorillonite Dosage

To investigate the impact of montmorillonite dosage on DOX removal, different concentrations of montmorillonite were added to remove DOX in a solution at pH 7.0. The adsorption time was 24 h, with an initial DOX concentration of 11.25 µM. As the montmorillonite dosage increased from 100 to 600 mg/L, DOX removal by adsorption increased from 7.58 to 12.57%. Meanwhile, the absolute value of the zeta potential gradually increased from 15 to 27 mV, indicating that montmorillonite was stable and dispersed in the solution, resulting in an increase in DOX adsorption (Figure 5).

3.2.2. Impact of pH

In order to explore the effect of pH on the DOX adsorption efficiency by montmorillonite, the initial pH of the experimental solution was set at 2, 4, 6, 7, 8 and 10. All experiments were conducted at room temperature, the reaction time was 24 h, the initial concentration of DOX was 5 mg/L and the concentration of montmorillonite was 100 mg/L. With the increase in pH from 2.0 to 10.0, DOX removal decreased (Figure 6).

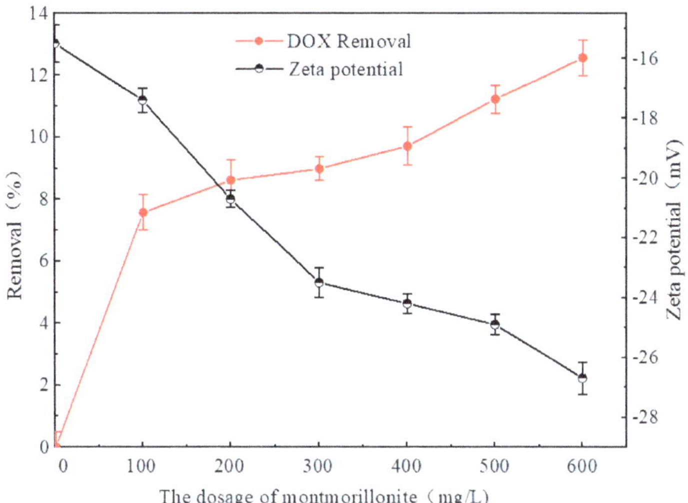

Figure 5. Effect of different dosages of montmorillonite on DOX removal.

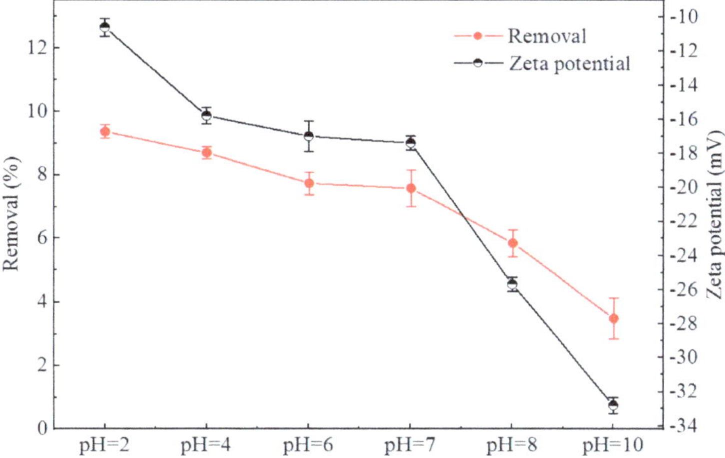

Figure 6. Effect of different pH values of montmorillonite on DOX removal.

Montmorillonite had the best effect on DOX removal (9.36%) at pH 7.0. Its removal under different pH conditions were: 8.7 (pH 4.0), 7.74 (pH 6.0), 7.58 (pH 7.0), 5.85 (pH 8.0) and 3.49% (pH 10.0). The adsorption of DOX by montmorillonite under acidic conditions was significantly higher than under alkaline conditions. For this result, one reasonable explanation is that the adsorption of DOX is influenced by DOX dissociation (pK_{a1} = 3.02, pK_{a2} = 7.97 and pK_{a3} = 9.15) [25]. The dominant species of DOX was ionized DOX^+ at pH 3.0. Concurrently, the zeta potential of the solution containing montmorillonite was negative, indicating that DOX could have been adsorbed via charge neutralization on the negatively charged surface of montmorillonite. Therefore, DOX adsorption by montmorillonite was the highest at pH 2. However, the dominant DOX species gradually turned from ionized DOX^+ to deprotonated DOX^- or DOX^{2-} as pH increased and the Zeta potential of montmorillonite surface become more negative (Figure 6) [32]. On account of the principle of heterogeneous charge repulsion, DOX adsorption by montmorillonite gradually decreased.

3.2.3. Impact of Adsorption Time

Adsorption time is one of the main factors affecting adsorption efficiency. In this experiment, 100 mg/L of montmorillonite was placed into a solution containing DOX (5 mg/L, 11.25 μM) at pH 7.0. Additionally, samples at predetermined times—0, 0.5, 1, 2, 4, 6, 10, 12, 14, 18 and 24 h—were analyzed (Figure 7).

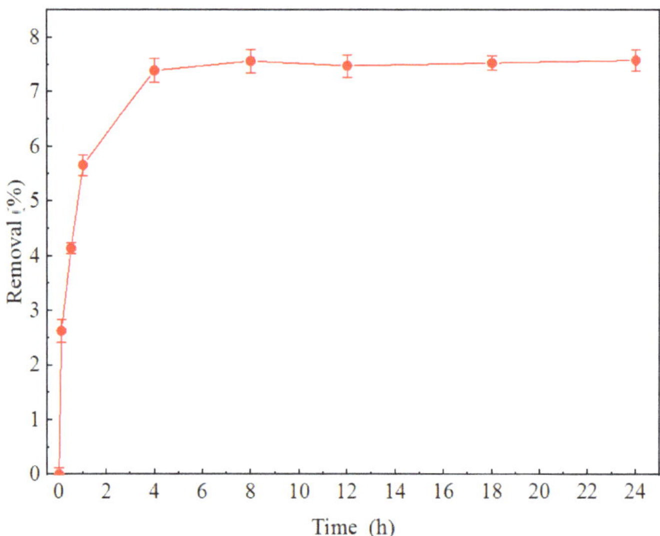

Figure 7. Effect of time on DOX removal by montmorillonite.

DOX adsorption by montmorillonite reached an equilibrium at 4 h, and the adsorption efficiency was very high in the early stages. The adsorption efficiency at 10 min was 2.63%, and with the extension of the adsorption time, the adsorption efficiency gradually increased to 4.14% at 30 min, 5.66% after 1 h and then 7.39% at 4 h. Afterwards, although the DOX concentration in the solution fluctuated slightly, it remained roughly unchanged. Therefore, there were two adsorption stages of DOX by montmorillonite, namely, an early rapid adsorption stage (0~4 h) and a later slow adsorption stage (4~24 h).

3.2.4. Effect of Adsorption Temperature

As temperature strongly affects adsorption, the influence of temperature on DOX adsorption by montmorillonite was investigated (Figure 8). In this study, 100 mg/L of montmorillonite was added to a solution containing DOX with concentrations ranging from 11.25 μM to 675.01 μM at pH 7. When the initial DOX concentration increased to 337.51 μM, adsorption equilibriums at different temperatures were observed: 20.38 (25 °C), 26.40 (35 °C) and 29.43 mg/g (45 °C), indicating that the temperature increase was beneficial to DOX adsorption by montmorillonite. The reason for this result may be that the ionized DOX$^-$ under neutral conditions was predominantly adsorbed by montmorillonite via van der Waals forces, which increased with faster molecular diffusion following a temperature rise [33].

3.3. Removal of DOX by Fe(VI) and Montmorillonite

To investigate the synergistic effect of oxidation and adsorption, Fe(VI) and montmorillonite were dosed in three different sequences under neutral conditions: (i) oxidation first and adsorption; (ii) adsorption first and then oxidation; (iii) simultaneous oxidation and adsorption. The above experimental results are shown in Figure 9. Additionally, the results of DOX removal by Fe(VI) and montmorillonite alone at pH 7.0 are shown in Figure 10. At pH 7.0, DOX removal (5 mg/L, 11.25 μM) by Fe(VI) with a molar ratio of 5:1 ([Fe(VI)]:[DOX]) was 82.07% after 60 min. Under the same experimental conditions, DOX adsorption by 100 mg/L montmorillonite after adsorption for 24 h was 7.58% (Figure S4).

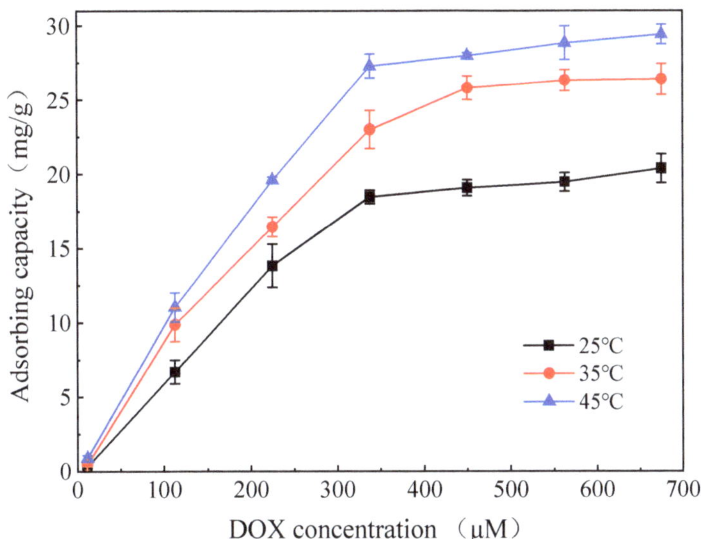

Figure 8. Effect of temperature on DOX removal by montmorillonite.

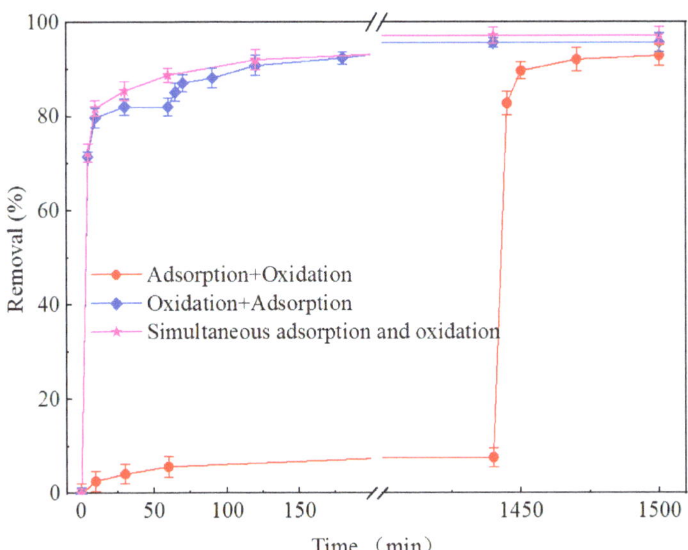

Figure 9. Comparison of different dosing sequences of Fe(VI) and montmorillonite on DOX removal.

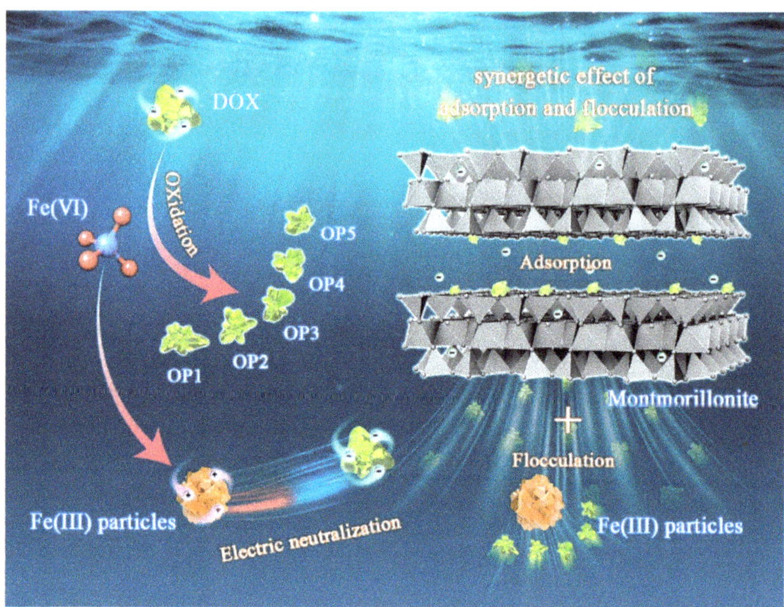

Figure 10. The synergy mechanism of DOX removal by Fe(VI) and montmorillonite.

The results of many experiments showed that DOX removal can be improved by combining Fe(VI) with montmorillonite. The synergy mechanism is shown in Figure 10. All three dosing sequences (Fe(VI) oxidation first and then adsorption, adsorption first and then Fe(VI) oxidation, simultaneous oxidation with adsorption) contribute to DOX removal (Figure 9). The optimal DOX removal efficiency (97.18%) was obtained under the synergetic effect of simultaneous Fe(VI) oxidation and montmorillonite adsorption. There may be two reasonable explanations for this. First, the highly positively charged intermediate product generated during Fe(VI) decomposition was neutralized with negatively charged montmorillonite. Additionally, the oxidation characteristics of Fe(VI) made DOX react with parts of the organic surface coating of colloidal particles, thus breaking through the organic coating, enhancing the electric neutralization of $Fe(OH)_3$ coagulants and colloidal particles [34]. This may reduce the repulsion between particles and enable larger flocs to agglomerate, improving the settling performance of flocs and the turbidity removal effect. Secondly, in the simultaneous adsorption experiment, montmorillonite not only adsorbed DOX and its degradation products, but it can also promote adsorption bridging, net flapping and sweeping and electric neutralization with negatively charged in situ Fe(III) particles, enhancing adsorption and flocculation. Furthermore, the turbidity caused by montmorillonite can be settled by Fe(III) flocculation.

3.4. Characterization of Fe(III)-Montmorillonite Composite and Montmorillonite

In order to evaluate the pore characteristic (specific surface, pore volume and pore size) of the Fe(VI)-montmorillonite composite, N_2 adsorption–desorption analysis was performed. Consistent with the IUPAC classification [35], both Fe(III)-montmorillonite composite and montmorillonite showed Type IVa isotherm pattern with a significant H3 type hysteresis loop (Figure S5), indicating the presence of mesopores [36], this can also be seen from BJH adsorption dV/dD pore volume(Figure S6). Due to the existence of the hysteresis curve, the pores are corrugated, resilient and expansible [37]. The Fe(III)-montmorillonite composite and montmorillonite surfaces were analyzed with a scanning electron microscope (SEM) and a transmission electron microscope (TEM). The results showed that montmorillonite has a layered structure (Figure S7a); after being combined with Fe(III) particles, it presents a non-geometric shape (Figure S7b,c), which indicated that Fe(III) and montmorillonite were successfully combined.

The BET surface area and BJH pore size of montmorillonite were 159.33 m^2/g and 6.42 nm, respectively. Comparatively, the Fe(III)-montmorillonite composite had a higher BET surface area and BJH pore size (195.37 m^2/g and 5.98 nm) (Figure S6 and Table S4), indicating the coordination of Fe(III) particle and montmorillonite promoted the adsorption capacity of sole montmorillonite. This result validated the synergistic effect of Fe(VI) and montmorillonite combination on DOX removal, due to the enhanced adsorption by formation of Fe(III)-montmorillonite composite.

3.5. Plausible DOX Degradation Pathway

GC–MS was used to identify the five intermediates of DOX when oxidized by Fe(VI). The degradation products detected at each reaction period are shown in Table S3 and Figure S8. Referring to the measured m/z value, the previous literature and considering the molecular structure of the target pollutants with the oxidation characteristics of Fe(VI) [30,31,38,39], the possible degradation pathway of DOX is proposed in Figure 11. The molecular structure of DOX contains a stable naphthol ring, so when Fe(VI) oxidizes DOX it predominantly attacks the hydroxyl, amino and methyl groups in the DOX structure. Combined with the TOC degradation results, it can be seen that it is difficult for Fe(VI) to open the naphthol ring and oxidize DOX into small molecules. From the TOC degradation effect of DOX (41.68%), it can be noted that, while the degradation efficiency of DOX by Fe(VI) was high, the degree of oxidation mineralization was insignificant, similar to previous research results concerning tetracycline hydrochloride removal by Fe(VI) [29,40,41]. As seen in Figure 10, when oxidized by Fe(VI), the phenolic hydroxyl groups and hydroxyl groups on the naphthol ring of DOX are first oxidized, and the phenolic hydroxyl groups are broken and oxidized into p-benzoquinone, while the C–O bond on the naphthol ring is broken, losing two hydroxyl groups to generate the intermediate product OP1. Subsequently, one C–N bond on the naphthol ring of OP1 is broken, losing one amino group and turning into a carboxyl group, while the other two C–N bonds break and lose two methyl groups to produce the intermediate OP2. Later, the intermediate OP2 continues to be oxidized, resulting in C–C bond breaking and the loss of a carboxyl group to form the intermediate OP3. At this point, the p-benzoquinone structure on the naphthol ring continues to lose one $H_2C=CH_2$ bond due to the oxidation of Fe(VI), producing the intermediate product OP4. Finally, OP4 continues to be oxidized, losing two carboxyl groups, and becomes OP5, the final product of DOX oxidation.

The FTIR band is captured for Fe(III)-montmorillonite at 400–4000 cm^{-1}, the FTIR spectra were performed to reveal the nature of the interfacial interactions between interlayers of Fe(III)-montmorillonite composite and montmorillonite after reaction (Figure 12). The bands located at 1043 cm^{-1} are assigned to NO_3^- symmetry stretching [42,43]. The spectrum also shows stretching modes at 3434 cm^{-1} [43], while the bending modes appear at 1635 cm^{-1} [44]. These samples show evidence for the presence of NH_4^+ [44]. It proves that the C-N bond on the naphthol ring of OP1 was broken, losing one amino group, and parts of them were oxidized to NO_3^- by Fe(VI).

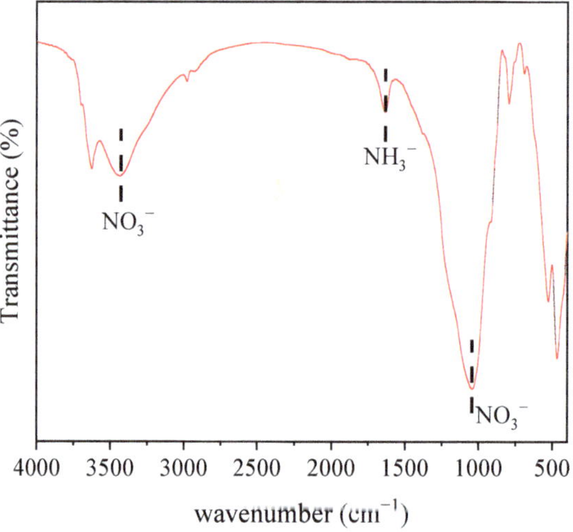

Figure 11. Proposed pathway of DOX degradation by Fe(VI).

Figure 12. FTIR spectra for the Fe(III) + montmorillonite composite.

4. Conclusions

Fe(VI) is highly oxidizing and DOX removal can be effectively improved by the incremental dosage of Fe(VI) and montmorillonite. Under acidic conditions, DOX oxidation by Fe(VI) and the adsorption of DOX by montmorillonite can be promoted in the initial reaction stage. The oxidation of DOX by Fe(VI) finishes within 30 min at an optimal pH of 9.

The adsorption of DOX by montmorillonite reaches an equilibrium at 4 h, and an increase in temperature promotes adsorption. In addition, DOX oxidation reactions by Fe(VI) follow second-order kinetics. Due to the stability of the structure of DOX, Fe(VI) cannot completely mineralize it. Although DOX was completely removed by Fe(VI), adding a small amount of montmorillonite to the system can promote the further degradation of DOX (DOX removal increased from 82.07 to 97.18%). The removal rate was observed to be higher than simple oxidation. For example, 82% of DOX was removed by the electrochemically generated Fe(VI) at the molar ratio of 5:1 [18]. The degradation ratios of DOX in the carbon black-activated peroxydisulfate and manganese–cobalt–tungsten composite oxidation process reached 87 and 80%, which was lower than Fe(VI)-montmorillonite synergetic system [31,38]. This indicated that the Fe(VI)-montmorillonite synergetic system was an efficient approach to the removal of DOX. In addition to the excellent degradation rate, the in situ Fe(III) particles formed by Fe(VI) decomposition acted as a good coagulant and flocculant, promoting the flocculation and precipitation of particles in solution, removing turbidity. Overall, this work demonstrates that DOX was almost eliminated via the synergy of Fe(VI) and montmorillonite, and provides insight into the combination of oxidants and adsorbents in water treatment.

Supplementary Materials: The following supporting information can be downloaded at: https://www.mdpi.com/article/10.3390/w15091758/s1, Text S1. The Fe(VI) determination by ABTS method; Figure S1: The recovery rate after filtering by 0.22 μm membrane; Figure S2: The standard curve of DOX; Figure S3: Concentration of Fe(VI) (a) and DOX (b) with the reaction time; Figure S4: Comparison of oxidation alone (left axis) and adsorption alone (right axis) on DOX removal; Figure S5: Isotherm Linear Plot: Montmorillonite (a), Fe(III)+montmorillonite (b); Figure S6: BJH Adsorption dV/dD Pore Volume; Figure S7: (a) SEM-montmorillonite; (b) SEM-Fe(III) with montmorillonite; (c) TEM-Fe(III)-montmorillonite particles; Figure S8: GC-MS spectra of DOX degraded (a) 5 min intermediates; (b) 30 min intermediates; (c) 45min intermediates; (d) 60 min intermediates; Table S1: The UPLC and GC–MS conditions; Table S2: Dissociation constants (pKa) of doxycycline [25,32]; Table S3: Detection products in DOX degradation by GC-MS; Table S4. Textural attributes of adsorbents.

Author Contributions: Conceptualization, H.Z.; methodology, H.Z. and S.W.; software, H.Z. and S.W.; validation, H.W.; formal analysis, J.S.; investigation, S.W. and H.Z.; data curation, S.W. and H.Z.; writing—original draft, H.Z. and S.W.; writing—review and editing, H.Z. and J.S.; visualization, S.W. and H.Z.; supervision, H.W.; project administration, H.W.; funding acquisition, H.W. and H.Z. All authors have read and agreed to the published version of the manuscript.

Funding: The authors express their gratitude to the National Natural Science Foundation of China (Grant Numbers: 22178323) and the Basic Scientific Research Business Fees of Provincial Colleges and Universities (grant number: FRF22QN004) for their financial support in conducting this work.

Data Availability Statement: Data is contained within the article or supplementary material. The data presented in this study are available in Supplementary Materials.

Conflicts of Interest: The authors declare no conflict of interest.

References

1. Cheng, Y.; Ding, T.; Qian, Y.; Li, M.; Li, J. Advances in Biodegradation of Pharmaceuticals and Personal Care Products. *Sheng Wu Gong Cheng Xue Bao* **2019**, *35*, 2151–2164. [PubMed]
2. Akinbowale, O.L.; Peng, H.; Barton, M.D. Diversity of Tetracycline Resistance Genes in Bacteria from Aquaculture Sources in Australia. *J. Appl. Microbiol.* **2007**, *103*, 2016–2025. [CrossRef] [PubMed]
3. Comeau, F.; Surette, C.; Brun, G.L.; Losier, R. The Occurrence of Acidic Drugs and Caffeine in Sewage Effluents and Receiving Waters from Three Coastal Watersheds in Atlantic Canada. *Sci. Total Environ.* **2008**, *396*, 132–146. [CrossRef] [PubMed]
4. Lishman, L.; Smyth, S.A.; Sarafin, K.; Kleywegt, S.; Toito, J.; Peart, T.; Lee, B.; Servos, M.; Beland, M.; Seto, P. Occurrence and Reductions of Pharmaceuticals and Personal Care Products and Estrogens by Municipal Wastewater Treatment Plants in Ontario, Canada. *Sci. Total Environ.* **2006**, *367*, 544–558. [CrossRef] [PubMed]
5. Wang, L.; Ying, G.G.; Zhao, J.L.; Yang, X.B.; Chen, F.; Tao, R.; Liu, S.; Zhou, L.J. Occurrence and Risk Assessment of Acidic Pharmaceuticals in the Yellow River, Hai River and Liao River of North China. *Sci. Total Environ.* **2010**, *408*, 3139–3147. [CrossRef]
6. Goswami, R.K.; Agrawal, K.; Verma, P. An Exploration of Natural Synergy Using Microalgae for the Remediation of Pharmaceuticals and Xenobiotics in Wastewater. *Algal Res.* **2022**, *64*, 102703. [CrossRef]

7. Hanna, N.; Sun, P.; Sun, Q.; Li, X.; Yang, X.; Ji, X.; Zou, H.; Ottoson, J.; Nilsson, L.E.; Berglund, B.; et al. Presence of Antibiotic Residues in Various Environmental Compartments of Shandong Province in Eastern China: Its Potential for Resistance Development and Ecological and Human Risk. *Environ. Int.* **2018**, *114*, 131–142. [CrossRef]
8. Ma, J.; Wang, Z.; Zhang, Z.; Liu, Q.; Li, L. Distribution Characteristics of 29 Antibiotics in Groundwater in Harbi. *Rock Miner. Anal.* **2021**, *40*, 944–953.
9. Rahman, N.; Raheem, A. Graphene Oxide/Mg-Zn-Al Layered Double Hydroxide for Efficient Removal of Doxycycline from Water: Taguchi Approach for Optimization. *J. Mol. Liq.* **2022**, *354*, 118899. [CrossRef]
10. Gatica, J.; Cytryn, E. Impact of Treated Wastewater Irrigation on Antibiotic Resistance in the Soil Microbiome. *Environ. Sci. Pollut. Res.* **2013**, *20*, 3529–3538. [CrossRef]
11. Wang, D.; Zeng, Z.; Zhang, H.; Zhang, J.; Bai, R. How Does PH Influence Ferrate(VI) Oxidation of Fluoroquinolone Antibiotics? *Chem. Eng. J.* **2022**, *431*, 133381. [CrossRef]
12. Acosta-Rangel, A.; Sánchez-Polo, M.; Rozalen, M.; Rivera-Utrilla, J.; Polo, A.M.S.; Berber-Mendoza, M.S.; López-Ramón, M.V. Oxidation of Sulfonamides by Ferrate(VI): Reaction Kinetics, Transformation Byproducts and Toxicity Assesment. *J. Environ. Manag.* **2020**, *255*, 109927. [CrossRef] [PubMed]
13. Shu, J.; Wang, K.; Sharma, V.K.; Xu, X.; Nesnas, N.; Wang, H. Efficient Micropollutants Degradation by Ferrate (VI)-Ti/Zn LDH Composite under Visible Light: Activation of Ferrate (VI) and Self-Formation of Fe (III) -LDH Heterojunction. *Chem. Eng. J.* **2023**, *456*, 141127. [CrossRef]
14. Lee, Y.; Zimmermann, S.G.; Kieu, A.T.; Von Gunten, U. Ferrate (Fe(VI)) Application for Municipal Wastewater Treatment: A Novel Process for Simultaneous Micropollutant Oxidation and Phosphate Removal. *Environ. Sci. Technol.* **2009**, *43*, 3831–3838. [CrossRef]
15. Sharma, V.K. Potassium Ferrate(VI): An Environmentally Friendly Oxidant. *Adv. Environ. Res.* **2002**, *6*, 143–156. [CrossRef]
16. Hörsing, M.; Ledin, A.; Grabic, R.; Fick, J.; Tysklind, M.; la Cour Jansen, J.; Andersen, H.R. Determination of Sorption of Seventy-Five Pharmaceuticals in Sewage Sludge. *Water Res.* **2011**, *45*, 4470–4482. [CrossRef]
17. Li, J.; Yin, C.; Ma, H.; Fen, X.; Cheng, G. Research Progress on Modification and Application of Montmorillonite. *Technol. Dev. Chem. Ind.* **2021**, *50*, 25–29.
18. Wang, K.M.; Shu, J.; Wang, S.J.; Hong, T.Y.; Xu, X.P.; Wang, H.Y. Efficient Electrochemical Generation of Ferrate(VI) by Iron Coil Anode Imposed with Square Alternating Current and Treatment of Antibiotics. *J. Hazard. Mater.* **2020**, *384*, 121458. [CrossRef]
19. Lee, Y.; Yoon, J.; Von Gunten, U. Spectrophotometric Determination of Ferrate (Fe(VI)) in Water by ABTS. *Water Res.* **2005**, *39*, 1946–1953. [CrossRef]
20. Yang, B.; Ying, G.G.; Zhang, L.J.; Zhou, L.J.; Liu, S.; Fang, Y.X. Kinetics Modeling and Reaction Mechanism of Ferrate(VI) Oxidation of Benzotriazoles. *Water Res.* **2011**, *45*, 2261–2269. [CrossRef]
21. Xu, N.; Dong, J.; Zhou, W.; Liu, Y.; Ai, X. Determination of Doxycycline, 4-Epidoxycycline, and 6-Epidoxycycline in Aquatic Animal Muscle Tissue by an Optimized Extraction Protocol and Ultra-Performance Performance Liquid Chromatography with Ultraviolet Detection. *Anal. Lett.* **2019**, *52*, 452–464. [CrossRef]
22. Wang, H.; Liu, Y.; Jiang, J.-Q. Reaction Kinetics and Oxidation Product Formation in the Degradation of Acetaminophen by Ferrate (VI). *Chemosphere* **2016**, *155*, 583–590. [CrossRef] [PubMed]
23. Feng, M.; Wang, X.; Chen, J.; Qu, R.; Sui, Y.; Leslie, C.; Wang, Z.; Sharma, V.K. Degradation of Fluoroquinolone Antibiotics by Ferrate(VI): Effects of Water Constituents and Oxidized Products. *Water Res.* **2016**, *103*, 48–57. [CrossRef] [PubMed]
24. Sharma, V.K.; Sohn, M.; Anquandah, G.A.K.; Nesnas, N. Kinetics of the Oxidation of Sucralose and Related Carbohydrates by Ferrate(VI). *Chemosphere* **2012**, *87*, 644–648. [CrossRef] [PubMed]
25. Qiang, Z.; Adams, C. Potentiometric Determination of Acid Dissociation Constants (PK a) for Human and Veterinary Antibiotics. *Water Res.* **2004**, *38*, 2874–2890. [CrossRef] [PubMed]
26. Wen, X.J.; Shen, C.H.; Niu, C.G.; Lai, D.C.; Zhu, M.S.; Sun, J.; Hu, Y.; Fei, Z.H. Attachment of Ag/AgCl Nanoparticles on CdMoO4 Microspheres for Effective Degradation of Doxycycline under Visible Light Irradiation: Degradation Pathways and Mineralization Activity. *J. Mol. Liq.* **2019**, *288*, 111063. [CrossRef]
27. Wen, X.J.; Niu, C.G.; Zhang, L.; Liang, C.; Zeng, G.M. A Novel Ag$_2$O/CeO$_2$ Heterojunction Photocatalysts for Photocatalytic Degradation of Enrofloxacin: Possible Degradation Pathways, Mineralization Activity and an in Depth Mechanism Insight. *Appl. Catal. B Environ.* **2018**, *221*, 701–714. [CrossRef]
28. Zhi, D.; Wang, J.; Zhou, Y.; Luo, Z.; Sun, Y.; Wan, Z.; Luo, L.; Tsang, D.C.W.; Dionysiou, D.D. Development of Ozonation and Reactive Electrochemical Membrane Coupled Process: Enhanced Tetracycline Mineralization and Toxicity Reduction. *Chem. Eng. J.* **2020**, *383*, 123149. [CrossRef]
29. Dalmázio, I.; Almeida, M.O.; Augusti, R.; Alves, T.M.A. Monitoring the Degradation of Tetracycline by Ozone in Aqueous Medium Via Atmospheric Pressure Ionization Mass Spectrometry. *J. Am. Soc. Mass Spectrom.* **2007**, *18*, 679–686. [CrossRef]
30. Park, J.A.; Pineda, M.; Peyot, M.L.; Yargeau, V. Degradation of Oxytetracycline and Doxycycline by Ozonation: Degradation Pathways and Toxicity Assessment. *Sci. Total Environ.* **2023**, *856*, 159076. [CrossRef]
31. Chen, Y.; Yin, R.; Zeng, L.; Guo, W.; Zhu, M. Insight into the Effects of Hydroxyl Groups on the Rates and Pathways of Tetracycline Antibiotics Degradation in the Carbon Black Activated Peroxydisulfate Oxidation Process. *J. Hazard. Mater.* **2021**, *412*, 125256. [CrossRef] [PubMed]

32. Zhang, X.; Bai, B.; Li, G.; Wang, H.; Suo, Y. Novel Sea Buckthorn Biocarbon SBC@β-FeOOH Composites: Efficient Removal of Doxycycline in Aqueous Solution in a Fixed-Bed through Synergistic Adsorption and Heterogeneous Fenton-like Reaction. *Chem. Eng. J.* **2016**, *284*, 698–707. [CrossRef]

33. Parsegian, V.A.; Ninham, B.W. *Temperature-Dependent Van Der Waals Forces*; Elsevier: Amsterdam, The Netherlands, 1970; Volume 10.

34. Ma, Y.; Gao, N.; Li, C. Degradation and Pathway of Tetracycline Hydrochloride in Aqueous Solution by Potassium Ferrate. *Environ. Eng. Sci.* **2012**, *29*, 357–362. [CrossRef]

35. Thommes, M.; Kaneko, K.; Neimark, A.V.; Olivier, J.P.; Rodriguez-Reinoso, F.; Rouquerol, J.; Sing, K.S.W. Physisorption of Gases, with Special Reference to the Evaluation of Surface Area and Pore Size Distribution (IUPAC Technical Report). *Pure Appl. Chem.* **2015**, *87*, 1051–1069. [CrossRef]

36. Li, B.; Zhao, Y.; Zhang, S.; Gao, W.; Wei, M. Visible-Light-Responsive Photocatalysts toward Water Oxidation Based on NiTi-Layered Double Hydroxide/Reduced Graphene Oxide Composite Materials. *ACS Appl. Mater. Interfaces* **2013**, *5*, 10233–10239. [CrossRef] [PubMed]

37. Xing, L.; Haddao, K.M.; Emami, N.; Nalchifard, F.; Hussain, W.; Jasem, H.; Dawood, A.H.; Toghraie, D.; Hekmatifar, M. Fabrication of HKUST-1/ZnO/SA Nanocomposite for Doxycycline and Naproxen Adsorption from Contaminated Water. *Sustain. Chem. Pharm.* **2022**, *29*, 100757. [CrossRef]

38. Luo, X.; You, Y.; Zhong, M.; Zhao, L.; Liu, Y.; Qiu, R.; Huang, Z. Green Synthesis of Manganese–Cobalt–Tungsten Composite Oxides for Degradation of Doxycycline via Efficient Activation of Peroxymonosulfate. *J. Hazard. Mater.* **2022**, *426*, 127803. [CrossRef] [PubMed]

39. Zhang, J.; Wu, H.; Zhang, D.; Zhang, L.; Zhu, C. Preparation of a Ruthenium-Modified Composite Electrode and Evaluation of the Degradation Process and Degradation Mechanism of Doxycycline at This Electrode. *J. Water Process Eng.* **2022**, *48*, 102904. [CrossRef]

40. Jiao, S.; Meng, S.; Yin, D.; Wang, L.; Chen, L. Aqueous Oxytetracycline Degradation and the Toxicity Change of Degradation Compounds in Photoirradiation Process. *J. Environ. Sci.* **2008**, *20*, 806–813. [CrossRef]

41. Jiao, S.; Zheng, S.; Yin, D.; Wang, L.; Chen, L. Aqueous Photolysis of Tetracycline and Toxicity of Photolytic Products to Luminescent Bacteria. *Chemosphere* **2008**, *73*, 377–382. [CrossRef]

42. Bei, W.; Junhua, J.; Ting, Z.; Zhonghui, L. Determination of Soluble Lons in Atmospheric Aerosol by FTTR. *J. Henan Mech. Electr. Eng. Coll.* **2010**, *18*, 15–17.

43. Socrates, G. *Infrared and Raman Characteristic Group Frequencies: Tables and Charts*; John Wiley & Sons, Inc.: Hoboken, NJ, USA, 2001; ISBN 978-0-470-09307-8.

44. Hakiri, R.; Ameur, I.; Abid, S.; Derbel, N. Synthesis, X-Ray Structural, Hirshfeld Surface Analysis, FTIR, MEP and NBO Analysis Using DFT Study of a 4-Chlorobenzylammonium Nitrate (C$_7$ClH$_9$N)+(NO$_3$)–. *J. Mol. Struct.* **2018**, *1164*, 486–492. [CrossRef]

Review

Fenton Reaction–Unique but Still Mysterious

Frantisek Kastanek [1], Marketa Spacilova [1,*], Pavel Krystynik [2], Martina Dlaskova [1] and Olga Solcova [1]

[1]   Institute of Chemical Process Fundamentals of the Czech Academy of Sciences, Rozvojova 1/135, 165 00 Prague, Czech Republic
[2]   Faculty of Environment, University of J. E. Purkyne in Usti nad Labem, Pasteurova 3632/15, 400 96 Usti nad Labem, Czech Republic
*   Correspondence: spacilova.marketa@icpf.cas.cz; Tel.: +420-220-390-280

Abstract: This study is devoted to the Fenton reaction, which, despite hundreds of reports in a number of scientific journals, provides opportunities for further investigation of its use as a method of advanced oxidation of organic macro- and micropollutants in its diverse variations and hybrid systems. It transpires that, for example, the choice of the concentrations and ratios of basic chemical substances, i.e., hydrogen peroxide and catalysts based on the Fe^{2+} ion or other transition metals in homogeneous and heterogeneous arrangements for reactions with various pollutants, is for now the result of the experimental determination of rather randomly selected quantities, requiring further optimizations. The research to date also shows the indispensability of the Fenton reaction related to environmental issues, as it represents the pillar of all advanced oxidation processes, regarding the idea of oxidative hydroxide radicals. This study tries to summarize not only the current knowledge of the Fenton process and identify its advantages, but also the problems that need to be solved. Based on these findings, we identified the necessary steps affecting its further development that need to be resolved and should be the focus of further research related to the Fenton process.

Keywords: Fenton reaction; Fenton-like reactions; advanced oxidation processes; wastewater treatment

Citation: Kastanek, F.; Spacilova, M.; Krystynik, P.; Dlaskova, M.; Solcova, O. Fenton Reaction–Unique but Still Mysterious. *Processes* **2023**, *11*, 432. https://doi.org/10.3390/pr11020432

Academic Editors: Gassan Hodaifa, Antonio Zuorro, Joaquín R. Dominguez, Juan García Rodríguez, José A. Peres, Zacharias Frontistis and Mha Albqmi

Received: 3 January 2023
Revised: 20 January 2023
Accepted: 30 January 2023
Published: 1 February 2023

1. Introduction

The Fenton reaction, discovered at the beginning of the 20th century [1], is primarily based on the idea of the formation of oxidizing radicals, which are created by the catalytic action of Fe^{2+} on the decomposition of H_2O_2, added in a certain amount and ratio to water, which also contains various organic substances. Oxidative radicals oxidize the present organic substances to varying degrees during the momentary period of their existence. This Fenton reaction is homogeneous because the catalyst (Fe^{2+} ion) is dissolved in water. However, the catalyst can also be heterogeneous. For example, the Fenton reaction is heterogeneous in the case of Fe particles, which under certain conditions release Fe ions into the solution.

The homogeneous Fenton reaction is highly popular, for instance, due to the decontamination of various organic pollutants in wastewater, both municipal and industrial. This reaction is capable of decomposing even more structurally complex organic substances, e.g., azo dyes orange II, tartrazine, nonsteroid antiphlogistics such as acetylsalicylic acid and most pharmaceuticals, as well as bacteriostatic antibiotics, e.g., tetracycline, 2,4-dichlorophene (precursor of the herbicide 2,4-dichlorophenoxyacetic acid), bisphenol A, various endocrine disruptors, e.g., estrogens, pesticides of all types, etc. On the contrary, it shows lower efficiency for the decomposition of simpler chlorinated substances, e.g., tri- and tetrachlorinated ethanes, chloroform, or tetrachloromethane [2]. The process of contaminant oxidation in an aqueous environment using the homogeneous Fenton reaction consists of several successive steps, namely the addition of reagents Fe^{2+} and H_2O_2 in appropriate concentrations and ratios to the contaminated water while adjusting the pH, usually to pH 3, the progress and control of the oxidation reaction, the alkalization of the solution after the reaction, and the coagulation–flocculation of the sludge and its separation.

The Fenton reaction is a pillar of the so-called advanced oxidation processes, which currently belong to the most important categories of chemical processes aimed at removing hazardous substances from the environment. To elucidate, a few selected interesting publications related to this reaction [3] are listed below. They are focused on the Fenton reaction in real wastewater from various industrial productions [4,5], wastewater from textile dyeing [6], leachate from landfills [7], fire extinguishing water containing phenol [8], and the removal of extracellular polymers in sewage sludge worsening its drainage [9]. There are already hundreds of publications dealing with the Fenton reaction, and its popularity continues even nowadays, whereby the attention of scientists addressing ecological issues is turning to previously poorly observed environmental pollutants, the so-called emerging pollutants. These previously unmonitored substances, contained in waters in concentrations of ng/L to µg/L, such as the so-called endocrine disruptors (hormones, surfactants, additives to plastics and cosmetics, phthalates, parabens, bisphenol A, nonylphenol) [10–12] or waste pharmaceuticals and drugs [13] are unwanted waste products of anthropogenic origin that negatively affect aquatic diversity and are also potentially dangerous for humans. This study summarizes the current knowledge of the Fenton process and identifies its advantages and also the problems that need to be solved, considering its historical role. Based on these findings, the necessary steps are mentioned, which can affect its further development and should be the focus of further research related to the Fenton process.

2. Advantages of the Fenton Process and Its Historical Role

The Fenton process is based on inexpensive reagents and the reactor arrangement is simple. Compared to other oxidation systems, e.g., gas–liquid (chlorination, ozonation), the reaction is not negatively affected by mass transfer. Concurrently, the system of Fenton components can also be bubbled with air or oxygen, similar to the heterogeneous Fenton reaction, where dissolved oxygen in water accelerates the formation of Fe^{2+} in the case of using solid Fe particles as their source. The reaction can be easily controlled, and the Fenton process is indispensable if simple and affordable chemicals together with a simple reactor equipment (essentially a stirred tank) are used.

Fenton's discovery of oxidizing radicals enabled the development of other new processes based on this phenomenon, in which various oxidizing radicals, e.g., singlet oxygen, are also generated by physicochemical processes, preferably by the action of UV radiation of various wavelengths, including solar radiation, ultrasound, cold plasma, and streams of electrons, from which dozens can be assembled by combination. This created a huge space for further theoretical and experimental studies of advanced oxidation processes and the verification of their effectiveness and the quality of products for the destruction of hundreds of different organic substances. It should be emphasized that the classic Fenton reaction is an excellent basis for the development of such advanced AOP modifications.

Essentially, AOPs are considered to be reactions where the formation of radicals is initiated by another application of energy intervention, such as ultraviolet radiation. Due to the more powerful generation of hydroxyl radicals, in most cases, AOPs also show a higher efficiency of decomposition of organic substances than the classical Fenton reaction, which is their greatest benefit. Some strongly recalcitrant substances (chloro and nitro derivatives of phenol, phthalates, polyaromatics, polychlorinated aromatics such as PCBs, dioxins and furans, pharmaceuticals, chlorinated fungicides and pesticides, alkyl benzyl sulfonates, etc.) can be significantly disrupted only by these new AOPs. However, they are also more energy intensive, more expensive, and their strong oxidizing power, in such situations where the mineralization of organic substances does not occur, can lead to the formation of oxidation byproducts, which appear more hazardous than the original organic substances. With the classical Fenton reaction, such danger is less serious.

Nevertheless, the simple version of the Fenton reaction is irreplaceable when the concentration of the target pollutant to be removed reaches the values given by regulations or standards due to its simplicity, incomparably lower costs, and the reduction in the risk of creating hazardous side- and endproducts, which happens frequently. This trend is evident not only from the great interest in studying the application of the Fenton reaction in developing countries, where this simple technology is experiencing a great renaissance, but also in economically developed ones, where it seems to be an ideal decontamination technique for the decontamination of water from various smaller brownfields.

Why Are These Promising Methods Not Applied in Industry More Commonly?

AOP is currently receiving extreme attention in literature. AOPs can be made both in the homogeneous variant (O_3/H_2O_2, H_2O_2/Fe^{2+} a H_2O_2/UV) and the heterogeneous one, e.g., $H_2O_2/TiO_2/UV$. However, the main drawback of the homogeneous system is the inhibition of radical propagation due to the presence of radical scavengers that are commonly present in water, such as carbonates, bicarbonates, and NOM. This is the main reason why homogeneous variants, including the classical Fenton method, are not implemented for the treatment of contaminated waters on an industrial scale, although the application of the Fenton method to various wastewaters is being addressed in laboratories. There have been a large number of similar contributions in the last decade, especially from developing countries, where methods of wastewater treatment from various local sources are beginning to be considered, e.g., from wastewater refineries, the textile industry, especially from dyeworks, tanneries, etc. [14–16]. However, Vega and Valdés [17] documented that the aforementioned limiting factor could be reduced by a heterogeneous arrangement.

3. Heterogenous Variant of Fenton Reaction

The simplest variant of the Fenton reaction is the application of metallic Fe particles, including nano-Fe, in which the formation of Fe^{2+} is applied under the oxidation conditions of dissolved oxygen in water and in the presence of metallic iron particles, e.g., as nano-Fe. Although the formation of radicals proceeds through a process similar to that of the homogeneous Fenton reaction, radicals are mainly formed in close proximity to the iron surface, where organic pollutants are also sorbed, which makes it easier for them to meet. Interfering radical absorbers present in the liquid volume have less opportunity to meet them. It is also possible to consider the addition of some heavy metal ions to Fe particles during the oxidation process, which would fulfill the ideal requirements for decontamination reactions based on the Fenton reaction as a universal process, during which organic pollutants would be decontaminated and some heavy metals extracted (sorbed) at the same time. However, how the sorption of heavy metal ions on nanoiron particles could affect the formation of oxidative radicals still remains unknown.

A heterogeneous catalyst is not represented only by particles of metallic Fe. It is also possible to anchor Fe on carriers (e.g., as Fe nanodots on alumina, activated carbon, and various biochars) when the adsorption of the pollutant on the active sites of the sorbent near Fe increases the opportunity for the reaction contact with the catalyst.

It appears that it is possible to incorporate metal ions into the mineralogical grid of clays (montmorillonite, bentonite, natural and synthetic zeolites), especially Fe, Cu, Ce, etc., forming structures known as pillared clays, which exhibit the properties of solid Fenton catalysts, e.g., Fe-ZSM5 [18]. The synthesis of the most common catalyst, Fe-ZSM5, is of great interest, e.g., [19–23], as it is possible to work with them under less acidic conditions. Additionally, they find application in the decontamination of a number of different types of wastewater, e.g., the decolorization of textile waters [24,25], and also pharmaceuticals [13,26,27]. Clays modified in this way act as successful catalysts for the destruction of organic pollutants, e.g., [28,29], (see Figure 1), either as Fenton process catalysts or as photocatalysts or wet oxidation catalysts.

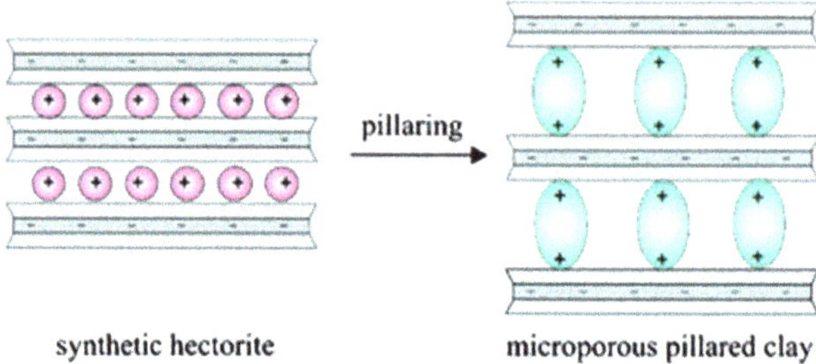

Figure 1. Pillaring, exchange of simple cations for polyoxo cations, e.g., Al-Fe polycation $(Al_{13-x}/Fe_x)^{7+}$, see Muñoz et al. [30]. Reproduced with permission from Stöcker et al., realization of truly microporous pillared clays, published by Chemical Communications, 2018 [31].

For this reason, research on heterogeneous catalysts in recent decades has mainly focused on two directions in which it is assumed that the inhibition of oxidative radicals is suppressed. For example, research focuses on catalysts applied in the so-called catalytic wet oxidation condition, which is a process used for the destruction of organic substances using oxygen as an oxidant while the reaction temperatures are characteristically high, usually over 200 °C. A number of solid catalysts have been applied based on various combinations of heavy metals and their oxides, especially transition metals and rare earths, including their various combinations, e.g., Mn/Ce, Co/Bi, Ru/Ce, or $CuO\text{-}ZnO\text{-}Al_2O$, etc. The reaction mechanism in the three-phase arrangement includes a number of interphase mass transfer steps, and the reaction usually takes place in a bubbled suspension reactor. High temperature and pressure contribute to the formation of oxidizing radical resistant to the possible presence of scavengers. Due to this, it is possible to obtain over 90% conversion of pollutants and mineralization of even complex organic substances into CO_2, ammonia and nitrogen, sulfates, phosphates, and chlorides. Wet oxidation and catalytic wet oxidation processes, which enable a significant reduction in temperatures and pressures, have also found industrial implementations for various types of wastewater (pharmaceutical and chemical industry) in the typical temperature range of 140–325 °C and 2–12 MPa. Concerning industrial designs, mainly Fe^{2+}/oxygen, Cu^{2+}/air or oxygen, Fe-Cu-Mn/H_2O_2, and Fe^{2+}/air/H_2O_2 catalysts were applied. Although the process is expensive and the corrosion of the metal parts of the reactors may cause a problem, it is reported that it can be cheaper than incineration with all the safeguards. All important aspects of this process are excellently described in Bhargava et al. [32]. It can be summarized that even this very promising process can be considered as a continuation of the heterogeneous Fenton reaction realized under high pressures.

Activated carbon, AC/H_2O_2 and AC/O_3, also plays a very important role. When in the presence of hydrogen peroxide, activated carbon acts as a catalyst because it can produce additional oxidative radicals, and it seems that the inhibition of radicals is suppressed in this system [17]. In the case of ozone, the presence of activated carbon generates oxidizing radicals via the decomposition of the ozone. However, the chemistry of AC surface groups plays a crucial role in radical production. Longer-term contact with hydrogen peroxide disrupts the composition of AC surface functional groups and reduces the formation of radicals due to the increased formation of acidic surface groups. Obviously, the presence of basic groups is mainly responsible for the catalytic decomposition of hydrogen peroxide, which leads to the production of radicals [17]. Nevertheless, other authors favor accelerating the formation of radicals with HNO_3 vapors [33]; thus, it would be appropriate to further examine the activated carbon including various biochars as catalysts for the decomposition of hydrogen peroxide. It is evident that not only the selection of suitable ACs and their surface treatment can reveal a longer duration of use in contact with peroxide. Additionally, Vega and Valdés [17] simultaneously showed that the presence of metal ions on the AC surface, mostly Fe, could play a significant role in the decomposition of hydrogen peroxide, thereby increasing the formation of radicals. Even this variant with AC and hydrogen peroxide can be considered as an $AC/H_2O_2/Fe^{2+}$ application of the Fenton process in a heterogeneous arrangement.

The presence of mesoporous-activated carbon particles is particularly advantageous in waters with a high TDS content (total dissolved soils), which inhibits the formation of radicals during a homogeneous Fenton reaction, as described in the work [34] focused on waters with a high TDS content of around 1%.

It is difficult to decide whether coal acts as a generator of radicals in the presence of dissolved oxygen, or only as a base with active sites for the contact of organic substances with radicals created by other processes than with the contribution of coal, or mainly as a TDS sorbent. Nevertheless, TDS can be sorbed onto the coal and stop interfering with the generation of radicals. Such situations, in the case of waters with a high TDS content (e.g., from tanneries), must be further studied. It is important for the heterogeneous Fenton reaction that the reaction takes place on the surface of the catalyst, and the adsorption of the organic pollutant is one of the important steps for the efficiency of oxidative degradation by radicals (see Figure 2). At different pHs, the isoelectric point of the catalyst would be different, and the electrical charge of the decontaminated target substance would be similarly affected. Simultaneously, the pH of the solution influences the electrostatic relations between the substance and the catalyst (i.e., repel or attract), which can explain the effect of pH on the greater success of the degradation of the target substance under either a homogeneous or heterogeneous Fenton reaction. However, even a pH closer to neutral can save neutralization costs, which are more favorable for heterogeneous Fenton reaction applications. Thus, further research in this direction is still open.

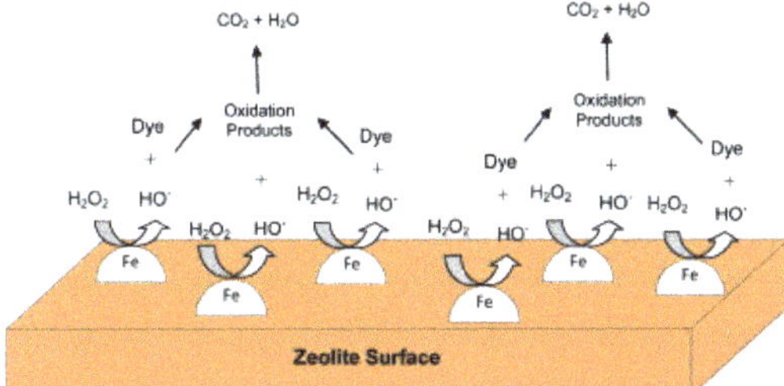

Figure 2. Formation of radicals on Fe-zeolite during oxidation of textile dyes. Reproduced with permission from Queirós et al., heterogeneous Fenton reaction oxidation using Fe/ZSM-5 as catalyst in a continuous stirred tank reactor, published by Separation and Purification Technology, 2015 [25].

It is evident that different heterogeneous variants of the Fenton reaction can enlarge AOP for water decontamination at industrially interesting scales and, at reasonable costs, result in useful solutions for environmental issues. For these reasons, considerable attention is currently focused on the development of heterogeneous catalysts with a high and long-term photocatalytic activity for the Fenton and photo-Fenton oxidation of wastewater. A promising heterogeneous variant can work in the future at a neutral pH (which is so far a practically unattainable wish of environmental engineers) and ambient temperature, without the need to neutralize the effluent after treatment. Illustrative examples are particles of magnetite, (Fe_3O_4), hematite (Fe_2O_3), goethite $(\alpha\text{-FeOOH})$, pyrite (FeS_2), and lepidocrocite $(\gamma\text{-FeOOH})$, although compared to the homogeneous variant, the oxidation reaction rate was usually lower due to the retarding mass transfer effect on the catalyst surface. Hematite and goethite have iron in the Fe^{3+} oxidation state, and for the formation of active radicals, it would be more appropriate to apply UV radiation, which formally follows from the theoretical relationships presented below. It was found that at the same time, they also act as photocatalysts and produce other necessary radicals due to the further reaction of the regenerated Fe^{2+} with peroxide (relation (2) below). Many possible synthetic and natural catalysts have already been described. Essentially, most solid substances containing transition metals, whether synthetic composites or natural minerals, when in contact with hydrogen peroxide, show catalytic abilities to generate hydroxide oxidation radicals, e.g., [35].

A number of works are devoted to the magnetic surface on powdered activated carbon, which is positive for the practical application, as it is relatively simple to produce a catalyst with magnetite nanoparticles. The coprecipitation reaction (addition of Fe^{2+} and Fe^{3+} salts to treated activated carbon by precipitation with ammonia) is used, see instructions, e.g., [36], while the range of possibilities to degrade various pollutants is really large.

However, the problem with these solid catalysts lies in the gradual leaching of the metal component into the solution and thus in the gradual transition to the conditions of the classic Fenton reaction. The solution to prevent this transition of the heterogeneous Fenton reaction lies in the use of stable nonmetallic catalysts such as graphene [37]. Graphene can act as an effective electrode in the electro-Fenton process [38], as well as graphene oxide, or graphene cornered on F_3O_4 particles, which are sacrificed during the reaction [39]. Related to this, it can significantly stabilize and regenerate Fe^{2+} formed by the reaction, i.e., prevent the premature formation of, e.g., Fe^{3+} sludge [40]. It turns out that in such constellation, Fe^{3+} is more active than Fe^{2+}; however, carbon nanotubes, diamond nanoparticles [41], or fullerenes [42] could also be used. These carbon species can act either as a separate catalyst or rather as a carrier with a large interfacial area for various combinations of transition metals, e.g., Ce-Fe-RGO (RGO-reduced graphene oxide) [43]. Simultaneously, it has been proved that the Fenton reaction can be operated over a much wider pH range. For example, Wan et al. [43] even stated the pH of 7. Nevertheless, the experimental verification of similarly optimistic data would be necessary.

Current experience with the application of nanocarbon catalysts to the Fenton process is provided by Xin et al. [42]. However, further studies are necessary to verify that the efforts associated with the preparation of these catalysts (e.g., the technique of laser bombardment of graphite, electrospinning of preprepared forms of nanocarbon particles, vapor deposition, etc.) will ensure the increased degradation efficiency, or optimally, the application of this reaction in a pH-neutral environment at reasonable costs. Commercially available activated carbon fibers saturated with Fe^{3+} can act competitively by simple immersion in chloride hexahydrate [44], and reduction, e.g., by borohydride in the case of nanoiron, as stated for the application of the natural fibers of palm biochar [45]. However, activated carbon impregnated with Fe ions can also be used. It was verified by Duarte et al. [46] and Messele et al. [47] for phenol degradation.

In particular, Soares et al. [48] showed that highly mesoporous supports (sol-gel-method-prepared Fe/Xerogel M) and supports with a large interfacial area (Fe/activated carbon M) with 2% Fe content and with increased N- functional groups as catalysts of the Fenton heterogeneous process had greater degradation capabilities than other forms of carbon catalysts in the degradation of p-nitrophenol, 3.6 mmol/L and H_2O_2 29, mmol/L compared to catalysts with carbon nanotubes, which greatly favor the Fenton reaction in a simple version with treated activated carbon (N- functional groups, e.g., after soaking in a solution of melamine or urea). It follows that the surface phenomena, which have not yet been completely clarified, are important at the application of the heterogeneous Fenton reaction. It is obvious that the combination of a high number of mesopores and a high interphase area together with the introduction of suitable functional groups (for *p*-nitrophenol, e.g., N groups) allow for a given wastewater to "tailor a catalyst" for optimal conversions. Therefore, research is also open to easily available activated carbon as a component of the Fenton reaction catalyst.

4. When Can the Original Homogeneous Fenton Reaction Be Used?

There are many examples in the literature where real wastewater from various technologies are applied. Unfortunately, the experiments were not carried out on a scale that would demonstrate the possibility of a continuous application of the Fenton reaction, which is essential for industrial use. Most of the published works were applied as demonstrations in glass on a laboratory scale, although experiments were conducted with real water samples collected in effluents from various technologies. Moreover, the Fenton reaction is currently very popular in the photo-Fenton variant, i.e., with the participation of UV radiation, which is another problem for scaling up.

However, the Fenton reaction is cited as the first option for reducing COD values in highly contaminated waters with a high content of dissolved substances, colloids, and NOM, especially strongly colored ones, e.g., for the removal of COD from wastewater after textile dyeing (from dozens of works, e.g., [49]) or even leachate from landfills of various wastes.

The Fenton reaction would be potentially usable as a pretreatment process followed by other processes such as bioremediation, adsorption, membrane separation, sterilization, ion exchange, etc. Another possibility could be its use as a part of a reduction–oxidation node (reduction of chlorinated substances by metal + Fenton reaction). Similarly, it is possible to include (electro)coagulation, filtration, or membrane separation before applying the Fenton process (e.g., Vergili and Gencdal [50], for removing pharmaceuticals). The achievement of a BOD5/COD ratio greater than 0.4 is often considered as a criterion for the suitability of subsequent biodegradation (see, e.g., Cortez et al. [7]). However, the Fenton reaction does not always reach this ratio, and therefore, it is necessary to apply other oxidation processes. If we abandon the use of UV as too expensive, then the combination of H_2O_2/ozone can be applied with a small concentration of peroxide, e.g., 400 mg/L [7].

Furthermore, it should be mentioned that reaching the expected final concentrations of the removed substances can be considered a success; however, the decisive factor is whether the resulting water is not ecotoxic after the treatment, i.e., particularly, if it is not carcinogenic and mutagenic. The negative result of the relevant ecotoxicological test, especially, for example, on lower organisms, is completely decisive for expressing the success of the method used. Considering this fact, it is necessary to emphasize the fundamental advantage of the Fenton reaction, namely that even if it usually does not reach the mineralization of dangerous organic substances, it generally does not provide ecotoxic intermediates or products, and related to this, it increases the possibility of the subsequent bioavailability of pollutants.

Simplified Theoretical Basis of the Fenton Reaction

Well-known relationships involve the principle of the oxidative radical formation and, simultaneously, the oxidation of the target substance R is indicated. The quantity and quality of oxidation radicals are apparently different. The equations are, for practical reasons, focused on the •OH radical, which reveals a very strong oxidizing potential and can be quantified by set methods.

$$Fe^{2+} + H_2O_2 \rightarrow Fe^{3+} + \bullet OH + OH^- \tag{1}$$

$$Fe^{3+} + H_2O_2 \rightarrow Fe^{2+} + HO_2\bullet + H^+ \tag{2}$$

Compared to (1), (2) is several orders of magnitude slower, Fe^{3+} at a pH higher than 5 forms solid oxyhydroxides, and the cycle is completed.

$$RH + \bullet OH \rightarrow R\bullet + H_2O \tag{3}$$

$$R\bullet + H_2O_2 \rightarrow ROH + \bullet OH \tag{4}$$

$$Fe^{3+} + HO\bullet_2 \rightarrow Fe^{2+} + O_2 + H^+ \tag{5}$$

$$R\bullet + O_2 \rightarrow ROO\bullet \tag{6}$$

These radicals, R• and ROO•, can further become disproportionate, or produce relatively stable molecules, or react with Fe ions (e.g., [51]). The produced organic intermediates can further react with hydroxyl radicals and with O_2 and thus lead to other decomposition products and, in the ideal case, to mineralized components (H_2O, CO_2, Cl^-, etc.), which is only sometimes completed. A number of similar simplified schemes can be found, as reported in detail, for example, by Umar et al. [52]. However, it is assumed that four main reaction pathways actually occur: the addition of radicals, abstraction of hydrogen, transfer of electrons, and combination of radicals.

However, it is important for practice that the Fenton reaction enables the oxidation and gradual destruction of a number of organic substances, although the exact theoretical background is still not clearly known.

5. Problem of the Fenton Process

The biggest problem with the Fenton reaction lies in the fact that the amount of hydrogen peroxide and ferrous ion is not optional and cannot be predicted for real wastewater. Moreover, the decontamination efficiency of various organic pollutants of the target components of interest depends on both the H_2O_2:R ratio and the H_2O_2:Fe^{2+} ratio. Furthermore, these ratios depend not only on the concentration, chemical composition, and structure of the target substance R but also, in real waters, on a variety of organic and inorganic substances, which are undefined in terms of both their composition and structure. These substances are defined by terms such as natural organic matter (NOM) or dissolved organic matter (DOM), the total amount of dissolved solids, usually salts, total dissolved solids (TDS), chemical oxygen demand (COD, an indicative measure of the amount of oxygen that will be consumed during the reaction), total organic carbon (TOC), or biological oxygen consumption (BOD5), which is the biochemical oxidation of organic or inorganic substances in water under given conditions. The mentioned group quantities are defined by the relevant standardized procedures. Regarding the mechanism of radical formation in AOP, these components consume oxidative radicals; however, they can also produce them under irradiation with light of different wavelengths. Some of them also act as radical absorbers, and heavy metals can even act as catalysts. In addition, these groups show very different values in real waters. For these reasons, to remove the same concentration of the target pollutant (e.g., 100 mg/L benzene) from different wastewaters, a different and difficult method to predict the amount of the substance is needed.

The existence of oxidizing radicals •OH is also very sensitive to both the amount of supplied H_2O_2 and Fe^{2+}, which can be formally expressed as relationships:

$$Fe^{2+} + \bullet OH \rightarrow Fe^{3+} + OH^- \tag{7}$$

$$H_2O_2 + \bullet OH \rightarrow H_2O + HO_2\bullet \tag{8}$$

$$HO\bullet_2 + \bullet OH \rightarrow H_2O + O_2 \tag{9}$$

However, the oxidizing power of both O_2 itself and other possibly present radicals is much smaller than that of the •OH radical.

It would be theoretically possible to estimate the H_2O_2:R ratio from the assumed oxidation schemes for the pure R component for pure water; nevertheless, it is impossible for real water. The ratio must be determined experimentally for each case as evident from data provided by various authors. The values and the ratio of H_2O_2:Fe^{2+} for diverse organic substances and real water with different concentrations of the target components R, COD, TOC, and BOD varied considerably, mostly from 2.5–40 mmol/L: 0.17–1.48 mmol/L, and the ratio was usually 20–27. However, the values of this ratio also reached 33 or 3 (at 4 mmol Fe^{2+} for landfill leachate with COD − 743 mg/L, see [7]) For instance, in Hasan et al. [53] and Diya'uddeen et al. [8], the optimal H_2O_2 concentration was chosen to be 40.3 mmol/L and for Fe^{2+} was 1.48 mmol/L in a molar ratio of 27.2 for real water containing COD and target phenol in measured concentrations of 2322 mg/L and 84.8 mg/L. However, for high COD values and complicated compositions, e.g., seepage waters containing a number of different micropollutants (phenols, PAHs, pesticides, phthalates, and nonylphenol in microgram quantities), hydrogen peroxide was applied in up to hundreds and thousands of mmol/L in order to achieve a reduction in COD below the recommended 1000 mg/L and such quality of the water, which would make it possible to lead it, for example, to the biological stage of the WWTP [54]. Even in the work devoted to the treatment of dyeing wastewater containing azo and anthraquinone dyes with a COD over 2500 mg/L, in Gulkaya et al. [55], the authors report a possible supply of 11,323 mmol/L H_2O_2 and a ratio of H_2O_2:Fe^{2+} (g/g) up to over 100 for 95% COD removal. Another example of pharmaceutical wastewater (COD = 18,000 mg/L) states 1000 mmol/L H_2O_2 and 50 mmol/L Fe^{2+}, i.e., the ratio of 20 [50], and a recommendation for hydrogen peroxide dosage in the range of 1–10 mg/L [5], which differs significantly from most designs. It is clear from the published results that there is a significant difference for the estimation of the basic Fenton equation parameters, whether it is the degradation of an organic component in water or in real water, where the synergistic oxidation of all present components, including NOM (usually expressed as COD), takes place. In this case, it is necessary to find the parameters experimentally.

The amount of hydrogen peroxide and the ratio to the catalyst is difficult to estimate even for modifications of the Fenton process. For example, the addition of ozone can reduce the dosage of hydrogen peroxide, while the pH is adjusted towards higher values within which ozone produces radicals. For very complex contaminated leachate from waste dumps (usually with COD above 2000 ng/kg and BOD5 below 100 mg/kg), the applied peroxide concentration was 50 mmol/L and the ratio to Fe^{2+} was 1 [56]. In the case of waters from the production of ammunition containing nitroaromatics and toxic azo components, even for the combination of Fenton with ultrasound, the inverted ratio $Fe^{2+}/H_2O_2 = 500$ (concentration in mg/L) was chosen [57]. Roudi et al. [58] mathematically determined the optimization for landfill leachate values of pH = 3, Fe^{2+} = 781.25 mg/L, $Fe^{2+}/H_2O_2 = 2$, which is in agreement with the value recommended by Cortez et al. [59].

A large number of similar articles can be found in the literature, mostly without a deeper attempt to correlate the obtained data. Closer connections with the content of COD and TOC were sought by Benatti et al. [60], who optimized these amounts for wastewater more generally as the ratios [COD]:[H_2O_2] = 1:9 and [H_2O_2]:[Fe^{2+}] = 4.5:1 (in mg/L, COD in mg/kg).

However, despite a certain chaos in the determination of the basic parameters of the Fenton process, this process has the preconditions to be applied in a real process, certainly as one of the rational pretreatments of heavily contaminated water. For orientation, cost estimates, e.g., to reduce COD minimally by 70% in wastewater from the production of vegetable juices, for [Fe^{2+}] = 20 mmol/L, [H_2O_2] = 100 mmol/L, pH = 3 a 4 h, the operating procedures were calculated at 4.38 € per m^3, see Amaral-Silva et al. [61].

5.1. COD and TOC, Important Criteria for Estimating the Inlet H_2O_2 Concentration and Ratio H_2O_2:Fe^{2+}

Turki et al. [62] searched for the optimum inlet concentration for highly polluted landfill leachate COD of 12,000 mg/L. They found the optimum at an 82 mmol/L H_2O_2 and H_2O_2/Fe^{2+} ratio at 4. However, the authors of [63], based on semitheoretical notions of the Fenton mechanism, estimated such quantities as 1.2 mg H_2O_2 per 1 mg COD input and 0.9 mg Fe^{2+} per 1 mg COD, which is a smaller molar ratio of the order. Similarly, concerning wastewater from a flax cleaning plant, Abou-Elela et al. [64] indicated a suitable molar ratio of H_2O_2/Fe^{2+} as 25 and a ratio, H_2O_2:COD = 0.75–1, for COD up to 6 g/kg, which is similar to Turki et al. [62]. However, for cork boiling waters with COD = 5000 mg/kg, they reported an optimal ratio of H_2O_2:COD of 2.2 (all in mg/L), an inlet concentration of 311 mmol/L, and an H_2O_2/Fe^{2+} ratio at 8.2.

The theoretical ratio of H_2O_2:COD was originally designed as stoichiometric: 1 g COD = 1 g O_2 = 0.03125 mol O_2 = 0.0625 mol H_2O_2 = 2.125 g H_2O_2 [65,66]. It was approximately valid, for example, for small values of COD = 964 mg/L, for the decontamination of wastewater from paper mills with the optimum for H_2O_2:COD = 0.52–1.04 and input values of H_2O_2 = 15–30 mmol/L [67]. Similarly, the determination of the most suitable ratio H_2O_2:COD = 2 for the decontamination of synthetic paint (COD = 421, H_2O_2 = 800 mg/L, Fe^{2+} = 150 mg/L, molar ratio H_2O_2:Fe^{2+} = 8.8) falls into the specified range [68]. However, these numbers are not generally valid, which is usual for Fenton reactions in the case of real water. For example, Talebi et al. [69] used 747 mg/L H_2O_2 for leachate from landfills with a COD of 3.511 g/L, whereas Saber et al. [70], exploring refinery wastewater, published the ratio of H_2O_2:COD = 10.03 and used a COD = 450.

Other authors suggest [53] that rather than relying on COD values, it seems more plausible to consider TOC. They reason that TOC measures carbon that is directly converted and, therefore, is affected neither by the oxidation state of the organic matter nor by organically bound elements such as nitrogen, hydrogen, and inorganic matter. In addition, COD inadequately reflects the actual oxygen requirements for organic pollutant oxidation, as it also includes the oxidation of other substances, e.g., ferrous ions, sulfides, etc. This may explain the reason why the used concentrations of the basic components of the Fenton process and their ratio can be difficult to correlate in real wastewater. However, in the case of a large difference in COD >> TOC concentration, it will be necessary to consider the different chemistry and structure of organic substances, and therefore increase the amount of peroxide. It is very likely that the dosing of chemicals will also affect the BOD and the BOD5:COD ratio and thus also the biodegradability, which opens up space for further research.

5.2. Effect of Organic Substances in Effluents (EfOM-Effluent Organic Matter)

The Fenton reaction is often applied to eluents from WWTPs, while effluents from treatment plants contain organic substances, which are, for example, substances resulting from the biological activity of microorganisms, known as so-called extracellular dissolved polymeric substances and soluble microbial products of microbial cell metabolism. These are the ingredients that make up the bulk of COD. Basically, they play the role of sensitizers in photochemical reactions, and therefore, they affect reactions related to UV and solar radiation; moreover, they can also affect the result of an ordinary Fenton reaction, carried out in dark or light environments. Apparently, all advanced oxidation processes are affected by the presence of EfOM, where the main oxidation process takes place with these components, with the micropollutants being degraded as a cometabolic process according to these ideas (see hypothesis in Giannakis et al. [71]). Humic substances absorb light and produce a hydroxyl radical; consequently, further reactions occur with the organic substances and dissolved oxygen present. The entire system, in the presence of hydrogen peroxide and Fe ions, further participates in the formation of oxidative radicals. It is clear that the oxidation process under the conditions of a simple Fenton reaction, but even Fenton under solar radiation (UVA, UVB) or hard UVC radiation, is very complicated, and some opinions of different researchers are mentioned in the publication Giannakis et al. [71].

Obviously, Fenton processes can contain organic sensitizers in different qualities and quantities in different real waters, especially in WWTP effluents; hence, the setting of the most suitable reaction conditions depends mainly on the EfOM content, which significantly varies. The presence of suspended particles would also have an impact, which would affect the penetration of light. Thus, the main process of advanced oxidation by radicals is the destruction of EfOM, while the oxidation of micropollutants is rather accidental and accompanying. Probably, micropollutants would be removed only after EfOM has been removed from these waters. Thus, the preliminary removal of these substances, for example by coagulation or membrane filtration, would be suitable for the deeper decontamination of micropollutants. It can be assumed for Fenton, solar Fenton, and UVB-photoFenton that degradation rates (e.g., pharmaceuticals) are rather slower than with UVC/H_2O_2 applications. However, the amount of micropollutants removed by the Fenton reaction can be significant for all reactions.

5.3. What Else Needs to Be Clarified concerning the Fenton Reaction?

Considering the fact that TOC and COD are usually associated with contaminated water, it is essential to quantify or realistically estimate the concentration of oxidizing radicals formed by the Fenton reaction necessary for the degradation of organic pollutants in order to achieve the permitted desired target values (concentration of the target component or final COD, e.g., 1000 mg/kg, zero ecotoxicity, etc.). The amount of oxidizing radicals can be quantified (e.g., [72–75]). Regrettably, the stoichiometric relationship between TOC and the amount of oxidizing radicals is usually not specified. The total amount of generated oxidizing radicals for different concentrations of H_2O_2 and Fe^{2+}, as well as for their H_2O_2/Fe^{2+} ratio (e.g., for pH 3), is not sufficiently known, which makes a more accurate estimation of the input composition of peroxide and Fe ions for individual cases of contaminated water impossible. Usually, these parameters are selected based on previous experience and are further optimized experimentally. For this reason, it is necessary to accept the situation that the setting of suitable concentrations of basic chemicals and their ratio in the Fenton reaction cannot be reliably determined in advance. Simultaneously, it should be taken into account that the vast majority of works on the Fenton reaction were carried out with simulated water, and their conclusions are unreliable for real water.

6. Disadvantages of the Basic Fenton Process and Possible Solutions

The basic disadvantage of the Fenton reaction is the low pH value around pH 3, above which the radicals are not stable. Moreover, at a pH below 3, there is a strong reduction in radical formation [13], and at a pH above 3, specifically around pH 5, insoluble $Fe(OH)_2$ is formed, and at a pH above 4, the low decomposition of peroxide, which preferentially decomposes into oxygen and water without the formation of radicals, appears. For these reasons, it is also necessary to focus on whether the target organic substances are soluble at all of the pH ranges suitable for the generation of radicals. In the case of heterogeneous catalysts, it is also essential to consider the isoelectric point and the surface charge of the catalyst as well as the target substance, i.e., whether the adsorption of the substance on the catalyst is favored at a given pH. It is one of the important control steps of radical oxidation on the catalyst. Treated water also needs to be neutralized before being released into the environment, which increases the cost of chemicals.

It appears that it is possible to apply an additive that "wraps" the Fe ion and prevents the precipitation of iron hydroxide (e.g., by adding resorcinol, see Romero et al. [76]). This direction becomes interesting for further research related to the effort to realize pollutant oxidation at a neutral pH, which is highly urgent.

The application of the Fenton reaction under desired neutral conditions is currently possible only with its modified hybrid version, e.g., with the catalyst ferric-nitrilotriacetate complex (Fe^{3+}-NTA) and under the influence of UVA radiation (0.178 mM Fe^{3+}-NTA (1:1), 4.54 mM H_2O_2, UVA intensity 4.05 mW/cm^2, hydraulic retention time (HRT) 2 h, influent pH 7.6 [77]). The application of various chelate complexes with Fe can also be considered as a variant of the homogeneous Fenton reaction, which is also possible in execution with UV. The generation of •OH radicals was also confirmed with the use of Fe^{3+}-ethylenediamine-N,N'-disuccinic acid (Fe^{3+}-EDDS), and regarding the degradation efficiency, it greatly surpassed other Fe^{3+} complexes, e.g., with citric, malic, oxalic, and wine acid.

Some authors solved the problem of reducing the high consumption of chemicals by treating effluents from treatment plants with extremely low values of iron and peroxide (from popular ratios such as H_2O_2/Fe^{2+} (in mmol/L) = 2/0.2 or 3/0.3. They choose the smallest one, perhaps 1.5/0.1), and the solution was doped with chelates (EDDS and citrate) when it was possible to work at an almost neutral pH, see Miralles-Cuevas et al. [78]. It was verified only for low values of drugs (carbamazepine, flumequine, ibuprofen, ofloxacin, and sulfa-methoxazole) around 15 μg/L and very low values of COD (30–40 mg/L). However, the result of the Fenton reaction, e.g., in the simultaneous presence of complex substances, can be different from the classic Fenton reaction, see Kuznetsova et al. [79], and unlike the classic Fenton reaction, the reaction can also take place at a higher pH, which would be highly favorable. Nevertheless, a more acidic environment would be more favorable even in the presence of chelates and with a heterogeneous Fenton reaction (which is shown by the reduction in •OH oxidation potential formation at a neutral pH (2.8 V at pH 3 vs. 1.9 V at pH 7) [25]).

Additionally, another disadvantage of this reaction is the necessity to subsequently rid the treated water of Fe ions or complexes. Further disadvantages of the homogeneous Fenton process, such as the formation of a large amount of Fe^{3+} and the limited pH range in the very acidic region, prompted the testing of the Fenton process in a heterogeneous form. Additionally, the application in real wastewater encounters the presence of various chelating substances and phosphates, which can react with Fe ions and, thus, interfere with the generation of oxidative radicals, as already mentioned.

7. Energy-Assisted Fenton Reactions

As already stated, the simple original Fenton reaction (Fe^{2+}/H_2O_2) can be simultaneously combined with UV radiation, ultrasound, or ozonation. Compared to the original classical technology based on simply mixing two common chemicals, such variations are now referred to as "advanced" and are a part of the AOP. The simplest advanced modification of the Fenton reaction is its execution under the application of UV radiation of different wavelengths, which is known as the photo-Fenton variation ($Fe^{2+}/H_2O_2/UV$).

The development of the knowledge of the theoretical foundations of the Fenton process has led to the study and optimization of the $Fe^{2+}:H_2O_2:UV$ ratio, and logically also to the idea of simplifying the scheme of the photo-Fenton arrangement, for example, by reducing the mass of the catalyst to a minimum, and studying simpler arrangements without the presence of a catalyst. It could even be a system where the amount of catalyst is zero, i.e., the H_2O_2/UV system, in which the oxidizing radicals are generated only by the action of UV radiation on hydrogen peroxide and water. This system is universally applicable to underground and surface wastewater, and it is the only one that enables the large-scale treatment of contaminated water for drinking purposes, perhaps with the exception of perfluorinated compounds. Some successes under strictly laboratory conditions—including photolysis and photo-Fenton for shorter perfluorinated compounds—are cited by Arvaniti and Stasinakis [80].

Photo-Fenton is already an independent process to which dozens of scientific reports are devoted [81–84]. Its simple variation could represent a promising water decontamination process; however, the decontamination products and intermediates may be more toxic than the parent contaminant, and hence there is space for further ecotoxicological research. This danger does not appear with the photocatalysis process. Instead of Fe^{2+} or particles of zero valent Fe in the heterogeneous Fenton variant, other catalysts can be applied, e.g., TiO_2, ZnO, i.e., chemical semiconductors which, when irradiated with UV light of certain wavelengths, produce radicals capable of oxidizing organic substances in water [85]. The combination with photocatalysis, when the full use of photogenerated electrons can be expected, which would increase the efficiency of the Fenton process for the application of various semiconductors (such as TiO_2, g-C_3N_4, graphene, $BiVO_4$, $ZnFeO_4$ and $BiFeO_3$, activated carbon/$CoFe_2O_4$ nanocomposites, etc.) is currently being thoroughly studied [86–88].

From an economic point of view, it would be interesting to replace UV radiation with solar radiation and simultaneously use ultrasounds such as F/US or ozone F/O_3. These modifications create different amounts of reactive oxidizing radicals, which are higher than their amount in the classic Fenton process [89]. Variations in the amount of peroxide and the ratio with Fe^{2+} were recorded, e.g., for leachate, 2000 mg/L H_2O_2 and 10 mg/L Fe^{2+} were used during aeration under a medium-pressure Hg lamp (125 W). The amount of COD dropped by 57% from 5200 mg/L within 60 min [90]. The decolorization of wastewater especially from the food industry is often preferred. The almost complete decolorization of strongly dark water from olive oil production, which is a significant environmental problem, e.g., for the Mediterranean countries of Europe, was achieved by the combination of H_2O_2/Fe (6/1 volumes) under UV radiation [91].

According to the targeted interest of most authors dealing with the oxidative decontamination of organic substances, photo-Fenton, which has been verified in the removal of various pesticides, antibiotics, or industrial waters [49,92,93] is currently preferred, as well as solar radiation at a neutral pH [94,95] or with heterogeneous catalysts [96,97].

It is evident that the application of heterogeneous variations, including photocatalysis, is more efficient and, due to the reduction in the difficult final sludge, more practical than the homogeneous Fenton reaction. Heterogeneous photo-Fenton with UV radiation could thus be a good alternative for solving a number of water decontamination problems. However, for highly polluted effluents, the penetration of UV radiation through a dense suspension will be strongly reduced. In this case, the solution could be a variation producing UV light using so-called electrodeless lamps with the application of microwaves, which are inserted directly into the catalyst suspension, and the radiation is thus in direct contact with the contaminant, e.g., heterogeneous Fenton + electrodeless lamp UV + microwaves. This combination, which has already been tested for the decontamination of highly polluted seepage waters [98] or without a catalyst for the removal of polybrominated substances by Kastanek et al. [99], could prove useful for point sources of limited volumes of wastewater, e.g., hospitals.

Another modification of the Fenton reaction is the electro-Fenton reaction, in which the two chemicals H_2O_2 and Fe^{2+} are produced electrochemically, while the generated radicals have an oxidative effect on the organic pollutants present (see Figure 3). It is an interesting process, suitable for laboratory verification and theoretical research with the potential of practical applications [100–103]. Hydrogen peroxide is formed on the cathode during bubbling with oxygen, which does not need to be added. Simultaneously, the Fe^{3+} ion formed by the Fenton reaction is reduced, which contributes to the renewal of the Fe^{2+} catalyst and the decrease in the ferric ion, thereby reducing the formation of iron sludge, which is one of the inconveniences of the classic Fenton reaction. However, the exact explanations for both radical generation and $Fe^{2+} \rightarrow Fe^{3+} \rightarrow Fe^{2+}$ transitions have not yet been found.

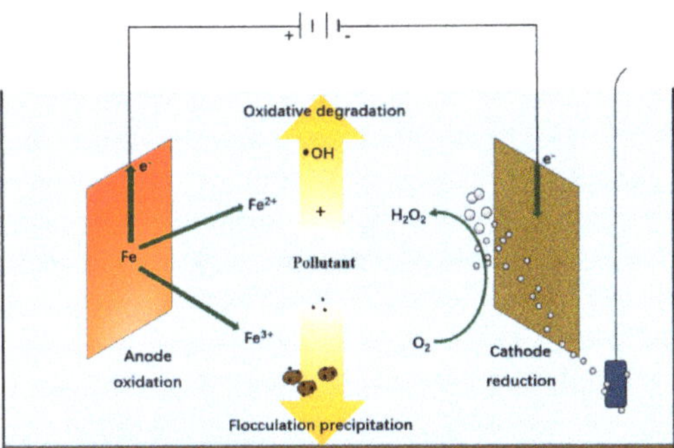

Figure 3. Electro-Fenton reaction mechanism diagram [42].

An important role is played by the design, arrangement, and material of the electrodes, especially the cathode, which affects the formation of peroxide [104]). For example, Wang et al. [105] recommends a carbon–polytetrafluoroethylene mixture as a base on a Ni-grid, Zhao et al. [106] recommends graphene, etc. Apparently, the biggest technical problem in practical use is solving the electrode distance for a scale up. Heidari et al. [107] showed that the optimal electrode distance was 3 cm (in a 400 mL beaker, diameter 8.8 cm) for the successful degradation of organics in water, specifically pentachlorophenol. For realistic wastewater flows, it would be necessary to construct a reactor as a multiple system of horizontally and vertically placed pairs of electrodes, which has not yet been tested.

However, the formation of insoluble hydroxide, which is inevitable during the course of the reaction and covers the cathode over time, which reduces the amount of generated hydrogen peroxide, is a problem. It will be difficult to estimate the ratio of peroxide to catalyst. Moreover, regarding real waters, a part of the Fe ions will react with humic substances and compounds of fulvic acid and NOM to solid flocs, which may dissolve during the reaction. It will be necessary to optimize whether it is more appropriate to separate the flocs (they retain significant amounts of organic substances and possibly heavy metals present) or to leave them in the reaction mixture as a source of catalyst. Coagulation–flocculation is, after all, a common pretreatment method for Fenton reactions, see Vedrenne et al. [108]. An electro-Fenton reaction would generally be suitable for wastewater with a high salt content, owing to the fact that there is no need to dope the water with an electrolyte (especially for the complex composition of, e.g., leachates wastewater from landfills, or wastewater from leather treatment, textile dyeing, etc.). Nevertheless, it is necessary to master the reaction arrangement in real large-volume conditions, namely the material of the electrodes, their size, distances, etc. For example, the so-called reticulated vitreous carbon RVC was applied as a cathode by El-Desoky et al. [109].

Another possibility for the generation of radicals is the use of ultrasound (US), which could be more economically advantageous than UV radiation. It is apparently possible to achieve good efficiencies with less chemical consumption and in a shorter time with the application of US [110]. This is an indisputable advantage that would significantly reduce the amount of sludge with Fe^{3+} [111]. Simultaneously, it is necessary to optimize power USs because increasing the power may cause more formations of cavitation bubbles, which generate oxidative free radicals OH•, and the supersaturation of the bubbles may even cause less implosion, resulting in less efficiency of organic matter destruction.

During the ultrasonic radiation process, there are three reaction degradation zones: hydrophilic substances are localized inside the solution, nonvolatile hydrophobic substances are located mainly at the bubble–water interface, and volatile substances are mostly inside the cavitation bubbles. Pyrolysis, as a degradation reaction, takes plays inside the cavitation bubbles. On the surface of the bubbles, the main reaction is the attack triggered by radicals arising during the implosion of the cavitation bubbles. Last but not least, in the volume of the solution, the reaction takes place with free radicals. Therefore, it is a very complicated process, where each substance is differently hydrophobic, and it is difficult to optimize the entire system [112]. Different ultrasonic variations in the Fenton reaction were again tested, e.g., the electro-Fenton reaction with US, in which hydrogen peroxide and Fe^{2+} were generated in an electrochemical cell with special electrodes (boron diamond doped electrodes) under the influence of US [113], or the completely reduced version without Fe ions (e.g., Rahdar et al. [114]), as an analogy to advanced UV/H_2O_2 oxidation. The use of US was also confirmed by Muñoz-Calderon et al. [115], who claimed that the correct dosing of the oxidant and catalyst, supported by ultrasound, could be a good alternative to the Fenton reaction.

8. Summarization

The Fenton reaction, by introducing the concept of oxidizing hydroxyl radicals, created the phenomenon of contemporary advanced oxidation processes. It is unique in its apparent simplicity and could significantly contribute to the healing of the environment in all its modifications. The Fenton reaction has been cited and experimentally verified for decades, and although hundreds of publications have been devoted to it, much remains unexplained. Even some very basic data, such as the primary determination of the optimal dosage of two single reaction quantities for the oxidation of the target component in real water, are the subject of rather randomly and/or traditionally chosen quanta, requiring laborious experimental optimizations for each individual case. However, most of the conclusions of a large number of studies on the application of the basic Fenton reaction to different real waters incline to the opinion that basically, regardless of the level of COD and the dosage of peroxide, COD removal efficiencies, even for complexly polluted waters, are on average around 70%. This fact is not only an interesting phenomenon of the Fenton reaction, which does not disappoint in terms of the effectiveness of decontamination even with a certain freedom in the selected amount of basic components, but also a confirmation that it is unique and difficult to be replaced. In addition, it is one of the most important processes for the remediation of water and soil, especially if the concentrations of pollutants are not high and the cost of the consumed H_2O_2 allows for a cost-effective treatment of the reaction medium.

It appears that layered and porous aluminosilicates and analogous double oxides containing iron, copper, or other transition metals exhibit a catalytic activity for the Fenton process. However, as in any catalytic process, the key point is the efficiency of using the reagent, in this case H_2O_2, to generate hydroxyl-free radicals. This issue has often been ignored, and mostly only model pollutant extinction and/or total organic carbon reduction has been studied. The optimization of H_2O_2 is also considered one of the important parameters in the evaluation of the efficiency of solid catalysts. The current situation in this area does not yet allow conclusions to be drawn regarding the relative catalytic activity of the various tested solids. In addition to the selective activity in H_2O_2 conversion, other parameters, such as catalyst stability, absence of leaching, and aging of catalytic sites, as well as operating conditions including pH, temperature, and catalyst amount, must be considered. It will also be necessary to confirm whether nonmetallic catalysts, e.g., graphene, have a high efficiency in generating hydroxyl radicals. Moreover, attention will need to be paid to surface modifications by introducing suitable functional groups on easily available activated carbon, perhaps also in the form of fibers, as a carrier of transit metals, mainly Fe for heterogeneous variations in the Fenton reaction.

Simultaneously, it can be assumed that the use of the Fenton reaction will increase in the near future, motivated by environmental problems and pollution remediation. Future developments in this area will lead to clarification of the current state, with some new materials emerging as major catalysts for the Fenton reaction. It already appears that various modifications of the Fenton method (especially photo-Fenton with UV and visible light) can, in many cases, be more effective for the decomposition of many recalcitrant pollutants than the classical Fenton method. One promising possibility is the verification of the Fenton/ultrasound variation, which can be operated in relatively large volumes, even continuously.

The growing contents of so-called emerging pollutants in waters of all kinds will probably, in the foreseeable future, inevitably lead to the need for their removal even on a real scale, and society will have to bear the necessary costs, which, however, must be rational. Deciding on which advanced oxidation process to use (there will probably be no doubt about its application in the necessary complex with pre- and post-treatment technologies) will be rational, and the choice will probably involve:

- Photolysis of the UV/H_2O_2 type, possibly with the application of various catalysts.
- Photocatalysis (most likely UV/TiO_2), possibly with the interaction of H_2O_2.
- The classic Fenton reaction (Fe^0, Fe^{2+}, Fe^{3+}, possibly in the form of a photo-Fenton reaction under UV exposure).
- Homogeneous Fenton reaction working at neutral pH, i.e., its so-called "green" form corresponding to current environmental trends.

Such a competition is not only still open but also indispensable, and the Fenton process seems to have a great chance. This is evidenced by the incessant publication of its various variants in very good scientific journals.

8.1. Future Directions

Research on the Fenton method will continue, and according to the current literature search, its focus should be directed on the following topics:

- Solar energy, including clarifying its real influence on the homogeneous and heterogeneous Fenton reaction, which could be interesting for subtropical and tropical regions, which currently usually suffer from high pollutant loads.
- An evaluation of the possibility of applying other components besides hydrogen peroxide, such as persulphates, percarbonates, or ferrate Fe^{4+}, which has been addressed so far by Lee et al. [116] and Karim et al. [117].

- A theoretical clarification of how to optimally dose the amount and ratio of catalyst (preferably Fe^{2+}) and H_2O_2 for different water compositions, i.e., COD, TOC concentrations, and the composition of various micropollutants, which still, after more than hundreds of years, remains unclear, left either to a random choice or supported by randomized experiments.
- Solving the use of sludge with Fe^{3+} from the point of view of waste policies, especially the use of sludge Fe as a catalyst, which would positively affect the economics of the Fenton process (see Xu et al. [3]).
- Finding Fe^{3+} chelates, which could shift the Fenton reaction to its "green" homogeneous variant, working at a normal pH, which would, among other things, open the Fenton reaction to other possibilities for removing, for example, emerging pollutants from wastewater, when their strong acidification and subsequent neutralization is difficult and expensive, e.g., Lekikot et al. [118], while the removal of the complex from the treated water including regeneration is an indispensable step [119].
- A rational evaluation of costs to achieve the required final values of concentrations and ecotoxicity for individual variants of the Fenton reaction, including hybrid ones, which is necessary for the real use.
- A verification of the Fenton reaction, at least in a pilot setting, on traditional pollutants from selected real producers of these wastes (PAU, BTEX) and on real wastewater from various industries (on paper mill water [67,120], on leachate from landfills [121], as well as emerging pollutants, which is mostly missing). It would make it possible to express a fundamental opinion based on the conviction that, regarding the situation in the issue of wastewater decontamination, the Fenton reaction is an irreplaceable environmental tool.

8.2. Conclusions

The more than 100-year-old principle of the Fenton reaction, which fundamentally influenced all subsequent advanced oxidation processes for the removal of pollutants from water and soil, is far from being exhausted. Under real conditions, the formation of oxidative radicals is strongly influenced not only by the presence of natural and dissolved organic substances such as humic and fulvic acids, natural surfactants, and the interaction of soil microorganisms, but also by the presence of natural chelates and other heavy metals. The biggest problem of the Fenton reaction application in the real soil and water environment is the necessity of a low pH of 3, which is ecologically harmful. However, the use of Fe chelates, which enable the reaction at a normal pH, is also not a solution. Various types of substances have been tried, from organic (based on EDTA, amino acids, or peptides) to inorganic (e.g., tripolyphosphates); regrettably, no one has any idea what happens to them in the real environment. The question also arises whether it would be wise to give up the beautiful simplicity of the Fenton reaction and substitute it for complex catalysts, which are more expensive and will eventually be connected with the same problems as simple Fe ions in real waters or soils. As part of the development of scientific knowledge, it is of course necessary to solve problems associated with the development of new catalysts, including nanoforms; nevertheless, it must be taken into consideration that the reality of contaminated soils and waters is complex, and thus experiments in simulated soil and water are rather misleading and will not solve this problem. There is a huge scope for experiments with real soils that would help shed light on the fates of oxidizing radicals under various specific conditions, which could significantly expand scientific knowledge.

In conclusion, it could be stated that there are still many challenging scientific and application issues associated with the Fenton reaction; therefore, we can look forward to the potential upcoming interesting solutions and results related to real applications.

Author Contributions: Conceptualization, F.K., O.S. and P.K.; methodology, F.K. and O.S.; validation, F.K., O.S., P.K., M.D. and M.S.; formal analysis, P.K., M.D. and M.S.; investigation, F.K., O.S., P.K., M.D. and M.S.; resources, F.K., O.S. and P.K.; data curation, F.K., O.S. and P.K.; writing—original draft preparation, F.K., O.S., P.K., M.D. and M.S.; writing—review and editing, M.D. and M.S.; visualization,

M.D. and M.S.; supervision, F.K. and O.S.; project administration, O.S.; funding acquisition, O.S. All authors have read and agreed to the published version of the manuscript.

Funding: This research was funded by the Ministry of Industry and Trade of the Czech Republic, project no. FV40126. Part of the work was financed by the research infrastructure NanoEnviCZ (LM2018124).

Data Availability Statement: Not applicable.

Acknowledgments: The financial support of the Ministry of Industry and Trade of the Czech Republic (project no. FV40126) is gratefully acknowledged. Part of the work was financed by the research infrastructure NanoEnviCZ (LM2018124).

Conflicts of Interest: The authors declare no conflict of interest.

References

1. Fenton, H.J.H.; Jones, H.O. The oxidation of organic acids in presence of ferrous iron. Part I. *J. Chem. Soc.* **1900**, *77*, 69–76. [CrossRef]
2. Deng, Y.; Engelhard, D.J. Treatment of landfill leachate by Fenton process. *Water Res.* **2006**, *40*, 3683–3694. [CrossRef] [PubMed]
3. Xu, M.; Wu, C.; Zhou, Y. Advancements in the Fenton Process for Wastewater Treatment. In *Advanced Oxidation Processes—Applications, Trends, and Prospects*; Bustillo-Lecompte, C., Ed.; IntechOpen: London, UK, 2020. [CrossRef]
4. Sindhi, Y.; Mehta, M. COD Removal of Different Industrial Wastewater by Fenton Oxidation Proces. *Int. J. Eng. Sci. Res. Technol.* **2014**, *3*, 1134–1139.
5. Pawar, V.; Gawande, S. An overview of the Fenton Process for Industrial Wastewater. *IOSR J. Mech. Civ. Eng.* **2015**, *2*, 127–136.
6. Rodrigues, C.S.D.; Madeira, L.M.; Boaventura, R.A.R. Decontamination of an Industrial Cotton Dyeing Wastewater by Chemical and Biological Processes. *Ind. Eng. Chem. Res.* **2014**, *53*, 2412–2421. [CrossRef]
7. Cortez, S.; Teixeira, P.; Oliveira, R.; Mota, M. Evaluation of Fenton and ozone-based advanced oxidation processes as mature landfill leachate pre-treatments. *J. Environ. Manag.* **2011**, *92*, 749–755. [CrossRef]
8. Diya'uddeen, B.H.; Aziz, A.R.A.; Ashri, W.D.W.M. Optimized Treatment of Phenol-Containing Fire Fighting Wastewater Using Fenton Oxidation. *J. Environ. Eng.* **2012**, *138*, 761–770. [CrossRef]
9. Neyens, E.; Baeyens, J. A review of classic Fenton's peroxidation as an advanced oxidation technique. *J. Hazard. Mater. B* **2003**, *98*, 33–50. [CrossRef]
10. Sarmento, A.P.; Borges, A.C.; Matos, A.T.d.; Romualdo, L.L. Sulfamethoxazole and Trimethoprim Degradation by Fenton and Fenton-Like Processes. *Water* **2020**, *12*, 1655. [CrossRef]
11. Wang, M.; Su, X.; Yang, S.; Li, L.; Li, C.; Sun, F.; Dong, J.; Rong, Y. Bisphenol A degradation by Fenton advanced oxidation process and operation parameters optimization. *E3S Web Conf.* **2020**, *144*, 01014. [CrossRef]
12. Guo, B.; Xu, T.; Zhang, L.; Li, S. A heterogeneous fenton-like system with green iron nanoparticles for the removal of bisphenol A: Performance, kinetics and transformation mechanism. *J. Environ. Manag.* **2020**, *272*, 111047. [CrossRef] [PubMed]
13. Mirzaei, A.; Chen, Z.; Haghighat, F.; Yerushalmi, L. Removal of pharmaceuticals from water by homo/heterogonous Fenton-type processes—A review. *Chemosphere* **2017**, *174*, 665–688. [CrossRef] [PubMed]
14. Adhami, S.; Jamshidi-Zanjani, A.; Darban, A.K. Phenanthrene removal from the contaminated soil using the electrokinetic-Fenton method and persulfate as an oxidizing agent. *Chemosphere* **2021**, *266*, 128988. [CrossRef] [PubMed]
15. Heydari, F.; Osfouri, S.; Abbasi, M.; Dianat, M.J.; Khodaveisi, J. Treatment of highly polluted grey waters using Fenton, UV/H_2O_2 and UV/TiO_2 processes. *Membr. Water Treat.* **2021**, *12*, 125–132.
16. Huong Le, T.X.; Drobek, M.; Bechelany, M.; Motuzas, J.; Julbe, A.; Cretin, M. Application of Fe-MFI zeolite catalyst in heterogeneous electro-Fenton process for water pollutants abatement. *Microporous Mesoporous Mater.* **2019**, *278*, 64–69. [CrossRef]
17. Vega, E.; Valdés, H. New evidence of the effect of the chemical structure of activated carbon on the activity to promote radical generation in an advanced oxidation process using hydrogen peroxide. *Microporous Mesoporous Mater.* **2018**, *259*, 1–8. [CrossRef]
18. Navalon, S.; Alvaro, M.; Garcia, H. Heterogeneous Fenton catalysts based on clays, silicas and zeolites. *Appl. Catal. B Environ.* **2010**, *99*, 1–26. [CrossRef]
19. Drumm, F.C.; Schumacher de Oliveira, J.; Foletto, E.L.; Dotto, G.L.; Flores, E.M.M.; Peters Enders, M.S.; Müller, E.I.; Janh, S.L. Response surface methodology approach for the optimization of tartrazine removal by heterogeneous photo-Fenton process using mesostructured Fe_2O_3-suppoted ZSM-5 prepared by chitin-templating. *Chem. Eng. Commun.* **2018**, *205*, 445–455. [CrossRef]
20. Mohan, N.; Cindrella, L. Direct synthesis of Fe-ZSM-5 zeolite and its prospects as effic electrode matcrial in methanol fuel cell. *Mater. Sci. Semicond. Proc.* **2015**, *40*, 361–368. [CrossRef]
21. Yuan, E.; Wu, G.; Dai, W.; Guan, N.; Li, L. One-pot construction of Fe/ZSM-5 zeolites for the selective catalytic reduction of nitrogen oxides by ammonia. *Catal. Sci. Technol.* **2017**, *7*, 3036–3044. [CrossRef]
22. Bian, K.; Zhang, A.; Yang, H.; Fan, B.; Xu, S.; Guo, X.; Song, C. Synthesis and Characterization of Fe-Substituted ZSM-5 Zeolite and Its Catalytic Performance for Alkylation of Benzene with Dilute Ethylene. *Ind. Eng. Chem. Res.* **2020**, *59*, 22413–22421. [CrossRef]

23. Hosseinpour, M.; Akizuki, M.; Yoko, A.; Oshima, Y.; Soltani, M. Novel synthesis and characterization of Fe-ZSM-5 nanocrystals in hot compressed water for selective catalytic reduction of NO with NH_3. *Microporous Mesoporous Mater.* **2020**, *292*, 109708. [CrossRef]

24. Rostamizadeh, M.; Abbas, J.; Soorena, G. High efficient decolorization of Reactive Red 120 azo dye over reusable Fe-ZSM-5 nanocatalyst in electro-Fenton reaction. *Sep. Purif. Technol.* **2018**, *192*, 340–347. [CrossRef]

25. Modarresi-Motlagh, S.; Bahadori, F.; Ghadiri, M.; Afghan, A. Enhancing Fenton-like oxidation of crystal violet over Fe/ZSM-5 in a plug flow reactor. *Reac. Kinet. Mech. Cat.* **2021**, *133*, 1061–1073. [CrossRef]

26. Adityosulindro, S.; Julcour-Lebigue, C.; Barthe, L. Heterogeneous Fenton oxidation using Fe-ZSM5 catalyst for removal of ibuprofen in wastewater. *J. Environ. Chem. Eng.* **2018**, *6*, 5920–5928. [CrossRef]

27. Su, R.; Chai, L.; Tang, C.; Li, B.; Yang, Z. Comparison of the degradation of molecular and ionic ibuprofen in a UV/H_2O_2 system. *Water Sci. Technol.* **2018**, *77*, 2174–2183. [CrossRef]

28. Reimbaeva, S.M.; Massalimova, B.K.; Kalmakhanova, M.S. New pillared clays prepared from different deposits of Kazakhstan. *Mater. Today Proc.* **2020**, *31*, 607–610. [CrossRef]

29. Baloyi, J.; Ntho, T.; Moma, J. Synthesis and application of pillared clay heterogeneous catalysts for wastewater treatment: A review. *RSC Adv.* **2018**, *8*, 5197–5211. [CrossRef]

30. Muñoz, H.-J.; Blanco, C.; Gil, A.; Vicente, M.-A.; Galeano, L.-A. Preparation of Al/Fe-Pillared Clays: Effect of the Starting Mineral. *Materials* **2017**, *10*, 1364. [CrossRef]

31. Stöcker, M.; Seidl, W.; Seyfarth, L.; Senker, J.; Breu, J. Realisation of truly microporous pillared clays. *Chem. Commun.* **2018**, *2008*, 629–631. [CrossRef]

32. Bhargava, S.K.; Tardio, J.; Prasad, J.; Föger, K.; Akolekar, D.B.; Grocott, S.C. Wet Oxidation and Catalytic Wet Oxidation. *Ind. Eng. Chem. Res.* **2006**, *45*, 1221–1258. [CrossRef]

33. Fang, G.-d.; Liu, C.; Gao, J.; Zhou, D.-m. New Insights into the Mechanism of the Catalytic Decomposition of Hydrogen Peroxide by Activated Carbon: Implications for Degradation of Diethyl Phthalate. *Ind. Eng. Chem. Res.* **2014**, *53*, 19925–19933. [CrossRef]

34. Sekaran, G.; Karthikeyan, S.; Boopathy, R.; Maharaja, P.; Gupta, V.K.; Anandan, C. Response surface modeling for optimization heterocatalytic Fenton oxidation of persistence organic pollution in high totaldissolved solid containing wastewater. *Environ. Sci. Pollut. Res. Int.* **2014**, *21*, 1489–1502. [CrossRef] [PubMed]

35. Nidheesh, P.V. Heterogeneous Fenton catalysts for the abatement of organic pollutants from aqueous solution: A review. *RCS Adv.* **2015**, *5*, 40552–40577. [CrossRef]

36. Badi, M.Y.; Azari, A.; Pasalari, H.; Esrafili, A.; Farzadkia, M. Modification of activated carbon with magnetic Fe_3O_4 nanoparticle composite for removal of ceftriaxone from aquatic solutions. *J. Mol. Liq.* **2018**, *261*, 146–154. [CrossRef]

37. Espinosa, J.C.; Navalon, S.; Primo, A.; Moral, M.; Fernándze Sanz, J.; Álvaro, M.; García, H. Graphenes as efficient metal-free Fenton catalyst. *Chemistry* **2015**, *21*, 11966–11971. [CrossRef]

38. Divyapriya, G.; Nidhees, P.V. Importance of graphene in electro-Fenton process. *ACS Omega* **2020**, *5*, 4725–4732. [CrossRef]

39. Li, W.; Wu, X.; Li, S.; Tang, W.; Chen, Y. Magnetic porous Fe_3O_4/carbon octahedra derived from iron-based metal-organic framework as heterogeneous Fenton-like catalyst. *Appl. Surf. Sci.* **2018**, *436*, 252–262. [CrossRef]

40. Zubir, N.A.; Yacou, C.; Motuzas, J.; Zhang, X.; Zhao, X.S.; Diniz da Costa, J.C. The sacrificial role of graphene oxide in stabilising a Fenton-like catalyst GO–Fe_3O_4. *Chem. Commun.* **2015**, *51*, 9291–9293. [CrossRef]

41. Navalon, S.; Dhakshinamoorthy, A.; Alvaro, M.; Gar, H. Heterogeneous Fenton Catalysts Based on Activated Carbon and Related Materials. *ChemSusChem* **2011**, *4*, 1712–1730. [CrossRef]

42. Xin, L.; Hu, J.; Xiang, Y.; Li, C.; Fu, L.; Li, Q.; Wei, X. Carbon-Based Nanocomposites as Fenton-Like Catalysts in Wastewater Treatment Applications: A Review. *Materials* **2021**, *14*, 2643. [CrossRef] [PubMed]

43. Wan, Z.; Hu, J.; Wang, J. Removal of sulfamethazine antibiotics using Ce-Fe-graphene nanocomposite as catalyst by Fenton-like process. *J. Environ. Manag.* **2016**, *182*, 284–291. [CrossRef] [PubMed]

44. Yao, Y.; Wang, L.; Sun, L.; Zhu, S.; Huang, Z.; Mao, Y.; Lu, W.; Chen, W. Efficient removal of dyes using heterogeneous Fenton catalysts based on activated carbon fibers with enhanced activity. *Chem. Eng. Sci.* **2013**, *101*, 424–431. [CrossRef]

45. Su, C.; Cao, G.; Lou, S.; Wang, R.; Yuan, F.; Yang, L.; Wang, Q. Treatment of Cutting Fluid Waste using Activated Carbon Fiber Supported Nanometer Iron as a Heterogeneous Fenton Catalyst. *Sci. Rep.* **2018**, *8*, 10650. [CrossRef] [PubMed]

46. Duarte, F.; Morais, V.; Maldonado-Hódar, F.J.; Madeira, L.M. Treatment of textile effluents by the heterogeneous Fenton process in a continuous packed-bed reactor using Fe/activated carbon as catalyst. *Chem. Eng. J.* **2013**, *232*, 34–41. [CrossRef]

47. Messele, S.A.; Stüber, F.; Bengoa, C.; Fortuny, A.; Fabregat, A.; Font, J. Phenol Degradation by Heterogeneous Fenton-Like Reaction Using Fe Supported Over Activated Carbon. *Procedia Eng.* **2012**, *42*, 1373–1377. [CrossRef]

48. Soares, S.G.P.; Rodrigues, C.S.D.; Madeira, L.M.; Pereira, M.F.R. Heterogeneous Fenton-Like Degradation of p-Nitrophenol over Tailored Carbon-Based Materials. *Catalysts* **2019**, *9*, 258. [CrossRef]

49. Ramos, M.D.N.; Lima, J.P.P.; Aquino, S.F.; Aguiar, A. A critical analysis of the alternative treatments applied to effluents from Brazilian textile industries. *J. Water Process. Eng.* **2021**, *43*, 102273. [CrossRef]

50. Vergili, I.; Gencdal, S. Applicability of combined Fenton oxidation and nanofiltration to pharmaceutical wastewater. *Desalination Water Treat.* **2015**, *56*, 3501–3509. [CrossRef]

51. Pignatello, J.J.; Oliveros, E.; MacKay, A. Advanced Oxidation Processes for Organic Contaminant Destruction Based on the Fenton Reaction and Related Chemistry. *Crit. Rev. Environ. Sci. Technol.* **2006**, *36*, 1–84. [CrossRef]

52. Umar, M.; Aziz, H.A.; Yusoff, M.S. Trends in the use of Fenton, electro-Fenton and photo-Fenton for the treatment of landfill leachate. *Waste Manag.* **2010**, *30*, 2113–2121. [CrossRef]

53. Hasan, D.B.; Raman, A.A.A.; WanDaud, W.M.A. Kinetic Modeling of a Heterogeneous Fenton OxidativeTreatment of Petroleum Refining Wastewater. *Sci. World J.* **2014**, *2014*, 252491. [CrossRef]

54. Ateş, H.; Argun, M.E. Advanced oxidation of landfill leachate: Removal of micropollutants and identification of by-products. *J. Hazard. Mater.* **2021**, *413*, 125326. [CrossRef]

55. Gulkaya, İ.; Surucu, G.A.; Dilek, F.B. Importance of H_2O_2/Fe^{2+} ratio in Fenton's treatment of a carpet dyeing wastewater. *J. Hazard. Mater.* **2006**, *136*, 763–769. [CrossRef]

56. Abu Amr, S.S.; Abdul Azis, H. New treatment of stbilized leachate by ozone/Fenton in the advanced oxidation process. *Waste Manag.* **2012**, *32*, 1692–1698. [CrossRef]

57. Li, Y.; Hsieh, W.-P.; Mahmudov, R.; Wei, X.; Huang, C.P. Combined ultrasound and Fenton (US Fenton) proces for the treatment of ammunition wastewater. *J. Hazard. Mater.* **2013**, *244–245*, 403–411. [CrossRef]

58. Roudi, A.M.; Chelliapan, S.; Mohtar, W.H.M.W.; Kamyab, H. Prediction and Optimization of the Fenton Process for the Treatment of Landfill Leachate Using an Artificial Neural Network. *Water* **2018**, *10*, 595. [CrossRef]

59. Cortez, S.; Teixeira, P.; Oliveira, R.; Mota, M. Fenton's oxidation as post-treatment of a mature municipal landfill leachate. *Int. J. Civ. Environ. Eng.* **2010**, *2*, 40–43. [CrossRef]

60. Benatti, C.T.; Tavares, C.R.; Guedes, T.A. Optimization of Fenton's oxidation of chemical laboratory wastewaters using the response surface methodology. *J. Environ. Manag.* **2006**, *80*, 66–74. [CrossRef]

61. Amaral-Silva, N.; Martins, R.C.; Castro-Silva, S.; Quinta-Ferreira, R.M. Fenton's treatment as an effective treatment for elderberry effluents: Economical evaluation. *Environ. Technol.* **2016**, *37*, 1208–1219. [CrossRef]

62. Turki, N.; Elghnijii, K.; Belhaj, D.; Bouzid, J. Effective degradation and detoxification of landfill leachates using a new combination process of coagulation/flocculation-Fenton and powder zeolite adsorption. *Desalination Water Treat.* **2015**, *55*, 151–162. [CrossRef]

63. Singh, S.K.; Tang, W.Z. Statistical analysis of optimum Fenton oxidation conditions for landfill leachate treatment. *Waste Manag.* **2013**, *33*, 81–88. [CrossRef]

64. Abou-Elela, S.I.; Ali, M.E.M.; Ibrahim, H.S. Combined treatment of retting flax wastewater using Fenton oxidation and granular activated carbon. *Arab. J. Chem.* **2016**, *9*, 511–517. [CrossRef]

65. Gernjak, W.; Krutzler, T.; Glaser, A.; Malato, S.; Caceres, J.; Bauer, R.; Fernández-Alba, A.R. Photo-Fenton treatment of water containing natural phenolic pollutants. *Chemosphere* **2003**, *50*, 71–78. [CrossRef]

66. Badawy, M.I.; Ghaly, M.Y.; Gad-Allah, T.A. Advanced oxidation processes for the removal of organophosphorus pesticides from wastewater. *Desalination* **2006**, *194*, 166–175. [CrossRef]

67. Tambosi, J.L.; Di Domenico, M.; Schirmer, W.N.; José, H.J.; de FPM Moreira, R. Treatment of paper and pulp wastewater and removal of odorous compounds by a Fenton-like proces at the pilot scale. *J. Chem. Technol. Biotechnol.* **2006**, *81*, 1426–1432. [CrossRef]

68. Ghaly, M.Y.; Hawash, S.I.; Mahdy, A.N.; El Marsafy, S.M.; Ahmed, E.M. Catalytic oxidation of Triactive Blue dye (TAB) by using fentons reagent. *J. Eng. Appl. Sci.* **2005**, *52*, 1245–1264.

69. Talebi, A.; Ismail, N.; Teng, T.T.; Alkarkhi, A.F.M. Optimzation of COD, apparent color, and turbidity reductions of landfill leachate by Fenton reagent. *Desalination Water Treat.* **2014**, *52*, 1524–1530. [CrossRef]

70. Saber, A.; Hasheminejad, H.; Taebi, A.; Ghaffari, G. Optimization of Fenton-based treatment of petroleum refinery wastewater with scrap iron using response surface methodology. *Appl. Water Sci.* **2014**, *4*, 283–290. [CrossRef]

71. Giannakis, S.; Vives, F.A.G.; Grandjean, D.; Magnet, A.; De Alencastro, L.F.; Pulgarin, C. Effect of advanced oxidation processes on the micropollutants and the effluent organic matter contained in municipal wastewater previously treated by three different secondary methods. *Water Res.* **2015**, *84*, 295–306. [CrossRef] [PubMed]

72. Biaglow, J.E.; Manevich, Y.; Uckun, F.; Held, K.D. Quantitation of hydroxyl radicals produced by radiation and copper-linked oxidation of ascorbate by 2-deoxy-D-ribose method. *Free Radic. Biol. Med.* **1997**, *22*, 1129–1138. [CrossRef]

73. Hu, Y.; Zhang, Z.; Yang, C. Measurement of hydroxyl radical production in ultrasonic aqueous solutions by a novel chemiluminescence method. *Ultrason. Sonochem.* **2008**, *15*, 665–672. [CrossRef]

74. Trojanowicz, M.; Bobrowski, K.; Szreder, T.; Bojanowska-Czajka, A. Chapter 9—Gamma-ray, X-ray and Electron Beam Based Processes. In *Advanced Oxidation Processes for Waste Water Treatment: Emerging Green Chemical Technology*; Ameta, S.C., Ameta, R., Eds.; Academic Press: London, UK, 2018; pp. 257–331. [CrossRef]

75. Prasad, M.N.V.; Vithanage, M.; Kapley, A. *Pharmaceuticals and Personal Care Products Waste Management and Treatment Technology: Emerging Contaminants and Micro Pollutants*; Elsevier Inc.: Oxford, UK, 2019.

76. Romero, V.; Acevedo, S.; Marco, P.; Giménez, J.; Esplugas, S. Enhancement of Fenton and photo-Fenton processes at initial circumneutral pH for the degradation of the β-blocker metoprolol. *Water Res.* **2016**, *88*, 449–457. [CrossRef]

77. Dong, W.; Jin, Y.; Zhou, K.; Sun, S.-P.; Li, Y.; Chen, X.D. Efficient degradation of pharmaceutical micropollutants in water and wastewater by Fe^{III}-NTA-catalyzed neutral photo-Fenton process. *Sci. Total Environ.* **2019**, *688*, 513–520. [CrossRef]

78. Miralles-Cuevas, S.; Oller, I.; Sanchez Perez, J.A.; Malato, S. Removal of pharmaceuticals from MWTP effluentby nanofiltration and solar photo-Fenton using twodifferent iron complexes at neutral pH. *Water Res.* **2014**, *64*, 23–32. [CrossRef]

79. Kuznetsova, E.V.; Savinov, E.N.; Vostrikova, L.A.; Parmon, V.N. Heterogeneous catalysis in the Fenton-type system FeZSM-5/H_2O_2. *Appl. Catal. B Environ.* **2004**, *51*, 165–170. [CrossRef]

80. Arvaniti, O.S.; Stasinakis, A.S. Review on the occurrence, fate and removal of perfluorinated compounds during wastewater treatment. *Sci. Total Environ.* **2015**, *524–525*, 81–92. [CrossRef]
81. Bao, S.; Shi, Y.; Zhang, Y.; He, L.; Yu, W.; Chen, Z.; Wu, Y.; Li, L. Study on the efficient removal of azo dyes by heterogeneous photo-Fenton process with 3D flower-like layered double hydroxide. *Water Sci. Technol.* **2020**, *81*, 2368–2380. [CrossRef]
82. De la Cruz, N.; Esquius, L.; Grandjean, D.; Magnet, A.; Tungler, A.; de Alencastro, L.F.; Pulgarín, C. Degradation of emergent contaminants by UV, UV/H$_2$O$_2$ and neutral photo-Fenton at pilot scale in a domestic wastewater treatment plant. *Water Res.* **2013**, *47*, 5836–5845. [CrossRef]
83. Behfar, R.; Davarnejad, R.; Heydari, R. Pharmaceutical Wastewater Chemical Oxygen Demand Reduction: Electro-Fenton, UV-enhanced Electro-Fenton and Activated Sludge. *Int. J. Eng.* **2019**, *32*, 1710–1715. [CrossRef]
84. Cuerda-Correa, E.M.; Alexandre-Franco, M.F.; Fernández-González, C. Advanced Oxidation Processes for the Removal of Antibiotics from Water. An Overview. *Water* **2020**, *12*, 102. [CrossRef]
85. Kar, P.; Shukla, K.; Jain, P.; Sathiyan, G.; Gupta, R.K. Semiconductor based photocatalysts for detoxification of emerging pharmaceutical pollutants from aquatic systems: A critical review. *Nano Mater. Sci.* **2021**, *3*, 25–46. [CrossRef]
86. Heidari, M.R.; Varma, R.S.; Ahmadian, M.; Pourkhosravani, M.; Asadzadeh, S.N.; Karimi, P.; Khatami, M. Photo-Fenton like Catalyst System: Activated Carbon/CoFe$_2$O$_4$ Nanocomposite for Reactive Dye Removal from Textile Wastewater. *Appl. Sci.* **2019**, *9*, 963. [CrossRef]
87. Dong, C.; Xing, M.; Zhang, J. Recent Progress of Photocatalytic Fenton-Like Process for Environmental Remediation. *Front. Environ. Chem.* **2020**, *1*, 8. [CrossRef]
88. Ren, G.; Han, H.; Wang, Y.; Liu, S.; Zhao, J.; Meng, X.; Li, Z. Recent Advances of Photocatalytic Application in Water Treatment: A Review. *Nanomaterials* **2021**, *11*, 1804. [CrossRef]
89. Tai, C.; Peng, J.-F.; Liu, J.-F.; Jiang, G.B.; Zou, H. Determination of hydroxyl radicals in advanced oxidation processes with dimethyl sulfoxide trapping and liquid chromatography. *Anal. Chim. Acta* **2014**, *527*, 73–80. [CrossRef]
90. De Morais, J.L.; Zamora, P.P. Use of advanced oxidation processes to improve the biodegradability of mature landfill leachates. *J. Hazard. Mater.* **2005**, *123*, 181–186. [CrossRef]
91. Ferreira, F.; Carvalho, L.; Pereira, R.; Antunes, S.C. Biological and photo-Fenton treatment of olive oil mill wastewater. *Glob. Nest J.* **2008**, *10*, 419–425. [CrossRef]
92. Çokay, E.; Eker, S. Pesticide Industry Wastewater Treatment with Photo-Fenton Process. *Environ. Sci. Proc.* **2022**, *18*, 15. [CrossRef]
93. Su, R.; Dai, X.; Wang, H.; Wang, Z.; Li, Z.; Chen, Y.; Luo, Y.; Ouyang, D. Metronidazole degradation by UV and UV/H$_2$O$_2$ advanced oxidation processes: Kinetics, mechanisms, and effects of natural water matrices. *Int. J. Environ. Res. Public Health* **2022**, *19*, 12354. [CrossRef]
94. Arzate, S.; Campos-Mañas, M.C.; Miralles-Cuevas, S.; Agüera, A.; García Sánchez, J.L.; Sánchez Pérez, J.A. Removal of contaminants of emerging concern by continuous flow solar photo-Fenton process at neutral pH in open reactors. *J. Environ. Manag.* **2020**, *261*, 110265. [CrossRef]
95. Oller, I.; Malato, S. Photo-Fenton applied to the removal of pharmaceutical and other pollutants of emerging concern. *Curr. Opin. Green Sustain. Chem.* **2021**, *29*, 100458. [CrossRef]
96. Molina, C.B.; Sanz-Santos, E.; Boukhemkhem, A.; Bedia, J.; Belver, C.; Rodriguez, J.J. Removal of emerging pollutants in aqueous phase by heterogeneous Fenton and photo-Fenton with Fe$_2$O$_3$-TiO$_2$-clay heterostructures. *Environ. Sci. Pollut. Res.* **2020**, *27*, 38434–38445. [CrossRef] [PubMed]
97. Liu, X.; Liu, H.-L.; Cui, K.-P.; Dai, Z.-L.; Wang, B.; Chen, X. Heterogeneous Photo-Fenton Removal of Methyl Orange Using the Sludge Generated in Dyeing Wastewater as Catalysts. *Water* **2022**, *14*, 629. [CrossRef]
98. Li, J.; Qin, L.; Zhao, L.; Wang, A.; Chen, Y.; Meng, L.; Zhang, Z.; Tian, X.; Zhou, Y. Removal of Refractory Organics from Biologically Treated Landfill Leachate by Microwave Discharge Electrodeless Lamp Assisted Fenton Process. *Int. J. Photoenergy* **2015**, *2015*, 643708. [CrossRef]
99. Kastanek, F.; Cirkva, V.; Kmentova, H.; Kastanek, P.; Maleterova, Y.; Solcova, O. Photodebromination of Opaque Slurries Using Titania-Coated Mercury Electrodeless Discharge Lamps. *J. Multidiscip. Eng. Sci. Technol.* **2016**, *3*, 5262–5267.
100. Zhang, H.; Choi, H.J.; Huang, C.-P. Optimization of Fenton process for the treatment of landfill leachate. *J. Hazard. Mater.* **2005**, *125*, 166–174. [CrossRef]
101. Zhang, H.; Zhang, D.; Zhou, J. Removal of COD from landfill leachate by electro-Fenton method. *J. Hazard. Mater.* **2006**, *135*, 106–111. [CrossRef]
102. Gadi, N.; Boelee, N.C.; Dewil, R. The Electro-Fenton Process for Caffeine Removal from Water and Granular Activated Carbon Regeneration. *Sustainability* **2022**, *14*, 14313. [CrossRef]
103. Marlina, E. Electro-Fenton for Industrial Wastewater Treatment: A Review. *E3S Web Conf.* **2019**, *125*, 03003. [CrossRef]
104. Nidheesh, P.V.; Gandhimathi, R. Trends in electro-Fenton process for water and wastewater treatment: An overview. *Desalination* **2012**, *299*, 1–15. [CrossRef]
105. Wang, Y.; Lia, X.; Zhen, L.; Zhang, H.; Zhang, Y.; Wang, C. Electro-Fenton treatment of concentrates generated in nanofiltration of biologically pretreated landfill leachate. *J. Hazard. Mater.* **2012**, *229–230*, 115–121. [CrossRef] [PubMed]
106. Zhao, X.; Liu, S.; Huang, Y. Removing organic contaminants by an electro-Fenton system constructed with graphene cathode. *Toxicol. Environ. Chem.* **2016**, *98*, 530–539. [CrossRef]

107. Heidari, Z.; Motevasel, M.; Jaafarzadeh, N.A. Application of Electro-Fenton (EF) Process to the Removal of Pentachlorophenol from Aqueous Solutions. *Iran. J. Oil Gas Sci. Technol.* **2015**, *4*, 76–87. [CrossRef]
108. Vedrenne, M.; Vasquez-Medrano, R.; Prato-Garcia, D.; Frontana-Uribe, B.A.; Ibanez, J.G. Characterization and detoxification of a mature landfill leachate using a combined coagulation–flocculation/photo Fenton treatment. *J. Hazrad. Mater.* **2012**, *205–206*, 208–215. [CrossRef]
109. El-Desoky, H.S.; Ghoneima, M.; El-Sheikhb, R.; Zidana, N.M. Oxidation of Levafix CA reactive azo-dyes in industrial wastewater of textile dyeing by electro-generated Fenton's reagent. *J. Hazard. Mater.* **2010**, *175*, 858–865. [CrossRef]
110. Cetinkaya, S.G.; Morcali, M.H.; Akarsu, S.A.; Ziba, C.A.; Dolaz, M. Comparison of classic Fenton with ultrasound Fenton processes on industrial textile wastewater. *Sustain. Environ. Res.* **2018**, *28*, 165–170. [CrossRef]
111. Yuan, D.; Zhou, X.; Jin, W.; Han, W.; Chi, H.; Ding, W.; Huang, Y.; He, Z.; Gao, S.; Wang, Q. Effects of the Combined Utilization of Ultrasonic/Hydrogen Peroxide on Excess Sludge Destruction. *Water* **2021**, *13*, 266. [CrossRef]
112. Camargo-Perea, A.L.; Rubio-Clemente, A.; Peñuela, G.A. Use of Ultrasound as an Advanced Oxidation Process for the Degradation of Emerging Pollutants in Water. *Water* **2020**, *12*, 1068. [CrossRef]
113. Dietrich, M.; Franke, M.; Stelter, M. Degradation of endocrine disruptor bisphenol A by ultrasound-assisted electrochemical oxidation in water. *Ultrason. Sonochem.* **2017**, *39*, 741–749. [CrossRef]
114. Rahdar, S.; Igwegbe, C.A.; Ghasemi, M.; Ahmadi, S. Degradation of aniline by the combined process of ultrasound and hydrogen peroxide (US/H_2O_2). *MethodsX* **2019**, *6*, 492–499. [CrossRef] [PubMed]
115. Muñoz-Calderón, A.; Zúñiga-Benítez, H.; Valencia, S.H.; Rubio-Clemente, A.; Aupegui, S.; Peñuela, G.A. Use of low frequency ultrasound for water treatment: Data on azithromycin removal. *Data Brief* **2020**, *31*, 105947. [CrossRef] [PubMed]
116. Lee, Y.; Cho, M.; Kim, J.Y.; Yoon, J. Chemistry of ferrate (Fe(VI)) in aqueous solution and its applications as a green chemical. *J. Ind. Eng. Chem.* **2004**, *10*, 161–171.
117. Karim, A.V.; Krishnan, S.; Pisharody, L.; Malhotra, M. Application of Ferrate for Advanced Water and Wastewater Treatment. In *Advanced Oxidation Processes—Applications, Trends, and Prospects*; IntechOpen: London, UK, 2020. [CrossRef]
118. Lekikot, B.; Mammer, L.; Talbi, K.; Benssassi, M.E.; Abdessemed, A.; Sehili, T. Homogeneous modified Fenton-like oxidation using FeIII-gallic acid complex for ibuprofen degradation at neutral pH. *Desalination* **2021**, *234*, 147–157. [CrossRef]
119. Zhu, Y.; Fan, W.; Zhou, T.; Li, X. Removal of chelated heavy metals from aqueous solution: A review of current methods and mechanisms. *Sci. Total Environ.* **2019**, *678*, 253–266. [CrossRef]
120. Ayoub, M. Fenton process for the treatment of wastewater effluent from the edible oil industry. *Water Sci. Technol.* **2022**, *86*, 1388–1401. [CrossRef]
121. Primo, O.; Rueda, A.; Rivero, M.J.; Ortiz, I. An Integrated Process, Fenton Reaction−Ultrafiltration, for the Treatment of Landfill Leachate: Pilot Plant Operation and Analysis. *Ind. Eng. Chem. Res.* **2008**, *47*, 946–952. [CrossRef]

 *catalysts*

Article

Vital Role of Synthesis Temperature in Co–Cu Layered Hydroxides and Their Fenton-like Activity for RhB Degradation

Ruixue Zhang, Yanping Liu, Xinke Jiang and Bo Meng *

Department of Engineering Research Center of Advanced Functional Material Manufacturing of Ministry of Education, Zhengzhou University, Zhengzhou 450001, China; zrx3066@163.com (R.Z.); lyp2423821591@gs.zzu.edu.cn (Y.L.); xinke7955@163.com (X.J.)
* Correspondence: mengbo8305@zzu.edu.cn

Abstract: Cu and Co have shown superior catalytic performance to other transitional elements, and layered double hydroxides (LDHs) have presented advantages over other heterogeneous Fenton catalysts. However, there have been few studies about Co–Cu LDHs as catalysts for organic degradation via the Fenton reaction. Here, we prepared a series of Co–Cu LDH catalysts by a co-precipitation method under different synthesis temperatures and set Rhodamine B (RhB) as the target compound. The structure-performance relationship and the influence of reaction parameters were explored. A study of the Fenton-like reaction was conducted over Co–Cu layered hydroxide catalysts, and the variation of synthesis temperature greatly influenced their Fenton-like catalytic performance. The Co–Cu$_{t=65°C}$ catalyst with the strongest LDH structure showed the highest RhB removal efficiency (99.3% within 30 min). The change of synthesis temperature induced bulk-phase transformation, structural distortion, and metal–oxygen (M–O) modification. An appropriate temperature improved LDH formation with defect sites and lengthened M–O bonds. Co–Cu LDH catalysts with a higher concentration of defect sites promoted surface hydroxide formation for H_2O_2 adsorption. These oxygen vacancies (Ovs) promoted electron transfer and H_2O_2 dissociation. Thus, the Co–Cu LDH catalyst is an attractive alternative organic pollutants treatment.

Keywords: Cu–Co LDH; oxygen vacancy; hydroxyl radicals; RhB; Fenton-like reaction

Citation: Zhang, R.; Liu, Y.; Jiang, X.; Meng, B. Vital Role of Synthesis Temperature in Co–Cu Layered Hydroxides and Their Fenton-like Activity for RhB Degradation. *Catalysts* **2022**, *12*, 646. https://doi.org/10.3390/catal12060646

Academic Editors: Gassan Hodaifa, Antonio Zuorro, Joaquín R. Dominguez, Juan García Rodríguez, José A. Peres and Zacharias Frontistis

Received: 5 May 2022
Accepted: 3 June 2022
Published: 13 June 2022

Publisher's Note: MDPI stays neutral with regard to jurisdictional claims in published maps and institutional affiliations.

1. Introduction

Rhodamine B (RhB) is a cationic xanthene dye with an aromatic structure and persistent stability. It is commonly used in the printing and dyeing industries [1]. However, it is harmful if swallowed by human beings and animals, and causes irritation to the skin, eyes, gastrointestinal tract, and respiratory tract [1]. Furthermore, it also causes phototoxic and photoallergic reactions. The carcinogenicity, reproductive and developmental toxicity, neurotoxicity, and chronic toxicity to humans and animals have been experimentally proven [2]. Therefore, the insufficient disposal of RhB-containing wastewater can lead to severe environmental problems and threaten human health. Many methods have been developed for dye wastewater treatment, such as adsorption [3,4], advanced oxidation [5–8], biological treatment [9,10], and photodegradation [11,12].

Among advanced oxidation processes (AOPs), the Fenton reaction has drawn great attention from the academic and industrial fields as an effective wastewater treatment owing to its low operating cost, low toxicity, and high degradation efficiency [13,14]. However, the homogeneous Fenton reaction has disadvantages, like a narrow working pH range, poor recyclability, and secondary pollution of residual sludge [15,16]. To overcome these drawbacks, the heterogeneous Fenton-like reaction with better stability and recyclability has been promoted as a promising alternative to the decomposition of refractory organics. Moreover, Cu and Co have exhibited superior catalytic performance to other transitional elements, so they have attracted a lot of research attention in the field of Fenton reaction [17–21].

Owing to their unique layered structure, layered double hydroxides (LDHs) have presented clear advantages over other heterogeneous Fenton catalysts, with negligible metal leaching during the reaction [22–24]. The LDH is a two-dimensional (2D) layered hydrotalcite formulated as $[M^{2+}_{1-x}M^{3+}_x(OH^-)_2]^{x+}(A^{n-})_{x/n}\cdot mH_2O$, in which M^{2+} and M^{3+} represent divalent and trivalent cations in the host layers, respectively, and A^{n-} stands for the compensative anion in the interlayer [25]. Moreover, abundant alkaline sites in the LDH structure can benefit the maintenance of a weak alkaline condition during reaction, which greatly depresses the leaching condition of catalysts [26,27]. Moreover, the metal component in the positive layer with low redox potential, such as Cu, Fe, Ni, and Co, can favor the electron transfer in reaction [28]. Furthermore, the oxygen vacancies (Ovs) with rich electrons can stretch the O–O bond in H_2O_2 after adsorption and thus facilitate the production of ·OH [29–31] with accelerated degradation of organics. In recent years, Cu-containing LDHs have drawn great attention among LDH catalysts for Fenton-like reactions. Wang et al. prepared the Cu1Ni2Sn0.75 LDH catalyst with almost total (97.8%) phenol mineralization with a neutral pH value [32]. Tao et al. fabricated Cu–Fe LDHs to treat methyl orange, with a nearly 100% removal in 13.5 min [33]. A Cu–Zn–Fe LDH was also developed for the decomposition of an acetaminophen-containing sample, with a degradation efficiency up to 100% within 24 h [34]. However, there have been few studies on Co–Cu LDHs as catalysts for organic degradation via the Fenton reaction.

In this work, we prepared a series of Co–Cu LDH catalysts by a co-precipitation method under different synthesis temperatures and set RhB as the target compound. In addition, the structure–performance relationship and the influence of reaction parameters were also explored.

2. Results

2.1. Bulk-Phase and Electronic Structure Characterization

Nitrogen adsorption–desorption isotherms of all catalysts are displayed in Figure 1a. The isotherms show that all catalysts possessed the typical IV type isotherms with obvious hysteresis loops at high relative pressures ($0.6 < P/P_0 < 1$), indicating the presence of mesopores. Meanwhile, there were almost no microspores in the samples, as there was little N_2 adsorption for all catalysts at low relative pressures ($P/P_0 < 0.1$), which can also be clearly observed from the pore size distribution curves (Figure 1b). In addition, as shown in Table 1, the specific surface area of the catalysts decreased with increased synthesis temperature, to 65 °C, yet it exhibited slightly increasing return when the synthesis temperature reached 75 °C and 85 °C. A larger specific surface area usually provides more active sites, yet it had little relationship with the catalytic activity in this case, as Co–Cu$_{t=65°C}$ exhibited the highest RhB removal with the smallest surface area. In addition, the catalysts had a higher specific surface area, which often exists in more active sites, and the specific surface area of the catalyst synthesized at 65 °C was the smallest. The result indicates that the structure of the catalyst is sensitive to the synthesis temperature. Furthermore, according to the experimental results of ICP and XPS (Table 1), the molar ratio of copper and cobalt elements in the catalyst is about 4.15 and 4.2, respectively, which were very close to the theoretical value of 4 for the molar ratio of copper and cobalt elements.

Table 1. BET surface area, BJH mean pore size, pore volume, and elemental composition of catalysts.

Sample	S_{BET} (m² g⁻¹)	Volume (cm³ g⁻¹)	Pore Size (nm)	Co:Cu Molar Ratio	Surface Co:Cu Molar Ratio
Cu–Co$_{t=45°C}$	65.99	0.32	13.96	4.12	4.4
Cu–Co$_{t=55°C}$	62.86	0.36	17.52	4.15	4.3

Table 1. *Cont.*

Sample	S_{BET} $(m^2\,g^{-1})$	Volume $(cm^3\,g^{-1})$	Pore Size (nm)	Co:Cu Molar Ratio	Surface Co:Cu Molar Ratio
Cu–Co$_{t=65°C}$	46.47	0.22	13.38	4.14	4.2
Cu–Co$_{t=75°C}$	57.33	0.32	17.19	4.14	4.1
Cu–Co$_{t=85°C}$	60.33	0.36	17.26	4.1	4.3

The Co:Cu molar ratio was analyzed by ICP, the surface Co:Cu Molar ratio was analyzed by XPS.

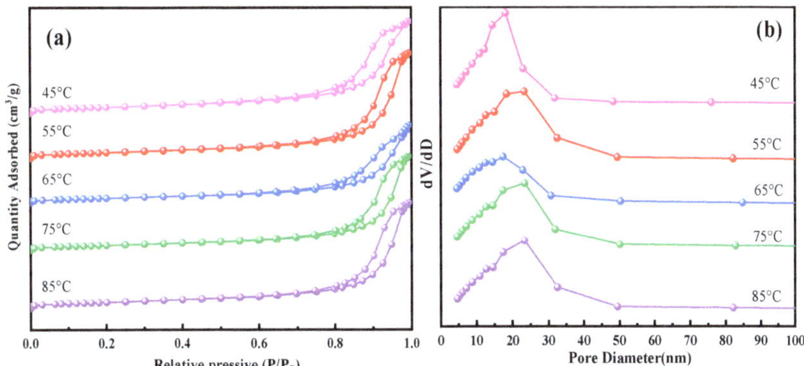

Figure 1. (**a**) N_2 adsorption/desorption isotherms and (**b**) pore size distribution curves of catalysts.

The bulk-phase structure of catalysts was further studied by XRD analysis. As shown in Figure 2a, when the synthesis temperature was in the range of 45–75 °C, all catalysts exhibited a group of characteristic diffractions corresponding to (003), (006), (012), and (015) crystalline planes of the hydrotalcite-like (JCPDS#35-0965) structure at 2θ of 11.7, 23.4, 33.7, and 39.1, respectively [35]. The basal spacing d(003) around 0.74 nm was in line with the presence of carbonate in the interlayer space, as in previous reports [25]. Moreover, a second phase was $Cu(OH)_2$ (JCPDS#80-0656) owing to the excessive Cu existence besides the necessary composition for LDH or the Jahn-Teller effect. This is because Cu^{2+} could be situated in near-lying octahedra with the formation of the copper compound with distorted octahedra, which is energetically preferred to the compound of LDH [25]. A higher synthesis temperature would enhance the crystallinity of the LDH structure, with an improved intensity of (003) reflection in the range of 45–65 °C. This may be highly related to the promoted oxidation of Co^{2+}, which resulted in the improved LDH structure with increasing temperature. However, the condition for Co–Cu$_{t=75°C}$ inversed for the decomposition of LDH framework under the circumstances. When the synthesis temperature was 85 °C, the LDH structure of the catalyst disappeared. Thus, the Co–Cu$_{t=65°C}$ catalyst had a better LDH structure. As shown in Figure 3, the catalyst presented sheet-like structures, as is typical for LDH materials.

FT-IR characterization was then carried out to investigate the special Co–Cu interaction, LDH property, and their influence on catalytic performance. The patterns in Figure 2b show some vibrational information of different catalyst structures. There are three types of O–H stretching modes in the catalysts. One is the ν_{O-H} vibration of the metal hydroxide located at about 3570 cm^{-1} and 3623 cm^{-1} for $Cu(OH)_2$ and $Co(OH)_2$, respectively [36]. Another broad band in the region of 3300–3500 cm^{-1} can be attributed to the O–H stretching of the adsorbed and interlayer water [25]. Furthermore, the weak band at 1634 cm^{-1} was associated with the hydroxyl deformation mode of the water molecules in the interlayer [37]. In addition, the peaks at c.a. 1361 cm^{-1}, 992 cm^{-1}, and 683 cm^{-1} can be attributed to different stretching modes of carbonate [25], indicating carbonates as the primary compensating anions in the LDH interlayer, which is in line with XRD results.

Moreover, the peaks located below 683 cm^{-1} were related to M–O vibrations (M = Co and/or Cu) [36], which varied across samples, indicating that the different M–O bonds may determine their corresponding catalytic performance.

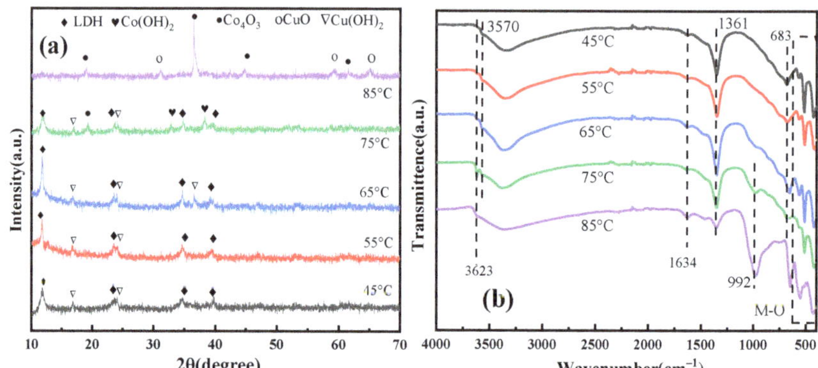

Figure 2. (**a**) XRD patterns of catalysts; (**b**) FT-IR spectra of catalysts.

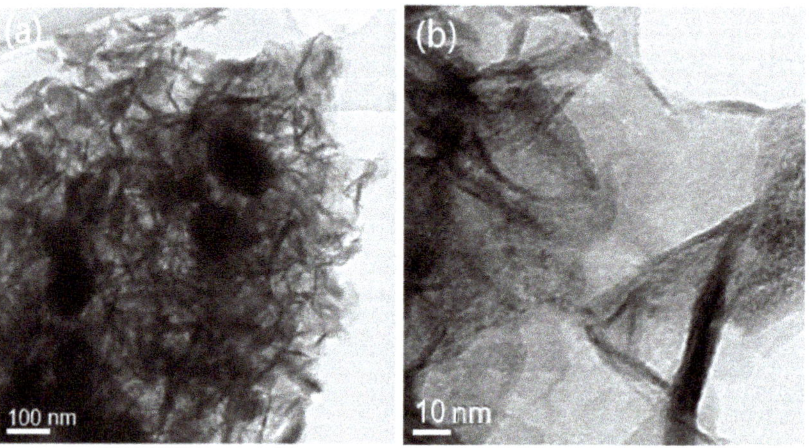

Figure 3. TEM images of Co–Cu$_{t=65°C}$ catalyst (**a**) lager-scale image (**b**) small-scale image.

Raman spectra analysis was further conducted to investigate these M–O bonds' properties. As shown in Figure 4a, the four vibrational modes exhibited at ca. 189, 461, 527, and 688 cm^{-1} all belonged to cobalt oxides corresponding to F_{2g}^1, E_g, F_{2g}^2, and A_{1g} species, respectively [38–40], and no Cu-related peaks were detected in all catalysts. In particular, the shifts for E_g suggested that this is probably related to the incorporation of Cu in the Co-related lattice. Specifically, the bands at 688 cm^{-1} (A1g) and 189 cm^{-1} (F_{2g}^1) could be assigned to the Raman vibration of Co^{3+}–O^{2-} at octahedral sites and Co^{2+}–O^{2-} at tetrahedral sites, respectively [41]. The bands at 461 cm^{-1} (E_g) and 527 cm^{-1} (F_{2g}^2) can be attributed to the combined vibrations in tetrahedral sites and octahedral oxygen motions [42]. Particularly, the band of 527 cm^{-1} can be ascribed to Co-related doubly occupied Ovs bound with donor defects [43]. According to our previous study, the band's peak intensity was positive to the concentration of surface Ovs. With the improved LDH crystallinity and peak intensity, shown in Figure 4a, F_{2g}^2 modes exhibited the same enhanced tendency, indicating an increasing Ov density.

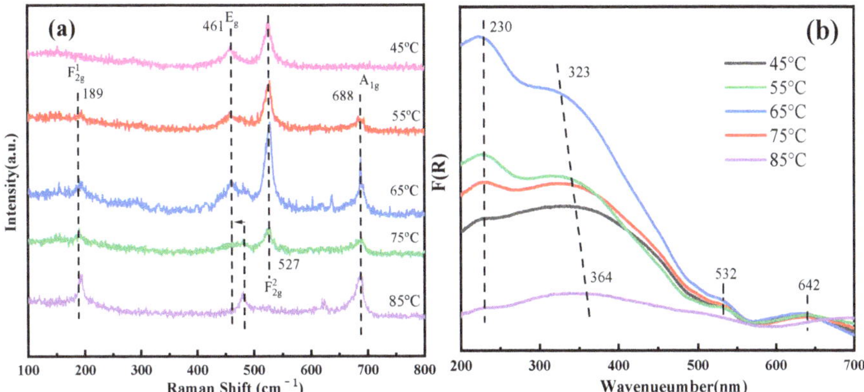

Figure 4. (**a**) Raman spectra of catalysts; (**b**) UV–Vis DRS spectra of catalysts.

As shown in Figure 5, the convolutions of O1s XPS spectra were carried out for detailed information, especially for oxygen defects. The core level spectra were fitted into three identified peaks. Peaks at 529.8 and 530.6 eV were attributed to oxygen atoms bound to lattice oxygen for metal oxides and hydroxyl species, respectively [31]. It is noted that the peak at ca. 533.4 eV can be ascribed to the presence of defect sites in the low oxygen-coordination [44]. In addition, Co–Cu$_{t=65°C}$ presented the highest defect density (Table 2), with its changing agreeing well with the Raman analysis.

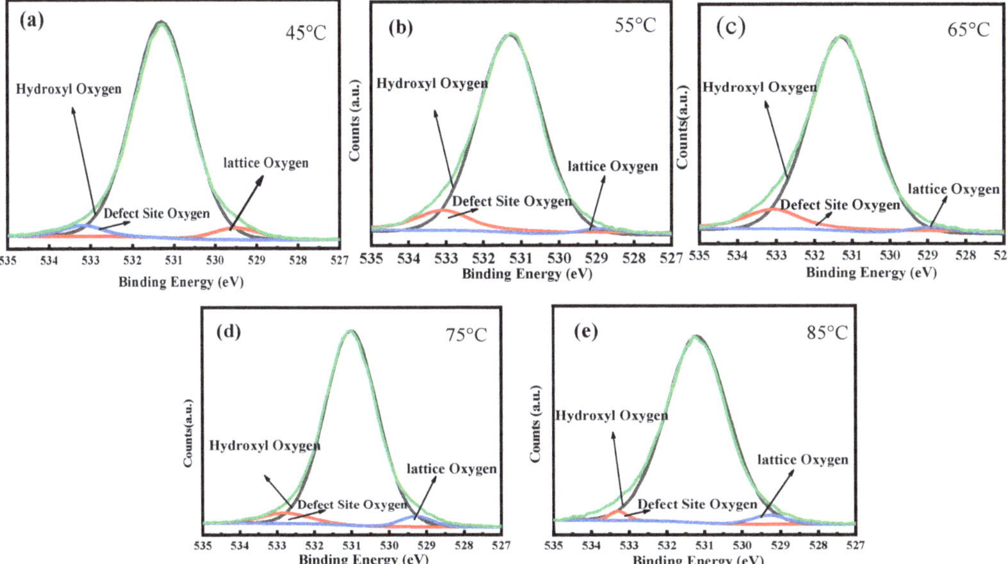

Figure 5. O1s XPS spectra of catalysts. (**a**) Co–Cu$_{t=45°C}$; (**b**) Co–Cu$_{t=55°C}$; (**c**) Co–Cu$_{t=65°C}$; (**d**) Co–Cu$_{t=75°C}$; (**e**) Co–Cu$_{t=85°C}$.

Table 2. Oxygen defect density of catalysts.

Sample	Oxygen Defect Density (%)	O1s BE for Defect Sites (eV)
$Cu–Co_{t=45°C}$	5.6	533.2
$Cu–Co_{t=55°C}$	7.8	533.1
$Cu–Co_{t=65°C}$	12.8	533.8
$Cu–Co_{t=75°C}$	9.2	533.7
$Cu–Co_{t=85°C}$	3.19	532.3

The density of defects was calculated by atomic ratio depending on the area percentage of defect site oxygen in total area for all peaks from XPS.

UV-Vis DRS spectra (Figure 4b) shed more light on the special electronic property of Ov and the LDH structure. The absorption peak observed at 230 nm can be attributed to ligand-to-metal charge-transfer excitations occurring in the MO_6 coordination [45]. Moreover, the peaks at c.a. 323–364 nm and 642 nm can be associated with $O^{2-} \rightarrow Co^{2+}$ and $O^{2-} \rightarrow Co^{3+}$, respectively [41,46,47]. The shifts of the peaks also depicted the different doping conditions of Cu in the LDH structure for different catalysts. In addition, the distinctive peak at 532 nm is related to the special metal–metal charge transfer of the Co–O–Cu oxo-bridge in the MO^6 environment for LDH, as it included the transitions of $d_z{}^2 \rightarrow d_x{}^2 - y^2$ for Cu^{2+}, $^1A_{1g} \rightarrow {}^1T_{1g}$ for Co^{3+}, and $^4T_{1g}(F) \rightarrow {}^4T_{1g}(P)$ for Co^{2+} with weak-field ligands [48–51], which is also related to the generation of surface Ov, according to previous publications [52]. A stronger peak intensity suggested a higher Ov concentration, with its changing agreeing well with the Raman analysis. Thus, it can be concluded from the above discussion that improved LDH structure is likely to possess more Co–O–Cu oxo–bridge interaction with more lattice disorder resulting from the incorporation of Cu in Co sites. This special Co–O–Cu structure can promote the generation of surface Ov, which is positively related to the catalytic performance [35,41].

2.2. Catalytic Performance

The catalytic activity for all catalysts with different synthesis temperatures are shown in Figure 6a. The RhB removal was raised from 82.5% to 99.3% as the fabricated temperature increased from 45 °C to 65 °C, respectively. However, a further increment in the synthesis temperature to 75 °C and 85 °C decreased the RhB degradation efficiency to 85.10% and 65.99%, respectively. Moreover, the adsorption only counted for less than 10% of the total removal without the addition of H_2O_2, indicating that self-decomposition and/or adsorption of RhB by catalysts can be excluded in this case, and the Fenton oxidation is the key factor for total organic elimination. The homogeneous Fenton test showed that the degradation resulting from metal leaching only contributed 9.3% of the total removal from the homogeneous part (Figure 6b). Furthermore, the quenching experiment showed that ·OH was the main reactive oxygen species of the reaction, as the RhB removal declined to 9.8% with the existence of a scavenger (Figure 6c). Thus, the RhB decomposition can be mostly ascribed to the ·OH generation over the catalyst surface rather than ·OH produced by leaching metals.

The results of the catalytic performance tests reveal that the synthesis temperature had a great impact on catalytic activity, as the optimum synthesis temperature of the catalyst was 65 °C. However, the $Co–Cu_{t=65°C}$ LDH catalyst had the smallest specific surface areas in the BET analysis, indicating that the catalyst performance is more structurally dependent than the surface area in the Fenton-like reaction. In addition, among other catalysts, the $Co–Cu_{t=65°C}$ LDH catalyst had the best performance, which can be attributed to its excellent LDH structure and high density of Ov based on the XRD and Raman analyses. According to a previous study [35], Ov is the main active site of the LDH structure, and it promotes H_2O_2 adsorption, electron transfer, and H_2O_2 dissociation during reaction, which benefit the generation of effective ·OH. In addition, the $Co–Cu_{t=65°C}$ LDH catalyst exhibited better RhB removal efficiency compared with other catalysts (Table 3). Therefore, we selected the $Co–Cu_{t=65°C}$ LDH catalyst to carry out the following experiments.

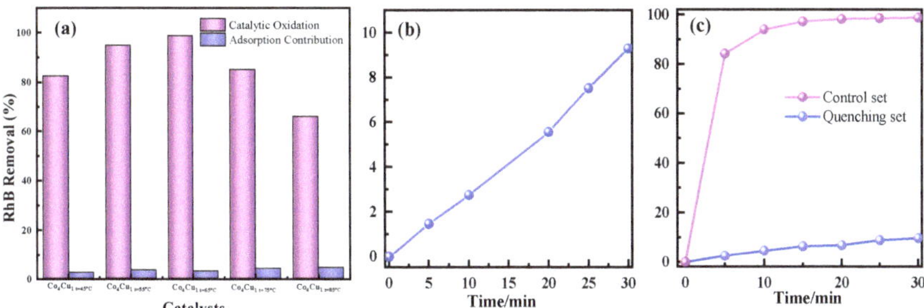

Figure 6. (**a**) Catalytic performance of catalysts; (**b**) homogeneous RhB degradation for Co–Cu$_{t=65°C}$ with C_{Cu2+} = 0.072 ppm and C_{Co2+} = 2.814 ppm according to metal leaching analysis; (**c**) quenching test for Co–Cu$_{t=65°C}$.

Table 3. Comparison of RhB removal among different catalysts.

Catalysts	Conditons	Removal	Ref.
Fe/MCM–41	1 g/L catalysts, 20 mM H$_2$O$_2$, 100 ppm RhB, pH = 4.0, 80 °C, 30 min.	99.1%	[53]
MgFe$_2$O$_4$	0.625 g/Lcatalysts, 1.00 vol%H$_2$O$_2$, 10 ppm RhB, 45 °C, pH = 6.44, 180 min	90.0%	[54]
Cu/Al$_2$O$_3$	1 g/Lcatalyst, 10 ppm RhB, 1000 ppm H$_2$O$_2$, pH = 5.14, 50 °C, 30 min	98.5%	[55]
Fe$_3$O$_4$/MI	0.5 g/Lcatalyst, 20 mM H$_2$O$_2$, 10 ppm RhB, pH = 7, 25 °C, 30 min	99.6%	[56]
Co–Cu LDH	0.1 g/Lcatalyst, 240 ppm H$_2$O$_2$, 10 ppm RhB, pH = 7, 40 °C, 30 min	99.3%	This work

2.3. Investigation of Reaction Parameters

To better understand the reaction process, we also performed condition experiments to investigate the influence of different reaction parameters such as H$_2$O$_2$ dosage, reaction temperature, and pH surroundings for RhB removal, and find out the optimum reaction parameters for the catalyst Co–Cu$_{t=65°C}$. All experiments were carried out in conditions of 100 mL, 10 ppm RHB, 10 mg catalyst, 240 ppm H$_2$O$_2$, and 50 °C under 700 rpm-stirring for 60 min with initial pH = 5.15 unless otherwise specified for investigations of each parameter. The kinetic study was based on the first-order reaction in the first 7.5 min to obtain a more satisfactory fitting model.

2.3.1. Effect of Initial H$_2$O$_2$ Dosage

The effect of the initial H$_2$O$_2$ dosage on degradation of RhB was investigated in the range of 160 to 960 ppm. As shown in Figure 7, as the H$_2$O$_2$ dosage rose from 160 to 480 ppm, the RhB removal also improved from 77.71 to 96.19% in 5 min, with its corresponding reaction rate constant k$_G$ (min^{-1}) surging from 0.1924 to 0.4938 min^{-1} (Table 4). This phenomenon can be ascribed to the fact that the adequate dose-up of initial H$_2$O$_2$ can directly promote the generation of ·OH with further improvement for RhB removal [57]. However, a further increase of the initial H$_2$O$_2$ dosage to 960 ppm brought about a drop of the catalytic performance, as excessive H$_2$O$_2$ and ·OH can quench ·OH and decrease the degradation efficiency [35,55,58].

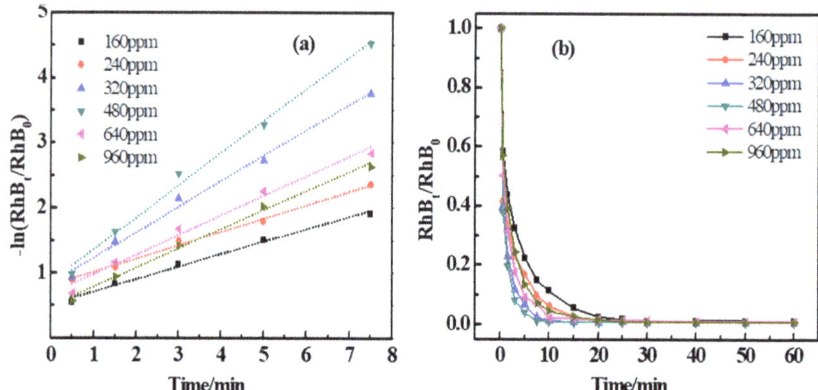

Figure 7. Catalytic degradation of RhB under different H_2O_2 concentration (**a**) plots of pseudo-first-order kinetics; (**b**) RhB concentration.

Table 4. The influence of H_2O_2 concentration on k_G for the first 7.5 min and RhB removal in the first 5 min.

H_2O_2 Concentration/ppm	k_G for First 7.5 min (min^{-1})	R_2	RhB Removal in First 5 min (%)
160	0.1924	0.9895	77.7
240	0.2070	0.9928	83.4
320	0.3930	0.9898	93.4
480	0.4938	0.9900	96.2
640	0.3019	0.9795	91.1
960	0.2943	0.9929	86.7

2.3.2. Effect of Reaction Temperature

Figure 8 shows the positive relationship between the reaction temperature and RhB. Specifically, the RhB removal in the first 5 min was 38.09%, 48.93%, 65.55%, and 84.07% with the reaction temperature of 27 °C, 33 °C, 40 °C, and 50 °C, respectively. In addition, the reaction rate constant k also increased with the rise of the reaction temperature (Table 5). This is attributable to the fact that a higher temperature promotes the collision between molecules, in which case more molecules obtain energy exceeding the activation barrier [59]. Thus, the reactions for the generation of ·OH and degradation of RhB were both improved simultaneously.

Table 5. The influence of reaction temperatures on k_G for the first 7.5 min and RhB removal in the first 5 min.

Reaction Temperature/°C	K_G for the First 7.5 min (min^{-1})	R_2	RhB Removal in the First 5 min (%)
27	0.0303	0.9841	38.1
33	0.0628	0.9967	48.9
40	0.1256	0.9915	65.5
50	0.2070	0.9929	84.1

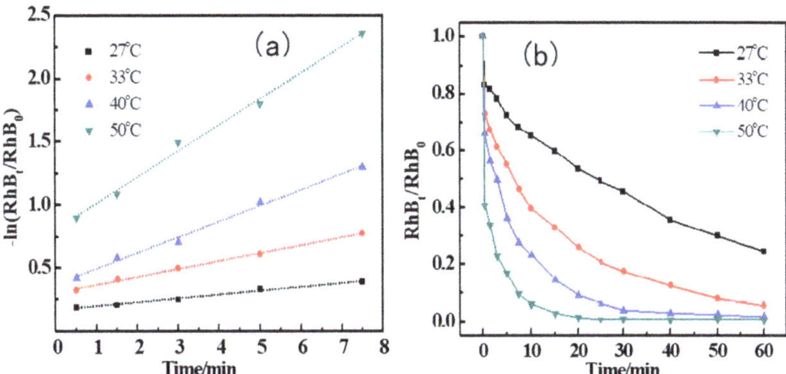

Figure 8. Catalytic degradation of RhB under different reaction temperatures (**a**) plots of pseudo-first-order kinetics; (**b**) RhB concentration.

2.3.3. Effect of pH Surroundings

It has been well established that the reaction surroundings, especially for the pH value, play an important role in Fenton or Fenton-like reactions. Thus, HNO_3 and NaOH were used to adjust the initial pH value in this work to better understand its influence on RhB removal. As shown in Figure 9, the catalyst presented excellent performance within 30 min over a wide pH range. Moreover, as the pH value increased from 5.2 to 7.9, the reaction rate constant k_G (Table 6) significantly improved within the first 5 min. In addition, the highest RhB removal of 97.5% was achieved at pH = 7.0. However, a further increase of the pH value had a negative impact on the reaction, with the removal in the first 5 min dropping to 61.7%.

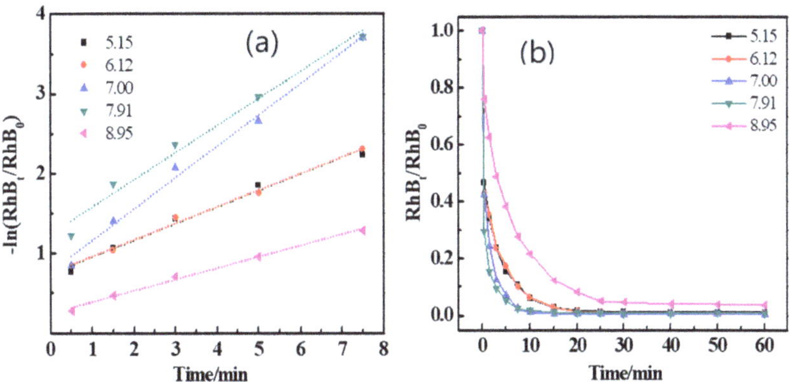

Figure 9. Catalytic degradation of RhB under different pH conditions (**a**) plots of pseudo-first-order kinetics; (**b**) RhB concentration.

Table 6. The influence of pH on k_G for the first 7.5 min and RhB removal in the first 5 min.

pH	K_G for the First 7.5 min (min^{-1})	R_2	RhB Removal in the First 5 min (%)
5.15	0.2075	0.9784	84.3
6.12	0.2070	0.9928	82.6
7.00	0.3930	0.9898	93.0
7.91	0.3405	0.9760	94.9
8.95	0.1418	0.9911	61.7

There are a few possible reasons for this phenomenon. In an acidic environment, the greater stability of H_2O_2 [60] and the metal leaching of catalysts lead to a decrease in catalytic activity. However, in strong alkaline solutions, the self-decomposition of H_2O_2 and the side reactions for $\cdot$OH scavenging are aggravated with a decline of the oxidation potential for $\cdot$OH/H_2O [61], which adversely affects the reaction.

2.3.4. The Reusability and Stability of Catalysts

To explore the stability of the Co–Cu$_{t=65°C}$ catalyst, repeated experiments of catalyst recycling and reuse were carried out under the same experimental conditions. Experiments were carried out in conditions of 100 mL, 10 ppm RHB, 10 mg catalyst, 240 ppm H_2O_2, 40 °C under 700 rpm-stirring for 60 min with initial pH = 5.15. As shown in Figure 10a, the efficiency of RhB dye degradation was only slightly decreased after being reused three times, but the removal rate of RhB could still exceed 85% after being reused three times. Hence, the catalysts showed good reusability in catalytic degradation. For the stability, the XRD spectra of Co–Cu$_{t=65°C}$ catalyst before and after the third run cycle are depicted in Figure 10b, where the spectra appear identical except for the slightly lower peak intensity of the (003) reflection at 2θ of 11.7, which presented a new $Cu(OH)_2$ phase after the recycling reaction. This can be attributed to the leaching of metal elements of the Co–Cu$_{t=65°C}$ catalyst after three cycles.

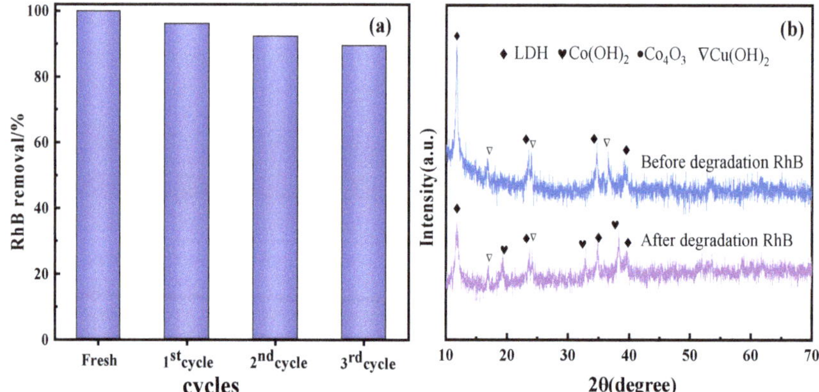

Figure 10. The reusability and stability of Cu–Co$_{t=65°C}$ catalyst (**a**); and XRD diffractograms before and after degradation (**b**).

2.4. Plausible Mechanism

According to the above experimental results, the possible heterogeneous Fenton-like reaction mechanisms on RhB degradation by the Cu–Co LDH/H_2O_2 were proposed in Equations (1)–(7). Irrespective of the existence of Oxygen vacancy (Ov), the events of Equations (1) and (2) occurred [62]. Surface active centers of Co and/or Cu are involved in one-electron oxidation to catalyze H_2O_2 to $\cdot$OH [30]. Concurrently, $M^{(n+1)+}$ (M = Co and Cu) was reduced by H_2O_2 according to the Haber-Weiss mechanism [63]. When oxygen vacancy co-existed, Equations (1) and (2) accelerated and the steps of Equations (3)–(6) were in effect [62]. Firstly, H_2O absorbed on the surface vacancy of Vo≡Mn' defect sites (Equation (3)) to form the surface hydroxyl group [64]. Secondly, through ligand exchange of the previous surface hydroxyl group which was prior to electron transfer, H_2O_2 adsorbed on the same defect sites [64] (Equation (4)). Thirdly, the existence of Ov stretched of the O–O bond, which facilitated activation of H_2O_2 to produce $\cdot$OH (Equation (5)) [21,65,66]. Finally, the reactive radicals ($\cdot$OH) readily attacked RHB that adsorbed nearby the active sites (Equation (7)):

$$\equiv M^{n+} + H_2O_2 \rightarrow \equiv M^{(n+1)+} - OH + \cdot OH \tag{1}$$

$$\equiv M^{(n+1)+} - OH + H_2O_2 \rightarrow \equiv M^{n+} + H_2O + \cdot O_2H \tag{2}$$

$$Ov\equiv M^{n+} + H_2O \rightarrow Ov\equiv M^{n+} - H_2O^* \tag{3}$$

$$Ov\equiv M^{n+} - H_2O^* + H_2O_2 \rightarrow H_2O_2^* - Ov\equiv M^{n+} - H_2O^* \tag{4}$$

$$H_2O_2^* - Ov\equiv M^{n+} - H_2O^* \rightarrow Ov\equiv M^{(n+1)+} - H_2O^* + \cdot OH \tag{5}$$

$$Ov\equiv M^{(n+1)+} - H_2O^* + H_2O_2 \rightarrow Ov\equiv M^{n+} - H_2O^* + \cdot O_2H + H^+ \tag{6}$$

$$\cdot OH + RHB \rightarrow \text{Intermediates} \rightarrow CO_2 + H_2O \tag{7}$$

Therefore, Ov plays a key role in the heterogeneous Fenton-like reaction. The higher the content of Ov in the catalyst, the faster the ·OH is generated, which in turn increases the degradation rate of RhB. It can be seen intuitively from Figure 11 that the Co–Cu$_{t=65°C}$ catalyst contains more Vo, and this is also the reason for the better performance of the Co–Cu$_{t=65°C}$ catalyst.

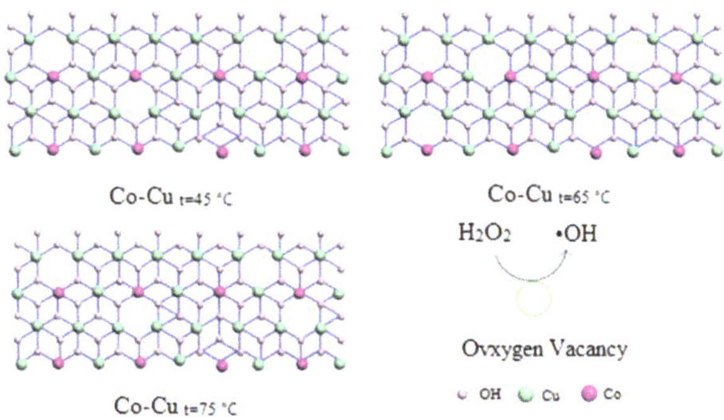

Figure 11. Structural model diagram of catalysts.

3. Materials and Methods

3.1. Catalyst Preparation

Cu–Co LDH catalysts were obtained by the co-precipitation method [25,35]. Specifically, mixed salts of Cu (NO$_3$)$_2$·3H$_2$O and Co (NO$_3$)$_2$·6H$_2$O with a Co/Cu molar ratio of 4:1 was dissolved in 150 mL deionized water. The above solution dropwise added a mixed alkaline of NaOH (0.4 M) and Na$_2$CO$_3$ (0.2 M) to adjust the pH to 12 under vigorous magnetic stirring at 25 °C for 1 h. The mixture was aged at different temperatures (45 °C, 55 °C, 65 °C, 75 °C, 85 °C) for 24 h with subsequent centrifugation after being washed by deionized water several times. The obtained pastes were then dried at 60 °C overnight and denoted as Cu–Co$_{t=x°C}$ as a final LDH catalyst.

3.2. Characterization

The BET surface area, pore volume, average pore size, and pore size distribution of samples were calculated from the isotherms of the N$_2$-physisorption, which is measured on a Micromeritics ASAP 2460 instrument by N$_2$ adsorption-desorption isotherms at −196 °C. All samples were dried at 50 °C for 72 h in a vacuum oven and then degassed at 50 °C for 72 h by nitrogen before measurements. The specific surface areas were calculated from the isotherms using Brunauer-Emmett-Teller (BET) method, and the pore size distributions were calculated by the Barrett-Joyner-Halenda (BJH) method.

Trans-mission electron microscopy (TEM) images were recorded on a TEM FEI Talos apparatus operating at acceleration voltage of 200 kV.

X-ray diffraction (XRD) was performed on a Bruker D8-Advance X-ray powder diffractometer with a Cu Kα ray source (λ = 0.154056 Å) at 40 kV and 40 mA. The intensity data were collected at room temperature in a 2θ range from 10° to 70° with a scan rate of 6°/min.

Fourier transform infrared spectra (FT-IR) experiments were carried out on a FT-IR/ATR spectrophotometer (Nicolet 50, Thermo Fisher, Waltham, MA, USA). The spectra were collected from 4000 to 500 cm^{-1} with 32 scans and a resolution of 4 cm^{-1}.

X-ray photoelectron spectroscopy (XPS, ESCALAB 250Xi, Thermo Fisher) was investigated with a source of Al-Kα radiation (1486.6 eV, a pass energy of 30.0 eV). C1s peak at 284.8 eV was used for calibration of all binding energies. The peaks were fitted according to references for O1s [31].

Laser Raman spectra (LRS) were collected on a confocal Raman (LabRam HR Evolution, Horiba, Kyoto, Japan) with a 50X objective under the excitation laser of 785 nm in ambient environment. Laser intensity was set as 3.2% ND Filter with 100 s acquisition time to ensure no damage of samples was caused during the experiment.

Diffuse reflectance UV-vis spectra in the range of 200–800 nm was recorded on an UV-3600 Plus (Shimadzu, Kyoto, Japan) spectrophotometer with $BaSO_4$ as reference.

3.3. Catalytic Performance Test

As for the catalytic activity, 10 mg of catalyst was dispersed in 100 mL of a 10-ppm RhB solution with a stirring rate of 700 rpm at 40 °C. The suspensions were first mixed for 1 h at a certain reaction temperature 40 °C to reach an adsorption-desorption equilibrium before H_2O_2 was added. Then, a 30 min Fenton-like reaction was activated by a dosage of 240 ppm H_2O_2 with an initial pH surrounding unless otherwise specified. Then, 3 mL of the liquid sample was taken from the reaction mixture at a given time and filtrated through a 0.22 μm Nafion membrane for the analysis of the RhB concentration by a UV-Vis spectrophotometer (UV-2600, Shimadzu, Kyoto, Japan) at an absorption wavelength of 554 nm. The RhB removal rate can be calculated by the following equation:

$$\text{RhB removal rate } (\%) = \frac{c_0 - c_t}{c_0} \times 100\% \tag{8}$$

where C_0 is the initial concentration of RhB, and C_t is the concentration of RhB at time t.

3.4. The Homogeneous Experiment

After the catalytic performance test, the metal ion concentration of the catalyst leached in the reaction solution was obtained by an atomic absorbance spectrometry (TAS-990F, Beijing Purkinje, Beijing, China) instrument. The conditions for the homogeneous experiment and the catalytic performance test were almost the same; the only difference is that equivalent metal nitrate salts were used as the catalyst in the homogeneous experiment.

3.5. The Quenching Test

The conditions for the quenching test and the catalytic performance test were also almost the same; the only difference is that an additional 50 mL of tert-butanol was added at the beginning of the experiment for the quenching test.

4. Conclusions

In this work, a novel Fenton-like catalyst, Co–Cu LDH, was prepared under different synthesis temperatures, and its best performance was achieved at a temperature of 65 °C. The superior catalytic activity can be attributed to the catalyst's structural defects, resulting from Cu doping, and Co–O–Cu in MO_6 surroundings enhanced the generation of Ov. Moreover, the high density of Ov under a strong Co–Cu interaction could improve the adsorption and dissociation of H_2O_2 as well as electron transfer to generate more ·OH, giving rise to increased RhB removal. Furthermore, the investigation of reaction parameters depicted the high catalytic activity of Co–Cu$_{t=65°C}$, which showed the highest removal, 96.2% RhB removal, in the first 5 min, with an H_2O_2 dosage of 480 ppm. The results indicate that the Co–Cu LDH catalyst is an attractive alternative organic pollutants treatment.

Author Contributions: Conceptualization and supervision, B.M.; methodology, validation, writing—original draft preparation, R.Z.; writing—review and editing, Y.L. and X.J.; All authors have read and agreed to the published version of the manuscript.

Funding: This research was funded by the Youth Program of the National Natural Science Foundation of China (Grant No. 21606209).

Data Availability Statement: Not applicable.

Conflicts of Interest: The authors declare no conflict of interest.

References

1. Rochat, J.; Demenge, P.; Rerat, J.C. Toxicologic study of a fluorescent tracer: Rhodamine B. *Toxicol. Eur. Res.* **1978**, *1*, 23–26. [PubMed]
2. Jain, R.; Mathur, M.; Sikarwar, S.; Mittal, A. Removal of the hazardous dye rhodamine B through photocatalytic and adsorption treatments. *J. Environ. Manag.* **2007**, *85*, 956–964. [CrossRef] [PubMed]
3. Namasivayam, C.; Sangeetha, D.; Gunasekaran, R. Removal of Anions, Heavy Metals, Organics and Dyes from Water by Adsorption onto a New Activated Carbon from Jatropha Husk, an Agro-Industrial Solid Waste. *Process Saf. Environ. Prot.* **2007**, *85*, 181–184. [CrossRef]
4. Purkait, M.K.; Maiti, A.; DasGupta, S.; De, S. Removal of congo red using activated carbon and its regeneration. *J. Hazard. Mater.* **2007**, *145*, 287–295. [CrossRef]
5. Javaid, R.; Qazi, U.Y.; Ikhlaq, A.; Zahid, M.; Alazmi, A. Subcritical and supercritical water oxidation for dye decomposition. *J. Environ. Manag.* **2021**, *290*, 112605. [CrossRef]
6. Javaid, R.; Qazi, U.Y.; Kawasaki, S. Highly efficient decomposition of Remazol Brilliant Blue R using tubular reactor coated with thin layer of PdO. *J. Environ. Manag.* **2016**, *180*, 551–556. [CrossRef]
7. Kim, S.C.; Lee, D.K. Preparation of Al-Cu pillared clay catalysts for the catalytic wet oxidation of reactive dyes. *Catal. Today* **2004**, *97*, 153–158. [CrossRef]
8. Hua, L.; Ma, H.R.; Zhang, L. Degradation process analysis of the azo dyes by catalytic wet air oxidation with catalyst CuO/gamma-Al$_2$O$_3$. *Chemosphere* **2013**, *90*, 143–149. [CrossRef]
9. Park, H.O.; Oh, S.; Bade, R.; Shin, W.S. Application of A2O moving-bed biofilm reactors for textile dyeing wastewater treatment. *Korean J. Chem. Eng.* **2010**, *27*, 893–899. [CrossRef]
10. Cripps, C.; Bumpus, J.A.; Aust, S.D. Biodegradation of Azo and Heterocyclic Dyes by Phanerochaete-Chrysosporium. *Appl. Environ. Microbiol.* **1990**, *56*, 1114–1118. [CrossRef]
11. Naeimi, A.; Sharifi, A.; Montazerghaem, L.; Abhari, A.R.; Mahmoodi, Z.; Bakr, Z.H.; Soldatov, A.V.; Ali, G.A.M. Transition metals doped WO3 photocatalyst towards high efficiency decolourization of azo dye. *J. Mol. Struct.* **2022**, *1250*, 131800. [CrossRef]
12. Solehudin, M.; Sirimahachai, U.; Ali, G.A.M.; Chong, K.F.; Wongnawa, S. One-pot synthesis of isotype heterojunction g-C3N4-MU photocatalyst for effective tetracycline hydrochloride antibiotic and reactive orange 16 dye removal. *Adv. Powder Technol.* **2020**, *31*, 1891–1902. [CrossRef]
13. Wu, Y.; Zhou, S.; Qin, F.; Zheng, K.; Ye, X. Modeling the oxidation kinetics of Fenton's process on the degradation of humic acid. *J. Hazard. Mater.* **2010**, *179*, 533–539. [CrossRef]
14. Zhang, L.; Nie, Y.; Hu, C.; Qu, J. Enhanced Fenton degradation of Rhodamine B over nanoscaled Cu-doped LaTiO3 perovskite. *Appl. Catal. B Environ.* **2012**, *125*, 418–424. [CrossRef]
15. Babuponnusami, A.; Muthukumar, K. A review on Fenton and improvements to the Fenton process for wastewater treatment. *J. Environ. Chem. Eng.* **2014**, *2*, 557–572. [CrossRef]
16. Lyu, L.; Yan, D.B.; Yu, G.F.; Cao, W.R.; Hu, C. Efficient Destruction of Pollutants in Water by a Dual-Reaction Center Fenton-like Process over Carbon Nitride Compounds-Complexed Cu(II)-CuAlO$_2$. *Environ. Sci. Technol.* **2018**, *52*, 4294–4304. [CrossRef]
17. Zhu, Y.P.; Ren, T.Z.; Yuan, Z.Y. Hollow cobalt phosphonate spherical hybrid as high-efficiency Fenton catalyst. *Nanoscale* **2014**, *6*, 11395–11402. [CrossRef]
18. Dai, C.; Tian, X.; Nie, Y.; Lin, H.M.; Yang, C.; Han, B.; Wang, Y. Surface Facet of CuFeO$_2$ Nanocatalyst: A Key Parameter for H$_2$O$_2$ Activation in Fenton-Like Reaction and Organic Pollutant Degradation. *Environ. Sci. Technol.* **2018**, *52*, 6518–6525. [CrossRef]
19. Khan, A.; Liao, Z.; Liu, Y.; Jawad, A.; Ifthikar, J.; Chen, Z. Synergistic degradation of phenols using peroxymonosulfate activated by CuO-Co$_3$O$_4$@MnO$_2$ nanocatalyst. *J. Hazard Mater.* **2017**, *329*, 262–271. [CrossRef]
20. Okamura, J.; Nishiyama, S.; Tsuruya, S.; Masai, M. Formation of Cu-supported mesoporous silicates and aluminosilicates and liquid-phase oxidation of benzene catalyzed by the Cu-mesoporous silicates and aluminosilicates. *J. Mol. Catal. A Chem.* **1998**, *135*, 133–142. [CrossRef]
21. Zhang, Y.; Liu, C.; Xu, B.; Qi, F.; Chu, W. Degradation of benzotriazole by a novel Fenton-like reaction with mesoporous Cu/MnO2: Combination of adsorption and catalysis oxidation. *Appl. Catal. B Environ.* **2016**, *199*, 447–457. [CrossRef]
22. Anantharaj, S.; Karthick, K.; Kundu, S. Evolution of layered double hydroxides (LDH) as high performance water oxidation electrocatalysts: A review with insights on structure, activity and mechanism. *Mater. Today Energy* **2017**, *6*, 1–26. [CrossRef]

23. Wang, Y.; Yan, D.; El Hankari, S.; Zou, Y.; Wang, S. Recent Progress on Layered Double Hydroxides and Their Derivatives for Electrocatalytic Water Splitting. *Adv. Sci.* **2018**, *5*, 1800064. [CrossRef] [PubMed]

24. Navalon, S.; Alvaro, M.; Garcia, H. Heterogeneous Fenton catalysts based on clays, silicas and zeolites. *Appl. Catal. B Environ.* **2010**, *99*, 1–26. [CrossRef]

25. Cavani, F.; Trifirò, F.; Vaccari, A. Hydrotalcite-type anionic clays: Preparation, properties and applications. *Catal. Today* **1991**, *11*, 173–301. [CrossRef]

26. Guan, Y.-H.; Ma, J.; Li, X.-C.; Fang, J.-Y.; Chen, L.-W. Influence of pH on the Formation of Sulfate and Hydroxyl Radicals in the UV/Peroxymonosulfate System. *Environ. Sci. Technol.* **2011**, *45*, 9308–9314. [CrossRef]

27. Liu, Y.; Yang, J.; Guan, Q.; Yang, L.; Liu, H.; Zhang, Y.; Wang, Y.; Wang, D.; Lang, J.; Yang, Y.; et al. Effect of annealing temperature on structure, magnetic properties and optical characteristics in $Zn_{0.97}Cr_{0.03}O$ nanoparticles. *Appl. Surf. Sci.* **2010**, *256*, 3559–3562. [CrossRef]

28. Nava-Andrade, K.; Carbajal-Arízaga, G.G.; Obregón, S.; Rodríguez-González, V. Layered double hydroxides and related hybrid materials for removal of pharmaceutical pollutants from water. *J. Environ. Manag.* **2021**, *288*, 112399. [CrossRef]

29. Li, H.; Shang, J.; Yang, Z.; Shen, W.; Ai, Z.; Zhang, L. Oxygen Vacancy Associated Surface Fenton Chemistry: Surface Structure Dependent Hydroxyl Radicals Generation and Substrate Dependent Reactivity. *Environ. Sci. Technol.* **2017**, *51*, 5685–5694. [CrossRef]

30. Jin, H.; Tian, X.; Nie, Y.; Zhou, Z.; Yang, C.; Li, Y.; Lu, L. Oxygen Vacancy Promoted Heterogeneous Fenton-like Degradation of Ofloxacin at pH 3.2–9.0 by Cu Substituted Magnetic Fe_3O_4@FeOOH Nanocomposite. *Environ. Sci. Technol.* **2017**, *51*, 12699–12706. [CrossRef]

31. Liu, P.F.; Yang, S.; Zhang, B.; Yang, H.G. Defect-Rich Ultrathin Cobalt–Iron Layered Double Hydroxide for Electrochemical Overall Water Splitting. *ACS Appl. Mater. Interfaces* **2016**, *8*, 34474–34481. [CrossRef]

32. Wang, H.; Zhang, Z.; Jing, M.; Tang, S.; Wu, Y.; Liu, W. Synthesis of CuNiSn LDHs as highly efficient Fenton catalysts for degradation of phenol. *Appl. Clay Sci.* **2020**, *186*, 105433. [CrossRef]

33. Tao, X.; Yang, C.; Wei, Z.; Huang, L.; Chen, J.; Cong, W.; Xie, R.; Xu, D. Synergy between Fenton process and DBD for methyl orange degradation. *Mater. Res. Bull.* **2019**, *120*, 110581. [CrossRef]

34. Lu, H.; Zhu, Z.; Zhang, H.; Zhu, J.; Qiu, Y.; Zhu, L.; Küppers, S. Fenton-Like Catalysis and Oxidation/Adsorption Performances of Acetaminophen and Arsenic Pollutants in Water on a Multimetal Cu-Zn-Fe-LDH. *ACS Appl. Mater. Interfaces* **2016**, *8*, 25343–25352. [CrossRef]

35. Guo, X.X.; Hu, T.T.; Meng, B.; Sun, Y.; Han, Y.-F. Catalytic degradation of anthraquinones-containing H_2O_2 production effluent over layered Co-Cu hydroxides: Defects facilitating hydroxyl radicals generation. *Appl. Catal. B Environ.* **2020**, *260*, 118157. [CrossRef]

36. Zeng, H.C.; Xu, Z.P.; Qian, M. Synthesis of Non-Al-Containing Hydrotalcite-like Compound $Mg_{0.3}CoII_{0.6}CoIII_{0.2}(OH)_2(NO_3)_{0.2}\cdot H_2O$. *Chem. Mater.* **1998**, *10*, 2277–2283. [CrossRef]

37. Zhao, X.; Niu, C.; Zhang, L.; Guo, H.; Wen, X.; Liang, C.; Zeng, G. Co-Mn layered double hydroxide as an effective heterogeneous catalyst for degradation of organic dyes by activation of peroxymonosulfate. *Chemosphere* **2018**, *204*, 11–21. [CrossRef]

38. Liu, Q.; Wang, L.-C.; Chen, M.; Cao, Y.; He, H.-Y.; Fan, K.-N. Dry citrate-precursor synthesized nanocrystalline cobalt oxide as highly active catalyst for total oxidation of propane. *J. Catal.* **2009**, *263*, 104–113. [CrossRef]

39. He, L.; Li, Z.; Zhang, Z. Rapid, low-temperature synthesis of single-crystalline Co_3O_4 nanorods on silicon substrates on a large scale. *Nanotechnology* **2008**, *19*, 4–15. [CrossRef]

40. Zhang, R.; Zhang, Y.-C.; Pan, L.; Shen, G.-Q.; Mahmood, N.; Ma, Y.-H.; Shi, Y.; Jia, W.; Wang, L.; Zhang, X.; et al. Engineering Cobalt Defects in Cobalt Oxide for Highly Efficient Electrocatalytic Oxygen Evolution. *ACS Catal.* **2018**, *8*, 3803–3811. [CrossRef]

41. Lou, Y.; Ma, J.; Cao, X.; Wang, L.; Dai, Q.; Zhao, Z.; Cai, Y.; Zhan, W.; Guo, Y.; Hu, P.; et al. Promoting Effects of In_2O_3 on Co_3O_4 for CO Oxidation: Tuning O2 Activation and CO Adsorption Strength Simultaneously. *ACS Catal.* **2014**, *4*, 4143–4152. [CrossRef]

42. Wang, X.F.; Xu, J.B.; Zhang, B.; Yu, H.G.; Wang, J.; Zhang, X.; Yu, J.G.; Li, Q. Signature of Intrinsic High-Temperature Ferromagnetism in Cobalt-Doped Zinc Oxide Nanocrystals. *Adv. Mater.* **2006**, *18*, 2476–2480. [CrossRef]

43. Gandhi, V.; Ganesan, R.; Abdulrahman Syedahamed, H.H.; Thaiyan, M. Effect of Cobalt Doping on Structural, Optical, and Magnetic Properties of ZnO Nanoparticles Synthesized by Coprecipitation Method. *J. Phys. Chem. C* **2014**, *118*, 9715–9725. [CrossRef]

44. Bao, J.; Zhang, X.; Fan, B.; Zhang, J.; Zhou, M.; Yang, W.; Hu, X.; Wang, H.; Pan, B.; Xie, Y. Ultrathin Spinel-Structured Nanosheets Rich in Oxygen Deficiencies for Enhanced Electrocatalytic Water Oxidation. *Angew. Chem. Int. Ed. Engl.* **2015**, *54*, 7399–7404. [CrossRef] [PubMed]

45. Parida, K.; Mohapatra, L.; Baliarsingh, N. Effect of Co2+Substitution in the Framework of Carbonate Intercalated Cu/Cr LDH on Structural, Electronic, Optical, and Photocatalytic Properties. *J. Phys. Chem. C* **2012**, *116*, 22417–22424. [CrossRef]

46. Godelitsas, A.; Charistos, D.; Tsipis, C.; Misaelides, P.; Filippidis, A.; Schindler, M. Heterostructures patterned on aluminosilicate microporous substrates: Crystallization of cobalt(III) tris(N,N-diethyldithiocarbamato) on the surface of a HEU-type zeolite. Microporous and Mesoporous. *Materials* **2003**, *61*, 69–77. [CrossRef]

47. Khassin, A.A.; Anufrienko, V.F.; Ikorskii, V.N.; Plyasova, L.M.; Kustova, G.N.; Larina, T.V.; Molina, I.Y.; Parmon, V.N. Physicochemical study on the state of cobalt in a precipitated cobalt-aluminum oxide system. *Phys. Chem. Chem. Phys.* **2002**, *4*, 4236–4243. [CrossRef]

48. Li, X.; Dai, Y.; Ma, Y.; Huang, B. Electronic and magnetic properties of honeycomb transition metal monolayers: First-principles insights. *Phys. Chem. Chem. Phys.* **2014**, *16*, 13383–13389. [CrossRef]
49. Lin, W.Y.; Frei, H. Anchored metal-to-metal charge-transfer chromophores in a mesoporous silicate sieve for visible-light activation of titanium centers. *J. Phys. Chem. B* **2005**, *109*, 4929–4935. [CrossRef]
50. Mohapatra, L.; Parida, K.M. Dramatic activities of vanadate intercalated bismuth doped LDH for solar light photocatalysis. *Phys. Chem. Chem. Phys.* **2014**, *16*, 16985–16996. [CrossRef]
51. Nakamura, R.; Okamoto, A.; Osawa, H.; Irie, H.; Hashimoto, K. Design of all-inorganic molecular-based photocatalysts sensitive to visible light: Ti(IV)-O-Ce(III) bimetallic assemblies on mesoporous silica. *J. Am. Chem. Soc.* **2007**, *129*, 9596. [CrossRef] [PubMed]
52. Nayak, S.; Mohapatra, L.; Parida, K. Visible light-driven novel g-C_3N_4/NiFe-LDH composite photocatalyst with enhanced photocatalytic activity towards water oxidation and reduction reaction. *J. Mater. Chem. A* **2015**, *3*, 18622–18635. [CrossRef]
53. Xu, D.; Sun, X.; Zhao, X.; Huang, L.; Qian, Y.; Tao, X.; Guo, Q. Heterogeneous Fenton Degradation of Rhodamine B in Aqueous Solution Using Fe-Loaded Mesoporous MCM-41 as Catalyst. *Water Air Soil Pollut.* **2018**, *229*, 317. [CrossRef]
54. Han, X.; Zhang, H.; Chen, T.; Zhang, M.; Guo, M. Facile synthesis of metal-doped magnesium ferrite from saprolite laterite as an effective heterogeneous Fenton-like catalyst. *J. Mol. Liq.* **2018**, *272*, 43–52. [CrossRef]
55. Sheng, Y.; Sun, Y.; Xu, J.; Zhang, J.; Han, Y.-F. Fenton-like degradation of rhodamine B over highly durable Cu-embedded alumina: Kinetics and mechanism. *AIChE J.* **2018**, *64*, 538–549. [CrossRef]
56. Phan, T.T.N.; Nikoloski, A.N.; Bahri, P.A.; Li, D. Enhanced removal of organic using LaFeO3-integrated modified natural zeolites via heterogeneous visible light photo-Fenton degradation. *J. Environ. Manag.* **2019**, *233*, 471–480. [CrossRef]
57. Vu, A.-T.; Xuan, T.N.; Lee, C.-H. Preparation of mesoporous $Fe_2O_3 \cdot SiO_2$ composite from rice husk as an efficient heterogeneous Fenton-like catalyst for degradation of organic dyes. *J. Water Process. Eng.* **2019**, *28*, 169–180. [CrossRef]
58. Abdullah, A.Z.; Ling, P.Y. Heat treatment effects on the characteristics and sonocatalytic performance of TiO_2 in the degradation of organic dyes in aqueous solution. *J. Hazard. Mater.* **2010**, *173*, 159–167. [CrossRef]
59. Sun, Y.; Yang, Z.; Tian, P.; Sheng, Y.; Xu, J.; Han, Y.-F. Oxidative degradation of nitrobenzene by a Fenton-like reaction with Fe-Cu bimetallic catalysts. *Appl. Catal. B Environ.* **2019**, *244*, 1–10. [CrossRef]
60. Herney-Ramirez, J.; Vicente, M.A.; Madeira, L.M. Heterogeneous photo-Fenton oxidation with pillared clay-based catalysts for wastewater treatment: A review. *Appl. Catal. B Environ.* **2010**, *98*, 10–26. [CrossRef]
61. Liu, Y.; Zhang, J.; Yin, Y. Removal of Hg0from flue gas using two homogeneous photo-fenton-like reactions. *AIChE J.* **2015**, *61*, 1322–1333. [CrossRef]
62. Ding, R.R.; Li, W.Q.; He, C.S.; Wang, Y.R.; Liu, X.C.; Zhou, G.N.; Mu, Y. Oxygen vacancy on hollow sphere $CuFe_2O_4$ as an efficient Fenton-like catalysis for organic pollutant degradation over a wide pH range. *Appl. Catal. B Environ.* **2021**, *291*, 120069. [CrossRef]
63. Pignatello, J.J.; Oliveros, E.; MacKay, A. Advanced oxidation processes for organic contaminant destruction based on the Fenton reaction and related chemistry (vol 36, pg 1, 2006). *Crit. Rev. Environ. Sci. Technol.* **2007**, *37*, 273–275.
64. Zhang, N.; Li, X.; Ye, H.; Chen, S.; Ju, H.; Liu, D.; Lin, Y.; Ye, W.; Wang, C.; Xu, Q.; et al. Oxide Defect Engineering Enables to Couple Solar Energy into Oxygen Activation. *J. Am. Chem. Soc.* **2016**, *138*, 8928–8935. [CrossRef]
65. Cahill, A.E.; Taube, H. The Use of Heavy Oxygen in the Study of Reactions of Hydrogen Peroxide. *J. Am. Chem. Soc.* **1952**, *74*, 2312–2318. [CrossRef]
66. Dixon, W.T.; Norman, R.O.C. Free radicals formed during the oxidation and reduction of peroxides. *Nature* **1962**, *196*, 891–892. [CrossRef]

Article

Degradation of Diazepam with Gamma Radiation, High Frequency Ultrasound and UV Radiation Intensified with H_2O_2 and Fenton Reagent

Michel Manduca Artiles [1], Susana Gómez González [2], María A. González Marín [1], Sarra Gaspard [3] and Ulises J. Jauregui Haza [4,*]

[1] Higher Institute of Technologies and Applied Sciences, University of Havana, Ave Salvador Allende # 1110 e/Infanta and Rancho Boyeros, A.P. 6163, Revolution Square, Havana 10400, Cuba; manduca@instec.cu (M.M.A.); magonzalez@instec.cu (M.A.G.M.)

[2] Pharmaceutical Laboratories Aica⁺, Ave 23 s/n e/268 y 270 San Agustín, La Lisa, Havana 17100, Cuba; susana1991@nauta.cu

[3] Laboratoire COVACHIM M2E, EA 3592, Université des Antilles, BP 250, CEDEX, Pointe-à-Pitre 97157, Guadeloupe; sarra.gaspard@univ-antilles.fr

[4] Instituto Tecnológico de Santo Domingo, Ave de los Próceres, Santo Domingo 10602, Dominican Republic

* Correspondence: ulises.jauregui@intec.edu.do

Citation: Manduca Artiles, M.; Gómez González, S.; González Marín, M.A.; Gaspard, S.; Jauregui Haza, U.J. Degradation of Diazepam with Gamma Radiation, High Frequency Ultrasound and UV Radiation Intensified with H_2O_2 and Fenton Reagent. *Processes* **2022**, *10*, 1263. https://doi.org/10.3390/pr10071263

Academic Editors: Joaquín R. Dominguez, José A. Peres, Juan García Rodríguez, Antonio Zuorro, Zacharias Frontistis and Gassan Hodaifa

Received: 16 April 2022
Accepted: 17 May 2022
Published: 27 June 2022

Abstract: A degradation study of diazepam (DZP) in aqueous media by gamma radiation, high frequency ultrasound, and UV radiation (artificial-solar), as well with each process intensified with oxidizing agents (H_2O_2 and Fenton reagent) was performed. The parameters that influence the degradation of diazepam such as potency and frequency, irradiation dose, pH and concentration of the oxidizing agents used were studied. Gamma radiation was performed in a ^{60}Co source irradiator; an 11 W lamp was used for artificial UV radiation, and sonification was performed at frequency values of 580 and 862 kHz with varying power values. In the radiolysis a 100% degradation was obtained at 2500 Gy. For the sonolysis, 28.3% degradation was achieved after 180 min at 862 kHz frequency and 30 W power. In artificial photolysis, a 38.2% degradation was obtained after 300 min of UV exposure. The intensification of each process with H_2O_2 increased the degradation of the drug. However, the best results were obtained by combining the processes with the Fenton reagent for optimum H_2O_2 and Fe^{2+} concentrations, respectively, of 2.95 mmol L^{-1} and of 0.06 mmol L^{-1}, achieving a 100% degradation in a shorter treatment time, with a dose value of 750 Gy in the case of gamma radiation thanks to increasing in the amount of free radicals in water. The optimized processes were evaluated in a real wastewater, with a total degradation at 10 min of reaction.

Keywords: advanced oxidation process; wastewater; diazepam; gamma radiation; high frequency ultrasound; UV radiation; Fenton reaction

1. Introduction

Contamination of surface water, groundwater, and wastewater has increased in recent years due to the presence of so-called "emerging pollutants", such as drugs and pesticides [1–5]. Many investigations report the inefficiency of conventional wastewater treatment plants for eliminating persistent pollutants, and as a result, the presence of contaminants in effluent from treatment plants, rivers, lakes, and to a lesser extent in groundwater [1]. Diazepam (DZP) is the most prescribed benzodiazepine for its hypnotic, tranquilizing, and anticonvulsive properties, with levels in water bodies varying from 10 ng L^{-1} to 1 µg L^{-1} [6–12]. The presence of benzodiazepines affects the ecosystems in different ways [13,14]. Due to benzodiazepines interaction with the GABAA receptor, they may affect the function of the nervous system of non-target species, such as aquatic organisms [13]. On the other hand, Subedi et al. showed that zebrafish (Danio rerio) larvae exposed to the mixtures of psychotic drug residues, including the benzodiazepines, had affected immune system and gene expression [14].

Considering the high impact of pharmaceutical products, it is very important to remove them from the wastewater before discharge. Several researches carried out in recent years point to the use of advanced oxidation processes (AOPs) as innovative technologies for the elimination of persistent pollutants [6,15–21].

For the degradation of DZP in aqueous medium, some studies report the use of ozone [18], ultraviolet (UV) radiation, and its combination with oxidizing agents [6,22–24]. However, the most studied AOPs for drug degradation in water use gamma radiation and high frequency ultrasound intensified with H_2O_2 and Fenton reagent [25,26].

In the present work, we studied the degradation of DZP in a synthetic matrix by radical attack ($HO^{\bullet}$, e^- (aq), $H^{\bullet}$, $HO_2^{\bullet}$) formed in the processes of radiolysis, sonolysis, and photolysis in aqueous media as well as its intensification with H_2O_2 and the Fenton reagent [26–28]. We hypothesize that the use of oxidizing agents will generate an additional amount of highly reactive $HO^{\bullet}$ radicals, allowing a greater efficiency in the degradation of the molecule. Different parameters were optimized in the degradation processes to increase the drug removal efficiency. These final conditions were applied in the analysis of samples of real residual water.

2. Materials and Methods

For the DZP degradation studies, the three AOPs were carried out separately at an initial concentration of 20 mg L^{-1} DZP (Sigma Aldrich, St. Louis, MO, USA) in acidified distilled water, at pH values between 2 and 3, adjusted with a 10% solution of concentrated H_2SO_4 (95–97% purity; Merck, Darmstadt, Germany).

Acetonitrile (HPLC grade purchased from Sigma-Aldrich) and orthophosphoric acid (85% supplied by Merck) were used for the analysis by High Resolution Liquid Chromatography (HPLC). Hydrogen peroxide (35%) was obtained from Fluka. 99.5% potassium iodide, 99.5% purity heptahydrate iron sulfate, and 98% sodium sulfite were purchased from Merck.

For gamma irradiation of the samples, an ISOGamma-LLCo irradiator, equipped with a ^{60}Co source was used. The irradiation doses used were 0.1, 0.25, 0.5, 0.75, 1.0, 2.5, and 5.0 kGy at a dose rate of 3.73 kGy h^{-1}. All experiments were performed at a temperature of $30 \pm 2\,°C$. In the experiments carried out with the gamma irradiation process, the aqueous solutions of DZP were packed in 50 mL bottles with screw cap, which reached a thermal equilibrium at room temperature and atmospheric pressure.

A high-frequency ultrasound horn (Meinhardtultraschall Technik, Leipzig, Germany) with flat transducer was used for the studies of sonolysis. Frequencies values of 580 and 862 kHz, with varying power levels, were evaluated. The degradation was carried out in a 1.5 L glass reactor at a controlled temperature ($25 \pm 2\,°C$) and constant stirring at 300 m^{-1}.

The laboratory-scale artificial photolysis was performed with a UV lamp of 254 nm and a power of 11 watts. The degradation was performed in a 500 mL beaker with irradiated area of 78.5 cm^2 and a constant stirring at 300 m^{-1}. The solar UV degradation scaling process was carried out in a 5 L open channel flat reactor of (irradiated area of 1600 cm^2) connected to a 20 L recirculation tank at the rate of 1 L min^{-1}.

For the processing of the liquid samples treated in the three processes, a Shimadzu High Resolution Liquid Chromatograph was used, with a LC-20AD double channel pump, an automated SIL-10AL injector, and a UV detector with SPD M20A type arrangement. The column used was a Merck reverse phase with modified C18 (100 mm × 4.6 mm, 5 μm) spherical silica packed in an isocratic regime with 1 mL min^{-1} flow, and a ratio of acetonitrile: water acidified to pH 2 with H_3PO_4: methanol of 40:40:20 v/v. The injection volume was 10 μL and the wavelength used was 230 nm.

The mineralization analysis of the samples was carried out using a Total Organic Carbon Analyzer Shimadzu (TOC-V CSN), equipped with an infrared detector of the non dispersive type. The sample was injected at 50 μL and the combustion process was carried out in a quartz tube at 680 °C with a platinum catalyst. An oxygen flow of entrainment of the vapors was used at a rate of 150 mL min^{-1}. The mean relative error in each result was less than 6% and 2% for TOC and HPLC, respectively.

The degradation and mineralization values of the experiments were calculated by Equations (1) and (2).

$$Degradation\ (\%) = \frac{C_i - C_f}{C_i} \times 100 \tag{1}$$

where C_i is the initial concentration of diazepam and C_f is the final concentration of diazepam at a point other than the initial value.

$$Mineralization\ (\%) = \frac{TOC_i - TOC_f}{TOC_i} \times 100 \tag{2}$$

where TOC_i is the initial value of the total organic carbon and TOC_f is the total organic carbon at the end of the reaction.

In the characterization of the radiolytic transformations of the solvent we used the concept of radiolytic performance (*G-value*), referred to the number of molecules, free radicals, ions, excited particles, among others, that form or decompose when the system absorbs 100 eV of energy from ionizing radiation [29,30]. This factor is calculated according to Equation (3).

$$G - value = \frac{RNa}{D(6.24 \times 10^{16})} \tag{3}$$

where R is the change in concentration of diazepam (M), Na is the Avogadro number, D is the absorbed dose (Gy), and 6.24×10^{16} is the conversion factor of Gy to eV/L. For the conversion of the *G-value* to mol J^{-1} multiplies by 1.04×10^{-7} [31].

3. Results and Discussion

3.1. Effect of Operational Conditions on the Degradation of DZP by Radiolysis, Sonolysis and Photolysis

Figure 1 shows the best conditions achieved in the process of degradation of DZP in aqueous solutions at an initial concentration of 20 mg L^{-1} by sonolysis and radiolysis. In the sonolysis process, three power values were tested for each frequency value used. For the 580 kHz frequency, the electric power outputs were of 1.4, 8.7, and 21.8 Watt and for 862 kHz of 2.1, 10.4, and 30.6 Watt. In the photolytic degradation, a unidirectional 11 W lamp was used. Figure 1 shows the results of DZP degradation by gamma radiation. It is observed that at doses lower than 500 Gy, the degradation of DZP is very low. However, for doses higher than 2500 Gy, a 100% elimination of the drug is obtained.

For doses below 500 Gy the *G-value* is low. In this case, the amount of DZP that is degraded is less than 2% which may be associated with weak collisions between the molecule and the radicals involved in the degradation process.

The drug degradation increases with adsorbed dose. At 1000 Gy an 83.4% elimination of the DZP corresponding to a maximum *G-value* of 0.015 mol J^{-1} is reached. At dose values above this point the *G-value* decays again since the degradation of the DZP reaches 100%, decreasing the probability that molecules are formed or destroyed in the system.

In the degradation of DZP by sonolysis a maximum at 180 min of 28.27% is obtained at values of 862 kHz frequency and 30.6 watts of power. This is explained due to the fact that at higher powers, cavitation bubbles are formed with high rupture energies which generate a greater amount of HO$^\bullet$ radicals, thus increasing the probability of interaction with the molecule and its degradation.

In the degradation of DZP by photolysis a maximum of 37.97% was reached at 300 min of irradiation. Although this degradation value is higher than that obtained by sonolysis, the exposure time of the molecule to the radiation increases by 66%.

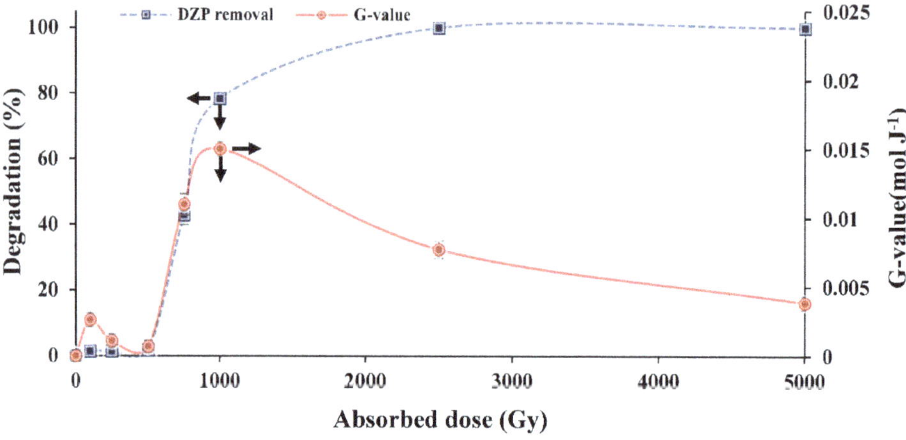

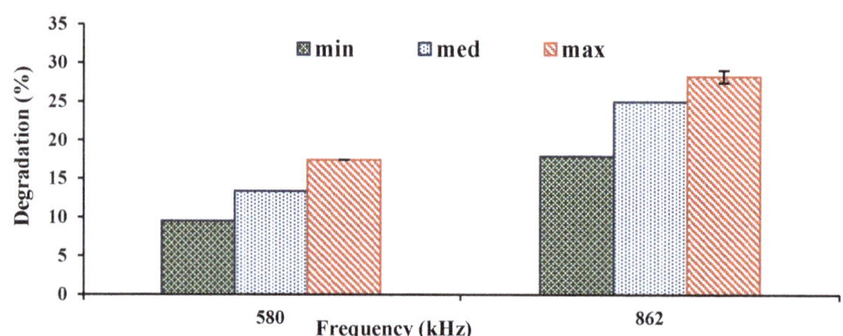

Figure 1. Degradation of DZP by gamma and utrasonic radiation. **Upper**: Influence of absorbed dose on DZP degradation and radiolytic yield (initial concentration: 20 mg L^{-1}, pH: 2.5, dose rate: 3.73 kGy h^{-1}); **Lower**: Influence of ultrasonic power and frequency on DZP degradation (initial concentration: 20 mg L^{-1}, pH: 2.5). Min, med, and max refer to the minimal, medium and maximal value of the ultrasonic power for each studied value of ultrasonic frequency.

3.2. Initial Effect of pH on Degradation of DZP

The initial pH is an important factor to study since it influences the chemical and physical conditions of the solution. The effect of pH on the radiolytic, photolytic, and sonolytic degradation of DZP (20 mg L^{-1}) at doses of 750 Gy, with a mercury lamp of 11 W, with a frequency of 862 kHz, and a power of 30.6 W.

The Figure 2 shows the influence of pH on DZP degradation. For radiolysis, drug degradation decreases with increasing pH, whereas for sonolysis at pH values 3, 5, and 7, degradation has a similar value (decreasing for pH 2.5 and 9).

Study of pH influence on sonolytic degradation shows that at pH 3, 5, and 7, the best degradation values are achieved. This result is closely related to the pKa = 3.4 value of the DZP corresponding to the carbonyl group present in its structure. This is explained by the fact that in acidic medium the molecule is dissociated in its protonated form, where the species with net positive charge, has an electrostatic interaction with the negative charges that are present in the periphery of the cavitation bubble [32], facilitating the process of drug degradation. In contrast, at pH less than 3, the degradation process is deprived, and hydrogen peroxide molecules formed (Equations (4)–(7)) can protonate and form more stable species such as H$_3$O$_2$$^+$, and or radicals HO$^{\bullet}$ can be attacked by the H$^+$ limiting the degradation process [32]. On the other hand, in basic medium (pH 9), there is practically no presence of charged species in the dissociation equilibrium, so the interaction with the cavitation bubbles is much smaller [32]. Similar results have been obtained by other authors who report that the ultrasonic degradation of different compounds is higher in acid medium, which is also influenced by the fact that in this medium the HO$^\bullet$ radicals present a higher oxidation potential [33,34].

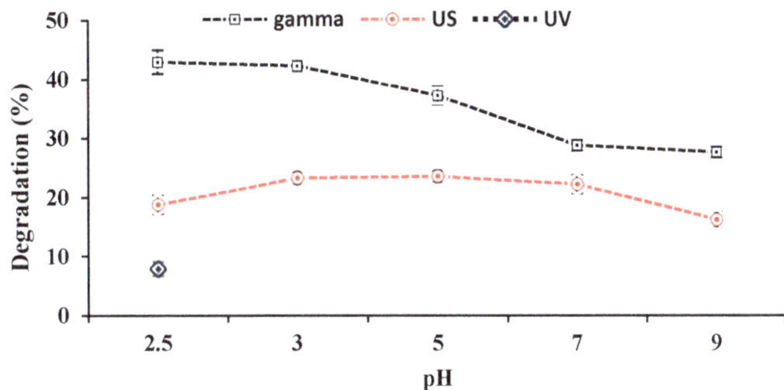

Figure 2. Influence of the pH of the solution on the sonolytic, radiolytic and photolytic degradation of the DZP (c (DZP) = 20 mg L^{-1}, T = 25 °C, US: v (agitation) = 300 m^{-1}; f = 862 kHz, t =180 min Gamma: D = 750 Gy, $\dot{D}$ = 3.73 kGy h^{-1}, UV: P = 11 W, λ = 254 nm, v (agitation) = 300 m^{-1}, t = 300 min).

$$H_2O + US \rightarrow H^\bullet + HO^\bullet \tag{4}$$

$$HO^\bullet + OH^- \rightarrow H_2O + O^- \qquad k = 1.2 \times 10^{10} \ M^{-1}s^{-1} \tag{5}$$

$$HO^\bullet + H_2O \rightarrow H_2O_2 + H^\bullet \tag{6}$$

$$HO^\bullet_{ac} + HO^\bullet_{ac} \rightarrow H_2O_{2\ (ac)} \qquad k = 5.5 \times 10^9 \ M^{-1}s^{-1} \tag{7}$$

For radiolysis, increasing pH leads to a decrease in the degradation of DZP. On the other hand, in acidic medium the hydrated electron is more likely to react with the H$^+$ and form the radical H$^+$ by Equation (9) favoring the recombination reaction (Equation (8)). In the alkaline medium, the dissociation of the HO$^\bullet$ radical occurs, so degradation of the drug is not favored (Equations (8)–(10)). Different authors report the same behavior in the radiolytic degradation of other drugs with increasing pH [35–37].

$$H^\bullet + HO^\bullet \rightarrow H_2O \qquad k = 7.0 \times 10^{10} \ M^{-1}s^{-1} \tag{8}$$

$$e^-_{(ac)} + H_3O^+_{(ac)} \rightarrow H^\bullet + H_2O \qquad k = 2.3 \times 10^{10} \ M^{-1}s^{-1} \tag{9}$$

$$HO^\bullet \leftrightarrow H^+_{(ac)} + O^\bullet_{(ac)} \qquad pKa = 11.9 \tag{10}$$

In the case of UV radiation, the effect of the initial pH was studied only for the value 2.5. In a previous publication, this studied was carried out for a similar lamp to the one used here, showing that best degradation results are obtained for pH value of 2.5 [38].

3.3. Effect of Hydrogen Peroxide on the Degradation of DZP by Sonolysis, Radiolysis and Photolysis

In the combined process of photolysis and sonolysis with H$_2$O$_2$, an increase in the formation of the hydroxyl radical occurs (Equation (11)), which is the main responsible for the degradation of the molecule. By radiolysis the formation of HO$^\bullet$ radicals is intensified, due to the interaction of H$_2$O$_2$ with the solvated electron and the hydronium radical present in the water radiolysis (Equations (9) and (11)) [31,39].

$$H^\bullet + H_2O_2 \rightarrow HO^\bullet + H_2O \qquad k = 9.0 \times 10^7 \ M^{-1}s^{-1} \tag{11}$$

Likewise, excess $OH^{\bullet}$ radicals in the medium can cause a decrease in the removal efficiency of the compound in the system. This is due to the fact that radicals tend to recombine at a very high reaction rate, as shown by Equations (7), (8) and (12)–(14) [31].

$$HO^{\bullet} + H_2O_2 \rightarrow HO_2^{\bullet} + H_2O \qquad\qquad k = 2.7 \times 10^8 \ M^{-1}s^{-1} \qquad (12)$$

$$HO_2^{\bullet} + HO^{\bullet} \rightarrow H_2O + O_2 \qquad k = 6.0 \times 10^9 \ M^{-1}s^{-1} \qquad (13)$$

$$HO^{\bullet} + e_{aq}^- \rightarrow OH^- \qquad\qquad k = 3.0 \times 10^{10} \ M^{-1}s^{-1} \qquad (14)$$

For DZP degradation study shown in Figure 3, a 20 mg L^{-1} solution of DZP at pH 2.5 for sonolysis, and pH 3 for photolysis and radiolysis was used. The power and working frequency was 30.6 W and 862 kHz, respectively, for the sonolysis. A 254 nm lamp with a power of 11 W was used for UV radiation, and the irradiation dose was 750 Gy for gamma radiation.

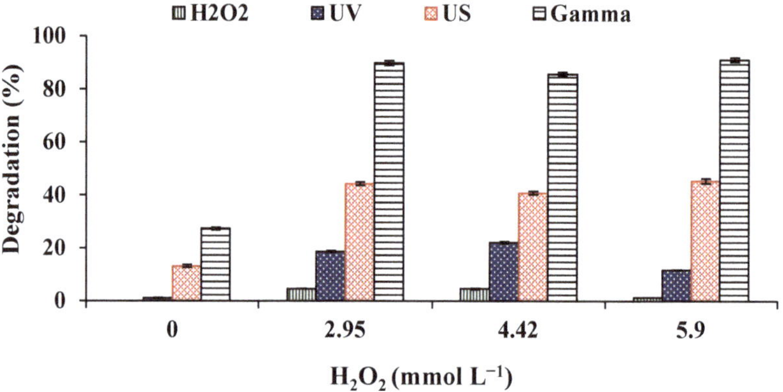

Figure 3. Influence of the addition of hydrogen peroxide on degradation of DZP (c (DZP) = 20 mg L^{-1}, T = 25 °C, t = 10 min US: v (agitation) = 300 m^{-1}; f= 862 kHz, pH = 3. Gamma: D = 750 Gy, $\dot{D}$ = 3.73 kGy h^{-1}, pH = 2.5, UV: v (agitation) = 300 m^{-1}, λ = 254 nm, P = 11 W, pH = 2.5).

In the three AOPs combined with H_2O_2 the DZP removal is intensified. For sonolysis and radiolysis, a maximum of 45.41% and 91.16% for a concentration of 5.9 mmol L^{-1} of H_2O_2 is reached. For the photolysis the maximum removal of 22.2% is achieved at 2.95 mmol L^{-1} of H_2O_2 decreasing for the higher concentration of oxidant used. This can be associated to processes of radical recombination, (Equations (3)–(5)) [21]. In general, the efficiency of the degradation process followed the order: AOPs with H_2O_2 > AOPs > H_2O_2.

The addition of H_2O_2 increases the efficiency of AOPs for the removal of contaminants in aqueous medium (You et al., 2021), obtaining the best results with the combination gamma/H_2O_2 as has been reported in other studies [31].

3.4. Effect of the Fenton Reagent on the Degradation of DZP by Sonolysis, Radiolysis and Photolysis

In Fenton processes, the $HO^{\bullet}$ radicals are generated by the catalytic decomposition of H_2O_2 using Fe^{3+} ions in acid medium at pH in the 2–4 range [40]. This method facilitates a high formation of $HO^{\bullet}$ (Equation (15)), however an excess of Fe^{2+} can trap them (Equation (15)), such as halogens, H_2O_2, or $HO_2^{\bullet}$ (Equation (16)) [31].

$$Fe^{2+} + H_2O_2 \rightarrow Fe^{3+} + OH^- + HO^{\bullet} \qquad k = 63 \ M^{-1}s^{-1} \qquad (15)$$

$$HO^{\bullet} + HO_2^{\bullet} \rightarrow H_2O + O_2 \quad k = 1.0 \cdot 10^{10} \ M^{-1}s^{-1} \qquad (16)$$

Table 1 shows the results of the degradation and mineralization of DZP by photolysis, sonolysis, and radiolysis combined with the Fenton reagent. Experiment 9 of each optimization series was repeated three times to verify the reproducibility of the experiments. In all three AOPs with the Fenton reagent, a 100% degradation is achieved for the H_2O_2/Fe^{2+} ratio and concentration values of 5.9 mmol L^{-1}. The best experimental condition in which the maximum mineralization for the three AOPs is reached corresponds to the H_2O_2/Fe^{2+} ratio of 29 with values of 4.42 mmol L^{-1} and 0.15 mmol L^{-1}, of hydrogen peroxide and ferrous salt, respectively.

Table 1. Evaluation of degradation of diazepam intensified with Fenton reagent (c (DZP) = 20 mg L^{-1}, T = 25 °C, % D at 10 min, % M at 30 min US: v (agitation) = 300 m^{-1} P = 30.6 W; f = 862 kHz; pH = 3. Gamma: D = 750 Gy; $\dot{D}$ = 3.73 kGy h^{-1}; pH = 2.5. UV: v (agitation) = 300 m^{-1}; λ = 254 nm; P = 11 W; pH = 2.5).

Run	H_2O_2 (mmol L^{-1})	Fe^{2+} (mmol L^{-1})	Fenton		Sono-Fenton		Gamma-Fenton		Photo-Fenton	
			% D	% M	% D	% M	% D	% M	% D	% M
1	-	-	13.2	27.4	1.1	-	-	-	13.2	27.4
2	5.90	0.59	98.0	7.3	100	14.3	100	37.5	100	10.6
3	5.90	0.20	78.2	4.3	80.8	17.9	94.9	51.2	86.7	14.8
4	5.90	0.12	71.1	2.8	74.8	17.4	92.8	50.1	46.5	14.5
5	2.95	0.29	89.6	1.9	100	14.4	96.1	48.7	79.9	14.0
6	2.95	0.10	73.2	2.1	74.3	19.4	96.3	58.4	64.3	16.9
7	2.95	0.06	51.3	3.0	58.7	18.3	96.4	56.7	35.8	16.4
8	4.42	0.44	97.2	9.6	100	15.0	96.5	53.0	98.1	15.2
9	4.42	0.15	71.6 ± 2.5	7.6 ± 0.8	65.9 ± 2.3	23.6 ± 1.1	95.3 ± 1.8	68.3 ± 2.7	47.7 ± 1.9	19.7 ± 0.7
10	4.42	0.09	69.1	4.9	59.0	22.5	95.3	66.0	37.0	19.3

The Fenton reagent alone can remove the DZP by 98% for the H_2O_2/Fe^{2+} ratio value of 10 with concentrations of 5.9 mmol L^{-1}. In spite of their high percentage of elimination of the drug, only 9.6% of mineralization is obtained for the ratio 10 H_2O_2/Fe^{2+} at concentrations of 4.44 mmol L^{-1} and 0.44 mmol L^{-1}, respectively.

A multiple regression analysis yielded a mathematical model to evaluate the influence of H_2O_2 and Fe^{2+} concentrations on DZP mineralization for each AOPs.

For the case of the US the percentage of mineralization responds to Equation (17) for a level of significance lower than 0.05 with a correlation coefficient of 0.89. The variable $(Fe^{2+})^2$ is not statistically significant and H_2O_2 concentration is proved to be the variable with the highest incidence in the mineralization process. Equation (18) represents the variation of mineralization for the photo-Fenton process.

$$\% \, M = -9.24 - 44.37\left[Fe^{2+}\right] + 15.92[H_2O_2] - 1.88[H_2O_2]^2 + 5.99\left[Fe^{2+}\right][H_2O_2] \quad (17)$$

$$\% \, M = -33.32 - 60.50\left[Fe^{2+}\right] + 49.12[H_2O_2] - 155.21\left[Fe^{2+}\right]^2 - 6.17[H_2O_2]^2 + 24.30\left[Fe^{2+}\right][H_2O_2] \quad (18)$$

where % M represents the percentage of mineralization, $[Fe^{2+}]$ the concentration of Fe^{2+}, and $[H_2O_2]$ the concentration of H_2O_2.

From the model the response surface and the isoline curves (Figure 4) of the mineralization were constructed as a function of H_2O_2 and Fe^{2+} concentrations. The mineralization reaches a maximum higher than 22.5% for values ranging from 3.6 mmol L^{-1} to 5.0 mmol L^{-1} for H_2O_2, progressively decreasing with the increase of the catalyst. This may be due to the fact that the amount of Fe^{2+} in the system begins to interact with the $HO^\bullet$ radicals, which decreases the process mineralization yield.

For the case of gamma radiation, the model adequately describes the experiment according to Equation (19) with a correlation coefficient of 0.99. All coefficients of the model are significant for a significance level of less than 0.05.

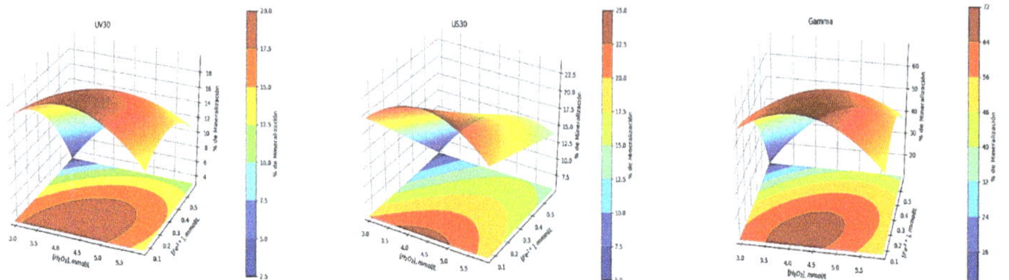

Figure 4. UV-Fenton, US-Fenton and gamma-Fenton processes. (DZP) = 20 mg L^{-1}, T = 25 °C, t = 30 min. v (agitation) = 300 m^{-1}, P = 30.6 W, f = 862 kHz, pH = 3. Gamma: D = 750 Gy, $\dot{D}$ = 3.73 kGy h^{-1}, pH = 2.5, v (agitation) = 300 m^{-1}, λ = 254 nm, P = 11 W, pH = 2.5.

$$\% M = -10.08 - 18.44\left[Fe^{2+}\right] + 14.49[H_2O_2] - 42.82\left[Fe^{2+}\right]^2 - 1.81[H_2O_2]^2 + 6.87\left[Fe^{2+}\right][H_2O_2] \tag{19}$$

Taking into account previous experiences using response surface methodology for optimizing AOPs [41], this procedure was used in this work. The response surface and the isoline curves were also constructed (Figure 4) where the mineralization reaches a maximum of 17.5% for H$_2$O$_2$ concentration values ranging from 3.3 mmol L^{-1} to 5.0 mmol L^{-1} and concentrations of Fe^{2+} lower than 0.3 mmol L^{-1} displaced towards an increase in the concentration of the former.

3.5. Evaluation of Energy Efficiency in the Degradation of DZP with AOPs

The estimation of energy consumption is an essential criterion for evaluating the most efficient AOPs in water treatment. A total of four criteria were used to estimate energy consumption in the degradation by radiolysis, photolysis, and DZP sonolysis. The first is the one proposed by Bolton and Carter in 1994, where the electric power is determined by order of magnitude (EE/O) (Equation (20)), which is expressed in kWh L^{-1} and represents the energy required to degrade a liter of polluted with the drug water [30,42]. The second is the DW (Equation (21)) which expresses the amount of energy needed to degrade a milligram of pollutant and its unit of measure is kWh mg^{-1} [31].

$$EE/O = \frac{Pt}{Vlog\left(\frac{C_o}{C_f}\right)} \tag{20}$$

where P is the consumption power at the facility (kW), t the irradiation time (h), V the volume of the treated water (L), C_o and C_f the initial and final concentrations in the working solution (mg L^{-1}).

$$DW = \frac{Pt}{\left(C_o - C_f\right)V} \tag{21}$$

where $(C_o - C_f)V$ is the mass of the degraded contaminant (mg).

Energy efficiency in mineralization was evaluated using the criteria (EE/O) and c (DW) through Equations (22) and (23). The (EE/O) c represents the amount of energy per carbon needed to treat a Liter of contaminant and the (DW) c amount of carbon energy needed to degrade one milligram of pollutant.

$$EE/O = \frac{Pt}{Vlog\left(\frac{TOC_o}{COT_f}\right)} \tag{22}$$

$$DW = \frac{Pt}{\left(TOC_o - TOC_f\right)V} \tag{23}$$

where P is the power of the equipment (kW), t the process energy consumption time (h), V is the effective volume (L), TOC_o the initial total organic carbon, and TOC_f the final total organic carbon.

In the study of energy efficiency, nine processes are compared in which three AOPs are used: radiolysis, photolysis, sonolysis, and their combination with oxidizing agents such as H_2O_2 and Fenton reactive. Experimental conditions were analyzed in which the best % degradation of DZP for each process and its combination with the oxidizing agents were reached.

The best results in the comparison of energy efficiency are found in the combination of AOPs with the Fenton reactive allowing to decrease irradiation times of 40 to 12 min for gamma-Fenton, 180 to 10 min in Fenton sleep and 300 to 10 min in photo-Fenton.

In Table 2, we report the best values obtained for energy efficiency for each studied process. The best results are obtained for the photo-Fenton reaction.

Table 2. Energy consumption in the degradation and mineralization (*) of DZP in sonolysis, photolysis and radiolysis. NA, not available.

Process	Time (min)	EE/O (kWh L^{-1})	DW (kWh mg^{-1})	* (EE/O) (kWh L^{-1})	* (DW) (kWh mg^{-1})
Gamma	40	1.2	2.9×10^{-1}	NA	NA
Gamma-H_2O_2	12	1.7	9.7×10^{-2}	NA	NA
Gamma-Fenton	12	3.3×10^{-1}	8.8×10^{-2}	17.8	1.0
Sonolysis	180	2.5	6.5×10^{-2}	NA	NA
Sono-H_2O_2	60	4.7×10^{-1}	1.3×10^{-2}	NA	NA
Sono-Fenton	10	3.9×10^{-3}	1.0×10^{-3}	0.5	2.0×10^{-2}
Photolysis	300	1.1	2.9×10^{-2}	NA	NA
Foto-H_2O_2	300	1.3×10^{-1}	1.1×10^{-2}	NA	NA
Foto-Fenton	10	1.4×10^{-3}	3.7×10^{-4}	0.2	8.0×10^{-3}

3.6. Study of the Photo-Fenton Process with Solar Radiation in a Real Wastewater

Based on the positive results obtained in the experiments on the degradation of DZP at laboratory scale using the photo-Fenton process, it was decided to scale-up this process to a flat open-channel reactor with a reaction volume of 20 L. The experiments used sunlight as the energy source, combined with Fenton's reagent. The wastewater sample was taken at the entrance of the "María del Carmen" wastewater treatment plant in Havana city, Cuba.

Three degradation experiments of DZP were carried out at a concentration of 20 mg L^{-1} for each molecule. The first experiment was carried out with technical water doped with the studied molecule, using the concentration of H_2O_2 and Fe^{2+} ions that were the most efficient in the photo-Fenton process with artificial UV light. For DZP, an H_2O_2 concentration of 5.9 mmol L^{-1} and an Fe^{2+} ion concentration of 0.59 mmol L^{-1} were used. The second experiment was carried out with wastewater doped with 20 mg L^{-1} of diazepam and the same concentrations of H_2O_2 and Fe^{2+} ions. In the third experiment, the COD of the wastewater sample was considered and the concentrations of H_2O_2 and Fe^{2+} ions to be used were recalculated, they were 11.6 mmol L^{-1} and 1.16 mmol L^{-1} respectively. The actual wastewater was previously characterized as established in NC 27:2012, Table 3.

As can be seen in Figure 5, 10 min after the reaction started a total degradation was achieved for the molecule in the technical water. This behavior was similar to that obtained in the photo-Fenton process with artificial UV at the same concentrations of H_2O_2 and Fe^{2+} ions. In the study, a total degradation of the DZP was obtained 30 min after the start of the reaction.

By increasing the concentration of the Fenton reagent, taking into account the COD of the wastewater, the total degradation of DZP was reached 10 min after the reaction started, a value that indicate the need of knowing the complexion of the matrix to improve the degradation process, and reaffirms the competitiveness of the organic molecules present in the wastewater for the HO$^\bullet$ radicals formed in the process.

Table 3. Analysis of wastewater before and after photo-Fenton treatment with sunlight.

Parameters	Waste Water	Treated Water	Efficiency (%)	NC 27: 2012
Temperature (°C)	27	26	-	<40
pH	7.65	7.32	-	6–9
Conductivity (μS cm^{-1})	1085 ± 1	87 ± 1	91.9	<2000
CO (mg of O$_2$ L^{-1})	188 ± 30	61.4 ± 6	67.3	<90
BOD$_5$ (mg of O$_2$ L^{-1})	91 ± 1	36.5 ± 1	59.8	<40
Settleable solids (mL L^{-1})	2.5 ± 0.1	0	100	<2
Floating material	present	absent	-	-
Iron (mg L^{-1})	0.91	0	100	-
TOC (mg de C L^{-1})	86.4	32.9 * 23.7 **	61.9 72.6	-

* TOC value at 30 min; ** TOC value at 120 min.

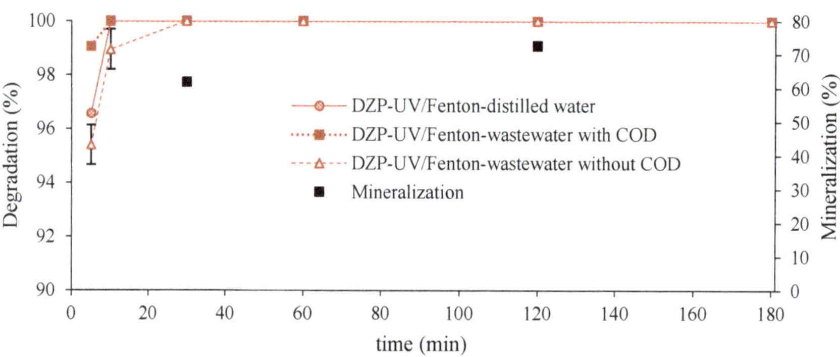

Figure 5. DZP degradation by photo-Fenton of technical water, wastewater without taking into account COD, and wastewater taking into account COD. [DZP] = 20 mg L^{-1}; pH 3.

The diazepam highest mineralization values, 61.9% and 72.6%, were reached at 30 and 120 min from the start of the reaction respectively, witnessing an increase of 14.6% between the intervals measured 90 min apart. This result is of great importance since although total degradation is obtained for DZP, there is a substantial increase in the per cent of mineralization at 120 min with respect to that achieved in technical water with the application of Fenton's reagent alone (12%), with a difference in the mineralization reached of 60.6% in both processes.

Taking in account the possible industrial application of this process in combination with other conventional treatments for wastewater, photo-Fenton is applicable before biological treatment. This is highly favorable since the microorganisms responsible for the remission of contaminants are not capable of eliminating these molecules. It is also important to note that the photo-Fenton process guarantees a satisfactory value of efficiency in the reduction of BOD$_5$ (59.8%) and COD (67.3%), that reduces the times of residence of the water in the system and increase the volume to be treated each day.

4. Conclusions

1. The three processes of advanced oxidation, sonolysis, radiolysis, and photolysis combined with the Fenton reagent guarantee the total degradation of diazepam in synthetic matrices and more than 80% in a real matrix with partial mineralization.
2. Diazepam sonolysis guarantees 28% degradation of the drug at 30.6 W and 862 kHz at 3 h of experimentation, while with photolysis 38% is achieved after 5 h. With the increase of the dose of irradiation increases the degradation of the drug for doses greater than 500 Gy, reaching the total degradation for a dose of 2500 Gy. In all cases, the process is favored in acidic conditions, with the best results for the ultrasonic and photolytic study at pH 2.5 and for the radiolytic at pH 3.

3. The combination of the processes with hydrogen peroxide guarantees the intensification of the same to the 10 min with an increase in more than 30%, 63%, and 21% for the sonolytic, radiolytic, and photochemical degradation, respectively.

4. Integration of the processes with Fenton reagent guarantees the total elimination of the drugs at 10 min in a synthetic matrix, and an increase in mineralization by more than 16%, 60%, and 12% for sonolytic, radiolytic degradation, and photochemistry, respectively.

5. All of the criteria for the evaluation of the energy consumption agree that the photo-Fenton process constitutes the one with the lowest energy consumption for the degradation of diazepam in the water matrix.

6. The degradation of the DZP in real residual water gave the best results in the experiments where the COD was taken into account to adjust the H_2O_2, and Fe^{2+} concentrations. The photo-Fenton process guarantees total degradation using solar radiation as a source of energy after 10 min. A decrease in COD, and BOD_5 of waste water was achieved below the limits required by NC-27-2012 for classification B. The gamma-Fenton process guarantees maximum efficiency in decreasing COD, BOD_5, and TOC with 89.2%, 82.1%, and 88.1%, respectively.

Author Contributions: Conceptualization, U.J.J.H., M.M.A. and S.G.; methodology, U.J.J.H. and M.M.A.; experimental work, M.M.A., S.G.G. and M.A.G.M.; formal analysis, U.J.J.H. and M.M.A.; resources, U.J.J.H., M.M.A. and S.G.; data curation, M.M.A., S.G.G. and M.A.G.M.; writing—original draft preparation, U.J.J.H., M.M.A. and S.G.G.; writing—review and editing, U.J.J.H. and S.G.G.; visualization, M.M.A.; supervision, U.J.J.H. and S.G.G.; project administration, U.J.J.H.; funding acquisition, U.J.J.H. and S.G.G. All authors have read and agreed to the published version of the manuscript.

Funding: This work was supported by the projects TATARCOP, InSTEC, University of Havana, and CAPES-Dolé: CAPES-2020, Guadeloupe.

Institutional Review Board Statement: Not applicable.

Informed Consent Statement: Informed consent was obtained from all subjects involved in this study.

Data Availability Statement: The data presented in this study are available on request from the corresponding author.

Conflicts of Interest: The authors declare no conflict of interest.

References

1. Serna-Galvis, E.A.; Silva-Agredo, J.; Botero-Coy, A.M.; Moncayo-Lasso, A.; Hernández, F.; Torres-Palma, R.A. Effective elimination of fifteen relevant pharmaceuticals in hospital wastewater from Colombia by combination of a biological system with a sonochemical process. *Sci. Total Environ.* **2019**, *670*, 623–632. [CrossRef] [PubMed]

2. Montgomery, A.B.; Bowers, I.; Subedi, B. Trends in Substance Use in Two United States Communities during Early COVID-19 Lockdowns Based on Wastewater Analysis. *Environ. Sci. Technol. Lett.* **2021**, *8*, 890–896. [CrossRef]

3. Nieto-Juárez, J.I.; Torres-Palma, R.A.; Botero-Coy, A.M.; Hernández, F. Pharmaceuticals and environmental risk assessment in municipal wastewater treatment plants and rivers from Peru. *Environ. Int.* **2021**, *155*, 106674. [CrossRef] [PubMed]

4. Liu, Z.; Hu, W.; Zhang, H.; Wang, H.; Sun, P. Enhanced Degradation of Sulfonamide Antibiotics by UV Irradiation Combined with Persulfate. *Processes* **2021**, *9*, 226. [CrossRef]

5. Tejeda, C.; Quiñonez, E.; Peña, M. Contaminantes emergentes en aguas: Metabolitos de fármacos. *Rev. Fac. Cienc. Básicas* **2014**, *10*, 80–101. [CrossRef]

6. Ellis, J.B. Pharmaceutical and personal care products (PPCPs) in urban receiving waters. *Environ. Pollut.* **2006**, *144*, 184–189. [CrossRef]

7. Bade, R.; Ghetia, M.; White, J.M.; Gerber, C. Determination of prescribed and designer benzodiazepines and metabolites in influent wastewater. *Anal. Methods* **2020**, *12*, 3637–3644. [CrossRef]

8. de Araujo, F.G.; Bauerfeldt, G.F.; Marques, M.; Martins, E.M. Development and Validation of an Analytical Method for the Detection and Quantification of Bromazepam, Clonazepam and Diazepam by UPLC-MS/MS in Surface Water. *Bull. Environ. Contam. Toxicol.* **2019**, *103*, 362–366. [CrossRef]

9. Fernández-Rubio, J.; Rodríguez-Gil, J.L.; Postigo, C.; Mastroianni, N.; de Alda, M.L.; Barceló, D.; Valcárcel, Y. Psychoactive pharmaceuticals and illicit drugs in coastal waters of North-Western Spain: Environmental exposure and risk assessment. *Chemosphere* **2019**, *224*, 379–389. [CrossRef]

10. Zeyuan, W.; Siyue, G.; Qingying, D.; Meirong, Z.; Fangixing, Y. Occurrence and risk assessment of psychoactive substances in tap water from China. *Environ. Pollut.* **2020**, *261*, 114163. [CrossRef]

11. Lei, H.-J.; Yang, B.; Ye, P.; Yang, Y.-Y.; Zhao, J.-L.; Liu, Y.-S.; Xie, L.; Ying, G.-G. Occurrence, fate and mass loading of benzodiazepines and their transformation products in eleven wastewater treatment plants in Guangdong province, China. *Sci. Total Environ.* **2021**, *755*, 142648. [CrossRef] [PubMed]

12. Ye, P.; You, W.-D.; Yang, B.; Chen, Y.; Wang, L.-G.; Zhao, J.-L.; Ying, G.-G. Pollution characteristics and removal of typical pharmaceuticals in hospital wastewater and municipal wastewater treatment plants. *Huanjing Kexue/Environ. Sci.* **2021**, *42*, 2928–2936.

13. Lebreton, M.; Malgouyres, J.-M.; Carayon, J.-L.; Bonnafé, E.; Géret, F. Effects of the anxiolytic benzodiazepine oxazepam on freshwater gastropod reproduction: A prospective study. *Ecotoxicology* **2021**, *30*, 1880–1892. [CrossRef] [PubMed]

14. Subedi, B.; Anderson, S.; Croft, T.L.; Rouchka, E.C.; Zhang, M.; Hammond-Weinberger, D.R. Gene alteration in zebrafish exposed to a mixture of substances of abuse. *Environ. Pollut.* **2021**, *278*, 116777. [CrossRef] [PubMed]

15. INCB. *Psychotropic Substances: Statistics for 2013*; Assessments of Annual Medical and Scientific Requirements for Substances in Schedules II, III and IV of the Convention on Psychotropic Substances of 19711 (E/INCB/2014/3) United Nations; INCB: New York, NY, USA, 2014.

16. Chaichi, M.J.; Alijanpour, S.O. A new chemiluminescence method for determination of clonazepam and diazepam based on 1-ethyl-3-methylimidazolium ethylsulfate/copper as catalyst. Spectrochimica Acta Part A. *Mol. Biomol. Spectrosc.* **2014**, *118*, 36–41. [CrossRef] [PubMed]

17. Caban, M.; Kumirska, J.; Bialk-Bielińska, A.; Stepnowski, P. Current issues in pharmaceutical residues in drinking water. *Curr. Anal. Chem.* **2015**, *12*, 249–257. [CrossRef]

18. Cunha, D.L.; da Silva, A.S.; Coutinho, R.; Marques, M. Optimization of Ozonation Process to Remove Psychoactive Drugs from Two Municipal Wastewater Treatment Plants. *Water Air Soil Pollut.* **2022**, *233*, 67. [CrossRef]

19. Mergenbayeva, S.P.; Poulopoulos, S.G. Comparative Study on UV-AOPs for Efficient Continuous Flow Removal of 4-tert-Butylphenol. *Processes* **2022**, *10*, 8. [CrossRef]

20. Serna-Galvis, E.A.; Porras, J.; Torres-Palma, R.A. A critical review on the sonochemical degradation of organic pollutants in urine, seawater, and mineral water. *Ultrason. Sonochem.* **2022**, *82*, 105861. [CrossRef]

21. Wang, J.; Chu, L. Irradiation treatment of pharmaceutical and personal care products (PPCPs) in water and wastewater: An overview. *Radiat. Phys. Chem.* **2016**, *125*, 56–64. [CrossRef]

22. Bagheri, H.; Akhami, A.; Noroozi, A. Removal of pharmaceutical compounds from hospital wastewaters using nanomaterials. A Review. *Anal. Bioanal. Chem. Res.* **2016**, *3*, 1–18.

23. Domènech, X.; Jardim, W.F.; Litter, M.I. Procesos avanzados de oxidación para la eliminación de contaminantes. In *Eliminación de Contaminantes por Fotocatálisis Heterogénea*; CYTED: La Plata, Argentina, 2001.

24. You, W.-D.; Ye, P.; Yang, B.; Luo, X.; Fang, J.; Mai, Z.-T.; Sun, J.-L. Degradation of 17 benzodiazepines by the UV/H_2O_2 Treatment. *Front. Environ. Sci.* **2021**, *9*, 764841. [CrossRef]

25. Ribeiro, A.R.; Nunes, O.C.; Pereira, M.; Silva, A. An overview on the advanced oxidation processes applied for the treatment of water pollutants defined in the recently launched Directive 2013/39/EU. *Environ. Int.* **2015**, *75*, 3351. [CrossRef] [PubMed]

26. Expósito, A.J.; Patterson, D.A.; Monteagudo, J.M.; Durán, A. Sono-photodegradation of carbamazepine in a thin falling film reactor: Operation costs in pilot plant. *Ultrason. Sonochem.* **2017**, *34*, 496–503. [CrossRef]

27. Alalm, M.G.; Tawfik, A.; Ookawara, S. Degradation of four pharmaceuticals by solar photo-Fenton process: Kinetics and costs estimation. *J. Environ. Chem. Eng.* **2015**, *3*, 46–51. [CrossRef]

28. Gong, Y.; Li, J.; Zhang, Y.; Zhang, M.; Tian, X.; Wang, A. Partial degradation of levofloxacin for biodegradability improvement by electro-Fenton process using an activated carbon fiber felt cathode. *J. Hazard. Mater.* **2016**, *304*, 320328. [CrossRef]

29. Xu, L. Degradation of Refractory Contaminants in Water by Chemical-Free Radicals Generated by Ultrasound and UV Irradiation. Ph.D. Thesis, The Hong Kong Polytechnic University, Hongkong, China, 2014.

30. Kim, H.Y.; Lee, O.M.; Kim, T.H.; Yu, S. Enhanced biodegradability of pharmaceuticals and personal care products by ionizing radiation. *Water Environ. Res.* **2015**, *87*, 321–325. [CrossRef]

31. Rivas, I.O.; Cruz, G.G.; Lastre, A.A.; Manduca, M.A.; Rapado, M.P.; Chávez, A.A.; Silva, A.C.; Jáuregui, U.H. Optimization of radiolytic degradation of sulfadiazine by combining Fenton and gamma irradiation processes. *J. Radioanal. Nucl. Chem.* **2017**, *314*, 2597–2607. [CrossRef]

32. Lastre, A.M.; Cruz, G.G.; Nuevas, P.L.; Jáuregui, H.U.; Silva, A.C. Ultrasonic degradation of sulfadiazine in aqueous solutions. *Environ. Sci. Pollut. Res.* **2015**, *22*, 918–925. [CrossRef]

33. Durán, A.; Monteagudo, J.A.; Martín, I.S. Operation costs of the solar photo-catalytic degradation of pharmaceuticals in water: A mini-review. *Chemosphere* **2018**, *211*, 482–488. [CrossRef]

34. Kanakaraju, D.; Glass, B.D.; Oelgemoller, M. Advanced oxidation process-mediated removal of pharmaceuticals from water: A review. *J. Environ. Manag.* **2018**, *219*, 189–207. [CrossRef] [PubMed]

35. Torun, M.; Gültekin, Ö.; Solpan, D.; Güven, O. Mineralization of paracetamol in aqueous solution with advanced oxidation processes. *Environ. Technol.* **2014**, *36*, 970–982. [CrossRef] [PubMed]

36. Quijano, P.D.; Orozco, D.J.; Holguín, H. Conocimientos y prácticas de pacientes sobre disposición de medicamentos no consumidos. Aproximación a la ecofarmacovigilancia. *Salud Pública* **2016**, *18*, 61–71. [CrossRef] [PubMed]

37. Rizzo, L.; Della, A.; Fiorentino, A.; Li Puma, G. Disinfection of urban wastewater by solar driven and UV lamp e TiO2 photocatalysis: Effect on a multi drug resistant Escherichia coli strain. *Water Res.* **2014**, *53*, 145–152. [CrossRef] [PubMed]
38. Rossi-Bautitz, I.; Pupo-Nogueira, R. Photodegradation of lincomycin and diazepam in sewage treatment plant effluent by photo-Fenton process. *Catal. Today* **2010**, *151*, 94–99. [CrossRef]
39. Monteiro, M.A.; Spisso, B.F.; dos Santos JR, M.P.; da Costa, R.P.; Ferreira, R.G.; Pereira, M.U.; Miranda, T.S.; de Andrade BR, G.; d'Avila, L.A. Occurrence of antimicrobials in river water samples from rural region of the state of Rio de Janeiro, Brazil. *J. Environ. Prot.* **2016**, *7*, 230–241. [CrossRef]
40. Matongo, S.; Birungi, G.; Moodley, B.; Ndungu, P. Pharmaceuticals residues in water and sediment of Msunduzi river, KwaZulu-Natal, South Africa. *Chemosphere* **2015**, *134*, 133–140. [CrossRef]
41. Roudi, A.M.; Salem, S.; Abedini, M.; Maslahati, A.; Imran, M. Response Surface Methodology (RSM)-Based Prediction and Optimization of the Fenton Process in Landfill Leachate Decolorization. *Processes* **2021**, *9*, 2284. [CrossRef]
42. Zhang, Y.; Xiao, Y.; Zhang, J.; Chang, V.; Lim, T.T. Degradation of cyclophosphamide and 5-fluorouracil in water using UV and UV/H2O2: Kinetics investigation, pathways and energetic analysis. *J. Environ. Chem. Eng.* **2017**, *5*, 1133–1139. [CrossRef]

water

Article

Efficient Removal of Rhodamine B in Wastewater via Activation of Persulfate by MnO$_2$ with Different Morphologies

Xinyi Zhang [1,2], Xinrui Gan [1], Shihu Cao [1], Jiangwei Shang [1,2] and Xiuwen Cheng [1,2,*]

[1] Key Laboratory of Pollutant Chemistry and Environmental Treatment, College of Chemistry and Environmental Science, Yili Normal University, Yining 835000, China
[2] Key Laboratory for Environmental Pollution Prediction and Control, College of Earth and Environmental Sciences, Lanzhou University, Lanzhou 730000, China
* Correspondence: chengxw@lzu.edu.cn

Abstract: In recent years, typical organic pollutants were frequently found in aquatic environments. Among them, synthetic dyes were widely used in many industries, which resulting in a large amount of wastewater contained dyes. Because of the characteristic of complex components, poor biodegradability and high toxicity, this kind of wastewater brought lots of harm to the ecological environment and organism. In this study, three different types of manganese dioxide (MnO$_2$) with the rod-like, needle-like and mixed morphologies respectively were successfully fabricated by hydrothermal method with changing the preparation conditions and addition of the metal ions, and utilized as activator of persulfate (PS) to remove the dyes aqueous. Subsequently, these MnO$_2$ nanocomposite was characterized by X-ray diffraction (XRD) and scanning electron microscope (SEM) measurements. In addition, Rhodamine B (Rh B), as a representative substance of xanthene dyes was chosen as the target degradants to researched and compared the efficiency of removal via PS activated by different MnO$_2$. By exploring the influences of different reaction parameters on the result of removal, results indicated that PS activated by the acicular MnO$_2$ (α-MnO$_2$) can remove 97.41% of Rh B over 60 min, with the optimal catalyst/PS ratio of 2:1 (the concentration of the α-MnO$_2$ and PS were 1.2 g/L and 0.6 g/L, respectively), pH value of 3, at the temperature of 20 °C. Meanwhile, the probable degradation mechanism was also proposed. At last, as the catalyst was reused for four times, the degradation rate of target degradants decreased less than 10%.

Keywords: PS; MnO$_2$; morphologies; SO$_4^{\bullet-}$; mechanism

Citation: Zhang, X.; Gan, X.; Cao, S.; Shang, J.; Cheng, X. Efficient Removal of Rhodamine B in Wastewater via Activation of Persulfate by MnO$_2$ with Different Morphologies. *Water* **2023**, *15*, 735. https://doi.org/10.3390/w15040735

Academic Editors: Gassan Hodaifa, Antonio Zuorro, Joaquín R. Dominguez, Juan García Rodríguez, José A. Peres, Zacharias Frontistis and Mha Albqmi

Received: 25 December 2022
Revised: 2 February 2023
Accepted: 8 February 2023
Published: 13 February 2023

1. Introduction

In recent decades, with the rapid development of industrialization and urbanization, the environmental problems caused by refractory organic wastewater are becoming increasingly serious and need to be effectively treated [1]. Due to the widespread usage of synthetic dyes in printing, dyeing, papermaking, textile, cosmetics, leather and other industries [2], a large number of highly toxic, complex ingredients, and difficult to biodegradable dye wastewater have been produced, and many organic dyes are difficult to metabolize in the body, gradual accumulation causes great harm to human health and ecological environment [3]. Therefore, it is crucial to find a more efficient and green treatment method to degrade organic dye wastewater. The rhodamine series of dyes are 3′,6′ diaminated xanthene-like fluorescents with excellent pH stability and a variety of structures, and different structures offer fluorescence with different wavelengths and colors. Rhodamine B (Rh B), as a representative substance of xanthene dyes, is most common and most frequently used, similar to diphenylmethane and triphenylmethane derivatives, generally used in industries such as colored glass, paper, textiles, fireworks, and etc. It is difficult to degrade in the environment, highly toxic, persistent and carcinogenic [4]. Herein, Rh B was selected as target pollutants in this study. It is hoped to provide some methods and references for the degradation of such organic dyes.

The usual methods for treating dyeing and printing wastewater include physical, biological and chemical methods. The commonly used physical methods generally include: adsorption, coagulation, membrane treatment, reverse osmosis, ion exchange and etc. However, physical methods face many application limitations: (1) adsorption materials are not suitable for hydrophobic dyes, and the regeneration cost is high and the loss is large; (2) coagulation method is not effective for hydrophilic dyes, and it is easy to produce sludge that needs further treatment; (3) membrane treatment and reverse osmosis membrane can treat printing and dyeing wastewater well, but the cost is too high, the membrane is easy to clog, and the treatment and disposal of waste membrane is also a problem; (4) ion exchange method is only suitable for treating ionic dye wastewater, and it needs to be regenerated continuously, which results in high loss. In addition, the biological treatment method is not ideal for dye wastewater because most of these dyes are poorly biochemical and difficult to biodegrade, among which it has been confirmed that reactive dyes are the most difficult to be degraded in dyeing wastewater [5,6]. Chemical methods refer to the oxidation of organic pollutants using oxidants, and the commonly used oxidants are H_2O_2, $NaClO$, O_3 and etc. The chemical method has the characteristics of good treatment effect, fast reaction rate, complete degradation of pollutants, high treatment efficiency and wide application range. Commonly used chemical methods are through the oxidant degradation of pollutants, and will use photocatalysis [7], electrocatalysis [8], high temperature and other means to catalyze the oxidant to produce a large number of hydroxyl radicals to improve the degradation effect. However, for some hard-to-degrade organic substances, traditional chemical methods still cannot effectively degrade them.

Advanced oxidation processes (AOPs), which have been developed in recent years, are currently attracting much attention and can make most organic pollutants to be effectively degraded by producing free radicals as the main active substance to eliminate pollutants completely [9]. It is mainly represented by AOPs based on sulfate radical ($SO_4^{\bullet-}$) and hydroxyl radical ($^{\bullet}OH$). By contrast, $SO_4^{\bullet-}$ possesses a higher redox potential than $^{\bullet}OH$ (redox potential of $^{\bullet}OH$ is 2.8 eV, while the $SO_4^{\bullet-}$ is 2.5–3.1 eV) [10]; $SO_4^{\bullet-}$ has better selectivity through electron transfer, dehydrogenation, addition reaction to degrade organic matter [11]; $SO_4^{\bullet-}$ is more stable than $^{\bullet}OH$ due to its shorter half-life [12]; the way to produce $SO_4^{\bullet-}$ is also easier and the origin of which shows a higher stability [13]. Currently, AOPs based on $SO_4^{\bullet-}$ radicals generally generate $SO_4^{\bullet-}$ radicals by activating persulfates, including peroxydisulfate (PS)and peroxymonosulfate (PMS). Among them, PS is a solid that exists stably at room temperature, is easily preserved for transport and transfer, dissolves easily in water, and is relatively inexpensive. The PS itself, as a strong oxidizing agent, can also produce $SO_4^{\bullet-}$ radicals in aqueous solution without a catalyst. In general, PS is more stable and the above reaction rates are extremely low. After activation by applied energy or catalyst, the peroxygen bond (O-O bond) in the PS breaks and a large number of $SO_4^{\bullet-}$ radicals can be efficiently produced with strong oxidizing properties. Herein, $SO_4^{\bullet-}$ radicals can degrade difficult-to-treat organic pollutants in wastewater into biodegradable pollutants or even directly mineralize to carbon dioxide and water. In this process, $SO_4^{\bullet-}$ radicals will be reduced to sulfate (SO_4^{2-}), and no toxic and harmful substances will be produced, which is an environmentally friendly oxidant.

Therefore, the activation technology of PS has become the focus of current research. The most commonly used PS activation methods are thermal activation, ultrasound, photoactivation, and transition metal activation [14]. The reactions involved are as follows:

$$S_2O_8^{2-} + \text{heat} \rightarrow 2\,SO_4^{\bullet-} \tag{1}$$

adjust the pH in the reaction parameter influence tests. All the experiments were carried out for three times to minimize the errors

$$\tag{2}$$

$$M^{n+} + S_2O_8^{2-} \rightarrow M^{(n+1)+} + SO_4^{\bullet-} + SO_4^{2-} \tag{3}$$

$$M^{n+} + HSO_5^- \rightarrow M^{(n+1)+} + SO_4^{\bullet-} + {}^{\bullet}OH \tag{4}$$

However, some of these methods require a large number of high-cost oxidants, high reaction conditions, or consume large amounts of energy [15]; likewise, some metal oxides or metal ions as catalysts are prone to secondary pollution, such as Co^{2+}, Co_3O_4, CuO, etc. [16–18]. Thus, the exploration and development of green and efficient persulfate activator has become a hot spot and focus now.

Manganese oxides are considered as common mineral oxides found in sediments and soils [19]. Among the various catalyst materials, MnO_2 has become a very promising candidate because of its natural abundance, wide source, low cost, high activation, and low toxicity [20]. Because of its superior oxygen mobility, manganese oxides can induce a wider range of redox reactions [21]. More importantly, the reactivity of manganese-based oxides is relatively high. Meanwhile, as a promising alternative to iron minerals, manganese oxides are often applied in Fenton or Fenton-like reactions [22]. In recent reports, MnO_2 have attracted considerable attention as activator for PS [23], especially, multi-component magnetic material compounded with it [19,24]. In this paper, three different morphologies of MnO_2 activated PS were prepared to degrade organic pollutants, which can provide a certain reference for the technology of activation of PS with MnO_2 and research of its multielement composite.

Herein, the main works of this study are: (1) Preparation of catalyst materials: three different morphologies of MnO_2 were fabricated by hydrothermal method by changing the reaction conditions and adding metal ions; (2) Characterization of catalysts' crystallographic structure and micromorphology: X-ray diffraction (XRD) was employed to analyze the materials' composition and structure of atoms or molecules inside. Scanning electron microscope (SEM) was used to observe the microscopic surface morphology of the MnO_2 with different morphologies; (3) Study on the catalytic performance: Rhodamine B was chosen as target pollutants to evaluate the catalytic activity of MnO_2 on PS activation. The tests of exploring the influences of different reaction parameters on the result of removal were also conducted; (4) Stability evaluation: the repeated degradation experiments were carried out for 4 times. The main objective of this study was to compare and select the most dominant species among three different morphologies of MnO_2 catalysts and to reveal the relationship between the microscopic morphology of the materials and their performance in activating PS for Rh B degradation. Ultimately, this research was expected to provide an important reference for the study of the activation of PS by MnO_2 and their multicomponent complexes and the development of new nanomaterials for the efficient removal of organic dyes from wastewater.

2. Materials and Methods

2.1. Chemicals

Potassium permanganate ($KMnO_4$), manganese sulfate monohydrate ($MnSO_4 \bullet H_2O$), manganese chloride tetrahydrate ($MnCl_2 \bullet 4H_2O$), ferric chloride hexahydrate ($FeCl_3 \bullet 6H_2O$), potassium persulfate ($K_2S_2O_8$), hydrochloric acid (HCl), absolute ethanol (C_2H_5OH), sulfuric acid (H_2SO_4), potassium hydroxide (KOH), Rhodamine B ($C_{28}H_{31}ClN_2O_3$, Rh B) were purchased from Kermel Chemical Reagent Co. Ltd. (Tianjin, China). All the chemicals above were directly used when received without further purification. Noted, deionized (DI) water was used throughout this experiment.

2.2. Fabrication of MnO_2 with Different Morphologies

In this experiment, MnO_2 was prepared by hydrothermal method. By changing the preparation conditions, the reactant ratio and trying to add metal ions as auxiliary agents, three different morphologies of MnO_2 were obtained.

2.2.1. Fabrication of Rod-Shaped MnO_2

8 mmol (1.3520 g) of $MnSO_4 \bullet H_2O$ and 8 mmol (1.2640 g) of $KMnO_4$ dissolved in 50 mL of deionized water, and performed magnetic stirring for 30 min to dissolve fully. The mixed solution was then transferred to a 100 mL Teflon-lined stainless-steel autoclave,

sealed and kept at 140 °C for 12 h. After cooling to room temperature, the precipitate was centrifuged and washed with deionized water and absolute ethanol alternately three times, finally dried in a vacuum oven at 120 °C for 12 h. The sample was ground and stored for following use [25].

2.2.2. Fabrication of Acicular MnO_2

The fabrication conditions of the above rod-shaped MnO_2 were changed to some extent, instead of changing the dosing ratio of $MnSO_4 \bullet H_2O$ and $KMnO_4$ (1:1), an appropriate amount of iron ions (Fe^{3+}) was added to the reaction system. In this test, 1 mmol (0.2702 g) of $FeCl_3 \bullet 6H_2O$ was added to the mixed solution, followed by magnetic stirring for 30 min to dissolve thoroughly. The mixed solution was then transferred to a 100 mL Teflon-lined stainless-steel autoclave, sealed and placed in a vacuum oven at 140 °C for 12 h. After cooling to room temperature, subsequent processing was consistent with the rod-shaped MnO_2 above. After grinding, sample was stored for use [26].

2.2.3. Fabrication of Mixed MnO_2

To obtain the mixed MnO_2, 1.5 mmol (0.2970 g) of $MnCl_2 \bullet 4H_2O$ and 2.5 mmol (0.3950 g) of $KMnO_4$ were dissolved in 50 mL of deionized water, and 0.45 mmol (0.1216 g) of $FeCl_3 \bullet 6H_2O$ were added to the mixture, which was sufficiently dissolved by vigorously stirring. The mixed solution was poured into a Teflon-lined stainless-steel autoclave, sealed and kept at 150 °C for 24 h. After cooling to room temperature, the treatment of the precipitate simulated the preparation process of the first two types of MnO_2 [27].

2.3. Characterizations

Powder X-ray diffraction (XRD) spectrograms of the samples were obtained on Panalytical multifunctional powder X-ray diffractometer (XRD D8 Advance, Rigaku Corporation, Matsumoto, Japan) with a graphite monochromatic Cu Kα radiation (λ = 0.15418 nm). The morphology and microstructure of three types of MnO_2 were determined by field-emission scanning electron microscopy (SEM, Japan Electron Optics Laboratory Co. Ltd., Mitaka, Tokyo, JSM-6701F) at 20 kV.

2.4. Catalytic Activity Tests

Prepared a 10 mg/L Rh B solution of the target substance for use and took 50 mL each time into 100 mL Erlenmeyer flasks, added a certain amount of catalyst and PS, then marked according to different reaction conditions. At the same time, set a blank group and pure adsorption group to contrast. Degradation experiments were conducted at the temperature of 20 °C, on the shaking table with the rocking speed kept 180 rpm. After the given time intervals, 5 mL of mixed solution was aspirated with a syringe and filtered through a 0.22 μm to remove the catalyst, and immediately measured at maximum absorption wavelengths of the Rh B (549 nm) in a T6 UV–vis spectrophotometer (Evolution 300, Thermo Fisher Scientific Inc., Shanghai, China) [28]. Besides, the ratio of the measured absorbance of the sample to the absorbance of the initial target solution (A_t/A_0) was regarded as the analysis index. Where A_t was the absorbance measured for each sampled sample and A_0 was the absorbance of the initial 10 mg/L solution of the target pollutant. Meanwhile, in the following equation, C_0 was the initial concentration of the target degradant and C_t was the concentration of the target degradant in the solution at time t. The degradation rate of this experiment was calculated using the following Equation (5) [29]. Additionally, H_2SO_4 (pH = 2) and KOH (pH = 11) were employed to adjust the pH in the reaction parameter influence tests. All the experiments were carried out for three times to minimize the errors.

$$\eta\% = [(C_0 - C_t)/C_0] \times 100\% = [(C_0 - C_t)/C_0] \times 100\% = (1 - A_t/A_0) \times 100\% \quad (5)$$

3. Results and Discussion

3.1. XRD and SEM Analysis

In order to obtain the crystallographic structure of different types of the as-prepared MnO_2, X-ray diffraction (XRD) was carried out and results were displayed in Figure 1. As depicted in Figure 1a, lines appearing at 28.7°, 37.3°, 40.9°, 42.8°, 56.7°, 59,3°, 64.9°, 67.3° and 72.3° can be appreciably consistent with characteristic lines of β-MnO_2 (JCPDS: 24-0735), corresponding to the (110), (101), (200), (111), (211), (220), (002), (310) and (301) planes of tetragonal β-MnO_2 crystals [30]. Meanwhile, several diffraction lines at 12.6°, 18.0°, 28.7°, 37.4°, 41.8°, 49.7°, 56.4°, 59.9°, 65.0° and 69.2° displayed in Figure 1b can be indexed to the hexagonal phase of $K_{1.33}Mn_8O_{16}$ (JCPDS: 77-1796) with the exposure of (110), (200), (130), (330), (240), (140), (251), (620), (002) and (541) planes. In addition, lines at 12.7°, 18.0°, 25.7°, 28.7°, 37.5°, 42.0°, 49.7°, 56.3°, 60.0°, 65.3°, 69.2° and 72.8° shown in Figure 1c, were attributed to (110), (200), (220), (310), (211), (301), (411), (600), (260), (002), (541) and (321) crystal facets of tetragonal α-MnO_2 (JCPDS: 44-0141) [31]. These phenomena could rationally verify that three types of MnO_2 were supposed to be pure β-MnO_2, mixed $K_{1.33}Mn_8O_{16}$ and pure α-MnO_2, respectively. In addition, all three MnO_2 catalysts prepared in this study exhibited an excellent crystallinity and no obvious spurious lines appeared in characteristic XRD patterns, indicating that pure MnO_2 materials were successfully synthesized.

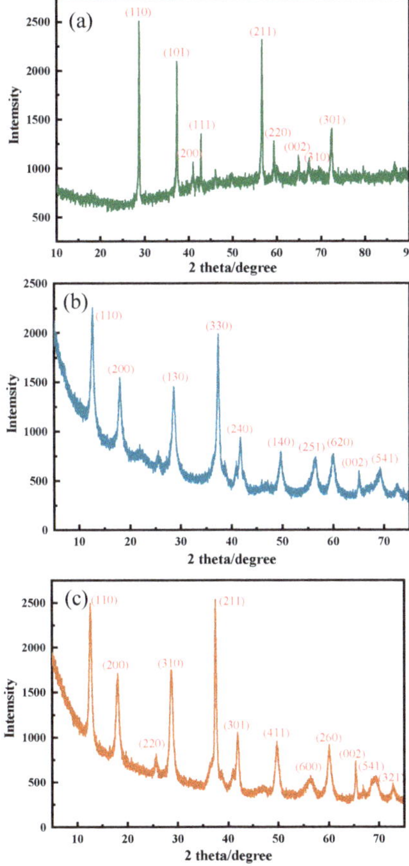

Figure 1. XRD patterns of β-MnO_2 (**a**), $K_{1.33}Mn_8O_{16}$ (**b**) and α-MnO_2 (**c**).

Besides, to further analyzed the surface morphology of β-MnO$_2$, K$_{1.33}$Mn$_8$O$_{16}$ and α-MnO$_2$, SEM consequences of different samples were shown in Figure 2. As exhibited in Figure 2a, as-prepared smooth and evenly distributed rod-like β-MnO$_2$ with an average diameter of about 80 nm showed a number of aggregated states. Moreover, the combination between MnO$_2$ nanoparticles and MnO$_2$ nanorods observed in Figure 2b, were matched with K$_{1.33}$Mn$_8$O$_{16}$, a kind of mixed MnO$_2$. While, α-MnO$_2$ displayed acicular-like appearance in Figure 2c were agglomerated with an average diameter of about 60 nm. In addition, rod-like β-MnO$_2$ and acicular-like α-MnO$_2$ are essentially MnO$_2$ nanorods, and the mixed type is a combination one. However, the shorter length and finer diameter of α-MnO$_2$ leaded to larger specific surface area in comparing with β-MnO$_2$ and K$_{1.33}$Mn$_8$O$_{16}$, which could theoretically provide more reaction and activation sites to better activate PS [32]. This inference was proved by the subsequent experiments.

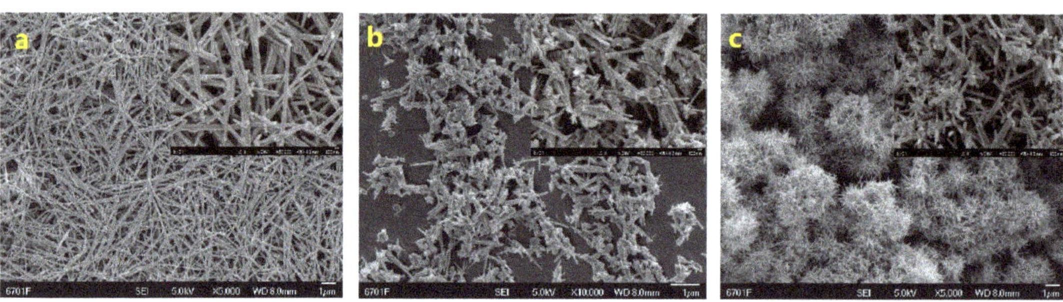

Figure 2. SEM images of β-MnO$_2$ (**a**), K$_{1.33}$Mn$_8$O$_{16}$ (**b**) and α-MnO$_2$ (**c**).

3.2. Organic Pollutants Degradation Performance

Herein, the degradation experiments of Rh B were carried out in various systems under the unadjusted pH (4.86) to evaluate and compare the organic pollutants degradation performance of three types of MnO$_2$. As presented in Figure 3, only about 4.01%, 7.13% and 9.46% of Rh B could be removed with absorption of β-MnO$_2$, K$_{1.33}$Mn$_8$O$_{16}$ and α-MnO$_2$, respectively, which revealed the decolorization of Rh B had little to do with adsorption. Besides, the degradation efficiency of Rh B was less than 45% over 60 min with the single PS addition, which was ascribed to the weak direct oxidation of PS (E$_0$ = 2.01 V) [33], This result suggested that chemical oxidation alone could not efficiently oxidized organic pollutants. Noted, for as-synthesized β-MnO$_2$, K$_{1.33}$Mn$_8$O$_{16}$ and α-MnO$_2$, the Rh B removal efficiency was markedly improved in presence of each type of MnO$_2$ and PS simultaneously, indicated that the activation effect on PS of catalysts in aqueous solution play a vital role in the degradation of organic pollutants [34]. Among them, the efficiency of PS activation on Rh B degradation with different catalysts was α-MnO$_2$ > K$_{1.33}$Mn$_8$O$_{16}$ > β-MnO$_2$, which was 78.43%, 67.48% an 47.62%, respectively. The result was consistent with those inference obtained by SEM analysis, which was mainly caused by larger specific surface area α-MnO$_2$ possessed. In detail, the specific surface area of β-MnO$_2$ with a rod-like structure was approximately 7.9~13.2 m^2/g, which was much lower than that of α-MnO$_2$ with a needle-like structure (44.4~76.5 m^2/g) [32,35], while the specific surface area of hybrid MnO$_2$ should be somewhere in between. Thus, α-MnO$_2$ enhanced the activation of PS with its ability to expose more active sites, facilitating the elimination of Rh B in water.

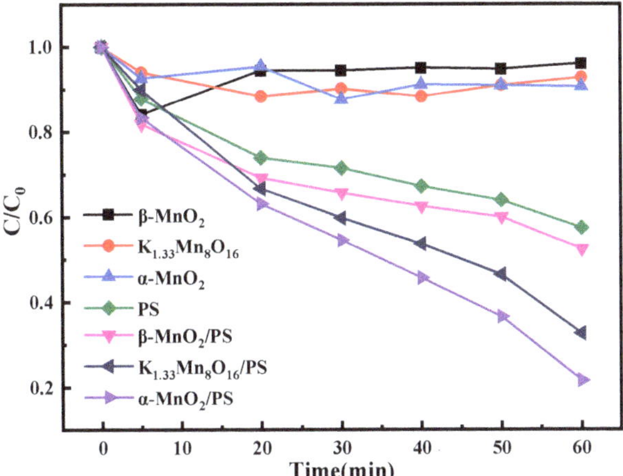

Figure 3. Time course of Rh B absorption and degradation via PS activation by different types of catalysts. Experiment conditions: PS concentration = 0.6 g/L, catalyst dosage = 0.8 g/L, Rh B concentration = 10 mg/L.

3.3. Optimization of Reaction Parameters

Generally speaking, as we all know that the addition of catalyst dosage, PS dosage and catalyst/PS dosing ratio play important roles in degradation experiments by PS activation. Moreover, detecting the influence of initial pH is great significance to prove the adaptability of α-MnO$_2$ in complicated practical water with various pH in the different places [36]. Therefore, a serious of tests were conducted to optimize reaction conditions. The effect of α-MnO$_2$ dosage (0.2 g/L, 0.4 g/L, 0.6 g/L, 0.8 g/L and 1.2 g/L) on the Rh B degradation was displayed in Figure 4a. Obviously, the removal rate of Rh B promoted (43.38%, 62.71%, 68.89%, 78.61%, 84.49% and 93.80%) with the increased concentration of the α-MnO$_2$. Likewise, as illustrated in Figure 4b, the decoloration efficiency of Rh B promoted with the increased dosage of PS, loading from 0.2 to 1.2 g/L corresponding to degradation efficiency of 38.31%, 59.85%, 75.25%, 81.39%, 85.61% and 86.71%. These phenomena might imply that with the increase of the amount of catalyst and PS, the available active metal ions increased, which accelerated the activation of PS and generated more active substance to degrade Rh B more efficiently [28]. In consideration of economics and inhibition caused by interaction between the catalyst and PS, reasonable catalyst/PS dosing ratio needed to be determined experimentally. It can be seen from Figure 4c that when the ratio of α-MnO$_2$/PS was 2:1 (the concentration of the α-MnO$_2$ and PS were 1.2 g/L and 0.6 g/L, respectively), the effect of activating PS to degrade Rh B was best. The effects of initial pH were studied at different values of 3, 5, 7, 9 and 11, as shown from Figure 4d, the degradation rate decreases significantly with the pH value increasing. 97.41% of Rh B could be removed for 60 min at the pH of 3, while the pH was increased to 5, 7, or 9, the degradation efficiency decreased by 18%, 25, or 27% over 60 min. In addition, when the pH was increased to 11, only 28.53% of Rh B was removed. The reasons of this result were as follow: (1) The zero point charge of MnO$_2$ and its complexes was reported to generally occur in the pH range of 3 to 5 [35,37,38]. Therefore, when the pH was sufficiently low, the crystalline surface became positively charged, which was more favorable for electrostatic adsorption between the electron-rich contaminants, PS and α-MnO$_2$; (2) When the pH was too high, the large amount of SO$_4{}^{2-}$ produced by the reactivity of SO$_4{}^{\bullet-}$ and OH$^-$ (Equation (6)), caused the loss of SO$_4{}^{\bullet-}$ [39].

$$\mathrm{SO_4}^{\bullet-} + \mathrm{OH}^- \rightarrow {}^{\bullet}\mathrm{OH} + \mathrm{SO_4}^{2-} \tag{6}$$

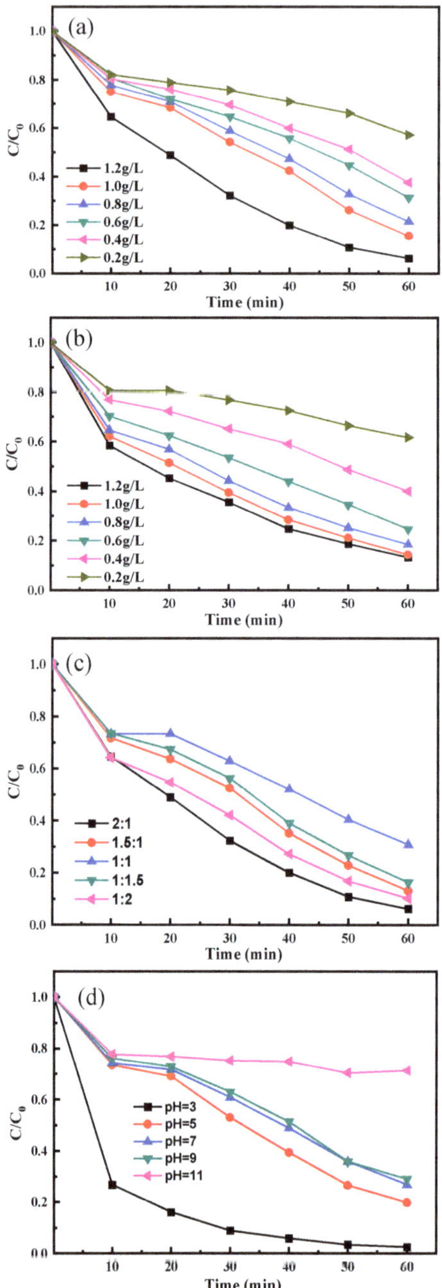

Figure 4. Effect of reaction parameters on the degradation of Rh B: catalyst dosage (**a**), persulfate concentration (**b**), catalyst/PS dosing ratio (**c**) and initial pH value (**d**). Experiment conditions: PS concentration = 0.6 g/L, catalyst dosage = 0.6 g/L, catalyst/PS dosing ratio = 2:1, Rh B concentration = 10 mg/L, and initial pH value = 4.86 (unadjusted).

3.4. Mechanism Analysis

Based on the previous study [19,24,28], we proposed a possible mechanism in α-MnO_2/PS reaction system. As shown in Figure 5, continuous conversion of Mn (IV) and Mn (III), which occurred on the surface of α-MnO_2 were depicted in Equations (7) and (8), a great number of $SO_4^{\bullet-}$ produced via the activation of PS. Moreover, moiety $SO_4^{\bullet-}$ could further react with H_2O to induce $\cdot OH$ (Equation (9)). Therefore, the mineralization of Rh B was mainly ascribed to strong active species, SRs ($SO_4^{\bullet-}$), hydroxyl radicals ($^{\bullet}OH$) and even persulfate (PS). $SO_4^{\bullet-}$ radicals and $^{\bullet}OH$ radicals would preferentially attack the central carbon ring position of Rh B to decolorize the dye and further degrade it through a ring opening process. Rh B would be converted into complex xanthene benzene radicals and produce three different forms of phthalic acid (phthalic acid, isophthalic acid and p-phthalic acid) [40]. In addition, an alternative degradation pathway for Rh B under $SO_4^{\bullet-}$ radicals and $^{\bullet}OH$ radicals attacking has been proposed. The main degradation products were malonic acid, oxalic acid, glycine, m-aminophenol, adipic acid, 2-hydroxybenzoic acid and trans-crotonic acid [41], as long as they were converted to simpler organic acids and inorganic small molecule compounds by stepwise oxidation. Most of the above low molecular weight acids were environmentally friendly and may also be eventually converted to CO_2, H_2O and inorganic compounds.

$$Mn(IV) + S_2O_8^{2-} \rightarrow Mn(III) + S_2O_8^{\bullet-} \tag{7}$$

$$Mn(III) + S_2O_8^{2-} \rightarrow Mn(IV) + SO_4^{\bullet-} + SO_4^{2-} \tag{8}$$

$$SO_4^{\bullet-} + H_2O \rightarrow {}^{\bullet}OH + H^+ + SO_4^{2-} \tag{9}$$

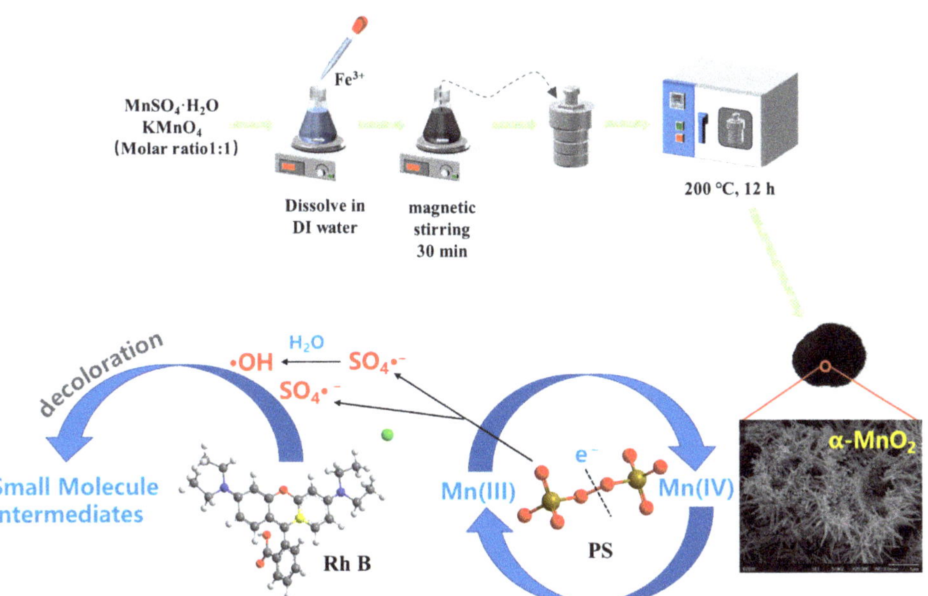

Figure 5. Proposed mechanism of Rh B degradation under α-MnO_2/PS system.

3.5. Reusability of α-MnO_2

Generally speaking, stability of catalysts is one of the most important characteristics in all evaluation indicators. Considering practical application potential of α-MnO_2, the repeated degradation experiment was performed for four runs. As exhibited in Figure 6, the degradation efficiency got slightly decreased after each cycle, from 97.41% to 89.76%, indicating that α-MnO$_2$ had expectation of putting into practical application in the organic pollutants' degradation due to its good repeatability. In addition, there were several reasons could explain this reduction of removal rate along the Rh B recycle degradation: (1) the slight leaching of metal ions on the catalyst [42,43]; (2) the absorption of degradation intermediate or remaining Rh B occupied the active sites on the surface of the catalyst in the previous cycle [43]; (3) the by-production produced during the several cycles would compete for active species with the Rh B molecules [44].

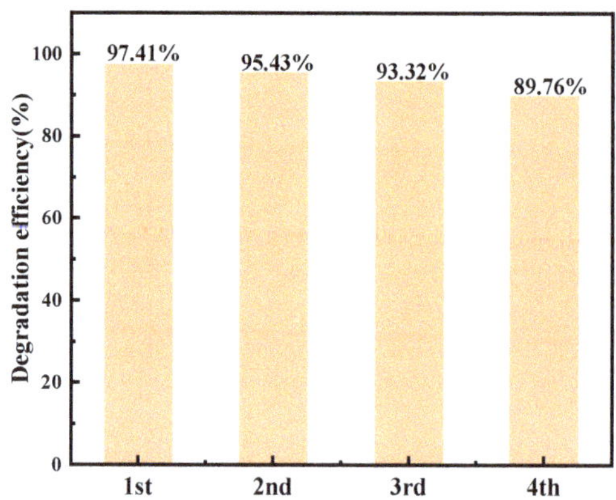

Figure 6. Performance of α-MnO$_2$ upon 4 successive uses for Rh B degradation. Experimental conditions: PS concentration = 0.6 g/L, catalyst dosage = 1.2 g/L, Rh B concentration = 10 mg/L.

4. Conclusions

All in all, rod-shaped, acicular and mixed MnO$_2$ were successfully fabricated by hydrothermal method with controlling the conditions of reaction and addition of the metal ions. From the results of surface morphology and crystallographic structure analyzed by scanning electron microscope (SEM) and X-ray diffraction (XRD), three different morphologies of MnO$_2$ were corresponding to β-MnO$_2$, mixed K$_{1.33}$Mn$_8$O$_{16}$ and α-MnO$_2$, respectively. In the microscopic state, the α-MnO$_2$ nanorods were more dispersed, and their nanorods were shorter in length and smaller in diameter. The mixed type, K$_{1.33}$Mn$_8$O$_{16}$ was a combination of manganese dioxide nanorods and manganese dioxide particles. In contrast, the β-MnO$_2$ mostly showed a polymerized state. Thus, the α-MnO$_2$ with needle-shaped microscopic morphology had a larger specific surface area and could provide more activation sites. As-prepared α-MnO$_2$ had a superior activity for Rh B degradation with the addition of PS, and 97.41% of Rh B could be removed within 60 min. More importantly, the reaction parameters are also optimized in this study, when the catalyst/PS ratio was 2:1 (the concentration of the α-MnO$_2$ and PS were 1.2 g/L and 0.6 g/L, respectively), pH value was 3, and the best catalytic efficiency for PS activation was obtained at the temperature of 20 °C. Meanwhile, the probable degradation mechanism was also proposed. What's more, the catalyst was reused 4 times, and the degradation rate decreased by less than 10%. In a word, this study provided an important reference for study on activation of PS by manganese oxides and their multicomponent complexes and exploitation of new nano materials for efficient removal of organic dyes in wastewater. In future research, the preparation of composite materials from α-MnO$_2$ with needle-shaped microscopic morphology together with other substances as raw materials could be explored in depth. For example, magnetic composites can be prepared to facilitate the recovery of the catalyst. Or other reactants can be added to obtain composites with better activation and regeneration performance.

Author Contributions: X.Z. contributed to the Data curation, writing—original draft preparation, X.G. and S.C. contributed to the writing—review and editing; J.S. and X.C. contributed to the supervision, funding acquisition. All authors have read and agreed to the published version of the manuscript.

Funding: This research received no external funding.

Data Availability Statement: Data available in a publicly accessible repository.

Acknowledgments: This work was supported by Research and Innovation Team Cultivation Program of Yili Normal University (CXZK2021004) and Natural Science Foundation of Xinjiang Uygur Autonomous Region (2021D01C461).

Conflicts of Interest: The authors declare no conflict of interest.

References

1. Zhang, X.; Wang, J.; Dong, X.-X.; Lv, Y.-K. Functionalized metal-organic frameworks for photocatalytic degradation of organic pollutants in environment. *Chemosphere* **2020**, *242*, 125144. [CrossRef] [PubMed]
2. Zhang, W.; Tay, H.L.; Lim, S.S.; Wang, Y.; Zhong, Z.; Xu, R. Supported cobalt oxide on MgO: Highly efficient catalysts for degradation of organic dyes in dilute solutions. *Appl. Catal. B Environ.* **2010**, *95*, 93–99. [CrossRef]
3. Guo, N.; Liu, H.; Fu, Y.; Hu, J. Preparation of Fe_2O_3 nanoparticles doped with In_2O_3 and photocatalytic degradation property for rhodamine B. *Optik* **2020**, *201*, 163537. [CrossRef]
4. Ji, R.; Zhao, Z.; Yu, X.; Chen, M. Determination of rhodamine B in capsicol using the first derivative absorption spectrum. *Optik* **2019**, *181*, 796–801. [CrossRef]
5. Pearce, C., Jr.; Lloyd, J.; Guthrie, J. The removal of colour from textile wastewater using whole bacterial cells: A review. *Dye. Pigment.* **2003**, *58*, 179–196. [CrossRef]
6. Banat, I.M.; Nigam, P.; Singh, D.; Marchant, R. Microbial decolorization of textile-dyecontaining effluents: A review. *Bioresour. Technol.* **1996**, *58*, 217–227. [CrossRef]
7. Foti, L.; Coviello, D.; Zuorro, A.; Lelario, F.; Bufo, S.A.; Scrano, L.; Sauvetre, A.; Chiron, S.; Brienza, M. Comparison of Sunlight-Aops for Levofloxacin Removal: Kinetics, Transformation Products, and Toxicity Assay on Escherichia Coli and Micrococcus Flavus. *Environ. Sci. Pollut. Res.* **2022**, *29*, 58201–58211. [CrossRef]
8. Ennouri, R.; Lavecchia, R.; Zuorro, A.; Elaoud, S.C.; Petrucci, E. Degradation of chloramphenicol in water by oxidation on a boron-doped diamond electrode under UV irradiation. *J. Water Process. Eng.* **2021**, *41*, 101995. [CrossRef]
9. Huixuan, Z.; Nengzi, L.-C.; Wang, Z.; Zhang, X.; Li, B.; Cheng, X. Construction of Bi_2O_3/Cunife Ldhs Composite and Its Enhanced Photocatalytic Degradation of Lomefloxacin with Persulfate under Simulated Sunlight. *J. Hazard. Mater.* **2020**, *383*, 121236.
10. Pirsaheb, M.; Hossaini, H.; Janjani, H. Reclamation of hospital secondary treatment effluent by sulfate radicals based–advanced oxidation processes (SR-AOPs) for removal of antibiotics. *Microchem. J.* **2019**, *153*, 104430. [CrossRef]
11. Lei, Z.; Yang, X.; Ji, Y.; Wei, J. Sulfate Radical-Based Oxidation of the Antibiotics Sulfamethoxazole, Sulfisoxazole, Sulfathiazole, and Sulfamethizole: The Role of Five-Membered Heterocyclic Rings. *Sci. Total Environ.* **2019**, *692*, 201–208.
12. Chen, C.; Feng, H.; Deng, Y. Re-evaluation of sulfate radical based–advanced oxidation processes (SR-AOPs) for treatment of raw municipal landfill leachate. *Water Res.* **2019**, *153*, 100–107. [CrossRef] [PubMed]
13. Wang, J.; Wang, S. Activation of persulfate (PS) and peroxymonosulfate (PMS) and application for the degradation of emerging contaminants. *Chem. Eng. J.* **2018**, *334*, 1502–1517. [CrossRef]
14. Yang, L.; He, L.; Xue, J.; Ma, Y.; Xie, Z.; Wu, L.; Huang, M.; Zhang, Z. Persulfate-based degradation of perfluorooctanoic acid (PFOA) and perfluorooctane sulfonate (PFOS) in aqueous solution: Review on influences, mechanisms and prospective. *J. Hazard. Mater.* **2020**, *393*, 122405. [CrossRef]
15. Matzek, L.W.; Carter, K.E. Activated persulfate for organic chemical degradation: A review. *Chemosphere* **2016**, *151*, 178–188. [CrossRef]
16. Zhang, M.; Chen, X.; Zhou, H.; Murugananthan, M.; Zhang, Y. Degradation of p-nitrophenol by heat and metal ions co-activated persulfate. *Chem. Eng. J.* **2015**, *264*, 39–47. [CrossRef]
17. Ruonan, G.; Meng, Q.; Zhang, H.; Zhang, X.; Li, B.; Cheng, Q.; Cheng, X. Construction of Fe_2O_3/Co_3O_4/Exfoliated Graphite Composite and Its High Efficient Treatment of Landfill Leachate by Activation of Potassium Persulfate. *Chem. Eng. J.* **2019**, *355*, 952–962.
18. Xing, S.; Li, W.; Liu, B.; Wu, Y.; Gao, Y. Removal of Ciprofloxacin by Persulfate Activation with Cuo: A Ph-Dependent Mechanism. *Chem. Eng. J.* **2020**, *382*, 122837. [CrossRef]
19. Junjing, L.; Guo, R.; Ma, Q.; Nengzi, L.-C.; Cheng, X. Efficient Removal of Organic Contaminant via Activation of Potassium Persulfate by Γ-Fe_2O_3/A-MnO_2 Nanocomposite. *Sep. Purif. Technol.* **2019**, *227*, 115669.
20. Miao, L.; Wang, J.; Zhang, P. Review on manganese dioxide for catalytic oxidation of airborne formaldehyde. *Appl. Surf. Sci.* **2019**, *466*, 441–453. [CrossRef]
21. Wang, Y.; Chen, S. Droplets impact on textured surfaces: Mesoscopic simulation of spreading dynamics. *Appl. Surf. Sci.* **2015**, *327*, 159–167. [CrossRef]

22. Ning, S.; Duan, Y.; Jiao, X.; Chen, D. Large-Scale Preparation and Catalytic Properties of One-Dimensional A/B-MnO_2 Nanostructures. *J. Phys. Chem. C* **2009**, *113*, 8560–8565.

23. Xu, Y.; Lin, H.; Li, Y.; Zhang, H. The mechanism and efficiency of MnO_2 activated persulfate process coupled with electrolysis. *Sci. Total. Environ.* **2017**, *609*, 644–654. [CrossRef]

24. Zhengyu, D.; Zhang, Q.; Chen, B.-Y.; Hong, J. Oxidation of Bisphenol a by Persulfate via Fe_3O_4-A-MnO_2 Nanoflower-Like Catalyst: Mechanism and Efficiency. *Chem. Eng. J.* **2019**, *357*, 337–347.

25. Deshan, Z.; Sun, S.; Fan, W.; Yu, H.; Fan, C.; Cao, G.; Yin, Z.; Song, X. One-Step Preparation of Single-Crystalline B-MnO_2 Nanotubes. *J. Phys. Chem. B* **2005**, *109*, 16439–16443.

26. Subramanian, V.; Zhu, H.; Vajtai, R.; Ajayan, P.M.; Wei, B. Hydrothermal Synthesis and Pseudocapacitance Properties of MnO_2 Nanostructures. *J. Phys. Chem. B* **2005**, *109*, 20207–20214. [CrossRef] [PubMed]

27. Jiechao, G.; Zhuo, L.; Yang, F.; Tang, B.; Wu, L.; Tung, C. One-Dimensional Hierarchical Layered $K_X MnO_2$ (X < 0.3) Nanoarchitectures: Synthesis, Characterization, and Their Magnetic Properties. *J. Phys. Chem. B* **2006**, *110*, 17854–17859.

28. Gong, C.; Zhang, X.; Gao, Y.; Zhu, G.; Cheng, Q.; Cheng, X. Novel Magnetic MnO_2/$MnFe_2O_4$ Nanocomposite as a Heterogeneous Catalyst for Activation of Peroxymonosulfate (Pms) toward Oxidation of Organic Pollutants. *Sep. Purif. Technol.* **2019**, *213*, 456–464.

29. Lavecchia, R.; Pugliese, A.; Zuorro, A. Removal of lead from aqueous solutions by spent tea leaves. *Chem. Eng. Trans.* **2010**, *19*, 73–78.

30. Jain, N.; Roy, A. Phase & morphology engineered surface reducibility of MnO_2 nano-heterostructures: Implications on catalytic activity towards CO oxidation. *Mater. Res. Bull.* **2019**, *121*, 110615.

31. Rao, T.P.; Kumar, A.; Naik, V.M.; Naik, R. Effect of Carbon Nanofibers on Electrode Performance of Symmetric Supercapcitors with Composite A-MnO_2 Nanorods. *J. Alloys Compd.* **2019**, *789*, 518–527. [CrossRef]

32. Yang, W.; Su, Z.; Xu, Z.; Yang, W.; Peng, Y.; Li, J. Comparative Study of A-, B-, Γ- and Δ-MnO_2 on Toluene Oxidation: Oxygen Vacancies and Reaction Intermediates. *Appl. Catal. B Environ.* **2020**, *260*, 118150. [CrossRef]

33. Ma, Q.; Zhang, H.; Zhang, X.; Li, B.; Guo, R.; Cheng, Q.; Cheng, X. Synthesis of magnetic CuO/$MnFe_2O_4$ nanocompisite and its high activity for degradation of levofloxacin by activation of persulfate. *Chem. Eng. J.* **2018**, *360*, 848–860. [CrossRef]

34. Huixuan, Z.; Song, Y.; Nengzi, L.-c.; Gou, J.; Li, B.; Cheng, X. Activation of Persulfate by a Novel Magnetic $CuFe_2O_4$/Bi_2O_3 Composite for Lomefloxacin Degradation. *Chem. Eng. J.* **2020**, *379*, 122362.

35. Zeming, W.; Wang, Z.; Li, W.; Lan, Y.; Chen, C. Performance Comparison and Mechanism Investigation of Co_3O_4-Modified Different Crystallographic MnO_2 (A, B, Γ, and Δ) as an Activator of Peroxymonosulfate (Pms) for Sulfisoxazole Degradation. *Chem. Eng. J.* **2022**, *427*, 130888.

36. Gong, C.; Nengzi, L.-C.; Gao, Y.; Zhu, G.; Gou, J.; Cheng, X. Degradation of Tartrazine by Peroxymonosulfate through Magnetic Fe_2O_3/Mn_2O_3 Composites Activation. *Chin. Chem. Lett.* **2020**, *31*, 2730–2736.

37. Hirakendu, B.; Singh, S.; Venkatesh, M.; Pimple, M.V.; Singhal, R.K. Graphene Oxide-MnO_2-Goethite Microsphere Impregnated Alginate: A Novel Hybrid Nanosorbent for as (Iii) and as (V) Removal from Groundwater. *J. Water Process Eng.* **2021**, *42*, 102129.

38. Shijun, Z.; Li, H.; Wang, L.; Cai, Z.; Wang, Q.; Shen, S.; Li, X.; Deng, J. Oxygen Vacancies-Rich A@Δ-MnO_2 Mediated Activation of Peroxymonosulfate for the Degradation of Cip: The Role of Electron Transfer Process on the Surface. *Chem. Eng. J.* **2023**, *458*, 141415.

39. Lin, X.; Ma, Y.; Wan, J.; Wang, Y.; Li, Y. Efficient degradation of Orange G with persulfate activated by recyclable $FeMoO_4$. *Chemosphere* **2019**, *214*, 642–650. [CrossRef]

40. Meng, F.; Song, M.; Song, B.; Wei, Y.; Cao, Q.; Cao, Y. Enhanced degradation of Rhodamine B via α-Fe_2O_3 microspheres induced persulfate to generate reactive oxidizing species. *Chemosphere* **2019**, *243*, 125322. [CrossRef]

41. Moutusi, D.; Bhattacharyya, K.G. Oxidation of Rhodamine B in Aqueous Medium in Ambient Conditions with Raw and Acid-Activated MnO_2, Nio, Zno as Catalysts. *J. Mol. Catal. A Chem.* **2014**, *391*, 121–129.

42. Huixuan, Z.; Wang, J.; Zhang, X.; Li, B.; Cheng, X. Enhanced Removal of Lomefloxacin Based on Peroxymonosulfate Activation by Co_3O_4/Δ-Feooh Composite. *Chem. Eng. J.* **2019**, *369*, 834–844.

43. Zhongjuan, W.; Nengzi, L.-C.; Zhang, X.; Zhao, Z.; Cheng, X. Novel $NiCo_2S_4$/Cs Membranes as Efficient Catalysts for Activating Persulfate and Its High Activity for Degradation of Nimesulide. *Chem. Eng. J.* **2020**, *381*, 122517.

44. Huixuan, Z.; Nengzi, L.-C.; Li, X.; Wang, Z.; Li, B.; Liu, L.; Cheng, X. Construction of $CuBi_2O_4$/MnO_2 Composite as Z-Scheme Photoactivator of Peroxymonosulfate for Degradation of Antibiotics. *Chem. Eng. J.* **2020**, *386*, 124011.

catalysts

Review

The Influence of Synthesis Methods and Experimental Conditions on the Photocatalytic Properties of SnO$_2$: A Review

Jéssica Luisa Alves do Nascimento [1], Lais Chantelle [2], Iêda Maria Garcia dos Santos [2], André Luiz Menezes de Oliveira [2,*] and Mary Cristina Ferreira Alves [1,*]

[1] Laboratório de Síntese Inorgânica e Quimiometria (LABSIQ), Departamento de Química, CCT, Universidade Estadual da Paraíba, Campus I, Campina Grande 58429-500, PB, Brazil; jessica_alvesn@yahoo.com

[2] Núcleo de Pesquisa e Extensão: Laboratório de Combustíveis e Materiais (NPE/LACOM), Departamento de Química, Universidade Federal da Paraíba, Campus I, João Pessoa 58051-900, PB, Brazil; lais.chantele@hotmail.com (L.C.); ieda@quimica.ufpb.br (I.M.G.d.S.)

* Correspondence: andrel_ltm@hotmail.com (A.L.M.d.O.); mary.alves@cct.uepb.edu.br (M.C.F.A.)

Abstract: Semiconductors based on transition metal oxides represent an important class of materials used in emerging technologies. For this, the performance of these materials strongly depends on the size and morphology of particles, surface charge characteristics, and the presence of bulk and surface defects that are influenced by the synthesis method and the experimental conditions the materials are prepared. In this context, the present review aims to report the importance of choosing the synthesis methods and experimental conditions to modify structural, morphological, and electronic characteristics of semiconductors, more specifically, tin oxide (SnO$_2$), since these parameters may be a determinant for better performance in various applications, including photocatalysis. SnO$_2$ is an n-type semiconductor with a band gap between 3.6 and 4.0 eV, whose intrinsic characteristics are responsible for its electrical conductivity, good optical characteristics, high thermal stability, and other qualities. Such characteristics have provided excellent results in advanced oxidative processes, i.e., heterogeneous photocatalysis applications. This process involves semiconductors in the production of hydroxyl radicals via activation by light absorption, and it is considered as an emerging and promising technology for domestic-industrial wastewater treatment. In our review article, we focused on the photodegradation of different organic dyes and types of persistent organic pollutants using SnO$_2$-based photocatalysts, and how the efficiency of these materials can be impacted by synthesis methods and experimental conditions employed to prepare them.

Keywords: synthesis method; experimental conditions; tin oxide; heterogeneous photocatalysis; persistent organic pollutants; dyes

Citation: do Nascimento, J.L.A.; Chantelle, L.; dos Santos, I.M.G.; Menezes de Oliveira, A.L.; Alves, M.C.F. The Influence of Synthesis Methods and Experimental Conditions on the Photocatalytic Properties of SnO$_2$: A Review. *Catalysts* **2022**, *12*, 428. https://doi.org/10.3390/catal12040428

Academic Editors: Gassan Hodaifa, Antonio Zuorro, Joaquín R. Dominguez, Juan García Rodríguez, José A. Peres and Zacharias Frontistis

Received: 18 February 2022
Accepted: 4 April 2022
Published: 11 April 2022

Publisher's Note: MDPI stays neutral with regard to jurisdictional claims in published maps and institutional affiliations.

1. Introduction

For decades, the world has been affected by serious environmental problems, especially water resources that have been polluted over the years. In particular, the disposal of domestic-industrial effluents in water bodies has been a major problem for modern society [1–5]. In this sense, the research community, public agencies, and environmentalists have been very concerned about this issue. Among the contaminants commonly present in wasters, persistent organic pollutants (POPs) are considered one of the main substances. POPs are heterogeneous synthetic compounds based on carbon from natural or from anthropogenic origins, with a high chemical resistance to degradation in the environment. These substances are characterized by a high level of toxicity and can be disruptive, neurotoxic, or immunosuppressive [6–9]. The excessive presence and accumulation of POPs in food chains has a major impact on human health and well-being. For instance, Mansouri and Reggabi [10] reported that the exposure to endocrine disruptors, and especially POPs, may contribute to the development of type 2 diabetes mellitus. According to these authors, diabetic subjects had higher plasma concentrations of POPs than non-diabetics, and environmental exposure to some POPs is associated with an increased risk of type 2 diabetes in the studied samples. In addition, authors have stated that the exposure to some other POPs may increase health problem risks [11,12].

The elimination, restriction, and control of POPs took place first in 2001, after the Stockholm Convention [13]. Currently, more than 35 (thirty-five) POPs are reported in literature, such as polychlorinated biphenyls (PCBs), dichlorodiphenyltrichloroethane (DDTs), and hexachlorocyclohexane isomers (HCHs), which are general types of drugs and medicine, pesticides, and synthetic dyes [14–17]. In this context, the search for new methodologies to promote full POP removal and degradation from the environment has attracted too much attention of the scientific community as well as public agencies, but it has also been a challenge to be faced today [18]. Different physical, chemical, and biological methods are used for degrading and removing pollutants present in water, air, and soil, among which adsorption [19,20], electrocoagulation [20], and advanced oxidative processes (AOPs) [21,22].

Among the diversity of water treatment methods, AOPs, particularly, represent a group of techniques characterized by the generation of highly oxidizing agents as free radicals. AOPs stand out due to their high efficiency in the degradation of numerous organic compounds, such as pesticides, surfactants, chlorophenols, and benzene, among others. APOs can promote mineralization of pollutants by transforming them into carbon dioxide, water, and inorganic anions [23–25]. AOPs are divided into homogeneous and heterogeneous, as well as photochemical and non-photochemical, processes [26]. Particularly, the heterogeneous photochemical process known as heterogeneous photocatalysis is one of the most important AOPs techniques used nowadays for being considered a promising technique for the treatment of domestic-industrial effluents and environmental decontamination [23,24,27–29]. In this process, semiconductors are irradiated by a light source (UV and visible light or even sunlight) with energy greater than or equal to its bang gap energy (Eg)—the minimum energy required to excite electrons from the lower energy band (valence band, VB) to a higher energy band (conduction band, CB) of a semiconductor. Thus, the energy absorption results in electronic excitation from VB to the CB, which leads to the formation of photoinduced charge pairs (e^-/h^+). Electron–hole pairs present very positive electrochemical potentials, which, when generated, give the semiconductor its redox properties [27,30].

During photocatalysis, photogenerated e^-/h^+ can migrate to the surface of the material and interact with an adsorbed species; in addition, they can be captured in intermediate energy states, or undergo recombination. When e^- are in the BC, one of the very important reactions can occur, which is the reduction of the adsorbed O_2 on the catalyst surface to superoxide ($^\bullet O_2{}^-$) radicals, avoiding recombination of electrons and the hole. This results in the accumulation of an oxygen radical species that can participate in photocatalytic reactions. On the other hand, when h^+ from the the VB migrates to the catalyst surface, it can react with adsorbed H_2O to generate hydroxyl ($^\bullet OH$) radicals, reacting with the pollutants to be degraded. It is worth mentioning that electronic recombination does not favor photocatalysis, because if the e^- and h^+ recombine, the formation of the photogenerated e^-/h^+ pairs fails to participate in the oxidation-reduction process and formation of free radicals in the process [30–32]. An illustration of the general mechanism involved in photocatalysis is given in Figure 1, as also reported in [33].

A great variety of semiconductors have been playing an important role in the heterogeneous photocatalysis for energy production from water splitting [34–36] and for environmental remediation toward the degradation of different organic pollutants such as POPs [36–38]. Within the broad range of oxide semiconductors applied in photocatalysis, wide band gap ones, such as SnO_2 [39,40], TiO_2 [41,42], ZnO [43], NiO [44], and WO_3 [45–47], have been extensively explored because they are considered good semiconductors and have showed promising results. Although most studies have been dedicated to titanium oxide (TiO_2), tin oxide (SnO_2) is also worth mentioning, as indicated by the number of published research articles listed in Table 1. SnO_2 is an n-type semiconductor with a large band gap value between 3.6 and 4.0 eV [47,48]. Its intrinsic characteristics are responsible for the material's conductivity, good optical and electrical characteristics, and high thermal stability, among other qualities. Furthermore, SnO_2 presents a high oxidation potential and chemical inertness, as well as corrosion resistance and non-toxicity [47]. Thus, SnO_2 has been successfully used in heterogeneous photocatalysis and showed to be very efficient in the degradation of different organic pollutants, especially dyes [49–51].

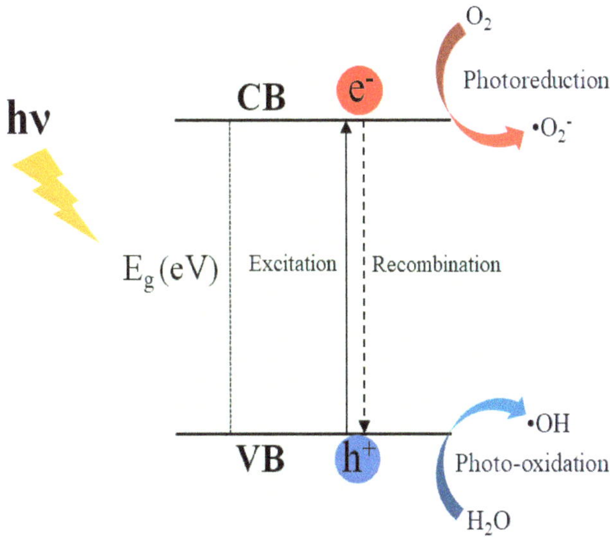

Figure 1. General mechanism of the heterogeneous photocatalysis using a hypothetical semiconductor.

In respect to the synthesis of SnO_2 that can affect the material properties and applications, the photocatalytic ones can be easily modified according to the synthesis route used to prepare the material. In addition, although most SnO_2-based photocatalysts are explored in their powder form, their separation and reuse after the process is difficult, as it needs filtration to recover the catalyst. Thus, other researchers have prepared SnO_2-based catalysts in their film form, which is considered an effective approach to overcome the filtration issues after a photocatalytic reaction, allowing the reusability of the catalyst without any material loss. These factors are important to design new and efficient photocatalysts.

Some review articles have dealt with the applications of SnO_2 materials in photocatalysis. However, to our knowledge, none of them draw attention to particular characteristics of synthesis methods and experimental conditions used to prepare the different SnO_2-based photocatalysts and how these directly impacted the sample characteristics, properties, and photocatalytic efficiency for organic dyes degradation. For instance, in the review published by Al-Hamdi et al. [47], results from studies conducted up to 2017 were shown, which focused on the applications of SnO_2 in an advanced oxidative process for the degradation of organic pollutants, such as phenols, phthalates, and other toxins in water. In addition, authors listed relevant investigations on fundamental aspects related to SnO_2, such as structure and properties, the charge transfer mechanism involved, and parameters that affect the photodegradation of pollutants in aqueous solutions, such as the catalyst load, concentration of the contaminant, and pH. In a recent review, Sun et al. [48] discussed the importance of employing SnO_2-based photocatalysis in environmental science and energy fields. The authors focused several studies dealing with strategies used to enhance the SnO_2 properties for photocatalysis. The effect of doping, formation of solid solutions, stoichiometry, particle size and morphology, besides the formation of hierarchical, porous, and heterojunctions structures, were demonstrated. Apart from the applications in water splitting and organic dyes photodegradation, the authors discussion gave examples of SnO_2 applications in Cr(VI) and CO_2 reduction. The authors also mentioned the complexities of applying photocatalysts in a large scale to simulate the real scope of the industry since, although significant progress has been achieved in the improvement of the photocatalytic efficiency of SnO_2, there are still major challenges to be faced.

Table 1. Metal oxides applied in heterogeneous photocatalysis of dyes over the last 5 years. Data collected from different research literature databases on 20 January 2022.

Oxide	Publications Attributed to Photocatalysis of POPs	Database	Publications Attributed to Photocatalysis of Dyes	Database
TiO_2	9.103	Scopus	2.132	Scopus
	1.369	Web of Science	3155	Web of Science
	2.492	ScienceDirect	3.933	ScienceDirect
	1.364	ACS Journal Search	1679	ACS Journal Search
ZnO	2.862	Scopus	1251	Scopus
	252	Web of Science	1759	Web of Science
	4.558	ScienceDirect	7.659	ScienceDirect
	542	ACS Journal Search	806	ACS Journal Search
NiO	346	Scopus	93	Scopus
	27	Web of Science	138	Web of Science
	1.164	ScienceDirect	1.500	ScienceDirect
	171	ACS Journal Search	174	ACS Journal Search
SiO_2	582	Scopus	149	Scopus
	41	Web of Science	150	Web of Science
	228	ScienceDirect	343	ScienceDirect
	438	ACS Journal Search	439	ACS Journal Search
WO_3	951	Scopus	135	Scopus
	10	Web of Science	191	Web of Science
	2.593	ScienceDirect	341	ScienceDirect
	2.570	ACS Journal Search	312	ACS Journal Search
SnO_2	425	Scopus	181	Scopus
	21	Web of Science	239	Web of Science
	173	ScienceDirect	330	ScienceDirect
	191	ACS Journal Search	261	ACS Journal Search

Based on what was mentioned, the present work aims to address important aspects about recent and relevant achievements of how the choice of the appropriate synthesis method and experimental parameters may influence the morphology, particle size, bandgap energy, and, consequently, the impact on the photocatalytic properties of SnO_2 for the degradation of different organic dyes as targets of persistent organic pollutants (POPs) molecules. Apart from the applications of pure SnO_2 toward the degradation of dyes, studies concerning photocatalysis using metal and nonmental-doped SnO_2, as well as SnO_2-based composites, are also summarized. We believe that this literature review may provide important aspects for the better development of SnO_2-based catalysts, and understanding the limitations, in order to use them in practical devices.

2. Persistent Organic Pollutants (POPs), Dyes, and Photocatalysts

Environmental problems, and especially water resources, have been a major problem for the population, researchers, and environmentalists [1,2]. There are several contaminants that are usually disposed into the environment, and can be classified into different types, depending on their purpose and origin: persistent organic pollutants (POPs), pharmaceuticals and personal care products (PPCPs), endocrine disrupting chemicals (EDCs), and agricultural chemicals (pesticides, herbicides) [51–53]. Particularly, persistent organic pollutants (POPs)—which are organic chemicals that are highly toxic to humans—animals, and vegetation stand out among these pollutants. The specific effects of POPs can include allergies, cancer, and damage to the central and peripheral nervous system and reproductive, endocrine, and immune systems [13,14,54].

According to the Stockholm Global Conference held in 2001 that aimed to discuss the consequences of the environmental degradation, twelve (12) chemical substances were included into the POPs class. However, thirty-five (35) POP substances are listed as most harmful [13,15,21,55–59]. Amongst the variety of POPs, drugs, pesticides, hormones, dyes, synthetic textile products, artificial sweeteners, some micro-organisms, and algae toxins have extensively been introduced on a large scale into the environment, which can strongly affect aquatic life and human beings [21,53]. Particularly, synthetic organic dyes stand out for being considerably used over the years in several technological sectors, such as textile industries, in medicine, and in the manufacturing process of pens, among others.

Synthetic dyes are classified as POPs because of their difficulty to be degraded and removed from the environment [55]. The disposal of effluents from textile industries is one of the main sources of pollutants responsible for changes in the quality of receiving water. The different dyes used in textile industries generally have a high organic load, and this can be very harmful to the environment if inappropriately discharged. There is an estimate that about 20% of the dyes used are disposed in effluents because of losses during the dyeing process [36,59,60]. In this context, several methods (chemical, physical, and biological) have been used for dealing with the treatment of contaminated effluents with POPs (Figure 2). However, great efforts have been dedicated to studies related to photodegradation and mineralization of organic dyes using photocatalytic materials under UV-visible light, as well as sunlight irradiation.

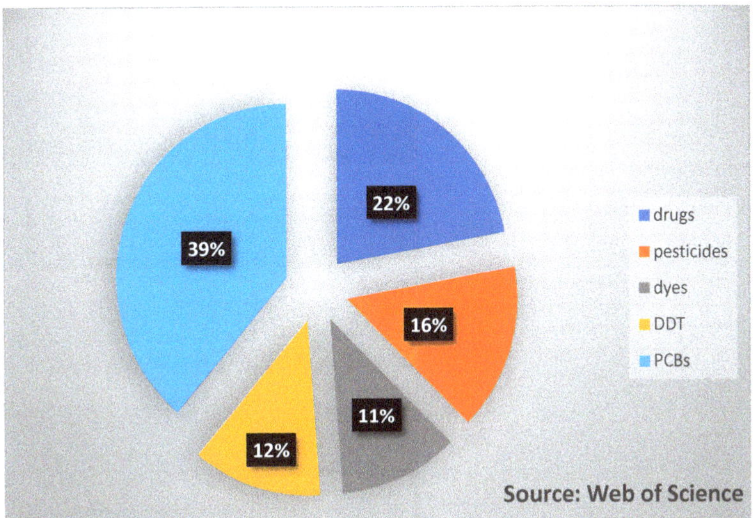

Figure 2. Published articles referring to the treatment of different persistent organic pollutants. Data collected from Web of Science on 20 January 2022.

As aforementioned, different materials have been applied in heterogeneous photocatalysis, especially for photodegradation of POPs. As this review aimed to discuss important works involving the photodegradation of organic dyes and types of POPs, the article's search was done considering different databases, as listed in Table 1. It is important to highlight that most of the publications have devoted to the application of TiO_2-based photocatalysts. However, different metal oxides have been explored for such purposes because of their unique and prominent properties, due to their structure, particle size, and morphology [61]. From this search, SnO_2 appears in 6th place among the six (6) oxide-based photocatalysts used in photocatalysis of organic dyes. It is still important to mention that scientific research based on tin materials has been increased over the last years.

It is important to highlight that the search was done using the following words for articles dealing with photodegradation of POPS, such as for SnO_2: "SnO_2, photocatalysis and POPs" and "tin oxide, photocatalysis and POPs"; while for works concerning photodegradation of dyes included, "SnO_2, photocatalysis and dyes" and "tin oxide, photocatalysis and dyes". A similar search has been completed for each metal oxide listed in Table 1. Compound words such as "photocatalytic properties", "photocatalytic efficiency", and "organic dyes" were not placed in quotes for the search of papers related to photocatalysis of dyes using these oxides.

Considering organic dyes, additional studies dealing with these pollutants are in need. For instance, the emission of effluents from textile industries is one of the main things re-sponsible for alterations in water bodies' quality. Organic dyes have a high organic load, generating strong impacts on the environment when inappropriately discharged. Depending on the type of dye and the place where it is discarded, they can react with other substances present in the environment, originating mutagenic and carcinogenic by-products [62,63]. In dye molecules, two important organic groups are present: the chromophore one, which is responsible for the color by light absorption, and the auxochrome group that is responsible for dye fixation on fibers and for color intensification. These groups are constituted by conjugated systems with double bonds and functional groups (auxochromes), which are electron donors or acceptors [64]. Such characteristics have made dyes one of the great villains in the process of polluting water. In this context, a great amount of research has explored the photocatalysis of different organic dyes using wide band gap semiconductors, such as SnO_2. These works are discussed as follows for SnO_2, which has been gaining prominence in this type of application over the years, as shown in Figure 3.

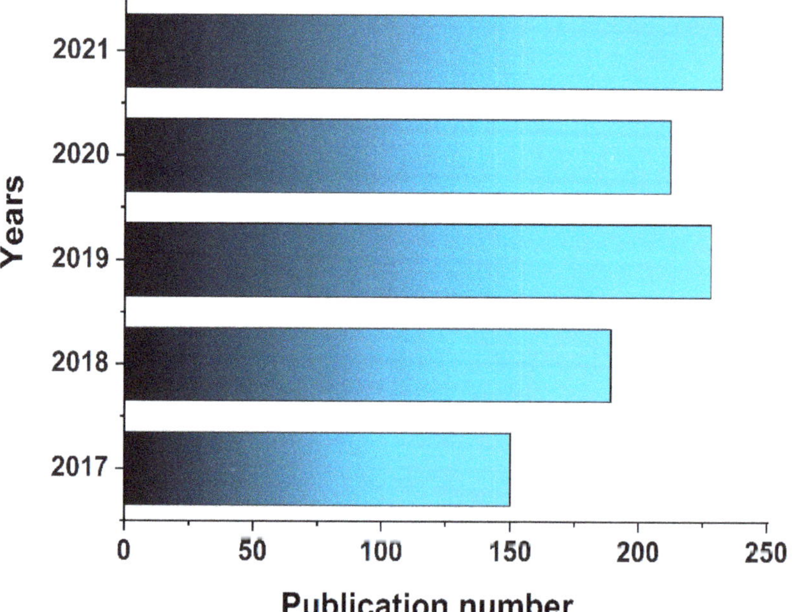

Figure 3. Number of published articles concerning the photodegradation of organic dyes using SnO_2-based photocatalysts. Search performed on 20 January 2022, using different databases. The period of 2017–2021 was considered for this search.

3. Tin-Oxide-Synthesis, Structure, Properties, and Applications as Heterogeneous Photocatalysts

Tin dioxide (SnO_2), obtained by combining Sn^{4+} and O^{2-}, is a ceramic material that has been used in a wide variety of applications, such as gas sensors, photovoltaic energy converters, and photocatalysts [65–67]. The success for multiple applications of SnO_2 is due to its intrinsic characteristics, such as: n-type conductivity, which is responsible for the conductivity of the material, and in addition to optical and electrical characteristics, high thermal stability, and high surface area [36,49,68]. Electrical conductivity can be described in terms of the movement of negatively charged electrons, and this gives rise to an n-type semiconductor [36,49,50,69–71].

As for the crystal structure of SnO_2, at room temperature, it adopts a rutile-type tetragonal structure (cassiterite) with *P42/mnm* space group, as illustrated in Figure 4. This structure is formed by a tetragonal unit cell defined by three (3) parameters: the *a* and *c* lattice parameters and the internal parameter, *u*, which defines the oxygen position (*u*, *u*, and 0). At room temperature, the theoretical lattice parameters for SnO_2 are *a* = 4.7374 Å, *c* = 3.1864 Å [72], and *u* = 0.3056 Å. It has been reported that SnO_2 can also adopt an orthorhombic, $CaCl_2$-type (*Pnnm*) structure, besides existing in an orthorhombic α-PbO_2-type (*Pbcn*), a cubic pyrite-type (*Pa-3*), an orthorhombic ZrO_2-type (*Pbca*), and a cotunnite-type (*Pnma*) structure. However, these structures are metastable at ambient conditions, and it is hard to follow these phase transitions through traditional methods under low pressure and temperatures [63,64].

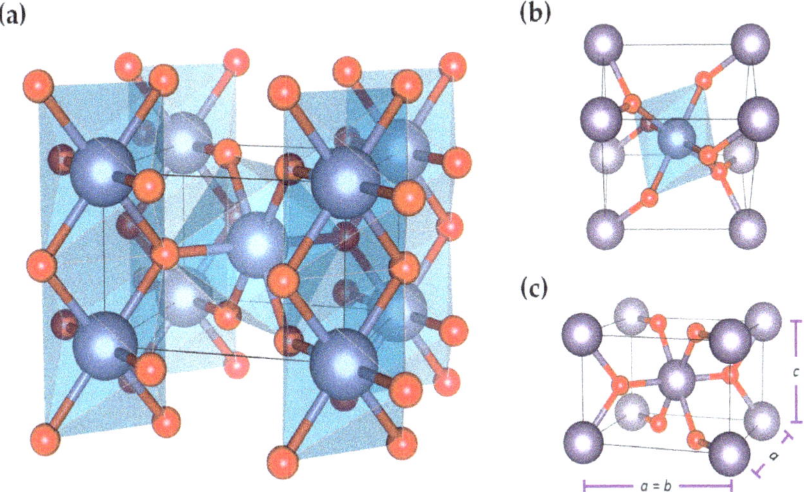

Figure 4. (**a**) A *P42/mnm* tetragonal crystal structure of SnO_2 showing (**b**) environment of distorted SnO_6 octahedron and (**c**) lattice parameters, obtained using VESTA software and structural parameters reported in [73].

The structure of SnO_2 consists of chains of SnO_6 octahedra in which each Sn atom is octahedral, surrounded by six oxygen atoms, while each oxygen is surrounded by three Sn atoms arranged at the corners of an equilateral triangle. The structure's Sn:O coordination is 6:3, with each octahedron not being regular, showing a slight orthorhombic distortion [72,73].

The literature has reported that SnO_2 can present nano- and micro-structured characteristics, however, most researchers have been dedicated to the synthesis of SnO_2 in the form of nanoparticles. Such characteristics have offered good opportunities to explore new physical and chemical applications. Recently, SnO_2 has been applied to the degradation of antibiotics [74], fuel cell catalysts [73], gas sensors [75], sodium ion batteries [76], light-emitting diodes [77], heterogeneous photocatalysis, and antimicrobial activity [78,79]. There are several synthesis routes and experimental conditions to obtain SnO_2 in powder form, the choice of which may influence the particles' morphology and texture, besides their structural, optical, and electronic properties, which will consequently impact the efficiency and final application of the material. In relation to SnO_2-based thin films, chemical deposition methods have appeared to be the most used [62,80–84].

With respect to the applications in photocatalysis, SnO_2 catalysts with a tetragonal, rutile-type structure are the main phases investigated, but a recent work reported that the coexistence of mixed tetragonal–orthorhombic phases affected the photocatalytic efficiency of SnO_2 [85]. All these factors are described below.

4. Synthesis Methods and the Influence of Experimental Parameters on the Characteristics and Photocatalytic Properties of SnO_2-Based Materials

One of the main aspects this review refers to is the approach of how different synthesis methods and experimental parameters affect the photocatalytic properties of SnO_2-based materials.

Researchers have employed different methods to produce new materials with specific properties and applications. The choice of an appropriate condition to prepare a material is crucial to modifying its physicochemical characteristics, such as crystal structure and morphological and texture characteristics of the particles (size, shape, surface area, and surface charge characteristics), as well as the electronic and optical properties. The variation on these characteristics may change the applicability of the material. In this sense, different researchers have devoted to obtaining SnO_2-based materials with specific characteristics and efficient photocatalytic behaviors. Among the great variety of methods known in the literature, SnO_2-based photocatalysts have been obtained, in their powder form, by a solvothermal reaction [86,87], the microwave-assisted hydrothermal method [88], chemical precipitation [68], the polymeric precursor method [89], and sol-gel [14], all which are mostly used.

In relation to the photocatalytic properties of SnO_2 particles, although there is a variety of organic pollutants, organic dyes such as Methylene blue (MB), Rhodamine B (RhB), and Methyl orange (MO) have been the most used as target molecules to evaluate the photocatalytic efficiency of SnO_2. However, other organic dyes such as Congo red (CR), Malachite green (MG) and Eriochrome black T (EBT) have also been explored for such purposes. In this context, the photodegradation of these organic dyes is highlighted in this review. Apart from powdered SnO_2 materials, SnO_2-based films have also been used as photocatalysts for dye photodegradation [90–93]. For the preparation of films, chemical [64,80,84,86,89] and physical-based deposition methods [64] are used. In the following sections, photocatalytic applications of pure and doped SnO_2-based catalysts, as well as the composites based on SnO_2 in powdered and film forms prepared by different techniques, are also discussed.

4.1. Pure SnO_2-Based Photocatalysts

Besides being synthesized in its pure form, SnO_2 has been prepared in its doped form with metals or nonmetal ions, as well as in the form of composites with other semiconductor materials, or even impregnated in inert, non-active photocatalytic materials. It is well known that the generation of electrons/holes (e^-/h^+) pairs by absorption of a photon of equal energy, to or higher than the band gap energy induced by light, is a basic prerequisite for a semiconductor to be used in photocatalysis. Because of its wide band gap of SnO_2 (3.6 eV), no absorption response to the visible light would be achieved, and this is the main disadvantage of this material, which restricts its application in practical devices. A wide variety of pure semiconductor materials, particularly SnO_2, have been investigated regarding the photocatalytic properties, but only few of them are considered effective photocatalysts.

Besides the value of the forbidden band energy corresponding to absorption in the visible region, it is required that the energy levels of the conduction and valence bands are suited to the redox potential of the water, so as to produce reactive agents to promote breaking of organic pollutants molecules. Therefore, different authors have employed various synthesis methods and varied experimental conditions in order to change particle characteristics, which include size, morphology, and texture, as well as the crystallinity and the presence of defects, in order to design efficient photocatalysts under UV irradiation. These parameters can play an important role during photocatalysis, as they can affect the adsorptive and photo-absorption capacity of the catalyst, besides reducing photogenerated changes of recombination during photoexcitation. In addition, some authors also explored the reaction mechanisms involved in the photodegradation of the dyes and how the reactive species act for the photocatalysis to occur. These points are discussed throughout the review.

For instance, considering pure tin oxide as a photocatalyst, Akram et al. [90] prepared SnO_2 nanoparticles by the continuous microwave flow synthesis (CMFS) method in a domestic microwave oven operating at 600 W, using tin chloride pentahydrate ($SnCl_4 \cdot 5H_2O$), sodium hydroxide (NaOH), ethanolic solutions. The solutions were pumped through the microwave with the aid of peristaltic pumps to attain 10 min of retention time inside the device. The resulting suspension was filtered, washed, and dried at 80 °C for 12 h, followed by heating at 200 °C for 2 h to obtain SnO_2 samples. The photocatalytic properties of the nanoparticles were investigated toward the photodegradation of Methylene blue (MB) dye under UV irradiation (365 nm). In their study, the authors also investigated the effect of the concentration of the reacting $SnCl_4 \cdot 5H_2O$ and NaOH materials on the crystallinity, particle size, and morphology, as well as the photocatalytic behavior of the SnO_2 nanoparticles. The authors evidenced that crystalline SnO_2 samples with tetragonal, rutile-type structures were obtained only after heating at 200 °C. The increase in the concentration of $SnCl_4 \cdot 5H_2O$ (from 0.25 to 0.75 M) and NaOH (from 1 to 3 M) provoked an increase in sample crystallinity and an average particles size from 4.33 to 8.56 nm, with no meaningful change in morphology. This phenomenon was attributed to the increased number of nuclei sites formed by the reacting species, as well as to the microwave irradiation that favors better nucleation and crystal growth. Surprisingly, a decrease of the band gap (*Eg*) values from 3.33 to 3.19 eV was also observed as a function of the precursor's concentration, and it was associated to the increase of the particle size. In relation to the photocatalytic property of the samples, that sample was prepared using the lower concentration of the reacting species, which presented the lowest degree of crystallinity (63%), smallest particle size (4.43 nm), largest surface area (153.57 $m^2 g^{-1}$), and widest band gap (3.33 eV), and this was the most efficient in the photodegradation of MB dye, reaching up to 93% of degradation in 240 min. According to the authors, the higher efficiency observed for that sample was mainly due to the large surface area and higher concentration of defects.

Still considering pure tin oxide photocatalysts, Abdelkader et al. [92] synthesized SnO_2 nanoparticles via a sol-gel method and calcined it at different temperatures (80, 450, and 650 °C) for 4 h, in order to achieve different crystallinity and particle morphology. The authors investigated the photocatalytic efficiency of the synthesized samples toward Congo red (CR) dye degradation under UVA irradiation. For the synthesis of the nanoparticles, tin chloride ($SnCl_2 \, 2H_2O$) was dissolved in 250 mL of deionized water to obtain a white suspension of a 0.4 M Sn (II) concentration. The suspension was stirred for 1h at room temperature and oxalic acid was added dropwise to the aqueous solution as a chelating agent with a molar ration of 1:1 (oxalic acid: tin cations). The obtained suspension was centrifuged, filtered, and washed several times to eliminate chloride ions. The washed precipitate was dried at 80 °C/24 h and calcined at 450 to 650 °C for 4 h. According to the results, the sample obtained after drying at 80° for 24 h (SnO_2-80) crystallizes in pure tin oxalate (SnC_2O_4) with a monoclinic structure. On the other hand, pure SnO_2 with a *P42/mnm* tetragonal, rutile-type structure was obtained after calcinations at 450 (SnO_2-450) and 650 °C (SnO_2-650). The authors draw attention to the use of oxalic acid as a chelating agent, and it affected the particle characteristics for controlling the nucleation and crystal orientation, as well as the calcination temperature in the control of the crystallinity. The authors evidenced that samples calcined at higher temperatures are more aggregated with a foamed aspect because of the smallest particle size of the powders. The samples calcined at different temperatures also presented a specific surface area that varied from 66.41 to 37.54 $m^2 g^{-1}$, and with band gap values from 3.35 to 3.49 eV for the SnO_2-450 and SnO_2-650 samples, respectively.

Curiously, regarding the photocatalytic behavior of the samples prepared by Abdelkader et al. [92], the highest efficiency in the degradation of CR dye (catalyst/CR dye concentration of 0.5 g L^{-1}) was achieved using SnO_2-650 that presented a lower surface area. This sample presented a dye photodegradation efficiency of 61.53% after 100 min under irradiation. The authors pointed out that the highest efficiency observed for this sample is especially due to its greater crystallinity that gives less surface defects as well as particle aggregation. The lower density of surface defects (Sn^{2+} and oxygen vacancies) could reduce recombination of the electron–hole (e^-/h^+) pairs. Because of this fact, the authors proposed the photocatalytic mechanism involved in the CR photodegradation. The reactions mechanism can be represented by Equations (1)–(6). According to the authors, under UVA light, electrons are excited from the VB to the CB of SnO_2 (Equation (1)) and, simultaneously, holes are created in the VB. The photoinduced e^- in the CB can directly reduce Sn^{4+} to Sn^{2+} (Equation (2)). However, as Sn^{4+} can act as a scavenger of e^-, Sn^{2+} can influence the photo-reactivity by altering a e^-/h^+ recombination (Equation (3)). The h^+ in the VB is captured by H_2O generating hydroxyl $^\bullet OH$ radicals (Equation (4)). In addition, according to the band energy position, the authors stated that the h^+ in the VB of SnO_2 (+3.50 eV/NHE) is more positive than that of the H_2O/OH couple (+1.9 eV/NHE), which is required for organic pollutant decomposition $R/R^{\bullet+}$ (+1 V/NHE), indicating that the photoinduced holes in the VB can oxidize the adsorbed CR dye and H_2O molecules on the SnO_2 surface. Thus, the formation of organic cation-radicals ($R^{\bullet+}$) is formed (Equation (5)). As a result, all the $O_2^{\bullet-}$, $^\bullet OH$, and $R^{\bullet+}$ radicals participate in redox reactions responsible for decomposing CR dye (Equation (6)).

$$SnO_2 + UVA(365\ nm) \rightarrow n - SnO_2\ (h^+{}_{(VB)} + e^-{}_{(CB)}) \tag{1}$$

$$Sn^{4+} + e^-{}_{(CB)} \rightarrow Sn^{2+} \tag{2}$$

$$Sn^{2+} + 2O_2 \rightarrow 2O_2{}^{\bullet-} + Sn^{4+} \tag{3}$$

$$H_2O + h^+{}_{(VB)} \rightarrow {}^\bullet OH \tag{4}$$

$$R + h^+{}_{(VB)} \rightarrow R^{\bullet+} \tag{5}$$

$$(O_2{}^{\bullet-}, {}^\bullet OH, R^{\bullet+}) + CR\ dye \rightarrow intermediate\ products \rightarrow Dye\ degradation \tag{6}$$

Although most studies concerning photocatalytic degradation of dyes using pure SnO_2 have been performed under UV light, some authors investigated the photocatalytic properties of SnO_2 under visible light and sunlight. For instance, Kumar et al. [94] reported the use of SnO_2 particles prepared by a simple, eco-friendly, and low-cost biosynthesis process using guava (*Psidium guajava*) leaf extract in the photodegradation of Reactive yellow 186 (RY186) dye under sunlight. The authors prepared the samples by mixing a 2.1 M $SnCl_4$ solution with the extracts in a ratio of 1:1, and kept under stirring at 60 °C for 4 h, followed by calcination at 400 °C for 4 h to obtain SnO_2 nanoparticles. The authors evidenced that the SnO_2 single-phase nanoparticles with a size of 8–10 nm showed a high photocatalytic efficiency, degrading 90% of RY186 dye in 180 min. For the photocatalysis, a concentration of 1 g L^{-1} of SnO_2 sample was used. According to the authors, superoxide ($O_2{}^{\bullet-}$) and hydroxyl ($^\bullet OH$) radicals are responsible for the photodegradation of dye. The efficiency of the photocatalyst in the degradation and mineralization of RY186 was confirmed by CO_2 evolution during the photocatalysis, which was analyzed by GC analysis. The authors evidenced the complete mineralization of dye led to CO_2 (0.8 μmol) and H_2O. As the photostability and reusability of the photocatalyst are important aspects, the authors evaluated them by performing five consecutive photocatalytic cycles. The photoactivity of the catalyst remained about constant even up to five experiments, although gradual losses in activity were expected. It is important to highlight that the use of powdered samples in photocatalysis might be disadvantageous due to the loss in the amount of the catalyst after centrifugation and filtration processes, which leads to a loss in photoactivity. However, the authors reinforced that the synthesized catalyst is easily separable from the solution.

Like Kumar et al. [94], Haq et al. [78] synthesized tin dioxide (SnO_2) nanoparticles by an eco-friendly process using leaves extracts. However, in the study conducted by Haq et al. [78] *Daphne mucronata* leaf extract was used as a capping and reducing agent in order to control particle size and morphology. The authors evaluated the photocatalytic performance of the samples toward Rhodamine 6G (R6G) dye degradation using 0.4 g L^{-1} of the catalyst. For the materials synthesis, a mixture of a 0.003M $SnCl_4 \cdot 5H_2O$ solution with 20mL of the leaf extract was kept under 500 rpm and stirring at 55 °C. A greenish gel was obtained after 40 min and aged for 24 h, which was subsequently washed with hot water, filtered, washed with ethanol, and finally dried at 100 °C for 6 h to obtain a fine, colorless powder. The authors obtained nanoparticle samples with an average particle size of 64 nm and specific surface area around $147 \text{ m}^2 \text{ g}^{-1}$. A maximum of 99.70% of the dye degradation was observed after 390 min under simulated sunlight. The charge generation and photodegradation mechanism reported by Haq et al. [78] is the same as the one described by Kumar et al. [94] for RY186 dye [94].

Kumar et al. [94] synthesized SnO_2 nanocrystals by a solution-phase growth technique, and the structural, optical, and photocatalytic properties of the nanostructures were investigated. The effects of reaction temperature (180 and 200 °C), time (24 and 30 h), the use of CTAB (CetylTrimethylAmmonium Bromide) surfactant on the particle size, morphology, and band gap energy were evaluated. According to the authors, the SEM images showed that samples synthesized at 180 °C for 24 (Sample 1) and 30 h without the surfactant (Sample 2) presented particles with a rod-like morphology and crystallite size of 17.12 and 26.51 nm, respectively. When the sample was synthesized at 200 °C for 24 h without surfactant (Sample 3), SnO_2 nanostructures with a nanoflower-like morphology, with crystallite size at 28.25 nm were obtained. On the other hand, the samples prepared at 180 (Sample 4) and 200 °C for 24 h with the addition of CTAB (Sample 5) showed particles with a nanosphere-type morphology with sizes of 23.80 and 32.14 nm, respectively. The band gap (*Eg*) values estimated for sample 1 (nanorods), Sample 2 (nanorods), Sample 3 (nanoflowers), Sample 4 (nanospheres), and sample 5 (nanospheres) were 4.05, 3.88, 3.84, 3.95, and 3.76 eV, respectively. The authors associated the variation of the *Eg* values mainly to the temperature and time conditions of which the samples were prepared, that directly affected the particle size and morphology and, therefore, impacted photocatalytic activity of the samples. Despite the wide band gap of the samples, the authors evaluated their photocatalytic activity using 0.5 g L^{-1} of the catalysts in the degradation of Rhodamine B (RB) dye under direct sunlight irradiation. Surprisingly, the authors observed a high photodegradation of dye under sunlight, and the efficiency of the SnO_2 nanostructures was strongly related to the particles' morphology. According to the authors, even presenting a *Eg* = 3.76 eV, sunlight was enough to promote photoexcitation in Sample 5 to degrade 91.7% of the dye after 2 h. As expected, Sample 1 (SnO_2 nanorods with *Eg* = 4.05 eV) displayed the lowest dye degradation efficiency in 2 h (76% of the dye is degraded). The authors established that size, morphology, specific surface area, and dispersion of the catalysts played key role in the photodegradation of the dye.

Assis et al. [58] used a polymeric precursors method to prepare SnO_2 particles at different temperatures (700, 800, and 900 °C). After being prepared, the powders were impregnated in polystyrene foams in order to increase surface area due to the porous characteristic of polystyrene, besides a favoring for the recovery of the material after use. The photocatalytic property of the samples was investigated in the degradation of RhB dye under UV irradiation with a catalyst/dye concentration of 0.4 g L^{-1}. The authors observed, using high-resolution transmission electron microscopy (HRTEM), that the SnO_2 samples present nanoparticles with sizes ranging between 20 and 80 nm. Moreover, the formation of agglomerates was observed in the samples calcined at higher temperatures (800 and 900 °C). The oxide obtained at lower temperatures presented a smaller particle size and a larger surface area, which resulted in a greater photocatalytic activity, degrading 98.2% of degradation of the rhodamine RhB after 70 min.

A very conventional synthesis procedure for obtaining oxide-based catalysts is the so-called sol-gel method. Thus, Najjar et al. [93] synthesized SnO_2 nanoparticles by a green sol-gel method, using chitosan as a polymerizing agent, and calcined it at different temperatures (500, 700, 800, and 1000 °C). The authors also draw attention to the use of chitosan that may increase particle stability, prevent particles aggregation, and reduce the particles' toxicity. According to the TGA-DTA analysis, the temperature of 700 °C (namely, SnO_2-NPs at 700 °C) proved to be more adequate to prepare the desirable SnO_2 catalyst. The material calcined at this temperature presented a spherical particle morphology with an average size of 10 nm, as observed by TEM analyses. The authors evaluated the photocatalytic properties of SnO_2-NPs at 700 °C toward the photodegradation of Eriochrome black T (EBT), an azo-type anionic dye. The photocatalytic tests were carried out by adding 21.1 mg L^{-1} of the catalyst in 100 mL of EBT dye solution (10^{-5} M) and kept at a constant stirring and UV irradiation (Hg vapor lamp, 500 W) for 270 min. Regarding the photocatalytic activity of the prepared SnO_2-NPs, a photodegradation efficiency of 77% was obtained after 270 min. In order to investigate the best conditions for optimum dye degradation using SnO_2-NPs, Najjar et al. [93] also investigated the influence of the catalyst concentration (8.7, 21.1, and 43.2 mg L^{-1}) and the solution pH (3.5, 5, 7, and 9). The authors observed an increase of the photodegradation rate by increasing the concentration of the photocatalyst from 8.7 to 21.1 mg L^{-1}, decreasing afterwards. According to the authors, this decrease in photocatalytic efficiency of SnO_2-NPs in a higher concentration is due to the fact of accumulation of nanoparticles that lead to a decrease in the generation of reactive radicals, such as hydroxyls ($^\bullet$OH). With respect to variation of the solution pH, the highest photodegradation efficiency (77%) was attained at the isoelectric point of SnO_2-NPs at pH 3.5 (Zeta potential = 0 eV). Curiously, the photodegradation of the anionic EBT dye at positive Zeta potential values (at pH 2) was not as expressive as that observed at the isoelectric point of the catalyst.

According to Najjar et al. [93] the general mechanism involved in the dye photodegradation using SnO_2-NPs is summarized by Equations (7)–(12), that are similar to those reactions displayed in Figure 1 for a hypothetical catalyst.

$$SnO_2 + h\upsilon \rightarrow SnO_2 + (h^+_{(VB)} + e^-_{(CB)}) \tag{7}$$

$$SnO_2\ (h^+_{(CB)}) + H_2O \rightarrow SnO_2 + H^+ + {}^\bullet OH \tag{8}$$

$$SnO_2\ (e^-_{(CB)}) + O_2 \rightarrow SnO_2 + O_2^{-\bullet} \tag{9}$$

$$O_2^{-\bullet} + H^+ \rightarrow HO_2^{\bullet} \tag{10}$$

$$HO_2^{\bullet} + H_2O \rightarrow H_2O_2 + {}^\bullet OH \tag{11}$$

$$Dye + {}^\bullet OH \rightarrow degradation\ dye \tag{12}$$

As the photostability of the catalyst is an important factor for its reuse in consecutive photocatalytic tests, cyclic experiments of EBT photodegradation were carried out for the SnO_2-NPs under the optimal conditions established in the work. Thus, Najjar et al. [93] evidenced that the degradation rate of EBT remained over 74% after five cycles. In addition, using FTIR, XRD, TEM, and FESEM analysis, the authors showed that no visible changes were observed in the samples after the fifth cycle, which confirms the high photostability of the synthesized catalyst.

Recently, Luque et al. [39] synthesized SnO_2 nanoparticles (SnO_2 NPs) by a green synthesis using *Citrus x paradisi* extract as a stabilizing capping agent. There were different concentrations of the extract (1, 2, and 4% in relation to the aqueous medium—SnO_2

NPs-1, 2, and 4%). It is important to highlight that a heating treatment at 400 °C for 1 h was completed to obtain crystallized SnO_2 NPs. The authors obtained crystalline SnO_2 nanoparticles with average sizes of 9.1, 5.1, and 4.7 nm when 1, 2, and 4% of the capping agent was used in the synthesis, respectively. These samples also presented band gap values (*Eg*) of 3.28, 2.77, and 2.69 eV, respectively, which were smaller than those *Eg* values reported by Mahmood et al. [95]. This confirmed the role of this capping agent in controlling the particles' size, as well as in the modification of optical band gap properties of SnO_2 NPs. The photocatalytic properties of the samples were then investigated under both solar and UV irradiation using a SnO_2 NPs/dye concentration of 1.0 g L^{-1}. Furthermore, Methyl orange (MO), Methylene blue (MB) and Rhodamine B (RhB) were used as target dyes. Regarding the photocatalytic efficiency of the SnO_2 NPs, SnO_2 NPs-4% presented the highest efficiency in the degradation of the dyes, degrading 100% of MO after 180 and 20 min under solar and UV irradiation, respectively. In relation to other dyes, SnO_2-NPs-4% degraded 100% of MB and RhB after 60 min under UV irradiation. The efficiency of this photocatalyst in degrading MO, MB, and RhB dyes under these conditions was confirmed by Turnover number (TON) and Turnover frequency (TOF) analysis. The authors associated the superior photocatalytic efficiency of SnO_2-NPs-4% to the smaller particle size, larger surface area, and the increased number of active sites present on the surface when compared to the other samples. Finally, the authors investigated the involvement in the degradation of dyes, and they evidenced that $^{\bullet}$OH radicals are the main species responsible for degradation.

Apart from the above-mentioned SnO_2-based photocatalysts prepared by various synthesis methods and the experimental conditions, the search for different photocatalytic materials with the desired efficiency is still a challenge. In this context, different authors have prepared SnO_2 catalysts, owing to the flexibility of applications, including photocatalysis. Compared to powdered materials, the use of films in photocatalysis has some advantages, especially for being easily recovered and reused in different batches. For instance, Bezzerouk et al. [80] deposited SnO_2 thin films on glass substrates at 450 °C by an ultrasonic spray pyrolysis technique, and desired polycrystalline SnO_2 films were obtained with a band gap of 3.80 eV, greater than that for bulk SnO_2 (*Eg* = 3.6 eV). The authors evaluated the photocatalytic property of the films toward Methylene blue (MB) degradation under UV-LEDs (340–400 nm, 7 W) and ultrasound (US) transducer (40 KHz). Different degradation processes were investigated, such as: photolysis (UV), photocatalysis (SnO_2 + UV), the sonolysis process (US), and sonocatalysis (SnO_2 + US) as well as sonophotolysis (US + UV) and sono-photocatalysis (SnO_2 + UV + US). Curiously, SnO_2 film did not show meaningful activity in the degradation of MB dye under UV irradiation (photocatalysis). However, when US was employed, a pronounced increase in the dye degradation was observed, reaching to 88.33%, 94.31%, 97.28%, and 98.25% of efficiency when sono-photocatalysis (SnO_2 + UV + US), sonolysis (US), sonophotolysis (US + UV) and sonocatalysis (SnO_2 + US) processes were used, respectively. The authors associated the highest efficiency of dye degradation using sonocatalysis to the production of acoustic cavitation in the water that can favor the dissociation of water and the formation of an important quantity of $^{\bullet}$OH radicals that participate in the degradation of dye. Finally, the authors associated the lower efficiency of the processes under UV irradiation to the rapid recombination of the electron–holes during SnO_2 photoexcitation. Therefore, authors showed different processes used to improve dye degradation using SnO_2-based material.

As one could see, several methods and experimental conditions were employed to synthesize undoped SnO_2 materials (in powder and film forms) with different characteristics and properties. It has been shown that these characteristics can directly impact the photocatalytic activity toward the degradation of dyes under UV-visible light, as well as under direct or simulated sunlight or even coupled with ultrasound irradiation. A summary of some important works concerning the photocatalytic applications of different undoped SnO_2 materials obtained by different methods is listed in Table 2.

Table 2. Summary of works related to the synthesis of pure SnO_2 materials by different methods and their application as photocatalysts for organic dye degradation.

Catalyst Type	Synthesis Method	Synthesis Conditions (Temperature/Time of Calcination)	Dye Solution Concentration (mg L^{-1})	Photocatalyst Concentration (g L^{-1})	Pollutant *	Irradiation	Efficiency/Time
SnO_2 NPs [94]	Green solution synthesis	400 °C/4 h	40	1.0	RY186	Sunlight	90%/3 h
SnO_2 NPs [90]	Continuous microwave flow synthesis (CMFS)	200 °C/2 h	50	-	MB	UV	93%/4 h
SnO_2 NPs [92]	Sol-gel	650 °C/4 h	20	0.5	CR	UVA	61.53%/1.66 h
SnO_2 NPs [40]	Green solution synthesis	400 °C/1 h	0.75	1.0	MO MB RhB	UV	(MO) 00%/0.33 h (MB) 100%/1 h (RhB) 100%/1 h
SnO_2 NPs [93]	Green sol-gel	700 °C/2 h	4.61	0.21	EBT	UV	77%/4.5 h
SnO_2 NPs [91]	Green solution synthesis using *Tinospora Cordifolia* extracts	400 °C/2 h	20	2.0	RhB	UV	99.9%/0.75 h
SnO_2 NPs [78]	Green solution synthesis	100 °C/6 h	15	0.4	R6G	Simulated sunlight	99.7/6.5 h
SnO_2 microflowers [95]	One-pot hydrothermal	190 °C/24 h	10	1.0	RhB	UV	99%/2 h
SnO_2 nanocrystals [96]	Solution phase growth technique	200 °C/24 h	5	0.5	RhB	UV-visible	91.7%/2 h
SnO_2 NPs [97]	Electrospinning by precursor solution	120 °C/48 h	5	1.33	MO MB	UV	(MO) 92%/4 h (MB) 95%/4 h
SnO_2 nanorods [85]	Chemical precipitation/hydrothermal	550 °C/4 h	20	0.6	MO	UV	52%/1 h
SnO_2 multilayered films [93]	Microwave hydrothermal	800 °C/2 h	10	-	RhB	UV	100%/4 h
SnO_2 thin films [88]	SPD using microscopy	500 °C/6 h	3.99	-	MB MO	UVA/H_2O_2	(MB) 42%/6 h (MO) 26%/6 h
SnO_2 thin films [80]	Ultrasonic spray pyrolysis technique	450 °C	-	-	MB	UV	98%/0.66 h

* Pollutants: RY186—Reactive yellow 186; MB—Methylene blue; MO—Methyl orange; MG—Malachite green; MR—Methylene red; CR—Congo red; RhB—Rhodamine B; R6G—Rhodamine 6G; EBT- Eriochrome Black T.

It is known that, to design an efficient system, photocatalysts usually need to meet some requirements, such as appropriate band gaps for light absorption, effective charge of carriers' separation, and appropriate VB and CB edge potentials. However, it is difficult for pure SnO_2-based photocatalysts to satisfy all of them. In this sense, doping SnO_2 with different cations and the formation of SnO_2-based composites with other materials have attracted interest, as they drive other possibilities of photocatalytic studies. Therefore, discussion of different works concerning the photocatalytic properties of doped SnO_2 and SnO_2-based composite is given in the following sections.

4.2. Doped SnO_2 Photocatalysts

Although pure SnO_2 nanoparticles with a different morphology have shown efficiency in the degradation of dyes under irradiation, different authors have developed strategies to overcome the low photoactivity of SnO_2 under visible light exposure. For instance, doping SnO_2 with different foreign ions has shown to be an efficient way to shorten its band gap and enhance its photoactivity.

Based on this fact, N. Mala et al. [98] synthesized SnO_2 nanoparticles doped with $Mg^{2+} + Co^{3+}$ cations by a low-cost chemical solution method and investigated the antibacterial activity and photocatalytic efficiency toward the degradation of Methylene blue (MB) and Malachite green (MG) dyes. The authors revealed that the samples presented a tetragonal crystalline phase, with an average crystallite size of 24 and 25 nm for pure SnO_2 and SnO_2-Mg:Co, respectively. The authors suggested that this slightl increase of the crystallite size after doping was due to local distortions in the SnO_2 lattice induced by the presence of dopants. A nanorod-like morphology was confirmed through SEM images, with a reduction in the crystal length and in the average diameter after doping. Surprisingly, an increase in the band gap energy estimated for SnO_2 (3.52 eV) and SnO_2-Mg:Co (4.22 eV) was observed. This behavior is attributed to the quantum confinement effect that normally happens when the nanoparticle size decreases. However, no meaningful variation was observed in the particle size for pure and doped SnO_2 samples. Regarding the photocatalytic activity of SnO_2 and SnO_2:Mg:Co nanoparticles, the authors observed that SnO_2 presented an efficiency of 82 and 86%, while SnO_2-Mg:Co displayed 89 and 92% efficiency toward MB and MG dyes degradation, respectively, under visible light after 60 min. The authors explained that three factors are responsible for the increase in the photocatalytic efficiency of doped SnO_2, which are: prevention of the recombination of electron–hole pairs photogenerated by surface defects, generation of greater number of oxidative species ($^{\bullet}OH$, $O_2{}^-$, and H_2O_2), and particle size reduction.

Chu et al. [99] synthesized Bi^{3+}-doped SnO_2 by the hydrothermal method at 180 °C for 24 h, with a variation of bismuth molar content (3, 5 and 7%). The Rhodamine B (RhB) and Ciprofloxacin hydrochloride (CIP) were used as target molecules to evaluate the photocatalytic activity of the synthesized materials under simulated sunlight. XRD analysis confirmed the cassiterite tetragonal phase for all the samples, with no secondary phases, confirming that Bi^{3+} is dissolved into the oxide crystal lattice by replacing Sn^{4+} during the synthesis. The Bi^{3+}/Sn^{4+} replacement in SnO_2 was confirmed by HR-TEM, UV–vis DRS, and XPS measurements. The average crystallite sizes decreased as a function of doping from 5.3 nm in SnO_2 to 3.3 nm in Bi-SnO_2(7%). In addition, the band gap (Eg) values showed a subtle variation of 3.72, 3.75, and 3.78 eV for Bi-SnO_2(3%), Bi-SnO_2(5%), and Bi-SnO_2(7%) samples, against 3.86 eV for the pure one. The authors state that the introduction of new levels in the band gap of materials can act as a trap center for electron and hole, reducing charge recombination, which is beneficial to improve photocatalytic activity. By using PL spectroscopy, authors confirmed the lower recombination charge rate in the Bi-SnO_2(5%) sample for presenting the lowest PL emission among all samples. Regarding the photocatalytic activity, Bi-SnO_2(5%) showed an efficiency of 98.28% of RhB dye degradation after 100 min and 92.13% of CIP degradation after 90 min under irradiation. The excellent photodegradation efficiency of the doped samples was due to the increase in light absorption, as well as the effective separation and migration of photogenerated charge carriers. All the results also indicated that there is an ideal amount of Bi^{3+} doping to optimize the mentioned characteristics in order to enhance the SnO_2 material functionality.

Although doped SnO_2 is most prepared in powder, thin films based on doped SnO_2 have also been studied in photocatalysis. For instance, S. Vadivel and G. Rajaraja et al. [84] prepared magnesium-doped SnO_2 films by the chemical bath deposition method, varying Mg^{2+} molar concentrations (1, 5, and 10%). The films were deposited on glass, and after deposition they were annealed at 500 °C for 5 h in air to promote crystallization. From XRD analysis, the tetragonal rutile phase was confirmed in all films. Atomic force microscopy (AFM) images revealed that the surface roughness decreases with increasing dopant concentration. The optical band gap energy for pure SnO_2 was 3.63 eV, decreasing to 3.42 eV for the film doped with 10% Mg. The photocatalytic activities of the films were evaluated by the degradation of Methylene blue (MB) and Rhodamine B (RhB) dyes under UV irradiation. The maximum photodegradation of the dyes was reached for 10% Mg-doped SnO_2 film, degrading 80% of MB and 90% of RhB after 120 min. Fast electron transfer and high efficiency in electron–hole pairs separation led to a significant improvement of photocatalytic activity in the doped sample.

In the work conducted by Haya et al. [82] films of pure SnO_2 and doped with 2, 4, 6, and 8% of Sr^{2+} were prepared by a chemical solution deposition method using the sol-gel method to deposit the solution coating on a glass substrate. The effect of doping on the structural, optical, morphological, and photocatalytic properties of the films were studied. According to the results, the increase in Sr^{2+} doping promotes a decrease in crystallite size and an increase in the lattice distortion. These effects generate a greater number of defects, such as grain boundaries, micro-stresses, and displacements in the thin film lattice. The average crystallite size decreased from 7.61 nm for undoped SnO_2 to 3.80 nm for 8% Sr-SnO_2. It was also observed by UV-visible analysis that the presence of dopants introduced new intermediate levels in the semiconductor band gap (*Eg*), decreasing *Eg* from 3.86 eV for pure SnO_2 film to 3.76 eV for Sr-richer SnO_2 film. Additionally, the morphology of the films was analyzed by AFM, and a smaller grain size was observed for 8% Sr-SnO_2 (4.96 nm). Consequently, it showed the lower surface roughness when compared to the other films. Concerning the photocatalytic activity, the greatest efficiency in the degradation of MB dye under irradiation was attained for 8% Sr-SnO_2 film, which was attributed to smaller grain sizes and surface roughness, as well as the introduction of new energy levels below the conduction band of the pure material, resulting from the Sr doping.

Using a non-conventional method to prepare thin films, Loyola Poul Raj et al. [83] prepared SnO_2 thin films doped with 3 and 6 mol% of Tb^{3+} on a glass substrate by the spray nebulized pyrolysis (NSP) method and calcined them at 400 °C to crystallize the materials. The authors investigated the photocatalytic property of the films in the degradation of MB dye under UV irradiation. According to the authors, doping SnO_2 films with up to 6% of Tb^{3+} cations induces a decrease in the grain size from 80 to 56 nm, and the band gap from 3.51 to 3.36 eV, which directly impacts photocatalysis, as reported by other authors. Indeed, a maximum of 85% of dye degradation was observed after 120 min under UV irradiation using SnO_2 film doped with the 6 mol% Tb^{3+}. Using PL spectroscopy, the authors reveal that Tb doping leads to the creation of more defects that act as reactive sites for catalyzed reactions. As a consequence of the study, the authors concluded that Tb doping favored photocatalytic reactions by reducing particle size, and therefore increasing the surface area and the number of reactive sites on the surface, which allows the dye adsorption. In addition, Tb doping induces a decrease in the band gap of the materials, which favors photo-absorption, aiming to potentialize charge carriers to participate in photocatalytic reactions.

Other studies based on the synthesis of doped SnO_2 catalysts and their applications in the photodegradation of different organic dyes are listed in Table 3.

Table 3. Summary of the studies for doped SnO_2-based photocatalysts applied in the photodegradation of different organic dyes.

Catalyst Type	Synthesis Method	Synthesis Conditions (Temperature/Time of Calcination)	Dye Solution Concentration ($mg\ L^{-1}$)	Photocatalyst Concentration ($g\ L^{-1}$)	Pollutant *	Irradiation	Efficiency/Time
SnO_2:Mg^{2+}Co^{3+} NPs [84]	Wet chemical method	550 °C/2 h	-	0.5	MB MG	Near UV	(MB) 89%/1 h (MG) 92%/1 h
SnO_2:Sr^{2+} thin films [82]	Sol–gel method using a dip-coating technique	500 °C/2 h	5	-	MB	UV	37.90%/2 h
SnO_2:Tb^{3+} thin films [83]	Nebulized spray pyrolysis (NSP) technique	400 °C	3.2	-	MB	UV	85%/2.08 h
SnO2:Bi^{3+} quantum dots [99]	One-step hydrothermal	180 °C/24 h	20	1.0	RhB	UV	98.58%/1.66 h
SnO_2:Mg^{2+} thin films [84]	Chemical bath deposition	500 °C/2 h	15	-	MB	UV	(MB) 80%/2 h
SnO_2:Co^{3+} [100]	Chemical solution	400 °C/2 h	60	0.2	MB	UV-visible	95.38%/2 h
SnO_2:Ni^{2+} NPs [101]	Chemical precipitation	410 °C/2 h	5	0.15	BG	UV	97.54%/1.75 h
SnO_2:Zn^{2+} NPs [102]	Combustion	600 °C/8 h	10	1.0	MB MO	UV	1.6%/1 h 40%/1 h
SnO_2:Zr^{4+} NPs [103]	Co-precipitation	200 °C/24 h	20	-	MO	UV	89.6%/3 h
SnO_2:Mn^{2+} nanowires [104]	Co-precipitation	400 °C/4 h	61	1.0	NBB	Solar irradiation	99%/3 h

* Pollutant: BG—Brilliant green; NBB—Naphthol blue black; MB—Methylene blue; MO—Methyl orange; MG—Malachite green; CR—Congo red; RhB—Rhodamine B.

4.3. SnO$_2$-Based Composite Photocatalysts

As one can see from studies discussed in the sections above, pure and doped SnO$_2$ particles and films have been well explored. However, SnO$_2$ has also been combined with other different semiconductors in order to reduce recombination of the photoinduced charge carriers, to therefore improve photocatalytic activity.

Considering this fact, Abdel-Messih et al. [105] synthesized SnO$_2$/TiO$_2$ nanoparticles with a spherical mesoporous morphology synthesized by the sol-gel process, using polymethylmethacrylate as a template. The amount of SnO$_2$ (0–25%) in relation to the mass of pure TiO$_2$ was varied to obtain composites with different compositions. The samples were calcined at 800 °C for 3 h to ensure complete organic polymer decomposition. In relation to the photocatalytic property of the materials, photodegradation of Rhodamine B (RhB) dye was performed under UV irradiation using a catalyst/dye concentration of 1 g L^{-1}. The photodegradation efficiency of the composites increased with the increase of tin oxide content up to 10% (about 92% of the dye degraded after 3 h). However, the sample with 25 mol% of SnO$_2$ showed the lowest efficiency, which was attributed to the loss of the titanium anatase phase. The authors concluded that there is an optimal amount of SnO$_2$ to achieve the maximum efficiency. In addition, the remarkable reduction in particle size by the existence of SnO$_2$ in the composites enhanced the oxidizing power and extended the photoinduced charge separation, and these were the main reasons for the increase in the catalytic activity of the samples.

Das et al. [36] prepared Sn/SnO$_2$ nanocomposites by the precipitation method, followed by carbothermal reduction and calcination at 800 °C for 2 h. The authors investigated the photocatalytic property of Sn/SnO$_2$ composites in the degradation of methylene blue under UV irradiation using a catalyst concentration of 0.5 g L^{-1}. It was found that there was a maximum efficiency of 41% for pure SnO$_2$ after 210 min under irradiation, while the Sn/SnO$_2$ composite showed a higher photocatalytic activity of 99%. The highest efficiency observed for the composite was related to the role of Sn on the surface of SnO$_2$ nanoparticles. As the Fermi energy level of Sn is higher than that observed for SnO$_2$ due to its lower work function, when metallic Sn is bound on the surface of SnO$_2$ nanoparticles, electrons migrate from Sn to SnO$_2$ to reach Fermi-level equilibrium. The effect of the pH solution on the photocatalytic efficiency of the composites was also evaluated. The pH had a direct influence on the photocatalytic process, being the neutral pH favorable for the degradation of MB dye. Finally, the authors investigated the reusability of the composites after three cycles and confirmed that the photocatalyst is stable, but gradual loss in efficiency was observed due to the loss of the material during recovery processes.

Li et al. [85] synthesized carbon-coated, mixed-phase (tetragonal/orthorhombic) SnO$_2$ (i.e., tetragonal/orthorhombic) nanorods photocatalysts by a combined chemical precipitation and hydrothermal method at 180 °C for 6 h. The SnO$_2$-C composite with a SnO$_2$ tetragonal phase was obtained after calcination of the hydrothermal products at 550 °C for 4 h. The photocatalytic activities of the samples were investigated toward the degradation of Methyl orange (MO) dye under UV irradiation using a catalyst/dye concentration of 0.6 g L^{-1}. The as-prepared mixed-phase SnO$_2$ showed a photodegradation activity of 52% against an efficiency of 39% for pure SnO$_2$ with a tetragonal phase after 60 min under irradiation, indicating the influence of different phases on the junction formation to tune photocatalytic activity. Coating mixed-phase SnO$_2$ nanorods with carbon provided a degradation activity of 98%. The tetragonal/orthorhombic-SnO$_2$ material exhibits very high stability after three cycles, remaining about constant without apparent deactivation. Photocatalytic activity was not primarily attributed to the narrower band gap or visible light absorption tail. By demonstrating that the transfer and separation of photogenerated electron–hole pairs are improved by the introduction of a carbon layer in interparticle space. To understand the photocatalytic mechanism, different scavengers were used in the study—triethanolamine (TEOA), tert-butyl alcohol (TBA), and benzoquinone (BQ). The results indicated that the species h$^+$, O$_2^{\bullet-}$, and HO$^\bullet$ played important roles in the degradation of MO.

Constantino et al. [97] synthesized a porous composite based on SnO_2/cellulose acetate with the electrospinning method and calcined it at 120 °C for 48 h. The photocatalytic activity of the nanocomposites was studied toward the degradation of MO and MB dyes under UV irradiation using a catalyst/dye concentration of 1.3 g L^{-1}. The photocatalytic efficiency of SnO_2/cellulose was approximately 92% of MO degradation after 210 min and 95% of MB degradation after 240 min. It is worth mentioning that the photodegradation process did not alter the average diameter and morphology of the fibers as well as their surface chemistry. From TOC analysis, the authors evidenced that only 54% and 79% of MO and MB dye are mineralized after the photocatalytic process. However, presence of other compounds as by-product of dye degradation was confirmed by LC–MS.

Silva et al. [37] synthesized spherical nanoparticles and microrods of Ag_3PO_4/SnO_2 composites, by the in situ coprecipitation method, with various molar ratios of 5, 10, 15, and 20% of SnO_2 in relation to the mass of pure Ag_3PO_4, followed by calcination at 350 °C for 2 h. The photocatalytic performance of the samples was investigated by the degradation of Rhodamine B (RhB) dye under visible light irradiation using a catalyst/dye concentration of 0.6 g L^{-1}. The authors observed superior photocatalytic activity for all the composites when compared to pure Ag_3PO_4. The authors evidenced that the excess of SnO_2 damaged the interfacial contact between the Ag_3PO_4 and SnO_2, which was due to the high degree of particle agglomeration. The photocatalytic mechanism involved in the photodegradation of the dye was also investigated for pure Ag_3PO_4 and Ag_3PO_4/SnO_2-15%. It has been confirmed that the photogenerated holes participated in the direct degradation of RhB when Ag_3PO_4 was a photocatalyst. On the other hand, there was a significant participation of $O_2^{\bullet-}$ radicals when Ag_3PO_4/SnO_2-15% is used. The highest photodegradation efficiency presented by the Ag_3PO_4/SnO_2-15% composite was confirmed by total organic carbon (TOC) analysis. Reusability tests were also performed for Ag_3PO_4/SnO_2-15% and a loss of 43.2% of its photocatalytic efficiency was observed after the third cycle, which was similar to that observed for pure Ag_3PO_4. Using the X-ray diffraction technique, the presence of Ag in the composition of the samples was observed after the photocatalysts were used in photocatalysis.

Apart from the above-mentioned studies performed using composite particles, SnO_2-based composite photocatalysts have also been explored as films. For instance, porous SnO_2/TiO_2 films were prepared using Ar-assisted, modified thermal evaporation, followed by the atomic layer deposition (ALD) technique at 300 °C in the reaction chamber [62]. To prepare SnO_2/TiO_2 films, TiO_2 layers were deposited on porous SnO_2 nanofoam, with those previous deposited on 2 × 2 cm^2 Si (100) wafers or ITO substrates, with variation in deposition cycles of TiO_2 (10, 25, 50, and 100 cycles) by ALD. The samples were denoted SnO_2/TiO_2-10, SnO_2/TiO_2-25, SnO_2/TiO_2-50, and SnO_2/TiO_2-100, respectively. After the TiO_2 deposition, the material was calcined at 700 °C for 1 h. The photocatalytic properties of the films were evaluated by the degradation of MB at a concentration of 1.2 mg L^{-1} under UV irradiation. The nanofoam heterostructures showed higher photocatalytic activity when compared to the porous SnO_2 nanofoam. SnO_2/TiO_2-50 nanofoam, which exhibited the highest efficiency, reached to 99% of MB degradation after 300 min. The authors correlated this fact to being due to a synergistic effect occurring between SnO_2 and TiO_2, and due to the sparse deposition of the TiO_2 layer on porous SnO_2. Separation of charge carriers due to the potential difference between SnO_2 and TiO_2 increases the lifetime of the charge and improves the interfacial charge transfer to the species adsorbed on the surface. This phenomenon, along with the strong oxidant $^{\bullet}OH$ radicals formed in the VB of the TiO_2 layer, improves photocatalytic efficiency of the SnO_2/TiO_2 heterostructure.

Other important works reporting the activity of SnO_2-based composite photocatalysts for the degradation of dyes are summarized in Table 4.

Table 4. SnO$_2$-based composites formed with different semiconductors for photodegradation of organic dyes.

Catalyst Type	Synthesis Method	Synthesis Conditions (Temperature/Time of Calcination)	Dye Solution Concentration (mg L^{-1})	Photocatalyst Concentration (mg L^{-1})	Pollutant *	Irradiation	Efficiency/Time
CrO$_4$-SnO$_2$ spherical NPs [106]	Coprecipitation	150, 300 and 450 °C/3 h	1 × 10^{-4} mol L^{-1}	1.6 g L^{-1}	TB	Sunlight	80%/1 h
g-C$_3$N$_4$/SnO$_2$ nanosheets decorated with NPs [107]	Microwave-assisted hydrothermal method	550 °C/2 h	10 mg L^{-1}	0.5 g L^{-1}	RhB	UV and visible lights	100%/4 h (UV) and 98.5% (visible light)
Zn$_2$SnO$_4$–SnO$_4$ and Zn$_2$SnO$_4$–Sn spherical NPs [108]	Solid state reaction	800 °C/1 h	20 mg L^{-1}	0.2 g L^{-1}	MB	UV	92%/2 h
Polyaniline SnO$_2$ nanoneedles and nanograins [109]	Chemical oxidation polymerization method using aniline monomer	500 °C/2 h	50 mg L^{-1}	1 g L^{-1}	RY	UV	96%/1 h
ZnO-SnO$_2$ thin films [81]	Sol-gel	550 °C/1 h	16 mg L^{-1}	-	MeG	UV	42%/45 min
ZnO-SnO$_2$/NPs hexagonal nanopillar [110]	Sol-gel	600 °C/2 h	10 mg L^{-1}	0.025 g L^{-1}	MO MB CR	UV	91.78%/3 h (MO) 93.21%/3 h (MB) 85.14%/3 h (CR)
CuCr$_2$O$_4$/SnO$_2$ [111]	Sol-gel/solid state reaction	900 °C/6 h and 600 °C	15 mg L^{-1}	1 g L^{-1}	CV	Sunlight	100%/1.5 h
SnO$_2$/GO spherical NPs [112]	Sonochemical method	180 °C/6 h and 100 °C/6 h	1 × 10^{-5} mol L^{-1}	0.5 g L^{-1}	RhB TDW	Sunlight	95%/2 h (RhB) 100%/2.5 h (TDW)
SnO$_2$/Zn$_2$SnO$_4$ cube-like NPs [113]	Hydrothermal	150 °C/12 h and 700 °C/2 h	10^{-5} mol L^{-1}	1 g L^{-1}	MB MO EBT	Simulated sunlight	97.1%/2.5 h (MB) 93.7%/3 h (MO) 87.9%/3.5 h (EBT)
SnO$_2$-MoS$_2$ spherical NPs [114]	Sonochemical liquid exfoliation method	80 °C/2 h	100 ppm	1 mL of SnO$_2$-MoS$_2$ added in 50 mL of dye solution	MR MB	Visible light	94.0%/2 h (MR) 58.5%/2 h (MB)
SnO$_2$-WO$_3$ NPs [115]	Green combustion method	500 °C/1 h	5 ppm	0.8 g L^{-1}	MB	Visible light	70%/3 h

* Pollutant: CV—Crystal violet; CR—Congo red; EBT—Eriochrome black T; MB—Methylene blue; MO—Methyl orange; MeG—Methyl green; MR—Methylene red; RhB—Rhodamine B; RY—Remazol yellow; TB—Trypan Blue; TDW—Textile dye wastewater.

Most of the authors reported that the number of materials for the formation of the SnO_2 with a different particle size and morphology, besides doped SnO_2 with an appropriate amount and type of dopant, and also the formation of the composite with SnO_2, plays an important role in improving the photocatalytic activity of the SnO_2 material. In relation to composites, the excess of both species can be harmful to the contact surface between the phases, mainly due to the high degree of particle agglomeration. Tests using scavengers, such as p-benzoquinone (BZ, $C_6H_4O_2$), isopropanol (ISO, $(CH_3)_2CHOH$), and ammonium oxalate monohydrate (AO, $(NH_4)_2C_2O_4 \cdot H_2O$) indicate $\bullet OH$ is the main species in most photocatalytic mechanisms. However, to obtain more insights about the photocatalytic mechanism involved in composite materials, we seek to understand the charge transfer between the phases from the band structures of each individual material. Structural and electronic defects can also generate energy levels between the VB and CB, and, therefore, modify the photocatalytic mechanism of composites. The creation of different interfaces between the phases may reduce charge carriers' recombination, leading to the formation of a great number of free radicals to improve photocatalysis. In addition, several other parameters can impact the photocatalytic efficiency of composites, such as phase composition, surface area, morphology, particle size, pore structure, electron–hole recombination rate, and band gap energy of the individual components. Some authors showed that the high surface area and the presence of pores are more effective parameters that affect dye degradation since the existence of several active sites, responsible for the adsorption of molecules, is crucial for the photocatalysis to occur.

Based on the findings above, it can be concluded that to design a new photocatalytic material with specific characteristic, one has to consider optimizing type and amount of dopants and interface characteristics between materials, or even the nature of the desired product (powder, film, etc.), besides the microstructure of the material (particle size and morphology), and by a choice of specific synthesis methodology and appropriate experimental conditions.

5. Conclusions and Final Remarks

From this review work, it was possible to evaluate the importance of synthesis methods and experimental parameters in obtaining tin-oxide-based materials with high performance in heterogenous photocatalysis of persistent organic pollutants, more specifically, organic dyes. The search for new materials and methodologies that provide efficient results for the remediation of such pollutants has been one of the great challenges for the scientific community. Among the studied promising materials, SnO_2 has shown excellent results as a catalyst in heterogeneous photocatalysis processes due to its intrinsic characteristics, which have been responsible for the material's conductivity, optical and electrical properties, and high thermal stability. As a consequence of the choice of the synthesis method and experimental conditions, it was possible to evidence different morphology, particle size, surface area, structural modifications, optical bandgap energy, and surface and bulk defects, and, therefore, obtain excellent results in the application of pure and modified SnO_2 toward the degradation of persistent organic pollutants (POPs). In general, SnO_2-based photocatalysts have shown promising efficiency for degrading a series of different organic dyes.

Considering pure SnO_2 catalysts, synthesis conditions may especially influence particle size and morphology, specific surface area, crystallinity, and the presence of electronic defects on surface and bulk of the materials, which are important parameters to change photocatalytic efficiency under both UV and visible irradiation. With respect to doped SnO_2, the type and number of dopants may introduce different intermediate levels within the gap, decreasing band gap energy to improve the photo-absorption in visible light. Finally, it has been demonstrated that the composite formed with SnO_2 is responsible for band structure alignment and improvement of charge separation that led to an increased photocatalytic activity when compared to the individual components. It is still important to highlight that the study of the reaction mechanism involved in the dye degradation is an important aspect, which allows the design of new and efficient SnO_2-based photocatalysts, and an understanding of their laminations in order to use them in practical devices.

Author Contributions: Conceptualization, M.C.F.A. and A.L.M.d.O.; methodology, J.L.A.d.N., A.L.M.d.O. and M.C.F.A.; validation, A.L.M.d.O. and M.C.F.A.; investigation, J.L.A.d.N.; formal analysis, J.L.A.d.N., L.C., A.L.M.d.O. and M.C.F.A.; data curation: J.L.A.d.N.; writing—original draft preparation, J.L.A.d.N.; writing—review and editing, L.C., A.L.M.d.O. and M.C.F.A.; visualization, I.M.G.d.S.; supervision, A.L.M.d.O. and M.C.F.A.; project administration, M.C.F.A. All authors have read and agreed to the published version of the manuscript.

Funding: This research was funded by CAPES-Brazil (Grant 88887.497404/2020-00).

Institutional Review Board Statement: Not applicable.

Informed Consent Statement: Not applicable.

Acknowledgments: The authors thank PRPGP/UEPB, PPGQ/UEPB and FAPESQ/PRONEX, for financial support. A.L.M.d.O thanks CAPES-Brazil for the PNPD funding (Grant 88882.317938/2019-01).

Conflicts of Interest: The authors declare that they have no conflict of interest regarding the publication of this article, financial and/or otherwise.

References

1. Anastopoulos, I.; Pashalidis, I.; Orfanos, A.G.; Manariotis, I.D.; Tatarchuk, T.; Sellaoui, L.; Bonilla-petriciolet, A.; Mittal, A.; Núnez-delgado, A. Removal of caffeine, nicotine and amoxicillin from (waste)waters by various adsorbents. A review. *J. Environ. Manag.* **2020**, *261*, 110236. [CrossRef] [PubMed]
2. Beltrame, T.F.; Lhamby, A.R.; Beltrame, A. Wastewater, solid waste and environmental education: A discussion about the subject. *Rev. Eletrônica Gestão Educ. Tecnol. Ambient.* **2016**, *20*, 351–362.
3. Kurade, M.B.; Há, Y.; Xiong, J.; Govindwar, S.P.; Jang, M.; Jeon, B. Phytoremediation as a green biotechnology tool for emerging environmental pollution: A step forward towards sustainable rehabilitation of the environment. *Chem. Eng. J.* **2021**, *415*, 129040. [CrossRef]
4. Tao, Y.; Wu, Y.; Zhou, J.; Wu, M.; Wang, S.; Zhang, L.; Xu, C. How to realize the effect of air pollution control? A hybrid decision framework under the fuzzy environment. *J. Clean. Prod.* **2021**, *305*, 127093. [CrossRef]
5. Shi, Z.; She, Z.; Chiu, Y.; Shijiong, Q.S.; Zhang, L. Assessment and improvement analysis of economic production, water pollution, and sewage treatment efficiency in China. *Socio-Econ. Plan. Sci.* **2021**, *74*, 100956. [CrossRef]
6. Kim, A.Y.; Park, J.B.; Woo, M.S.; Lee, S.Y.; Kim, H.Y.; Yoo, Y.H. Review: Persistent Organic Pollutant-Mediated Insulin Resistance. *Int. J. Environ. Res. Pub. Health* **2019**, *16*, 448. [CrossRef]
7. Tyutikov, S.F. Migration and Biogeochemical Indication of Persistent Organic Pollutants. *Geochem. Int.* **2018**, *56*, 1028–1035. [CrossRef]
8. Arthur, R.B.; Ahern, J.C.; Patterson, H.H. Review Application of BiOX Photocatalysts in Remediation of Persistent Organic Pollutants. *Catalysts* **2018**, *8*, 604. [CrossRef]
9. Li, C.; Yanga, L.; Shic, M.; Liu, G. Persistent organic pollutants in typical lake ecosystems. *Ecotoxicol. Environ. Saf.* **2019**, *180*, 668–678. [CrossRef]
10. Mansouri, E.H.; Reggabi, M. Association between type 2 diabetes and exposure to chlorinated persistent organic pollutants in Algeria: A case-control study. *Chemosphere* **2021**, *264*, 128596. [CrossRef]
11. Ulutaş, O.K.; Çok, I.; Darendeliler, F.; Aydin, B.; Çoban, A.; Henkelmann, B.; Schramm, K. Blood concentrations and risk assessment of persistent organochlorine compounds in newborn boys in Turkey. *A pilot study. Environ. Sci. Pollut. Res.* **2015**, *22*, 19896–19904. [CrossRef] [PubMed]
12. Nascimento, F.P.; Kuno, R.; Lemes, V.G.R.; Kussumi, T.A.; Nakano, V.E.; Rocha, S.B.; Oliveira, M.C.C.; Kimura, I.A.; Gouveia, N. Organochlorine pesticides levels and associated factors in a group of blood donors in São Paulo, Brazil. *Environ. Monit. Assess.* **2017**, *189*, 380. [CrossRef] [PubMed]
13. Stockholm Convention Secretariat United Nations Environment. *An Introduction to the Chemicals Added to the Stockholm Convention as Persistent Organic Pollutants by the Conference of the Parties;* International Environmental House: Vernier, Switzerland, 2017; pp. 2–25.
14. Varakina, Y.; Lahmanov, D.; Aksenov, A.; Trofimova, A.; Korobitsyna, R.; Belova, N.; Sobolev, N.; Kotsur, D.; Sorokina, T.; Grjibovsk, A.M.; et al. Concentrations of Persistent Organic Pollutants in Women's Serum in the European Arctic Russia. *Toxics* **2021**, *9*, 6. [CrossRef] [PubMed]
15. Wagner, M.; Lin, K.A.; Oh, W.; Lisak, G. Metal-organic frameworks for pesticidal persistent organic pollutants detection and adsorption—A mini review. *J. Hazard. Mater.* **2021**, *413*, 125325. [CrossRef]

16. Wang, S.; Hu, C.; Lu, A.; Wang, Y.; Cao, L.; Wu, W.; Li, H.; Wu, M.; Yan, C. Association between prenatal exposure to persistent organic pollutants and neurodevelopment in early life: A mother-child cohort (Shanghai, China). *Ecotoxicol. Environ. Saf.* **2021**, *208*, 111479. [CrossRef]

17. Chen, Y.; Zhi, D.; Zhou, Y.; Huang, A.; Wu, S.; Yao, B.; Tang, Y.; Sun, C. Electrokinetic techniques, their enhancement techniques and composite techniques with other processes for persistent organic pollutants remediation in soil: A review. *J. Ind. Eng. Chem.* **2021**, *97*, 163–172. [CrossRef]

18. Junior, J.C.A.P.; Albuquerque, L.S.; Delfino, N.M.; Muniz, E.P.; Rocha, S.M.S.; Porto, P.S.S. Treatment of synthetic methylene blue dye effluent by electrofloculation. *Braz. J. Prod. Eng.* **2017**, *3*, 105–113.

19. Li, M.; Zhao, H.; Lu, Z. Porphyrin-based porous organic polymer, Py-POP, as a multifunctional platform for efficient selective adsorption and photocatalytic degradation of cationic dyes. *Microporous Mesoporous Mater.* **2020**, *292*, 109774. [CrossRef]

20. Zhao, L.; Houb, H.; Iwasaki, K.; Terada, A.; Hosomi, M. Utilization of recycled charcoal as a thermal source and adsorbent for the treatment of PCDD/Fs contaminated sediment. *J. Hazard. Mater.* **2012**, *225*, 182–188. [CrossRef]

21. Titchou, F.E.; Zazou, H.; Afanga, H.; Gaayda, J.E.; Akbour, R.A.; Hamdani, M. Review Removal of Persistent Organic Pollutants (POPs) from water and wastewater by adsorption and electrocoagulation process. *Groundw. Sustain. Dev.* **2021**, *13*, 100575. [CrossRef]

22. Ji, S.; Yang, Y.; Zhou, Z.; Li, X.; Liu, Y. Photocatalysis-Fenton of Fe-doped g-C3N4 catalyst and its excellent degradation performance towards RhB. *J. Water Process Eng.* **2021**, *40*, 101804. [CrossRef]

23. Ajmal, A.; Majeedb, I.; Malika, R.N.; Iqbal, M.; Nadeemb, M.A.; Hussain, I.; Yousaf, S.; Mustafa, G.; Zafara, M.I.; Nadeem, M.A. Photocatalytic degradation of textile dyes on Cu_2O CuO/TiO_2 anatase powders. *J. Environ. Chem. Eng.* **2016**, *4*, 2138–2146. [CrossRef]

24. Waki, M.; Shirai, S.; Yamanaka, K.; Maegawa, Y.; Inagaki, S. Heterogeneous water oxidation photocatalysis based on periodic mesoporous organosilica immobilizing a tris(2,20-bipyridine)ruthenium sensitizer. *RSC Adv.* **2020**, *10*, 13960. [CrossRef]

25. Lima, A.S.; Rocha, R.D.C.; Pereira, E.C.; Sikora, M.S. Photodegradation of Ciprofoxacin antibiotic over TiO_2 grown by PEO: Ecotoxicity response in *Lactuca sativa* L. and Lemna minor. *Int. J. Environ. Sci. Technol.* **2021**, *19*, 2771–2780. [CrossRef]

26. Araújo, K.S.; Antonelli, R.; Gaydeczka, B.; Granato, A.C.; Malpass, G.R.P. Processos oxidativos avançados: Uma revisão de fundamentos e aplicações no tratamento de águas residuais urbanas e efluentes industriais. *Rev. Ambient. Água* **2016**, *11*, 387–401.

27. Honorio, I.M.C.; Santos, M.V.B.; Filho, E.C.S.; Osajima, J.A.; Maia, A.S.; Santos, I.M.G. Alkaline earth stannates applied in photocatalysis: Prospection and review of literature. *Cerâmica* **2018**, *64*, 559–569. [CrossRef]

28. Cornejo, O.M.; Murrieta, M.F.; Castañeda, L.F.; Nava, J.L. Characterization of the reaction environment in flow reactors fitted with BDD electrodes for use in electrochemical advanced oxidation processes: A critical review. *Electrochim. Acta* **2020**, *331*, 135373. [CrossRef]

29. Moradi, M.; Vasseghian, Y.; Khataee, A.; Kobya, M.; Arabzade, H.; Dragoi, E. Service life and stability of electrodes applied in electrochemical advanced oxidation processes: A comprehensive review. *J. Ind. Eng. Chem.* **2020**, *87*, 18–39. [CrossRef]

30. Honorio, L.M.C.; Oliveira, A.L.M.; Silva-Filho, E.C.; Osajima, J.A.; Hakki, A.; Macphee, D.E.; Santos, I.M.G. Supporting the photocatalysts on ZrO_2: An effective way to enhance the photocatalytic activity of $SrSnO_3$. *Appl. Surf. Sci.* **2020**, *528*, 146991. [CrossRef]

31. Sun, H.; Qin, P.; Wu, Z.; Liao, C.; Guo, J.; Luo, S.; Chai, Y. Visible light-driven photocatalytic degradation of organic pollutants by a novel Ag_3VO_4/Ag_2CO_3 pen heterojunction photocatalyst: Mechanistic insight and degradation pathways. *J. Alloys Compd.* **2020**, *834*, 155211. [CrossRef]

32. Riente, P.; Fianchini, M.; Llanes, P.; Pericàs, M.A.; Noël, T. Shedding light on the nature of the catalytically active species in photocatalytic reactions using Bi_2O_3 semiconductor. *Nat. Commun.* **2021**, *12*, 625. [CrossRef] [PubMed]

33. Hu, W.; Quang, N.D.; Majumder, S.; Jeong, M.J.; Park, J.H.; Cho, Y.J.; Kim, S.B.; Lee, K.; Kim, D.; Chang, H.S. Three-dimensional nanoporous SnO_2/CdS heterojunction for high-performance photoelectrochemical water splitting. *Appl. Surf. Sci.* **2021**, *560*, 149904. [CrossRef]

34. Li, Y.; Wang, J.; Sun, H.; Hua, W.; Liu, X. Heterostructured SnS_2/SnO_2 nanotubes with enhanced charge separation and excellent photocatalytic hydrogen production. *Int. J. Hydrog. Energy* **2018**, *43*, 14121–14129. [CrossRef]

35. Chen, S.; Yang, J.; Wu, J. Three-Dimensional Undoped Crystalline SnO2 Nanodendrite Arrays Enable Efficient Charge Separation in $BiVO_4/SnO_2$ Heterojunction Photoanodes for Photoelectrochemical Water Splitting. *ACS Appl. Energy Mater.* **2018**, *1*, 2143–2149. [CrossRef]

36. Das, O.R.; Uddin, M.T.; Rahman, M.M.; Bhoumick, M.C. Highly active carbon supported Sn/SnO_2 photocatalysts for degrading organic dyes. *J. Phys.* **2018**, *1086*, 012011. [CrossRef]

37. Silva, G.N.; Martins, T.A.; Nogueira, I.C.; Santos, R.K.; Li, M.S.; Longo, E.; Botelho, G. Synthesis of Ag_3PO_4/SnO_2 composite photocatalyst for improvements in photocatalytic activity under visible light. *Mater. Sci. Semicond. Process.* **2021**, *135*, 106064. [CrossRef]

38. Xua, L.; Xiana, F.; Zhanga, Y.; Wang, W.; Qiua, K.; Xu, J. Synthesis of ZnO-decorated SnO_2 nanopowder with enhanced photocatalytic performance. *Opt. Int. J. Light Electron. Opt.* **2019**, *194*, 162965. [CrossRef]

39. Luque, P.A.; Garrafa-Galvez, H.E.; Nava, O.; Olivas, A.; Martínez-Rosas, M.E.; Vilchis-Nestor, A.R.; Villegas-Fuentes, A.; Chinchillas-Chinchillas, M.J. Efficient sunlight and UV photocatalytic degradation of Methyl Orange, Methylene Blue and Rhodamine B, using Citrus×paradisi synthesized SnO_2 semiconductor nanoparticles. *Ceram. Int.* **2021**, *5*, 91. [CrossRef]

40. Santos, J.E.L.; Moura, D.C.; Cerro-López, M.; Quiroz, M.A.; Martínez-Huitle, C.A. Electro- and photo-electrooxidation of 2,4,5-trichlorophenoxiacetic acid (2,4,5-T) in aqueous media with PbO$_2$, Sb-doped SnO$_2$, BDD and TiO$_2$-NTs anodes: A comparative study. *J. Electroanal. Chem.* **2020**, *873*, 114438. [CrossRef]

41. La Fourni'ere, E.M.; Meichtry, J.M.; Gautier, E.A.; Leyva, A.G.; Litter, M.I. Treatment of ethylmercury chloride by heterogeneous photocatalysis with TiO$_2$. *J. Photochem. Photobiol. A Chem.* **2021**, *411*, 113205. [CrossRef]

42. Furtado, R.X.S.; Sabatini, C.A.; Zaiat, M.; Azevedo, E.B. Perfluorooctane sulfonic acid (PFOS) degradation by optimized heterogeneous photocatalysis (TiO$_2$/UV) using the response surface methodology (RSM). *J. Water Process Eng.* **2021**, *41*, 10198.

43. Yanga, D.; Gondalb, M.A.; Yamani, Z.H.; Baig, U.; Qiao, Q.; Liua, G.; Xud, Q.; Xiange, D.; Maof, J.; Shen, K. 532 nm nanosecond pulse laser triggered synthesis of ZnO$_2$ nanoparticles via a fast ablation technique in liquid and their photocatalytic performance. *Mater. Sci. Semicond. Process.* **2017**, *57*, 124–131. [CrossRef]

44. Khatri, A.; Rana, P.S. Visible light assisted photocatalysis of Methylene Blue and Rose Bengal dyes by iron doped NiO nanoparticles prepared via chemical co-precipitation. *Phys. B Phys. Condens. Matter.* **2020**, *579*, 411905. [CrossRef]

45. Mohite, S.V.; Ganbavle, V.V.; Rajpure, K.Y. Photoelectrocatalytic activity of immobilized Yb doped WO$_3$ photocatalyst for degradation of methyl orange dye. *J. Energy Chem.* **2017**, *26*, 440–447. [CrossRef]

46. Omrani, N.; Nezamzadeh-Ejhieh, A. BiVO$_4$/WO$_3$ nano-composite: Characterization and designing the experiments in photodegradation of sulfasalazine. *Environ. Sci. Pollut. Res.* **2020**, *27*, 44292–44305. [CrossRef]

47. Al-Hamdi, A.M.; Rinner, U.; Sillanpää, M. Tin dioxide as a photocatalyst for water treatment: A review. *Process Saf. Environ. Prot.* **2017**, *107*, 190–205. [CrossRef]

48. Sun, C.; Yang, J.; Xu, M.; Cui, Y.; Ren, W.; Zhang, J.; Zhao, H.; Liang, B. Recent intensification strategies of SnO$_2$-based photocatalysts: A review. *Chem. Eng. J.* **2022**, *427*, 131564. [CrossRef]

49. Lavanya, N.; Fazio, E.; Neri, F.; Bonavita, A.; Leonardi, S.G.; Neri, G.; Sekar, C. Simulta neous electrochemical determination of epinephrine and uric acid in the presence of ascorbic acid using SnO$_2$/graphene nanocomposite modified glassy carbon electrode. *Sens. Actuators B Chem.* **2015**, *221*, 1412–1422. [CrossRef]

50. Sakthiraj, K.; Balachandrakumar, K. Influence of Ti addition on the room temperature ferromagnetism of tin oxide (SnO$_2$) nanocrystal. *J. Magn. Magn. Mater.* **2015**, *395*, 205–212. [CrossRef]

51. Xiong, L.; Guo, Y.; Wen, J.; Liu, H.; Yang, G.; Qin, P.; Fang, G. Review on the Application of SnO$_2$ in Perovskite Solar Cells. *Adv. Funct. Mater.* **2018**, *28*, 1802757. [CrossRef]

52. Mouly, T.A.; Toms, L.L. Breast cancer and persistent organic pollutants (excluding DDT): A systematic literature review. *Environ. Sci. Pollut. Res.* **2016**, *23*, 22385–22407. [CrossRef] [PubMed]

53. Gopinath, K.P.; Madhav, N.V.; Krishnan, A.; Malolan, R.; Rangarajan, G. Present applications of titanium dioxide for the photocatalytic removal of pollutants from water: A review. *J. Environ. Manag.* **2020**, *270*, 110906. [CrossRef] [PubMed]

54. Tang, S.H.; Zaini, M.A.A. Isotherm studies of malachite green removal by yarn processing sludge-based activated carbon. *Chem. Didact. Ecol. Metrol.* **2019**, *24*, 127–134. [CrossRef]

55. Adithya, S.; Jayaraman, R.S.; Krishnan, A.; Malolan, R.; Gopinath, K.P.; Arun, J.; Kim, W.; Govarthanan, M. A critical review on the formation, fate and degradation of the persistent organic pollutant hexachlorocyclohexane in water systems and waste streams. *Chemosphere* **2021**, *271*, 129866. [CrossRef]

56. Bilal, M.; Adeel, M.; Rasheed, T.; Zhao, Y.; Iqbal, H.M.N. Emerging contaminants of high concern and their enzyme-assisted biodegradation—A review. *Environ. Int.* **2019**, *124*, 336–353. [CrossRef]

57. Hernandez-Vargas, G.; Sosa-Hernández, J.E.; Hernandez, S.; Villalba-Rodríguez, A.M.; Parra-Saldivar, R.; Iqbal, H.M.N. Electrochemical biosensors: A solution to pollution detection with reference to environmental contaminants. *Biosensors* **2018**, *8*, 29. [CrossRef]

58. Assis, G.C.; Skovroinski, E.; Leite, V.D.; Rodrigues, M.O.; Galembeck, A.; Alves, M.C.F.; Eastoe, J.; Oliveira, R.J. Conversion of "Waste Plastic" into Photocatalytic Nanofoams for Environmental Remediation. *Appl. Mater. Interfaces* **2018**, *10*, 8077–8085. [CrossRef]

59. Aziz, A.; Ali, N.; Khan, A.; Bilal, M.; Malik, S.; Khan, H. Chitosan-zinc sulfide nanoparticles, characterization and their photocatalytic degradation efficiency for azo dyes. *Int. J. Biol. Macromol.* **2020**, *153*, 502–512. [CrossRef]

60. Rawat, D.; Mishra, V.; Sharma, R. Detoxification of azo dyes in the context of environmental processes. *Chemosphere* **2016**, *155*, 591–605. [CrossRef]

61. Chu, D.; Zhu, S.; Wang, L.; Wang, G.; Zhang, N. Hydrothermal synthesis of hierarchical flower-like Zn-doped SnO$_2$ architectures with enhanced photocatalytic activity. *Mater. Lett.* **2018**, *224*, 92–95. [CrossRef]

62. Kim, S.; Kwang, H.-K.C.; Kim, B.; Hyun-Jong, K.; Lee, H.-N.; Park, T.J.; Park, Y.M. Highly Porous SnO$_2$/TiO$_2$ Heterojunction Thin-Film Photocatalyst Using Gas-Flow Thermal Evaporation and Atomic Layer Deposition. *Catalysts* **2021**, *11*, 1144. [CrossRef]

63. Ono, S. High-pressure phase transitions in SnO$_2$. *J. Appl. Phys.* **2005**, *97*, 073523. [CrossRef]

64. Lian, Y.; Huang, X.; Yu, J.; Tang, T.B.; Zhang, W.; Gu, M. Characterization of scrutinyite SnO$_2$ and investigation of the transformation with 119Sn NMR and complex impedance method. *AIP Adv.* **2018**, *8*, 125226. [CrossRef]

65. Lin, S.S.; Tsai, Y.; Bai, K. Structural and physical properties of tin oxide thin films for optoelectronic applications. *Appl. Surf. Sci.* **2016**, *380*, 203–209. [CrossRef]

66. Feng, B.; Feng, Y.; Qin, J.; Wang, Z.; Zhang, Y.; Du, F.; Zhao, Y.; Wei, J. Self-template synthesis of spherical mesoporous tin dioxide from tin-polyphenol-formaldehyde polymers for conductometric ethanol gas sensing. *Sens. Actuators B Chem.* **2021**, *341*, 12995. [CrossRef]
67. Noha, M.F.M.; Soha, M.F.; Tehb, C.H.; Lim, E.L.; Yap, C.C.; Ibrahima, M.A.; Ludina, N.A.; Teridi, M.A.M. Effect of temperature on the properties of SnO$_2$ layer fabricated via AACVD and its application in photoelectrochemical cells and organic photovoltaic devices. *Sol. Energy* **2017**, *158*, 474–482. [CrossRef]
68. Das, M.; Roy, S. Preparation, Characterization and Properties of Newly Synthesized SnO$_2$-Polycarbazole Nanocomposite via Room Temperature Solution Phase Synthesis Process. *Mater. Today Proc.* **2019**, *18*, 5438–5446. [CrossRef]
69. Muthusamy, S.; Charles, J. In situ synthesis of ternary prussian blue, hierarchical SnO$_2$ and polypyrrole by chemical oxidative polymerization and their sensing properties to volatile organic compounds. *Optik* **2021**, *241*, 166968. [CrossRef]
70. Mallik, A.; Roy, I.; Chalapathi, D.; Narayana, C.; Das, T.D.; Bhattacharya, A.; Bera, S.; Bhattacharya, S.; De, S.; Das, B.; et al. Single step synthesis of reduced graphene oxide/SnO$_2$ nanocomposites for potential optical and semiconductor applications. *Mater. Sci. Eng. B* **2021**, *264*, 114938. [CrossRef]
71. Housecroft, C.E.; Sharpe, A.G. *Química Inorgânica*, 4th ed.; LTC, 2013; Volume 1, p. 6. Available online: https://www.ifsc.edu.br/documents/35977/1670122/PPC+do+CST+em+Processos+Qu%C3%ADmicos/db5f1bdb-c035-471f-aa4f-8c8557f9a917 (accessed on 1 February 2022).
72. Stöwe, K.; Weber, M. Niobium, tantalum, and tungsten doped tin dioxides as potentialn support materials for fuel cell cata-lyst applications. *Z. Anorg. Allg. Chem.* **2020**, *646*, 1470–1480. [CrossRef]
73. Bolzan, A.A.; Fong, C.; Kennedy, B.; Howard, J.C. Structural Studies of Rutile-Type Metal Dioxides. *Acta Crystallogr. B* **1995**, *53*, 373–380. [CrossRef]
74. Yang, C.; Fan, Y.; Li, P.; Gu, Q.; Li, X. Freestanding 3-dimensional macro-porous SnO$_2$ electrodes for efficient electrochemical degradation of antibiotics in wastewater. *Chem. Eng. J.* **2021**, *422*, 130032. [CrossRef]
75. Liu, B.; Li, K.; Luo, Y.; Gao, L.; Duan, G. Sulfur spillover driven by charge transfer between AuPd alloys and SnO$_2$ allows high selectivity for dimethyl disulfide gas sensing. *Chem. Eng. J.* **2021**, *420*, 129881. [CrossRef]
76. Qiua, H.; Zhenga, H.; Jin, Y.; Yuana, Q.; Zhanga, X.; Zhao, C.; Wangb, H.; Jia, M. Mesoporous cubic SnO$_2$-CoO nanoparticles deposited on graphene as anode materials for sodium ion batteries. *J. Alloys Compd.* **2021**, *874*, 159967. [CrossRef]
77. Kim, M.J.; Kim, T.G. Fabrication of Metal-Deposited Indium Tin Oxides: Its Applications to 385 nm Light-Emitting Diodes. *Appl. Mater. Interfaces* **2016**, *8*, 5453–5457. [CrossRef]
78. Haq, S.; Rehman, W.; Waseem, M.; Shah, A.; Khan, A.R.; Rehman, M.; Ahmad, P.; Khan, B.; Ali, G. Green synthesis and characterization of tin dioxide nanoparticles for photocatalytic and antimicrobial studies. *Mater. Res. Express* **2020**, *7*, 025012. [CrossRef]
79. Hojamberdiev, M.; Czech, B.; Goktas, A.C.; Yubuta, K.; Kadirova, Z.C. SnO$_2$@ZnS photocatalyst with enhanced photocatalytic activity for the degradation of selected pharmaceuticals and personal care products in model wastewater. *J. Alloys Compd.* **2020**, *827*, 154339. [CrossRef]
80. Bezzerrouk, M.A.; Bousmaha, M.; Akriche, A.; Kharroubi, B.; M'hamed, G. Hybrid structure comprised of SnO$_2$, ZnO and Cu$_2$S thin film semiconductors with controlled optoelectric and photocatalytic properties. *Thin Solid Films* **2013**, *542*, 31–37.
81. Han, K.; Peng, X.-L.; Li, F.; Yao, M.-M. SnO$_2$ Composite Films for Enhanced Photocatalytic Activities. *Catalysts* **2018**, *8*, 453. [CrossRef]
82. Haya, S.; Brahmia, O.; Halimi, O.; Sebais, M.; Boudine, B. Sol–gel synthesis of Sr-doped SnO$_2$ thin films and their photocatalytic properties. *Mater. Res. Express* **2017**, *4*, 106406. [CrossRef]
83. Raj, I.L.P.; Revathy, M.S.; Christy, A.J.; Chidhambaram, N.; Ganesh, V.; AlFaify, S. Study on the synergistic effect of terbium-doped SnO$_2$ thin film photocatalysts for dye degradation. *J. Nanopart. Res.* **2020**, *22*, 359. [CrossRef]
84. Va-divel, S.; Rajarajan, G. Effect of Mg doping on structural, optical and photocatalytic activity of SnO$_2$ nanostructure thin films. *J. Mater. Sci. Mater. Electron.* **2015**, *26*, 3155–3162. [CrossRef]
85. Li, Q.; Zhao, H.; Sun, H.; Zhao, X.; Fan, W. Doubling the photocatalytic performance of SnO$_2$ by carbon coating mixed-phase particles. *RSC Adv.* **2018**, *8*, 30366–30373. [CrossRef]
86. Hermawan, A.; Asakura, Y.; Inada, M.; Yin, S. One-step synthesis of micro-/mesoporous SnO$_2$ spheres by solvothermal method for toluene gas sensor. *Ceram. Int.* **2019**, *45*, 15435–15444. [CrossRef]
87. Hermawan, A.; Asakura, Y.; Inada, M.; Yin, S. A facile method for preparation of uniformly decorated-spherical SnO$_2$ by CuO nanoparticles for highly responsive toluene detection at high temperature. *J. Mater. Sci. Technol.* **2020**, *51*, 119–129. [CrossRef]
88. Wang, X.; Fan, H.; Ren, P.; Li, M. Homogeneous SnO$_2$ core–shell microspheres: Microwave-assisted hydrothermal synthesis, morphology control and photocatalytic properties. *Mater. Res. Bull.* **2014**, *50*, 191–196. [CrossRef]
89. Rodrigues, E.C.P.E.; Olivi, P. Preparation and characterization of Sb-doped SnO$_2$ films with controlled stoichiometry from polymeric precursors. *J. Phys. Chem. Solids* **2003**, *64*, 1105–1112. [CrossRef]
90. Akram, M.; Saleh, A.T.; Ibrahim, W.A.W.; Awan, A.S.; Hussain, R. Continuous microwave flow synthesis (CMFS) of nano-sized tin oxide: Effect of precursor concentration. *Ceram. Int.* **2016**, *42*, 8613–8619. [CrossRef]
91. Fatimah, I.; Purwiandono, G.; Jauhari, H.M.; Aisyah, A.A.; Sagadevan, P.; Oh, W.-C.; Doong, R.-A. Synthesis and control of the morphology of SnO$_2$ nanoparticles via various concentrations of Tinospora cordifolia stem extract and reduction method. *Arab. J. Chem.* **2022**, *15*, 103738. [CrossRef]

92. Abdelkader, E.; Nadjia, L.; Naceur, B.; Noureddine, B. SnO$_2$ foam grain-shaped nanoparticles: Synthesis, characterization and UVA light induced photocatalysis. *J. Alloys Compd.* **2016**, *679*, 408–419. [CrossRef]
93. Najjar, M.; Hosseini, H.A.; Masoudi, A.; Sabouri, Z.; Mostafapour, A.; Khatami, M.; Darroudi, M. Green chemical approach for the synthesis of SnO$_2$ nanoparticles and its application in photocatalytic degradation of Eriochrome Black T dye. *Optik* **2021**, *242*, 167152. [CrossRef]
94. Kumar, M.; Mehta, A.; Mishra, A.; Singh, J.; Rawat, M.; Basu, S. Biosynthesis of tin oxide nanoparticles using Psidium Guajava leave extract for photocatalytic dye degradation under sunlight. *Mater. Lett.* **2018**, *215*, 121–124. [CrossRef]
95. Mahmood, H.; Khan, M.A.; Mohuddin, B.; Iqbal, T. Solution-phase growth of tin oxide (SnO$_2$) nanostructures: Structural, optical and photocatalytic properties. *Mater. Sci. Eng. B* **2020**, *258*, 114568. [CrossRef]
96. Wang, J.; Fan, H.-Q.; Yu, H.-W. Synthesis of Hierarchical SnO$_2$ Microflowers Assembled by Nanosheets and Their Enhanced Photocatalytic Properties. *Mater. Trans.* **2015**, *56*, 1911–1914. [CrossRef]
97. Costantino, F.; Armirotti, A.; Carzino, R.; Gavioli, L.; Athanassiou, A.; Fragouli, D. In situ formation of SnO$_2$ nanoparticles on cellulose acetate fibrous membranes for the photocatalytic degradation of organic dyes. *J. Photochem. Photobiol. A Chem.* **2020**, *398*, 112599. [CrossRef]
98. Mala, N.; Ravichandran, K.; Pandiarajan, S.; Srinivasan, N.; Ravikumar, B.; Nithiyadevi, K. Enhanced antibacterial and photocatalytic activity of (Mg+Co) doped tin oxide nanopowders synthesised using wet chemical route. *Mater. Technol.* **2017**, *32*, 1328082. [CrossRef]
99. Chu, L.; Fangfang Duo, F.; Zhang, M.; Wu, Z.; Sun, Y.; Wang, C.; Donga, S.; Sun, J. Doping induced enhanced photocatalytic performance of SnO$_2$:Bi^{3+}quantum dots toward organic pollutants. *Colloids Surf. A* **2020**, *589*, 124416. [CrossRef]
100. Ragupathy, S.; Manikandan, V.; Devanesan, S.; Ahmed, M.; Ramamoorthy, M.; Priyadharsan, A. Enhanced sun light driven photocatalytic activity of Co doped SnO$_2$ loaded corn cob activated carbon for methylene blue dye degradation. *Chemosphere* **2022**, *295*, 133848. [CrossRef]
101. Ragupathy, S.; Sathya, T. Photocatalytic decolorization of brilliant green by Ni doped SnO$_2$ nanoparticles. *J. Mater. Sci. Mater. Electron.* **2018**, *29*, 8710–8719. [CrossRef]
102. Soltan, W.B.; Ammar, S.; Olivie, C.; Toupance, T.; Soltan, W.B. Influence of zinc doping on the photocatalytic activity of nanocrystalline SnO$_2$ particles synthesized by the polyol method for enhanced degradation of organic dyes. *J. Alloys Compd.* **2017**, *729*, 638–647. [CrossRef]
103. Baig, A.; Ali Baig, A.; Rathinam, V.; Palaninathan, J. Fabrication of Zr-doped SnO$_2$ nanoparticles with synergistic influence for improved visible-light photocatalytic action and antibacterial performance. *Appl. Water Sci.* **2020**, *10*, 54. [CrossRef]
104. Borker, P.; Salker, A.; Gaokar, R.D. Sunlight driven improved photocatalytic activity of Mn doped SnO$_2$ nanowires. *Mater. Chem. Phys.* **2021**, *270*, 124797. [CrossRef]
105. Abdel-Messih, M.F.; Ahmed, M.A.; El-Sayed, A.S. Photocatalytic decolorization of Rhodamine B dye using novel mesoporous SnO$_2$–TiO$_2$ nano mixed oxides prepared by sol–gel method. *J. Photochem. Photobiol. A Chem.* **2013**, *260*, 1–8. [CrossRef]
106. Subashri, K.; Santhi, N. Synthesis and characterization of CrO$_4$-SnO$_2$ nanocomposites and their industrial and electrochemical applications. *Mater. Today Proc.* **2022**, *48*, 443–453. [CrossRef]
107. Wang, X.; He, Y.; Xu, L.; Xia, Y.; Gang, R. SnO$_2$ particles as efficient photocatalysts for organic dye degradation grown in-situ on g-C$_3$N$_4$ nanosheets by microwave-assisted hydrothermal method. *Mater. Sci. Semicond. Process.* **2021**, *121*, 105298. [CrossRef]
108. Onwudiwe, D.C.; Oyewo, O.A. Facile synthesis and structural characterization of zinc stannate/tin oxide and zinc stannate/tin composites for the removal of methylene blue from water. *Mater. Res. Express* **2019**, *6*, 125025. [CrossRef]
109. Akti, F. Photocatalytic degradation of remazol yellow using polyaniline–doped tin oxide hybrid photocatalysts with diatomite support. *Appl. Surf. Sci.* **2018**, *455*, 931–939. [CrossRef]
110. Palai, A.; Panda, N.R.; Sahu, D. Novel ZnO blended SnO$_2$ nanocatalysts exhibiting superior degradation of hazardous pollutants and enhanced visible photoemission properties. *J. Mol. Struct.* **2021**, *1244*, 131245. [CrossRef]
111. Lahmar, H.; Benamira, M.; Douafer, S.; Akika, F.Z.; Hamdi, M.; Avramova, I.; Trari, M. Photocatalytic degradation of Crystal Violet dye on the novel CuCr$_2$O$_4$/SnO$_2$ hetero-system under sunlight. *Optik* **2020**, *219*, 165042. [CrossRef]
112. Steplinpaulselvin, S.; Rajaram, R.; Silambarasan, T.S.; Chen, Y. Survival assessment of simple food webs for dye wastewater after photocatalytic degradation using SnO$_2$/GO nanocomposites under sunlight irradiation. *Sci. Total Environ.* **2020**, *721*, 137805.
113. Li, Z.; Zhao, Y.; Hua, L.; Bi, D.; Xie, J.; Zhou, Y. Hollow SnO$_2$/Zn$_2$SnO$_4$ cubes with porous shells towards n-butylamine sensing and photocatalytic applications. *Vacuum* **2020**, *182*, 109693. [CrossRef]
114. Rani, A.; Singh, K.; Patel, A.S.; Chakraborti, A.; Kumar, S.; Ghosh, K.; Sharma, P. Visible light driven photocatalysis of organic dyes using SnO$_2$ decorated MoS$_2$ nanocomposites. *Chem. Phys. Lett.* **2020**, *738*, 136874. [CrossRef]
115. Manjunatha, A.S.; Pavithra, N.S.; Shivanna, M.; Nagaraju, G.; Ravikumar, C.R. Synthesis of Citrus Limon mediated SnO$_2$-WO$_3$ nanocomposite: Applications to photocatalytic activity and electrochemical sensor. *J. Environ. Chem. Eng.* **2020**, *8*, 104500. [CrossRef]

Article

Methodology for Simultaneous Analysis of Photocatalytic deNO$_x$ Products

Jan Suchanek, Eva Vaneckova, Michal Dostal, Eliska Mikyskova, Libor Brabec, Radek Zouzelka and Jiri Rathousky *

Center for Innovations in the Field of Nanomaterials and Nanotechnologies, J. Heyrovsky Institute of Physical Chemistry of the CAS, Dolejskova 2155/3, 182 23 Prague, Czech Republic; jan.suchanek@jh-inst.cas.cz (J.S.); eva.vaneckova@jh-inst.cas.cz (E.V.); michal.dostal@jh-inst.cas.cz (M.D.); eliska.mikyskova@jh-inst.cas.cz (E.M.); libor.brabec@jh-inst.cas.cz (L.B.); radek.zouzelka@jh-inst.cas.cz (R.Z.)
* Correspondence: jiri.rathousky@jh-inst.cas.cz

Abstract: The ISO standard 22197-1:2016 used for the evaluation of the photocatalytic nitric oxide removal has a main drawback, which allows only the decrease of nitric oxide to be determined specifically. The remaining amount, expressed as "NO$_2$", is considered as a sum of HNO$_3$, HONO NO$_2$, and other nitrogen-containing species, which can be potentially formed during the photocatalytic reaction. Therefore, we developed a new methodology combining our custom-made analyzers, which can accurately determine the true NO$_2$ and HONO species, with the conventional NO one. Their function was validated via a photocatalytic experiment in which 100 ppbv of either NO or NO$_2$ dispersed in air passed over (3 L min^{-1}) an Aeroxide© TiO$_2$ P25 surface. The gas-phase analysis was complemented with the spectrophotometric determination of nitrates (NO_3^-) and/or nitrites (NO_2^-) deposited on the P25 layer. Importantly, an almost perfect mass balance (94%) of the photocatalytic NOx abatement was achieved. The use of custom-made analyzers enables to obtain (i) no interference, (ii) high sensitivity, (iii) good linearity in the relevant concentration range, (iv) rapid response, and (v) long-term stability. Therefore, our approach enables to reveal the reaction complexity and is highly recommended for the photocatalytic NOx testing.

Keywords: photocatalysis; air purification; NOx; HONO; NO$_2$; mass balance

Citation: Suchanek, J.; Vaneckova, E.; Dostal, M.; Mikyskova, E.; Brabec, L.; Zouzelka, R.; Rathousky, J. Methodology for Simultaneous Analysis of Photocatalytic deNO$_x$ Products. *Catalysts* **2022**, *12*, 661. https://doi.org/10.3390/catal12060661

Academic Editors: Gassan Hodaifa, Juan García Rodríguez, Antonio Zuorro, Joaquín R. Dominguez, José A. Peres and Zacharias Frontistis

Received: 9 May 2022
Accepted: 14 June 2022
Published: 16 June 2022

Publisher's Note: MDPI stays neutral with regard to jurisdictional claims in published maps and institutional affiliations.

1. Introduction

Nitrogen oxides (NOx) present in the air are considered to be toxic gases. There exist several strategies for NOx removal, focusing either on the emission control and prevention of their formation, or dealing with NOx conversion to N$_2$ or HNO$_3$ using additional chemical reagents [1]. While these methods cannot be used for NOx concentration of parts per billion (ppbv), heterogeneous photocatalysis offers such a possibility [2–5].

Because of the need to compare the performance of photocatalysts and assess their environmental applicability, the standard tests for the air-purification performance of semiconductor photocatalytic materials were developed, namely ISO 22197-1:2016, CEN (CEN/TS 16980-1, 2016), and UNI standards for nitric oxide (UNI 11247, 2010 and UNI 114874, 2013), further referred as ISO, CEN, and UNI.

Although these methods are well designed and allow the interlaboratory comparison of results in terms of photocatalytic activity, they are less meaningful for the prediction of the actual environmental impact under real-world conditions. The high input concentration of 500–1000 ppbv NO is one order of magnitude higher than that in heavily polluted air. Only NO is used as the model pollutant (except UNI 11247, 2010), although NO$_2$ is more toxic (its amount in air is restricted by specific limit). While the NOx analyzer (chemiluminescent) specified in ISO 7996 measures only NO selectively, NO$_2$ measurement suffers from the interference of other N-containing gases, including HONO and HNO$_3$.

Thus, to evaluate the real impact of NO_x photocatalytic abatement technology, a complete analysis of the degradation intermediates and products even at their very low concentration is of utmost importance. Such analytical data will enable to proceed with a complete balance of NO_x compounds. This is necessary for the realistic assessment of the performance of photocatalytic technology.

To address this issue, several strategies were reported (Table S1). Ifang et al. [6] suggested testing the photocatalytic activity under relevant atmospheric conditions and low reactant concentrations (<100 ppbv), and strongly recommended NO_2 as a tested reactant. According to the authors, the yield of nitrate formed by the photooxidation of NO_2 should be quantified by ion chromatography to exclude or confirm the formation of other N-containing products. The interference issues of chemiluminescence analyzers can be solved by using carbonate denuder as a trap for HONO [7].

Besides the common detection of NO and NO_2, several papers have dealt with the detection of other N-containing species. A chemiluminescence method for the detection of HONO was reported [8]). Alternatively, HONO was detected by employing the long-path absorption photometer (LOPAP) method [9,10] developed by Heland et al. [11]. The detection of NO_2, HONO, and N_2O by the combination of selective catalytic conversion measurement, LOPAP, and the gas chromatography/electron-capture detector, respectively, was published [12]. Furthermore, these studies also focused on the determination of nitrate (NO_3^-) and nitrite (NO_2^-) anions adsorbed on the surfaces using ion chromatography. N_2O detected by GC-MS with cryotrapping was found to be the main product from the NO reactant; however, the NO concentration was 100 ppmv, i.e., very high [13].

The analysis of the all-important N-compounds (NO, NO_2, HONO, HNO_3, and N_2O) needed for the overall evaluation of photocatalyst performance is still exceptional within the research community. Dedicated research dealing with possible solutions of complete NO_x degradation monitoring and photocatalytic activity assessment is of great value.

Therefore, we developed a methodology for the continuous monitoring of N-species participating in photocatalytic NO_x abatement. Our novel complex apparatus is partly based on the ISO standard (NO analyzer, bed-flow reactor, total reactor flow, light intensity, humidity) in order to enable interlaboratory results; however, we have supplemented it with very sensitive and selective NO_2 and HONO analyzers. After optimization procedures, the determination of the full balance of N-products ($N^{+2,3,4,5}$) was carried out, including those adsorbed in the photocatalyst (HNO_3 and HONO). This enabled us to assess more realistically the possible environmental impact of the air purification by means of photocatalysts.

2. Results and Discussion

2.1. Requirements to the Apparatus

There are a number of requirements which should be fulfilled. We show that our analytical system used in the developed experimental set-up fulfills following requirements:

1. No interference with each other;
2. Sufficient sensitivity achieving low detection limit;
3. Good linearity of the response in the needed concentration range;
4. Rapid response;
5. Long-term stability–suppressed zero drift;
6. Reasonable investment and operating costs.

Ad 1: Analyzers connected in parallel must not interfere with each other. Specifically, the analyzers must not exhibit chemical interferences towards other N-containing species. It is desirable to test independently the proper function of each analyzer and perform their calibration. This is realized by a system of three-way valves, as specified below. The whole apparatus tightness is crucial due the toxic gases used, and any leaks must be revealed easily. Therefore, the system should be well-arranged and sufficiently easy to operate. Apparatus materials have to be inert for all the used chemicals, and the tubes should be as short as possible to minimize the sorption of the gases. Chemiluminescence analyzers

(HONO and NO_2) were tuned to be interference-free [14–16]. The NO_2 analyzer showed no detectable response to 1.0 ppmv of NO, and the HONO analyzer showed no detectable response both to 100 ppbv NO_2 and NO.

Ad 2: Due to the very low concentrations to be monitored, a high sensitivity of analyzers is essential. Techniques based on chemiluminescence detection fulfill this requirement very well. The advantages of chemiluminescence techniques lie in their selectivity as well as very good sensitivity. The sensitivity (and the dynamic range) can be optimized by the composition of the chemiluminescence solutions as well as by the gas flow through the analyzer.

The detection limit ought to be in units of ppbv, as it usually occurs in natural air. The nominal detection limits of the NO_2 and HONO analyzers according to the producers are much lower than is needed for our application [14–16]. However, the detection limits determined in our laboratory with the analyzers connected to the measurement system were much higher: 6.0 ppbv for NO_2 and 1.0 ppbv for HONO. Meaningful reasons may be the additional instabilities caused by parallel sampling and the additional adsorption of gases within the tubes. The limit of detection was estimated as the concentration at which the signal would be equal three-times the standard deviation of the background signal.

Ad 3 and 4: For the possible application in kinetic studies, it is crucial that the analyzers respond rapidly with a sufficient frequency of data collection at least in one-second intervals. Furthermore, the response needs to be linear over the range of expected concentrations ranged from 1.0 to 1000 ppbv. For the HONO analyzer, a good linearity was found in the tested range 2.0–300 ppbv. The NO_2 analyzer performed a linear trend for concentrations ranging from 10 to 1000 ppbv. Below 10 ppbv, nonlinearity was observed, as was previously described in the literature [15].

Ad 5: A critical parameter for the applicability of the selected techniques is their long-term stability. For the chemiluminescence custom-made analyzers, it was tested with following concentrations: 100 ppbv for NO_2 and 34 µg L^{-1} for NO_2^- (corresponding to 12 ppbv of HONO in the air under given experimental conditions). The dosing of the chemicals by peristaltic pumps may be unstable for longer time periods. Therefore, for NO_2, the stability was evaluated by observing the averages of the signals by periods of 10 min (122 points) within 10 h, with the standard deviation of those averages being less than 1%. For HONO, the signal was averaged for 5 min before and after 8 h of the photocatalytic experiment (signal obtained from standard solution of NO_2^-), the difference being 1.5%. Both custom-made analyzers showed acceptable stability during a time period of at least 8 h, which is sufficient to ensure a reliable assessment of the photocatalytic activity.

Ad 6: The investment needed for assembling the custom-made analyzers is given mainly by the cost of the photomultiplier and the pumps (for liquid reagents and gas sampling). From current prices of these components, the estimated cost is below EUR 4000. This is roughly five times lower than the cost of a commercial NO_2 analyzer. Currently, commercial analyzers for HONO have not been developed.

2.2. Photocatalytic Reactor and Analytical System

Our self-developed set-up consisted of three main parts: a gas supply manifold, a photoreactor, and an analytical system (Figure 1). A set of mass flow controllers (Bronkhorst, The Netherlands) was used to dilute the pollutant (50 ppmv NO or NO_2 in N_2, Messer Technogas, Prague, Czech Republic) to the desired concentrations of 100 and 1000 ppbv. The carrier gas (synthetic air) was humidified using a Drechsel bottle. Stream of the gas passed through the photoreactor (ISO 22197-1, 2016) over a photocatalyst layer. Specifically, the following testing conditions have been retained: total flow rate 3 L min^{-1}, relative humidity 50% at 25 °C, radiation intensity 1.0 mW cm^{-2} of UVA light (TL-D 15W BLB, Philips, Prague, Czech Republic) with a dominant wavelength at 365 nm, a geometric area of the layer 5×10 cm^2, and the space height of 0.5 cm above the layer for the flowing gas. The output of the reactor was divided into branches leading to the analyzers for HONO,

NO$_2$, and NO$_x$/NO monitoring (Figure 1). These are described in the Supplementary Materials (Figures S1 and S2).

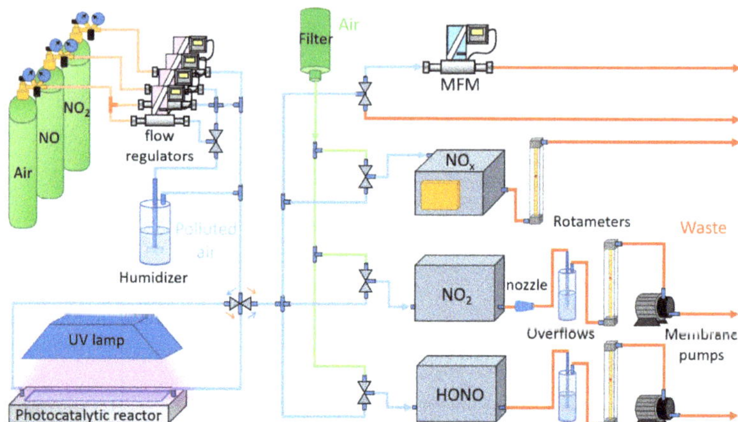

Figure 1. Set-up for testing the performance of photocatalysts for NOx abatement equipped with NO/NO$_x$, NO$_2$, and HONO analyzers.

Downstream of the reactor, the gas is simultaneously distributed to the analyzers. They actively suck the gas by pumps, which are protected by liquid overflow bottles placed in front of the pumps. The exhaust tubes downstream of the pumps are equipped by one-way valves, preventing the return of the exhaust gases. The inlet to each analyzer can be switched by a three-way ball valve between the analyzed gas and air from the lab drawn through a filter. The analyzers can thus be tested individually. The excess gas goes either directly to waste or can be passed by a 3-way ball valve through a capillary mass-flow meter (MFM) to test the leakages in the segment of the system, including the photoreactor. The bubbler has to be first bypassed to avoid damaging the MFM by water vapor.

2.3. Determination of Products in Gaseous Phase

The deNOx curves were integrated based on the formula in Equation (1), as follows:

$$deNOx = \frac{p(qx_{NOx\ in}t - \int_{t=0}^{t} qx_{NOx}dt)}{RT} \tag{1}$$

where p is normal gas pressure (101.325 kPa), q is the total gas flow (m^3 s^{-1}), t is the time of photocatalysis (s), $qx_{NOx\ in}$ is a molar fraction of the pollutant in the input gas mixture, x_{NOx} is the measured instantaneous NO$_x$ molar fraction in the gas mixture behind the photoreactor, R is the molar gas constant (8.314 J.K^{-1} mol^{-1}), and T is temperature (273.15 K). The gas conditions are given from the mass flow controllers, where the controlled volume flow is given under normal conditions.

2.4. Quantification of NO$_2^-$ and NO$_3^-$ Anions (HONO and HNO$_3$) Adsorbed on the Photocatalyst

Nitrite and nitrate anions accumulated on the layer during the photocatalytic reaction were determined by means of an ultralow-range colorimeter (HI764 Nitrite Checker, Hanna Instruments, Prague, Czech Republic) and corresponding reagents. After a 24 h extraction of the anions to an exact volume (60 or 80 mL) of deionized water, a proper dose of diazotation reagents (Hanna Instruments, Prague, Czech Republic) was added to the solution and stirred carefully. In the presence of NO$_2^-$ anions, the solution became a shade of pink, depending on the nitrite concentration. NO$_3^-$ anions were first reduced to NO$_2^-$ by cadmium (reagents NitraVer6, Hach Lange, GmbH, Berlin, Germany) and, afterwards, analyzed by the equal procedure as for NO_2^-. Concentration values obtained for the nitrites were determined using the known $NaNO_2$ solution concentration, and those for nitrates were adjusted according to the measurement of the known KNO_3 solution concentration. In several cases, the solutions with the extracted nitrites/nitrates were analyzed spectrophotometrically in a commercial lab for comparison, and the results were in the reasonable agreement. Always fresh layers were used for these experiments.

2.5. Modified ISO Test Conditions

Unlike the standardized methods, we focused on the following experimental conditions. First, the inlet concentration of the model NO or NO_2 pollutant was adjusted to 100 ppbv, which corresponds to realistic environmental conditions. Due to simple reaction stoichiometry, the sum of the concentrations of all assumed gaseous N-compounds (NO, NO_2, HONO, HNO_3, but not N_2O) present in the air flow was expected to be the same as the initial concentration of reactant. The standard total gas flow (3 L min^{-1}) through the reactor maintained unmodified.

Second, the duration of the photocatalytic test was longer. Our experiments showed that for highly active samples, the steady-state was not reached even after five hours. Third, three analyzers were used, in contrast to the only one in the ISO standard.

3. Validation

3.1. Determination of Products in Gaseous Phase

To verify the proper functioning of the apparatus, the photocatalytic reaction of NO_2 was tested on the TiO_2 (P25) layer (49.1 mg). A model NO_2 pollutant with a concentration of 100 ppbv in the air with 50% relative humidity flowed through the reactor at a flow rate of 3 L min^{-1}. After purging the reactor and stabilizing the input concentration, the layer was irradiated with a UV lamp of an irradiation intensity of 1.0 mW cm^{-2}. At the very beginning of the photocatalytic reaction, the NO_2 concentration immediately sharply dropped to half of its initial value, while neither HONO nor NO were formed in the gas phase (Figure 2A). We can deduce that all the removed NO_2 was adsorbed onto the surface of the photocatalyst, mainly in the form of NO_2 and HNO_3. This explains the subsequent partial deactivation of photocatalyst in the first one hour [17]. After this period, the reaction achieved a steady-state character with a negligible change of NO_2 concentration (0.3 ppbv per hour).

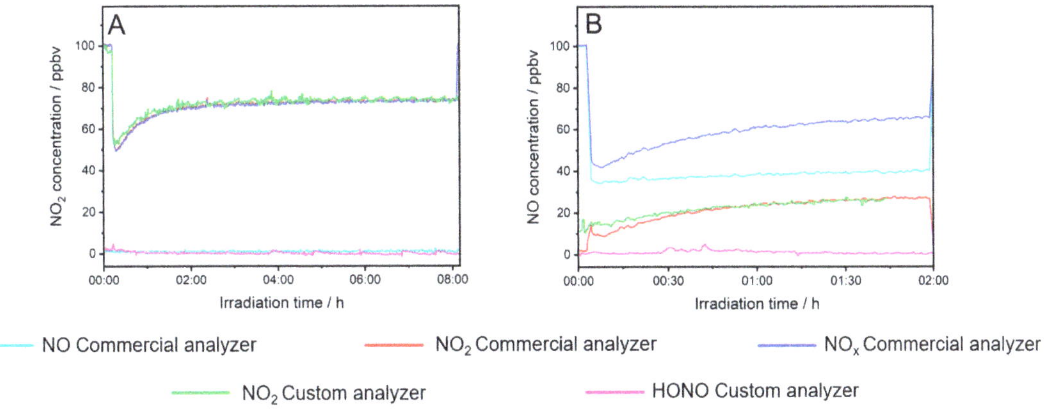

Figure 2. Set-up for testing the performance of photocatalysts for NOx abatement equipped with NO/NO$_x$, NO$_2$, and HONO analyzers. A 100 ppbv inlet concentration of NO$_2$ (**A**) and NO (**B**), respectively.

Measurement via a commercial analyzer (APNA-370, Horiba, Tokyo, Japan) was complemented with the selective measurement of NO_2 and HONO using our custom-made analyzers. In this case, the NO_2 measurements are in excellent agreement with the values obtained by the commercial APNA-370 (Figure 2A). This agreement can be associated with the negligible formation of gaseous HONO and HNO_3, which would otherwise contribute to the NO_2 and NOx signals of the APNA-370 analyzer.

Similar experiments were performed for 100 ppbv of NO while maintaining the same reaction conditions. The decrease in the NO concentration in the air to 40 ppbv caused by oxidation was accompanied by the gradual rise of the NO_2 concentration (Figure 2B). The NO_2 concentration measured by the custom-made NO_2 analyzer (Figure 2B, green line) is again in very good agreement with the NO_2 values obtained from the commercial one (Figure 2B, red line). A tiny discrepancy observed at the beginning of the experiment (up to 30 min) can be explained by the calibration of the custom-made NO_2 analyzer below 10 ppbv due to a nonlinear behavior, as well as to the detection limit (6.0 ppbv). Moreover, a negligible concentration of gaseous HONO was detected during the photocatalytic abatement of NO.

3.2. Determination of Products in Solid Phase

HNO_3 and HONO formation and their presence on the P25 layer were confirmed by the extraction, described in the Experimental section. Amounts of HONO were at least ten times smaller than those of HNO_3, since nitrite anions were mostly oxidized to nitrate during the layer transfer from the reactor and following the one-day extraction of anions into water. Commercial analysis by spectrophotometry found the nitrite anions to be three orders of magnitude less than the nitrate ones.

3.3. NOx Balance

Table 1 shows the comparison of the amount of NO_x removed from the gas stream and the HNO_3 deposited on the photocatalyst surface. From these data, the percentage imbalance p between the inlet, outlet, and deposited amount was evaluated according to Equation (2):

$$p = \left[\frac{(O + D) - I}{(O + D)} \right] \times 100 \ [\%] \tag{2}$$

where O is outlet, D is deposited, and I is the inlet amount.

Table 1. A comparison of the amount of NO_x removed from the gas stream and HNO_3 deposited on the photocatalyst surface.

Pollutant	Irradiation Intensity/ mW cm^{-2}	Inlet Concentration/ ppbv	Total NO_x Passed over Photocatalyst [1]/ μmol	NO_x Removed from Gas Phase [2]/ μmol	HNO_3 Deposited on Photocatalyst [3]/ μmol
NO	0	100	3.68	0	<0.35
NO	1.0	100	3.68	0.65	0.90
NO$_2$	1.0	100	3.68	1.10	1.70
NO	1.0	1000	36.8	1.20	1.30

[1] in 5 h. [2] determined via integration of analyzer signal (after 5 h on stream). [3] determined via spectrophotometry of solid phase extraction (after 5 h on stream).

We observed that the amount of NO_x removed from the flowing gas determined by the integration of the analyzer signal is slightly lower than the amount of HNO_3 deposits (Table 1). For 100 ppbv of NO, and especially for 1000 ppbv NO, this imbalance is low, being 6 per cent, and practically negligible for the former and latter case, respectively. However, for 100 ppbv NO_2 in the inlet gas, the imbalance is higher, achieving about 14 per cent.

There exist several potential reasons for this difference. First, a shift of the analyzer output (e.g., due to a baseline instability). However, extensive tests showed that this is not the case, as the performance of the analyzer is stable and reproducible. Moreover, the tightness of the set-up was checked by a helium test, and potential leaks can be completely excluded. Further, the sorption of NO_x on the walls of the set-up did not have any effect on the outcomes of the experiments, as the tests themselves were started after an equilibrium was achieved. At the same time, a potential content of nitrates or nitrites in the purchased TiO_2 powder was excluded by analytical measurement.

Second, the reason for the higher nitrate amount deposited on the photocatalyst may be the formation of nitrate by the oxidation of nitrogen in the carrier air on TiO_2. This speculation is supported by Yuan et al. [18], who observed such oxidation even by sunlight. Interestingly, we tested the possibility of a heterogeneous reaction of NO with the surface photocatalyst in the darkness at an inlet concentration of 100 ppbv; however, no formation of nitrates was detected.

Regarding the selection of either NO or NO_2 for the standard testing of the photocatalyst performance, an interesting conclusion follows. NO_2 seems more suitable due to its higher toxicity, and especially as a pollutant whose concentration is monitored according to relevant national limits and WHO guidelines. However, the imbalance given above shows that it is a rather problematic compound due to its chemical properties, such as complex sorption on the surface of not only the photocatalyst but also of other parts of the experimental set-up in contact with the polluted air. These complex relations were observed by Araña et al. [19], who determined the role of its sorption or disproportionation in producing NO. Another observed process was the formation of stable $[(NO_3^-)\text{-}(H_2O)\text{-}NO_2]$ complexes on the surface of the photocatalyst.

4. Preparation of TiO$_2$ Layers

For photocatalytic testing the commercial Aeroxide P25 (Evonik Industries, Essen, Germany, batch No. 616000298) was chosen. The surface of 5×10 cm^2 glass substrate was abraded by silicon carbide of grits 220 (Top Geo Mineralienhandel GmbH, Satteldorf, Germany) to increase the adhesion of ordinary soda-lime glass. Water suspension with 10 wt.% TiO_2 was homogenized by dispersing instrument T25 (IKA, Staufen, Germany) for 5 min at 20,000 rpm. This suspension was applied on the glass plate using an airbrush (PME, Enfield, UK) in multiple layers to achieve the required sample weight (50 mg). After spraying, each individual layer was dried with a blow dryer. All layers were stored in a dark chamber under the nitrogen atmosphere. Figure 3 confirms the homogeneity of deposited layers.

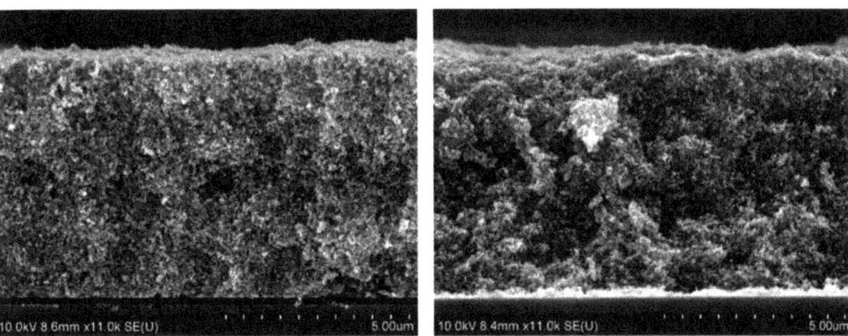

Figure 3. SEM micrographs of multiple P25 layer profiles. Left side–fivefold spraying. Right side–elevenfold spraying. In both, the same amount of TiO_2 deposited, and consequently the thickness of the multilayer is the same.

5. Conclusions

We have developed a new methodology for a broader assessment of the photocatalytic removal of NOx species in the air. Our custom-made analyzers, complemented with the conventional one, exhibited high performance in the simultaneous detection of the true NO_2, HONO, and NO species during the photocatalytic NOx degradation. This innovative set-up enables to reliably determine a mass balance of the NO_x abatement when the analysis of deposits on the photocatalyst surface is involved. By our approach, almost perfect mass balance for NO (94%) and NO_2 (86%) removal was achieved. Furthermore, our study provides important suggestions concerning the suitability of NO and NO_2 as test molecules, with NO being more suitable. As a next step of our research, the upgrade of the photocatalytic apparatus by the implementation of a custom-made HNO_3 analyzer will follow.

Supplementary Materials: The following are available online at https://www.mdpi.com/article/10.3390/catal12060661/s1, Figure S1: Custom-made analytical system for HONO detection in air, designed and assembled by Mikuska and co-workers, Figure S2: Custom-made analytical system for NO_2 detection in air, designed and assembled by Mikuska and co-workers. Table S1: Overview of the analytical systems used to determine gaseous nitrogen-based species and ozone. Ref. [20] is cited in Supplementary Materials.

Author Contributions: Conceptualization, J.R. and E.V.; methodology, J.S., E.V. and R.Z.; validation, L.B. and E.M.; investigation, J.S., E.V. and E.M.; resources, J.R.; writing—original draft preparation, J.S., E.V., M.D. and L.B.; writing—review and editing, J.R. and R.Z.; supervision, J.R. and L.B.; project administration, L.B.; funding acquisition, J.R. All authors have read and agreed to the published version of the manuscript.

Funding: This research was funded by the Czech Science Foundation, project no. 19-12109S.

Data Availability Statement: All raw- and metadata are archived and are available on request.

Acknowledgments: The authors acknowledge the assistance provided by the Research Infrastructures NanoEnviCz (project no. LM2018124) supported by the Ministry of Education, Youth and Sports of the Czech Republic, and the project Pro-NanoEnviCz (reg. no. CZ.02.1.01/0.0/0.0/16_013/0001821) supported by the Ministry of Education, Youth and Sports of the Czech Republic and the European Union–European Structural and Investments Funds in the frame of the Operational Programme Research Development and Education. Further, the research was supported by the Academy of Sciences of the Czech Republic within the program Strategy AV21 no. 23—"The City as a Laboratory of Change; buildings, cultural heritage and an environment for a safe and valuable life". Finally, R. Zouzelka thanks the Academy of Sciences of the Czech Republic for the financial support (project no. MSM200402101).

Conflicts of Interest: The authors declare no conflict of interest. The funders had no role in the design of the study; in the collection, analyses, or interpretation of data; in the writing of the manuscript, or in the decision to publish the results.

References

1. Skalska, K.; Miller, J.S.; Ledakowicz, S. Trends in NOx abatement: A review. *Sci. Total Environ.* **2010**, *408*, 3976–3989. [CrossRef] [PubMed]
2. Muscetta, M.; Russo, D. Photocatalytic applications in wastewater and air treatment: A patent review (2010–2020). *Catalysts* **2021**, *11*, 834. [CrossRef]
3. Chen, H.; Nanayakkara, C.E.; Grassian, V.H. Titanium dioxide photocatalysis in atmospheric chemistry. *Chem. Rev.* **2012**, *112*, 5919–5948. [CrossRef] [PubMed]
4. He, F.; Jeon, W.; Choi, W. Photocatalytic air purification mimicking the self-cleaning process of the atmosphere. *Nat. Commun.* **2021**, *12*, 10–13. [CrossRef] [PubMed]
5. Jiang, Z.; Yu, X. Impact of Visible-Solar-Light-Driven photocatalytic pavement on air quality improvement. *Transp. Res. Part D Transp. Environ.* **2020**, *84*, 102341. [CrossRef]
6. Ifang, S.; Gallus, M.; Liedtke, S.; Kurtenbach, R.; Wiesen, P.; Kleffmann, J. Standardization methods for testing photo-catalytic air remediation materials: Problems and solution. *Atmos. Environ.* **2014**, *91*, 154–161. [CrossRef]
7. Monge, M.E.; D'Anna, B.; George, C. Nitrogen dioxide removal and nitrous acid formation on titanium oxide surfaces—an air quality remediation process? *Phys. Chem. Chem. Phys.* **2010**, *12*, 8991. [CrossRef] [PubMed]
8. George, C.; Beeldens, A.; Barmpas, F.; Doussin, J.F.; Manganelli, G.; Herrmann, H.; Kleffmann, J.; Mellouki, A. Impact of photocatalytic remediation of pollutants on urban air quality. *Front. Environ. Sci. Eng.* **2016**, *10*, 2. [CrossRef]
9. Gandolfo, A.; Bartolomei, V.; Gomez Alvarez, E.; Tlili, S.; Gligorovski, S.; Kleffmann, J.; Wortham, H. The effectiveness of indoor photocatalytic paints on NOx and HONO levels. *Appl. Catal. B Environ.* **2015**, *166–167*, 84–90. [CrossRef]

10. Mothes, F.; Ifang, S.; Gallus, M.; Golly, B.; Boréave, A.; Kurtenbach, R.; Kleffmann, J.; George, C.; Herrmann, H. Bed flow photoreactor experiments to assess the photocatalytic nitrogen oxides abatement under simulated atmospheric conditions. *Appl. Catal. B Environ.* **2018**, *231*, 161–172. [CrossRef]
11. Heland, J.; Kleffmann, J.; Kurtenbach, R.; Wiesen, P. A new instrument to measure gaseous nitrous acid (HONO) in the atmosphere. *Environ. Sci. Technol.* **2001**, *35*, 3207–3212. [CrossRef] [PubMed]
12. Laufs, S.; Burgeth, G.; Duttlinger, W.; Kurtenbach, R.; Maban, M.; Thomas, C.; Wiesen, P.; Kleffmann, J. Conversion of nitrogen oxides on commercial photocatalytic dispersion paints. *Atmos. Environ.* **2010**, *44*, 2341–2349. [CrossRef]
13. Lu, W.; Olaitan, A.D.; Brantley, M.R.; Zekavat, B.; Erdogan, D.A.; Ozensoy, E.; Solouki, T. Photocatalytic Conversion of Nitric Oxide on Titanium Dioxide: Cryotrapping of Reaction Products for Online Monitoring by Mass Spectrometry. *J. Phys. Chem. C* **2016**, *120*, 8056–8067. [CrossRef]
14. Mikuska, P.; Motyka, K.; Vecera, Z. Determination of nitrous acid in air using wet effluent diffusion denuder–FIA technique. *Talanta* **2008**, *77*, 635–641. [CrossRef]
15. Mikuška, P.; Večeřa, Z. Effect of complexones and tensides on selectivity of nitrogen dioxide determination in air with a chemiluminescence aerosol detector. *Anal. Chim. Acta* **2000**, *410*, 159–165. [CrossRef]
16. Mikuska, P.; Vecera, Z. Determination of nitrogen dioxide with a chemiluminescent aerosol detector. *Anal. Chem.* **1992**, *64*, 2187–2191. [CrossRef]
17. Han, R.; Andrews, R.; O'Rourke, C.; Hodgen, S.; Mills, A. Photocatalytic air purification: Effect of HNO_3 accumulation on NOx and VOC removal. *Catal. Today* **2021**, *380*, 105–113. [CrossRef]
18. Yuan, S.-J.; Chen, J.-J.; Lin, Z.-Q.; Li, W.-W.; Sheng, G.-P.; Yu, H.-Q. Nitrate formation from atmospheric nitrogen and oxygen photocatalysed by nano-sized titanium dioxide. *Nat. Commun.* **2013**, *4*, 2249. [CrossRef] [PubMed]
19. Araña, J.; Sousa, D.G.; Díaz, O.G.; Melián, E.P.; Rodríguez, J.D. Effect of NO_2 and NO_3^-/HNO_3 adsorption on NO photocatalytic conversion. *Appl. Catal. B* **2019**, *244*, 660–670. [CrossRef]
20. Schiff, H.I.; Mackay, G.I.; Castledine, C.; Harris, G.W.; Tran, Q. Atmospheric measurements of nitrogen dioxide with a sensitive luminol instrument. In *Acidic Precipitation*; Springer: Berlin/Heidelberg, Germany, 1986; pp. 105–114.

water

Article

Arsenite to Arsenate Oxidation and Water Disinfection via Solar Heterogeneous Photocatalysis: A Kinetic and Statistical Approach

Felipe de J. Silerio-Vázquez [1], **Cynthia M. Núñez-Núñez** [2], **José B. Proal-Nájera** [3,*] **and María T. Alarcón-Herrera** [1,*]

[1]   Departamento de Ingeniería Sustentable, Centro de Investigación en Materiales Avanzados, S.C. Calle CIMAV 110, Colonia 15 de Mayo, Durango 34147, Mexico

[2]   Ingeniería en Tecnología Ambiental, Universidad Politécnica de Durango, Carretera Durango-México km 9.5, Durango 34300, Mexico

[3]   CIIDIR-Durango, Instituto Politécnico Nacional, Calle Sigma 119, Fraccionamiento 20 de Noviembre II, Durango 34220, Mexico

*   Correspondence: jproal@ipn.mx (J.B.P.-N.); teresa.alarcon@cimav.edu.mx (M.T.A.-H.);
     Tel.: +52-618-1341781 (J.B.P.-N.); +52-614-4394896 (M.T.A.-H.)

Citation: Silerio-Vázquez, F.d.J.; Núñez-Núñez, C.M.; Proal-Nájera, J.B.; Alarcón-Herrera, M.T. Arsenite to Arsenate Oxidation and Water Disinfection via Solar Heterogeneous Photocatalysis: A Kinetic and Statistical Approach. *Water* **2022**, *14*, 2450. https://doi.org/10.3390/w14152450

Academic Editors: Antonio Zuorro, Zacharias Frontistis, José A. Peres, Gassan Hodaifa, Joaquín R. Dominguez and Juan García Rodríguez

Received: 6 July 2022
Accepted: 5 August 2022
Published: 8 August 2022

Publisher's Note: MDPI stays neutral with regard to jurisdictional claims in published maps and institutional affiliations.

Abstract: Arsenic (As) poses a threat to human health. In 2014, more than 200 million people faced arsenic exposure through drinking water, as estimated by the World Health Organization. Additionally, it is estimated that drinking water with proper microbiological quality is unavailable for more than 1 billion people. The present work analyzed a solar heterogeneous photocatalytic (HP) process for arsenite (As^{III}) oxidation and coliform disinfection from a real groundwater matrix employing two reactors, a flat plate reactor (FPR) and a compound parabolic collector (CPC), with and without added hydrogen peroxide (H_2O_2). The pseudo first-order reaction model fitted well to the As oxidation data. The treatments FPR–HP + H_2O_2 and CPC–HP + H_2O_2 yielded the best oxidation rates, which were over 90%. These treatments also exhibited the highest reaction rate constants, 6.7×10^{-3} min^{-1} and 6.8×10^{-3} min^{-1}, respectively. The arsenic removal rates via chemical precipitation reached 98.6% and 98.7% for these treatments. Additionally, no coliforms were detected at the end of the process. The collector area per order (A_{CO}) for HP treatments was on average 75% more efficient than photooxidation (PO) treatments. The effects of the process independent variables, H_2O_2 addition, and light irradiation were statistically significant for the As^{III} oxidation reaction rate ($p < 0.05$).

Keywords: groundwater; real water matrix; reactor prototype; arsenic speciation; collector area per order

1. Introduction

The abundance of arsenic (As) in the Earth's crust is around 1.5–3 mg/kg, making it the 20th most abundant element and a component of more than 245 minerals [1]. When the groundwater's pH and bicarbonate anion (HCO_3^-) concentration are high, As is easily dissolved and passes to the groundwater cycle [2]. As toxicity affects all body systems, causing both acute and chronic poisoning [3]. Acute exposure is rare and happens mostly with exposure to arsenite (As^{III}) than arsenate (As^V) [4]. Long-term exposure leads to a variety of illnesses known as arsenicosis [5], which includes skin, bladder, kidney, and lung cancer, along with black foot disease [6]. As^{III} is more toxic than As^V [7] and also harder to remove from water [8].

With nearly 1 billion people exposed to arsenic by food, and more than 200 million people exposed to it via drinking water [9–11], As is a serious threat to the physical, social, and economic well-being of people, affecting especially the population of developing countries [12]. Countries affected by high arsenic concentrations in groundwater include

Argentina, Australia, Bangladesh, China, Chile, Mexico, India, New Zealand, Nepal, USA, Vietnam, and Taiwan [13].

In addition, the typical and harmful pollutants in developing countries are of biological origin, as diseases present in almost 50% of their populations are associated with water, both for supply and sanitation [14]. Worldwide, according to estimations, safe drinking water is unavailable for 1.1 billion people; water scarcity is suffered by 2.7 billion people, and 5 million people die each year due to waterborne infections [15,16]. These infections can be caused by viruses, bacteria, or protozoa [17].

The World Health Organization (WHO) recommends that the As amount in water should not surpass 10 μg/L [18]; as for microbiological water quality, the WHO recommends the use of the following microbial water quality indicators: total coliforms, thermotolerant coliforms, *Escherichia coli*, intestinal enterococci, enteric virus, and coliphage virus (none of which should be detected in drinking water) [19].

The growing population and climate change are two of the main factors that are increasing the demand for drinking water; it is then a priority to research drinking water treatments to improve processes in terms of reliability, efficiency, safeness, and ease of implementation [20]. Additionally, economic feasibility, technical viability, and environmental safeness must be complied with for a technology to be considered sustainable [21].

Recently, the scientific community has shown extensive interest in advanced oxidation processes (AOPs), considering them as the most promising technologies for the potabilization of water and the treatment of wastewater on account of the nonselectivity of reactive oxidizing species (ROS), enabling AOPs to remove pollutants, including microbes and organic and inorganic contaminants [22].

Heterogeneous photocatalysis (HP) with semiconductors (or photocatalysts) is an AOP developed in 1972 with several advantages, including its ability to use solar light and that fact that is environmentally friendly and has a relatively low cost [23–25]. When the photocatalyst is irradiated with light whose energy is higher than the photocatalyst bandgap energy level, electrons in the valence band (VB) migrate to the conduction band (CB), generating a positive hole (h^+) and an extra electron (e^-) in the VB and CB, respectively [26]. Oxygen adsorbed on the photocatalyst surface can react with e^- to form superoxide radicals ($O_2^{\bullet-}$), while water can react with h^+ to generate hydroxyl radicals ($HO^\bullet$) [27].

HP with titanium dioxide (TiO_2) has been investigated for As oxidation, which is a good approach as As^{III} is found as a neutral charge oxyanion in a wide pH range, and, in contrast to other oxyanions, adsorption onto metal oxides or clays is inefficient, and precipitation at near neutral pH barely occurs [28]. Although As^V is a triprotic acid and can be found in several forms depending of the medium pH, its removal from water is easier with processes such as chemical precipitation [29] and adsorption [30].

For the last two decades, HP has also been widely investigated for water disinfection, showing potential for treatment through oxidative stress caused by ROS, Gram positive and negative bacteria, DNA and RNA viruses, and even algae [31]. ROS can attack cell membrane components, altering cell integrity, which results in a cytoplasm leakage [32]; they can inhibit required cell activities such as protein synthesis [33]; they can also break organic covalent bonds present in biomolecules [34]. Many factors affect the efficiency of disinfection via HP, including the chemical nature and concentration of the microorganisms, time of treatment, light intensity, water matrix, deficiency of atomic ligands, surface energy level, photocatalyst properties, and solution pH [22].

One of the main drawbacks limiting commercial and industrial HP application is the lack of reactor designs efficient enough to handle large volumes of water [35]. Many types of reactors have been studied and developed, but standard procedures for scale up are still lacking; HP technology readiness level (TRL), which ranges from TRL = 1 (proof-of-concept stage) to TRL = 9 (full operational scale stage), is between TRL = 4 (lab scale validation) and TRL = 5 (ongoing pilot scale applications) [36]. Other relevant issues concerning reactor design, such as reducing photon and mass transfer limitations [37] or a thorough understanding of heat, mass, and light transfer in the system [38], are a current research interest as are operation conditions, such as analyzing reactor performance with real water matrixes instead of synthetic water matrixes [39].

In this work, heterogeneous photocatalytic arsenic oxidation and water disinfection were explored in two types of solar reactors, a compound parabolic collector reactor (CPC) and a flat plate reactor (FPR); a real groundwater matrix was used, and coliform disinfection was also analyzed. As removal via chemical precipitation with ferric chloride ($FeCl_3$) was explored, following oxidation. Estimations of the collector area per order (A_{CO}) (m^2/m^3-order) were performed for the evaluation of the area or energy requirements by every reactor. The results were also analyzed from a reaction kinetic and statistical standpoint.

2. Materials and Methods

2.1. Location and Operation Timeframe for Experiments

Experiments were carried out in the city of Durango (25.613238° N, 103.435395° W, 1900 m above sea level; Durango, Mexico), and they were performed during the summer, autumn, and winter seasons at solar noon. The city is located within the planet's sunbelt, which receives an average solar radiation in the range of 5.5 kWh/m^2/day to 6.5 kWh/m^2/day [40].

2.2. Photocatalytic Reactors

The experiments were performed in a CPC tubular reactor set above a compound parabolic collector; the tube was made of 3 mm thick polymethylmethacrylate (PMMA), with an inside diameter of 4.48 cm and a longitude of 90 cm; 3 mm thick PMMA plates (which served as the TiO_2 support) were placed within the tube. Galvanized iron sheet was used to make the compound parabolic reflector; commercial reflective adhesive paper was used to cover the reflector. A submergible water pump was used to feed water to the reactor from a reservoir into one end of the tube and sent it back to the reservoir using a tubing arrangement (Figure 1).

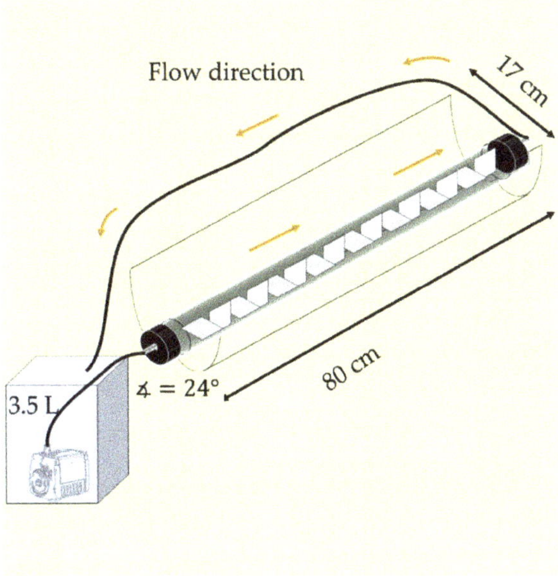

Figure 1. CPC reactor scheme.

The FPR reactor has been described in previous works [41–43], but briefly, it consisted of a PMMA compartment, with a feeding tube (with small holes separated every 0.5 cm) made of polyvinyl chloride set at the compartment's top, plus a drainage located at the compartment's bottom. The PMMA compartment was placed onto an inclinable metal assembly; reactor's drainage led to the water reservoir, which, with the help of a submergible water pump (Model H-331 BioPro, Jiangsu, China), was fed to the reactor (Figure 2). Table 1 shows additional reactor operational parameters, as reported in previous work [44].

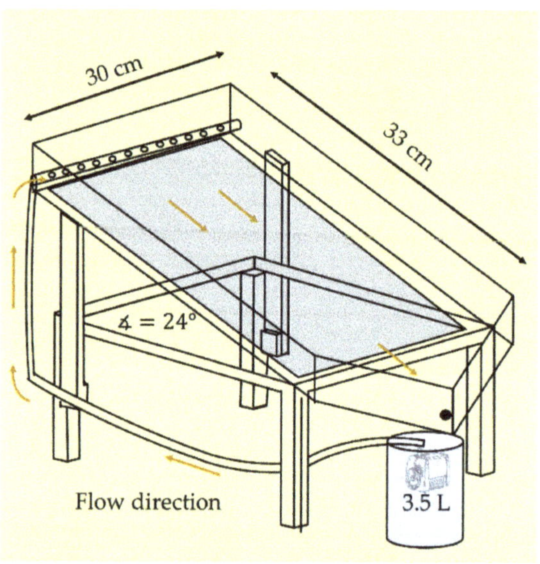

Figure 2. FPR reactor scheme.

Table 1. Operational parameters differing between reactors.

Reactor	Illuminated Net Area (m^2)	Photocatalyst Covered Area (m^2)	Volumetric Flow (m^3/h)
FPR	0.10 [a]	0.10 [a]	0.18
CPC	1.40 [b]	0.07 [c]	1.50

Note: [a] Glass plate area, equal to area covered by photocatalyst; [b] Reactor reflector area; [c] PMMA plate area covered by photocatalyst.

In the CPC, TiO_2 (Degussa-Evonik, Essen, Germany. CAS: 13463-67-7) was immobilized on a 4.5 cm wide and 5 cm long PMMA plate, on both plate sides; sixteen plates were used. While in the FPR, a 33 cm wide and 30 cm long frosted glass plate was used as photocatalyst support. The immobilization methodology has been reported elsewhere [44].

2.3. Experimental Conditions

Each experimental run was carried out with 3.5 L of a 300 μg/L As^{III} solution prepared using sodium meta-arsenite ($NaAsO_2$; J.T. Baker, Radnor, PA, USA. CAS: 7784-46-5) with groundwater from a local well, which already had 46.06 μg/L of As^{III} and 5.46 μg/L of As^{V}; groundwater physicochemical characterization is presented in Table 2. The solution was also spiked with water from a municipal wastewater treatment plant (MWTE) to analyze As oxidation in the presence of coliforms; the initial most probable number (MPN) per 100 mL was >2419. A total of 0.358 mL from a 30% H_2O_2 (Fermont, Guadalajara, Mexico. CAS: 7722-84-1) solution was added to the water to analyze H_2O_2 addition effect. Experiments in the dark were carried out for each experimental condition as control experiments, as well as photooxidation experiments without TiO_2.

Table 2. Groundwater physicochemical characterization.

pH	8.52
Electrical Conductivity	548.25 µS/cm
Major ions (mg/L)	
Na^+	4.38 mg/L
K^+	53.10 mg/L
Ca^{+2}	9.87 mg/L
Mg^{+2}	60.55 mg/L
F^-	1.54 mg/L
$NO_3{}_-$	38.35 mg/L
$NO_2{}^-$	1.83 mg/L
Cl^-	26.39 mg/L
$HCO_3{}^-$	148.50 mg/L
$SO_4{}^{-2}$	59.75 mg/L
Arsenic (µg/L)	
As^{III}	46.06
As^V	5.46

Experiments lasted 300 min (which began 150 min before solar midday); 20 mL samples were collected at 0, 30, 60, 100, 150, 220, and 300 min to measure As^{III} oxidation during the course of the experiment. To separate As^{III} and As^V, for each sample, an As speciation cartridge (MetalSoft Centre, Buena Park, CA, USA) was used, according to the manufacturer's specifications [45]. A radiometer (CUV5, Kipp & Zonen, Delft, The Netherlands) was used to measure UV irradiance during experiments.

As removal was also explored. After solar treatment was over, 0.5 L samples [46] were collected to carry out chemical precipitation with an optimized dose (2.19 mg/L) of $FeCl_3$ (Merck, Darmstadt, Germany. CAS: 7705-08-0; determination included in Supplementary Materials), and after adding the $FeCl_3$, pH was adjusted to 6.5 using hydrochloric acid (HCl; Merck, Darmstadt, Germany. CAS: 7647-01-0), and samples were taken to a jar tester machine with a rapid mixing phase of 400 rpm for 1 min, a slow mixing phase of 20 rpm for 10 min, and a settling phase for 20 min [47]. Samples were taken carefully with a micropipette from the supernatant for As quantification. Figure 3 shows a flow diagram of the process.

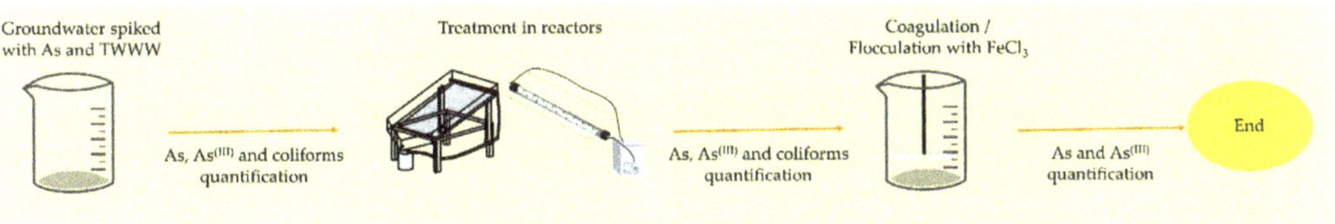

Figure 3. Process flow diagram.

Graphite furnace atomic absorption spectroscopy (GFAAS) (Avanta GBC model XplorAA) was carried out to quantify As.

For the experiments exploring disinfection, coliform bacteria inactivation was assessed by MPN using the Quanty-Tray Method with defined substrate Colilert (IDEXX Laboratories, Westbrook, ME, USA) [48], which is approved by the USEPA [49]. Colilert media was added to 100 mL samples (collected at the start and end of each experiment), mixed until properly dissolved, and mixed solutions were then poured into a Quanty-Tray/2000, sealed using the Quanty-Tray sealer, then put at 35 °C for 24 h for incubation [50]. This method is based on the enzymatic activity of coliform bacteria, which metabolizes the nutrient-indicating molecule ortho-nitrophenyl-β-D-galacto-pyranoside that turns colorless water yellow [51].

2.4. Kinetic Analysis

Scientific reports indicate that As^{III} heterogeneous photocatalytic oxidation employing supported TiO_2 follows a pseudo first-order reaction rate [52,53], which is useful to describe reactions limited by adsorption, as has been reported for heterogeneous photocatalytic reactions with TiO_2 [54,55]. As^{III} data were used to determine the K_{phC}, which was obtained by fitting the data to the pseudo first-order reaction model equation, which was obtained by established and conventional kinetics [56], as Equation (1) shows:

$$\left[As^{III}\right]_t = \left[As^{III}\right]_0 e^{-K_{phC}\, t},\tag{1}$$

where $\left[As^{III}\right]_0$ is As^{III} at t = 0, $\left[As^{III}\right]_t$ at time t, and K_{phC} is the photocatalytic reaction rate constant. Assumptions, such as laminar flow with a Reynolds number below 1000, steady state operation, and constant fluid viscosity and density must be considered [57].

2.5. Fluence Analysis

Fluence (Q_{UV}) was calculated as indicated elsewhere [58–61] and is shown in Equation (2):

$$Q_{UV} = Q_{UV,n-1} + UV_n \cdot (t_n - t_{n-1}) \cdot \frac{A_i}{V_t},\tag{2}$$

where the fluence is represented by Q_{UV}, the UV irradiance by UV_n, the time by t_n, the net irradiated area by A_i, and the volume by V_t. $\left[As^{III}\right]_t$ can be calculated as a function of Q_{UV} [59] employing Equation (3):

$$\left[As^{III}\right]_t = \left[As^{III}\right]_0 e^{-k_{UV}Q_{UV}},\tag{3}$$

where K_{UV} denotes the rate constant calculated for Q_{UV}.

2.6. Collector Area per Order Determination

The collector area per order (A_{CO}) is the net area of irradiated photocatalyst needed, considering an average solar UV irradiance, to decrease a contaminant's concentration by one order of magnitude, within a unit of volume [40,62]. It can be calculated by Equation (4):

$$A_{CO} = \frac{A_c \overline{E}_s t}{V_t \log\left(\left[As^{III}\right]_0 / \left[As^{III}\right]_t\right)};\tag{4}$$

where the collector area per order is represented by A_{CO}, the photocatalyst covered area by A_c, the As concentration by $[As^{III}]$, the volume by V_t, the time by t, and the average solar UV irradiance by $\overline{E}_s$.

2.7. Experimental Design and Statistical Analysis

A full factorial 2^5 experimental design was used to test the effects of 5 independent variables, each variable with 2 levels (for a total of 32 different experimental runs, carried out in triplicate), which can be seen in Table 3.

Table 3. Independent variables and their levels.

AOP	H_2O_2	Reactor	MWTE Spike	Irradiation
Photooxidation	0 mM	CPC	0 mL	No irradiation (dark control)
Heterogeneous photocatalysis	1 mM	FPR	10 mL	Solar UV

An analysis of variance (ANOVA) was performed to determine if the effect of each one of the independent variables was significant for the result of the dependent variables [63]. Equation (5) shows the linear model used:

$$x_{ijklmn} = \overline{\mu} + AOP_i + H_2O_{2j} + R_k + MWTE_l + DC_m + Q_{UVn} + \varepsilon_{ijklmn}, \tag{5}$$

$\overline{\mu}$ is the general mean; AOP_i represents the effect of the AOP; H_2O_{2j} is H_2O_2 addition; R_k denotes the type of reactor employed; $MWTE_l$ indicates MWTE spike added; DC_m stands for DC treatment; Q_{UVn} signifies QUV, and ε_{ijklmn} is the linear model error. To perform this analysis, SAS Studio 3.8 (SAS Institute Inc., Cary, NC, USA), which is a statistical software, was used.

3. Results and Discussion

3.1. Arsenic Oxidation

Figure 4 shows As^{III} photooxidation during the experiment on both CPC and FPR. The initial As^{III} concentration was 350 ± 2 µg/L. DC experiments showed very little oxidation, which can be attributed to the effect of water and oxygen [64]. Solar photooxidation experiments showed less than 25% As^{III} oxidation. It is known that solar irradiation accelerates oxidation, which can be attributed to UV and visible light, and, as it has been reported, they also promote ROS formation in water without the need of a photocatalyst [65].

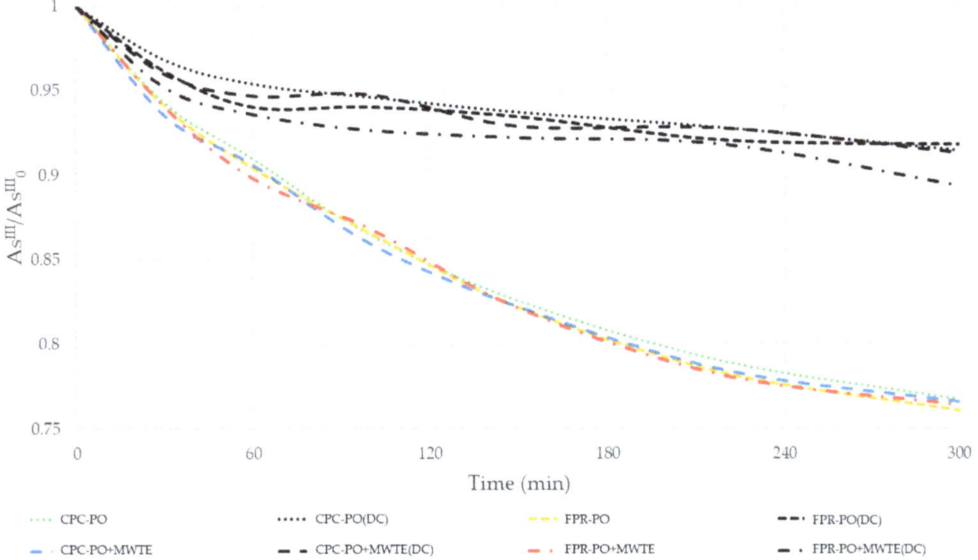

Figure 4. As^{III} photooxidation (PO) experiments carried out both in CPC and FPR, with and without MWTE spike. Experiments performed in the dark as control experiments (DC) are included as well.

Figure 5 shows also photooxidation experiments in the same treatments discussed above, but with H_2O_2 addition. The accelerated oxidation fostered by H_2O_2 observed in all cases was expected, as it has been reported that H_2O_2 is an effective oxidant in alkali conditions [66] (working solution pH = 8.5). Experiments under solar irradiation yielded an increased oxidation, as UV light causes H_2O_2 to undergo homolytic cleavage, generating $HO^\bullet$ [27], whose oxidative potential of 2.73–2.8 V is only surpassed by that of fluorine [67]. Both $HO^\bullet$ and H_2O_2 are able to promote As^{III} oxidation [68].

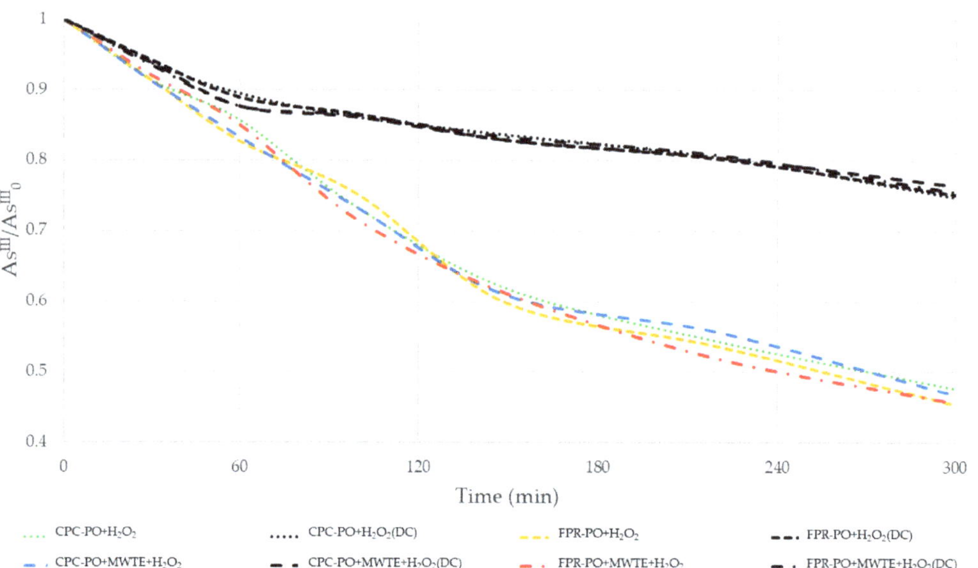

Figure 5. As^{III} photooxidation (PO) experiments carried out both in CPC and FPR, H_2O_2 added, and with and without MWTE spike. Experiments performed in the dark as control experiments (DC) are included as well.

Figure 6 shows As^{III} oxidation via solar HP; DC experiments apparently showed an increased oxidation, but most likely As^{III} adsorption onto TiO_2 surface is the main reason for this observation [69], as at pH = 8.5, As is mostly found as arsenious acid, which has a neutral charge [70]. The higher oxidation observed in HP experiments is clearly due to the generated ROS by HP, which also includes $O_2{}^{\bullet-}$ [71].

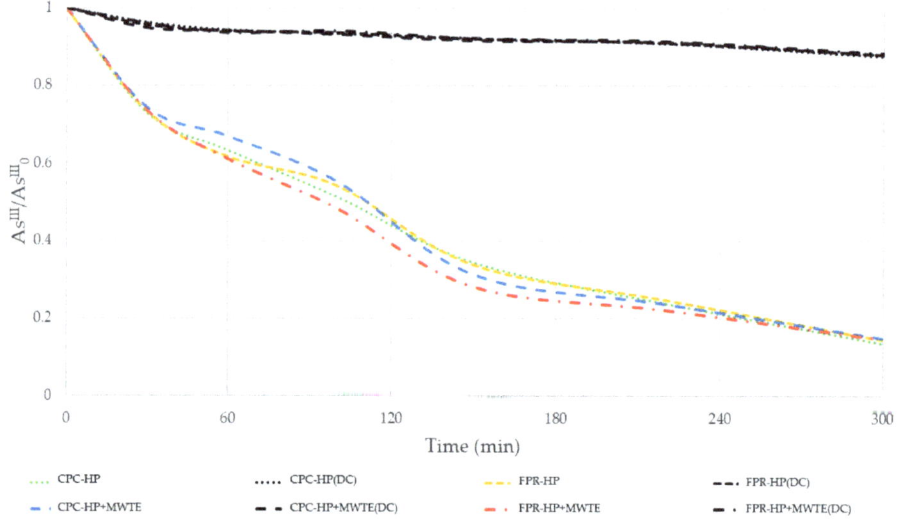

Figure 6. As^{III} heterogeneous photocatalytic (HP) oxidation experiments carried out both in CPC and FPR, and with and without MWTE spike. Experiments performed in the dark as control experiments (DC) are included as well.

In Figure 7, As^{III} oxidation via HP+ H_2O_2 is shown. The highest oxidation was achieved in this round of experiments, which conjoins the effects of the previous treatments. The main ROS driving As^{III} oxidation has been debated [72], but for cases in which H_2O_2 is added, it has been recently proposed that a nonradical species, surface complexes Ti-peroxo (Ti–OOH), would be the main oxidative species [73,74]. This fact can be theorized as an As^{III} oxidation experiment using cerium dioxide (CeO_2), with cerium being in the same periodic group as titanium, found in the presence of Ce-peroxo surface complexes [75]. Additionally, Ti–OOH (and Ce–OOH) has been reported as the main oxidative species in antimony (which is located in the same periodic group as As) oxidation experiments employing H_2O_2 over TiO_2/CeO_2 [76].

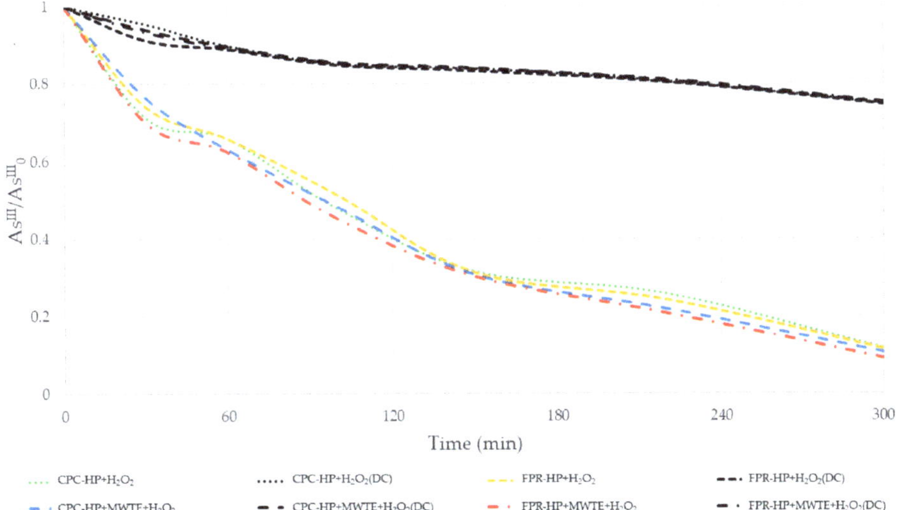

Figure 7. As^{III} heterogeneous photocatalytic (HP) oxidation experiments carried out both in CPC and FPR, H_2O_2 added, and with and without MWTE spike. Experiments performed in the dark as control experiments (DC) are included as well.

Since the groundwater matrix used in the experiments contained several cations and anions, other reactions could take place in some of the treatments; $HO^{\bullet}$ can react with CO_3^{2-} and Cl^- to form carbonate and chloride radicals, respectively ($CO_3^{\bullet-}$ and $Cl^{\bullet}$); NO_3^- can give place to nitrite radical ($NO_3^{\bullet}$) [77]. Although the aim of this research is not to study the effect of said species in the outcome of As^{III} oxidation, it is worth mentioning that they may play a role in the whole process.

The pH did not remain constant throughout the experiments; however, the change was negligible (final pH > 8.0). Since As^{III} (in the form of arsenious acid) pKa = 9.29 [78] and TiO_2 zero charge point is at pH = 6.3 [79], the pH change should not have a significant effect on the outcome of the experiment, not even after As^{III} oxidation, as the As^V pKa values are pKa_1 = 2.2, pKa_2 = 7.08, and pKa_3 = 11.5 [80].

3.2. Arsenic Removal

Water As concentration after chemical precipitation with $FeCl_3$ is reported in Table 4. Although more than 90% removal was achieved with every treatment, HP treatments were the only ones to attain an As concentration that complies with the WHO guidelines. Although it is well known that As^{III} is harder to remove than As^V due to its neutral charge [81,82], chemical precipitation is not a stoichiometric process and is also able to remove As [83]. The increased removal in DC samples involving H_2O_2 can be explained due to As^{III} oxidation caused by the Fenton reagent that is mostly carried out under acidic conditions of pH < 3, but can also take place at higher pH values, at which point sedimentation will occur as well [84].

3.3. Coliform Disinfection

Total coliform disinfection results are presented in Table 5. DC experiments did not offer enough stress as to cause a noticeable effect in coliform MPN. Partial disinfection was observed when H_2O_2 was added, which can be attributed to both internal and external cell damage caused by the ROS [85]. Coliforms were not detected in solar experiments, which is expected as SODIS is a well-known point of use technology for water disinfection, which is effective for several microorganisms species [86]. It is inferred that disinfection could have happened faster in the treatments involving TiO_2 and H_2O_2, as has been reported in a previous experiment [42].

As HP is nonselective, it can be used to deal with an assorted variety of contaminants [87] such as inactivating coliforms and oxidize arsenic as in the present case. This characteristic makes it attractive for further research and additional development to keep improving its technical feasibility. It is important to have options when it comes to water treatment and potabilization, as most of the times, one solution cannot fit to all cases.

3.4. Kinetic Analysis

The calculated rate constants, both photooxidative (k_{po}) and photocatalytic (K_{phC}), are shown in Table 6. The reaction rate constants for DC experiments were much lower than their irradiated counterparts in every case. H_2O_2 treatments also showed a higher reaction rate constant than those of treatments without H_2O_2. The k_{po} values were lower than the K_{phC} values, and the higher reaction rate constants were observed in experimental runs with higher As^{III} oxidation. The coefficient of determination (R^2) for DC treatments was 0.87 on average, while for irradiated treatments it was 0.96, which is an acceptable value to fit the data to the pseudo first-order reaction model that has been reported to describe As^{III} heterogeneous photocatalytic oxidation [52,53].

3.5. Fluence Analysis

Table 7 presents the fluence and reaction rate constants as a function of Q_{UV} for each treatment. Fluence was lower in FPR treatments than in CPC treatments as the total irradiated surface is roughly 40% larger in the CPC than in the FPR. The proposed reaction rate constant in the function of Q_{UV} has been used to describe the kinetics of a reaction in function of cumulative UV dose as an alternative to kinetics in function of time [59]. The average R^2 value was 0.96, which is almost identical to the R^2 obtained when fitting the data to the pseudo first-order reaction model. It can be inferred that K_{UV} is suitable for explaining As^{III} oxidation via HP as well.

Table 4. Arsenic concentration after chemical precipitation with $FeCl_3$.

Treatment	As (µg/L)	As Removed (%)	Treatment	As (µg/L)	As Removed (%)
CPC–PO	23.4 ± 2.9	93.0 ± 0.9%	CPC–HP	8.9 ± 2.2	98.1 ± 0.6%
CPC–PO(DC)	25.1 ± 1.4	92.3 ± 0.4%	CPC–HP(DC)	22.7 ± 0.3	93.5 ± 0.1%
FPR–PO	20.72 ± 0.8	93.9 ± 0.3%	FPR–HP	8.2 ± 1.3	93.3 ± 0.4%
FPR–PO(DC)	24.3 ± 1.1	92.8 ± 0.3%	FPR–HP(DC)	16.2 ± 1.4	92.5 ± 1.1%
CPC–PO + MWTE	23.9 ± 0.8	93.3 ± 0.2%	CPC–HP + MWTE	9.2 ± 1.4	97.2 ± 0.4%
CPC–PO + MWTE(DC)	24.9 ± 2.1	92.5 ± 0.6%	CPC–HP + MWTE(DC)	21.1 ± 1.2	94 ± 0.4%
FPR–PO + MWTE	22.3 ± 2.5	92.8 ± 0.7%	FPR–HP + MWTE	8.3 ± 0.9	97.5 ± 0.3%
FPR–PO + MWTE(DC)	24.9 ± 2.1	92.6 ± 0.6%	FPR–HP + MWTE(DC)	23.3 ± 0.1	93.3 ± 0.1%
CPC–PO + H_2O_2	16.5 ± 1.7	95.8 ± 0.5%	CPC–HP + H_2O_2	5.3 ± 1.1	98.7 ± 0.3%
CPC–PO + H_2O_2(DC)	18.7 ± 0.5	94.5 ± 0.2%	CPC–HP + H_2O_2(DC)	10.2 ± 1.2	96.8 ± 0.4%
FPR–PO + H_2O_2	16.5 ± 0.3	95.3 ± 0.1%	FPR–HP + H_2O_2	5.1 ± 0.1	98.6 ± 0.0%
FPR–PO + H_2O_2(DC)	19.5 ± 0.7	94.2 ± 0.2%	FPR–HP + H_2O_2(DC)	10.2 ± 2.7	96.6 ± 0.8%
CPC–PO + MWTE + H_2O_2	16.9 ± 0.8	94.9 ± 0.2%	CPC–HP + MWTE + H_2O_2	1.6 ± 0.5	99.7 ± 0.1%
CPC–PO + MWTE+ H_2O_2(DC)	18.7 ± 0.5	94.5 ± 0.2%	CPC–HP + MWTE + H_2O_2(DC)	11.8 ± 0.8	94.9 ± 0.2%
FPR–PO + MWTE + H_2O_2	16.2 ± 0.2	95.5 ± 0.2%	FPR–HP + MWTE + H_2O_2	1.7 ± 0.2	99.5 ± 0.1%
FPR–PO + MWTE + H_2O_2(DC)	18.7 ± 0.5	94.5 ± 0.2%	FPR–HP + MWTE + H_2O_2(DC)	11.5 ± 1.5	94.8 ± 0.4%

Table 5. Most probable number of coliforms in samples at the beginning and end of each treatment.

Treatment	0 min (MPN/100 mL)	300 min (MPN/100 mL)	Treatment	0 min (MPN/100 mL)	300 min (MPN/100 mL)
CPC–PO + MWTE	>2419	N.D.[a]	CPC–HP + MWTE	>2419	N.D.
CPC–PO + MWTE (DC)	>2419	>2419	CPC–HP + MWTE (DC)	>2419	>2419
FPR–PO + MWTE	>2419	N.D.	FPR–HP + MWTE	>2419	N.D.
FPR–PO + MWTE (DC)	>2419	>2419	FPR–HP + MWTE (DC)	>2419	>2419
CPC–PO + MWTE + H_2O_2	>2419	N.D.	CPC–HP + MWTE + H_2O_2	>2419	N.D.
CPC–PO + MWTE + H_2O_2 (DC)	>2419	574–727	CPC–HP + MWTE + H_2O_2 (DC)	>2419	658–755
FPR–PO + MWTE + H_2O_2	>2419	N.D.	FPR–HP + MWTE + H_2O_2	>2419	N.D.
FPR–PO + MWTE + H_2O_2 (DC)	>2419	629–686	FPR–HP + MWTE + H_2O_2 (DC)	>2419	613–689

Note: N.D. [a] = Not detectable.

Table 6. Calculated photooxidative and photocatalytic reaction rate constants for arsenic oxidation.

Treatment	k_{po} ($\times 10^{-3}$ min^{-1})	Treatment	k_{po} ($\times 10^{-3}$ min^{-1})	Treatment	K_{phC} ($\times 10^{-3}$ min^{-1})	Treatment	K_{phC} ($\times 10^{-3}$ min^{-1})
CPC–PO	0.8	CPC–PO + H_2O_2	2.6	CPC–HP	6.2	CPC–HP + H_2O_2	6.8
CPC–PO (DC)	0.2	CPC–PO + H_2O_2 (DC)	0.8	CPC–HP (DC)	0.3	CPC–HP + H_2O_2 (DC)	0.8
FPR–PO	0.8	FPR–PO + H_2O_2	2.6	FPR–HP	6.0	FPR–HP + H_2O_2	6.7
FPR–PO (DC)	0.2	FPR–PO + H_2O_2 (DC)	0.8	FPR–HP (DC)	0.3	FPR–HP + H_2O_2 (DC)	0.8

Table 7. Fluence and calculated reaction rate constants in function of fluence.

Treatment	Q_{UV} (kJ L^{-1})	K_{UV} ($\times 10^{-3}$ kJ^{-1} L)	Treatment	Q_{UV} (kJ L^{-1})	K_{UV} ($\times 10^{-3}$ kJ^{-1} L)
CPC–PO	354.99	0.7	FPR–PO	251.03	1.0
CPC–PO + H$_2$O$_2$	339.45	2.2	FPR–PO + H$_2$O$_2$	240.04	3.1
CPC–HP	366.79	5.1	FPR–HP	259.38	7.0
CPC–HP + H$_2$O$_2$	361.57	5.6	FPR–HP + H$_2$O$_2$	255.69	7.9

3.6. Collector Area per Order

Table 8 shows the A_{CO} value for each one of the treatments. A smaller Aco is generally associated with a more efficient process [44]. On average, CPC needed 30% less photocatalyst covered area than the FPR, which is in accordance with the improved optical efficiency provided by the reflector [52], accounting for an improved use of the photocatalyst-covered area.

Table 8. Estimation of A_{CO} for each treatment and comparative efficiency between reactors, $\varepsilon = [(A_{CO_{FPR}} - A_{CO_{CPC}})/(A_{CO_{FPR}})] \times 100$, for arsenic oxidation, considering batch operation and a first-order rate reaction.

Treatment	A_{CO} (m$^2\cdot$m^{-3}-Order)	Treatment	A_{CO} (m$^2\cdot$m^{-3}-Order)	Efficiency (ε)
CPC–PO	83.3	FPR–PO	121.7	31.5
CPC–PO + H$_2$O$_2$	28.4	FPR–PO + H$_2$O$_2$	39.2	27.4
CPC–HP	11.3	FPR–HP	15.3	26.4
CPC–HP + H$_2$O$_2$	10.2	FPR–HP + H$_2$O$_2$	15.0	32.0

The A_{CO} for HP treatments was on average 75% more efficient than PO treatments $(\varepsilon = [(A_{CO_{PO}} - A_{CO_{HP}})/(A_{CO_{PO}})] \times 100)$, while H$_2O_2$ treatments were 38% more efficient than treatments without H$_2$O$_2$ ($\varepsilon = [(A_{CO_{0mMH2O2}} - A_{CO_{1mMH2O2}})/(A_{CO_{0mMH2O2}})] \times 100$).

3.7. ANOVA Results

The summary of the ANOVA is shown in Table 9. The F-value for the model was statistically significant ($p < 0.05$); hence, it can be assumed that the linear model was appropriate for the analysis. The R^2 value was 0.76, leaving enough room for improvement for a better linear model that better explains variance [58].

Table 9. Summary of ANOVA for As^{III} oxidation.

Source	DF	Sum of Squares ($\times 10^4$)	Mean Square ($\times 10^4$)	F-Value	*p*-Value Probr > F
Model	6	4.7	0.7	49.61	<0.0001
AOP	1	1.3	1.3	88.53	<0.0001
H$_2$O$_2$ addition	1	0.2	0.2	13.96	0.0003
MWTE spike	1	0.0	0.0	0.01	0.9355
Reactor employed	1	0.0	0.0	0.01	0.9410
Irradiation exposure	1	3.0	3.0	195.13	<0.0001
Q_{UV}	1	0.0	0.0	0.00	0.9662

The results obtained through the ANOVA support the observations made previously in Figures 4–7. The AOP effect was statistically significant; for PO, the UV light was the main driver for As^{III} oxidation, which is favored in alkaline conditions [88], but it is not as fast as HP, which promotes ROS generation to accelerate As^{III} oxidation [89]. H$_2$O$_2$ addition accelerated oxidation due to ROS generation [90] or Ti-peroxo species formation (in HP treatments only) [91]; hence, its effect being statically significant is within expectations.

Solar light exposure (whether or not as in DC treatments) was the most significant effect for As^{III} oxidation ($p < 0.05$ in all three cases).

The MWTE spike did not have a statistically significant effect on the As^{III} oxidation reaction rate, which is within the expectations of AOPs being nonselective [92]; it is evident that not every generated ROS is directed to As^{III}. The effect of the used reactor was not statistically significant either, despite the fact that their geometry allows for a completely different operation (light distribution, light propagation, surface area to volume ratio, etc. [93]). This finding supports the fact that the outcome of an HP operation might depend on a plethora of variables and the interactions between them, such as the photocatalyst properties, the source of light, the intensity of the light, the pH, the concentration of the pollutant, and the temperature of the geometry of the reactor [44]. Finally, the effect of the covariable, Q_{UV}, was not statistically significant, possibly because of uniformity in the timeframe of the operation, which translated into minimal Q_{UV} variations (solar noon operation and only on sunny days) [41,44] ($p < 0.05$ in all three cases).

3.8. Perspectives and Outlook

The present work results are in agreement with the HP TRL ongoing pilot scale applications. An As concentration of 300 µg/L is extremely high, and the obtained results provide a notion of the resources needed to treat water with such characteristics. Disinfection can be achieved with solar light alone, but As^{III} oxidation was greatly improved when using HP, and even more with H_2O_2. Both TiO_2 and H_2O_2 pose no threat to the environment [94,95] and could be used readily. The present work analyzed only a volume of 3.5 L, and it took a considerable amount of time (300 min) to oxidize most of the As^{III}, so an optimization in function of time and volume treated with respect to As^{III} concentration should be followed, as proper experimental conditions have been found in the present work. Photocatalytic reactors are not easily scaled up due to the intrinsic nature of light; however, if space is available, scaling out is always possible [96].

TiO_2 is only active when irradiated with UV light, which limits its efficiency for solar applications, as UV light accounts for less than 5% of the received solar energy [97]; a direct comparison against visible light-active photocatalysts should also be explored. TiO_2 exhibits visible light activity when doped with some elements, for example nitrogen or silver [98].

4. Conclusions

The treatments CPC–HP + H_2O_2 and FPR–HP + H_2O_2 yielded the best oxidation for As^{III}, with rates around 90%. These treatments also exhibited the highest oxidation reaction rate constants, with 6.8×10^{-3} min^{-1} and 6.7×10^{-3} min^{-1}, respectively.

As removal rates achieved via chemical precipitation for the aforementioned treatments were 98.7% and 98.6%, reaching the As concentration level recommended by the WHO, which is below 10 µg/L.

Additionally, no coliforms were detected in the irradiated treatments, which adds up to the advantages of HP as a potential and promising technology for water potabilization and wastewater treatment.

The determination of A_{CO} showed that CPC was on average 30% more efficient than the FPR, requiring less photocatalyst-covered area.

The effects of AOP, H_2O_2 addition, and light irradiation were statistically significant for the As^{III} oxidation reaction rate, while the type of reactor utilized, spiking with MWTE, or fluence were not ($p < 0.05$), as found out with an ANOVA.

Supplementary Materials: The following supporting information can be downloaded at: https://www.mdpi.com/article/10.3390/w14152450/s1, Table S1: FeCl$_3$ dose and As concentration in the supernatant. References [47,99–101] are cited in the supplementary materials.

Author Contributions: Conceptualization, F.d.J.S.-V. and J.B.P.-N.; Data curation, F.d.J.S.-V.; Formal analysis, F.d.J.S.-V. and C.M.N.-N.; Investigation, F.d.J.S.-V.; Methodology, F.d.J.S.-V. and J.B.P.-N.; Resources, J.B.P.-N. and M.T.A.-H.; Supervision, J.B.P.-N. and M.T.A.-H.; Visualization, C.M.N.-N., J.B.P.-N. and M.T.A.-H.; Writing—original draft, F.d.J.S.-V.; Writing—review and editing, C.M.N.-N., J.B.P.-N. and M.T.A.-H. All authors have read and agreed to the published version of the manuscript.

Funding: The research was internally financed by Centro de Investigación en Materiales Avanzados and Instituto Politécnico Nacional (SIP project: 20210507). The first author is grateful to the Consejo Nacional de Ciencia y Tecnología (CONACyT) through the doctorate scholarship granted to him, student identification DCTA1001002.

Institutional Review Board Statement: Not applicable.

Informed Consent Statement: Not applicable.

Data Availability Statement: All data and materials have been provided within the manuscript.

Acknowledgments: The authors would like to thank academic technicians Luis Arturo Torres Castañón and José Rafael Irigoyen Campuzano for their valuable collaboration in the analytical determinations.

Conflicts of Interest: The authors declare no conflict of interest to disclose. Funders played no role in study design; data collection, analysis, and interpretation; manuscript writing, and results publication.

References

1. Mandal, B. Arsenic Round the World: A Review. *Talanta* **2002**, *58*, 201–235. [CrossRef]
2. Urseler, N.; Bachetti, R.; Morgante, V.; Agostini, E.; Morgante, C. Groundwater Quality and Vulnerability in Farms from Agricultural-Dairy Basin of the Argentine Pampas. *Environ. Sci. Pollut. Res.* **2022**. [CrossRef]
3. Jain, N.; Chandramani, S. Arsenic Poisoning- An Overview. *Indian J. Med. Spec.* **2018**, *9*, 143–145. [CrossRef]
4. Flora, S.J.S. Arsenic and Dichlorvos: Possible Interaction between Two Environmental Contaminants. *J. Trace Elem. Med. Biol.* **2016**, *35*, 43–60. [CrossRef]
5. Bjørklund, G.; Oliinyk, P.; Lysiuk, R.; Rahaman, M.S.; Antonyak, H.; Lozynska, I.; Lenchyk, L.; Peana, M. Arsenic Intoxication: General Aspects and Chelating Agents. *Arch. Toxicol.* **2020**, *94*, 1879–1897. [CrossRef]
6. Anand, V.; Kaur, J.; Srivastava, S.; Bist, V.; Singh, P.; Srivastava, S. Arsenotrophy: A Pragmatic Approach for Arsenic Bioremediation. *J. Environ. Chem. Eng.* **2022**, *10*, 107528. [CrossRef]
7. Yu, S.; Li, L.-H.; Lee, C.-H.; Jeyakannu, P.; Wang, J.-J.; Hong, C.-H. Arsenic Leads to Autophagy of Keratinocytes by Increasing Aquaporin 3 Expression. *Sci. Rep.* **2021**, *11*, 17523. [CrossRef]
8. García-Rosales, G.; Longoria-Gándara, L.C.; Cruz-Cruz, G.J.; Olayo-González, M.G.; Mejía-Cuero, R.; Pérez, P.Á. Fe-TiOx Nanoparticles on Pineapple Peel: Synthesis, Characterization and As(V) Sorption. *Environ. Nanotechnol. Monit. Manag.* **2018**, *9*, 112–121. [CrossRef]
9. Khan, M.I.; Ahmad, M.F.; Ahmad, I.; Ashfaq, F.; Wahab, S.; Alsayegh, A.A.; Kumar, S.; Hakeem, K.R. Arsenic Exposure through Dietary Intake and Associated Health Hazards in the Middle East. *Nutrients* **2022**, *14*, 2136. [CrossRef]
10. Guglielmi, G. Arsenic in Drinking Water Threatens up to 60 Million in Pakistan. *Science* **2017**, *14*, 2136. [CrossRef]
11. World Health Organization. *Guidelines for Drinking-Water Quality: Fourth Edition Incorporating the First Addendum*; World Health Organization: Geneva, Switzerland, 2017.
12. Fatoki, J.O.; Badmus, J.A. Arsenic as an Environmental and Human Health Antagonist: A Review of Its Toxicity and Disease Initiation. *J. Hazard. Mater. Adv.* **2022**, *5*, 100052. [CrossRef]
13. Raju, N.J. Arsenic in the Geo-Environment: A Review of Sources, Geochemical Processes, Toxicity and Removal Technologies. *Environ. Res.* **2022**, *203*, 111782. [CrossRef]
14. Taviani, E.; Pedro, O. Impact of the Aquatic Pathobiome in Low-Income and Middle-Income Countries (LMICs) Quest for Safe Water and Sanitation Practices. *Curr. Opin. Biotechnol.* **2022**, *73*, 220–224. [CrossRef]
15. Hanif, Z.; Tariq, M.Z.; Khan, Z.A.; La, M.; Choi, D.; Park, S.J. Polypyrrole-Coated Nanocellulose for Solar Steam Generation: A Multi-Surface Photothermal Ink with Antibacterial and Antifouling Properties. *Carbohydr. Polym.* **2022**, *292*, 119701. [CrossRef]
16. Javaid, M.; Qasim, H.; Zia, H.Z.; Dasliu, M.A.; Syeda Amber Hameed, A.Q.; Samiullah, K.; Hashem, M.; Morsy, K.; Dajem, S.B.; Muhammad, T.; et al. Bacteriological Composition of Groundwater and Its Role in Human Health. *J. King Saud Univ. Sci.* **2022**, *34*, 102128. [CrossRef]
17. Adelodun, B.; Ajibade, F.O.; Ighalo, J.O.; Odey, G.; Ibrahim, R.G.; Kareem, K.Y.; Bakare, H.O.; Tiamiyu, A.O.; Ajibade, T.F.; Abdulkadir, T.S.; et al. Assessment of Socioeconomic Inequality Based on Virus-Contaminated Water Usage in Developing Countries: A Review. *Environ. Res.* **2021**, *192*, 110309. [CrossRef]
18. Ahmad, A.; Bhattacharya, P. Arsenic in Drinking Water: Is 10 Mg/L a Safe Limit? *Curr. Pollut. Rep.* **2019**, *5*, 1–3. [CrossRef]
19. Wen, X.; Chen, F.; Lin, Y.; Zhu, H.; Yuan, F.; Kuang, D.; Jia, Z.; Yuan, Z. Microbial Indicators and Their Use for Monitoring Drinking Water Quality—A Review. *Sustainability* **2020**, *12*, 2249. [CrossRef]

20. Valdiviezo Gonzales, L.G.; García Ávila, F.F.; Cabello Torres, R.J.; Castañeda Olivera, C.A.; Alfaro Paredes, E.A. Scientometric Study of Drinking Water Treatments Technologies: Present and Future Challenges. *Cogent Eng.* **2021**, *8*, 1929046. [CrossRef]
21. Sundar, K.P.; Kanmani, S. Progression of Photocatalytic Reactors and It's Comparison: A Review. *Chem. Eng. Res. Des.* **2020**, *154*, 135–150. [CrossRef]
22. Jabbar, Z.H.; Esmail Ebrahim, S. Recent Advances in Nano-Semiconductors Photocatalysis for Degrading Organic Contaminants and Microbial Disinfection in Wastewater: A Comprehensive Review. *Environ. Nanotechnol. Monit. Manag.* **2022**, *17*, 100666. [CrossRef]
23. Sibhatu, A.K.; Weldegebrieal, G.K.; Sagadevan, S.; Tran, N.N.; Hessel, V. Photocatalytic Activity of CuO Nanoparticles for Organic and Inorganic Pollutants Removal in Wastewater Remediation. *Chemosphere* **2022**, *300*, 134623. [CrossRef]
24. Wang, H.; Li, X.; Zhao, X.; Li, C.; Song, X.; Zhang, P.; Huo, P.; Li, X. A Review on Heterogeneous Photocatalysis for Environmental Remediation: From Semiconductors to Modification Strategies. *Chin. J. Catal.* **2022**, *43*, 178–214. [CrossRef]
25. Xie, L.; Hao, J.G.; Chen, H.Q.; Li, Z.X.; Ge, S.Y.; Mi, Y.; Yang, K.; Lu, K.Q. Recent Advances of Nickel Hydroxide-Based Cocatalysts in Heterogeneous Photocatalysis. *Catal. Commun.* **2022**, *162*, 106371. [CrossRef]
26. Karim, A.V.; Krishnan, S.; Shriwastav, A. An Overview of Heterogeneous Photocatalysis for the Degradation of Organic Compounds: A Special Emphasis on Photocorrosion and Reusability. *J. Indian Chem. Soc.* **2022**, *99*, 100480. [CrossRef]
27. Sharma, A.; Ahmad, J.; Flora, S.J.S. Application of Advanced Oxidation Processes and Toxicity Assessment of Transformation Products. *Environ. Res.* **2018**, *167*, 223–233. [CrossRef]
28. Marinho, B.A.; Cristóvão, R.O.; Boaventura, R.A.R.; Vilar, V.J.P. As(III) and Cr(VI) Oxyanion Removal from Water by Advanced Oxidation/Reduction Processes—A Review. *Environ. Sci. Pollut. Res.* **2019**, *26*, 2203–2227. [CrossRef]
29. García, F.E.; Litter, M.I.; Sora, I.N. Assessment of the Arsenic Removal From Water Using Lanthanum Ferrite. *ChemistryOpen* **2021**, *10*, 790–797. [CrossRef]
30. Deng, Y.; Li, Y.; Li, X.; Sun, Y.; Ma, J.; Lei, M.; Weng, L. Influence of Calcium and Phosphate on PH Dependency of Arsenite and Arsenate Adsorption to Goethite. *Chemosphere* **2018**, *199*, 617–624. [CrossRef]
31. Hosseini, F.; Assadi, A.A.; Nguzen-Tri, P.; Ali, I.; Rtimi, S. Titanium-Based Photocatalytic Coatings for Bacterial Disinfection: The Shift from Suspended Powders to Catalytic Interfaces. *Surf. Interfaces* **2022**, *32*, 102078. [CrossRef]
32. John, D.; Jose, J.; Bhat, S.G.; Achari, V.S. Integration of Heterogeneous Photocatalysis and Persulfate Based Oxidation Using TiO_2-Reduced Graphene Oxide for Water Decontamination and Disinfection. *Heliyon* **2021**, *7*, e07451. [CrossRef] [PubMed]
33. Byrne, J.; Dunlop, P.; Hamilton, J.; Fernández-Ibáñez, P.; Polo-López, I.; Sharma, P.; Vennard, A. A Review of Heterogeneous Photocatalysis for Water and Surface Disinfection. *Molecules* **2015**, *20*, 5574–5615. [CrossRef] [PubMed]
34. Xu, Y.; Liu, Q.; Liu, C.; Zhai, Y.; Xie, M.; Huang, L.; Xu, H.; Li, H.; Jing, J. Visible-Light-Driven Ag/AgBr/$ZnFe_2O_4$ Composites with Excellent Photocatalytic Activity for *E. coli* Disinfection and Organic Pollutant Degradation. *J. Colloid Interface Sci.* **2018**, *512*, 555–566. [CrossRef] [PubMed]
35. Serrà, A.; Philippe, L.; Perreault, F.; Garcia-Segura, S. Photocatalytic Treatment of Natural Waters. Reality or Hype? The Case of Cyanotoxins Remediation. *Water Res.* **2021**, *188*, 116543. [CrossRef] [PubMed]
36. Antonopoulou, M.; Kosma, C.; Albanis, T.; Konstantinou, I. An Overview of Homogeneous and Heterogeneous Photocatalysis Applications for the Removal of Pharmaceutical Compounds from Real or Synthetic Hospital Wastewaters under Lab or Pilot Scale. *Sci. Total Environ.* **2021**, *765*, 144163. [CrossRef]
37. Espíndola, J.C.; Vilar, V.J.P. Innovative Light-Driven Chemical/Catalytic Reactors towards Contaminants of Emerging Concern Mitigation: A Review. *Chem. Eng. J.* **2020**, *394*, 124865. [CrossRef]
38. Chen, L.; Tang, J.; Song, L.-N.; Chen, P.; He, J.; Au, C.-T.; Yin, S.-F. Heterogeneous Photocatalysis for Selective Oxidation of Alcohols and Hydrocarbons. *Appl. Catal. B Environ.* **2019**, *242*, 379–388. [CrossRef]
39. Binjhade, R.; Mondal, R.; Mondal, S. Continuous Photocatalytic Reactor: Critical Review on the Design and Performance. *J. Environ. Chem. Eng.* **2022**, *10*, 107746. [CrossRef]
40. Zaruma-Arias, P.E.; Núñez-Núñez, C.M.; González-Burciaga, L.A.; Proal-Nájera, J.B. Solar Heterogenous Photocatalytic Degradation of Methylthionine Chloride on a Flat Plate Reactor: Effect of PH and H_2O_2 Addition. *Catalysts* **2022**, *12*, 132. [CrossRef]
41. Silerio-Vázquez, F.; Alarcón-Herrera, M.T.; Proal-Nájera, J.B. Solar Heterogeneous Photocatalytic Degradation of Phenol on TiO_2/Quartz and TiO_2/Calcite: A Statistical and Kinetic Approach on Comparative Efficiencies towards a TiO_2/Glass System. *Environ. Sci. Pollut. Res.* **2022**, *29*, 42319–42330. [CrossRef]
42. Núñez-Núñez, C.M.; Osorio-Revilla, G.I.; Villanueva-Fierro, I.; Antileo, C.; Proal-Nájera, J.B. Solar Fecal Coliform Disinfection in a Wastewater Treatment Plant by Oxidation Processes: Kinetic Analysis as a Function of Solar Radiation. *Water* **2020**, *12*, 639. [CrossRef]
43. González-Burciaga, L.A.; Núñez-Núñez, C.M.; Morones-Esquivel, M.M.; Avila-Santos, M.; Lemus-Santana, A.; Proal-Nájera, J.B. Characterization and Comparative Performance of TiO_2 Photocatalysts on 6-Mercaptopurine Degradation by Solar Heterogeneous Photocatalysis. *Catalysts* **2020**, *10*, 118. [CrossRef]
44. Silerio-Vázquez, F.d.J.; Núñez-Núñez, C.M.; Alarcón-Herrera, M.T.; Proal-Nájera, J.B. Comparative Efficiencies for Phenol Degradation on Solar Heterogeneous Photocatalytic Reactors: Flat Plate and Compound Parabolic Collector. *Catalysts* **2022**, *12*, 575. [CrossRef]

45. Leiva-Aravena, E.; Vera, M.A.; Nerenberg, R.; Leiva, E.D.; Vargas, I.T. Biofilm Formation of *Ancylobacter* Sp. TS-1 on Different Granular Materials and Its Ability for Chemolithoautotrophic As(III)-Oxidation at High Concentrations. *J. Hazard. Mater.* **2022**, *421*, 126733. [CrossRef] [PubMed]

46. Vieira, B.R.C.; Pintor, A.M.A.; Boaventura, R.A.R.; Botelho, C.M.S.; Santos, S.C.R. Arsenic Removal from Water Using Iron-Coated Seaweeds. *J. Environ. Manag.* **2017**, *192*, 224–233. [CrossRef] [PubMed]

47. Laky, D.; Licskó, I. Arsenic Removal by Ferric-Chloride Coagulation—Effect of Phosphate, Bicarbonate and Silicate. *Water Sci. Technol.* **2011**, *64*, 1046–1055. [CrossRef]

48. Cravo, A.; Barbosa, A.B.; Correia, C.; Matos, A.; Caetano, S.; Lima, M.J.; Jacob, J. Unravelling the Effects of Treated Wastewater Discharges on the Water Quality in a Coastal Lagoon System (Ria Formosa, South Portugal): Relevance of Hydrodynamic Conditions. *Mar. Pollut. Bull.* **2022**, *174*, 113296. [CrossRef]

49. Luvhimbi, N.; Tshitangano, T.G.; Mabunda, J.T.; Olaniyi, F.C.; Edokpayi, J.N. Water Quality Assessment and Evaluation of Human Health Risk of Drinking Water from Source to Point of Use at Thulamela Municipality, Limpopo Province. *Sci. Rep.* **2022**, *12*, 6059. [CrossRef]

50. Hile, T.D.; Dunbar, S.G.; Sinclair, R.G. Microbial Contamination of Drinking Water from Vending Machines of Eastern Coachella Valley. *Water Supply* **2021**, *21*, 1618–1628. [CrossRef]

51. Horváth, E.; Gabathuler, J.; Bourdiec, G.; Vidal-Revel, E.; Benthem Muñiz, M.; Gaal, M.; Grandjean, D.; Breider, F.; Rossi, L.; Sienkiewicz, A.; et al. Solar Water Purification with Photocatalytic Nanocomposite Filter Based on TiO_2 Nanowires and Carbon Nanotubes. *npj Clean Water* **2022**, *5*, 10. [CrossRef]

52. García, A.; Rosales, M.; Thomas, M.; Golemme, G. Arsenic Photocatalytic Oxidation over TiO_2-Loaded SBA-15. *J. Environ. Chem. Eng.* **2021**, *9*, 106443. [CrossRef]

53. Wei, Z.; Fang, Y.; Wang, Z.; Liu, Y.; Wu, Y.; Liang, K.; Yan, J.; Pan, Z.; Hu, G. PH Effects of the Arsenite Photocatalytic Oxidation Reaction on Different Anatase TiO_2 Facets. *Chemosphere* **2019**, *225*, 434–442. [CrossRef] [PubMed]

54. Kanth, N.; Xu, W.; Prasad, U.; Ravichandran, D.; Kannan, A.M.; Song, K. PMMA-TiO_2 Fibers for the Photocatalytic Degradation of Water Pollutants. *Nanomaterials* **2020**, *10*, 1279. [CrossRef] [PubMed]

55. Li, Z.; Zhang, Z.; Dong, Z.; Wu, Y.; Zhu, X.; Cheng, Z.; Liu, Y.; Wang, Y.; Zheng, Z.; Cao, X.; et al. CuS/TiO_2 Nanotube Arrays Heterojunction for the Photoreduction of Uranium (VI). *J. Solid State Chem.* **2021**, *303*, 122499. [CrossRef]

56. Kuhn, H.; Försterling, H.-D.; Waldeck, D.H. Chemical Kinetics. In *Principles of Physical Chemistry*; John Wiley & Sons, LTD: West Sussex, UK, 2000; pp. 735–794, ISBN 978-0-470-08964-4.

57. Giménez, J.; Curcó, D.; Queral, M. Photocatalytic Treatment of Phenol and 2,4-Dichlorophenol in a Solar Plant in the Way to Scaling-Up. *Catal. Today* **1999**, *54*, 229–243. [CrossRef]

58. Gutiérrez-Alfaro, S.; Acevedo, A.; Rodríguez, J.; Carpio, E.A.; Manzano, M.A. Solar Photocatalytic Water Disinfection of *Escherichia coli*, *Enterococcus* spp. and *Clostridium perfringens* Using Different Low-Cost Devices. *J. Chem. Technol. Biotechnol.* **2016**, *91*, 2026–2037. [CrossRef]

59. Mac Mahon, J.; Pillai, S.C.; Kelly, J.M.; Gill, L.W. Solar Photocatalytic Disinfection of *E. Coli* and Bacteriophages MS2, ΦX174 and PR772 Using TiO_2, ZnO and Ruthenium Based Complexes in a Continuous Flow System. *J. Photochem. Photobiol. B Biol.* **2017**, *170*, 79–90. [CrossRef]

60. Pereira, J.H.O.S.; Vilar, V.J.P.; Borges, M.T.; González, O.; Esplugas, S.; Boaventura, R.A.R. Photocatalytic Degradation of Oxytetracycline Using TiO_2 under Natural and Simulated Solar Radiation. *Sol. Energy* **2011**, *85*, 2732–2740. [CrossRef]

61. Rincón, A.-G.; Pulgarin, C. Field Solar *E. Coli* Inactivation in the Absence and Presence of TiO_2: Is UV Solar Dose an Appropriate Parameter for Standardization of Water Solar Disinfection? *Sol. Energy* **2004**, *77*, 635–648. [CrossRef]

62. Bolton, J.R.; Bircher, K.G.; Tumas, W.; Tolman, C.A. Figures-of-Merit for the Technical Development and Application of Advanced Oxidation Technologies for Both Electric- and Solar-Driven Systems (IUPAC Technical Report). *Pure Appl. Chem.* **2001**, *73*, 627–637. [CrossRef]

63. Faul, F.; Erdfelder, E.; Buchner, A.; Lang, A.-G. Statistical Power Analyses Using G*Power 3.1: Tests for Correlation and Regression Analyses. *Behav. Res. Methods* **2009**, *41*, 1149–1160. [CrossRef]

64. Manjiao, C.; Zhengfu, Z.; Xinjun, H.; Jianping, T.; Jingsong, W.; Rundong, W.; Xian, Z.; Xinjun, Z.; PeiLun, S.; Dianwen, L. Oxidation Mechanism of the Arsenopyrite Surface by Oxygen with and without Water: Experimental and Theoretical Analysis. *Appl. Surf. Sci.* **2022**, *573*, 151574. [CrossRef]

65. Heiba, H.F.; Bullen, J.C.; Kafizas, A.; Petit, C.; Skinner, S.J.; Weiss, D. The Determination of Oxidation Rates and Quantum Yields during the Photocatalytic Oxidation of As(III) over TiO_2. *J. Photochem. Photobiol. A Chem.* **2022**, *424*, 113628. [CrossRef]

66. Su, J.; Lyu, T.; Cooper, M.; Mortimer, R.J.G.; Pan, G. Efficient Arsenic Removal by a Bifunctional Heterogeneous Catalyst through Simultaneous Hydrogen Peroxide (H_2O_2) Catalytic Oxidation and Adsorption. *J. Clean. Prod.* **2021**, *325*, 129329. [CrossRef]

67. Wang, X.; Chen, J.; Bu, Z.; Wang, H.; Wang, W.; Li, W.; Sun, T. Accelerated C-Face Polishing of Silicon Carbide by Alkaline Polishing Slurries with Fe_3O_4 Catalysts. *J. Environ. Chem. Eng.* **2021**, *9*, 106863. [CrossRef]

68. Hong, J.; Liu, L.; Ning, Z.; Liu, C.; Qiu, G. Synergistic Oxidation of Dissolved As(III) and Arsenopyrite in the Presence of Oxygen: Formation and Function of Reactive Oxygen Species. *Water Res.* **2021**, *202*, 117416. [CrossRef]

69. Song, J.; Yan, L.; Duan, J.; Jing, C. TiO_2 Crystal Facet-Dependent Antimony Adsorption and Photocatalytic Oxidation. *J. Colloid Interface Sci.* **2017**, *496*, 522–530. [CrossRef]

70. Ning, R.Y. Arsenic Removal by Reverse Osmosis. *Desalination* **2002**, *143*, 237–241. [CrossRef]

71. Litter, M.I. Last Advances on TiO_2-Photocatalytic Removal of Chromium, Uranium and Arsenic. *Curr. Opin. Green Sustain. Chem.* **2017**, *6*, 150–158. [CrossRef]
72. Silerio-Vázquez, F.; Proal Nájera, J.B.; Bundschuh, J.; Alarcon-Herrera, M.T. Photocatalysis for Arsenic Removal from Water: Considerations for Solar Photocatalytic Reactors. *Environ. Sci. Pollut. Res.* **2021**. [CrossRef]
73. Meng, F.; Zhang, S.; Zeng, Y.; Zhang, M.; Zou, H.; Zhong, Q.; Li, Y. Promotional Effect of Surface Fluorine on TiO_2: Catalytic Conversion of O_3 and H_2O_2 into ·OH and ·O_2^- Radicals for High-Efficiency NO Oxidation. *Chem. Eng. J.* **2021**, *424*, 130358. [CrossRef]
74. Naniwa, S.; Yamamoto, A.; Yoshida, H. Visible Light-Induced Minisci Reaction through Photoexcitation of Surface Ti-Peroxo Species. *Catal. Sci. Technol.* **2021**, *11*, 3376–3384. [CrossRef]
75. Shan, C.; Liu, Y.; Huang, Y.; Pan, B. Non-Radical Pathway Dominated Catalytic Oxidation of As(III) with Stoichiometric H_2O_2 over Nanoceria. *Environ. Int.* **2019**, *124*, 393–399. [CrossRef] [PubMed]
76. Ren, Y.; Liu, Y.; Liu, F.; Li, F.; Shen, C.; Wu, Z. Extremely Efficient Electro-Fenton-like Sb(III) Detoxification Using Nanoscale Ti-Ce Binary Oxide: An Effective Design to Boost Catalytic Activity via Non-Radical Pathway. *Chin. Chem. Lett.* **2021**, *32*, 2519–2523. [CrossRef]
77. Ghanbari, F.; Giannakis, S.; Lin, K.-Y.A.; Wu, J.; Madihi-Bidgoli, S. Acetaminophen Degradation by a Synergistic Peracetic Acid/UVC-LED/Fe(II) Advanced Oxidation Process: Kinetic Assessment, Process Feasibility and Mechanistic Considerations. *Chemosphere* **2021**, *263*, 128119. [CrossRef]
78. Zhang, X.; Dong, Q.; Wang, Y.; Zhu, Z.; Guo, Z.; Li, J.; Lv, Y.; Chow, Y.T.; Wang, X.; Zhu, L.; et al. Water-Stable Metal–Organic Framework (UiO-66) Supported on Zirconia Nanofibers Membrane for the Dynamic Removal of Tetracycline and Arsenic from Water. *Appl. Surf. Sci.* **2022**, *596*, 153559. [CrossRef]
79. Lou, T.; Song, S.; Gao, X.; Qian, W.; Chen, X.; Li, Q. Sub-20-Nm Anatase TiO_2 Anchored on Hollow Carbon Spheres for Enhanced Photocatalytic Degradation of Reactive Red 195. *J. Colloid Interface Sci.* **2022**, *617*, 663–672. [CrossRef]
80. Dudek, S.; Kołodyńska, D. Arsenic(V) Removal on the Lanthanum-Modified Ion Exchanger with Quaternary Ammonium Groups Based on Iron Oxide. *J. Mol. Liq.* **2022**, *347*, 117985. [CrossRef]
81. Chi, Z.; Zhu, Y.; Liu, W.; Huang, H.; Li, H. Selective Removal of As(III) Using Magnetic Graphene Oxide Ion-Imprinted Polymer in Porous Media: Potential Effect of External Magnetic Field. *J. Environ. Chem. Eng.* **2021**, *9*, 105671. [CrossRef]
82. Zeng, H.; Zhai, L.; Qiao, T.; Yu, Y.; Zhang, J.; Li, D. Efficient Removal of As(V) from Aqueous Media by Magnetic Nanoparticles Prepared with Iron-Containing Water Treatment Residuals. *Sci. Rep.* **2020**, *10*, 9335. [CrossRef]
83. Weerasundara, L.; Ok, Y.-S.; Bundschuh, J. Selective Removal of Arsenic in Water: A Critical Review. *Environ. Pollut.* **2021**, *268*, 115668. [CrossRef]
84. Hug, S.J.; Leupin, O. Iron-Catalyzed Oxidation of Arsenic (III) by Oxygen and by Hydrogen Peroxide: pH-Dependent Formation of Oxidants in the Fenton Reaction. *Environ. Sci. Technol.* **2003**, *37*, 2734–2742. [CrossRef]
85. García-Gil, Á.; Feng, L.; Moreno-SanSegundo, J.; Giannakis, S.; Pulgarín, C.; Marugán, J. Mechanistic Modelling of Solar Disinfection (SODIS) Kinetics of *Escherichia coli*, Enhanced with H_2O_2—Part 2: Shine on You, Crazy Peroxide. *Chem. Eng. J.* **2022**, *439*, 135783. [CrossRef]
86. Cowie, B.E.; Porley, V.; Robertson, N. Solar Disinfection (SODIS) Provides a Much Underexploited Opportunity for Researchers in Photocatalytic Water Treatment (PWT). *ACS Catal.* **2020**, *10*, 11779–11782. [CrossRef]
87. Oturan, M.A.; Aaron, J.-J.J. Advanced Oxidation Processes in Water/Wastewater Treatment: Principles and Applications. A Review. *Crit. Rev. Environ. Sci. Technol.* **2014**, *44*, 2577–2641. [CrossRef]
88. Amyot, M.; Bélanger, D.; Simon, D.F.; Chételat, J.; Palmer, M.; Ariya, P. Photooxidation of Arsenic in Pristine and Mine-Impacted Canadian Subarctic Freshwater Systems. *J. Hazard. Mater. Adv.* **2021**, *2*, 100006. [CrossRef]
89. Saleh, S.; Mohammadnejad, S.; Khorgooei, H.; Otadi, M. Photooxidation/Adsorption of Arsenic (III) in Aqueous Solution over Bentonite/Chitosan/TiO_2 Heterostructured Catalyst. *Chemosphere* **2021**, *280*, 130583. [CrossRef]
90. Shen, J.; Yu, H.; Shu, Y.; Ma, M.; Chen, H. A Robust ROS Generation Strategy for Enhanced Chemodynamic/Photodynamic Therapy via H_2O_2/O_2 Self-Supply and Ca^{2+} Overloading. *Adv. Funct. Mater.* **2021**, *31*, 2106106. [CrossRef]
91. Huang, L.; Liu, S.; Li, X.; Peng, X.; Liu, D. Controllable High-Efficiency Transformation of H_2O_2 to Reactive Oxygen Species via Electroactivation of Ti-Peroxo Complexes. *Sep. Purif. Technol.* **2022**, *289*, 120747. [CrossRef]
92. Pretali, L.; Fasani, E.; Sturini, M. Current Advances on the Photocatalytic Degradation of Fluoroquinolones: Photoreaction Mechanism and Environmental Application. *Photochem. Photobiol. Sci.* **2022**, *21*, 899–912. [CrossRef]
93. Abdel-Maksoud, Y.K.; Imam, E.; Ramadan, A.R. TiO_2 Water-Bell Photoreactor for Wastewater Treatment. *Sol. Energy* **2018**, *170*, 323–335. [CrossRef]
94. Ding, Y.; Zhou, W.; Gao, J.; Sun, F.; Zhao, G. H_2O_2 Electrogeneration from O_2 Electroreduction by N-Doped Carbon Materials: A Mini-Review on Preparation Methods, Selectivity of N Sites, and Prospects. *Adv. Mater. Interfaces* **2021**, *8*, 2002091. [CrossRef]
95. Dharma, H.N.C.; Jaafar, J.; Widiastuti, N.; Matsuyama, H.; Rajabsadeh, S.; Othman, M.H.D.; Rahman, M.A.; Jafri, N.N.M.; Suhaimin, N.S.; Nasir, A.M.; et al. A Review of Titanium Dioxide (TiO_2)-Based Photocatalyst for Oilfield-Produced Water Treatment. *Membranes* **2022**, *12*, 345. [CrossRef] [PubMed]

96. Zhang, J.; Mo, Y. A Scalable Light-Diffusing Photochemical Reactor for Continuous Processing of Photoredox Reactions. *Chem. Eng. J.* **2022**, *435*, 134889. [CrossRef]
97. Kanakaraju, D.; anak Kutiang, F.D.; Lim, Y.C.; Goh, P.S. Recent Progress of Ag/TiO$_2$ Photocatalyst for Wastewater Treatment: Doping, Co-Doping, and Green Materials Functionalization. *Appl. Mater. Today* **2022**, *27*, 101500. [CrossRef]
98. Nur, A.S.M.; Sultana, M.; Mondal, A.; Islam, S.; Robel, F.N.; Islam, A.; Sumi, M.S.A. A Review on the Development of Elemental and Codoped TiO$_2$ Photocatalysts for Enhanced Dye Degradation under UV–Vis Irradiation. *J. Water Process Eng.* **2022**, *47*, 102728. [CrossRef]
99. Laky, D. Predictive model for drinking water treatment technology design – the efficiency of arsenic removal by in-situ formed ferric-hydroxide. *Period. Polytech. Civ. Eng.* **2010**, *54*, 45. [CrossRef]
100. Tshukudu, T.; Zheng, H.; Yang, J. Optimization of Coagulation with PFS-PDADMAC Composite Coagulants Using the Response Surface Methodology Experimental Design Technique. *Water Environ. Res.* **2013**, *85*, 456–465. [CrossRef]
101. Corral Bobadilla, M.; Lorza, R.; Escribano García, E.; Somovilla Gómez, F.; Vergara González, E. Coagulation: Determination of Key Operating Parameters by Multi-Response Surface Methodology Using Desirability Functions. *Water* **2019**, *11*, 398. [CrossRef]

International Journal of
*Environmental Research
and Public Health*

Article

White Light-Photolysis for the Removal of Polycyclic Aromatic Hydrocarbons from Proximity Firefighting Protective Clothing

Aline Marcelino Arouca [1,*], Victor Emmanuel Delfino Aleixo [2], Maurício Leite Vieira [3], Márcio Talhavini [3] and Ingrid Távora Weber [2]

1   Federal Institute of Education Science and Technology of Brasilia—IFB, Subcentro Leste—Complexo Boca da Mata, 02, Samambaia Sul, Brasilia 72302-300, Brazil
2   Laboratory of Inorganic and Materials (LIMA), Chemistry Institute, University of Brasilia—UNB, Brasilia 70904-970 , Brazil
3   National Institute of Criminalistics, Brazilian Federal Police, SAIS Quadra 07 Lote 23, Brasilia 70610-200, Brazil
*   Correspondence: aline.arouca@ifb.edu.br; Tel.: +55-61-99955-5220

Abstract: The presence of polycyclic aromatic hydrocarbons (PAHs) on firefighters' personal protective equipment is a concern. One form of preventing from these compounds is to decontaminate proximity firefighting protective clothing (PFPC). Traditional decontamination methods do not promote total removal of pollutants and alter the properties of PFPC. The objective of this work was to evaluate the effectiveness of white light-photolysis (WLP), an advanced oxidation process (AOP), for removing PAHs from PFPC, while maintaining the integrity of the fabric fibers. Experiments were carried out, varying reaction time and concentration of H_2O_2. With WLP (without H_2O_2), it was possible to remove more than 73% of the PAHs tested from the outer layer of PFPC in 3 days. The WLP provided the greatest removal of PAHs, compared with the most common mechanical decontamination techniques (laundering and wet-soap brushing). The fibers' integrity after exposure to the white light was evaluated with infrared spectroscopy and scanning electron microscopy/energy dispersive X-ray spectrometry. In addition, a tearing strength test was performed. No remarkable fabric degradation was observed, indicating a possible, routine-compatible, simple, and inexpensive method of decontamination of PFPC, based on photolysis, which is effective in the degradation of PAHs and maintains the integrity of fabric fibers.

Keywords: photolysis; polycyclic aromatic hydrocarbons (PAHs); firefighters; personal protective equipment (PPE); proximity firefighting protective clothing (PFPC); advanced oxidation process (AOP)

Citation: Arouca, A.M.; Aleixo, V.E.D.; Vieira, M.L.; Talhavini, M.; Weber, I.T. White Light-Photolysis for the Removal of Polycyclic Aromatic Hydrocarbons from Proximity Firefighting Protective Clothing. *Int. J. Environ. Res. Public Health* **2022**, *19*, 10054. https://doi.org/10.3390/ijerph191610054

Academic Editors: Gassan Hodaifa, Antonio Zuorro, Joaquín R. Dominguez, Juan García Rodríguez, José A. Peres and Zacharias Frontistis

Received: 6 June 2022
Accepted: 23 June 2022
Published: 15 August 2022

Publisher's Note: MDPI stays neutral with regard to jurisdictional claims in published maps and institutional affiliations.

1. Introduction

Firefighting is a high-risk activity, affecting the physical and mental health of these professionals. The International Agency for Research on Cancer (IARC) has classified firefighters' exposure to toxic materials as Group 2B, that is, possibly carcinogenic to humans [1]. As well as the obvious effects of combustion of hazardous materials, the contamination of firefighters' proximity protective clothing may promote danger to their health as well.

One of the main compounds that is formed during the combustion and pyrolysis process of a material are the polycyclic aromatic hydrocarbons (PAHs) [2,3]. PAHs represent a class of complex organic chemicals with more than 100 compounds containing two or more aromatic rings condensed in different ways in their structures [4]. PAHs can be formed by different pathways, as in an oil spill or oil seepage. The incomplete combustion of organic matter and fossil fuels is the most prominent source of PAHs in the environment [5]. The concern about the presence of these compounds in fire residues is due to the fact that many PAHs are considered carcinogenic, mutagenic, and teratogenic [6,7].

In the case of firefighters, the main routes of exposure are by inhalation and dermal absorption. One way to reduce these contaminations is through the use of a self-contained breathing apparatus (SCBA) and the use of proximity firefighting protective clothing (PFPC) [8]. For dermal contamination, exposure can occur through the deposit of the contaminants directly on the skin or through cross-contamination. This can be evidenced by the fact that even when full personal protective equipment (PPE) is being used, including SCBA, the PAHs level in firefighters is higher in post-combat situations than pre-combat situations [8–10]. This is due to the cross-contamination that happens after the firefighters' contact with a contaminated suit or equipment [11].

In order to reduce cross-contamination, an effective decontamination of all personal protective equipment must be carried out. A number of studies have been carried out to develop a safe and effective method of mechanical decontamination of proximity firefighting protective clothing. Most decontamination (decon) methods involve laundering, brushing, and using compressed air. Some of these mechanical decon techniques are not sufficient to remove PAHs completely [12–15]. Another important fact is that the PFPC is an expensive piece of equipment. In a 2017 bid made by the Military Firefighting Corps (Corpo de Bombeiros Militar) of the Federal district of Brasilia, the purchase price of a unit cost €1419.58 [16]. Considering this high purchase price per unit, suit conservation is extremely important, so any decon method must not damage it. Laundering, however, may alter the properties of fabric of the firefighting turnout gear [17–19].

An alternative to mechanical decon is chemical decontamination, such as advanced oxidation processes (AOPs). Lucena et al. [20] evaluated the ozonolysis decon of model PAHs (pyrene and 9-methylanthracene) in pieces of impregnated turnout firefighters' fabrics. Despite being able to partially decontaminate the sample, the result was unsatisfactory (due to the low degradation rate obtained and to the formation of compounds with similar or greater toxicity than parent compounds, such as phenanthrene-derived compounds).

Considering positive results of using photolysis, an effective and less expensive AOP used in the degradation of PAHs in liquid [21–24] and/or solid [25–27] samples, this represents a possible way to decontaminate firefighters' suits. This approach is favored because most PAHs absorb in the range of 210–386 nm [28,29]. In addition, photolysis is a simple, easy-to-perform, and low-cost method which a priori tends to not damage fabrics.

The goal of this paper is to evaluate the AOP chemical decon of firefighting protective clothing impregnated with PAHs using white light photolysis as an alternative to the mechanical decon methods already used. Besides studying the effect of reaction time on PAHs removal, we also evaluated the addition of hydrogen peroxide, an oxidizing agent that favors radical oxidation reactions. The aim of the study was to maximize the removal of polycyclic aromatic hydrocarbon pollutants while minimizing damage to the fabric fibers, and to maintain integrity of the fibers as well as the physical properties of the turnout gear.

2. Materials and Methods

2.1. Chemicals and Reagents

The analytes 9-methylanthracene (9MA, 98%) and pyrene (PYR, 98%) were purchased from Sigma-Aldrich (São Paulo, Brazil) and used without additional purification. Acetonitrile, suitable for HPLC (>99.9%, Exodo and JT Baker, São Paulo, Brazil), hydrogen peroxide (H_2O_2 35% w/v, Synth, São Paulo, Brazil), and deionized water (≥ 10 MΩ cm^{-1}), produced in a Milli-Q purification system, were used in all experiments.

All glassware was washed with a neutral detergent, and rinsed with deionized water, then re-rinsed with ethanol (Synth), ethyl acetate (Synth), and dichloromethane (Merk, São Paulo, Brazil), all HPLC grade.

2.2. White Light-Photolysis Experiments

All experiments were performed in a photoreactor equipped with a 250 W mercury vapor and tungsten filament white light (emission spectrum in Supplementary Figure S1), and two coolers to promote cooling and airflow, as shown in Supplementary Figure S3.

PAHs (9-methylanthracene and pyrene) were deposited onto a piece of fabric from the outer shell of the firefighting suits. The fabric had a nominal composition of 58% para-aramid (Kevlar®), 40% meta-aramid (Nomex®), and 2% carbon from Unishell® with a water repellant coating of fluorocarbon (Teflon®). The samples of fabric were provided by Santanense Workwear and were used in the manufacturing of Brazilian firefighters' protective clothing. Pieces of fabric were cut into 3×3 cm^2 squares (Supplementary Figure S4).

A 0.005 M pyrene and 9-methylanthracene stock solution was prepared in acetonitrile. Then, 200 µL was dripped onto the fabric samples. The solvent solution was air-dried for 10 min at room temperature. After 10 min, 200 µL of H_2O_2 was also deposited onto the fabric surface containing PAHs. H_2O_2 concentrations of 0%, 0.35%, and 3.5% of H_2O_2 in Milli-Q water were evaluated. Photolysis (WLP) was carried out inside the photoreactor for periods of 0, 1, and 3 days. A similar experiment was conducted without light irradiation (BLK), as reference.

After the photolysis reaction took place, the fabric samples were extracted with 10 mL of acetonitrile in a test tube. Each tube was placed into an ultrasonic bath for 20 min at room temperature. After this procedure, the extracts were stored in amber flasks and stored in a freezer. The extracts were analyzed by UV/VIS spectroscopy.

The kinetics were evaluated. Tests were conducted on the best condition of white light-photolysis reaction (0% of H_2O_2) and in different reaction times (0, 1, 3, 6, 9, 12, and 15 days). The extracts were obtained as previously described and stored in a freezer for further analysis. These samples were analyzed with a gas chromatogram coupled with mass spectrometer (GC/MS) in order to determine the PAHs concentration and to detect possible photolysis by-products.

2.3. Three-Layer Fabric White Light-Photolysis Decontamination

In order to analyze the effect of white light-photolysis decon within a more realistic scenario, the decontamination procedure was carried out with real samples of the three-layer fabric from used Brazilian firefighters' protective clothing.

The fabrics from the turnout gear were cut into 10 cm diameter disks and sewn together in the same order that is found on the Brazilian firefighters' gear (Supplementary Figure S5). Each disk was contaminated with 1.8 mL of the PAHs stock solution, then was air-dried for 10 min. The best reaction conditions found in the previous tests (reaction time and quantity of H_2O_2) were reproduced. For the PAHs extraction, 80 mL of acetonitrile was used with the fabric disks in an ultrasonic bath for 20 min at room temperature.

2.4. Three-Layer Fabric Mechanical Decontamination

To compare the white light-photolysis decon with mechanical decontamination techniques, a series of experiments were carried out on the three-layer fabric samples.

For the wet-soap brushing decon, the procedure described by Fent, K.W., et al. [13] was adapted. A neutral soap (0.5 mL of soap in 380 mL of water) solution was sprayed onto a pre-soaked three-layer sample using a spray bottle. Then, the disk was scrubbed 10 times with a plastic bristle brush and rinsed quickly, preventing water from penetrating the bottom layer of the sample.

Another common decon procedure which was evaluated was laundering. To evaluate the efficiency of this technique, an experiment using a bucket and a mechanical stirrer was carried out, simulating a washing machine. This adaptation was necessary given the small size (10 cm) of the disks (instead the whole suit).

The simulated laundering decon was adapted following the methodology described by NFPA 1851 [30], described in the Supplementary Material (Supplementary Figure S6). Two washing cycles (20 min then 10 min) were completed, using 1 mL of a commercial liquid laundry detergent. Then, three rinse cycles were completed: one for 10 min and two more for 5 min each. In all cycles, 4 L of water was used, and after each cycle time the water was completely drained.

2.5. Extract Analysis

2.5.1. UV-VIS Spectroscopy

After the decontamination experiments, absorption spectra of extracts (Varian ultraviolet/visible spectrophotometer, model Cary 5000) were obtained to determine the concentration of each target compound. For this, the extract was diluted in a volume of 1:1 (outer shell fabric experiments) and 1:2 (three-layer fabric experiments) for better analysis. The spectra were obtained in the 220–400 nm range.

Quantification of the analytes was conducted with an analytical curve (9-methylanthracene and pyrene). The curve and the ANOVA table are shown in Supplementary Material. The absorbance of samples was subtracted from blank samples (without the PAHs) for the concentration determination.

2.5.2. Gas Chromatography-Mass Spectrometry

The concentration of pyrene and 9-methylanthracene was determined by gas chromatograph (Agilent model 6890N) coupled with mass spectrometer (Agilent model 5973 inert), in samples from the kinetics evaluation. The ASTM method 8270E [31] was used with a Rxi$^{®}$-1 ms stationary phase capillary column, with 100% methylpolysiloxane, dimensions 25 m × 0.20 mm × 0.33 µm (RESTEK).

The chromatographic operating conditions were the following. The injector temperature was maintained at 280 °C, in Splitless mode with 1.3 µL injection. The column was maintained with a constant flow of helium at 0.6 mL.min^{-1}. The chromatographic oven programming had an initial temperature of 40 °C, held for 4 min, heating at a rate of 10 °C.min^{-1} to 320 °C, and remaining at this temperature for 2 min. The total time of analysis was 34 min. Solvent delay was used for 4.00 min and a gain factor of 20.

The GC/MS interface was maintained at 280 °C, and the mass spectrometer was operated in scan mode in the scan range from 35 to 500 m/z, with HiSense.u. The mass spectra obtained were analyzed using the Chemstation Data Analysis program and the NIST Search program (version 2.3). Quantification was conducted with external analytical curves of 9-methylanthracene and pyrene. All the data is available in the Supplementary Material (Figures S8, S9 and S21 and Tables S2–S4).

2.6. Fabric Evaluation

To assess whether the white light-photolysis decon would deteriorate the outer shell fabric used in firefighting protective clothing, Unishell$^{®}$ samples were analyzed. Samples of fabric without PAHs previous contamination were exposed for 3 days to white light (WLP) and without it (BLK), using 0%, 0.35%, and 3.5% H_2O_2. Additionally, a long exposure (30 days) test was performed with the best white light-photolysis condition.

2.6.1. Attenuated Total Reflectance Fourier Transform Infrared (ATR-FTIR) Spectroscopy

Infrared analysis was carried out using a Thermo Scientific Nicolet iS10 FTIR Spectrometer with The Thermo Scientific Smart iTX ATR sampling accessory. Spectra were collected in the range of 460–4000 cm^{-1} with a DTGS detector, KBr beam splitter, and a HeNe laser.

For each fabric sample, 10 FTIR spectra were collected at different locations on the sample and the average spectrum was obtained. Spectral baseline correction and normalization was carried out using Origin 2021 program.

2.6.2. Tearing Strength

The sample submitted to photolysis best condition for 30 days (long exposure) was chosen to be submitted to a tear resistance test. This was performed in accordance with ASTM D2261:2013 (2017)e1 [32], Standard Test Method for Tearing Strength of Fabrics by the Tongue (Single Rip) Procedure (Constant-Rate-of-Extension Tensile Testing Machine). The experimental conditions are described in the Table 1 below.

Table 1. Tearing strength experiment conditions.

Distance between the Jaws	75 mm
Speed	50 mm/min
Dynamometer Type	CRT
Calculation Methodology	Average of 5 peaks
Software Used	Bluehill 3
Dimensions of each jaw	Front: 2.5 mm × 7.5 mm Back: 2.5 mm × 7.5 mm
Tear Direction	Parallel to the warp and to the weft

2.6.3. Scanning Electron Microscopy with Energy Dispersive X-ray Spectrometry SEM/EDS

SEM/EDS images were obtained from fabric fibers after the treatment using a Zeiss, model EVO 15, Scanning Electron Microscope with EDS Oxford UltimMax 40. Fiber diameters were obtained using the Gwyddion® 2.60 program: 10 points were measured in 5 images for each sample (total of 50 measurements/sample). The images were obtained using a backscattered electron detector and low vacuum, without conducting coating. For the compositional maps, a 40 mm square silicon detector was used and AzTech software.

2.7. Statistical Evaluation

All experiments were conducted in triplicate ($n = 3$). All data obtained were evaluated using ANOVA test and paired-sample t test to assess the statistically significant differences among mean values.

3. Results and Discussion

3.1. White Light-Photolysis Experiments

The spectrum of the white lamp has several bands, both in lower wavelengths (283 nm, 322 nm, 353 nm) and in the longer wavelength region (465 nm, 496 nm, 498 nm, 545 nm, and 577 nm). Comparing the absorption spectra of the model PAHs with emission spectra, we can see a superposition of spectra, indicating that this lamp was suitable for the PAHs photolysis. All spectra are available in Supplementary Material (Figures S1 and S2).

Before analysis of the effect of H_2O_2 concentration and reaction time, an experiment was conducted in the dark (BLK) without using any energy source. After 3 days, the longest reaction time tested, approximately $20 \pm 1\%$ of PAHs removal had been obtained (Figure 1a). This removal is probably related to the carryover of the analytes due to the airflow maintained inside the photoreactor and due to the compound volatilization. The addition of an oxidizing agent (H_2O_2,) did not affect PAHs concentration (Figure 1b,c). About $23 \pm 6\%$ and $26 \pm 8\%$ of PAHs removal was obtained with 0.35% and 3.5% of H_2O_2, respectively. This may be related to the fact that hydroxyl radical formation is favored in an aqueous medium and light is necessary to break the O-O bond in H_2O_2 [24].

Determining the basal level of PAHs removal, the same experiment was conducted using the white light as an energy source. Again, the H_2O_2 content varied from 0 to 3.5% w/v. The best result was obtained without H_2O_2 (0% H_2O_2), as shown in Figure 1d. As observed in the BLK experiments, there was no change by adding an oxidizing agent. About $65 \pm 9\%$ and $83 \pm 12\%$ of removal was obtained with 1 and 3 days of light exposition. With the addition of 0.35 or 3.5% of H_2O_2, very similar results were achieved ($81 \pm 8\%$ and $76 \pm 5\%$, respectively, removal after 3 days). This result strongly suggests that PAHs tested are more susceptible to photoinduced degradation than to hydroxyl radical attacks. This is not surprising, as the reaction was not carried out in an aqueous media (in which analytes are insoluble). Moreover, considering potential applications, this can be seen as quite an encouraging result, since very mild conditions were used. Comparing the PAHs removal (Figure 2), the best condition was determined as 3 days of irradiation of with light without H_2O_2 (about 83% removal of PAHs).

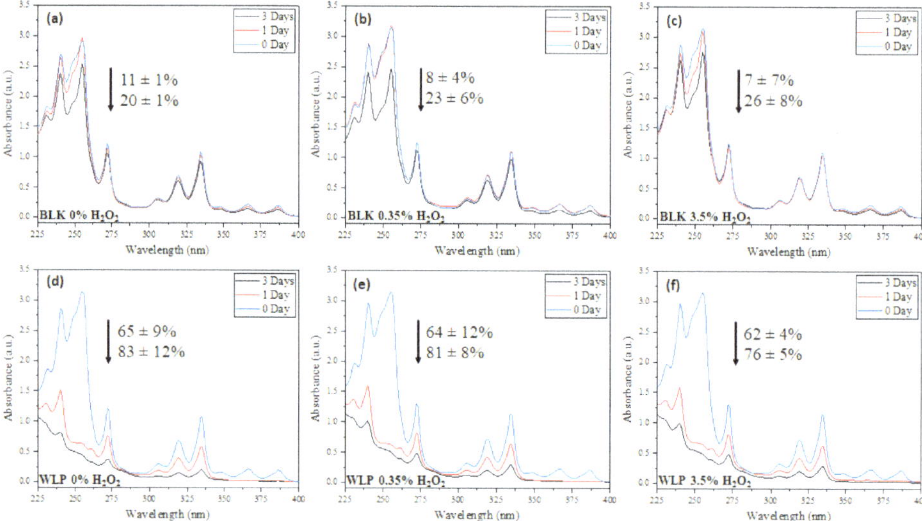

Figure 1. UV/VIS spectra of experiments performed in 1 and 3 days with (**a**) BLK 0% of H_2O_2, (**b**) BLK 0.35% of H_2O_2, (**c**) BLK 3.5% of H_2O_2, (**d**) WLP 0% of H_2O_2, (**e**) WLP 0.35% of H_2O_2, and (**f**) WLP 3.5% of H_2O_2.

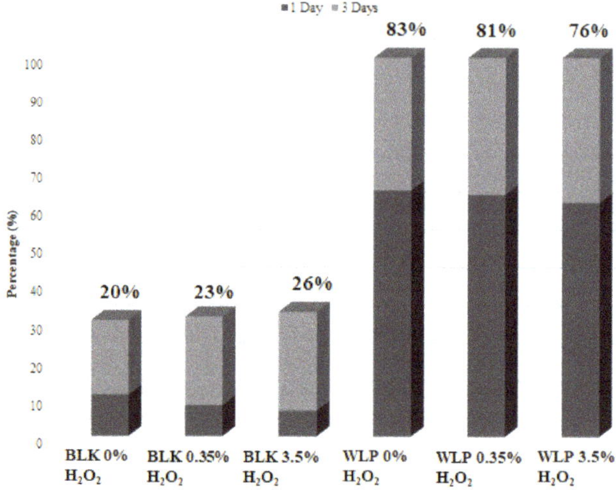

Figure 2. Comparative graph of PAHs removal in all tested conditions: BLK 0% of H_2O_2, BLK 0.35% of H_2O_2, BLK 3.5% of H_2O_2, WLP 0% of H_2O_2, WLP 0.35% of H_2O_2, and WLP 3.5% of H_2O_2.

3.2. White Light-Photolysis Kinetics

For determination of the degradation kinetics of each PAHs, the samples were submitted to longer exposure and the extracts obtained after the WLP (without H_2O_2) were analyzed by GC/MS. Both 9-methylanthracene and pyrene follow a pseudo-second-order kinetics. The plot of 1/[PAH] versus reaction time was used to determine the rate constant K (Figure 3).

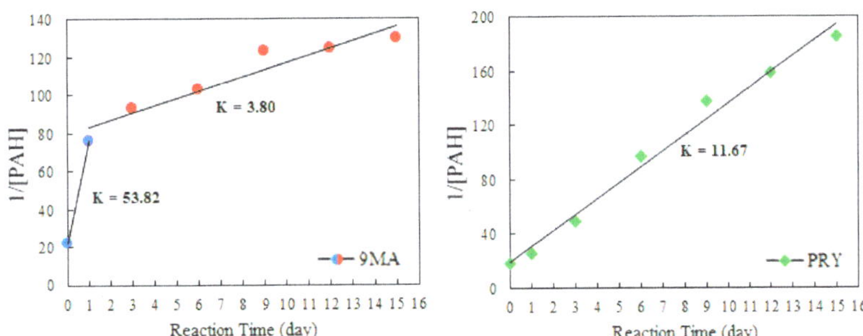

Figure 3. Pseudo second order rate constants for WLP (with 0% of H_2O_2) degradation reaction of 9-methylanthracene and pyrene.

The 9-methylanthracene curve draws attention for presenting two regions: initially a faster reaction, with a rate constant of 53.8, and then a slower kinetic, with K = 3.8. Pyrene, on the other hand, presents a well-behaved curve with a unique constant rate K = 11.67. The efficiencies obtained throughout the test are described in Figure 4.

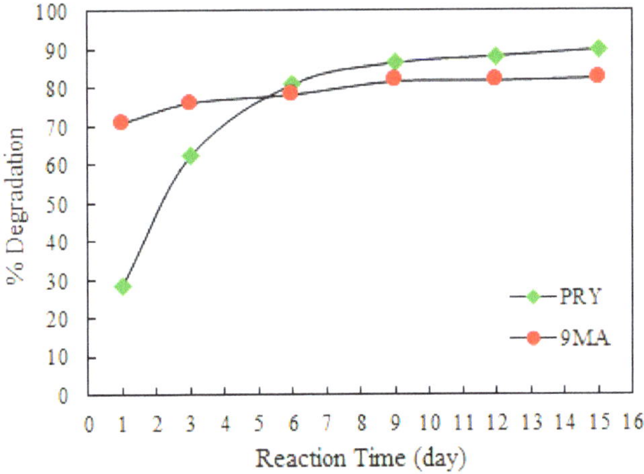

Figure 4. WLP (0% of H_2O_2) efficiency in degradation of PAHs obtained in 0, 1, 3, 6, 9, 12, and 15 days.

For 9MA, given that the first step reaction was relatively fast, 70% of the compound was removed after only one day. Considering the cost–benefit, this implies that the photolysis decon method could be performed in one day. With the slower kinetics of PRY, on the other hand, relevant removal is observed around the 6th day of exposure. The optimal results, considering both PAHs, were obtained between 6th and 9th days. In real situations, however, this exposure time is very long and works against optimization of the cost–benefit ratio.

Additionally, the mass spectra obtained in this experiment were evaluated and no PAHs derivatives more harmful than the original PAHs were formed.

3.3. Evaluation of Fiber Integrity

In order to assess fiber integrity after photolysis, Fourier Transform Infrared/Attenuated Total Reflection (FTIR/ATR) spectra and SEM/EDS images were acquired, and a tearing strength test was carried out.

3.3.1. FTIR/ATR

The fabric contained a mixture of Nomex® (meta-aramid) and Kevlar® (para-aramid) fibers, which can undergo oxidation forming acids, alcohols, and/or amides [33–35]. In the case of fiber degradation, it was possible to observe a reduction in the intensity of the amide bands and an increase in the intensity of the bands in the O-H and N-H regions. FTIR spectrum of untreated fabric was compared to treated samples. The spectra are shown in Figure 5.

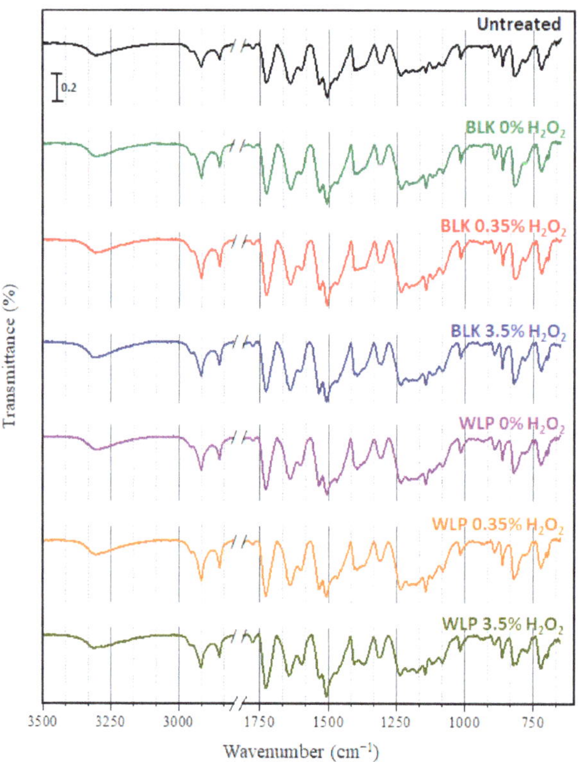

Figure 5. FTIR/RTA spectra of untreated sample and treated samples after exposure for 3 days: BLK 0% of H_2O_2, BLK 0.35% of H_2O_2, BLK 3.5% of H_2O_2, WLP 0% of H_2O_2, WLP 0.35% of H_2O_2, and WLP 3.5% of H_2O_2.

All the spectra are similar. No new band was observed nor even the widening or shifting of existing bands, suggesting that the fabric did not undergo chemical changes through treatment. The band at 3307 cm^{-1} refers to the N-H stretch of the amide of the para-aramid and meta-aramid compounds, whereas the band at 1724 cm^{-1} refers to the C=O stretch. The fibers have aromatic rings; it is possible to identify the C=C aromatic stretch at 1640 cm^{-1}. The region at 1017–820 cm^{-1} corresponds to the out-of-plane vibration of the C-H, indicating a meta and para substitution. Bands in the 2918–2850 cm^{-1} range are characteristic of the C-H sp^3 stretch. However, since fibers are not expected to have saturated carbons, the bands must have come from water repellant carbon-based coating, which seems to have been preserved during treatment. For the Unishell® clothing, the outer shell fabric has a fluorocarbon (Teflon®) coating, which is consistent with the C-H sp^3 stretch. The other attributions of the bands are described in Table 2 and are in accordance with what is described in the literature [33,35,36].

Table 2. Assignment of FTIR/ATR bands.

Band (cm^{-1})	Assignment
3307	N-H stretch
2918–2850	C-H sp^3 stretch
1724	C=O amide stretch
1640	C=C aromatic stretch
1537–1472	N-H deformation in plane and C-N stretch
1411 and 1304	C-N aromatic stretch
1017–820	C-H vibration out-of-plane

In addition to the spectra obtained after 3 days of decontamination, an extended experiment was carried out with WLP with 0% of H$_2$O$_2$ for 30 days (720 h). Even after the longer exposure (Figure 6), no changes in spectral profile were observed.

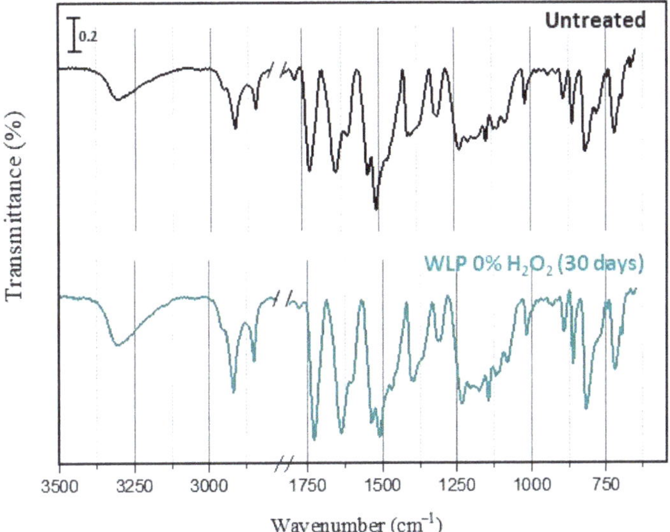

Figure 6. FTIR/ATR spectra of untreated sample and WLP 0% of H$_2$O$_2$ exposed for 30 days.

These results suggest the treatments based on photolysis degraded the deposited PAHs without altering the chemical structure of fibers. Even in prolonged experiments, no substantial differences were observed in the FTIR/ATR spectra.

Davis et al. [33] conducted a similar experiment, where outer shell fabrics used in firefighter jackets and pants were exposed to simulated ultraviolet sunlight at 50 °C and 50% relative humidity for different periods of time. The experiments were carried out in NIST SPHERE (Simulated Photodegradation via High Energy Radiant Exposure), a simulated photodegradation device. After 13 days exposure, the researchers observed a significant degradation of the water repellant coating. Additionally, changes in the fibers were identified by formation of compounds resulting from oxidation and break-up of the amides bonds. In our work, a lower energy irradiation source was used (white light), therefore there was less degradation of fabric than expected. In addition, the use of white light is cheaper and imposes less risk to the operator. The satisfactory result obtained using white light can be partially attributed to the match between the absorption of the PAHs and the emission of the used lamp.

3.3.2. Tearing Strength

As described in the NIST report 1751 [34], a way to assess changes in fabric constitution and possible wear from the decontamination process is by performing a tearing strength test. The NFPA 1971 [37] standard, which determines the necessary conditions for firefighting protective clothing, indicates that the fabrics of the outer layer of combat suits must present tear resistance higher than 100 N.

For the tearing strength test, only the sample treated for 30 days with WLP and 0% H_2O_2 was compared to the untreated fabric. For the Unishell® fabric, the tearing strengths were 260.97 ± 10.1 N and 218.39 ± 28.8 N for warp and weft, respectively. After the WLP decontamination process, tear strength values of 192.62 ± 31.8 N and 181.52 ± 22.9 N were obtained for warp and weft, respectively (Table 3).

Table 3. Tearing strength (N) values with expanded uncertainty with 95% confidence, performed in accordance with ASTM D2261 [32].

	Warp	Weft
Untreated sample of fabric	260.97 ± 10.1	218.39 ± 28.8
After 30 days WLP decon	192.62 ± 31.8	181.52 ± 22.9
ΔN (absolute)	68.35	36.87
ΔN (%)	26%	17%

There was a reduction of warp (26%) and weft (17%) values, which could be related to the breaking of the bonds in the fabric's water repellant coating, and not necessarily due to fiber degradation, as described by Davis et al. [33]. This fiber treatment makes the fabric more resistant, increasing the force needed to tear it. By reducing the repellant coating there is a reduction in the tearing resistance. This fact indicates that there is a reduction in the fabric's water repellant coating, not in the fabric's fibers bonds, that is, no alterations were observed in the FTIR/ATR spectra that would indicate a break-up of the para-aramid and meta-aramid molecule.

3.3.3. SEM/EDS

Micrographs of outer shell fabric were obtained to visualize the morphology of the fibers and investigate possible alterations. Images were obtained after decontamination for 3 and 30 days, as well as for the untreated sample, as shown in Figure 7.

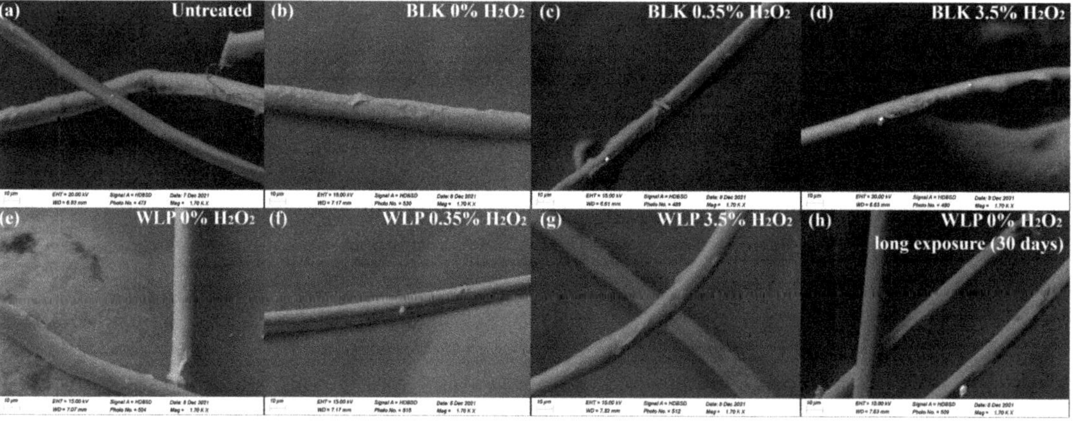

Figure 7. Micrographs of fibers before and after treatment with (**a**) untreated sample, (**b**) BLK 0% of H_2O_2, (**c**) BLK 0.35% of H_2O_2, (**d**) BLK 3.5% of H_2O_2, (**e**) WLP 0% of H_2O_2, (**f**) WLP 0.35% of H_2O_2, (**g**) WLP 3.5% of H_2O_2, and (**h**) WLP 0% of H_2O_2 for 30 days.

No change in fabric morphology was perceived and all fibers showed similar diameters within the standard deviation (Table 4). The results suggest no damage on fabric surface and the fibers maintained their structure, even for those undergoing photolysis with white light for 30 days.

Table 4. Fiber diameter.

Sample	Mean (μm)	Standard Deviation (μm)
Untreated	14.15	2.71
BLK 0%	13.90	1.50
BLK 0.35%	13.68	1.19
BLK 3.5%	14.35	1.62
WLP 0%	14.00	2.57
WLP 0.35%	14.36	1.26
WLP 3.5%	13.85	1.95
WLP 0% for 30 days	14.53	1.80

This result reinforces the conclusion that the reduction in tearing strength described in the previous item was associated with the Teflon® water repellant coating. Teflon® is a polytetrafluoroethylene (PTFE) polymer used as coating for various materials [38]. Due to the presence of a fluorine element, the presence of PTFE coating can be analyzed by scanning electron microscopy (SEM) with energy dispersive X-ray spectrometry (EDS). For this, elemental mapping of several points on untreated and treated fabric for 30 days with WLP 0% H_2O_2 by SEM/EDS were obtained. (Figure 8). The untreated sample showed a higher density of fluorine, indicating the presence of the protective PTFE coating layer. The 30 days treated sample showed a reduction in fluorine distribution, which indicates the partial removal of PTFE coating in the outer shell samples, corroborating the idea that polyamides remain intact after treatment and reduction in tearing strength can be attributed to PTFE partial removal. It is important to point out that PTFE coating removal is expected and occurs naturally with the use of the turnout gear and needs to be replaced after some period. All maps obtained are available in Supplementary Material.

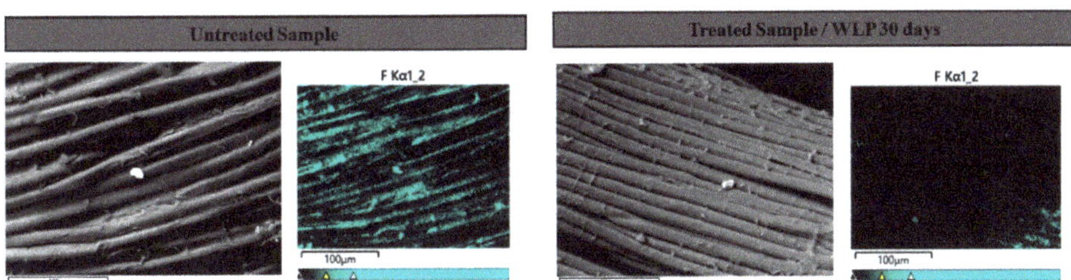

Figure 8. Compositional maps for fluorine for the untreated sample and for the treated sample with WLP and 0% H_2O_2 for 30 days.

3.4. Three-Layer Fabric Decontamination: Mechanical Decon vs. White Light-Photolysis Decon

To compare the WLP with traditional mechanical decon, pieces of three-layer fabric were contaminated with pyrene and 9-methylanthracene. The photolysis was carried out according to previous tests carried out for 3 days. Wet soap brushing and laundering was performed as described by Fent et al. [13].

After performing the tests, the samples were extracted with acetonitrile and the extracts analyzed by UV/VIS for PAHs quantification. Degradation rates and spectra related to each decontamination are depicted in Figure 9.

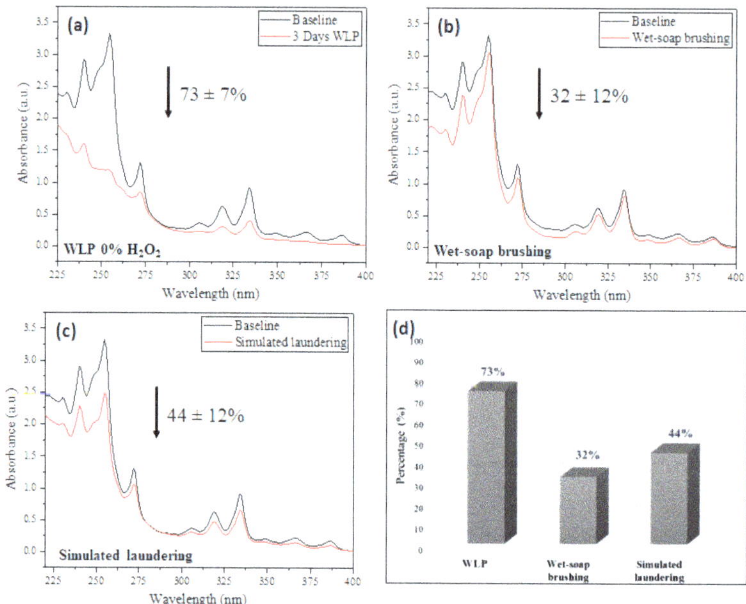

Figure 9. UV/VIS spectra of decontamination procedure performed by (**a**) WLP for 3 days without H_2O_2; (**b**) simulated laundering; and (**c**) wet-soap brushing. (**d**) Summary of percentage of PAHs removal.

The simulated laundering reduced the PAHs contamination by $44 \pm 12\%$ of PAHs, while the wet-soap brushing reduced it by $32 \pm 12\%$. Decontamination via photolysis gave the best result, $73 \pm 7\%$. This result demonstrates that photolysis with white light is efficient in removing the PAHs deposited on firefighters' protective clothing up to three layers of fabric.

One of the main complaints related to laundering and wet-soap brushing is the fact that the turnout gear remains damp after decontamination. This increases the risk of burns, in addition to the discomfort and extra weight [39] or the need to wait a drying interval before using the suit. The WLP, being a dry method, provides an advantage, in addition to being relatively fast and efficient.

Another advantage of WLP is that there is a lower risk of cross contamination. The firefighter ends up having less contact with the contaminated protective clothing when compared with mechanical decontamination methods. When laundering is done at home (the most common in Brazil), there is a greater risk of contaminating of other pieces of clothes and even other residents, due to the use of the washing machine.

Another important fact is the durability of the turnout gear. The washing process can promote further fiber degradation. Horn et al. [17] identified a reduction in tearing strength tests after laundering. The group evaluated the three layers of fabric and three methods of decontamination: laundering in a washing machine, wet-soap brushing, and dry brushing. Laundering had lower trap tear strength than the other treatments evaluated. After 10 washes, there was a reduction of 1.2–9.1%. With 20 washes the reduction was 9.3–19.6%, and with 40 washes it was 24.1–41.1%. Moreover, when fabric was submitted to 40 times in laundering machine, tear strength dropped below 100 N, the minimum requirement following NFPA 1971 [37].

The results discussed above could have been affected by experimental details and, therefore, present discrepancies between studies conducted by different groups. In this study, laundering was conducted on a small scale, using an adapted experimental set up to simulate a washing machine. The bucket adaptation does not promote the friction that facilitates washing, generated in the conventional washing machine. This may have disfavored PAHs removal and caused the lowest decontamination rate. Some studies have published more efficient results using mechanical decon methods in the removal of analytes than those described in this work.

Fent et al. [13] obtained wet-soap decon as the most effective technique in reducing PAHs contamination, with an 85% reduction. It is important to note that those authors did not use any chemical means for decontamination. In another study, performed by Banks, et al. [12], laundering decon method failed to promote the total removal of the PAHs and other compounds deposited on firefighting clothing after a fire event. In fact, in some cases, they found a higher concentration, indication a cross-contamination.

4. Conclusions

The goal of this research was to promote a chemical method for decontamination of polycyclic aromatic hydrocarbons deposited on firefighting turnout gear, using white light photolysis. The method provides an alternative to mechanical decon processes already used by firefighters.

The WLP decon promoted the removal of pyrene and 9-methylanthracene analytes deposited on fabrics of the outer layer of firefighter turnout gear without the need to add an oxidizing agent. After 3 days, the method promoted the removal of $81 \pm 8\%$ of both PAHs following a pseudo-second-order rate. The proposed decontamination route was compared with commonly applied techniques, laundering and wet-soap brushing, on three-layer fabric. Photolysis showed a removal of 73% of the deposited PAHs, while laundering removed 44% and wet-soap brushing 32%.

Regarding the fabric analysis, the protective clothing outer shell fabrics submitted to 3 and 30 days of photolysis were evaluated to assess whether the white light-photolysis decon promoted deterioration of the fibers. The chemical structure of the fibers was evaluated with FTIR/ATR; no changes were obtained in the spectra that could indicate the breakdown of fiber polymers. The tearing strength test presented a reduction in tearing strength by 17–26% after 30 days of light exposure. This reduction is not necessarily due to fiber degradation but related to the partial removal of fabric's water repellant coating. The treated sample showed a reduction on fluorine density, indicating the suggesting that the Teflon was degraded, but still was present. No change was observed in the fibers' morphology.

In view of these results, the 3 days WLP without H_2O_2 can be considered an efficient chemical decon method for the removal of PAHs deposited on proximity firefighting protective clothing, maintaining fabric's integrity and properties. The WLP decon has several advantages in comparison to what is currently used. Besides promoting greater removal of PAHs than the mechanical decon techniques, it is not necessary with WLP to use water on the turnout gear, which can result in burns if the suit is used while still damp. In addition, the technique developed is simple, cheap, environmentally friendly, and safer for the firefighter, as it reduces contact with contaminants.

Supplementary Materials: The following supporting information can be downloaded at: https://www.mdpi.com/article/10.3390/ijerph191610054/s1, Figure S1: Emission spectrum of 250 W mercury vapor and tungsten filament white light; Figure S2: spectrum of 250 W mercury vapor and tungsten filament white light; Figure S3: Photoreactor equipped white light; Figure S4: Samples of fabric from the outer shell of the firefighting uniform with composition of 58% para-aramid (kevlar®), 40% meta-aramid (nomex®) and 2% carbon, from Unishell®; Figure S5: Samples of three-layer fabrics from Brazilian firefighters' protective clothing. (a) jacket pieces, (b) three-layer fabrics samples (c) disk with the three layers sewn together; Figure S6: Scheme of three-layer fabric laundering decontamination; Figure S7: UV/VIS (a) calibration curve for PAHs solution and (b) residual graph; Table S1: UV/VIS ANOVA table for PAHs solution; Figure S8: GC/MS (a) calibration curve for PAHs solution and (b) residual graph for 9-methylanthracene; Table S2: GC/MS ANOVA table for 9-methylanthracene; Figure S9: GC/MS (a) calibration curve for PAHs solution and (b) residual graph for pyrene; Table S3: GC/MS ANOVA table for pyrene; Table S4. Limits of detection (LD) and quantification (LQ), in mmol.L^{-1}, obtained by the analytical curves for the analytes using GC/MS;

Figure S10: UV/VIS spectra of experiments performed in 0, 1 and 3 days with BLK 0% of H_2O_2; Figure S11: UV/VIS spectra of experiments performed in 0, 1 and 3 days with BLK 0.35% of H_2O_2; Figure S12: UV/VIS spectra of experiments performed in 0, 1 and 3 days with BLK 3.5% of H_2O_2; Figure S13: UV/VIS spectra of experiments performed in 0, 1 and 3 days with WLP 0% of H_2O_2; Figure S14: UV/VIS spectra of experiments performed in 0, 1 and 3 days with WLP 0.35% of H_2O_2; Figure S15: UV/VIS spectra of experiments performed in 0, 1 and 3 days with WLP 3.5% of H_2O_2; Figure S16: UV/VIS spectra of decontamination procedure performed for WLP decon for 3 days with 0% of H_2O_2; Figure S17: UV/VIS spectra of decontamination procedure performed for simulated laundering decon; Figure S18: UV/VIS spectra of decontamination procedure performed for wet-soap brushing decon; Figure S19: Pseudo first order rate constants for WLP 0% of H_2O_2 reaction of (a) 9-methylanthracene and (b) pyrene degradation; Figure S20: Pseudo second order rate constants for WLP 0% of H_2O_2 reaction of (a) 9-methylanthracene and (b) pyrene degradation; Figure S21: Chromatograms replicas of WLP 0% H_2O_2 experiments for the analyte 9-methylanthracene (a–c) obtained in the retention time of 23.673 min and for the analyte pyrene (d–f) obtained in the retention time of 25.166 min; Figure S22: Micrographs of fibers before treatment; Figure S23: Micrographs of fibers after treatment with BLK 0% of H_2O_2; Figure S24: Micrographs of fibers after treatment with BLK 0.35% of H_2O_2; Figure S25: Micrographs of fibers after treatment with BLK 3.5% of H_2O_2; Figure S26: Micrographs of fibers after treatment with WLP 0% of H_2O_2; Figure S27; Micrographs of fibers after treatment with WLP 0.35% of H_2O_2: Figure S28: Micrographs of fibers after treatment with WLP 3.5% of H_2O_2; Figure S29; Micrographs of fibers after treatment with WLP 0% of H_2O_2 for 30 days; Figure S30: Compositional maps for fluorine for the untreated sample and for the treated sample with WLP and 0% H_2O_2 for 30 days.

Author Contributions: Conceptualization, A.M.A., I.T.W. and M.T.; methodology, A.M.A., M.L.V. and I.T.W.; experimental analysis, A.M.A. and V.E.D.A.; investigation, all authors; data analysis, A.M.A., M.L.V. and V.E.D.A.; writing—original draft preparation, A.M.A.—review and editing, all coauthors; supervision, I.T.W.; project administration, I.T.W. All authors have read and agreed to the published version of the manuscript.

Funding: This research was funded by Conselho Nacional de Desenvolvimento Científico e Tecnológico (CNPq), by Fundação de Apoio à Pesquisa do Distrito Federal (FAPDF), by Coordenação de Aperfeiçoamento de Pessoal de Nível Superior (CAPES/PROFORENSE NEQUIFOR 3509/2014) and Universidade de Brasília (DPG/UnB 001/2021).

Institutional Review Board Statement: Not applicable.

Informed Consent Statement: Not applicable.

Data Availability Statement: Please, find the Supplementary Material. For the moment, no additional datasets are publicly available online.

Acknowledgments: We thank the Brazilian Federal Police and experts from SEPLAB-PF for research collaboration. Also, we thank Santanense Workwear for fabric samples. This work was supported by CNPq, CAPES PROFORENSE NEQUIFOR 3509/2014 and DPG/UnB. The English text of this paper has been revised by Sidney Pratt, Canadian, MAT (The Johns Hopkins University), RSAdip—TESL (Cambridge University).

Conflicts of Interest: The authors declare no conflict of interest.

References

1. World Health Organization; International Agency for Research on Cancer. *Painting, Firefighting, and Shiftwork*; IARC Press: Lyon, France, 2010; Volume 98, ISBN 978-92-832-1298-0.
2. Abdel-Shafy, H.I., Mansour, M.S.M. A Review on Polycyclic Aromatic Hydrocarbons: Source, Environmental Impact, Effect on Human Health and Remediation. *Egypt. J. Pet.* **2016**, *25*, 107–123. [CrossRef]
3. Oanh, N.T.K.; Reutergårdh, L.B.; Dung, N.T. Emission of Polycyclic Aromatic Hydrocarbons and Particulate Matter from Domestic Combustion of Selected Fuels. *Environ. Sci. Technol.* **1999**, *33*, 2703–2709. [CrossRef]
4. Meire, R.O.; Azeredo, A.; Torres, J.P.M. Aspectos Ecotoxicológicos de Hidrocarbonetos Policíclicos Aromáticos. *Oecologia Bras.* **2007**, *11*, 188–201. [CrossRef]
5. Lima, A.L.C.; Farrington, J.W.; Reddy, C.M. Combustion-Derived Polycyclic Aromatic Hydrocarbons in the Environment—A Review. *Environ. Forensics* **2005**, *6*, 109–131. [CrossRef]

6. Peng, N.; Li, Y.; Liu, Z.; Liu, T.; Gai, C. Emission, Distribution and Toxicity of Polycyclic Aromatic Hydrocarbons (PAHs) during Municipal Solid Waste (MSW) and Coal Co-Combustion. *Sci. Total Environ.* **2016**, *565*, 1201–1207. [CrossRef]
7. World Health Organization; International Agency for Research on Cancer. *Polynuclear Aromatic Compounds, Part I, Chemical, Environmental and Experimental Data*; IARC Press: Lyon, France, 1983; Volume 32, ISBN 92 832 12320.
8. Fent, K.W.; Eisenberg, J.; Snawder, J.; Sammons, D.; Pleil, J.D.; Stiegel, M.A.; Mueller, C.; Horn, G.P.; Dalton, J. Systemic Exposure to Pahs and Benzene in Firefighters Suppressing Controlled Structure Fires. *Ann. Occup. Hyg.* **2014**, *58*, 830–845. [CrossRef]
9. Kirk, K.M.; Logan, M.B. Firefighting Instructors Exposures to Polycyclic Aromatic Hydrocarbons during Live Fire Training Scenarios. *J. Occup. Environ. Hyg.* **2015**, *12*, 227–234. [CrossRef]
10. Stec, A.A.; Dickens, K.E.; Salden, M.; Hewitt, F.E.; Watts, D.P.; Houldsworth, P.E.; Martin, F.L. Occupational Exposure to Polycyclic Aromatic Hydrocarbons and Elevated Cancer Incidence in Firefighters. *Sci. Rep.* **2018**, *8*, 2476. [CrossRef]
11. Fent, K.W.; Toennis, C.; Sammons, D.; Robertson, S.; Bertke, S.; Calafat, A.M.; Pleil, J.D.; Geer Wallace, M.A.; Kerber, S.; Smith, D.L.; et al. Firefighters' and Instructors' Absorption of PAHs and Benzene during Training Exercises. *Int. J. Hyg. Environ. Health* **2019**, *222*, 991–1000. [CrossRef]
12. Banks, A.P.W.; Wang, X.; Engelsman, M.; He, C.; Osorio, A.F.; Mueller, J.F. Assessing Decontamination and Laundering Processes for the Removal of Polycyclic Aromatic Hydrocarbons and Flame Retardants from Firefighting Uniforms. *Environ. Res.* **2021**, *194*, 110616. [CrossRef]
13. Fent, K.W.; Alexander, B.; Roberts, J.; Robertson, S.; Toennis, C.; Sammons, D.; Bertke, S.; Kerber, S.; Smith, D.; Horn, G. Contamination of Firefighter Personal Protective Equipment and Skin and the Effectiveness of Decontamination Procedures. *J. Occup. Environ. Hyg.* **2017**, *14*, 801–814. [CrossRef] [PubMed]
14. Mayer, A.C.; Fent, K.W.; Bertke, S.; Horn, G.P.; Smith, D.L.; Kerber, S.; La Guardia, M.J. Firefighter Hood Contamination: Efficiency of Laundering to Remove PAHs and FRs. *J. Occup. Environ. Hyg.* **2019**, *16*, 129–140. [CrossRef] [PubMed]
15. Calvillo, A.; Haynes, E.; Burkle, J.; Schroeder, K.; Calvillo, A.; Reese, J.; Reponen, T. Pilot Study on the Efficiency of Water-Only Decontamination for Firefighters' Turnout Gear. *J. Occup. Environ. Hyg.* **2019**, *16*, 199–205. [CrossRef]
16. Corpo de Bombeiros Militar Do Distrito Federal Bid. Available online: https://www4.cbm.df.gov.br/2016-06-24-19-5004/licitacoes-cbmdf?task=document.viewdoc&id=13139 (accessed on 5 September 2020).
17. Horn, G.P.; Kerber, S.; Andrews, J.; Kesler, R.M.; Newman, H.; Stewart, J.W.; Fent, K.W.; Smith, D.L. Impact of Repeated Exposure and Cleaning on Protective Properties of Structural Firefighting Turnout Gear. *Fire Technol.* **2020**, *57*, 791–813. [CrossRef]
18. Mayer, A.C.; Horn, G.P.; Fent, K.W.; Bertke, S.J.; Kerber, S.; Kesler, R.M.; Newman, H.; Smith, D.L. Impact of Select PPE Design Elements and Repeated Laundering in Firefighter Protection from Smoke Exposure. *J. Occup. Environ. Hyg.* **2020**, *17*, 505–514. [CrossRef] [PubMed]
19. Stull, J.O.; Crown, E. *Evaluation of the Cleaning Effectiveness and Impact of Esporta and Industrial Cleaning Techniques on Firefighter Protective Clothing: Technical Report*; International Personnel Protection, Incorporated: Albuquerque, NM, USA, 2006.
20. de Lucena, M.A.M.; Zapata, F.; Mauricio, F.G.M.; Ortega-Ojeda, F.E.; Quintanilla-López, M.G.; Weber, I.T.; Montalvo, G. Evaluation of an Ozone Chamber as a Routine Method to Decontaminate Firefighters' Ppe. *Int. J. Environ. Res. Public Health* **2021**, *18*, 10587. [CrossRef] [PubMed]
21. Engwall, M.A.; Pignatello, J.J.; Grasso, D. Degradation and Detoxification of the Wood Preservatives Creosote and Pentachlorophenol in Water by the Photo-Fenton Reaction. *Water Res.* **1999**, *33*, 1151–1158. [CrossRef]
22. Mojiri, A.; Zhou, J.L.; Ohashi, A.; Ozaki, N.; Kindaichi, T. Comprehensive Review of Polycyclic Aromatic Hydrocarbons in Water Sources, Their Effects and Treatments. *Sci. Total Environ.* **2019**, *696*, 133971. [CrossRef]
23. Shemer, H.; Linden, K.G. Aqueous Photodegradation and Toxicity of the Polycyclic Aromatic Hydrocarbons Fluorene, Dibenzofuran, and Dibenzothiophene. *Water Res.* **2007**, *41*, 853–861. [CrossRef]
24. Jacobs, L.E.; Weavers, L.K.; Chin, Y.P. Direct and Indirect Photolysis of Polycyclic Aromatic Hydrocarbons in Nitrate-Rich Surface Waters. *Environ. Toxicol. Chem.* **2008**, *27*, 1643–1648. [CrossRef]
25. Eker, G.; Sengul, B. Removal of Polycyclic Aromatic Hydrocarbons (PAHs) from Industrial Soil with Solar and UV Light. *Polycycl. Aromat. Compd.* **2020**, *40*, 1238–1250. [CrossRef]
26. Zhang, L.; Xu, C.; Chen, Z.; Li, X.; Li, P. Photodegradation of Pyrene on Soil Surfaces under UV Light Irradiation. *J. Hazard. Mater.* **2010**, *173*, 168–172. [CrossRef] [PubMed]
27. Xu, C.; Dong, D.; Meng, X.; Su, X.; Zheng, X.; Li, Y. Photolysis of Polycyclic Aromatic Hydrocarbons on Soil Surfaces under UV Irradiation. *J. Environ. Sci.* **2013**, *25*, 569–575. [CrossRef]
28. Sanches, S.; Leitão, C.; Penetra, A.; Cardoso, V.V.; Ferreira, E.; Benoliel, M.J.; Crespo, M.T.B.; Pereira, V.J. Direct Photolysis of Polycyclic Aromatic Hydrocarbons in Drinking Water Sources. *J. Hazard. Mater.* **2011**, *192*, 1458–1465. [CrossRef]
29. Miller, J.S.; Olejnik, D. Photolysis of Polycyclic Aromatic Hydrocarbons in Water. *Water Res.* **2001**, *35*, 233–243. [CrossRef]
30. *NFPA 1851*; Standard on Selection, Care, and Maintenance of Protective Ensembles for Structural Fire Fighting and Proximity Fire Fighting. National Fire Protection Association (NFPA): Quincy, MA, USA, 2020.
31. *Method 8270E (SW-846)*; Semivolatile Organic Compounds by Gas Chromatography/Mass Spectrometry (GC/MS). U.S. EPA Test Methods; EPA: Washington, DC, USA, 2018; p. 64.
32. *ASTM D2261*; Standard Test Method for Tearing Strength of Woven Fabrics by the Tongue (Single Rip) Method (Constant-Rate-of-Traverse Tensile Testing Machine). American Society for Testing & Mater (ASTM): West Conshohocken, PA, USA, 2017.

33. Davis, R.; Chin, J.; Lin, C.C.; Petit, S. Accelerated Weathering of Polyaramid and Polybenzimidazole Firefighter Protective Clothing Fabrics. *Polym. Degrad. Stab.* **2010**, *95*, 1642–1654. [CrossRef]
34. Nazaré, S.; Flynn, S.; Davis, R.; Chin, J. *Accelerated Weathering of Firefighter Protective Clothing Containing Melamine Fiber Blends*; NIST Technical Note 1751; US Department of Commerce; National Institute of Standards and Technology: Washington, DC, USA, 2012. [CrossRef]
35. Villar-Rodil, S.; Paredes, J.I.; Martinez-Alonso, A.; Tascón, J.M.D. Atomic Force Microscopy and Infrared Spectroscopy Studies of the Thermal Degradation of Nomex Aramid Fibers. *Chem. Mater.* **2001**, *13*, 4297–4304. [CrossRef]
36. Mosquera, M.E.G.; Jamond, M.; Martínez-Alonso, A.; Tascón, J.M.D. Thermal Transformations of Kevlar Aramid Fibers During Pyrolysis: Infrared and Thermal Analysis Studies. *Chem. Mater.* **1994**, *6*, 1918–1924. [CrossRef]
37. *NFPA 1971*; Standard on Protective Ensembles for Structural Fire Fighting and Proximity Fire Fighting. National Fire Protection Association (NFPA): Quincy, MA, USA, 2018.
38. The TeflonTM Brand and Continuous Innovation. Available online: https://www.chemours.com/en/industries-applications/application-development/continuous-innovation (accessed on 1 June 2021).
39. Military Fire Department of the Federal District (CBMDF). Fire Fighting Techniques. Firef. Instr. Man. 2009, p. 234. Available online: https://www.cbm.df.gov.br/manuais-operacionais-prevencao-e-combate-a-incendio/ (accessed on 5 January 2022).

International Journal of
Environmental Research and Public Health

MDPI

Article

GO-TiO$_2$ as a Highly Performant Photocatalyst Maximized by Proper Parameters Selection

Aida M. Díez [1,2,*], Marta Pazos [2], M. Ángeles Sanromán [2] and Yury V. Kolen'ko [1]

[1] Nanochemistry Research Group, International Iberian Nanotechnology Laboratory, Avenida Mestre José Veiga s/n, 4715-330 Braga, Portugal

[2] CINTECX, Grupo de Bioingeniería y Procesos Sostenibles, Departamento de Ingeniería Química, Campus Lagoas-Marcosende, Universidade de Vigo, 36310 Vigo, Spain

* Correspondence: adiez@uvigo.es

Abstract: The synthesis and characterization of novel graphene oxide coupled to TiO$_2$ (GO-TiO$_2$) was carried out in order to better understand the performance of this photocatalyst, when compared to well-known TiO$_2$ (P25) from Degussa. Thus, its physical-chemical characterization (FTIR, XRD, N$_2$ isotherms and electrochemical measurements) describes high porosity, suitable charge and high electron mobility, which enhance pollutant degradation. In addition, the importance of the reactor set up was highlighted, testing the effect of both the irradiated area and distance between lamp and bulb solution. Under optimal conditions, the model drug methylthioninium chloride (MC) was degraded and several parameters were assessed, such as the water matrix and the catalyst reutilization, a possibility given the addition of H$_2$O$_2$. The results in terms of energy consumption compete with those attained for the treatment of this model pollutant, opening a path for further research.

Keywords: titanium dioxide doping; graphene oxide; methylthioninium chloride; electrochemical characterization measurements; photodegradation; process and catalyst characterization

Citation: Díez, A.M.; Pazos, M.; Sanromán, M.Á.; Kolen'ko, Y.V. GO-TiO$_2$ as a Highly Performant Photocatalyst Maximized by Proper Parameters Selection. *Int. J. Environ. Res. Public Health* **2022**, *19*, 11874. https://doi.org/10.3390/ijerph191911874

Academic Editors: José A. Peres, Joaquín R. Dominguez, Gassan Hodaifa, Antonio Zuorro, Juan García Rodríguez, Zacharias Frontistis and Mha Albqmi

Received: 11 July 2022
Accepted: 14 September 2022
Published: 20 September 2022

Publisher's Note: MDPI stays neutral with regard to jurisdictional claims in published maps and institutional affiliations.

1. Introduction

Photocatalysis has been regarded lately as a wastewater treatment alternative. It has several advantages, namely the non-selective degradation of pollutants and the possibility of working with sunlight [1].

Nevertheless, in order to achieve acceptable results when compared to other alternatives, such as absorption, researchers are focused on UVA-photocatalysis, where high degradation rates can be achieved by the utilization of the high performant TiO$_2$ [2].

However, the activity of TiO$_2$ seems to be not enough for scale-up applications, as little output has been found in real applications. This lack of real usability can be explained by the drawbacks of TiO$_2$, such as its activation only under UVA radiation [3] and its insufficient absorption capacities [1].

In order to overcome this, some modifications have been performed to commercial TiO$_2$ (P25 from Degussa), such as doping with elements to reduce its band gap, which enhances its activation under wavelengths higher than UVA. For instance, Chakinala et al. [3] added Bi and Ag to TiO$_2$ to more efficiently degrade contaminants such as methylthioninium chloride (MC) or rhodamine B. Likewise, the addition of porous compounds, such as SiO$_2$, has been essayed, favoring a reduction in the agglomeration between photocatalyst particles and promoting absorption due to an increase in photocatalyst global porosity [1]. In this regard, other authors have proposed the addition of GO [4,5], although the efficiencies attained can be widely improved with more sophisticated reactor set-ups.

With these alternatives for efficiency enhancement, photocatalysis processes could compete with other less expensive alternatives, such as absorption, which has to cope with spent absorbent disposal. Another approach that facilitates the real application of photo-based processes is to synthesize photocatalysts with dual activity under UVA and visible radiation [3,5]. With this, the treatment times are not subject to either sunlight or lamp operability.

Nevertheless, few researchers [1,3,6] have focused on the synthesis of dual activity photocatalysts, although increasing attention on the topic has been noticed (Figure S1-SM: Supplementary Material). Moreover, in those cases, the catalysts were expensive, either due to the presence of costly elements, such as Ag [3], or due to the high temperature increase requirement in the synthesis processes [1,6].

This study proposes the utilization of GO-TiO$_2$ as a dual activity photocatalyst for the degradation of a model drug (MC). For this, research was carried out under UVA radiation and validated afterwards with simulated solar radiation. Moreover, the photocatalyst was characterized by traditional measurements, such as X-ray diffraction (XRD), Fourier transform infrared spectroscopy (FTIR), N$_2$ isotherm and bang gap calculation, and by poorly studied electrochemical measurements, such as impedance or electrochemical active surface area (ECSA). The unstudied photocatalyst reusability was also assessed.

2. Materials and Methods

2.1. Reagents

MC of high purity was purchased from Alfa Aesar. GO (>99%) was bought from ACS Material. H$_3$BO$_3$ (>99.5%), (NH$_4$)$_2$TiF$_6$ (>99.9%) and ethanol (>99.8%) were acquired from Sigma. TiO$_2$ was acquired from Evonik (so-called P25), which contains 77.1% anatase, 15.9% rutile and 7% of amorphous TiO$_2$. Milli-Q water was attained from an Advantage A10 system (Millipore).

2.2. Catalyst Synthesis

The catalyst was synthesized using a previously reported procedure for the treatment of industrial dyes [7], so that this study can deeply evaluate the photocatalyst performance with the study of the drug MC. Briefly, it consists of the dispersion of GO (10 mg/mL) in 10 mL of ethanol. Then, 0.1 M of (NH$_4$)$_2$TiF$_6$ is added dropwise under stirring, as well as H$_3$BO$_3$ 0.3 M. This was kept for 1 h under vigorous stirring. Then, the prepared mixture is placed on a hydrothermal autoclave at 60 °C during 2 h. Then, the product is placed on the furnace at 200 °C for 2 h so 1GO-TiO$_2$ can be attained. The dosage of GO was studied and consequently, $\frac{1}{2}$GO-TiO$_2$ and 3-GO-TiO$_2$ were synthesized.

2.3. Catalyst Characterization

The characterization of TiO$_2$ was compared to the best performant GO-TiO$_2$ catalyst for MC photodegradation. Room-temperature XRD was carried out on a Rigaku diffractometer with an X PERT PRO MRD (Pananalytical, USA), using Cu-Kα radiation (λ = 1.54187 Å). Each pattern was recorded in the 2θ range from 20 to 80° with a step of 0.03° and the total collection time of 2 h. The analysis of XRD patterns was carried out with Higscore software 3.0 (Pananalytical, Westborough, MA, USA).

FTIR measurements were performed using the NBI—Bruker Vertex 80 v vacuum FTIR of Bruker within 400 and 4000 nm under N$_2$ conditions.

The UV–Vis spectra of the photocatalyst were measured in a 1 wt % ethanol matrix using a spectrophotometer (V-2550, Shimadzu, Kyoto, Japan).

The N$_2$ isotherm was carried out with Autosorb IQ2 (quantachrome) by degassing the sample at 303 K for 4.5 h and introducing N$_2$ at 77.35 K.

The point of zero charge (PZC) was evaluated by plotting the set pH (with HCl or NaOH) vs. the difference in pH after 24 h stirring, following reported procedures [5].

Electrochemical impedance spectroscopy (EIS) and ECSA were measured using the Autolab PGSTAT302N of Metrohm (Herisau, Switzerland). A three-electrode system was used, with Pt as the counter electrode, calomel electrode as the reference and Ni-foam (1 cm^2) as the working electrode, where the catalysts were impregnated. Then, 3 mg of the catalysts were dissolved in 630 µL of a mixture of Nafion:EtOH (1:20). The experiments were carried out in a 0.5 M Na$_2$SO$_4$ solution. EIS was measured on a frequency range of

$0.1\text{--}10^5$ Hz with a sinusoidal perturbation of 10 mV. ECSA was measured by measuring cyclic voltammetries (CVs) in a range of 0.1 V, so it crossed the 0 current. These CVs were measured at scan rates between 10 and 200 mV s^{-1}. Then, a linear trend was obtained by plotting the scan rate vs. half the difference in current density between the anodic and cathodic sweeps. The linear fitting slope of this graph provides the geometric double-layer capacitance (CDL). ECSA is calculated following equation 1, where Cs is the specific capacitance and has a value of 40 µF cm^{-2} [8].

$$ECSA = CDL/Cs \tag{1}$$

2.4. Reactor Set-Up

Taking into account the porosity provided by GO addition, the equilibrium under stirring on dark conditions was maintained for 30 min. The photocatalytic performance was evaluated on two reactors. (I) The wide-open-reactor was cylindrical, with a diameter of 10 cm and a height of 1.5 cm, in order to promote photocatalyst activation by the lamp placed at 2 cm above (Figure 1), although a 6 cm gap was also assessed. The UVA-LED lamp used was acquired from Seoulviosys model 78 CMF-AR-A03 (365 nm, 4.8 W). (II) In order to test the efficiency of the high radiated surface area, a test that switched the wide-open reactor for a narrow-open reactor was carried out (Figure 1B). This reactor has a 5 cm diameter and is 10 cm high. (III) A test with simulated solar radiation was carried out with a 400–740 nm emitting lamp (600 W, from Toplanet) placed at 2 cm next to the narrow-open reactor, so a high surface area was irradiated (providing that glass is transparent to visible radiation) (Figure 1C).

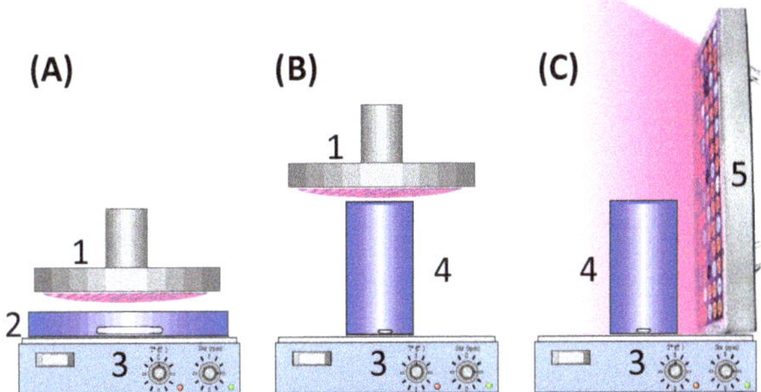

Figure 1. Proposed reactors: (**A**) wide-open-UVA reactor, (**B**) narrow-open-UVA reactor, (**C**) visible reactor, where 1: UVA LED lamp, 2: 10 cm ø, 1.5 cm high cylindrical reactor, 3: magnetic stirring, 4: 6 cm ø, 10 cm high cylindrical reactor, 5: visible lamp.

All preliminary experiments were carried out using Milli-Q water (Sartorious) as the working matrix, with a MC concentration of 20 mg/L. However, validation tests were performed with real wastewater, kindly donated by WTP (wastewater treatment plant) of Guillarei, Tui, Galicia, which remediates the municipal effluents by a primary process (so-called physically treated water) and subsequent secondary treatment (physically + biologically treated water). The characteristics of those effluents are provided in the supplementary material (Table S1-SM).

2.5. Process Efficiency Analysis

Decoloration of the samples was followed using Equation (2), where A$_i$ is the initial absorbance and A$_t$ is the absorbance at a time t. Absorbance was measured using the spectrophotometer (V-2550, Shimadzu, Kyoto, Japan). A calibration curve was produced in order to assess the exact remaining MC concentration with time.

$$\text{Decoloration} = (A_i - A_t / A_i) \times 100 \qquad (2)$$

Energy consumption (EC) was measured in order to evaluate the efficiency of the processes and to compare the results. For that, Equation (3) was used, where L is the lamp power (W), t is the treatment time (h) and C is the amount of degraded MC (mg).

$$\text{EC (Wh/mg)} = L \times t/C \qquad (3)$$

3. Results

3.1. Preliminary Tests

3.1.1. GO Dosage

Initially, in order to select the catalyst to work with, the influence of GO dosage was studied, as the amount of porous agents added to TiO_2 has been shown to have an effect on photo-activity [1,4]. In fact, some differences regarding dark absorption and photodegradation were detected (Figure 2). The MC degradation enhancement is remarkable when comparing well-known TiO_2 to TiO_2 with any GO dosage. Consequently, the synthesized 1GO-TiO_2 was prepared with half ($\frac{1}{2}$GO-TiO_2) and three-fold (3GO-TiO_2) the amount of GO. In addition, $\frac{1}{2}$GO-TiO_2 was the best candidate in terms of photocatalysis performance, and it was thereafter called GO-TiO_2.

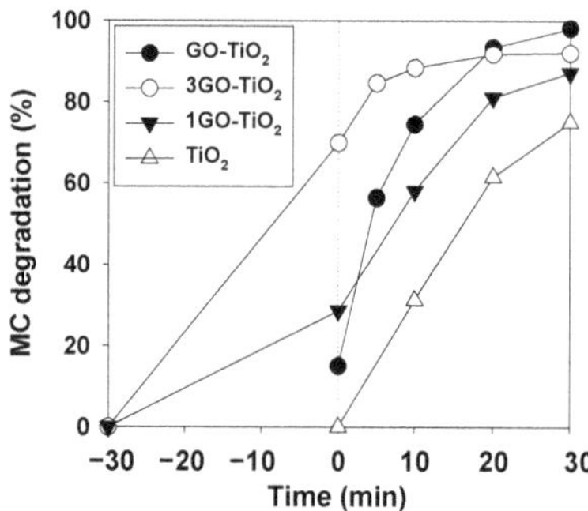

Figure 2. MC photodegradation (20 mg/L) at catalyst concentration of 800 mg/L under UV radiation. TiO_2 (white triangles), $\frac{1}{2}$GO-TiO_2 (so-called GO-TiO_2) (black circles), 1GO-TiO_2 (black triangles), and 3GO-TiO_2 (white circles).

3.1.2. Reactor Set-Up

In order to confirm the suitability of the proposed reactor (Figure 1A), the comparison between the aforementioned reactors was carried out. The results provided in Figure 3 demonstrate the importance of irradiation surface and intensity. In fact, EC was calculated for both the wide-open-reactor, with the lamp placed at 2 and 6 cm, and for the narrow-open-reactor and provided EC values of 0.12, 0.21 and 0.18 Wh/mg, respectively. Consequently, the open-wide-reactor with the lamp placed at 2 cm was used thereafter.

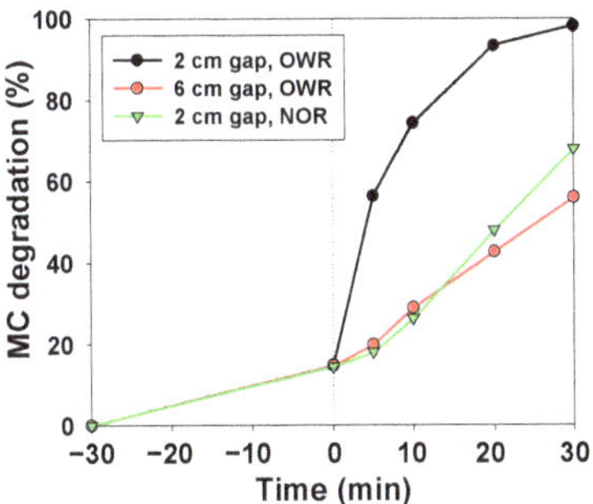

Figure 3. MB degradation on the open-wide-reactor (OWR) and narrow-open-reactor (NOR), placing the UV-lamp at different gaps.

3.2. Catalyst Characterization

3.2.1. XRD

The XRD data of both TiO_2 and GO-TiO_2 demonstrate that the former is stable in the synthesis process (Figure S2-SM), possibly due to the mild conditions during the GO addition.

3.2.2. FTIR

The FTIR spectra (Figure S3-SM) show that GO-TiO_2 has characteristic peaks of both GO and TiO_2, demonstrating successful synthesis.

3.2.3. N_2 Isotherms

Figure 4a depicts the N_2 isotherm of both photocatalysts that fit a III type isotherm, which indicates the presence of macropores and a weak interaction photocatalyst-MC [9]. GO-TiO_2 showed significantly higher absorbance of N_2, which may be related to its higher absorption capacity. Surface area, pore diameter and volume are depicted in Table S2-SM.

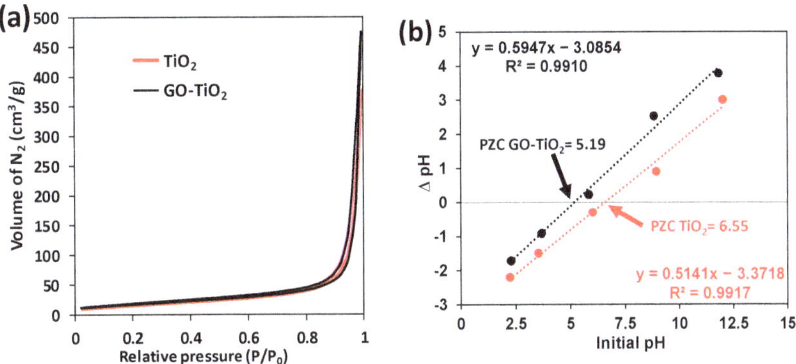

Figure 4. Mechanisms explaining the absorption behavior: (**a**) N_2 isotherm and (**b**) point of zero charge of TiO_2 and GO-TiO_2.

3.2.4. PZC

The PZC was calculated, and some differences were detected between TiO_2 and GO-TiO_2 (Figure 4b). Indeed, the PZC was more than 1 pH unit of acid than the commercial TiO_2. The TiO_2 PZC is in concordance with other studies [10].

3.2.5. UV Spectra

Both catalyst dispersions in ethanol (1% wt) were measured between 190 and 900 nm and the resulted profile is shown in Figure S4A-SM, where a slight augmentation in the absorbance response can be observed, which can be related to higher photo-activity [1]. Tauc plots were calculated following previous protocols [1,2] (Figure S4B-SM), and band gap values were measured, being 3.2 and 2.2 eV for TiO_2 and GO-TiO_2, respectively.

3.2.6. Electrochemical Measurements

The relationship between photo-activity and electrochemical activity has been demonstrated previously [11,12]. Consequently, the electrochemical impedance spectroscopy with the attainment of Nyquist graphs, and the calculation of the ECSA were carried out (Figure 5), in which the data can be fitted to equivalent circuits (Figure S5-SM). The series resistance represents the electrode, which is practically constant for both catalysts, considering the usage of the same set-up. In the case of the parallel resistance, it is related to the layer resistance of the material, with a 4-fold smaller resistance in the case of GO-TiO_2 when compared to TiO_2.

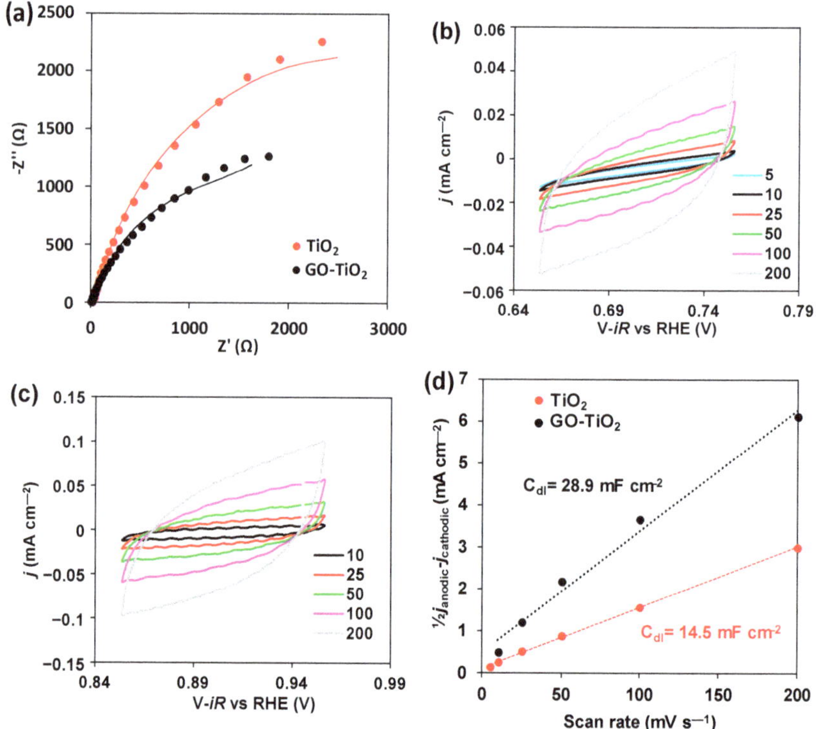

Figure 5. Electrochemical measurements: (**a**) Nyquist graph, where the lines represent the equivalent circuit adjustment; (**b**,**c**) are CVs at different scan rates (in mV s^{-1}) for TiO_2 (**b**) and TiO_2 GO (**c**). CDL representation of both samples (**d**).

Regarding the Nyquist graphs (Figure 5a), the semicircle arc is smaller in the case of GO-TiO_2. It can be also noticed, when comparing CVs from TiO_2 and GO-TiO_2 (Figure 5b,c, respectively), that the broadening of the CVs is practically twice in the case of GO-TiO_2 (Figure 5c), obtaining a range of 0.2 mA compared to the 0.11 mA for TiO_2 at 200 mV s^{-1}.

Regarding the CDL results, GO-TiO$_2$ provided a more than double CDL value, demonstrating that the increase in electroactive surface area has been accomplished, which is in concordance with the N$_2$ isotherm results (Figure 4a).

3.3. MC Degradation

3.3.1. Duality UVA–Vis

To our knowledge, the number of studies centered on dual activity UVA–Vis is much smaller than those on the independent evaluation of the radiation source (Figure S1-SM). Although, the trend is increasing with time, due to the obvious advantage of the possibility of working with daylight and with low consumption lamps during the night. Consequently, the synthesis and study of dual activity photocatalysts should be carried out in order to facilitate real usages of photodegradation processes.

The photo activity of GO-TiO$_2$ turned out to be higher than TiO$_2$ (Figure 6); thus, GO-TiO$_2$ favors MC degradation not only under UVA, but also under visible radiation when compared to TiO$_2$.

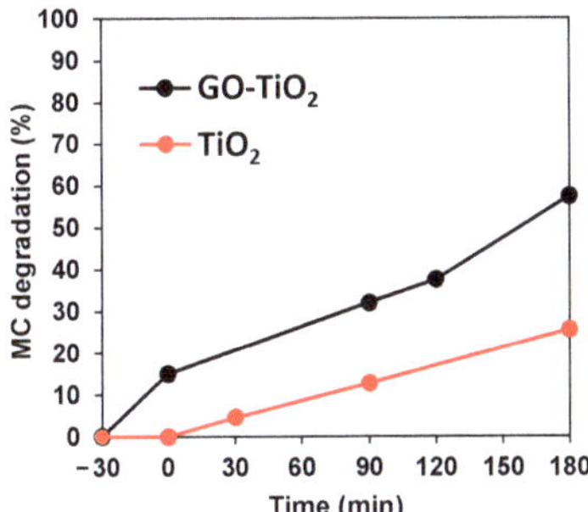

Figure 6. MC photodegradation (20 mg/L) at the catalyst concentration of 800 mg/L with TiO$_2$ and GO-TiO$_2$ under simulated solar radiation.

3.3.2. Catalyst Reuse

Catalyst reuse directly affects the process cost [3]; thus, it was evaluated. Lower MC degradation rates were reported (Figure 7a) for the 1st cycle (100%) to 2nd (70%) and 3rd cycles (38%). As a solution, H$_2$O$_2$ (0.66 mg/mL) was added, as some references have tested the enhancement of TiO$_2$ photocatalyst performance with H$_2$O$_2$ addition [13]. In this case, the efficiency was kept constant during the three cycles (Figure 7b).

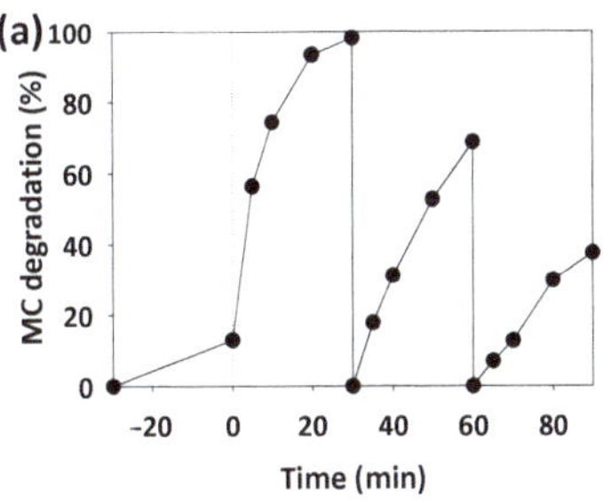
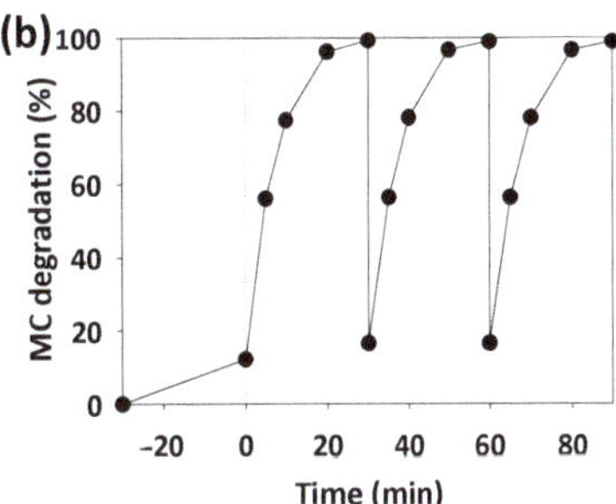

Figure 7. MC photodegradation under UV radiation (20 mg/L each cycle), using the same photocatalyst (at a concentration of 800 mg/L) during three cycles (**a**) and with the addition of 0.66 mg/mL of H$_2$O$_2$ in each cycle (**b**).

3.3.3. Matrix Effect

The working matrix is known to have an impact on photocatalysis degradation [14,15]; thus, its effect was validated. In fact, Figure 8 demonstrates that it was not only the degradation that diminished when using real effluents, but also it was a slight reduction in the absorption.. Indeed, a degradation decrease of 27% in the case of the physical and biologically treated effluent and of 38% in the case of the physically treated effluent was observed, when compared to the MC treatment with Milli-Q water.

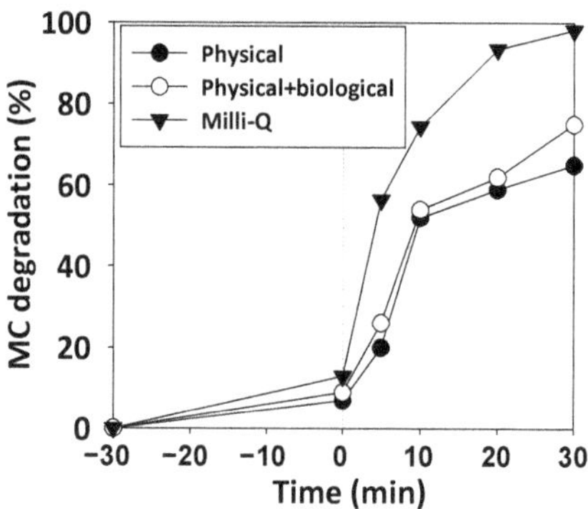

Figure 8. MC photodegradation (20 mg/L) at catalyst concentration of 800 mg/L under UV radiation. Matrix composed of real wastewater that received a physical treatment (black circles), matrix that received a physical and biological treatment (white circles), and milli-Q water (black triangles).

4. Discussion

4.1. Preliminary Tests

4.1.1. GO Dosage

Slight differences in MC degradation were detected when the ratio of GO varied from 1/2 to 3 (Figure 2), although variations were noticed on the adsorption behavior (at time 0). The higher the GO dosage, the higher initial absorption in dark conditions, which is in concordance with previous studies and reflects the porous characteristics of GO [4,5]. However, the photocatalytic behavior dramatically changed; thus, 3GO-TiO$_2$ caused 67% absorption in darkness, increasing only to 83% MC degradation after 30 min of illumination. On the other hand, $\frac{1}{2}$GO-TiO$_2$ caused 15% absorption and MC photodegradation reached 100% after illumination. This demonstrated that the lower the GO quantity, the lower the absorption within the GO porous structure and the better the photocatalytic activity. This also demonstrates the fact that too much MC absorption within the photocatalyst structure hinders the photocatalytic activity, due to the blockage of active sites [5], which also explains the decrease during the catalyst reuse process, due to the adsorption of MC on the photocatalyst (Figure 7a).

Additionally, $\frac{1}{2}$GO-TiO$_2$ (called from now on GO-TiO$_2$) surpasses commercial TiO$_2$ performance by 30%. This may be caused by the enhancement of porosity and hydrophobicity caused by GO addition [5]. These results are in concordance with the work of Mahanta et al. [1], who noticed that combining TiO$_2$ with porous SiO$_2$ enhanced TiO$_2$ activity, but the lowest concentration was the most optimal. Additionally, GO-TiO$_2$ improvement may be caused by easier photocatalyst activation, given the smaller band gap of GO-TiO$_2$ (Figure S4-SM).

4.1.2. Reactor Set-Up

The proposed wide-open reactor has a high surface area (10 cm) and the radiation source is placed extremely close to the bulb solution (2 cm), which enhances the photodegradation process. Indeed, this statement was corroborated by the results depicted in Figure 3. Thus, when placing the open-wide reactor at 6 cm instead to 2 cm, the degradation performance was diminished by more than 40%. This is in concordance with other studies, which have highlighted the decrease in the light intensity with distance. For instance, Anku et al. [10] accomplished 95% methyl orange degradation with a xenon lamp placed at 10 cm, whereas only 55% was achieved when the lamp was placed at 13 cm. This is in concordance with the work of Casado et al. [16], who reported that the radiation W that reached the reactor surface reduced with distance. For instance, when placing an LED lamp at 4 cm from the solution, 15 W was reported. In contrast, if they placed the lamp at 8 cm, only 5 W was recorded. Moreover, MC degradation suffered a decrease of 30% when reducing the illuminated surface area from 10 to 5 cm (when changing the open-wide reactor to the narrow-open reactor). In this case, this is in concordance with the intensity of radiation by the illuminated surface, and although the intensity is higher on the very center, extra photons can reach the photocatalyst [16] in the case of the open-wide reactor.

In both cases, the decrease is caused by the reduction in light intensity, reducing the number of photons that can reach the surface of the catalyst [10]. When focusing on the degradation behavior with time, one can notice that until the first 10 min, the performance of MC degradation by the narrow-open reactor is worse than with the open-wide-reactor placed at 6 cm; this may be caused by the light scattering of the concentrated MC solution and the depth of the effluent in the narrow-open reactor. As time passes, the darkness of the solution is diminished and with that, light scattering is reduced. Consequently, after 10 min of reaction, the MC degradation performance is higher with the narrow-open reactor placed at 2 cm than with the open-wide reactor placed at 6 cm.

4.2. Characterization

4.2.1. XRD

Regarding XRD, the samples showed almost the same profile (Figure S2-SM), where anatase and rutile phases are depicted. This is concordance with other authors, which noticed that TiO_2 was stable during superficial doping, demonstrating the high stability of this material. For instance, Kurniawan et al. [4] added from 0.05 to 50% of GO to TiO_2 in order to enhance MC degradation and the XRD did not significantly change, as GO has low crystallinity.

4.2.2. FTIR

FTIR measurements (Figure S3-SM) demonstrate the broadening of the peaks due to the bonds generated within $GO\text{-}TiO_2$. This is in concordance with the work of Kurniawan et al. [4], who labelled the peak at around 1400 cm^{-1} as a H-O typical of TiO_2 samples and reported that the new peaks between 1600 and 1700 cm^{-1} were caused by C = C and C = O bonds, which are found in the GO sample. Indeed, Figure S3-SM demonstrates the presence of several functional groups, which are known to promote worthy photocatalytic activity [5], promoting charge transfer.

4.2.3. N_2 Isotherms

Both TiO_2 and $GO\text{-}TiO_2$ showed type III isotherms related to the presence of macropores [9]. $GO\text{-}TiO_2$ showed higher N_2 absorption (Figure 4a), which is related to a 19% surface area increase in $GO\text{-}TiO_2$ when compared to TiO_2 (Table S2-SM), for which the pore diameter and volume are higher due to surface clogging with GO and the increase in the particle due to process synthesis [9,17].

These results are in concordance with the fact that $GO\text{-}TiO_2$ showed a 15% absorption rate (Figure 2); thus, this absorption may be caused by the physical trapping of MC due to the higher surface area of the synthesized photocatalyst. Indeed, other authors have experienced higher MC absorption on their photocatalysts when the surface area was higher [18].

4.2.4. PZC

The PZC is a way of measuring the possible electrostatic interactions between MC and the photocatalysts. Thus, the absorption of MC within GO-TiO$_2$ may be caused not only by the physical trapping due to the higher surface area but also due to charge attraction.

Indeed, the PZC of TiO$_2$ is 6.55 and PZC of GO-TiO$_2$ is 5.19 (Figure 4b); this means that at a pH below this value, the absorbent would be positively charged and at a pH above the PZC, the absorbent would be negatively charged.

Thus, considering the MC structure where the large organic part is positive, negatively charged absorbents would attract more of this pollutant. Considering that the PZC of GO-TiO$_2$ is lower than TiO$_2$, the aforementioned negative charge is more common in the former and with MC interaction. In fact, the pH of the MC 20 mg/L solution is 6.67, so it is clear that TiO$_2$ would almost not be negatively charged (PZC = 6.55), whereas GO-TiO$_2$ would be negatively charged, as there is a 1.5 pH unit difference. Thus, the interaction between MC-GO-TiO$_2$ is not only physical, but also electrostatic.

4.2.5. UV Spectra

The UV–Vis response of GO-TiO$_2$ was evaluated and compared to that of TiO$_2$ and the results (Figure S4-SM) demonstrate the much better activity of the synthesized GO-TiO$_2$. Thus, GO-TiO$_2$, with a band gap of 2.2 eV, would be more easily activated under radiation than TiO$_2$ (3.2 eV, which is in concordance with previously reported data) [1,2]. Moreover, the GO-TiO$_2$ band gap means that it is also activated under visible radiation (Section 3.3.1), due to the smaller electron–hole recombination rate [4].

4.2.6. Electrochemical Measurements

The Nyquist graph (Figure 5a) reflects a higher electron mobility in the case of GO-TiO$_2$, as it presents a smaller arc radius [2]. This higher electron mobility has been highlighted as a reason for photocatalyst performance improvement in the GO-TiO$_2$ samples [5]. Thus, GO addition favors the movement of electrons and foments the charge carrier transference from TiO$_2$ to GO, avoiding electron–hole recombination. This fact is demonstrated by the better performance of GO-TiO$_2$ when compared to TiO$_2$ (Figure 2). Indeed, the smaller value of the parallel resistance (Figure S5-SM) (17.5 kΩ vs 75.2 kΩ) indicates a lower charge-transfer resistance, which enhances the photocatalytic activity [2].

Moreover, increasing the surface area (BET) (Table S2-SM) and ECSA (Figure 5d) also has a positive effect on the degradation performance, due to the increase in the surface reactions of photogenerated carriers [5].

4.3. MC Degradation

4.3.1. Duality UVA–Vis

The results provided in Figure 6 demonstrated the activity of the GO-TiO$_2$ photocatalyst under visible radiation. This is explained by the smaller band gap (from 3.2 of TiO$_2$ to 2.2 eV of GO-TiO$_2$) when adding GO to TiO$_2$ [4]. Thus, even though this photocatalyst adsorbs 15% more MC than TiO$_2$, the degradation performance is doubled with this synthesized catalyst. Other authors have demonstrated the dual UV and visible activity of their synthesized catalysts, such as V/Mo-TiO$_2$ [6] or TiO$_2$-SiO$_2$ [1], although GO-TiO$_2$ may be an easier-to-apply option. For instance, Chakinala et al. [3] had already demonstrated the duality of a catalyst that was useful for both UV and visible radiation. Nevertheless, their catalyst was more complex and expensive (Di/Ag-TiO$_2$). Other authors have reported that TiO$_2$ can experience a shift to visible radiation with the addition of compounds such as GO [1].

Nevertheless, regarding the UV–Vis spectra, no significant shift was detected as in the case of Mahanta et al. [1] for their TiO$_2$-SiO$_2$ photocatalyst. These authors explained the dual activity of UVA-visible radiation by the small fraction of wavelengths in the visible range, which are enough to activate the photocatalyst due to SiO$_2$ addition in their case, or GO in this study.

4.3.2. Catalyst Reuse

The efficiency of the photolysis process is diminished after the second cycle (Figure 7a). This is caused by the blocking of the active sites with MC molecules and also by the generation of more stable intermediates [3]. In both cases, the addition of an extra oxidant, such as H_2O_2, is a good alternative (Figure 7b), by helping with the cleaning of the active sites and favoring the degradation of stable intermediates. In fact, the combination of TiO_2 and H_2O_2 promotes the generation of peroxy-titanium complexes (-TiOOH), which promote the generation of HO· radicals [13]. Moreover, extra HO· are generated by the H_2O_2 reaction with the electrons from the conduction band of GO-TiO_2 and also with the superoxide radicals that are produced by the photodegradation process [19].

The presence of macropores (Section 4.2.3) makes desorption difficult and this can explain the fact that in the photocatalyst reuse process, the second and third cycles did not show absorption because the MC within the structure was not degraded during the first cycle (Figure 7a), demonstrated by the blueish color of the photocatalyst. In the case of H_2O_2 addition the oxidants generated are enough to degrade the MC absorbed, a fact demonstrated by the clean photocatalyst after reuse and the restored 15% adsorption between cycles, as shown in Figure 7b. These results are superior to previous data, where the performance of the catalyst was also reduced over the cycles [1,3]. These authors could have considered H_2O_2 addition in order to reuse the photocatalyst. This opens a path for real usages, where photocatalyst reuse could reduce waste disposal and simplify plant operation.

4.3.3. Matrix Effect

Figure 8 demonstrates the negative effect of the usage of real wastewater as matrixes. This is caused by the presence of organic and inorganic matter in those effluents (Table S1-SM). Those compounds can act as scavengers and reduce the rate of the degradation process [14]. Moreover, even absorption under dark conditions was slightly reduced; this can be caused by the absorption of the other organic matter present in the real effluents, which cause an increase in COD (Table S1-SM), and by the clogging of the active pores, due to the presence of suspended solids. Indeed, other authors [20] have noticed a reduction in compound absorption under dark conditions with salts such as NaCl or $NaHCO_3$, or when treating real water, such as river water and treated wastewater. This can be explained by a modification in the electrostatic attraction between the photocatalyst and the contaminant due to the salt content variation [20]. Thus, in real effluents, the ionic strength and the organic matter content vary, leading to particle agglomeration [15].

4.3.4. Comparison with Previous Research

The achieved results demonstrate the importance of suitable photocatalyst selection, synthesis procedure and the importance of the reactor set-up to maximize its utilization. Indeed, bibliography research has been carried out in order to compare similar photodegradation processes with the attained results (Table 1). The term EC (Equation (3)) has been used to normalize the results, regardless of the initial MC concentration, the treatment time or the lamp power. Thus, the difference in EC is caused by either photocatalyst efficiency or reactor set-up.

In relation to photocatalyst suitability, the current results are superior to those of other authors, who have tried the addition of other compounds to TiO_2, such as SiO_2 [1], V and Mo [6] or even natural zeolite [21]. This highlights the suitability of the combination of GO-TiO_2, where GO addition favors the attraction between the drug and the photocatalyst, accelerating the degradation process [4] due to the acceleration of surface reactions [5]. This increased attraction is caused by the higher acidity and surface area when compared to raw TiO_2, as it was demonstrated in Figure 4. Moreover, the GO addition increases the electron transfer capacities of GO-TiO_2 [5], as it was demonstrated by the electrochemical impedance measurements (Figure 5).

Nevertheless, the efficiency has been improved when compared to other GO-TiO$_2$-based photocatalysts (Table 1). In this case, the reactor set-up may be the differentiating aspect, as for instance, Wafi et al. [5] placed the lamp at 10 cm from the solution, when it is well known that the radiation intensity decreases with distance [2,22]. This fact has been proven when comparing the MC degradation by placing the wide-open reactor either at 2 or 6 cm from the UVA lamp (Figure 3). The results demonstrated more than 30% improvement when the lamp was placed closer. This is in concordance with the work of Sun et al. [22], who degraded 10% more MC within a film with their N-TiO$_2$ photocatalyst when the lamp was placed at 0.5 cm from the film than when it was placed at 3 cm, caused by, in the latter case, lower photo current intensity due to the distance [2]. Moreover, the synthesis process was much easier in this presented study. Thus, Wafi et al.'s experiment [5] not only required calcination at 550 °C, but also the addition of dangerous reagents, such as NaBH$_4$. Other authors' experiments required long synthesis processes [4,23], instead of the brief GO-TiO$_2$ process proposed in this work (Section 2.2).

Table 1. Recent studies on MC degradation with UVA photocatalysis.

MC (mg/L)	Catalyst (mg/L)	Lamp (nm, w)	Time (min)	Degradation (%)	EC (W·h/mg)	Reference
20	GO-TiO$_2$ (800)	360–365 nm, 4.8 W	30	100	0.12	This study
10	TiO$_2$-SiO$_2$ (1000)	8 W	30	85	0.47	[1]
10	TiO$_2$ nanoparticles (60)	UV, 300 W, >420 nm	24	99	12.12	[2]
10	Bi-Ag-TiO$_2$ (15)	Vis, 250 W	120	90	111.1	[3]
5	GO-TiO$_2$ (200)	500 W	60	92	108.70	[4]
6.4	Reduced GO-TiO$_2$ (1000)	6 W, 365 nm	60	86	1.09	[5]
3.2	V/Mo-TiO$_2$ (1000)	365 nm, 8 W	60	86.7	2.88	[6]
10	Zeolite/TiO$_2$ (50)	UV, 16 W	120	100	3.20	[21]
73.57	Graphene-TiO$_2$ (50)	360 nm, 17 W	480	87	2.12	[23]

Additionally, the EC of previous studies (Table 1) is much higher than the attained EC in this study, which demonstrates the efficiency of the reactor set-up. Indeed, it was not only the aforementioned reduction in the distance between the bulb solution and lamp that provided a much higher performance, but also the increase in the radiated surface area. In fact, this reality is also proven in Figure 3, where it can be noticed that increasing the irradiated area from 19.6 to 78.5 cm^2 (for the narrow-open reactor and the wide-open reactor, respectively) caused an augmentation of around 40% in MC degradation. This was caused by an enhancement of the catalyst exposure to radiation and by the uniform irradiance of the photocatalyst through the small reactor depth [24].

Moreover, out of the selected references, no authors have validated their process with real wastewater effluents (Figure 8). Furthermore, only Mahata et al. [1] and Chakinala et al. [3] validated the reusability of the catalyst, although neither of them considered the utilization of an extra oxidant and even though Chakinala et al. [3] faced an almost 40% decrease in pollutant degradation. H$_2$O$_2$ addition may have increased the lifetime of their Bi-Ag-TiO$_2$ photocatalyst.

5. Conclusions

GO-TiO$_2$ was successfully synthesized and characterized in order to assess its performance as a dual (UV–Vis) photocatalyst. MC was used as a model pollutant in order to understand the photodegradation process. Thus, characterization and MC degradation unite to offer conclusions when compared to highly performant TiO$_2$. For instance, GO-TiO$_2$ resulted in a higher surface area, which experimentally implied more MC absorption under dark conditions, and a higher degradation rate due to better interactions of the drug-photocatalyst when compared to TiO$_2$. Regarding the higher electron transfer of GO-TiO$_2$, measured by Nyquist graphs, it was clear that the electron–hole recombination was diminished in contrast to TiO$_2$, which also explained the better photo-activity of the former. Moreover, the results were validated by reusing the several photocatalyst cycles by the assistance of H$_2$O$_2$ addition, by treating real wastewaters and by the comparison to previous data. Indeed, the EC was 0.12 Wh/mg, which has not been achieved previously; this makes room for further studies regarding the usage of the GO-TiO$_2$ catalyst and similar reactor set-ups.

Supplementary Materials: The following supporting information can be downloaded at: https://www.mdpi.com/article/10.3390/ijerph191911874/s1, Figure S1: Number of publications indexed on Scopus by searching for photocatalysis + UV, photocatalysis + Visible or photocatalysis + Visible and UV; Figure S2: XRD spectra of TiO$_2$ (black) and GO-TiO$_2$ (red) catalysts; Figure S3: FTIR spectra of TiO$_2$ (black), GO (green) and GO-TiO$_2$ (red) compounds. The insert figure is the zoom in of the spectra between 800 and 3700 nm. Sym. str. = symmetric stretching; Figure S4: A: UV-Vis spectra of TiO$_2$ and GO-TiO$_2$.B. Tauc plots of both TiO$_2$ and GO-TiO$_2$ photocatalysts; Figure S5: Fitted reactors to the Nyquist results for TiO$_2$ (a) and GO-TiO$_2$ (B) catalysts. Table S1: Characteristics of the real working matrixes on the study; Table S2: Porous characteristics of TiO$_2$ and GO-TiO$_2$.

Author Contributions: Conceptualization, A.M.D.; Methodology, A.M.D.; Software, A.M.D.; Validation, A.M.D.; Formal Analysis, A.M.D.; Investigation, A.M.D.; Resources, M.P., M.Á.S. and Y.V.K.; Data Curation, A.M.D.; Writing—Original Draft Preparation, A.M.D.; Writing—Review & Editing, M.P., M.Á.S. and Y.V.K.; Visualization, A.M.D.; Supervision, M.P., M.Á.S. and Y.V.K.; Project Administration, M.P., M.Á.S. and Y.V.K.; Funding Acquisition, A.M.D., M.P., M.Á.S. and Y.V.K. All authors have read and agreed to the published version of the manuscript.

Funding: This research was funded by Xunta de Galicia grant number [ED481B 2019/091]. This research was also funded by Project PID2020-113667GB-I00 464 funded by MCIN/AEI /10.13039/501100 011033.

Institutional Review Board Statement: Not applicable.

Informed Consent Statement: Not applicable.

Conflicts of Interest: The authors declare no conflict of interest.

References

1. Mahanta, U.; Khandelwal, M.; Deshpande, A.S. TiO$_2$@SiO$_2$ nanoparticles for methylene blue removal and photocatalytic degradation under natural sunlight and low-power UV light. *Appl. Surf. Sci.* **2022**, *576*, 151745. [CrossRef]
2. Niu, L.; Zhao, X.; Tang, Z.; Lv, H.; Wu, F.; Wang, X.; Zhao, T.; Wang, J.; Wu, A.; Giesy, J. Difference in performance and mechanism for methylene blue when TiO$_2$ nanoparticles are converted to nanotubes. *J. Clean. Prod.* **2021**, *297*, 126498. [CrossRef]
3. Chakinala, N.; Gogate, R.R.; Chakinala, G. Highly efficient bi-metallic bismuth-silver doped TiO$_2$ photocatalyst for dye degradation. *Korean J. Chem. Eng.* **2021**, *38*, 2468–2478. [CrossRef]
4. Kurniawan, T.A.; Mengting, Z.; Fu, D.; Yeap, S.K.; Othman, M.H.D.; Avtar, R.; Ouyang, T. Functionalizing TiO$_2$ with graphene oxide for enhancing photocatalytic degradation of methylene blue (MB) in contaminated wastewater. *J. Environ. Manag.* **2020**, *270*, 110871. [CrossRef] [PubMed]
5. Wafi, M.A.E.; Ahmed, M.A.; Abdel-Samad, H.S.; Medien, H.A.A. Exceptional removal of methylene blue and p-aminophenol dye over novel TiO$_2$/RGO nanocomposites by tandem adsorption-photocatalytic processes. *Mater. Sci. Energy Technol.* **2022**, *5*, 217–231. [CrossRef]
6. Zhang, X.; Chen, W.; Bahmanrokh, G.; Kumar, V.; Ho, N.; Koshy, P.; Sorrell, C.C. Synthesis of V- and Mo-doped/codoped TiO$_2$ powders for photocatalytic degradation of methylene blue. *Nano-Struct. Nano-Objects* **2020**, *24*, 100557. [CrossRef]

7. Ojha, A.; Thareja, P. Graphene-based nanostructures for enhanced photocatalytic degradation of industrial dyes. *Emerg. Mater.* **2020**, *3*, 169–180. [CrossRef]
8. Coleman, N.; Lovander, M.D.; Leddy, J.; Gillan, E.G. Phosphorus-Rich Metal Phosphides: Direct and Tin Flux-Assisted Synthesis and Evaluation as Hydrogen Evolution Electrocatalysts. *Inorg. Chem.* **2019**, *58*, 5013–5024. [CrossRef]
9. Abd Elkodous, M.S.; El-Sayyad, G.; Abdel Maksoud, M.I.A.; Kumar, R.; Maegawa, K.; Kawamura, G.; Tan, W.K.; Matsuda, A. Nanocomposite matrix conjugated with carbon nanomaterials for photocatalytic wastewater treatment. *J. Hazard. Mater.* **2021**, *410*, 124657. [CrossRef] [PubMed]
10. Anku, W.; Oppong, S.O.; Shukla, S.K.; Govender, P.P. Comparative photocatalytic degradation of monoazo and diazo dyes under simulated visible light using Fe3 /C/S doped-TiO$_2$ nanoparticles. *Acta Chim. Slov.* **2016**, *63*, 380–391. [CrossRef]
11. Zheng, H.; Meng, X.; Chen, J.; Que, M.; Wang, W.; Liu, X.; Yang, L.; Zhao, Y. In situ phase evolution of TiO$_2$/Ti$_3$C$_2$Tx heterojunction for enhancing adsorption and photocatalytic degradation. *Appl. Surf. Sci.* **2021**, *545*, 149031. [CrossRef]
12. Pang, X.; Xue, S.; Zhou, T.; Xu, Q.; Lei, W. 2D/2D nanohybrid of Ti$_3$C$_2$ MXene/WO$_3$ photocatalytic membranes for efficient water purification. *Ceram. Int.* **2022**, *48*, 3659–3668. [CrossRef]
13. Chinh, P.D.; Duyen, C.T.M.; Cuong, N.M.; Dong, N.T.; Hoang, N.V.; Hoi, B.V.; Nguyen, T.T.T. Integrating photocatalysis and microfiltration for methylene blue degradation: Kinetic and cost estimation. *Chem. Eng. Technol.* **2022**, *45*, 1748–1758. [CrossRef]
14. Kocijan, M.; Ćurković, L.; Radošević, T.; Podlogar, M. Enhanced Photocatalytic Activity of Hybrid rGO@TiO$_2$/CN Nanocomposite for Organic Pollutant Degradation under Solar Light Irradiation. *Catalysts* **2021**, *11*, 1023. [CrossRef]
15. Castanheira, B.; Otubo, L.; Oliveira, C.L.P.; Montes, R.; Quintana, J.B.; Rodil, R.; Brochsztain, S.; Vilar, V.J.; Teixeira, A.C.S. Functionalized mesoporous silicas SBA-15 for heterogeneous photocatalysis towards CECs removal from secondary urban wastewater. *Chemosphere* **2022**, *287*, 132023. [CrossRef] [PubMed]
16. Casado, C.; Timmers, R.; Sergejevs, A.; Clarke, C.T.; Allsopp, D.W.E.; Bowen, C.R.; van Grieken, R.; Marugán, J. Design and validation of a LED-based high intensity photocatalytic reactor for quantifying activity measurements. *Chem. Eng. J.* **2017**, *327*, 1043–1055. [CrossRef]
17. Diamantopoulou, A.; Sakellis, E.; Romanos, G.E.; Gardelis, S.; Ioannidis, N.; Boukos, N.; Falaras, P.; Likodimos, V. Titania photonic crystal photocatalysts functionalized by graphene oxide nanocolloids. *Appl. Catal. B Environ.* **2019**, *240*, 277–290. [CrossRef]
18. Alosaimi, E.H.; Alsohaimi, I.H.; Dahan, T.E.; Chen, Q.; Younes, A.A.; El-Gammal, B.; Melhi, S. Photocatalytic Degradation of Methylene Blue and Antibacterial Activity of Mesoporous TiO$_2$-SBA-15 Nanocomposite Based on Rice Husk. *Adsorpt. Sci. Technol.* **2021**, *2021*, 9290644. [CrossRef]
19. Danyliuk, N.; Tatarchuk, T.; Kannan, K.; Shyichuk, A. Optimization of TiO$_2$-P25 photocatalyst dose and H$_2$O$_2$ concentration for advanced photo-oxidation using smartphone-based colorimetry. *Water Sci. Technol.* **2021**, *84*, 469–483. [CrossRef]
20. Náfrádi, M.; Hernadi, K.; Kónya, Z.; Alapi, T. Investigation of the efficiency of BiOI/BiOCl composite photocatalysts using UV, cool and warm white LED light sources—Photon efficiency, toxicity, reusability, matrix effect, and energy consumption. *Chemosphere* **2021**, *280*, 130636. [CrossRef]
21. Badvi, K.; Javanbakht, V. Enhanced photocatalytic degradation of dye contaminants with TiO$_2$ immobilized on ZSM-5 zeolite modified with nickel nanoparticles. *J. Clean. Prod.* **2021**, *280*, 124518. [CrossRef]
22. Sun, X.; Xu, K.; Chatzitakis, A.; Norby, T. Photocatalytic generation of gas phase reactive oxygen species from adsorbed water: Remote action and electrochemical detection. *J. Environ. Chem. Eng.* **2021**, *9*, 104809. [CrossRef]
23. Acosta-Esparza, M.A.; Rivera, L.P.; Pérez-Centeno, A.; Zamudio-Ojeda, A.; González, D.R.; Chávez-Chávez, A.; Santana-Aranda, M.A.; Santos-Cruz, J.; Quiñones-Galván, J. UV and Visible light photodegradation of methylene blue with graphene decorated titanium dioxide. *Mater. Res. Express* **2020**, *7*, 035504. [CrossRef]
24. Constantino, D.S.M.; Dias, M.M.; Silva, A.M.T.; Faria, J.L.; Silva, C.G. Intensification strategies for improving the performance of photocatalytic processes: A review. *J. Clean. Prod.* **2022**, *340*, 130800. [CrossRef]

Article

Checking the Efficiency of a Magnetic Graphene Oxide–Titania Material for Catalytic and Photocatalytic Ozonation Reactions in Water

Manuel Checa , Vicente Montes, Javier Rivas and Fernando J. Beltrán *

Departamento de Ingeniería Química y Química Física, Instituto Universitario de Investigación del Agua, Cambio Climático y Sostenibilidad (IACYS), Universidad de Extremadura, 06006 Badajoz, Spain
* Correspondence: fbeltran@unex.es

Abstract: An easily recoverable photo-catalyst in solid form has been synthesized and applied in catalytic ozonation in the presence of primidone. Maghemite, graphene oxide and titania (FeGOTi) constituted the solid. Additionally, titania (TiO_2) and graphene oxide–titania (GOTi) catalysts were also tested for comparative reasons. The main characteristics of FeGOTi were 144 m^2/g of surface area; a 1.29 Raman D and G band intensity ratio; a 26-emu g^{-1} magnetic moment; maghemite, anatase and brookite main crystalline forms; and a 1.83 eV band gap so the catalyst can absorb up to the visible red region (677 nm). Single ozonation, photolysis, photolytic ozonation (PhOz), catalytic ozonation (CatOz) and photocatalytic ozonation (PhCatOz) were applied to remove primidone. In the presence of ozone, the complete removal of primidone was experienced in less than 15 min. In terms of mineralization, the best catalyst was GOTi in the PhCatOz processes (100% mineralization in 2 h). Meanwhile, the FeGOTi catalyst was the most efficient in CatOz. FeGOTi led, in all cases, to the highest formation of HO radicals and the lowest ozone demand. The reuse of the FeGOTi catalyst led to some loss of mineralization efficacy after four runs, likely due to C deposition, the small lixiviation of graphene oxide and Fe oxidation.

Keywords: magnetic catalyst; graphene oxide; titania; primidone; catalytic ozonation; photocatalytic ozonation; water treatment

Citation: Checa, M.; Montes, V.; Rivas, J.; Beltrán, F.J. Checking the Efficiency of a Magnetic Graphene Oxide–Titania Material for Catalytic and Photocatalytic Ozonation Reactions in Water. *Catalysts* **2022**, *12*, 1587. https://doi.org/10.3390/catal12121587

Academic Editors: Gassan Hodaifa, Antonio Zuorro, Joaquín R. Dominguez, Juan García Rodríguez, José A. Peres, Zacharias Frontistis and Mha Albqmi

Received: 27 October 2022
Accepted: 4 December 2022
Published: 6 December 2022

Publisher's Note: MDPI stays neutral with regard to jurisdictional claims in published maps and institutional affiliations.

1. Introduction

Catalytic ozonation processes in the absence and presence of radiation are applied to increase the efficiency of single ozonation to remove contaminants from water [1,2]. The application of ozone in water treatment is a well-known established technology that can be used to fulfil different objectives such as disinfection, the elimination of common by-products generated during drinking water chlorination [3] or emerging contaminant removal, such as pharmaceuticals, from urban wastewater [4]. Ozone reacts selectively with certain organics containing nucleophilic points in their molecular structure (i.e., phenolic structures containing activating substituents of electrophilic reactions or unsaturated moieties [5]). In advanced oxidation processes, ozone treatments are characterized by the generation of highly reacting free radicals, such as the hydroxyl radical, that unselectively react with organics in water [6]. This latter oxidizing route can be enhanced to a significant level by combining ozone with one or more other agents such as hydrogen peroxide, UV radiation or catalysts [6–8]. In this sense, the literature reports a huge variety of catalytic materials and, among them, solids with a carbonaceous base are one of the most used catalysts [9]. Hence, the synthesis and application of graphene-derived catalysts are raising increasing research interest recently. Graphene, discovered in 2004, is a 2D monolayer solid with a graphite-like structure of a high surface area, mechanical strength, significant electrical conductivity and optical properties [10,11]. This material has been studied recently in several applications [10,11] including water treatments as catalytic ozonation (CatOz) or photocatalytic ozonation (PhCatOz) for pollutant removal [12,13].

In any case, most of these works and others using different materials present important drawbacks that have limited the possibilities of real application. One of the main problems described is that most of these catalysts have been tested in powder form or are in nanometer size, so their separation from water, once they have been used, represents a technological and economically costly process. A possible solution to this problem is supporting these catalysts on porous materials such as alumina or ceramic foams [14,15], but a significant decrease in surface area is noticed. Another important drawback involves the radiation employed and is the fact that the main photoactive solids, such as titania, present a high band gap that makes visible radiation useless to activate the catalyst. Consequently, the application of solar or visible light as a radiation source is not recommended. These two important disadvantages could be eliminated or reduced when materials such as graphene oxide, iron oxides and titania are linked into one unique catalyst. Iron and graphene oxide supply magnetic properties and the capacity of absorbing visible light, respectively, whereas titania has the capability of creating electron-hole pairs that trigger the formation of hydroxyl radicals. Then, the use of this type of catalyst could be a good strategy to improve catalytic or photocatalytic processes, mainly when ozone is also present [16].

Some similar catalysts have been reported in the literature with different GO loadings in their composition. Thus, the following can be cited: Chang et al. [17] with 1 to 3% of TiO_2, Nada et al. [18] with 10, 20 and 30% of GO with magnetite or Guskos et al. [19] with 8% of GO. However, when GO linked to TiO_2 is used in photocatalytic oxidation, the previous works have highlighted that the GO percentage should be limited to values lower than 4% [20,21] to achieve the maximum removal of pollutants. This is a convenient fact as GO is an expensive material which should be minimized in large-scale processes, such as water decontamination. For this reason, in this work, the GO percentage in the prepared catalyst has been limited to 1%.

Accordingly, in this work, a magnetic graphene oxide titania (FeGOTi) catalyst was synthesized, characterized and applied to check for any possible enhancement of primidone removal from water. This compound was chosen because its presence is usual in urban wastewater [22], and its direct reaction with ozone is relatively low [23].

2. Results and Discussion

2.1. Catalyst Characterization

The main textural and electronic properties of the catalysts are summarized in Table 1 and discussed throughout the present section. From Table 1, it can be observed that titania synthesis, using the sol-gel methodology in the presence of commercial GO, led to GOTi composites with similar $S_{B.E.T.}$ values. However, after the third-phase addition, the $S_{B.E.T.}$ values decrease to 144.3 m^2 g^{-1} in the final FeGOTi catalyst. The analogous materials GO-Fe_3O_4 and Fe_3O_4-TiO_2 were discarded from this study due to the strong lixiviation of Fe in ultrapure water (measured using an Fe test in dissolution), reaching nearly total dissolution that contrasts with the null lixiviation detected for the FeGOTi. Similar effects were reported in other heterojunction systems where the graphene-like structure would increase the semiconductor stability [24,25].

Powder X-ray diffraction (XRD) patterns of bare TiO_2, GOTi and FeGOTi are depicted in Figure 1a. In all cases, the catalyst preparation procedure led to a TiO_2 characterized by the presence of an anatase (PDF 01-075-2547) and brookite (PDF 01-076-1934) phase mixture. Regarding GOTi as the FeGOTi precursor, the XRD patterns confirmed that the TiO_2 crystallographic phase remains constant after Fe_3O_4 deposition. Nevertheless, the wide signal around 12° present in GOTi, typical in GO materials [26], disappears in FeGOTi, probably due to a final GO content (c.a. 0.4 wt%) too low to be appreciated in the diffractograms or a partial reduction induced using the catalyst synthesis procedure. The XRD pattern of the FeGOTi catalyst exhibited some high intensity and sharp signals associated with the presence of magnetite (PDF 01-075-0449), titanium maghemite (TiMag,

Fe(Fe$_{0.92}$Ti$_{0.62}$)O$_4$; PDF 01-071-6450) or even other iron oxide mixtures (Fe$_{21.34}$O$_{32}$; PDF 01-080-2186). Fe and Ti were homogeneously distributed as evidenced using SEM-EDX mapping (Figure S1), and C was not possible to follow due to its low percentage. Similarly, using XPS, C1s did not show significant differences between TiO$_2$, GOTi and FeGOTi (see Figures S2–S4).

Table 1. Main characterization results concerning the photocatalyst studied: reference TiO$_2$, 1% GO-TiO$_2$ (GOTi) and 63.0% Fe$_3$O$_4$−0.4% GO–36.6% TiO$_2$ (FeGOTi).

Catalyst	Band Gap (eV)	$S_{B.E.T.}$ m^2 g^{-1}	TGA (% C)	XRD (nm) *	Crystallinity Degree	Raman (I_D/I_G)
TiO$_2$	3.02	195.0	-	6.9 (A)	69.7	-
GOTi	2.86	184.2	1.24	7.3 (A)	70.0	1.31
FeGOTi	1.83	144.3	0.49	38.6 (M); 8.7 (A)	84.0	1.29

* Particle size determined using Scherrer equation method from XRD main signal: (A) = anatase; (M) = maghemite/magnetite; TGA = thermogravimetric analysis; XRD = X-ray diffraction; I_D/I_G = D and G band ratio intensity.

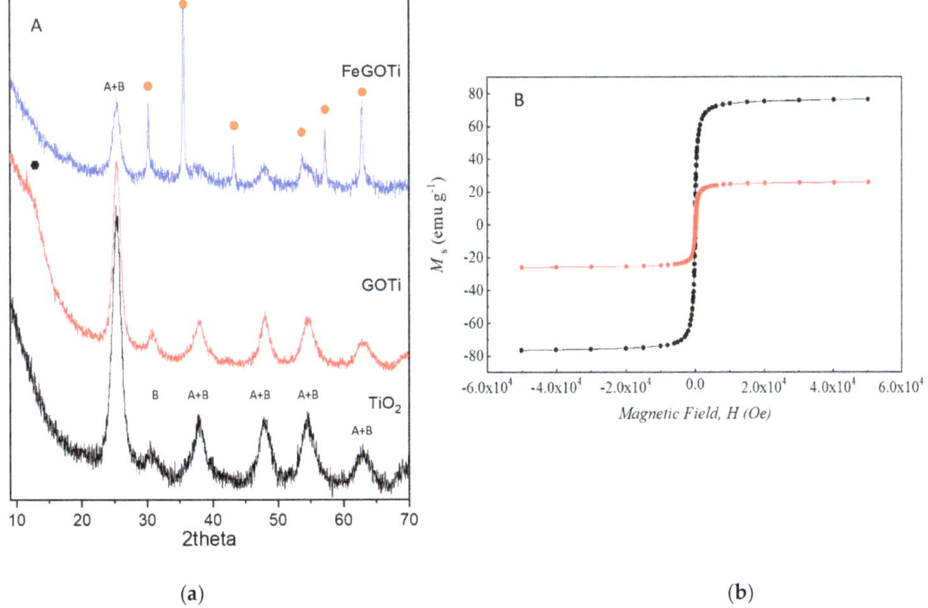

Figure 1. (a) XRD patterns of the solids. Identified phases were anatase (A), brookite (B), graphene oxide (⬢) and maghemite/magnetite (🟠). (b) Magnetic moment, Ms, versus applied magnetic field, H, for calcined magnetite (black) and FeGOTi catalyst (red).

The determination of the FeGOTi magnetic properties was performed by means of the SQUID technique. The main results can be found in Figure 1b. In general, the magnetic moment values, M$_s$, depend on the particle size. In the case of pure magnetite, a range of values from 60 to 92 emu g^{-1} is commonly reported for pure and bulk material, respectively [16,27,28]. In contrast, the M$_s$ value for the FeGOTi catalyst was estimated around 25 emu g^{-1} which is quite far from the magnetite values, demonstrating that the synthesis procedure affects the iron-containing phase. This fact allows a much more correct assignation of the phase detected in XRD as maghemite that presents similar magnetic moment values in the bibliography [29,30].

The Raman spectra collected between 1000 and 2000 cm^{-1} provides valuable information regarding the graphene structure. In this Raman shift range, carbonaceous materials "D" and "G" bands could be observed around 1300 and 1600 cm^{-1}, respectively. In the bibliography, D bands are associated with sp^{3}C presence and, in the case of GO, indicate the presence of oxygenated groups that disrupt the sp^2 graphitic structure, with G bands as a measure of the latter. The ratio between both band intensities can be interpreted as a measure related to the presence of GO in the material. As observed in Table 1 and Figure S5, catalysts containing GO, such as GOTi and FeGOTi exhibited I_D/I_G ratios of 1.31 and 1.29, respectively, within the range of the previously reported values for GO (1.2–1.4) [31].

A TGA-DTA-MS analysis of the catalyst allows the estimation of the carbon content and can be observed in Figure S6. The mass losses associated with carbonaceous material combustion exhibit an exothermic peak in the DTA analysis, whereas the endothermic peaks are associated with gas desorption. The quantification would be carried out from the measurement of CO and CO_2 when an exothermic peak is detected. In this way, the C content was determined as presented in Table 1, with 1.24% in GOTi and 0.49% in FeGOTi.

Concerning the band-gap measurements, a narrowing tendency was observed. Firstly, after GO incorporation, a light narrowing from 3.02 to 2.86 eV made the catalyst able to absorb photons of wavelengths up to 433 nm within the violet region (380–450 nm). Iron oxides have been studied in photocatalysis and usually exhibit band-gap values around 2.3 eV [32,33]. Synergism in the ternary catalyst between the maghemite–GO–titania phases leads to an important band-gap narrowing to 1.83 eV, meaning that FeGOTi can harvest photons of wavelengths up to the red region wavelength (677 nm in this case). In Figure S7, the normalized absorbance spectra of the three catalysts (TiO$_2$, GOTi and FeGOTi) and the normalized irradiance collected for the light source employed in this work are shown.

2.2. Activity Tests

Considering the ozone-involving experiments, different processes are studied. The simplest process to remove the pollutant is the direct treatment of the sample with ozone in the absence of light, which is called simple ozonation. The radical species of this treatment can be intensified by the incorporation of a radiation source with the PhOz process or in darkness by the incorporation of a catalyst that enhances the decomposition of ozone in the reaction media (CatOz). Finally, the combination of ozone with both a catalyst and light (PhCaTOz) exponentially increases the production of radical species responsible for performing organic matter mineralization. The catalyst activity for bare TiO$_2$, GOTi and FeGOTi was evaluated at different reaction conditions including catalytic oxidation (Cat), photocatalytic oxidation (PhCat), catalytic ozonation (CatOz) and photocatalytic ozonation (PhCatOz) as well as in photolysis and single ozonation conducted as blank experiments. The main results can be found in Figure 2, Figure 3 and Figure S8.

As can be observed in Figure 2, Figure 3 and Figure S8, ozonation, regardless of the presence of a catalyst and light, leads to fast and total primidone removal [34,35] in less than 30 min. Under photolysis conditions, primidone remains unchanged throughout the reaction time, exhibiting less than 10% removal and negligible changes in the TOC after 3 h of reaction. In the absence of ozone, the synergy between TiO$_2$ and GO in photocatalysis was revealed when the final primidone conversions after 3 h of treatment were compared (Figure 2). After 3 h, complete primidone removal took place in the presence of GOTi, whereas only about 40% could be eliminated using TiO$_2$. Surprisingly, despite the lower band gap of the FeGOTi catalyst (1.8 eV, Table 1), less than 20% of primidone removal was detected under photocatalysis conditions. This fact is important and can be a consequence of the catalyst's internal composition. The TOC measurements indicate that ozone-free photocatalytic treatments only achieved 30% mineralization when the most active catalyst (GOTi, see Figure S8) was used, so the addition of ozone is still necessary for deeper mineralization.

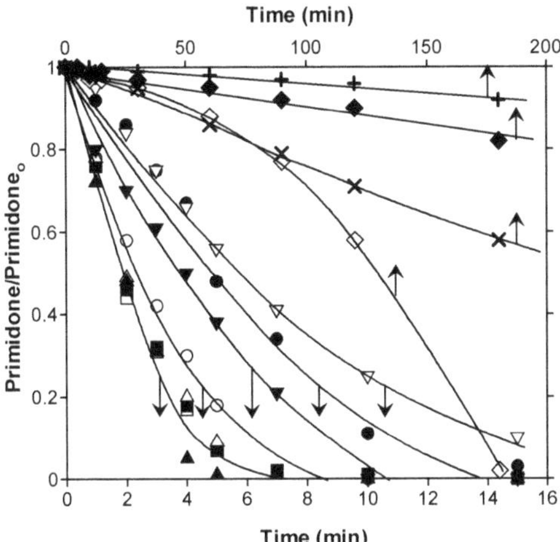

Figure 2. Variation in normalized primidone-remaining concentration (C/C_0) with time in experiments: without catalyst—ozonation (▼), photolytic ozonation (▽) and photolysis (+); with bare TiO_2, CatOz (▲), PhCatOz (Δ) and PhCat (x); with GOTi, CatOz (■), PhCatOz (*) and PhCat (◊); and with FeGOTi, CatOz (○), PhCatOz (●) and PhCat (♦). Conditions: gas flow rate, 35 L h^{-1}; inlet ozone gas concentration, 10 mg L^{-1}; primidone initial concentration, C_0 = 10 mg L^{-1}; catalyst dosage, 0.25 g L^{-1}; agitation speed, 700 rpm.

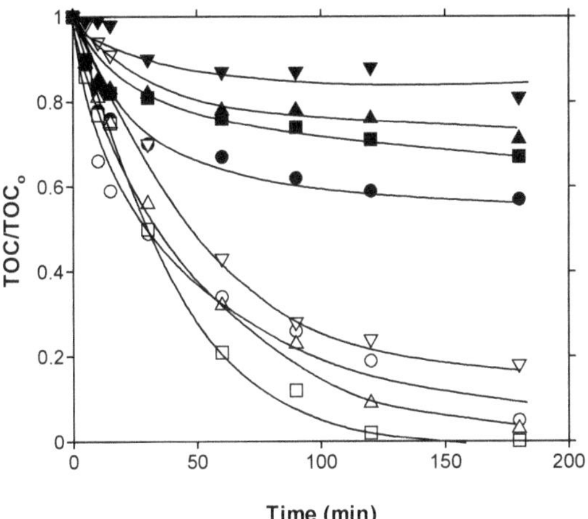

Figure 3. Variation in normalized TOC/TOC_0 with time from primidone ozone advanced oxidation experiments: without catalyst—ozonation (▼) and photolytic ozonation (▽); with bare TiO_2, CatOz (▲) and PhCatOz (Δ); with GOTi, CatOz (■) and PhCatOz (□); and with FeGOTi CatOz (●) and PhCatOz (○). Conditions: gas flow rate, 35 L h^{-1}; inlet ozone gas concentration, 10 mg L^{-1}; initial TOC = 6.6 mg L^{-1}; catalyst dosage, 0.25 g L^{-1}; agitation speed, 700 rpm.

Due to the similar ionic radius of Fe^{3+} and Ti^{4+}, the heterojunction of both semiconductors allowed the facilitating of the loading of Fe^{3+} species over TiO_2, acting as a nucleation point and generating an intimate union between both phases [33,36]. This fact has been evidenced in the FeGOTi XRD analysis (Figure 1a), where TiMag was identified as a possible iron-containing phase and using the different magnetic moment measured using SQUID. These latter values were closer to those reported for maghemite than for magnetite. The fact that TiMag is the dominant phase in the catalyst, explains the low band-gap value observed in FeGOTi (1.83 eV) when compared to the reported values for Fe_2O_3 (2.3–2.4 eV), although the low activity in the photocatalysis processes remains unclear [37]. The difference between the TiO_2 conduction band energy and O_2/O_2^- reaction potential facilitates the spontaneous generation of superoxide ion radicals in the titania surface as deduced by comparing the semiconductor band energies with that of NHE for the photocatalytic initiation reaction (see Figure S9). When Fe is introduced, it can first be admitted that the system acts as a heterostructure with a TiMag interface that facilitates electron mobility among the different phases and, second, as the conduction band of the Fe phase has lower potential energy with respect to titania, the Fe-containing phase would act as an electron sink, trapping the photogenerated electrons in a conduction band unable to generate O_2^- radicals, inhibiting this initiation route [38–40].

In the ozonation and PhCatOz processes, primidone was removed in the first 15 min of reaction (see Figure 2), but different trends could be observed in the TOC (Figure 3). The GOTi catalyst exhibited similar final TOC removal in the catalytic ozonation as in the photocatalytic process, whereas FeGOTi showed the best results in CatOz. A possible explanation is that the smaller particle size of the catalyst facilitates the ozone decomposition [41]. Nevertheless, in this case, TiO_2 and GOTi exhibited similar particle size, so this factor explains their similar TOC removal in CatOz. Regarding FeGOTi, the particle size influence could not be the only explanation as the maghemite phase exhibited particle sizes of around 38.4 nm. According to the bibliography, transition metal oxides with several oxidation states, such as Ti and Fe, generate oxygen vacancies at the catalyst surface that act as driving forces enhancing ozone decomposition [42–45]. Oxygen vacancy (V_o) generation implies a structural defect where a superficial oxygen atom is removed from the metal oxide crystalline structure. As a consequence, the neighboring metal requires a change in the oxidation state to maintain the charge neutrality of the material, generating the Ti^{3+}/Ti^{4+} or Fe^{2+}/Fe^{3+} pairs, respectively. These pairs are visible using XPS, Ti^{3+}/Ti^{4+} in fresh TiO_2 and GOTi, and Fe^{2+}/Fe^{3+} in FeGOTi (see Figures S2–S4). As the amounts of V_o in the surface are proportional to the particle size [46], the fact that FeGOTi exhibits a better behavior in catalytic ozonation seems to be sustained. Recently, Liu et al. [47] reported that O_3 decomposition over oxygen vacancies on TiO_2-P25 facilitates water dissociation and generates Ti-OH surface groups, enhancing the pollutant degradation process. This effect can also be increased by the presence of a light source for pure TiO_2. However, as discussed with the photocatalysis data, when FeGOTi is considered as a coupled semiconductor system with Fe phases acting as the electron sink, the recombination of the e–h^+ pair is favored to the detriment of the photodegradation process.

In PhCatOz, after 3 h of treatment, all of the catalysts tested exhibited nearly total mineralization (>90% TOC removal, see Figure 3). When GOTi was used, total mineralization was reached in the first 2 h of treatment. Due to the O_3 electron affinity, photogenerated electrons in the solid can be collected more efficiently than in the presence of O_2, enhancing the radical production and TOC removal and revealing the synergism between the processes. These processes follow pseudo-first-order kinetics, as can be observed from Figure S10. Despite apparent pseudo-first-order rate constants, k_{ap} has no rigorous kinetic basis, and they could only be used as a comparison tool to classify the catalysts according to their activity. In this sense, Li et al. [48] formulated a synergy factor (SF), calculated through Equation (1):

$$SF = \frac{k_{ap}(\text{photocatalytic ozonation})}{k_{ap}(\text{photolytic ozonation}) + k_{ap}(\text{photocatalytic oxidation})} \tag{1}$$

From the SF values, a synergism estimation could be experienced, and the catalyst efficiency could be compared [48,49].

As observed from Table 2, for the PhCatOz processes, the order of catalysts regarding the TOC removal efficiency was GOTi > TiO_2 > FeGOTi. This order is the same as that observed in PhCat. The lower synergy in FeGOTi could be a consequence of the synthesis procedure that uses GOTi as a nucleation point for FeO_x deposition. Nevertheless, the possibility of catalyst recovery and a similar mineralization tendency may be worthy of further research on FeGOTi.

Table 2. Values of k_{ap} and synergy factor of TOC removal for the catalysts in PhCat, CatOz and PhCatOz processes.

	PhCat		PhCatOz		SF
	k_{ap} (min^{-1})	R^2	k_{ap} (min$^{-1)}$	R^2	
No catalyst	0 [a]	-	0.0079 [b]	0.98	-
FeGOTi	0	-	0.0095	0.97	1.20
GOTi	0.0016	0.95	0.0405	0.90	4.23
TiO$_2$	0.0007	0.95	0.0176	0.98	2.03

[a] It corresponds to photolysis. [b] It corresponds to photolytic ozonation.

Regarding the possible catalytic reaction mechanism, the bibliography reports several hypotheses. Firstly, the mechanism can be analyzed considering each phase as a single component or as an individual catalyst. Secondly, the mechanistic study can be addressed considering the heterojunction of the components by pairs, for instance, TiO_2 and graphene or Fe_3O_4 and graphene. And finally, the third possibility implies an understanding of the system as a three-component heterojunction. As there is no scientific agreement about what kind of perspective is the most probable, especially when graphene is involved, it is very difficult to propose an accurate and specific reaction mechanism for the process. These difficulties were also found in other research work [50–55].

2.3. pH Stability Range and FeGOTi Reusage

Iron leaching is a common limitation for FeO_x-based catalyst aqueous applications. In ozone processes, H_2O_2 is usually formed from direct ozone reactions or hydroxyl ion ozone decomposition [56–58]. At low pH values, in PhCatOz, the presence of Fe-soluble species could lead to a second catalytic oxidation way due to the homogeneous Fenton and photo-Fenton processes in the presence of H_2O_2 (also detected in this work) [59–61]. Under unbuffered reaction conditions with FeGOTi, pH evolves from 7.5–8.0 to 4.8–5.0 at the end of the reaction, likely due to the generation of short-chain organic acids intermediates [34]. In this work, however, no Fe ions were detected in the aqueous phase. To delimit a possible pH application range, FeGOTi was tested in buffered PhCatOz experiments at pH values of 4, 6 and 8, with an $HPO_4^{-2}/H_2PO_4^-$ 50 mM buffer solution. This buffer was selected due to the high presence of PO_4^{3-} ions in urban wastewater. For pH values lower than 4, Fenton and photo-Fenton processes are viable, so an additional reaction was performed at pH 3.8 where Fe begins to dissolve. As observed in Table 3, the presence of PO_4^{3-} species from the buffer solution negatively affected the catalyst performance, reaching mineralization levels lower than those found in ultrapure water and simultaneously inducing some Fe leaching. PO_4^{3-} ions can be adsorbed on the solid inhibiting $HO^{\cdot}$ radical production on the catalyst surface [62,63]. When the pH decreases below 4, Fe leaching increases, and the levels of Fe in solution are high enough to induce an additional photo-Fenton process, leading to almost total mineralization. On the opposite hand, pH values above 8 would induce the leaching of non-active aquo Fe-complex species in the photo-Fenton process, leading to the lowest mineralization value observed (in the photolytic ozonation range).

Table 3. Effect of pH on PhCatOz final mineralization values and Fe leaching detected with FeGOTi. Conditions: gas flow rate, 35 L h^{-1}; inlet ozone gas concentration, 10 mg L^{-1}; primidone, C$_0$ = 10 mg L^{-1}; TOC$_0$ = 6.6 mgL^{-1}; catalyst dosage, 0.25 g L^{-1}; agitation speed, 700 rpm. HPO$_4$$^{-2}$/H$_2PO_4$$^-$ 50 mM buffer solution.

pH	% TOC Removal	% Fe Leached
3.8	99.2	2.64
4.0	74.6	0.12
6.0	77.5	0.16
8.0	68.0	1.68

Reusability tests were conducted using the spent FeGOTi catalyst. The solid was recovered with a magnet, washed with 3 × 50 mL of ultrapure water and dried overnight at 110 °C before being reused. As inferred from Figure 4, the activity of the catalyst experiences a slight decay after the first use, maintaining a similar mineralization conversion in the following reuses. Except for the fresh catalyst, mineralization conversion reached a stable value after 2 h of treatment, exhibiting similar values as the photolytic process after 180 min. No soluble Fe species in the reaction media were detected until the third and fourth run (0.06–0.11% leaching); this fact, however, does not explain the activity decay.

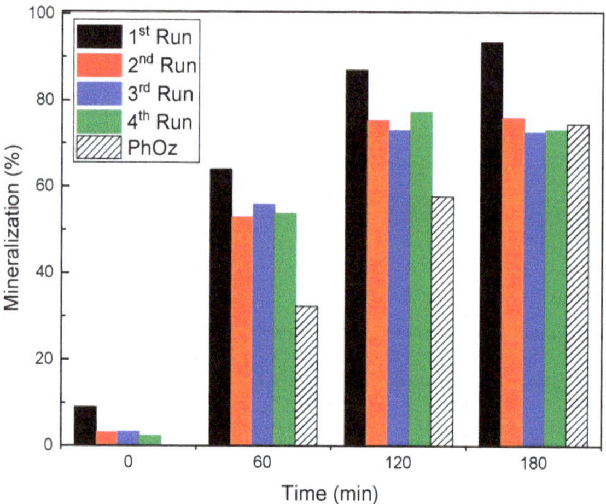

Figure 4. TOC mineralization at different reaction times for the FeGOTi reusability test under PhCatOZ conditions and photolytic ozonation as reference (PhOz). Conditions: gas flow rate, 35 L h^{-1}; inlet ozone gas concentration, 10 mg L-1; primidone, C$_0$ = 10 mg L^{-1}; TOC$_0$ = 6.6 mgL^{-1}; catalyst dosage, 0.25 g L^{-1}; agitation speed, 700 rpm. Results at 0 min correspond to dark adsorption for a 30 min period.

After the fourth run, the catalyst was recovered and characterized. In terms of optical properties, the used catalyst exhibited a band-gap value of 1.80 eV, close to the fresh catalyst value reported in Table 1. This suggests that the reaction conditions do not affect the light absorption capacity of the catalyst despite Fe leaching. On the contrary, the textural properties were affected. A decrease in the surface area from 144 to 123 m^2 g^{-1} was detected after the reaction, sustaining a clear effect of the reaction conditions on the catalyst structure. When XRD was performed, the crystalline phase detected was the same in both the fresh and used catalyst (Figure S11A). Despite the TiMag and anatase particle size exhibiting similar values to the fresh catalyst (39.3 and 10.7 nm for the TiMag and anatase phases, respectively), an increase in the FeGOTi crystalline grade from 84.0 to 93.1% was detected. This increment indicates a partial recrystallization of the solid under PhCatOz conditions.

Special attention must be paid to the 10–15° 2θ value, where a broad and low signal can be appreciated in the used catalyst. As commented in Figure 1, in these diffraction angles, graphene oxide exhibits its characteristic signal. The fact that this diffraction signal could be not appreciated in the fresh catalyst indicates that the synthesis conditions gave rise to a partial reduction in GO leading to a reduced graphene oxide structure (rGO). The typical signal of rGO would be hindered by the anatase signal that also appears at the same diffraction angle. Under ozone reaction conditions, rGO would recover some oxygenated groups as it is seen in the 10–15° 2θ band of Figure S11A.

To study the potential causes of catalyst deactivation, TGA experiments with the recovered catalyst were performed (see Figure S11B). The main mass loss detected was associated with water desorption. The quantification of CO_2 and CO associated with the exothermic peaks and combustion signals indicates 0.32% of C content in the recovered catalyst, which could be due to carbon deposition and some graphene oxide lixiviation. Moreover, the XPS analysis evidenced a change in the Fe-oxidation state during the reactions (Figure S12). In the fresh catalyst, Fe2p presents three clear peaks, with maximums at 709.5, 711.2 and 713 eV, whereas after four runs, only two peaks were observed at 710 and 712.1 eV. The peak around 710 eV is typical for Fe oxides, such as magnetite (PDF 01-075-0449), titanium maghemite (TiMag, $Fe(Fe_{0.92}Ti_{0.62})O_4$; PDF 01-071-6450) or even other iron oxide mixtures ($Fe_{21.34}O_{32}$; PDF 01-080-2186) detected using XRD. The peaks at a higher eV evidenced the presence of species such as FeOOH, which, in the presence of ozone, can promote the generation of HO radicals [30], and thus, Fe is oxidized during the first run to a less active form.

In view of these facts, it can be concluded that the reaction conditions affect the catalyst structure, especially from the fourth use. Under these conditions, a recrystallization of the solid likely took place, and Fe leaching could be clearly appreciated. Despite the decrease in activity observed after the first use, the activity of the catalyst stabilizes in the following reuses, reaching similar mineralization levels as the photolytic ozonation in the first 120 min.

2.4. Catalysts' Efficiency Comparison

In previous works, the degradation mechanism of primidone was studied, and the role of HO˙ radicals as being responsible of the predominant elimination route was demonstrated [34,35,58]. Considering the mechanism of ozone processes, the generation of HO˙ radicals can be assumed as a consequence of ozone interaction with a photogenerated electron in the semiconductor conduction band or by direct reaction with hydrogen peroxide [58]. In this context, three parameters—transferred ozone dose (TOD), R_{ct} and R_{O3OH}—can be considered to compare catalyst performance. These parameters can be determined through Equations (2)–(4), respectively [64]. Firstly, the determination of the ozone consumed by the water matrix through the transferred ozone dose (TOD) can be used as a descriptive parameter of the catalyst [65,66]:

$$\text{TOD} = \frac{v_g}{V} \int_0^t \left(C_{O3gas_{in}} - C_{O3gas_{out}} \right) dt \qquad (2)$$

where v_g refers to the gas flow rate, V the liquid phase volume, t the reaction time and $C_{O3gasin}$ and $C_{O3gasout}$ the ozone concentration in the gas at the entrance and exit of the reactor, respectively. Considering the low concentration of dissolved ozone with respect to the TOD, the TOD can be considered as ozone demand. As observed in Figure 5, there is a relationship between the TOD and consumed TOC in catalytic and photocatalytic ozonation. Although FeGOTi was the catalyst that requires less TOD per mg of the TOC removed in catalytic ozonation, the synergy of GOTi under PhCatOz conditions completely surpasses the efficiency of FeGOTi, requiring less than 40 mg L^{-1} of ozone to reach total mineralization.

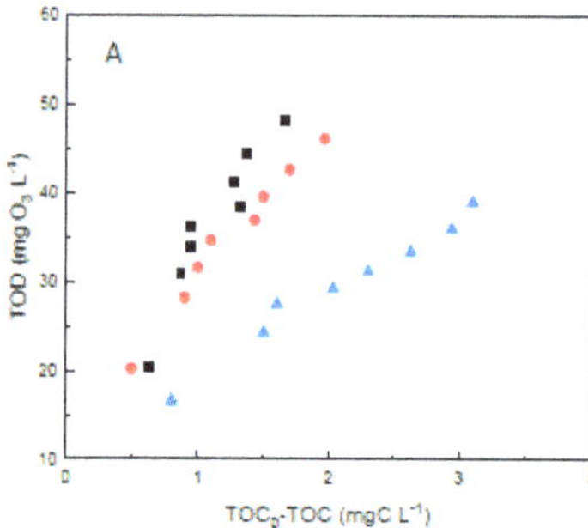

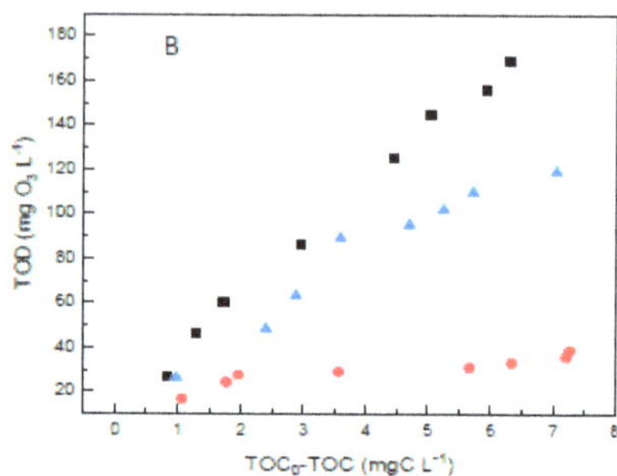

Figure 5. Changes in TOD with eliminated TOC during CatOz (**A**) and PhCatOz (**B**) for TiO_2 (■), GOTi (●) and FeGOTi (▲). Conditions: gas flow rate, 35 L h^{-1}; inlet ozone gas concentration, 10 mg L^{-1}; primidone, C_0 = 10 mg L^{-1}; TOC_0 = 6.6. mgL^{-1}; catalyst dosage, 0.25 g L^{-1}; agitation speed, 700 rpm.

Secondly, R_{CT}, considered as a tool for emerging contaminant oxidation modeling, provides a relationship between HO· and O_3 exposure [66], allowing the estimation of a catalyst's ability to generate HO· radicals from O_3 in the reaction media:

$$\mathrm{Ln}\left(\frac{TOC}{TOC_0}\right) = -R_{CT}k_{HO}\int_0^t C_{O_3}dt \tag{3}$$

where k_{OH} is the apparent rate constant of the reaction of TOC and HO· radicals that can be taken as 5×10^9 M^{-1}s^{-1} [67].

Finally, the R_{O3OH} parameter allows an estimation of HO· radicals regardless of the ozone concentration applied [68,69]:

$$\mathrm{Ln}\left(\frac{TOC}{TOC_0}\right) = -R_{O_3HO}k_{HO}\left(TOD - C_{O_3dis}\right) \tag{4}$$

The values of both R_{CT} and R_{O3HO} are shown in Table 4 for CatOz and PhCatOz experiments with different catalysts. As observed, the TiO_2 catalyst, similar to the TOD, is also the material with the lowest efficiency due to its lowest R_{CT} and R_{O3HO} values and, hence, the catalyst allows the formation of the lowest HO concentration. In the case of GOTi and FeGOTi, the corresponding values of R_{CT} and R_{O3HO} also confirmed the highest efficiency of FeGOTi in CatOz and GOTi in PhCatOz. The former leads to the highest HO concentration in the catalytic process, whereas the latter presents the best performance in producing HO radicals in the PhCatOz process.

Table 4. Values of R_{CT} and R_{O3HO} for PhCatOz and CatOz processes with different catalysts. Conditions as in Table 3.

	PhCatOz		CatOz	
Catalyst	R_{CT}	R_{O3HO}	R_{CT}	R_{O3HO}
FeGOTi	3.28×10^{-4}	5.52×10^{-9}	1.04×10^{-4}	2.25×10^{-9}
GOTi	1.26×10^{-3}	5.95×10^{8}	6.80×10^{-5}	1.38×10^{-9}
TiO$_2$	1.49×10^{-4}	2.72×10^{-9}	5.27×10^{-5}	1.02×10^{-9}

Finally, the results obtained in terms of primidone removal were compared with others previously reported, as shown in Table 5. Biological treatments such as biofiltration were not considered as primidone remains unaltered after the process [70], highlighting the importance of advanced oxidation processes for their removal.

Table 5. Main values reported for primidone AOP treatment.

Process	Radiation Source	Catalyst (Dosage)	Removal, % (Time, h) of:		Reference
			Primidone	TOC	
UV/Cl_2	Low Pressure Hg lamp	-	80–90 (0.17)	5 (1 h)	[71]
UV/Cl_2	Low Pressure Hg lamp	$FeCl_3$ (50 µmol/L)	100 (0.085)	6.8 (1 h)	[71]
Peroxymonosulfate	Solar radiation	PMS (5 mM)	95–100 (2)		[72]
UV/H_2O_2	Two G15T8 germicidal lamps	H_2O_2 (20 mg/L)	10–89 (0.33)	-	[73]
O_3/UV Semicontinuous	UV-LED	TiO_2 (0.5 g/L)	100 (0.25)	95 (1 h)	[74]
O_3/Vis	425 nm Vis-LED	TiO_2 (0.25 g/L)	100 (0.17)	90 (2 h)	[35]
O_3/Vis	425 nm Vis-LED	GO/TiO_2 (0.25 g/L)	100 (0.17)	80 (2 h)	[35]
O_3/UV-Vis	Solar radiation	TiO_2 (0.25 g/L)	100 (0.1)	95 (3 h)	This work
O_3/UV-Vis	Solar radiation	GO/TiO_2 (0.25 g/L)	100 (0.1)	100 (2 h)	This work
O_3/UV-Vis	Solar radiation	FeGOTi (0.25 g/L)	100 (0.25)	90 (3 h)	This work

3. Materials and Methods

3.1. Catalyst Preparation

Ternary Fe_3O_4-GO-TiO_2 catalyst was prepared following a sequential procedure employing commercial GO purchased from Graphenea®. Firstly, GO-TiO_2 was prepared using a sol-gel methodology adapted from previous works [35,75]. Briefly, the catalysts, under vigorous stirring, were prepared by diluting 21.8 mL of titanium (IV) butoxide (97%) in 6.4 mL of 2-propanol, adding, thereafter, 205 mL of distilled water at pH 2 (adjusted with 65% HNO_3). Titania sol was obtained after 75°C reflux for 24 h and transferred to a rotary evaporator to remove the alcohol using distillation at 80 °C under vacuum. Next, an amount of GO powder, calculated to obtain 1% of GO content in the final solid, was added and dispersed in the titania sol under ultrasonic treatment for 1 h. After evaporation to dryness under vacuum at 80 °C on a rotary evaporator, the solid was heated overnight in an oven at 100 °C. The obtained solid was milled and sieved. Reference TiO_2 was prepared following the same procedure in the absence of GO.

Fe_3O_4 incorporation was adapted from the method described in [76]. Under N_2 atmosphere, 0.5 g GO/TiO_2 was suspended in 350 mL of deionized water previously bubbled with N_2 during 30 min. Then, an aqueous solution (18 mL) containing 0.17 and 3.51 mmol of $FeCl_3 \cdot 6H_2O$ and $FeCl_2 \cdot 4H_2O$, respectively, was added, and the flask sealed. The mixture was vigorously stirred for 5 h under N_2 at room temperature. After the addition of 16 mL of 5 M NH_4OH, the temperature was increased to 65 °C, and reaction allowed to continue for 2.5 h. The product was collected using centrifugation, washed three times with water to remove excess ions and heated overnight in an oven at 100 °C. Finally, the solid obtained was milled and sieved. Theoretical catalyst weight composition was 0.4% GO, 36.6% TiO_2 and 63% of Fe_3O_4.

To increase the stability of the catalyst and remove possible NH_3 residues from the synthesis, the solids were calcined in a muffle furnace. After a temperature ramp of 10 °C min^{-1}, temperature was stabilized at 200 °C and maintained for 6 h. Then, the muffle was cooled down, and the solid was milled and sieved. The solids obtained were named as FeGOTi, GOTi and TiO_2, referencing the calcined composites of Fe_3O_4-GO-TiO_2, GO-TiO_2 and bare TiO_2, respectively.

3.2. Catalyst Characterization

Calcined material structure and morphology were characterized using a TEM Tecnai G20 Twin 200 kV transmission electron microscope (TEM. FEI Billerica, Billerica, MA, USA). All samples were mounted on 3 mm and 300 mesh lacey carbon film copper grids. The images were acquired with "TEM Imaging & analysis" 4.0 software.

A powder Bruker D8 Advance XRD diffractometer with CuKα radiation (λ= 0.1541 nm) and Ge 111 as monochromator was used to infer the crystalline phases present in the photocatalysts. The data were collected from 2θ = 20 to 80° at a scan rate of 0.02 s^{-1} and 1 s per point.

Thermogravimetric analysis was carried out in a SETSYS Evolution—16 (TGA. SETARAM—Scientific & Industrial Equipment, Lyon, France). An amount of 30 mg of sample was placed in an alumina crucible for TGA–DTA analysis and heated at temperatures from 50 to 900 °C at a rate of 10 °C/min under a stream of synthetic air at 40 mL min^{-1} in order to measure weight loss, heat flow and derivative weight loss.

Surface areas of the solids were determined from nitrogen adsorption–desorption isotherms obtained at liquid nitrogen temperature on an Autosorb iQ2-C Series, using the Brunnauer–Emmett–Teller ($S_{B.E.T.}$) method. All samples were degassed to 0.1 Pa at 120 °C and 12 h prior to measurement.

RAMAN analysis was performed in a Thermo Scientific Almega XR dispersive RAMAN spectrometer, equipped with two objectives for sample focus: MPlan 10× BD and MPlan 50× BD. A 630 nm laser beam as radiation source, 1288–92 cm^{-1} as spectral range. Fluorescence correction was used, and the spot size employed was 2.5 µm. All Raman spectra were registered and treated with OMNIC thermo scientific software.

UV-vis diffuse reflectance spectra were collected on a Cary 5000 (Agilent) UV-Vis-NIR spectrophotometer with an integrating sphere. The wavelength range studied was 200 to 800 nm. The resulting reflectance spectra were transformed into apparent absorption spectra using the Kubelka–Munk function (F(R)). The optical band gap of the materials was determined through the construction of Tauc plots by plotting (F(R)hν)n against (hν), with n = 1/2 as TiO_2 is an indirect semiconductor. The optical band gap was obtained by extrapolating the linear part of this plot to the energy axis.

For X-ray photoelectron spectroscopy (XPS) analysis, a Leibold–Heraeus LHS10 spectrometer (Bad Ragaz, Switzerland) capable of operating down to less than 2×10^{-9} Torr was used. This was equipped with an EA-200MCD hemispherical electron analyzer with a dual X-ray source, using AlKα (hν = 1486.6 eV) at 120 W, at 30 mA and utilizing C (1 s) as energy reference (284.6 eV). The spectra were recorded on 4 × 4 mm pellets 0.5 mm thick that were obtained by gently pressing the powdered materials following outgassing to a pressure below about 2×10^{-8} Torr at 150 °C in the instrument prechamber to remove chemisorbed volatile species.

Catalyst magnetic properties were determined using a Tesla superconducting quantum interference device (SQUID). Magnetic moment, M, was measured at 300 K as a function of applied magnetic field (from 0 to 7 T).

3.3. Photocatalytic Test

All tests were performed placing the reactor in an Atlas$^{®}$ simulated solar box SUNTEST CPS+, equipped with a Xe arc lamp as radiation source and magnetic stirring. The radiation fluency was kept constant at 500 Wm^{-2}, and incident radiation wavelength was >300 nm. Ozone was generated from pure oxygen with a 30/7 Sander laboratory ozonator type. In a typical experiment, 0.125 mg of catalyst and 0.5 L of an aqueous solution of primidone were placed in a borosilicate three-neck round-bottom reactor to have a final 10 mgL^{-1} concentration. In the absence of light, the reaction mixture was stirred for 30 min until adsorption equilibrium conditions were reached. Then, radiation source was connected in light-involving runs, and a 35 L h^{-1} gas flow carrying pure O_2 or O_3/O_2 mixture (10 mgL^{-1} ozone concentration) was bubbled in the reaction media. Steadily, samples were withdrawn from the reactor, filtered and analyzed.

3.4. Analytical Methods

As described in a previous work [35], primidone concentration was determined by means of HPLC-DAD performed in a Hitachi Elite LaChrom chromatograph equipped with a Phenomenex C-18 column (5 µm; 150 mm long and 3 mm diameter). The analysis was carried out following an isocratic method with acetonitrile/acidified water (20:80 *v/v*. 0.1% phosphoric acid) as mobile phase, a flow rate of 0.6 mL·min^{-1} and 215 nm as detection wavelength. Total organic carbon (TOC) in samples was determined in a Shimadzu TOC-VSCH analyzer. Ozone gas concentration in outlet gas was monitored with an Anseros GM-19 ozone detector. Dissolved ozone in water was determined using the Indigo method [77] with a Hach Lange 2800 Pro spectrophotometer set at 600 nm. Hydrogen peroxide was analyzed following Masschelein et al.'s method [78]. Total iron concentration was evaluated spectrophotometrically at 565 nm following the specification of Spectroquant® iron test (Merck 1.14761.0001, Kenilworth, NJ, USA).

4. Conclusions

The main conclusions reached in this work are:

The characterization results demonstrated a magnetically easily recoverable photocatalyst synthesized through a sequential sol-gel procedure with a loading of graphene oxide below 1%. This small loading clearly increased the efficiency in CatOz and PhCatOz. The main characteristics of FeGOTi were 144 m^2/g of surface area; a 1.29 Raman D and G band intensity ratio; a 26-emu g^{-1} magnetic moment; maghemite, anatase and brookite main crystalline forms; and a 1.83 eV band gap so the catalyst can absorb up to the visible red region (677 nm).

Regarding the activity of the catalysts, although the presence of ozone alone already leads to the significant removal of primidone (100% removal in 15 min), mineralization (the TOC reduction), as already reported previously, is less than 20% after a 3 h reaction. However, the presence of catalysts both in the CatOz and PhCatOz processes allows increasing removal efficiency both for primidone and, especially, mineralization. Thus, the three catalysts lead to total primidone removal in less than 8 min in the CatOz processes, and similar results are reached during PhCatOz with the TiO_2 and GOTi catalysts. As far as mineralization is concerned, the FeGOTi and GOTi catalysts lead to the highest mineralization in the CatOz and PhCatOz processes with best performances observed in the latter process with 100% and 95% mineralization reached in 2 h with GOTi and 3 h with FeGOTi, respectively.

The best pH values to perform PhCatOz were between 4 and 6 where low Fe leaching was observed (less than 0.16%). However, at pH 3.8 with 2.68% Fe leaching, 99.2% mineralization was observed in the FeGOTi process, likely due to the contribution of photo-Fenton oxidation.

The reuse of the FeGOTi catalyst in PhCatOz leads to a decrease in mineralization removal from 95 to about 78% after a 3 h reaction. This was likely due to some C deposition, lixiviation of graphene oxide and Fe oxidation during the first run.

Finally, the parameters, such as the TOD, RCT and R_{O3HO} that measure the amount of ozone needed to reach some mineralization degree or the formation of HO radicals, also confirm the best performance of GOTi during PhCatOz and FeGOTi during CatOz, respectively. Despite these differences between both GO-containing catalysts, it must be pointed out that FeGOTi allows an easy separation from water which results in lower costs for a possible real application. Then, the study of magnetic catalysts also containing GO and TiO_2 can be an attractive way of continuing the research of photocatalytic ozonation processes, especially with the use of solar light.

Supplementary Materials: The following supporting information can be downloaded at: https://www.mdpi.com/article/10.3390/catal12121587/s1, Figure S1. (a) SEM image of FeGOTi (b), oxygen-mapping using SEM-EDX (c), iron-mapping using SEM-EDX (d), titanium-mapping using SEM-EDX, Figure S2. XPS of synthesized TiO2 catalyst, Figure S3. XPS of synthesized GOTi catalyst, Figure S4. XPS of synthesized FeGOTi catalyst, Figure S5. Raman spectra of the catalyst, Figure S6. TGA and MS detected for the catalysts: TiO2 (A), GOTi (B) and FeGOTi (C), Figure S7. Comparison of normalized absorbance spectra (A; a.u.) of TiO_2 (black dots), GOTi (red dots) and FeGOTi (blue dots) and normalized irradiance collected for the light source employed in this work (E; a.u.), Figure S8. Variation of TOC/TOC_0 with time from ozone-free primidone photolysis and photocatalysis experiments, Figure S9. Schematic diagram of band positions relative to NHE, Figure S10. Checking apparent first order kinetics for PhOz (a) and PhCatOz (b) with TiO_2, (c) with GOTi and (d) with

FeGoTi, Figure S11. Comparison of XRD pattern of fresh and used FeGOTi (A), TGA and MS detected for the recovered catalyst (B), Figure S12. Fe2p XPS of fresh (a) and four times reused (b) FeGOTi catalyst [12,37,79].

Author Contributions: Conceptualization, F.J.B. and M.C.; methodology, M.C. and V.M.; validation, M.C. and V.M.; investigation, M.C.; writing—original draft preparation, J.R., M.C. and F.J.B.; writing—review and editing, J.R., M.C., V.M. and F.J.B.; project administration, F.J.B.; funding acquisition, F.J.B. and V.M. All authors have read and agreed to the published version of the manuscript.

Funding: This research was funded by the Agencia Estatal de Investigación of Spain (PID2019-104429RBI00/MCIN/AEI/10.13039/501100011033) and Junta de Extremadura and European Funds for Regional Development (FEDER) (IB20042 and TA18037).

Conflicts of Interest: The authors declare no conflict of interest.

References

1. Issaka, E.; AMU-Darko, J.N.O.; Yakubu, S.; Fapohunda, F.O.; Ali, N.; Bilal, M. Advanced catalytic ozonation for degradation of pharmaceutical pollutants—A review. *Chemosphere* **2022**, *289*, 133208. [CrossRef] [PubMed]
2. Mecha, A.C.; Chollom, M.N. Photocatalytic ozonation of wastewater: A review. *Environ. Chem. Lett.* **2020**, *18*, 1491–1507. [CrossRef]
3. Beltrán, F.J.; Rey, A.; Gimeno, O. The role of catalytic ozonation processes on the elimination of dbps and their precursors in drinking water treatment. *Catalysts* **2021**, *11*, 521. [CrossRef]
4. Gomes, J.; Costa, R.; Quinta-Ferreira, R.M.; Martins, R.C. Application of ozonation for pharmaceuticals and personal care products removal from water. *Sci. Total Environ.* **2017**, *586*, 265–283. [CrossRef] [PubMed]
5. Gunten, U. Von Ozonation of drinking water: Part I. Oxidation kinetics and product formation. *Water Res.* **2003**, *37*, 1443–1467. [CrossRef]
6. Glaze, W.H.; Kang, J.W. Advanced Oxidation Processes: Test of a Kinetic Model for the Oxidation of Organic Compounds with Ozone and Hydrogen Peroxide in a Semibatch Reactor. *Ind. Eng. Chem. Res.* **1989**, *28*, 1580–1587. [CrossRef]
7. Glaze, W.H.; Kang, J.-W.; Chapin, D.H. The Chemistry of Water Treatment Processes Involving Ozone, Hydrogen Peroxide and Ultraviolet Radiation. *Ozone Sci. Eng.* **1987**, *9*, 335–352. [CrossRef]
8. Nawrocki, J.; Kasprzyk-Hordern, B. The efficiency and mechanisms of catalytic ozonation. *Appl. Catal. B Environ.* **2010**, *99*, 27–42. [CrossRef]
9. Orge, C.A.; Sousa, J.P.S.; Gonçalves, F.; Freire, C.; Órfão, J.J.M.; Pereira, M.F.R. Development of novel mesoporous carbon materials for the catalytic ozonation of organic pollutants. *Catal. Lett.* **2009**, *132*, 1–9. [CrossRef]
10. Chen, H.; Chen, Z.; Yang, H.; Wen, L.; Yi, Z.; Zhou, Z.; Dai, B.; Zhang, J.; Wu, X.; Wu, P. Multi-mode surface plasmon resonance absorber based on dart-type single-layer graphene. *RSC Adv.* **2022**, *12*, 7821–7829. [CrossRef]
11. Shangguan, Q.; Chen, Z.; Yang, H.; Cheng, S.; Yang, W.; Yi, Z.; Wu, X.; Wang, S.; Yi, Y.; Wu, P. Design of Ultra-Narrow Band Graphene Refractive Index Sensor. *Sensors* **2022**, *22*, 6483. [CrossRef] [PubMed]
12. Li, X.; Yu, J.; Wageh, S.; Al-Ghamdi, A.A.; Xie, J. Graphene in Photocatalysis: A Review. *Small* **2016**, *12*, 6640–6696. [CrossRef] [PubMed]
13. Beltrán, F.J.; Álvarez, P.M.; Gimeno, O. Graphene-based catalysts for ozone processes to decontaminate water. *Molecules* **2019**, *24*, 3438. [CrossRef] [PubMed]
14. Rodríguez, E.M.; Rey, A.; Mena, E.; Beltrán, F.J. Application of solar photocatalytic ozonation in water treatment using supported TiO2. *Appl. Catal. B Environ.* **2019**, *254*, 237–245. [CrossRef]
15. Rosal, R.; Gonzalo, M.S.; Rodríguez, A.; García-Calvo, E. Catalytic ozonation of fenofibric acid over alumina-supported manganese oxide. *J. Hazard. Mater.* **2010**, *183*, 271–278. [CrossRef]
16. Chávez, A.M.; Solís, R.R.; Beltrán, F.J. Magnetic graphene TiO2-based photocatalyst for the removal of pollutants of emerging concern in water by simulated sunlight aided photocatalytic ozonation. *Appl. Catal. B Environ.* **2020**, *262*, 118275. [CrossRef]
17. Chang, Y.N.; Ou, X.M.; Zeng, G.M.; Gong, J.L.; Deng, C.H.; Jiang, Y.; Liang, J.; Yuan, G.Q.; Liu, H.Y.; He, X. Synthesis of magnetic graphene oxide-TiO 2 and their antibacterial properties under solar irradiation. *Appl. Surf. Sci.* **2015**, *343*, 1–10. [CrossRef]
18. Nada, A.A.; Tantawy, H.R.; Elsayed, M.A.; Bechelany, M.; Elmowafy, M.E. Elaboration of nano titania magnetic reduced graphene oxide for degradation of tartrazine dye in aqueous solution. *Solid State Sci.* **2018**, *78*, 116–125. [CrossRef]
19. Guskos, N.; Zolnierkiewicz, G.; Guskos, A.; Aidinis, K.; Wanag, A.; Kusiak-Nejman, E.; Narkiewicz, U.; Morawski, A.W. Magnetic moment centers in titanium dioxide photocatalysts loaded on reduced graphene oxide flakes. *Rev. Adv. Mater. Sci.* **2021**, *60*, 57–63. [CrossRef]
20. Pedrosa, M.; Pastrana-Martínez, L.M.; Pereira, M.F.; Faria, J.L.; Figueiredo, J.L.; Silva, A.M.T. N/S-doped graphene derivatives and TiO2 for catalytic ozonation and photocatalysis of water pollutants. *Chem. Eng. J.* **2018**, *348*, 888–897. [CrossRef]
21. Pastrana-Martínez, L.M.; Morales-Torres, S.; Likodimos, V.; Figueiredo, J.L.; Faria, J.L.; Falaras, P.; Silva, A.M.T. Advanced nanostructured photocatalysts based on reduced graphene oxide-TiO2 composites for degradation of diphenhydramine pharmaceutical and methyl orange dye. *Appl. Catal. B Environ.* **2012**, *123–124*, 241–256. [CrossRef]

22. Hass, U.; Duennbier, U.; Massmann, G. Occurrence and distribution of psychoactive compounds and their metabolites in the urban water cycle of Berlin (Germany). *Water Res.* **2012**, *46*, 6013–6022. [CrossRef]

23. Real, F.J.; Javier Benitez, F.; Acero, J.L.; Sagasti, J.J.P.; Casas, F. Kinetics of the chemical oxidation of the pharmaceuticals primidone, ketoprofen, and diatrizoate in ultrapure and natural waters. *Ind. Eng. Chem. Res.* **2009**, *48*, 3380–3388. [CrossRef]

24. Chen, P.; Liu, F.; Ding, H.; Chen, S.; Chen, L.; Li, Y.-J.; Au, C.-T.; Yin, S.-F. Porous double-shell CdS@C3N4 octahedron derived by in situ supramolecular self-assembly for enhanced photocatalytic activity. *Appl. Catal. B Environ.* **2019**, *252*, 33–40. [CrossRef]

25. Cao, W.; Lin, L.; Qi, H.; He, Q.; Wu, Z.; Wang, A.; Luo, W.; Zhang, T. In-situ synthesis of single-atom Ir by utilizing metal-organic frameworks: An acid-resistant catalyst for hydrogenation of levulinic acid to γ-valerolactone. *J. Catal.* **2019**, *373*, 161–172. [CrossRef]

26. Gupta, V.; Sharma, N.; Singh, U.; Arif, M.; Singh, A. Higher oxidation level in graphene oxide. *Optik* **2017**, *143*, 115–124. [CrossRef]

27. Quiñones, D.H.; Rey, A.; Álvarez, P.M.; Beltrán, F.J.; Plucinski, P.K. Enhanced activity and reusability of TiO2 loaded magnetic activated carbon for solar photocatalytic ozonation. *Appl. Catal. B Environ.* **2014**, *144*, 96–106. [CrossRef]

28. Nuñez, J.M.; Hettler, S.; Lima Jr, E.; Goya, G.F.; Arenal, R.; Zysler, R.D.; Aguirre, M.H.; Winkler, E.L. Onion-like Fe_3O_4/MgO/$CoFe_2O_4$ magnetic nanoparticles: New ways to control magnetic coupling between soft and hard magnetic phases. *J. Mater. Chem. C* **2022**, *10*, 15339–15352. [CrossRef]

29. Majidnia, Z.; Idris, A. Combination of maghemite and titanium oxide nanoparticles in polyvinyl alcohol-alginate encapsulated beads for cadmium ions removal. *Korean J. Chem. Eng.* **2015**, *32*, 1094–1100. [CrossRef]

30. Shokrollahi, H. A review of the magnetic properties, synthesis methods and applications of maghemite. *J. Magn. Magn. Mater.* **2017**, *426*, 74–81. [CrossRef]

31. Mei, X.; Meng, X.; Wu, F. Hydrothermal method for the production of reduced graphene oxide. *Phys. E Low-Dimens. Syst. Nanostruct.* **2015**, *68*, 81–86. [CrossRef]

32. Dai, X.; Lu, G.; Hu, Y.; Xie, X.; Wang, X.; Sun, J. Reversible redox behavior of Fe_2O_3/TiO_2 composites in the gaseous photodegradation process. *Ceram. Int.* **2019**, *45*, 13187–13192. [CrossRef]

33. Hitam, C.N.C.; Jalil, A.A. A review on exploration of Fe_2O_3 photocatalyst towards degradation of dyes and organic contaminants. *J. Environ. Manag.* **2020**, *258*, 110050. [CrossRef] [PubMed]

34. Figueredo, M.A.; Rodríguez, E.M.; Checa, M.; Beltran, F.J. Ozone-Based Advanced Oxidation Processes for Primidone Removal in Water using Simulated Solar Radiation and TiO_2 or WO_3 as Photocatalyst. *Molecules* **2019**, *24*, 1728. [CrossRef]

35. Checa, M.; Figueredo, M.; Aguinaco, A.; Beltrán, F.J. Graphene oxide/titania photocatalytic ozonation of primidone in a visible LED photoreactor. *J. Hazard. Mater.* **2019**, *369*, 70–78. [CrossRef]

36. Abbas, N.; Shao, G.N.; Haider, M.S.; Imran, S.M.; Park, S.S.; Kim, H.T. Sol-gel synthesis of TiO2-Fe2O3 systems: Effects of Fe2O3 content and their photocatalytic properties. *J. Ind. Eng. Chem.* **2016**, *39*, 112–120. [CrossRef]

37. Kurien, U.; Hu, Z.; Lee, H.; Dastoor, A.P.; Ariya, P.A. Radiation enhanced uptake of Hg0(g) on iron (oxyhydr)oxide nanoparticles. *RSC Adv.* **2017**, *7*, 45010–45021. [CrossRef]

38. Mishra, M.; Chun, D.M. α-Fe2O3 as a photocatalytic material: A review. *Appl. Catal. A Gen.* **2015**, *498*, 126–141. [CrossRef]

39. Kormann, C.; Bahnemann, D.W.; Hoffmann, M.R. Environmental photochemistry: Is iron oxide (hematite) an active photocatalyst? A comparative study: α-Fe_2O_3, ZnO, TiO_2. *J. Photochem. Photobiol. A Chem.* **1989**, *48*, 161–169. [CrossRef]

40. Palanisamy, B.; Babu, C.M.; Sundaravel, B.; Anandan, S.; Murugesan, V. Sol-gel synthesis of mesoporous mixed Fe_2O_3/TiO_2 photocatalyst: Application for degradation of 4-chlorophenol. *J. Hazard. Mater.* **2013**, *252–253*, 233–242. [CrossRef]

41. Beltrán, F.J.; Rivas, J.; Álvarez, P.; Montero-de-Espinosa, R. Kinetics of heterogeneous catalytic ozone decomposition in water on an activated carbon. *Ozone Sci. Eng.* **2002**, *24*, 227–237. [CrossRef]

42. Ding, Y.; Zhang, X.; Chen, L.; Wang, X.; Zhang, N.; Liu, Y.; Fang, Y. Oxygen vacancies enabled enhancement of catalytic property of Al reduced anatase TiO_2 in the decomposition of high concentration ozone. *J. Solid State Chem.* **2017**, *250*, 121–127. [CrossRef]

43. Yan, L.; Bing, J.; Wu, H. The behavior of ozone on different iron oxides surface sites in water. *Sci. Rep.* **2019**, *9*, 14752. [CrossRef] [PubMed]

44. Pinheiro Da Silva, M.F.; Soeira, L.S.; Daghastanli, K.R.P.; Martins, T.S.; Cuccovia, I.M.; Freire, R.S.; Isolani, P.C. CeO2-catalyzed ozonation of phenol: The role of cerium citrate as precursor of CeO_2. *J. Therm. Anal. Calorim.* **2010**, *102*, 907–913. [CrossRef]

45. Bing, J.; Hu, C.; Zhang, L. Enhanced mineralization of pharmaceuticals by surface oxidation over mesoporous γ-Ti-Al2O3 suspension with ozone. *Appl. Catal. B Environ.* **2017**, *202*, 118–126. [CrossRef]

46. Kongsuebchart, W.; Praserthdam, P.; Panpranot, J.; Sirisuk, A.; Supphasrirongjaroen, P.; Satayaprasert, C. Effect of crystallite size on the surface defect of nano-TiO_2 prepared via solvothermal synthesis. *J. Cryst. Growth* **2006**, *297*, 234–238. [CrossRef]

47. Liu, B.; Zhang, B.; Ji, J.; Li, K.; Cao, J.; Feng, Q.; Huang, H. Effective regulation of surface bridging hydroxyls on TiO_2 for superior photocatalytic activity via ozone treatment. *Appl. Catal. B Environ.* **2022**, *304*, 120952. [CrossRef]

48. Solís, R.R.; Rivas, F.J.; Gimeno, O.; Pérez-Bote, J.L. Photocatalytic ozonation of clopyralid, picloram and triclopyr. Kinetics, toxicity and influence of operational parameters. *J. Chem. Technol. Biotechnol.* **2016**, *91*, 51–58. [CrossRef]

49. Li, J.; Guan, W.; Yan, X.; Wu, Z.; Shi, W. Photocatalytic Ozonation of 2,4-Dichlorophenoxyacetic Acid using LaFeO3 Photocatalyst Under Visible Light Irradiation. *Catal. Letters* **2018**, *148*, 23–29. [CrossRef]

50. Kumar, P.; Joshi, C.; Barras, A.; Sieber, B.; Addad, A.; Boussekey, L.; Szunerits, S.; Boukherroub, R.; Jain, S.L. Core–shell structured reduced graphene oxide wrapped magnetically separable rGO@CuZnO@Fe_3O_4 microspheres as superior photocatalyst for CO_2 reduction under visible light. *Appl. Catal. B Environ.* **2017**, *205*, 654–665. [CrossRef]

51. Jo, W.K.; Kumar, S.; Isaacs, M.A.; Lee, A.F.; Karthikeyan, S. Cobalt promoted TiO_2/GO for the photocatalytic degradation of oxytetracycline and Congo Red. *Appl. Catal. B Environ.* **2017**, *201*, 159–168. [CrossRef]
52. Che, H.; Che, G.; Jiang, E.; Liu, C.; Dong, H.; Li, C. A novel Z-Scheme CdS/Bi_3O_4Cl heterostructure for photocatalytic degradation of antibiotics: Mineralization activity, degradation pathways and mechanism insight. *J. Taiwan Inst. Chem. Eng.* **2018**, *91*, 224–234. [CrossRef]
53. Kaus, N.H.M.; Rithwan, A.F.; Adnan, R.; Ibrahim, M.L.; Thongmee, S.; Yusoff, S.F.M. Effective strategies, mechanisms, and photocatalytic efficiency of semiconductor nanomaterials incorporating rgo for environmental contaminant degradation. *Catalysts* **2021**, *11*, 302. [CrossRef]
54. Ahn, Y.; Oh, H.; Yoon, Y.; Park, W.K.; Yang, W.S.; Kang, J.W. Effect of graphene oxidation degree on the catalytic activity of graphene for ozone catalysis. *J. Environ. Chem. Eng.* **2017**, *5*, 3882–3894. [CrossRef]
55. Yang, F.; Zhao, M.; Wang, Z.; Ji, H.; Zheng, B.; Xiao, D.; Wu, L.; Guo, Y. The role of ozone in the ozonation process of graphene oxide: Oxidation or decomposition? *RSC Adv.* **2014**, *4*, 58325–58328. [CrossRef]
56. Decoret, C.; Royer, J.; Legube, B.; Dore, M. Experimental and theoretical studies of the mechanism of the initial attack of ozone on some aromatics in aqueous medium. *Environ. Technol. Lett.* **1984**, *5*, 207–218. [CrossRef]
57. Staehelin, J.; Hoigne, J. Decomposition of ozone in water: Rate of initiation by hydroxide ions and hydrogen peroxide. *Environ. Sci. Technol.* **1982**, *16*, 676–681. [CrossRef]
58. Beltrán, F.J.; Checa, M.; Rivas, J.; García-Araya, J.F. Modeling the Mineralization Kinetics of Visible Led Graphene Oxide/Titania Photocatalytic Ozonation of an Urban Wastewater Containing Pharmaceutical Compounds. *Catalysts* **2020**, *10*, 1256. [CrossRef]
59. Kiwi, J.; Lopez, A.; Nadtochenko, V. Mechanism and kinetics of the OH-radical intervention during Fenton oxidation in the presence of a significant amount of radical scavenger (Cl-). *Environ. Sci. Technol.* **2000**, *34*, 2162–2168. [CrossRef]
60. Munoz, M.; de Pedro, Z.M.; Casas, J.A.; Rodriguez, J.J. Preparation of magnetite-based catalysts and their application in heterogeneous Fenton oxidation—A review. *Appl. Catal. B Environ.* **2015**, *176–177*, 249–265. [CrossRef]
61. Matafonova, G.; Batoev, V. Recent advances in application of UV light-emitting diodes for degrading organic pollutants in water through advanced oxidation processes: A review. *Water Res.* **2018**, *132*, 177–189. [CrossRef]
62. Rodríguez, E.M.; Márquez, G.; Tena, M.; Álvarez, P.M.; Beltrán, F.J. Determination of main species involved in the first steps of TiO_2 photocatalytic degradation of organics with the use of scavengers: The case of ofloxacin. *Appl. Catal. B Environ.* **2014**, *178*, 44–53. [CrossRef]
63. Wu, D.; Li, X.; Tang, Y.; Lu, P.; Chen, W.; Xu, X.; Li, L. Mechanism insight of PFOA degradation by ZnO assisted-photocatalytic ozonation: Efficiency and intermediates. *Chemosphere* **2017**, *180*, 247–252. [CrossRef] [PubMed]
64. Buffle, M.O.; Schumacher, J.; Meylan, S.; Jekel, M.; Von Gunten, U. Ozonation and advanced oxidation of wastewater: Effect of O3 dose, pH, DOM and HO.-scavengers on ozone decomposition and HO. generation. *Ozone Sci. Eng.* **2006**, *28*, 247–259. [CrossRef]
65. Xu, P.; Janex, M.L.; Savoye, P.; Cockx, A.; Lazarova, V. Wastewater disinfection by ozone: Main parameters for process design. *Water Res.* **2002**, *36*, 1043–1055. [CrossRef]
66. Elovitz, M.S.; Von Gunten, U. Hydroxyl radical/ozone ratios during ozonation processes. I. The R(ct) concept. *Ozone Sci. Eng.* **1999**, *21*, 239–260. [CrossRef]
67. Beltrán, F.J.; Aguinaco, A.; García-Araya, J.F. Kinetic modelling of TOC removal in the photocatalytic ozonation of diclofenac aqueous solutions. *Appl. Catal. B Environ.* **2010**, *100*, 289–298. [CrossRef]
68. Justo, A.; González, O.; Aceña, J.; Pérez, S.; Barceló, D.; Sans, C.; Esplugas, S. Pharmaceuticals and organic pollution mitigation in reclamation osmosis brines by UV/H_2O_2 and ozone. *J. Hazard. Mater.* **2013**, *263*, 268–274. [CrossRef]
69. Kwon, M.; Kye, H.; Jung, Y.; Yoon, Y.; Kang, J.W. Performance characterization and kinetic modeling of ozonation using a new method: ROH,O_3 concept. *Water Res.* **2017**, *122*, 172–182. [CrossRef] [PubMed]
70. Hermes, N.; Jewell, K.S.; Schulz, M.; Müller, J.; Hübner, U.; Wick, A.; Drewes, J.E.; Ternes, T.A. Elucidation of removal processes in sequential biofiltration (SBF) and soil aquifer treatment (SAT) by analysis of a broad range of trace organic chemicals (TOrCs) and their transformation products (TPs). *Water Res.* **2019**, *163*, 114857. [CrossRef] [PubMed]
71. Mao, Y.; Ding, X.; Li, M.; Wang, L.; Wang, Y. Degradation efficiency and mechanism of primidone by UV/chorine process. *Chinese J. Environ. Eng.* **2021**, *15*, 3524–3535. [CrossRef]
72. Solís, R.R.; Rivas, F.J.; Chávez, A.M.; Dionysiou, D.D. Peroxymonosulfate/solar radiation process for the removal of aqueous microcontaminants. Kinetic modeling, influence of variables and matrix constituents. *J. Hazard. Mater.* **2020**, *400*, 123118. [CrossRef] [PubMed]
73. Rosario-Ortiz, F.L.; Wert, E.C.; Snyder, S.A. Evaluation of UV/H_2O_2 treatment for the oxidation of pharmaceuticals in wastewater. *Water Res.* **2010**, *44*, 1440–1448. [CrossRef] [PubMed]
74. Figueredo, M.; Rodríguez, E.M.; Rivas, J.; Beltrán, F.J. Kinetic model basis of ozone/light-based advanced oxidation processes: A pseudoempirical approach. *Environ. Sci. Water Res. Technol.* **2020**, *6*, 1176–1185. [CrossRef]
75. Beltrán, F.J.; Checa, M. Comparison of graphene oxide titania catalysts for their use in photocatalytic ozonation of water contaminants: Application to oxalic acid removal. *Chem. Eng. J.* **2020**, *385*, 123922. [CrossRef]
76. Lin, Y.; Geng, Z.; Cai, H.; Ma, L.; Chen, J.; Zeng, J.; Pan, N.; Wang, X. Ternary graphene-TiO_2-Fe_3O_4 nanocomposite as a recollectable photocatalyst with enhanced durability. *Eur. J. Inorg. Chem.* **2012**, *2012*, 4439–4444. [CrossRef]
77. Bader, H.; Hoigné, J. Determination of ozone in water by the indigo method. *Water Res.* **1981**, *15*, 449–456. [CrossRef]

78. Masschelein, W.; Denis, M.; Ledent, R. Spectrophotometric determination of residual hydrogen peroxide. *Water Sew. Work.* **1977**, *124*, 69–72.
79. Wu, W.; Jiang, C.; Roy, V.A.L. Recent Progress in Magnetic Iron Oxide-Semiconductor Composite Nanomaterials as Promising Photocatalysts. *Nanoscale* **2015**, *7*, 38–58. [CrossRef]

International Journal of
*Environmental Research
and Public Health*

Article

Degradation of Agro-Industrial Wastewater Model Compound by UV-A-Fenton Process: Batch vs. Continuous Mode

Nuno Jorge [1,2], Ana R. Teixeira [2], José R. Fernandes [3], Ivo Oliveira [4], Marco S. Lucas [2] and José A. Peres [2,*]

[1] Escuela Internacional de Doctorado (EIDO), Campus da Auga, Campus Universitário de Ourense, Universidade de Vigo, As Lagoas, 32004 Ourense, Spain
[2] Centro de Química de Vila Real (CQVR), Departamento de Química, Universidade de Trás-os-Montes e Alto Douro (UTAD), Quinta de Prados, 5000-801 Vila Real, Portugal
[3] Centro de Química de Vila Real (CQVR), Departamento de Física, Universidade de Trás-os-Montes e Alto Douro (UTAD), Quinta de Prados, 5000-801 Vila Real, Portugal
[4] Centre for the Research and Technology of Agro-Environmental and Biological Sciences (CITAB), University of Trás-os-Montes and Alto Douro (UTAD), 5000-801 Vila Real, Portugal
* Correspondence: jperes@utad.pt

Abstract: The degradation of a model agro-industrial wastewater phenolic compound (caffeic acid, CA) by a UV-A-Fenton system was investigated in this work. Experiments were carried out in order to compare batch and continuous mode. Initially, batch experiments showed that UV-A-Fenton at pH 3.0 (pH of CA solution) achieved a higher generation of $HO^\bullet$, leading to high CA degradation (>99.5%). The influence of different operational conditions, such as H_2O_2 and Fe^{2+} concentrations, were evaluated. The results fit a pseudo first-order (PFO) kinetic model, and a high kinetic rate of CA removal was observed, with a $[CA] = 5.5 \times 10^{-4}$ mol/L, $[H_2O_2] = 2.2 \times 10^{-3}$ mol/L and $[Fe^{2+}] = 1.1 \times 10^{-4}$ mol/L ($k_{CA} = 0.694$ min^{-1}), with an electric energy per order (E_{EO}) of 7.23 kWh m^{-3} order^{-1}. Under the same operational conditions, experiments in continuous mode were performed under different flow rates. The results showed that CA achieved a steady state with higher space-times ($\theta = 0.04$) in comparison to dissolved organic carbon (DOC) removal ($\theta = 0$–0.020). The results showed that by increasing the flow rate (F) from 1 to 4 mL min^{-1}, the CA and DOC removal rate increased significantly ($k_{CA} = 0.468$ min^{-1}; $k_{DOC} = 0.00896$ min^{-1}). It is concluded that continuous modes are advantageous systems that can be adapted to wastewater treatment plants for the treatment of real agro-industrial wastewaters.

Keywords: caffeic acid; electric energy per order; environmental impact; photo-Fenton; UV-A LEDs; winery wastewater

Citation: Jorge, N.; Teixeira, A.R.; Fernandes, J.R.; Oliveira, I.; Lucas, M.S.; Peres, J.A. Degradation of Agro-Industrial Wastewater Model Compound by UV-A-Fenton Process: Batch vs. Continuous Mode. *Int. J. Environ. Res. Public Health* **2023**, *20*, 1276. https://doi.org/10.3390/ijerph20021276

Academic Editor: Paul B. Tchounwou

Received: 21 December 2022
Revised: 5 January 2023
Accepted: 6 January 2023
Published: 10 January 2023

1. Introduction

The rapid expansion of the agro-industry in both developed and developing countries is a major contributor of environmental pollution worldwide [1,2]. Agro-industrial wastewater characteristics are much more diverse than domestic wastewater, which is usually qualitatively and quantitatively similar in its composition. This industry produces large quantities of highly polluted wastewater containing toxic substances and organic and inorganic compounds such as: heavy metals, pesticides, phenols, and derivatives [3,4]. In the case of industries such as wine or olive oil production, large volumes of wastewaters are produced from different activities, namely tank washing, transfer, bottling, and filtration [5–7]. These wastewaters are characterized by an elevated content of suspended solids, low pH (3–5), and high organic load composed of phenolic compounds, sugars, organic acids, and esters [8,9]. Among these compounds, polyphenols are considered to be hazardous compounds because they are not mineralized by conventional biologic treatments [10,11], among which polyphenolic compounds are found to be the most abundant, including gallic, syringic, protocatechuic, vanillic, and caffeic acids [12–15]. Caffeic acid

(3,4-dihydroxycinnamic acid) is considered to be one of the most refractory phenolic compounds to biologic degradation, because it exhibits high toxicity and antibacterial activity and represents a risk to human health, reaching 100 mg L^{-1}, with a half-life ranging from 21.2 to 26.7 min in natural conditions [12,16,17].

Due to the limitations of conventional wastewater treatment technologies in removing recalcitrant pollutants such as caffeic acid, a more effective treatment is required to achieve a complete mineralization of these organic compounds, such as advanced oxidation processes (AOPs). In the AOPs, there is the generation of hydroxyl radicals ($HO^\bullet$), which react with organic compounds, oxidizing them to simpler intermediates and possibly to CO_2 and H_2O [18,19]. The $HO^\bullet$ radicals are advantageous because they (1) do not generate additional waste, (2) are not toxic and have a very short lifetime, (3) are not corrosive to pieces of equipment, (4) are usually produced by assemblies that are simple to manipulate, and (5) have a high oxidizing potential (E° = 2.80 V) regarding sulfate radical anion (E° = 2.60 V), ozone (E° = 2.08 V), hydrogen peroxide (E° = 1.76), and chlorine (E° = 1.36) [20,21]. Among the AOPs, the photo-Fenton process appears as a suitable treatment process for $HO^\bullet$ radical production [22]. It employs Fe^{2+} and H_2O_2, which are readily available, easy to handle, and environmentally benign [18] and "near-UV to visible region" of light, up to a wavelength of 600 nm [23] to improve the $HO^\bullet$ radical production and to rapidly reduce the Fe^{3+} back to Fe^{2+} [24,25]. Traditional mercury-based UV-C radiation lamps can become very unstable, due to their overheating, decreasing their efficiency and lifetime [26,27]. The UV-A LED lights are a good alternative to UV-C mercury lamps because they have longer lifetimes, lower energy consumption, higher efficiency, do not overheat, and are less harmful to the environment [28,29].

Although batch reactors have been shown to be efficient in contaminant degradation, they are expensive when upscaling the treatment of wastewater; however, to counteract this tendency, most industrial treatment facilities operate in continuous mode [30]. Therefore, the aim of this work was to evaluate the degradation of caffeic acid in batch and continuous mode, in a UV-A LED reactor with a wavelength of 365 nm. It is also intended to evaluate how the different variables (pH, concentration of H_2O_2, and Fe^{2+}) affect the kinetic rate of CA degradation and energy consumption. The novelty of this work lies in the application of a continuous system coupled with a UV-A reactor to degrade the CA phenolic, which has never been performed before.

2. Materials and Methods

2.1. Reagents

Caffeic acid (3,4-Dihydroxycinnamic acid) was acquired from Sigma-Aldrich, St. Louis, MO, and used as received without further purification. The molecular structure of caffeic acid in non-hydrolyzed form is illustrated in Table 1. Iron (II) sulfate heptahydrate ($FeSO_4 \bullet 7H_2O$) was acquired from Panreac, Barcelona, Spain, and hydrogen peroxide (H_2O_2 30% *w/w*) and titanium (IV) oxysulfate solution 1.9–2.1%, for determination of hydrogen peroxide, were acquired from Sigma-Aldrich, St. Louis, MO, USA. Sodium sulfite anhydrous (Na_2SO_3) was acquired from Merk, Darmstadt, Germany. Trifluoroacetic acid (HPLC grade, $\geq$99.0%) was acquired from Riedel-de Haën, Seelze, Germany, and acetonitrile (HPLC grade, $\geq$99.9%) was acquired from Chem-Lab, Zedelgem, Belgium. For pH adjustment, sodium hydroxide (NaOH) from Labkem, Barcelona, Spain, it was used, along with sulfuric acid (H_2SO_4, 95%) from Scharlau, Barcelona, Spain. Deionized water was used to prepare the respective solutions.

Table 1. Molecular formula, molecular structure, maximum absorption wavelength and molecular weight of caffeic acid (CA) [31].

Name	Molecular Formula	Molecular Structure	λ_{max} (nm)	Molecular Weight (g/mol)
Caffeic acid (CA)	$(HO)_2C_6H_3CH{-}CHCO_2H$		324	180.16

2.2. Analytical Determinations

Different parameters were measured in order to determine the effect of the treatments. The dissolved organic carbon (DOC) in mg C/L and total nitrogen (TN) in mg N/L were determined by direct injection of filtered samples into a Shimadzu TOC-LCSH analyzer (Shimadzu, Kyoto, Japan), equipped with an ASI-L autosampler, provided with an NDIR detector and calibrated with standard solutions of potassium phthalate. The hydrogen peroxide concentration was determined using titanium (IV) oxysulfate (DIN 38 402H15 method) at 410 nm using a portable spectrophotometer from Hach (Loveland, CO, USA), the pH and oxidation reduction potential (ORP) were measured by a 3510 pH meter (Jenway, Cole-Parmer, UK), and the iron concentrations were analyzed by atomic absorption spectroscopy (AAS) using a Thermo Scientific™ iCE™ 3000 Series (Thermo Fisher Scientific, Waltham, MA, USA).

The caffeic acid (CA) eluting peaks were monitored at 280 nm using software Chromeleon™ 7.2.9 (Thermo Fisher Scientific, Waltham, MA, USA). The CA concentration was monitored by a UHPLC Ultimate 3000 (Thermo Fisher Scientific, Waltham, MA, USA), using a C18 reverse phase column (250 × 4.6 mm, 5 µm) with a flowrate of 1 mL min^{-1} at 25.0 °C. The volume of injection was 10.00 µL, and the eluents used were ultrapure water/trifluoroacetic acid (99.9:0.1, *v/v*) (solvent A) and acetonitrile/trifluoroacetic acid (99.9:0.1, *v/v*) (solvent B) upon the linear gradient scheme (t in min; %B): 0, 0%B; 5, 20%B; 10, 100%B; 16, 0%B, 20, 0%B.

2.3. Fenton-Based Experimental Procedure

The Fenton-based batch experiments were performed in a self-designed lab-scale reactor with 500 mL capacity and a solution depth of 1.4 cm (Figure 1). The lab reactor had a rectangular shape with a bottom and mirror walls. The UV-A LEDs system was composed of 12 Indium Gallium Nitride (LnGaN) LEDs lamps (Roithner AP2C1-365E LEDs) with a λ_{max} = 365 nm. 250 mL of CA solution with a concentration of 5.5×10^{-4} mol L^{-1} (pH = 3.90 ± 0.19, DOC = 68.7 ± 2.1 mg C L^{-1}), which was constantly agitated (350 rpm) by a L32 Basic Hotplate Magnetic Stirre 20 L (Labinco, Breda, Netherlands) at ambient temperature (298 K) for 15 min. Initially, different AOPs were tested (H_2O_2, UV-A, Fe^{2+} + UV-A, H_2O_2 + UV-A, Fenton, and UV-A-Fenton), then the pH (3.0–7.0), H_2O_2 concentration (5.5×10^{-4}–8.8×10^{-3} mol L^{-1}), and Fe^{2+} concentration (0.18×10^{-4}–11×10^{-4} mol L^{-1}) were varied.

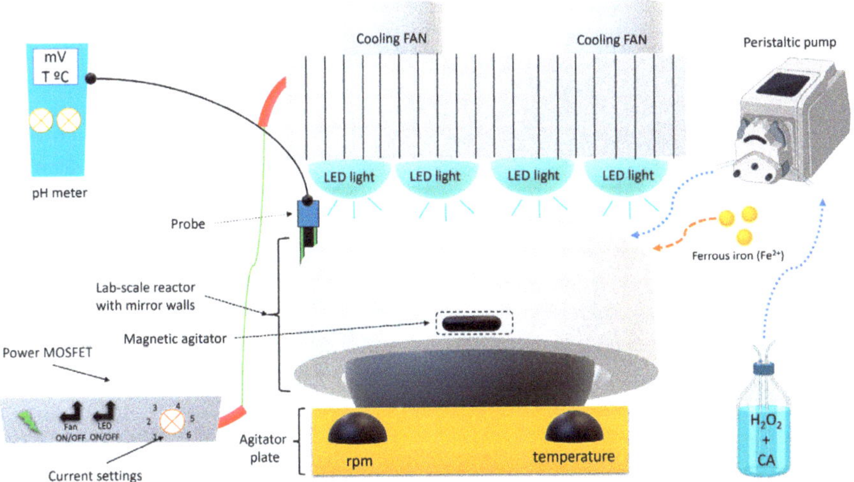

Figure 1. Schematic representation of UV-A LEDs lab-scale reactor with peristaltic pump.

For the Fenton-based continuous experiments, a reservoir filled with CA with a concentration of 5.5×10^{-4} mol L^{-1} and H_2O_2 was pumped by a peristaltic pump (Shenchen, Hvidovre, Denmark) into the UV-A reactor, in which different flow rates (1–4 mL min^{-1}) were applied to the aqueous mean, with a hydraulic retention time (HRT) of 15 min (Figure 1). The H_2O_2 was added in continuous mode to prevent radical scavenging and to enhance CA degradation.

To determine the CA and DOC removal, Equation (1) was applied [32].

$$\text{Removal } (\%) = \frac{C_0 - C_t}{C_0} \times 100 \tag{1}$$

where C_0 and C_t are the initial and final concentrations of CA and DOC.

2.4. Electrical Energy Determination

Assuming the degradation of CA as a pseudo-first order (PFO) kinetic model ($\ln[CA]_t = -kt + \ln[CA]_0$), the electrical energy per order (E_{EO}) can be determined by conversion of the units of the first order kinetic constants to min^{-1}, which results in Equation (2) [33]:

$$E_{EO} = \frac{38.4 \times 10^{-3} * P * 1000}{V * k_{obs}} \tag{2}$$

where P is the nominal power of the reactor (kW), V is the volume (m^3), and k_{obs} is the pseudo-first order kinetic observed (min^{-1}).

All the CA removal experiments were performed in triplicate, and the observed standard deviation was always less than 5% of the reported values. Differences among means were determined by analysis of variance (ANOVA) using OriginLab 2019 software (Northampton, MA, USA), and the Tukey's test was used for the comparison of means, which were considerate different when $p < 0.05$, and the data are presented as mean and standard deviation (mean $\pm$ SD).

3. Results

3.1. Chemical Degradability of Caffeic Acid

Considering the high content of impurities present in agro-industrial wastewaters, it is difficult to understand the degradation of the single compounds present in their constitution. Therefore, in this work caffeic acid (CA) was selected as a model compound because it exists in many types of agro-industrial wastewater. To evaluate the capacity of the UV-A reactor for the degradation of CA phenolic, several experiments were carried out: (1) CA+H_2O_2, (2) CA+UV-A, (3) CA+Fe^{2+}+UV-A, (4) CA+H_2O_2+UV-A, (5) CA+H_2O_2+Fe^{2+}, (6) CA+H_2O_2+Fe^{2+}+UV-A. In Figure 2a are represented the results of CA removal after each treatment. From the results, it was possible to observe that CA is resistant to oxidation with the application of H_2O_2, UV-A, Fe^{2+}+UV-A, and H_2O_2+UV-A, with 1.5, 4.7, 7.9, and 12.7%, respectively. With the application of H_2O_2, UV-A, and Fe^{2+}+UV-A, there is no generation of hydroxyl radicals (HO$^\bullet$) and CA shows to be resistant to degradation by radiation, H_2O_2, and iron. In accordance with these results, CA is harder to degrade regarding other phenolics, such as gallic acid [34]. The combination of H_2O_2 + UV-A increased the degradation of CA, due to the conversion of H_2O_2 into HO$^\bullet$ radicals (Equation (3)). The highest CA removals were observed with application of the Fenton and UV-A-Fenton processes, with 92.3 and 99.9% after 15 min of reaction. An analysis of the ORP values showed higher values with the application of Fenton and photo-Fenton, which could be linked to the production of HO$^\bullet$ radicals. To understand these results, the conversion of the H_2O_2 and the concentration of Fe^{2+} available in solution were studied (Figure 2b). With the application of H_2O_2 and H_2O_2 + UV-A, there was a low consumption of H_2O_2 (0.24 and 0.45 mM), thus a low generation of HO$^\bullet$ radicals occurred. With application of Fenton and UV-A-Fenton, the H_2O_2 consumption increased (0.80 and 1.20 mM). This increase could be due to the reaction of Fe^{2+} with H_2O_2 (Equation (4)). This difference in H_2O_2 consumption could be due to the Fe^{2+} available. Due to the reaction of Fe^{2+} with H_2O_2, the Fe^{2+} was oxidized to Fe^{3+}, precipitating and ferric hydroxide. The UV-A was able to regenerate the Fe^{3+} to Fe^{2+} (Equation (5)) [35,36], thus a higher $HO^\bullet$ radicals generation occurred.

$$H_2O_2 + UV \rightarrow 2HO^\bullet \tag{3}$$

$$Fe2^+ + H_2O_2 \rightarrow Fe3^+ + HO^\bullet + HO^- \tag{4}$$

$$Fe(HO)^{2+} + UV \rightarrow Fe2^+ + HO^\bullet \tag{5}$$

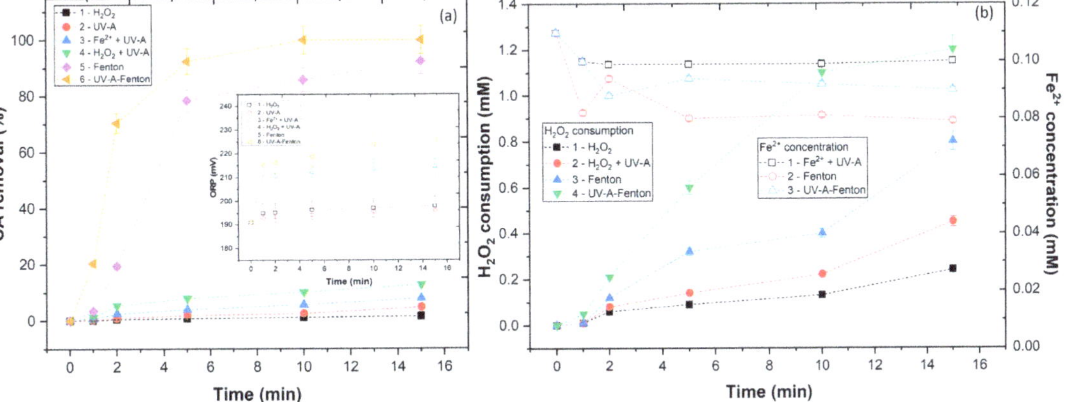

Figure 2. Assessment of different AOPs in (**a**) CA removal and ORP variation and (**b**) H_2O_2 consumption and Fe^{2+} concentration. Operational conditions: $[CA] = 5.5 \times 10^{-4}$ mol L^{-1}, $[H_2O_2] = 2.2 \times 10^{-3}$ mol L^{-1}, $[Fe^{2+}] = 1.1 \times 10^{-4}$ mol L^{-1}, pH = 3.0, agitation 150 rpm, temperature = 298 K, radiation UV-A, $I_{UV} = 32.7$ W m^{-2}, t = 15 min.

These results were in agreement with the work of Cruz et al. [37], who observed an efficient removal of Sulfamethoxazole by the combination of catalyst + H_2O_2 + UV radiation.

3.2. Effect of pH

The pH of the solution has an important effect in the degradation efficiency of CA. Previous authors observed that the downward trend of $HO^\bullet$ radical production increased with higher pH, because the Fenton reaction's optimal pH is around pH 2–4 [38]. The pH of the CA solution was varied (3.0–7.0), with results showing a CA removal of 99.9, 87.7, 43.0, and 41.3%, respectively, for pH 3.0, 4.0, 6.0, and 7.0 (Figure 3a). As the pH increased above pH > 3.0, the availability of the iron decreased (0.090, 0.079, 0.068, and 0.063 mM, respectively) (Figure 3b). At pH above 3.0, the Fe^{3+} produced by the oxidation of Fe^{2+} in the Fenton process began to precipitate in the form of amorphous $Fe(HO)_3$. This formation not only decreased the iron concentration, but also inhibited the regeneration of Fe^{2+} [39]. In accordance with Equations (6) and (7), the reaction between Fe^{3+} and H_2O_2 generates ferric-hydroperoxyl complexes that decomposes to produce $HO_2^\bullet$ radicals and Fe^{2+}, as shown in Equation (8); however, these reactions have a very slow reaction rate (2.7×10^{-3} M s^{-1}) [40]. Thus, at pH > 3.0, a lower concentration of H_2O_2 was converted to $HO^\bullet$ radicals, which decreased the efficiency of the UV-A-Fenton process.

$$Fe3^+ + H_2O_2 \rightarrow Fe(HO_2)^{2+} + H^+ \tag{6}$$

$$FeHO^{2+} + H_2O_2 \rightarrow Fe(HO)(HO_2)^+ + H^+ \tag{7}$$

$$Fe(HO)(HO_2)^+ \rightarrow Fe2^+ + HO^- + HO_2^\bullet \tag{8}$$

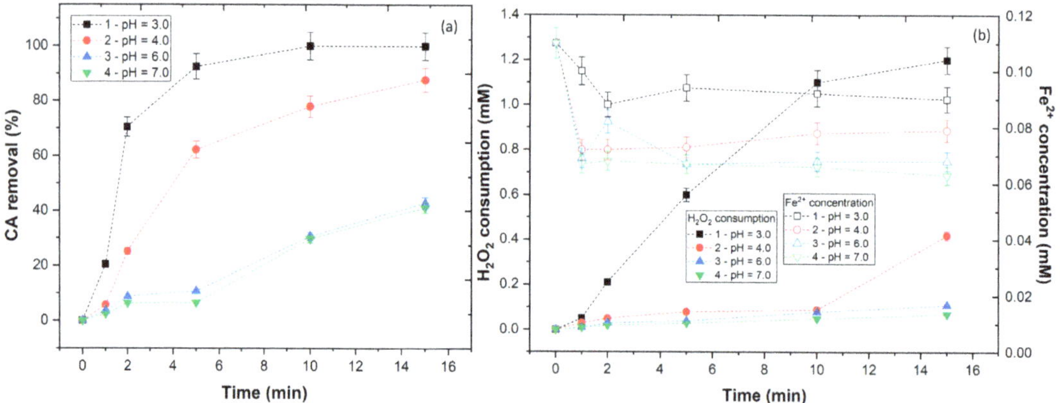

Figure 3. Effect of pH in (**a**) CA removal and (**b**) H_2O_2 consumption and Fe^{2+} concentration. Operational conditions: $[CA] = 5.5 \times 10^{-4}$ mol L^{-1}, $[H_2O_2] = 2.2 \times 10^{-3}$ mol L^{-1}, $[Fe^{2+}] = 1.1 \times 10{-4}$ mol L^{-1}, agitation 150 rpm, temperature = 298 K, radiation UV-A, $I_{UV} = 32.7$ W m^{-2}, t = 15 min.

A similar result was observed in the work of Zha et al. [41], in the degradation of landfill leachate by the photo-Fenton process. Considering that the pH of winery wastewaters is usually between 3 and 4, the application of photo-Fenton at pH 3.0 decreases the requirement of reagent consumption for pH adjustments.

3.3. Effect of H_2O_2 Concentration

In this section, the effect of the oxidant (H_2O_2) concentration in the efficiency of the UV-A-Fenton to degrade the CA phenolic was evaluated. To keep the UV-A-Fenton competitive with other processes, it is essential that this process represents a low-cost operation, thus the control of the H_2O_2 concentration is implied. In observation of Figure 4a, different H_2O_2:CA molar ratios (R) were tested, with results showing a tendency to increase the CA removal by increasing the molar rate. With application of an R = 1 and 2, a low CA removal was observed; however, when the R = 4 was applied, a near complete removal of CA was observed, reaching a plateau after 10 min of reaction. Above this ratio, no considerable CA removal was observed. To understand these results, the H_2O_2 consumption and Fe^{2+} concentration, along with the reaction, were studied (Figure 4b). With application of R = 1 and 2, the H_2O_2 consumption was low (0.10 and 0.30 mM, respectively), despite the high concentration of Fe^{2+} available (0.088 and 0.095 mM, respectively), which could mean that the H_2O_2 consumed was insufficient to produce $HO^\bullet$ radicals in a necessary amount to degrade the CA phenolic. With application of R = 4, 8, and 16, the iron concentrations remained similar (0.090, 0.087, and 0.086 mM, respectively); however, a higher H_2O_2 consumption was observed (1.2, 1.5, and 1.9 mM), thus a higher amount of $HO^\bullet$ radicals were generated, leading to the degradation of the CA phenolic. The consumption of H_2O_2 within R = 8 and 16 could be linked to the scavenging reactions between the excess of H_2O_2 with $HO^\bullet$ (Equations (9) and (10)), leading to the production of hydroperoxyl radicals ($HO_2^\bullet$) and superoxide anion radicals ($O_2^{\bullet-}$) with a lower oxidation potential ($E^\circ = 1.70$ and 0.40 V) than $HO^\bullet$ radicals ($E^\circ = 2.80$ V) [42].

$$H_2O_2 + HO^\bullet \;\rightarrow\; H_2O + HO_2^\bullet \tag{9}$$

$$H_2O_2 + HO^\bullet \;\rightarrow\; H_2O + H^+ + O_2^{\bullet-} \tag{10}$$

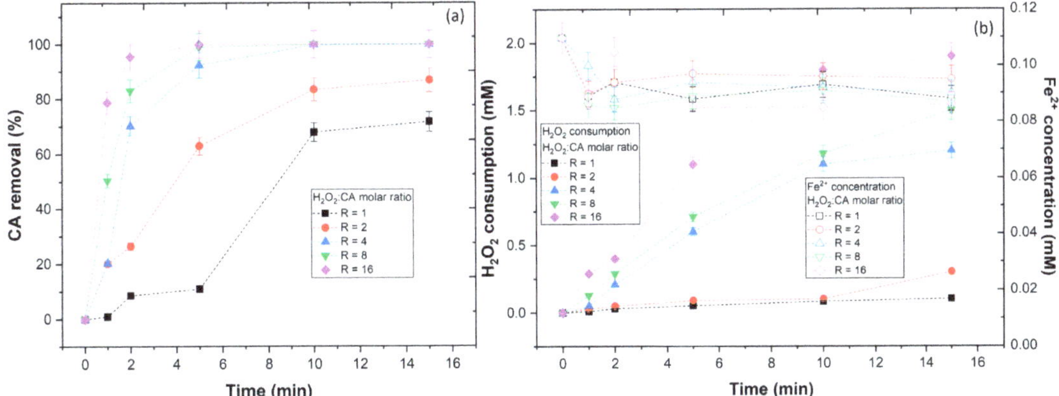

Figure 4. Evaluation of H_2O_2:CA molar ratio in (**a**) CA removal and (**b**) H_2O_2 consumption and Fe^{2+} concentration. Operational conditions: [CA] = 5.5×10^{-4} mol/L, [Fe^{2+}] = 1.1×10^{-4} mol L^{-1}, pH = 3.0, agitation 150 rpm, temperature = 298 K, radiation UV-A, I_{UV} = 32.7 W m^{-2}, t = 15 min.

To have a better understanding of the effect of the H_2O_2 in phenolic degradation, the results showed a good fitting to the pseudo-first order (PFO) kinetic model. The *k* values vs. [H_2O_2] were plotted and two different kinetic rates were separated (Figure 5a). In a first one, up to 2.2×10^{-3} mol L^{-1} H_2O_2, the PFO constant is directly proportional to the concentration of H_2O_2 applied, with a slope of 0.012 mol L^{-1} min^{-1}. In the second kinetic rate (up to 8.8×10^{-3} mol L^{-1}), the PFO constant increased linearly with the concentration of H_2O_2, with a slope of 0.489 mol L^{-1} min^{-1}. Based on these results, it can be assumed that the LED light fully penetrates the solution, independent of the concentration of H_2O_2 used, so the geometry of the UV-A reactor and the depth of the solution appeared to be adequate. These results were shown to be in agreement with the work of Li and Cheng [43], who observed a direct relation between the increase in the kinetic rate of malachite green degradation with higher H_2O_2 concentrations.

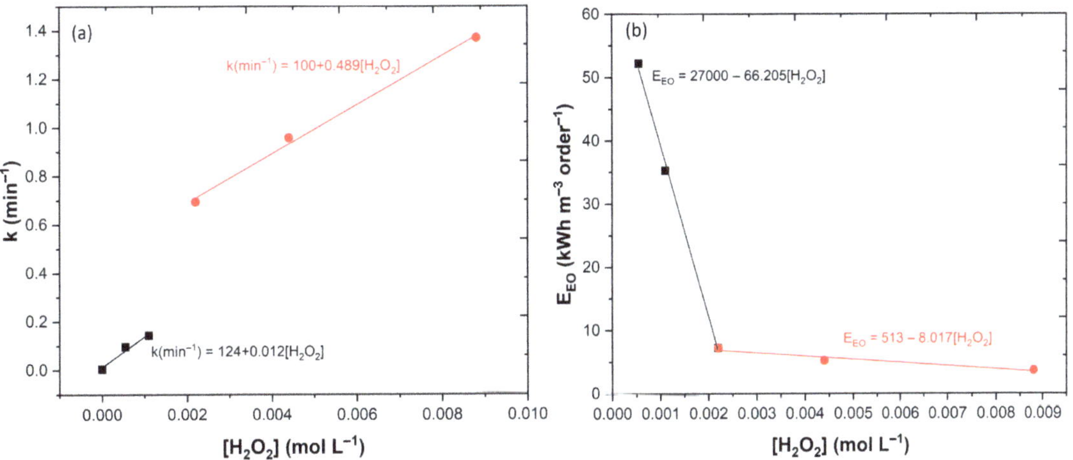

Figure 5. Outcome of several concentrations of H_2O_2 in CA degradation by the UV-A-Fenton process; (**a**) dependence on the kinetic constant rates; (**b**) electric energy per order (E_{EO}).

It is also necessary to study the how the concentration of H_2O_2 affects the energy consumption of the reactor, because the energy consumption determines part of the treatment costs. In Figure 5b, the E_{EO} values vs. the $[H_2O_2]$ were plotted, with results showing two distinct areas: (1) one for $[H_2O_2] \leq 2.2 \times 10^{-3}$ mol L^{-1} and (2) one for $[H_2O_2] \geq 2.2 \times 10^{-3}$ mol L^{-1}. For concentrations below 2.2×10^{-3} mol L^{-1}, the degradation of CA was too slow, which implied a higher energy consumption. For concentrations higher than 2.2×10^{-3} mol L^{-1}, only small gains were observed, which are not profitable, considering the high cost of H_2O_2 [44].

3.4. Effect of Fe^{2+} Concentration

As previously observed in Figure 2, the Fe^{2+} catalyst had a major effect regarding UV-A radiation in the conversion of H_2O_2 into $HO^\bullet$ radicals. Therefore, the $H_2O_2{:}Fe^{2+}$ molar ratio (R1) was varied to study the effect of the Fe^{2+} concentration in the UV-A-Fenton process (Figure 6a). The results showed that by increasing the catalyst concentration, the rate of the CA phenolic degradation was increased. Clearly, the application of an R1 from 120 to 40 was revealed to be insufficient to achieve a complete degradation of CA. Qualitatively, the results indicated that increasing the Fe^{2+} concentration in solution, for a constant H_2O_2 concentration, increased the degradation rate of CA. To understand the effect of the Fe^{2+} concentration, the H_2O_2 consumption and the concentration of iron that remained in solution were analyzed (Figure 6b). With the application of an R1 from 120 to 40, lower concentrations of H_2O_2 were consumed, regarding the values obtained within R1 from 20 to 2, thus more $HO^\bullet$ radicals were generated. Figure 6b shows that within the first 5 min a higher concentration of H_2O_2 was consumed with the application of R1 = 2, which is consistent with the higher concentration of iron present, thus more $HO^\bullet$ radicals were generated. However, from this point onwards, the H_2O_2 consumption decreased to values lower than those observed with the application of R1 = 4, which could be a strong indicator that Fe^{2+} began consuming the $HO^\bullet$ radicals. To confirm this theory, the Fe^{2+} present in the solution was determined. The results showed higher Fe^{2+} concentrations available within the application of R1 from 20 to 2 (Figure 6b). A similar behavior was observed in the work of Faggiano et al. [45], in which increasing the concentration of iron to a certain extent lead to a reduction of the H_2O_2 consumption.

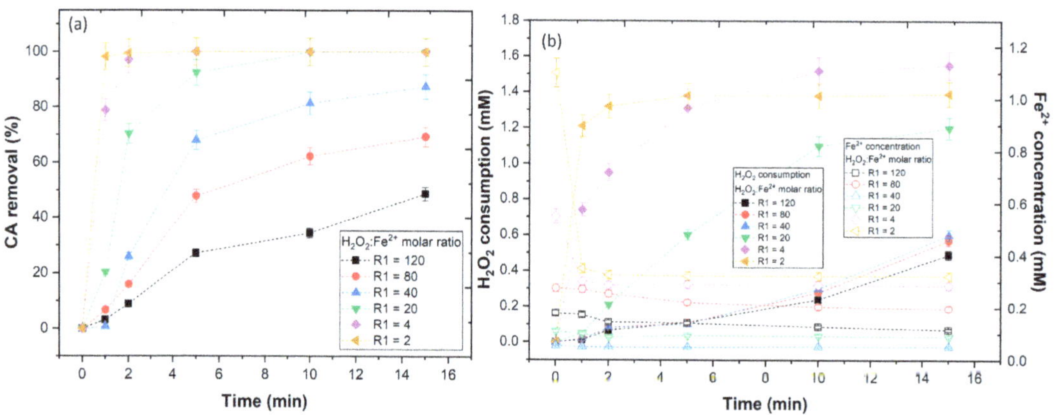

Figure 6. Assessment of the $H_2O_2{:}Fe^{2+}$ molar ratio in (a) CA degradation and (b) H_2O_2 consumption and Fe^{2+} concentration. Operational conditions: $[CA] = 5.5 \times 10^{-4}$ mol L^{-1}, $[H_2O_2] = 2.2 \times 10^{-3}$ mol L^{-1}, pH = 3.0, agitation 150 rpm, temperature = 298 K, radiation UV-A, $I_{UV} = 32.7$ W m^{-2}, t = 15 min.

The results obtained showed a good fitting to the PFO kinetic model, and by plotting the k values vs. $[Fe^{2+}]$, two different dominions are shown (Figure 7a). In a first dominion, up to 1.1×10^{-4} mol L^{-1}, the PFO kinetic constant is directly proportional to the concentration of Fe^{2+}, with a slope of 0.071 mol L^{-1} min^{-1}. A second dominion was observed from 1.1×10^{-4} to 11×10^{-4} mol L^{-1}, in which the PFO kinetic constant increased linearly with the concentration Fe^{2+}, with a slope of 0.749 mol L^{-1} min^{-1}. In a closer look, from 5.5×10^{-4} to 11×10^{-4} mol L^{-1}, the kinetic rate decreased. This could be attributed to scavenging reactions between the excess of iron present in solution and the $HO^{\bullet}$ radicals generated (Equation (11)) [46].

$$Fe2^{+} + HO^{\bullet} \rightarrow Fe3^{+} + HO^{-} \tag{11}$$

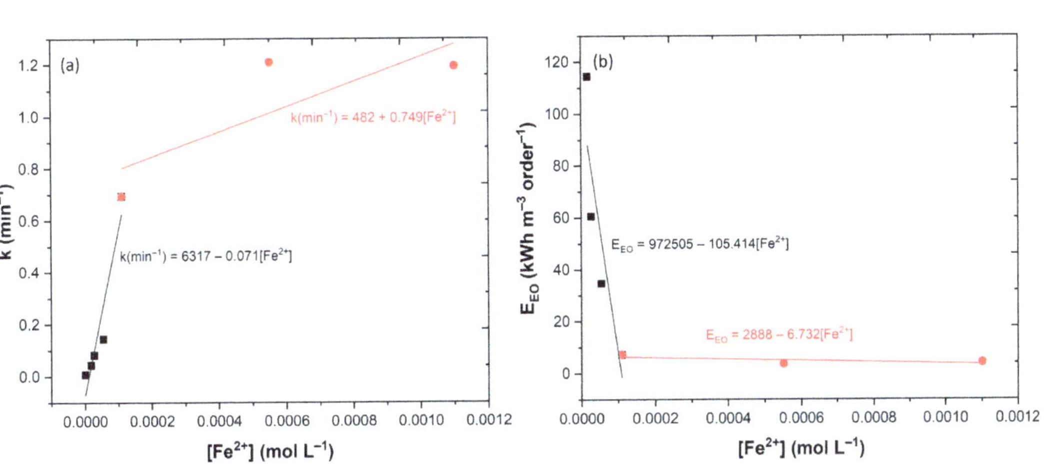

Figure 7. Outcome of several concentrations of Fe^{2+} in CA degradation by UV-A-Fenton process; (**a**) dependence on the kinetic constant rates; (**b**) electric energy per order (E_{EO}).

These results were in agreement with the work of Xavier et al. [47], who observed that the application of an excess of Fe^{2+} increased the scavenging reactions and decreased the efficiency of the photo-Fenton process in the degradation of Magenta MB. Parallel to the kinetic rate, the energy consumption was determined to evaluate the feasibility of the reactor. The results (Figure 7b) showed two different dominions, one for $[Fe^{2+}] \leq 1.1 \times 10^{-4}$ mol L^{-1} and one for $[Fe^{2+}] \geq 1.1 \times 10^{-4}$ mol L^{-1}. These results clearly stipulate that for concentrations below 1.1×10^{-4} mol L^{-1}, the reactions were too prolonged, implying larger electric consumptions, thus it could result in higher costs for treatment plants. For application of Fe^{2+} above 1.1×10^{-4} mol L^{-1}, the gains were very low and do not justify the application of those amounts of iron. The selection of the appropriate radiation type with the correct geometry influences the kinetic rate of degradation and the costs of treatment. This fact was observed in the work of Tapia-Tlatelpa et al. [48], who observed from testing UV lamps with different geometries, that LED lamps with a radial position (similar to this work) achieved the lowest energy consumption.

In Table 2, it is shown research that compares the E_{EO} values obtained by other treatment processes applied in the degradation of phenolic compounds, dyes, and compounds of emerging concern. When compared to the work of Yáñez et al. [12], the methodology created in this work was shown to be more efficient, reducing the energy required to degrade the CA phenolic. In a different work, the application of a low-pressure Hg lamp (8 W power, 250 nm) presented an ideal process to convert the H_2O_2 into $HO^{\bullet}$ radicals [49]; however, without a catalyst, this was observed to be a slow process with high energy consumption.

Table 2. Disclosed results of electric energy per order (E_{EO}) obtained after degradation of single compounds by different treatment processes.

Contaminants	AOP Processes	Conditions	E_{EO} (kWh.m^{-3} order^{-1})	References
Caffeic acid	UV-A LED/H$_2$O$_2$/Fe^{2+}	[CA] = 5.5 × 10^{-4} mol L^{-1}, [H$_2$O$_2$] = 2.2 × 10^{-3} mol L^{-1}, [Fe^{2+}] = 1.1 × 10^{-4} mol L^{-1}	7.23	Present work
Caffeic acid (CA)	UV-A-Fenton	[CA] = 10 mg L^{-1}, [H$_2$O$_2$] = 82.4 µmol L^{-1}, [Fe^{2+}] = 558.6 µmol L^{-1}	30	Yáñez et al. [12]
Winery wastewater	UV-A-Fenton	TOC = 1601 mg C L^{-1}, [Fe^{2+}] = 2.5 mM, [H$_2$O$_2$] = 225 mM, pH = 3.0, agitation = 350 rpm, t = 150 min	641	Jorge et al. [51]
Winery wastewater	UV-C-Fenton	TOC = 1601 mg C L^{-1}, [Fe^{2+}] = 2.5 mM, [H$_2$O$_2$] = 225 mM, pH = 3.0, agitation = 350 rpm, t = 150 min	170	Jorge et al. [51]
Poultry slaughterhouse wastewater	UV-C-Fenton	TOC = 68.66 mg C L^{-1}, [Fe^{2+}] = 20 mg L^{-1}, [H$_2$O$_2$] = 98 mM, pH = 3.3, agitation = 350 rpm, t = 150 min	248	Kanafin et al. [52]
Oxytetracycline (OTC)	UV-C/H$_2$O$_2$	[OTC] = 250 mg L^{-1}, [H$_2$O$_2$] = 375 mg L^{-1}	47.18	Rahmah et al. [49]
Sufamethoxazole (SMX)	UV-C	[SMX] = 30 mg L^{-1}, [H$_2$O$_2$] = 10 mM	1.50	Kim et al. [50]
Sufamethoxazole (SMX)	Ozone	[SMX] = 30 mg L^{-1}	27.53	Kim et al. [50]
Sufamethoxazole (SMX)	Electron beam	[SMX] = 30 mg L^{-1}	0.46	Kim et al. [50]
Acid Red 88 (AR88)	UV-A-Fenton	pH 3.0, [AR88] = 50 mg L^{-1}, [H$_2$O$_2$] = 4 mM, [Fe^{2+}] = 0.15 mM, [NTA] = 0.10 mM,	26	Teixeira et al. [53]
Orange PX-2R	UV-A LED/TiO$_2$	[OPX-2R] = 0.1 g/L, [TiO$_2$] = 1.0 g/L	119.04	Tapia-Tlatelpa et al. [48]
p-hydroxybenzoic acid (pHBA)	UV-A LED/TiO$_2$	[pHBA] = 50 mg L^{-1}, [TiO$_2$] = 1000 mg L^{-1}	115	Ferreira et al. [33]
Z-thiacloprid	UV/TiO$_2$	[Z-thiacloprid] = 1.0 × 10^{-4} mol L^{-1}, [TiO$_2$] = 1 g L^{-1}	80.0	Rózsa et al. [54]

In the work of Kim et al. [50], three different reactors were applied: (1) UV-C lamp (6 W power, 254 nm) with H$_2$O$_2$, (2) ozone batch reactor (1.12 mg O$_3$/min), and (3) E-beam (1 MeV and 40 kW). The results showed that the ozonation process achieved worse E_{EO} results in comparison to the UV-A reactor used in this work. Although the UV-C and E-beam reactors showed promising results, the concentration of the SMX degraded was much lower than the concentration of CA used in this work. The UV-A reactors with an emission wavelength of 365 nm were applied in photocatalytic degradation of dyes and phenolics [33,48], with results showing efficient removals; however, they are not revealed as being competitive in comparison to the results obtained in this work. In Table 2, several examples of the application of photo-Fenton process to treat real wastewaters are also shown, such as in winery wastewater [51] and poultry slaughterhouse wastewater [52]. It is shown that the demand for electric energy to treat real wastewaters is much higher in comparison to single compounds, as is the amount of reagents required to degrade the organic carbon.

3.5. Experiments in Continuous Mode

Previous sections have shown that batch reactors fitted with UV-A LED lights achieved a near complete removal of CA from the aqueous solution. It was also observed in the H$_2$O$_2$ and Fe^{2+} concentration variations sections that the degradation of CA occurred in two phases, a faster one within the first 5 min, followed by a slower rate, reaching a plateau. This occurs because, in the first period, the Fe^{2+} reacted very fast with the H$_2$O$_2$ (k = 78 mol^{-1} dm^3 s^{-1}), generating a high content of HO$^\bullet$ radicals. Following this period, the ferric ions produced earlier reacted with the H_2O_2 producing $HO_2^\bullet$, with a much lower kinetic rate ($k = 0.02$ mol^{-1} dm^3 s^{-1}) [55]. Therefore, these results indicate that the residence time (τ, min) within the continuous mode is an important parameter to be considered. In an ideal continuous reactor, the space-time is equal to the mean residence time, and is determined in accordance with Equation (12) [55].

$$\tau = \frac{V}{F} \tag{12}$$

where V is the volume of the reactor (mL) and *F* is the flow rate (mL min^{-1}). In Figure 8a are shown the removal results of CA and DOC as a function of dimensionless reaction time values ($\theta = t/\tau$), in which each θ is comparable to one τ [56]. The results showed that a CA steady state was achieved after 0.04 space-time values, while the mineralization of CA showed a steady state from 0 to 0.020 space-time values. By the analysis of the results obtained, the flow rate had a significant influence on the removal rate of CA and mineralization, in which, with *F* = 4 mL min^{-1}, higher CA and DOC removal was achieved (99.2 and 35.9%, respectively). The Fe^{2+} concentration analysis revealed that the availability of iron remaining in the solution was similar to that observed in batch experiments; therefore, the concentration of H_2O_2 was revealed to be the decisive factor. In observation of Figure 8, it can be seen that the application of *F* = 1 and 2 mL min^{-1} led to considerable lower additions of H_2O_2 regarding the application of *F* = 4 mL min^{-1}, thus less $HO^\bullet$ radicals were generated. These results were in agreement with the work of Rosales et al. [30], who observed that increasing the residence time enhanced the generation of $HO^\bullet$ radicals, increasing the discoloration of textile wastewaters.

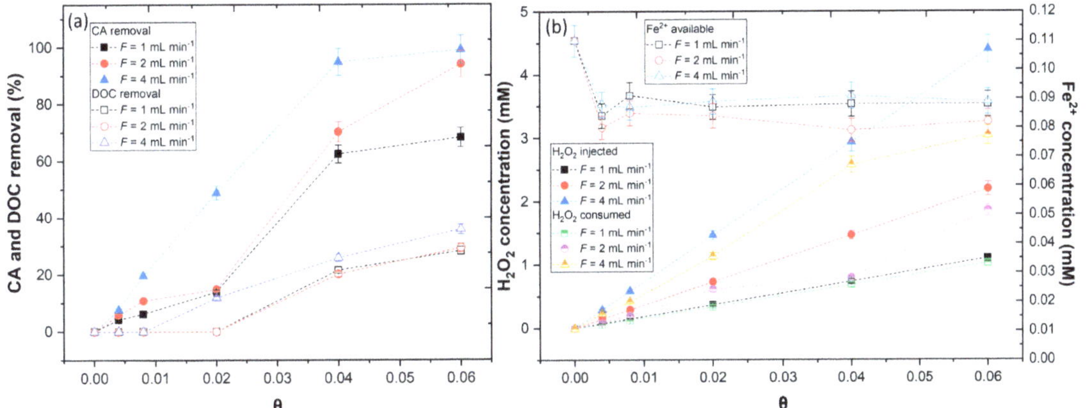

Figure 8. Assessment of different flow rates (1–4 mL min^{-1}) in (**a**) CA and DOC removal; (**b**) H_2O_2 and Fe^{2+} concentration available as function of θ. Operational conditions: [CA] = 5.5 × 10^{-4} mol L^{-1}, [Fe^{2+}] = 1.1 × 10^{-4} mol L^{-1}, pH = 3.0, agitation 150 rpm, temperature = 298 K, radiation UV-A, I_{UV} = 32.7 W m^{-2}, HRT = 15 min.

To understand the results obtained by the continuous mode, the kinetic constants were determined by application of a continuous mode simplified model, assuming a first order reaction Equation (13) [33].

$$k\tau = \frac{1 - \frac{C_{SS}}{C_0}}{\frac{C_{SS}}{C_0}} \tag{13}$$

where C_0 and C_{SS} are the initial and steady state concentrations of CA (mol L^{-1}) and DOC (mg C L^{-1}). In Table 3 are shown the kinetic rate constants obtained under different flow rates. The results are plotted in Figure 9, showing that as the flow rate increases, the kinetic rate of CA and DOC also increased, meaning that it achieved a good penetration of the UV-A radiation into the water. Considering this behavior, the continuous mode in association with the UV-A LEDs with horizontal geometry could be considered as a good system for phenolic degradation.

Table 3. Kinetic constants from CA and DOC removal observed for the assays with different flow rates (mL min^{-1}). Means in the same column with different letters represent significant differences ($p < 0.05$) within each parameter (k_{CA} and k_{DOC}) by comparing flow rates.

F mL min^{-1}	τ min	C$_{SS}$/C$_0$	DOC$_{SS}$/DOC$_0$	k$_{CA}$ min^{-1}	k$_{DOC}$ min^{-1}
1	250	0.318	0.718	$0.009 \pm 4.5 \times 10^{-4}$ [a]	$0.00157 \pm 7.8 \times 10^{-5}$ [a]
2	125	0.059	0.707	$0.063 \pm 3.2 \times 10^{-3}$ [b]	$0.00332 \pm 1.7 \times 10^{-4}$ [b]
4	62.5	0.008	0.641	$0.468 \pm 2.3 \times 10^{-2}$ [c]	$0.00896 \pm 4.5 \times 10^{-4}$ [c]

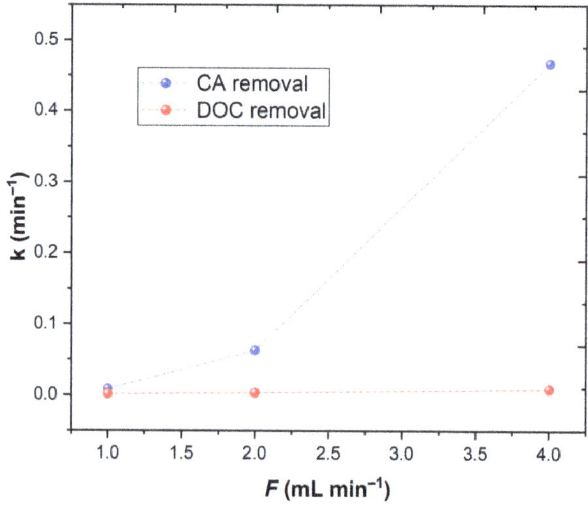

Figure 9. Dependence of the kinetic constant (k) to the flow rate (*F*) of the continuous mode reactor.

4. Conclusions

This work attempted to study the effect of several parameters in the kinetic rate and energy consumption of the hydroxyl radical-based oxidation processes. The experiments were carried out employing a UV-A LED system emitting at 365 nm, H_2O_2 as the oxidant, and Fe^{2+} as the catalyst. Based in the results, the UV-A-Fenton process at pH 3.0 achieved the highest CA degradation rate. It is concluded that increasing the H_2O_2 concentration increases the kinetic rate of CA degradation; however, above 2.2×10^{-3} mol L^{-1}, the reactor is not economically viable, because no significant E_{EO} reduction is observed. It is concluded that the Fe^{2+} achieves the best performance, with 1.1×10^{-4} mol L^{-1}, with an $E_{EO} = 7.23$ kWh m^{-3} order. Above this concentration, the kinetic rate increases; however, no significant gains in energy consumption were observed. When batch conditions are applied to continuous mode, it is concluded that the degradation of CA is dependent on the flow rate. It is also concluded that higher flow rates lead to higher CA and DOC removals ($k_{CA} = 0.468$ min^{-1}; $k_{DOC} = 0.00896$ min^{-1}).

Based in the results obtained from this work, a UV-A LED reactor adapted to continuous mode could provide an excellent process to reduce the soaring energy costs associated with these systems and make them viable for wastewater treatment plants.

Author Contributions: Conceptualization, N.J., A.R.T. and I.O.; methodology, N.J. and I.O.; software, N.J.; validation, N.J., J.R.F., M.S.L. and J.A.P.; formal analysis, N.J.; investigation, N.J.; resources, N.J.; data curation, N.J.; writing—original draft preparation, N.J.; writing—review and editing, N.J., J.R.F., M.S.L. and J.A.P.; visualization, N.J., J.R.F., M.S.L. and J.A.P.; supervision, M.S.L. and J.A.P.; project administration, J.A.P.; funding acquisition, M.S.L. and J.A.P. All authors have read and agreed to the published version of the manuscript.

Funding: The authors are grateful for the financial support of the Project AgriFood XXI NORTE-01-0145-FEDER-000041 and Fundação para a Ciência e a Tecnologia (FCT) to CQVR (UIDB/00616/2020). Ana R. Teixeira also thanks the FCT for the financial support provided through the doctoral scholarship UI/BD/150847/2020.

Institutional Review Board Statement: Not applicable.

Informed Consent Statement: Not applicable.

Data Availability Statement: Not applicable.

Acknowledgments: The authors are thankful to CITAB, supported by National Funds by the FCT—Portuguese Foundation for Science and Technology, under the project UIDB/04033/2020.

Conflicts of Interest: The authors declare no conflict of interest.

References

1. Alayu, E.; Yirgu, Z. Advanced technologies for the treatment of wastewaters from agro-processing industries and cogeneration of by-products: A case of slaughterhouse, dairy and beverage industries. *Int. J. Environ. Sci. Technol.* **2018**, *15*, 1581–1596. [CrossRef]
2. Rajagopal, R.; Saady, N.M.C.; Torrijos, M.; Thanikal, J.V.; Hung, Y.T. Sustainable agro-food industrial wastewater treatment using high rate anaerobic process. *Water* **2013**, *5*, 292–311. [CrossRef]
3. Amor, C.; Marchão, L.; Lucas, M.S.; Peres, J.A. Application of advanced oxidation processes for the treatment of recalcitrant agro-industrial wastewater: A review. *Water* **2019**, *11*, 205. [CrossRef]
4. Krzemińska, D.; Neczaj, E.; Borowski, G. Advanced oxidation processes for food industrial wastewater decontamination. *J. Ecol. Eng.* **2015**, *16*, 61–71. [CrossRef]
5. Spennati, E.; Casazza, A.A.; Converti, A. Winery wastewater treatment by microalgae to production purposes. *Energies* **2020**, *13*, 2490. [CrossRef]
6. Ioannou, L.A.; Puma, G.L.; Fatta-Kassinos, D. Treatment of winery wastewater by physicochemical, biological and advanced processes: A review. *J. Hazard. Mater.* **2015**, *286*, 343–368. [CrossRef]
7. Jorge, N.; Santos, C.; Teixeira, A.R.; Marchão, L.; Tavares, P.B.; Lucas, M.S.; Peres, J.A. Treatment of agro-industrial wastewaters by coagulation-flocculation-decantation and advanced oxidation processes—A literature review. *Eng. Proc.* **2022**, *19*, 33. [CrossRef]
8. Ferreira, R.; Gomes, J.; Martins, R.C.; Costa, R.; Quinta-Ferreira, R.M. Winery wastewater treatment by integrating Fenton's process with biofiltration by *Corbicula fluminea*. *J. Chem. Technol. Biotechnol.* **2018**, *93*, 333–339. [CrossRef]
9. Gernjak, W.; Maldonado, M.I.; Malato, S.; Caceres, J.; Krutzler, T.; Glaser, A.; Bauer, R. Pilot-Plant treatment of olive mill wastewater (OMW) by solar TiO_2 photocatalysis and solar photo-Fenton. *Sol. Energy* **2004**, *77*, 567–572. [CrossRef]
10. Vijayaraghavalu, S.; Prasad, H.K.; Kumar, M. Treatment and recycling of wastewater from winery. In *Advances in Biological Treatment of Industrial Waste Water and their Recycling for a Sustainable Future. Applied Environmental Science and Engineering for a Sustainable Future*; Singh, R.L., Singh, R.P., Eds.; Springer: Singapore, 2019; pp. 167–197.
11. Beltrán-Heredia, J.; Sánchez-Martín, J.; Muñoz-Serrano, A.; Peres, J.A. Towards overcoming TOC increase in wastewater treated with moringa oleifera seed extract. *Chem. Eng. J.* **2012**, *188*, 40–46. [CrossRef]
12. Yáñez, E.; Santander, P.; Contreras, D.; Yáñez, J.; Cornejo, L.; Mansilla, H.D. Homogeneous and heterogeneous degradation of caffeic acid using photocatalysis driven by UVA and solar light. *J. Environ. Sci. Health Part A* **2016**, *51*, 78–85. [CrossRef]
13. Filipe-Ribeiro, L.; Milheiro, J.; Matos, C.C.; Cosme, F.; Nunes, F.M. Data on changes in red wine phenolic compounds, headspace aroma compounds and sensory profile after treatment of red wines with activated carbons with different physicochemical characteristics. *Data Br.* **2017**, *12*, 188–202. [CrossRef] [PubMed]
14. Milheiro, J.; Filipe-Ribeiro, L.; Cosme, F.; Nunes, F.M. A simple, cheap and reliable method for control of 4-ethylphenol and 4-ethylguaiacol in red wines. screening of fining agents for reducing volatile phenols levels in red wines. *J. Chromatogr. B* **2017**, *1041–1042*, 183–190. [CrossRef] [PubMed]
15. Soto, M.L.; Moure, A.; Domínguez, H.; Parajó, J.C. Recovery, concentration and purification of phenolic compounds by adsorption: A review. *J. Food Eng.* **2011**, *105*, 1–27. [CrossRef]
16. Venditti, F.; Cuomo, F.; Ceglie, A.; Avino, P.; Russo, M.V.; Lopez, F. Visible light caffeic acid degradation by carbon-doped titanium dioxide. *Langmuir* **2015**, *31*, 3627–3634. [CrossRef]
17. Tolba, M.F.; Azab, S.S.; Khalifa, A.E.; Abdel-Rahman, S.Z.; Abdel-Naim, A.B. Caffeic acid phenethyl ester, a promising component of propolis with a plethora of biological activities: A review on its anti-inflammatory, neuroprotective, hepatoprotective, and cardioprotective effects. *IUBMB Life* **2013**, *65*, 699–709. [CrossRef]

18. Bello, M.M.; Raman, A.A.A.; Asghar, A. A review on approaches for addressing the limitations of Fenton oxidation for recalcitrant wastewater treatment. *Process Saf. Environ. Prot.* **2019**, *126*, 119–140. [CrossRef]
19. Neyens, E.; Baeyens, J. A Review of classic Fenton's peroxidation as an advanced oxidation technique. *J. Hazard. Mater.* **2003**, *98*, 33–50. [CrossRef]
20. Cuerda-Correa, E.M.; Alexandre-Franco, M.F.; Fernández-González, C. Advanced oxidation processes for the removal of antibiotics from water. An overview. *Water* **2020**, *12*, 102. [CrossRef]
21. Wang, J.; Chen, H. Catalytic ozonation for water and wastewater treatment: Recent advances and perspective. *Sci. Total Environ.* **2020**, *704*, 135249. [CrossRef]
22. Dowd, K.O.; Pillai, S.C. Photo-Fenton disinfection at near neutral pH: Process, parameter optimization and recent advances. *J. Environ. Chem. Eng.* **2020**, *8*, 104063. [CrossRef]
23. Chong, M.N.; Jin, B.; Chow, C.W.; Saint, C. Recent developments in photocatalytic water treatment technology: A review. *Water Res.* **2010**, *44*, 2997–3027. [CrossRef] [PubMed]
24. Umar, M.; Aziz, H.A.; Yusoff, M.S. Trends in the use of Fenton, electro-Fenton and photo-Fenton for the treatment of landfill leachate. *Waste Manag.* **2010**, *30*, 2113–2121. [CrossRef] [PubMed]
25. Wu, Y.; Passananti, M.; Brigante, M.; Dong, W.; Mailhot, G. Fe (III)–EDDS complex in Fenton and photo-Fenton processes: From the radical formation to the degradation of a target compound. *Environ. Sci. Pollut. Res.* **2014**, *21*, 12154–12162. [CrossRef]
26. Chen, J.; Loeb, S.; Kim, J. LED revolution: Fundamentals and prospects for UV disinfection applications. *Environ. Sci. Water Res. Technol.* **2017**, *3*, 188–202. [CrossRef]
27. Bronze, M.R.; Crespo, J.G.; Pereira, V.J.; Janssens, R.; Cristo, B.M.; Luis, P. Photocatalysis using UV-A and UV-C light sources for advanced oxidation of anti-cancer drugs spiked in laboratory-grade water and synthetic urine. *Ind. Eng. Chem. Res.* **2020**, *59*, 647–653. [CrossRef]
28. Natarajan, K.; Natarajan, T.S.; Bajaj, H.C.; Tayade, R.J. Photocatalytic Reactor Based on UV-LED/TiO$_2$ coated quartz tube for degradation of dyes. *Chem. Eng. J.* **2011**, *178*, 40–49. [CrossRef]
29. Rojviroon, T.; Laobuthee, A.; Sirivithayapakorn, S. Photocatalytic activity of toluene under UV-LED light with TiO$_2$ thin films. *Int. J. Photoenergy* **2012**, *2012*, 898464. [CrossRef]
30. Rosales, E.; Pazos, M.; Longo, M.A.; Sanromán, M.A. Electro-Fenton Decoloration of Dyes in a Continuous Reactor: A promising technology in colored wastewater treatment. *Chem. Eng. J.* **2009**, *155*, 62–67. [CrossRef]
31. Mudnic, I.; Modun, D.; Rastija, V.; Vukovic, J.; Brizic, I.; Katalinic, V.; Kozina, B.; Medic-Saric, M.; Boban, M. Antioxidative and vasodilatory effects of phenolic acids in wine. *Food Chem.* **2010**, *119*, 1205–1210. [CrossRef]
32. Jorge, N.; Teixeira, A.R.; Matos, C.C.; Lucas, M.S.; Peres, J.A. Combination of coagulation–flocculation–decantation and ozonation processes for winery wastewater treatment. *Int. J. Environ. Res. Public Health* **2021**, *18*, 8882. [CrossRef] [PubMed]
33. Ferreira, L.C.; Fernandes, J.R.; Rodriguez-Chueca, J.; Peres, J.A.; Lucas, M.S.; Tavares, P.B. Photocatalytic degradation of an agro-industrial wastewater model compound using a UV LEDs system: Kinetic study. *J. Environ. Manag.* **2020**, *269*, 110740. [CrossRef] [PubMed]
34. Benitez, F.J.; Real, F.J.; Acero, J.L.; Leal, A.I.; Garcia, C. Gallic acid degradation in aqueous solutions by UV/H$_2$O$_2$ treatment, fenton's reagent and the photo-fenton system. *J. Hazard. Mater.* **2005**, *126*, 31–39. [CrossRef] [PubMed]
35. Kaplan, F.; Hesenov, A.; Gözmen, B.; Erbatur, O. Degradations of model compounds representing some phenolics in olive mill wastewater via electro-Fenton and photoelectro-Fenton treatments. *Environ. Technol.* **2011**, *32*, 685–692. [CrossRef] [PubMed]
36. Jorge, N.; Teixeira, A.R.; Lucas, M.S.; Peres, J.A. Combined organic coagulants and photocatalytic processes for winery wastewater treatment. *J. Environ. Manag.* **2023**, *326*, 116819. [CrossRef] [PubMed]
37. Cruz, A.; Couto, L.; Esplugas, S.; Sans, C. Study of the contribution of homogeneous catalysis on heterogeneous Fe (III)/Alginate mediated photo-fenton process. *Chem. Eng. J.* **2017**, *318*, 272–280. [CrossRef]
38. Feng, X.; Wang, Z.; Chen, Y.; Tao, T.; Wu, F.; Zuo, Y. Effect of Fe (III)/Citrate concentrations and ratio on the photoproduction of hydroxyl radicals: Application on the degradation of diphenhydramine. *Ind. Eng. Chem. Res.* **2012**, *51*, 7007–7012. [CrossRef]
39. Masomboon, N.; Ratanatamskul, C.; Lu, M. Mineralization of 2, 6-dimethylaniline by photoelectro-Fenton process. *Appl. Catal. A Gen.* **2010**, *384*, 128–135. [CrossRef]
40. Gao, L.; Cao, Y.; Wang, L.; Li, S. A review on sustainable reuse applications of Fenton sludge during wastewater treatment. *Front. Environ. Sci. Eng.* **2022**, *16*, 77. [CrossRef]
41. Zha, F.G.; Yao, D.X.; Hu, Y.B.; Gao, L.M.; Wang, X.M. Integration of US/Fe^{2+} and photo-Fenton in sequencing for degradation of landfill leachate. *Water Sci. Technol.* **2016**, *73*, 260–266. [CrossRef] [PubMed]
42. Agrawal, S.; Niwan, N.; Chohadia, A. Degradation of acriflavine using environmentally benign process involving singlet-oxygen-photo-fenton: A comparative study. *J. Photochem. Photobiol. A Chem.* **2020**, *398*, 112547. [CrossRef]
43. Li, Y.; Cheng, H. Chemical kinetic modeling of organic pollutant degradation in Fenton and solar photo-Fenton processes. *J. Taiwan Inst. Chem. Eng.* **2021**, *123*, 175–184. [CrossRef]
44. Jorge, N.; Teixeira, A.R.; Lucas, M.S.; Peres, J.A. Agro-industrial wastewater treatment with *acacia dealbata* coagulation/flocculation and photo-Fenton-based processes. *Recycling* **2022**, *7*, 54. [CrossRef]
45. Faggiano, A.; De Carluccio, M.; Fiorentino, A.; Ricciardi, M.; Cucciniello, R.; Proto, A.; Rizzo, L. Photo-Fenton like process as polishing step of biologically co-treated olive mill wastewater for phenols removal. *Sep. Purif. Technol.* **2023**, *305*, 122525. [CrossRef]

46. Gupta, A.; Garg, A. Degradation of ciprofloxacin using Fenton's oxidation: Effect of operating parameters, identification of oxidized by-products and toxicity assessment. *Chemosphere* **2018**, *193*, 1181–1188. [CrossRef]
47. Xavier, S.; Gandhimathi, R.; Nidheesh, P.V.; Ramesh, S.T. Comparative removal of magenta mb from aqueous solution by homogeneous and heterogeneous photo-Fenton processes. *Desalin. Water Treat.* **2016**, *57*, 12832–12841. [CrossRef]
48. Tapia-tlatelpa, T.; Buscio, V.; Trull, J.; Sala, V. Performance analysis and methodology for replacing conventional lamps by optimized LED arrays for photocatalytic processes. *Chem. Eng. Res. Des.* **2020**, *156*, 456–468. [CrossRef]
49. Rahmah, A.U.; Harimurti, S.; Murugesan, T. Experimental investigation on the effect of wastewater matrix on oxytetracycline mineralization using UV/H$_2$O$_2$ system. *Int. J. Environ. Sci. Technol.* **2017**, *14*, 1225–1233. [CrossRef]
50. Kim, T.; Don, S.; Young, H.; Joo, S.; Lee, M.; Yu, S. Degradation and toxicity assessment of sulfamethoxazole and chlortetracycline using electron beam, ozone and UV. *J. Hazard. Mater.* **2012**, *227–228*, 237–242. [CrossRef] [PubMed]
51. Jorge, N.; Teixeira, A.R.; Marchão, L.; Lucas, M.S.; Peres, J.A. Application of combined coagulation–flocculation–decantation/photo-Fenton/adsorption process for winery wastewater treatment. *Eng. Proc.* **2022**, *19*, 22. [CrossRef]
52. Kanafin, Y.N.; Makhatova, A.; Meiramkulova, K.; Poulopoulos, S.G. Treatment of a poultry slaughterhouse wastewater using advanced oxidation processes. *J. Water Process Eng.* **2022**, *47*, 102694. [CrossRef]
53. Teixeira, A.R.; Jorge, N.; Fernandes, J.R.; Lucas, M.S.; Peres, J.A. Textile dye removal by *acacia dealbata* link. pollen adsorption combined with UV-A/NTA/Fenton process. *Top. Catal.* **2022**, *65*, 1045–1061. [CrossRef]
54. Rózsa, G.; Kozmér, Z.; Alapi, T.; Schrantz, K.; Takács, E.; Wojnárovits, L. Transformation of Z-thiacloprid by three advanced oxidation processes: Kinetics, intermediates and the role of reactive species. *Catal. Today* **2017**, *284*, 187–194. [CrossRef]
55. Esteves, B.M.; Rodrigues, C.S.D.; Madeira, L.M. Synthetic olive mill wastewater treatment by Fenton's process in batch and continuous reactors operation. *Environ. Sci. Pollut. Res.* **2018**, *25*, 34826–34838. [CrossRef] [PubMed]
56. Monteil, H.; Pechaud, Y.; Oturan, N.; Trellu, C.; Oturan, M.A. Pilot scale continuous reactor for water treatment by electrochemical advanced oxidation processes: Development of a new hydrodynamic/reactive combined model. *Chem. Eng. J.* **2021**, *404*, 127048. [CrossRef]

 *catalysts*

Review

Investigation of Advanced Oxidation Process in the Presence of TiO$_2$ Semiconductor as Photocatalyst: Property, Principle, Kinetic Analysis, and Photocatalytic Activity

Amir Hossein Navidpour [1], Sedigheh Abbasi [2], Donghao Li [3], Amin Mojiri [4] and John L. Zhou [1,*]

[1] Centre for Green Technology, School of Civil and Environmental Engineering, University of Technology Sydney, 15 Broadway, Sydney, NSW 2007, Australia
[2] Central Research Laboratory, Esfarayen University of Technology, Esfarayen 9661998195, North Khorasan, Iran
[3] Department of Chemistry, Yanbian University, Yanji 133002, China
[4] Department of Civil and Environmental Engineering, Graduate School of Advanced Science and Engineering, Hiroshima University, Higashihiroshima 739-8527, Japan
* Correspondence: junliang.zhou@uts.edu.au

Abstract: Water pollution is considered a serious threat to human life. An advanced oxidation process in the presence of semiconductor photocatalysts is a popular method for the effective decomposition of organic pollutants from wastewater. TiO$_2$ nanoparticles are widely used as photocatalysts due to their low cost, chemical stability, environmental compatibility and significant efficiency. The aim of this study is to review the photocatalytic processes and their mechanism, reaction kinetics, optical and electrical properties of semiconductors and unique characteristics of titanium as the most widely used photocatalyst; and to compare the photocatalytic activity between different titania phases (anatase, rutile, and brookite) and between colorful and white TiO$_2$ nanoparticles. Photocatalytic processes are based on the creation of electron–hole pairs. Therefore, increasing stability and separation of charge carriers could improve the photocatalytic activity. The synthesis method has a significant effect on the intensity of photocatalytic activity. The increase in the density of surface hydroxyls as well as the significant mobility of the electron–hole pairs in the anatase phase increases its photocatalytic activity compared to other phases. Electronic and structural changes lead to the synthesis of colored titania with different photocatalytic properties. Among colored titania materials, black TiO$_2$ showed promising photocatalytic activity due to the formation of surface defects including oxygen vacancies, increasing the interaction with the light irradiation and the lifetime of photogenerated electron–hole pairs. Among non-metal elements, nitrogen doping could be effectively used to drive visible light-activated TiO$_2$.

Keywords: black TiO$_2$; colored titania; photocatalytic activity; reaction kinetics; titania phases

Citation: Navidpour, A.H.; Abbasi, S.; Li, D.; Mojiri, A.; Zhou, J.L. Investigation of Advanced Oxidation Process in the Presence of TiO$_2$ Semiconductor as Photocatalyst: Property, Principle, Kinetic Analysis, and Photocatalytic Activity. *Catalysts* **2023**, *13*, 232. https://doi.org/10.3390/catal13020232

Academic Editors: Gassan Hodaifa, Antonio Zuorro, Joaquín R. Dominguez, Juan García Rodríguez, José A. Peres, Zacharias Frontistis and Mha Albqmi

Received: 16 December 2022
Revised: 12 January 2023
Accepted: 15 January 2023
Published: 19 January 2023

1. Introduction

Water scarcity is known as the greatest threat to natural ecosystems and human health, especially in arid areas of the world. Economic development and population growth have led to an increase in water demand and reduction of water availability consequently [1]. It is estimated that there will be one billion people living in the arid area suffering from absolute water scarcity by 2025. Thus, people living in such areas will be forced into a reduction of water consumption in agricultural sectors and transmitting water to other parts, which could reduce their domestic food production. Organic wastes, which originate from different industrial products such as dyes, plastics, pesticides, and detergents, are toxic pollutants that could lead to serious diseases in humans. Although wastewater treatment has been extensively used for the degradation of organic pollutants, persistent organic pollutants (POPs) are resistant to degradation and could remain in the treated water [2]. In addition, there are different indoor air pollutants, which could result in wide-ranging health problems.

Different methods have been used for the degradation of organic pollutants in water, including sonochemical degradation [3,4], advanced oxidation process (AOP) [5–10], adsorption [11,12], micellar-enhanced ultrafiltration (MEUF) [13,14], electrochemical oxidation (EO) [15,16], precipitation [17], coagulation [17,18], biodegradation [19], ion exchange [20,21] and forward/reverse osmosis [22,23]. It is notable that each of these processes has advantages and disadvantages. AOPs are divided into various techniques including photo-Fenton process, heterogeneous photocatalysis, ozonation, Fenton process and H_2O_2 photolysis [24]. In general, AOPs have been proposed for poorly biodegradable organic pollutants [2]. Extremely highly reactive hydroxyl radicals ($\bullet OH$), which are known as the (secondary) oxidant agents in these processes, are formed by primary oxidants (such as ozone or hydrogen peroxide) and are capable of oxidizing various organic pollutants. The hydroxyl radicals, which are the common aspect of all AOPs, are among the non-selective and powerful oxidants that could react through three mechanisms: the transfer of electrons, the abstraction of hydrogen and the addition of radicals. Thus, they can be potentially used for the degradation of complex chemical structures such as refractory species [25,26]. Although AOPs need input of chemicals and energy [27], they are becoming more efficient processes, e.g., photocatalysis. Around 60% of the expenses of photocatalytic reactors could be allocated to the energy consumption [28]. AOPs have a major advantage, as they are capable of partial or even complete degradation of pollutants. In comparison, coagulation, flocculation and membrane processes concentrate pollutants within one phase or transfer them from one phase to another [29].

Environmental photocatalysis has received remarkable attention since the discovery of the TiO_2 photocatalytic activity in 1972 [30]. The oil crisis was one of the reasons that encouraged research in non-fossil fuels in the early 1970s. Moreover, searching for renewable energy sources is another major driver [31]. After their first observation over water splitting by TiO_2, the photocatalytic reduction of CO_2 by different inorganic photocatalysts was reported by Fujishima et al. in 1979 [32]. SnO_2, TiO_2, and ZnO have been extensively used for photocatalytic decomposition of organic pollutants. Despite its several advantages, the removal efficiency of organic pollutants in the presence of a tin semiconductor is not that promising, limiting its use compared to main semiconductors [33–36], whereas TiO_2 and ZnO semiconductor nanoparticles are widely used for the photocatalytic removal of organic pollutants. In general, TiO_2 and ZnO might act as electron acceptors/donors when irradiated [37]. Because of its transparency, compatibility with the environment, and suitable band gap, ZnO has attracted the attention of many researchers for photocatalytic activities [38,39]. ZnO has three main crystallite structures including wurtzite, rocksalt, and zinc blend. Noteworthy, wurtzite is its most stable polymorph at ambient temperature and pressure [40]. Although ZnO has shown higher photo-absorption ability than TiO_2, it suffers from photocorrosion while exposed to UV irradiation [10]. Overall, TiO_2 (as an ideal semiconductor) has attracted the attention of many researchers due to its unique properties such as significant physicochemical stability, low cost, non-toxicity, electronic and optical properties, as well as abundance, in various fields such as pigments, catalysts, mineral membranes, dielectric materials and removal of pollutants [41,42]. TiO_2 nanoparticles have shown higher photocatalytic activity than bulk TiO_2 (due to larger surface area). The properties of TiO_2 nanoparticles, including morphology, crystal structure, composition of phases, degree of crystallinity, specific surface area, and band gap are influenced by the synthesis method [43–45]. TiO_2 nanoparticles have three major crystal structures including anatase, rutile, and brookite with unique band gap energies of 3.2, 3.0, and 3.1 eV, respectively. Notably, anatase phase with its wider band gap energy than other titania polymorphs has received more attention for photocatalytic activities [44]. The crystal structure of the main phases of titania, i.e., anatase, rutile, and brookite, could be explained by various arrangements of TiO_6 octahedra (Ti^{4+} surrounded by six O^{2-}). Hence, 3-D assembly of TiO_6 octahedra and different degrees of distortion are differences between these crystal structures. Crustal structures of anatase, rutile, and brookite TiO_2 are exhibited in Figure 1 [46]. Noteworthy, the blue and red balls represent Ti and O atoms, respectively.

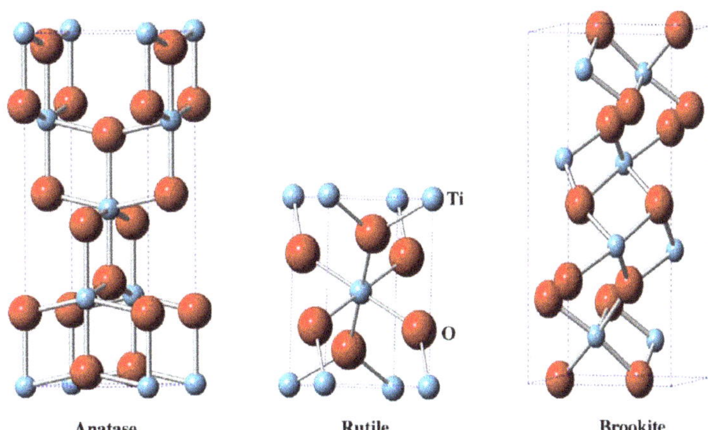

Figure 1. Crystal structures of brookite, rutile, and anatase TiO_2. Reprinted with permission from Ref. [46]. Copyright 2015, Elsevier.

Photocatalysis is a branch of research that uses photons' energy to initiate chemical reactions, which has been inspired by natural photosynthesis [31]. Photocatalytic (PC) and photosynthetic (PS) devices are different technologies that show sensitivity to charge-transfer kinetics, carrier mobility and specific surface area though at varying degrees [47].

From the viewpoint of thermodynamics, chemical reactions are divided into uphill and downhill categories. The schematic of these reactions is illustrated in Figure 2 (the '*' sign could represent the transition state). Uphill reactions that are powered by light could not be considered as photocatalysis [31]. Thermodynamically uphill reactions are nonspontaneous, and the use of a canonical catalyst could not initiate these reactions [31]. Usually, photocatalysis is considered as an appropriate terminology to describe a large number of spontaneous chemical reactions (downhill reactions) that are hindered by high activation energies. Thus, the introduction of light, as a source of energy, could enable these reactions [31,47]. Notably, there is another group of reactions that borrows the photocatalysis term. These reactions that use external potentials, usually, are referred to as photo-electrocatalytic/photo-electrochemical reactions [31]. Unlike photocatalysis, the photo-electrochemical water splitting is a thermodynamically uphill reaction [48–50]. Thus, during the process of photocatalytic water splitting, the energy of photons is converted to chemical energy (by using a photocatalytic material) [50].

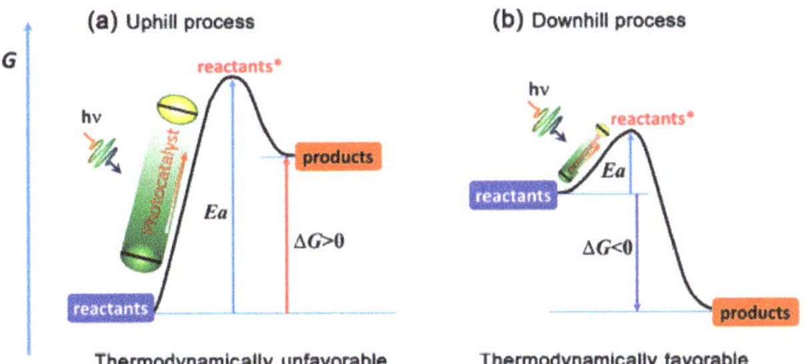

Figure 2. The schematic illustration of (**a**) uphill and (**b**) downhill reactions. Reprinted with permission from Ref. [51]. Copyright 2018, American Chemical Society.

The fundamental principles of photocatalysis and photocatalytic activity of various kinds of semiconducting materials have been widely evaluated by researchers, especially in recent years. Although this process is well-known as an efficient process for water and wastewater treatment, several key points should be further considered to facilitate the use of this process for practical applications. Firstly, the use of promising photocatalytic materials increasing the proficiency of this process is of high importance. Moreover, the expansion of the scope of reactions should receive more attention. For example, TiO_2 has found applications in building materials (for antifogging and self-cleaning properties) and antibacterial materials since the discovery of its super hydrophilicity [52]. Super hydrophilicity is an effect observed for materials whose surfaces display water contact angles (WCAs) below $10°$ [53]. Super hydrophilicity and photocatalysis are two simultaneous phenomena occurring upon the illumination of TiO_2 [54]. The attempt to degrade emerging pollutants plays an important role in the development and practical applications of this process. However, broad applications of this process would be unlikely without consideration of an appropriate method of photocatalyst immobilization. Another key point is that 3–6% of the solar spectrum is from the ultraviolet radiation [55], while TiO_2, as the most common photocatalytic material, has negligible visible-light adsorption. Since about 43% of the solar spectrum is allocated to visible light [56], attempts are intensifying currently to drive visible-light activated photocatalysts. In recent years, photocatalytic processes in the presence of semiconductors have attracted the attention of many researchers for the removal and decomposition of organic pollutants. Several factors such as photocatalyst type, structure, synthesis method and doping with metallic or non-metallic elements could affect the efficiency of semiconductors. Therefore, the main goal of this study is to review the kinetics of photocatalytic reactions and specifically clarify the photocatalytic properties of TiO_2 with different phases and colors.

2. Photocatalysis

2.1. Principles

Overall, semiconductor photocatalysis relies on four steps as follows [31]:

a. Absorption of photons with equal/larger energy than the bandgap energy of semi-conductor under irradiation of light that results in the photogeneration of electron and hole pairs;
b. Charge carrier separation;
c. Transfer of charge carriers to the surface of semiconducting material;
d. Redox reactions initiated by charge carriers.

The photogeneration of electron and hole pairs could be followed by the generation of active species [57,58]. These highly reactive oxygen species (ROSs), including hydroxyl radicals ($^{\bullet}OH$), superoxide radical ions ($^{\bullet}O_2^{-}$), hydrogen peroxide (H_2O_2), hydroperoxy radicals ($^{\bullet}HO_2$) and singlet oxygen ($^1O_2$), are responsible for further reactions [55,59–61]. The oxidation potential of hydroxyl radical ($E°(^{\bullet}OH/H_2O = 2.8$ V) is just below that of fluorine [62]. Notably, the photogenerated e^-/h^+ pairs could migrate through various pathways, as they could migrate to the surface of the semiconductor material. The electron acceptors (A) such as O_2 could receive the photogenerated electrons, and the photogenerated holes could result in the oxidation of donor species (D) under these circumstances. The redox potential levels of the adsorbed species and the position of conduction band (CB) and valence band (VB) play an important role in the rate and probability of the charge transfer processes for photogenerated e^-/h^+ pairs. The more is the lifetime of photogenerated e^-/h^+ pairs, the more quantity of ROS should be produced, which leads to the higher efficiency of photodegradation of organic contaminations in air and water. On the other hand, the recombination of photogenerated e^-/h^+ pairs, which could happen either at the surface or in the bulk of the semiconductor, prevents electron transfer processes and could result in the emission of light or release of heat [63,64]. To yield the highest photocatalytic activity, the recombination rate of charge carriers should be restricted. For such a purpose, several techniques such as heterojunction formation [65], metal deposition [66], creation of oxygen vacancies [67], and non-metal/metal ion doping [68] have been employed. It is noteworthy that surface reactions take place only if the reduction and oxidation potentials are more positive and negative than CB and VB levels, respectively [69].

The mechanism of an ideal heterogeneous photocatalytic process is based on the Equations (1)–(12) [70–72].

$$\text{Semiconductor} + h\upsilon \rightarrow \text{Semiconductor } (h^+{}_{VB} + e^-{}_{CB}) \tag{1}$$

$$h^+{}_{VB} + e^-{}_{CB} \rightarrow \text{heat} \tag{2}$$

$$OH^- + h^+{}_{VB} \rightarrow {}^\bullet OH \tag{3}$$

$$H_2O + h^+{}_{VB} \rightarrow {}^\bullet OH + H^+ \tag{4}$$

$${}^\bullet OH + {}^\bullet OH \rightarrow H_2O_2 \tag{5}$$

$$e^- + H^+ + O_2 \rightarrow {}^\bullet HO_2 \tag{6}$$

$${}^\bullet HO_2 + {}^\bullet HO_2 \rightarrow O_2 + H_2O_2 \tag{7}$$

$$O_2 + e^- \rightarrow {}^\bullet O_2{}^- \tag{8}$$

$${}^\bullet O_2{}^- + H^+ + {}^\bullet HO_2 \rightarrow O_2 + H_2O_2 \tag{9}$$

$$H_2O_2 + e^- \rightarrow {}^\bullet OH + OH^- \tag{10}$$

$$H_2O_2 + {}^\bullet OH \rightarrow H_2O + {}^\bullet HO_2 \tag{11}$$

$${}^\bullet OH + \text{organic substances} \rightarrow \text{intermediate} \rightarrow \text{product} \tag{12}$$

The photocatalytic process in water could be carried out as follows [30]:

a. Transfer of the pollutants from the liquid phase to the surface of catalyst;
b. Pollutants adsorption onto the surface of the activated catalyst;
c. Photogeneration of ROSs, including ${}^\bullet OH$, followed by pollutants degradation;
d. Desorption of intermediates from the surface of catalyst;
e. Transferring intermediates into the liquid phase.

Due to the crucial role of surface reactions in the photocatalysis mechanism, the adsorption of reactants on the surface of the semiconducting material is of great importance [31]. For instance, various mechanisms of oxidation have been observed for the degradation of phenol, benzene, and formic acid using TiO_2 [73]. Formic acid could be chemisorbed strongly on titania in water, which leads to direct oxidation by trapping photogenerated holes. In comparison, benzene is physisorbed on titania and has been photo-oxidized by an indirect transfer mechanism. When it comes to phenol, both chemisorption and physisorption processes could take place, and its photooxidation mechanism depends on the solvent used [31]. In addition to the reactants, the relative binding strength of products/intermediates to the surface of photocatalytic material is of high importance. For instance, the photocurrent decrease of WO_3 in $HClO_4$ has been allocated to the bond of photogenerated species on it, which could hinder the water oxidation of active sites [31]. Another key point is the probable interactions between water and the photocatalysts, which should be considered to determine the mechanism of photocatalytic processes.

2.2. Kinetic Analysis

The classical equation of Langmuir–Hinshelwood (L-H) has been widely used for the kinetic analysis of both liquid-phase and gas-phase photocatalysis [55,74–78]. Using this model, the relationship between the oxidation rate and the concentration of reactant is shown in Equation (13) (when the concentrations of oxygen and water remain constant) [77,78]:

$$r = -\frac{dD}{dt} = \frac{r_L K_L^{app}[D]}{1 + K_L^{app}[D]} \tag{13}$$

where r and [D] are the oxidation rate (mg m^{-3} min^{-1}) and the concentration of reactant (mg m^{-3}), respectively [78]. Langmuir–Hinshelwood parameters are K_L^{app} and r_L that are identified as the apparent Langmuir adsorption constant (m^3 min^{-1}) and the apparent L-H rate constant of the reaction, respectively [78,79].

Equation (13) can be transformed as Equation (14) [77]:

$$\frac{1}{r} = \frac{1}{r_L}\left(1 + \frac{1}{K_L^{app}[D]}\right) \tag{14}$$

The time dependence of the concentration of the adsorbed reactant is given in Equation (15) [77]

$$\frac{d[D_{ad}]}{d_t} = k_a[D]([S_D] - [D_{ad}]) - k_d[D_{ad}] - k_i I[D_{ad}] \tag{15}$$

where $[S_D]$, k_a, k_d, and k_i are the total concentration of the adsorption site for reactant, adsorption rate constant in the dark, desorption rate constant in the dark, and photoinduced desorption rate constant, respectively. There are some concerns for robotic application of L-H model to interpret the kinetic data [73,79]:

a. Adsorption–desorption equilibrium of substrate species is not disturbed under illumination (as a pre-assumption);
b. Ambiguous photon flow intervention (as an experimental parameter);
c. Issues in intervening physical meaning;
d. Chemical nature of the semiconductor surface does not change during photocatalysis (as a pre-assumption);
e. Disregarding the electronic interaction of surface with substrate species;
f. Considering that chemisorption of organic species onto the surface of catalyst is vital for photocatalysis.

The dark Langmuir adsorption constant (K_L) is not the same as K_L^{app} (usually $K_L^{app} \gg K_L$). It has been shown that k_{LH} is a function of absorbed light intensity (I^β) where β equals 1.0 (at low absorbed light intensity) or 0.5 (at high absorbed light intensity) [80]. After reevaluation of the semiconductor photo-assisted reactions, another simple mechanism has been suggested as follows [80]:

$$D(liquid) \longleftrightarrow D(ads) \tag{16}$$

$$D(ads) \xrightarrow{k_{LH}} products \tag{17}$$

It is assumed that the reaction of a surface hydroxyl radical and the absorbed reactant takes place in the latter step, Equation (17), which results in the formation of products. Based on the relationship between this step and absorbed light intensity (and the incident light intensity, consequently) correlation between k_{LH} and I^β has been proposed as follows [80]:

$$k_{LH} = \alpha I^\beta \tag{18}$$

In which α is proportionality constant and β equals 1.0 or 0.5 as discussed above. The surface coverage of the reactant (θ_D) is defined as follows by consideration of a pseudo-steady-state hypothesis [80]:

$$\theta_D = \frac{K_{ads}^{app}[D]}{1 + K_{ads}^{app}[D]} \tag{19}$$

where K_{ads}^{app} is given by Equation (20) as follows [80]:

$$K_{ads}^{app} = \frac{1}{K_{diss}^{app}} = \frac{k_1}{k_{-1} + \alpha I^\beta} \tag{20}$$

K_{diss}^{app} is defined as the apparent dissociation constant. Thus, the oxidation rate, r, is given as follows by considering that $-r = k_{LH}\theta_D$ [80]:

$$-r = \frac{k_{LH}[D]}{K_{diss}^{app} + [D]} \tag{21}$$

Boulamanti et al. have proposed the following equation by studying the photocatalytic activity of target gas and by considering the effect of water vapor [81]:

$$r = k\frac{K_{LH}D}{1 + K_{LH}D + K_w D_w} \tag{22}$$

where K_w is a Langmuir adsorption constant proportional to the adhered water molecules to the surface of the catalyst, and D_w is the concentration of water in the gas-phase.

3. Band Gap Estimation and Quantum Size Effect

Inorganic/organic semiconductors display various physical properties, and thus can be used for a variety of purposes [82]. The optical properties of semiconducting nanomaterials are their band gap energy (E_g) as well as absorption coefficient (K), potentially. The band gap energy of semiconductors is a crucial characteristic of their electronic structure, which determines their potential applications [83]. Although it could be affected by the synthesis/processing method, the exact determination of its value is a challenge in materials science and engineering. In direct band gap semiconductors, the recombination of e^-/h^+ pairs could be radiative, if the VB maximum and CB minimum are aligned in the momentum space, which results in photoluminescence [82]. When it comes to the indirect band gap semiconductors, phonons (lattice vibrations) that are involved in the emission and absorption of light allow the semiconductor to convert the light energy into mechanical (photoacoustic) or thermal (photothermal) responses [82]. When a semiconductor is exposed to photons of energy larger than its band gap, the transfer of electrons from the valence band to the conduction band leads to an abrupt increase in the absorbency of the semiconductor to the wavelength, which corresponds to the band gap energy. The relationship between the incidental photon energy and the absorption coefficient is dependent on the type of band gap (direct or indirect). The electronic properties of semiconductors are usually studied by conversion of DR spectra to pseudo-absorption spectra, $F(R_\infty)$, using the Kubelka–Munk function as follows [83,84]:

$$F(R_\infty) = \frac{K}{S} = \frac{(1 - R_\infty)^2}{2R_\infty} \tag{23}$$

$$R_\infty = \frac{R_{sample}}{R_{standard}} \tag{24}$$

where $F(R_\infty)$ is the Kubelka–Munk function and R_∞ is the diffused reflectance of the non-transparent material with infinite thickness. Thus, the supporting material is not contributed (in the case of semiconductors deposited on a substrate). The S and K parameters are the scattering and absorption K-M coefficients, respectively. It is notable that S does not depend on the wavelength for particles larger than 5 μm. Therefore, S could be considered as a constant and $F(R_\infty)$ could be considered as a pseudo-absorption function, consequently [83]. By far, a combination of absorption-based spectroscopic methods, including diffuse reflectance spectroscopy (DRS) on bulk samples or transmission measurements on thin films and coatings and the Tauc method, is the most common procedure of determination of band gap energy [83]. Since absorption-based spectroscopic methods do not confound bulk properties with surface effects, they are preferable for the measurement of bulk properties [83]. The Tauc method was proposed by Tauc et al. as a method for determination of band gap energy by plotting an optical absorption coefficient against energy while evaluating the electronic and optical properties of amorphous germanium.

The strength of optical absorption is dependent on the difference between the energy of photons and the band gap energy as follows [84]:

$$(\alpha h\nu)^{\frac{1}{n}} = A\left(E - E_g\right) \tag{25}$$

where ν is the photon/light frequency, α is the linear absorption coefficient, E_g is the band gap energy, h is the Planck's constant, A is a proportionality constant, and n is a constant that depends on the nature of the electronic transition [84]:

For direct allowed transitions: n = 1/2;
For indirect allowed transitions: n = 2;
For direct forbidden transitions: n = 3/2;
For indirect forbidden transitions: n = 3.

The 'E' term is the incident photon energy and is calculated as follows:

$$E(eV) = h\nu = \frac{hC}{\lambda(nm)} = \frac{1236}{\lambda(nm)} \tag{26}$$

It is notable that the majority of basic absorption processes are allowed transitions (n = 2 or n = 1/2 for indirect and direct transitions, respectively) [84]. Finally, the graph of $(\alpha h\nu)^{\frac{1}{n}}$ vs. $h\nu$ is plotted, and the band gap energy is easily extrapolated by the interception of the linear region to the X-axis. Usually, different values of n (usually 1, and 1/2) are used to identify the correct type of transition. As an example, the Tauc plot of TiO_2 is shown in Figure 3 [85]. Noteworthy, the straight blue line has been plotted to estimate E_g.

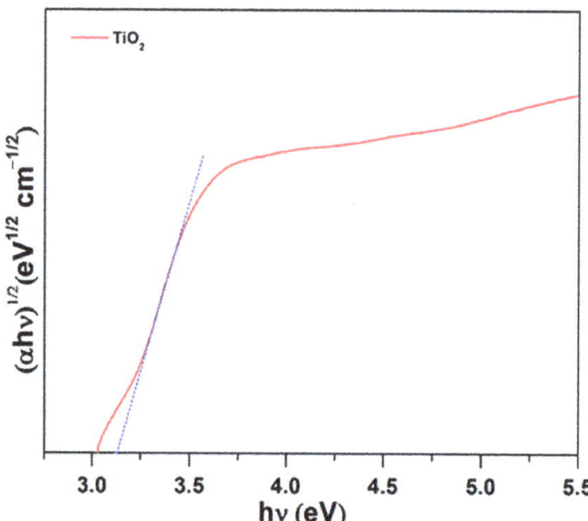

Figure 3. The Tauc plot of TiO_2. Reprinted with permission from Ref. [85]. Copyright 2015, Royal Society of Chemistry.

It is notable that the size of particles could change the band gap energy of semiconductors [55,86]. The planetary model of the Bohr hydrogen atom is generally used for description of the movement of the bonded electron–hole pair originated by photoexcitation. The exciton Bohr radius, a_B, is used for calculation of the region of delocalization of the electron hole pair as follows [86]:

$$a_B = \frac{\hbar^2 c}{e^2}\left[\frac{1}{m_e^* m_O} + \frac{1}{m_h^* m_O}\right] \tag{27}$$

where ε is the dielectric constant of the semiconductor, e is the charge of an electron, $\hbar$ is the reduced Planck's constant, m_e^* is the effective mass of electron, m_h^* is the effective mass of hole, and m_O is the rest mass of an electron. The steric limitation of photoexcited charges is observed in the bulk of a semiconductor whose nanocrystals are smaller than the region of delocalization of exciton. This could change some characteristic features of semiconductors. However, the degree of changes depends on the ratio of nanoparticle radius, R, and the value of a_B. In this case, two different modes could be considered [86]:

$a_B \leq$ R: There is a weak restriction in this region so that a similar electronic structure is observed for both the bulk crystal and the nanoparticle.

$a_B >$ R: There is a strong quantum confinement in the region so that a radical rearrangement is observed for the electronic structure of semiconductor nanoparticle. For example, energy of the exciton excitation increases. Moreover, discrete electronic levels are provided by gradual change in the energy bands of the semiconductor.

This phenomenon, in which the top of VB and the bottom of CB shift in positive and negative directions, respectively, is called a quantum-size effect and could result in the expansion of band gap energy. The Bohr radius of rutile and anatase TiO_2 is around 0.3 and 2.5 nm, respectively. Thus, interpreting the results by using quantum-size effect needs deep evaluations, in cases such as TiO_2, since synthesis of such a small size might be difficult [55]. It is noteworthy that the band gap energy of semiconducting nanomaterials could be highly dependent on their morphology. For instance, titania particles with nanoburrs or lateral organization could have lower band gap energies [87]. Based on the band gap energy, semiconductors are divided into two major categories: wide band gap semiconductors and narrow band gap semiconductors. Band gap energies of some major semiconductors are illustrated in Figure 4 [88]. Among different photocatalysts, TiO_2 is used as the most common semiconductor for photocatalytic applications and is discussed in the following part.

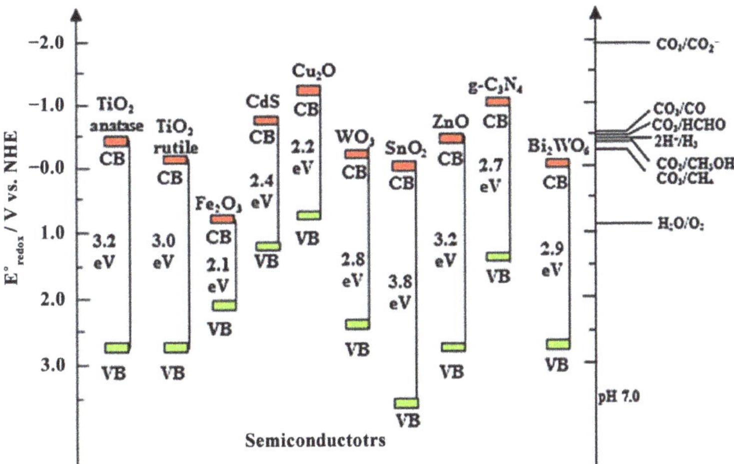

Figure 4. Band gap of major semiconductors (with respect to the redox potential of various chemical species measured at pH 7). Reprinted with permission from Ref. [88]. Copyright 2015, Elsevier.

4. TiO₂

4.1. General Properties and Applications

TiO_2 has found several applications in toothpaste, sunscreens, plastics, printing ink, condensers, paper, electronic components, leather, cosmetics, ceramics, food, and pigmentation since its commercial production [89–91]. High refractive indices of anatase and rutile phases, which lead to high reflectivity from surfaces, takes account for its several applications where white coloration is desired [42,92]. Since the discovery of the photocatalytic water splitting on a titanium oxide electrode under UV illumination by Fujishima and Honda, serious efforts have been allocated to the study of its efficiency in photocatalysis, photovoltaics, sensors, and photoelectrochemical applications, which can be generally divided into two discrete categories of environmental and energy [93,94]. TiO_2 is non-toxic and earth abundant, and has high photostability [94,95]. It is the most common semiconducting material used for photocatalytic applications and has been employed as a benchmark photocatalyst in many research works (especially commercial Degussa P-25 powder, which consists of both anatase and rutile phases) [96–98]. Although there are various semiconducting materials, an overwhelming majority of researchers (around 60%) have focused on different TiO_2 phases in photocatalytic works during the 2000s [99]. The position of the top valence band of TiO_2 is ca. 3 V vs. NHE (at pH 0). Therefore, the potential of its positive holes is remarkably more positive than oxidation potential of ordinary organic compounds. The valence bands of the majority of metal oxides are composed of the same orbitals (O2p). Thus, most of them possess the same potential of the top valence band and the same oxidation ability is expected for all of them [55]. In addition, the surface chemistry of water and oxygen is fundamental to important processes such as the oxygen reduction reaction (ORR) and photocatalytic water oxidation. These reactions are mediated by photoexcited electrons and holes, respectively. Water oxidation is a hole-mediated process as follows [100]:

$$2H_2O + 4h^+ \rightarrow 4H^+ + O_2 \tag{28}$$

The ORR is a process mediated by photoexcited electrons that is followed by reactions with coadsorbed water as follows [100]:

$$O_2 + e^- \rightarrow (O_2)^- \tag{29}$$

$$OOH^- + e^- \rightarrow OH^- + (O_{ad})^- \tag{30}$$

$$(O_{ad})^- + e^- + H_2O \rightarrow 2OH^- \tag{31}$$

$$H_2O + OOH^- \rightarrow OH^- + H_2O_2 \tag{32}$$

$$(O_2)^- + e^- + H_2O \rightarrow OH^- + OOH^- \tag{33}$$

$$H_2O_2 + 2e^- \rightarrow 2OH^- \tag{34}$$

Although both the ORR and water oxidation involve the same intermediate species (H_2O_2, OH^-, or OOH^-), the identification of the reaction intermediates is not simple (since they are usually short-lived) [100]. To illustrate this point, characteristic times for different photo-reactions in the heterogeneous photocatalysis on TiO_2 are shown in Table 1 [101].

Therefore, one of the probable reasons for high photocatalytic activity of TiO_2 could be its high reduction potential that is responsible for the injection of photogenerated electrons into the molecular O_2 on its surface [55]. Notably, equal numbers of positive holes and electrons must be consumed for completion of photocatalytic reactions. Thus, positive holes cannot be used for oxidation reactions, despite of their high potentials, unless the photogenerated electrons are consumed simultaneously. Most other metal oxides have high oxidation ability; however, their low reduction ability may result in their lower photocatalytic activity compared with TiO_2 [55].

Table 1. Characteristic times for heterogeneous photocatalysis on TiO_2: where e_{cb}^- is the conduction band electron, h_{vb}^+ is the valence band hole, e_{tr}^- is the trapped conduction band electron, $>TiOH$ shows the primary hydrated surface functionality of titania, $>Ti^{IV}OH\bullet+$ is the surface-bound hydroxyl radical, $>Ti^{III}OH$ is the surface-trapped conduction band electron, Red is the reductant, and OX is the oxidant [101].

Type of Primary Reaction	Primary Reaction	Characteristic Time (s)
Generation of charge carriers	$TiO_2 + h\upsilon \rightarrow e_{cb}^- + h_{vb}^+$	fs
Charge carrier trapping	$H_{vb}^+ + > Ti^{IV}OH \rightarrow > \{Ti^{IV}OH\}^+$	10 ns
• Deep trap (irreversible)	$e_{cb}^- + > Ti^{IV} \rightarrow >Ti^{III}$	10 ns
• Shallow trap (dynamic equilibrium)	$e_{cb}^- + > Ti^{IV}OH \leftrightarrow \{>Ti^{III}OH\}$	100 ps
Charge carrier recombination	$h_{vb}^+ + \{> Ti^{III}OH\} \rightarrow >Ti^{IV}OH$	10 ns
	$e_{cb}^- + > \{Ti^{IV}OH^\bullet\}^+ \rightarrow >Ti^{IV}OH$	100 ns
Interfacial charge transfer	$\{>Ti^{IV}OH^\bullet\}^+ + Red \rightarrow > Ti^{IV}OH + Red^{\bullet+}$	100 ns
	$e_{tr}^- + OX \rightarrow Ti^{IV}OH + OX^{\bullet-}$	ms

4.2. Optical and Electrical Properties of TiO_2

In general, rutile TiO_2 is more anisotropic than anatase TiO_2 in the range from infrared to visible spectra. On the other hand, an important anisotropy is observed for anatase phase in the band gap region [102]. Since titania is known as an n-type semiconductor, donor-type defects (e.g., titanium interstitials and oxygen vacancies) are responsible for its conductivity [103]. The excessive electrons in the solid, introduced by oxygen vacancies, could enhance the electrical conductivity [102]. However, oxidized titania may show p-type properties as a result of concurrent presence of acceptor-type defects including titanium vacancies. $Ti_{1-x}O_2$ or TiO_{2+x} could be applied, instead of TiO_2, where titanium vacancies could make a remarkable contribution to the defect disorder (under prolonged oxidation conditions) [103]. Doping is an efficient method that could be used to form both n-type and p-type titania. For example, manganese-doped TiO_2 has shown p-type electrical conduction [104,105]. In addition to manganese, iron, nickel, chromium and cobalt could act as electron acceptors. Thus, Fe-doped [106,107], Ni-doped [108], Cr-doped [109–111] and Co-doped [112,113] TiO_2 could show p-type electrical conduction. The ratio between the concentration of metal dopants and that of oxygen vacancies could be crucial to the reduction or augment of the electrical conductivity. In addition to the concentration of oxygen vacancies, their contribution should also be taken into account for evaluation of the electrical conductivity [102]. On the other hand, niobium and tantalum could act as electron donors. Therefore, W-doped [114], Ta-doped [115], and Nb-doped [116,117] TiO_2 could show n-type electrical conduction.

Various non-metal elements including N, P, S, B, and I have been extensively employed to drive visible-light-activated TiO_2. Among those, N-doping has received remarkable attention as a method efficiently used for driving visible-light activated TiO_2 [118–122]. It has been shown that nitrogen-doping could remarkably reduce the formation energy of oxygen vacancies as confirmed by DFT calculations. Thus, N-doped TiO_2 can take advantage of the presence of oxygen vacancies (especially on its surface) [123]. By far, various methods have been used for the synthesis of N-doped TiO_2 including sputtering, sol-gel, chemical vapor deposition, decomposition of nitrogen-containing metal-organic precursors, spray pyrolysis, pulsed laser deposition, combustion reaction, high-energy milling, and implantation. Schematic illustrations of models for doping of different nitrogen species including substitutional N-doping, interstitial N-doping, substitutional NO-doping, substitutional NO_2-doping, and interstitial NO-doping are shown in Figure 5.

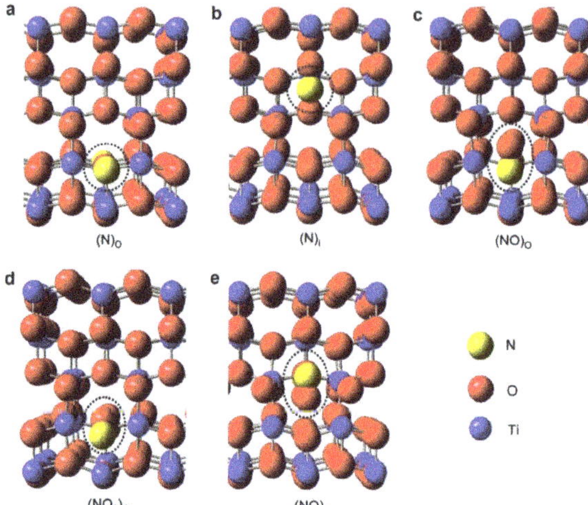

Figure 5. Schematic models for (**a**) substitutional N-doping, (**b**) interstitial N-doping, (**c**) substitutional NO-doping, (**d**) substitutional NO$_2$-doping, and (**e**) interstitial NO-doping. Reprinted with permission from Ref. [124]. Copyright 2007, Elsevier.

Thus, nitrogen doping into a TiO$_2$ lattice structure could be either interstitial or substitutional. Peng et al. synthesized both interstitial and substitutional N-doped TiO$_2$ for photocatalytic degradation of methyl orange and phenol. They suggested that although both interstitial and substitutional nitrogen doping could drive visible-light activated TiO$_2$, interstitial N-doped TiO$_2$ could show a higher photocatalytic activity than that of substitutional N-doped TiO$_2$ [125]. Zeng et al. compared the photocatalytic activity of visible-light activated interstitial and substitutional N-doped TiO$_2$ toward degradation of benzene under the same conditions (similar grain size, crystallinity, and specific surface area of the as-prepared samples) [126]. Unlike Peng et al., they concluded that substitutional N-doped TiO$_2$ could show higher photocatalytic activity than that of interstitial N-doped TiO$_2$. Lower recombination rate of the photogenerated e$^-$/h$^+$ pairs and higher proportion of surface hydroxyl groups have accounted for the higher photocatalytic activity of substitutional N-doped TiO$_2$. The higher photocatalytic activity of interstitial N-doped TiO$_2$ than that of substitutional N-doped TiO$_2$, as reported by Peng et al. [125], could originate from different methods used by Peng et al. for the synthesis of interstitial and substitutional N-doped TiO$_2$ [126]. It should be noted that the photocatalytic activity of semiconductors could be highly influenced by their grain/particle size [127–129], specific surface area [130–132], and crystallinity/polymorph [133–135]. Of note, nitrogen doping could be achieved through simultaneous interstitial and substitutional doping as confirmed by XPS analysis [136]. In substitutional doping, either oxygen atoms or titanium atoms can be substituted by nitrogen atoms. Therefore, TiO$_{2-x}$N$_x$ (titanium oxynitride) and Ti$_{1-y}$O$_{2-x}$N$_{x+y}$ are the results of the substitution of oxygen atoms and substitution of both oxygen and titanium atoms by nitrogen atoms, respectively. As suggested by Peng et al., Ti$_{1-y}$O$_{2-x}$N$_{x+y}$ could be formed at high concentrations of nitrogen (as high as 21 mol%) [137]. Valentin et al. suggested that nitrogen doping is likely accompanied by the formation of oxygen vacancies. Besides, they proposed that the abundance of nitrogen-doping species is dependent on the preparation conditions including the annealing temperature and the oxygen concentration (in the atmosphere) [138]. As for the substitutional N-doped TiO$_2$ (N-Ti-O and Ti-O-N), localized N 2p states (slightly above the VB of pure TiO$_2$) could be responsible for the visible-light response [139]. Migration of photo-excited electrons from VB to the N 2p states could reduce the band gap energy of TiO$_2$ [136]. Notably, NO bond with π-character that generates new localized states (slightly above the VB of pure TiO$_2$) accounts for the visible-light response of interstitial N-doped TiO_2 [139]. Valentin et al. suggested that the highest localized states for the interstitial and substitutional nitrogen species are 0.73 and 0.14 eV above the top of the valence band, respectively [140]. Not only could interstitial and substitutional nitrogen doping increase the visible-light harvest, but it could also intensify the absorption of UV-light, which originates from the appearance of defect energy levels in the band gap [126].

It has been shown that the photocatalytic efficiency of titania could be strongly correlated to its (micro) structure and semiconducting properties [103,141]. Additionally, the electrical properties of titania nanostructures depend on the crystallographic directions [102]. It is why the evaluation of the electrical properties of TiO_2 is of high importance for photocatalytic applications, as shown in Table 2.

Table 2. Comparison of the crystal structural, optical and electrical properties for TiO_2 nanostructures. Adapted with permission from Ref. [102]. Copyright 2018, Royal Society of Chemistry. This article is licensed under a Creative Commons Attribution-NonCommercial 3.0 Unported Licence (https://creativecommons.org/licenses/by-nc/3.0/) (Accessed on 1 December 2022).

Properties	TiO_2 Nanostructures		
	Rutile	Anatase	Brookite
Crystal Structure	Tetragonal	Tetragonal	Orthorhombic
Lattice constant (A)	a = 4.5936 c =2.9587	a = 3.784 c = 9.515	a = 9.184 b = 5.447 c = 5.154
Molecule (cell)	2	2	4
Volume/molecule ($A°3$)	31.21	34.061	32.172
Density ($g\ cm^{-3}$)	4.13	3.79	3.99
Ti–O bond length ($\overset{\circ}{A}$)	1.949 (4) 1.980 (2)	1.937 (4) 1.965 (2)	1.87–2.04
O–Ti–O bond angle	$81.2°$ $90°$	$77.7°$ $92.6°$	$77-105°$
Band gap at 10 K	3.051 eV	3.46 eV	
Static dielectric constant (ε_0, in MHz range)	173	48	
High frequency dielectric constant, ε_∞ ($\lambda = 600$ nm)	8.35	6.25	

When defects are electrically charged, defect disorder could largely affect the electrical properties. There are several reports on the investigation of the electrical properties of rutile TiO_2 at high temperatures and in nanocrystalline form [142–146]. Although evaluation of the electrical properties of anatase TiO_2 is of interest, several experimental problems, owing to its grain growth and partial phase transformation into rutile at elevated temperatures, has prevented its widespread studies [141]. The electrical properties of oxide semiconductors including titania are generally temperature dependent. This dependency is generally considered by the activation energy of electrical conductivity [103,147]. It is notable that the oxygen partial pressure could play a crucial role in the activation energy of electrical conductivity of single-crystal (SC) and polycrystalline (PC) TiO_2 [147]. The electrical conductivity of metals and n-type semiconductors including TiO_2 is the reciprocal of their resistivity and could be expressed as follows [147,148]:

$$\sigma = \frac{1}{\rho} = en\mu_n \tag{35}$$

in which σ is the electrical conductivity, ρ is the resistivity, n is the concentration of electrons, μ_n is the mobility of electrons, and e is the elementary charge [147]. Based on the four probe method, sample dimensions are required to determine the resistivity and/or electrical conductivity as follows [148]:

$$\rho = R\frac{L}{A} \tag{36}$$

where R, A, and L are resistance, surface area of the cross section, and distance between voltage electrodes, respectively [148]. The mobility is defined as follows [149]:

$$\mu_n = \frac{v}{\varepsilon} \tag{37}$$

The mobility is usually determined by scattering with photons (in an intrinsic semiconductor). The mobility's of electrons (μ_n) for some semiconductors such as Si, Ge, GaAs, GaN, InSb, InAs, InP, and ZnO are 1300, 4500, 8800, 300, 77,000, 33,000, 4600, and 230 cm^2/Vs, respectively, and the mobilities of holes (μ_p) for the same semiconductors are 500, 3500, 400, 180, 750, 460, 150, and 8 cm^2/Vs, respectively. Usually, semiconductors possess a much higher mobility than that of metals including Cu (i.e., 35 cm^2/Vs) [149]. In the case of TiO_2, the slope of log σ versus log p (O_2), which is dependent on the oxygen activity, is $-1/6$ and $-1/4$ in highly reduced conditions ($p(O_2) < 10^{-5}$ Pa) and in oxidized conditions ($p(O_2) > 10$ Pa), respectively (at elevated temperatures).

4.3. Promising Phases of TiO₂ for Photocatalytic Applications

Titanium dioxide, which is an n-type semiconductor owing to the presence of oxygen vacancies, crystalizes naturally in three major different phases including anatase (tetragonal), brookite (orthorhombic) and rutile (tetragonal). Among all TiO_2 polymorphs, anatase and rutile phases are generally used in photocatalytic applications [150]. However, brookite TiO_2 could also find applications in photocatalysis [151–154]. It has shown a higher photocatalytic activity than that of rutile TiO_2 in some cases. For instance, hydrothermally synthesized brookite TiO_2 has exhibited superior photocatalytic activity than that of hydrothermally synthesized rutile TiO_2 in degradation of some hazardous pharmaceuticals (ibuprofen, diatrizoic acid, and cinnamic acid) and phenol as model organic pollutant. The higher efficiency of the brookite phase (in this case study) has been devoted to the existence of highly active species, superoxide radicals and valence band holes, in the photocatalytic treatment [155]. In another study, titania nanoparticles (including pure anatase, anatase-rich, pure brookite, and brookite-rich phases) have been synthesized by Kandiel et al. for photocatalytic purposes.

Interestingly, higher photocatalytic activity was observed by using as-prepared brookite titania nanoparticles compared with as-prepared anatase TiO_2 nanoparticles. The surface area of the latter has been three times higher than that of the former. The higher photocatalytic efficiency of brookite titania nanoparticles for methanol photooxidation has been devoted to their higher crystallinity and/or to the cathodic shift of the CB of brookite TiO_2 compared with that of anatase TiO_2, which could facilitate the transfer of interfacial electrons to the molecular oxygen [156]. Although anatase and rutile TiO_2 are the most common titania polymorphs used for photocatalytic applications, it has been shown that brookite TiO_2 could exhibit markedly high photocatalytic activity, even higher than the anatase phase, in some cases (e.g., for degradation of ibuprofen, cinnamic acid, and phenol, and photooxidation of methanol) [152,156]. However, brookite titania has not received great attention due to some limitations such as its difficult synthesis procedure to generate high purities [157]. Of note, rutile TiO_2 is thermodynamically stable at ambient conditions [156] while anatase is a meta-stable phase. The band gap energy of rutile phase is lower than that of the anatase phase with the band gap energy of ~3.2 eV [158]. The phase transition between titania polymorphs depends on several factors. For instance, particle size, surface area, heating rate, particle shape (aspect ratio), volume of sample, atmosphere, measurement technique, soaking time, nature of sample container, and impurities are among various factors that affect the phase transformation of undoped anatase to rutile TiO_2 [92]. The transition of anatase to rutile is a reconstructive process in which the Ti-O bonds are ruptured. Afterwards, a structural rearrangement and formation of new Ti-O bonds occur which result in the formation of rutile phase. It is worth mentioning that the phase transformation of anatase to rutile is a nucleation and growth process [92]. However, it has been shown that this transition is limited by nucleation stage and not by growth. It should be considered that when it comes to the phase transition of nanocrystalline anatase, both surface and interface nucleation processes could be observed though at varying rates. The experiments have confirmed that interface nucleation is more rapid than surface nucleation [159]. Usually, rutile phase nucleates at (112) twin interfaces, in pure anatase, because of similar structure of these sites to rutile phase [92]. Upon heating, both meta-stable brookite and anatase phases could transform to rutile phase, irreversibly, though at different temperatures [92,158,160]. Notably, this transformation could take place at different temperatures and various attempts have been made to promote/inhibit it [92]. When it comes to the pure synthetic TiO_2, it could occur between 600–700 °C [160]. According to the report of Wu et al. [161], the effect of temperature on the phase stability and crystallite size of TiO_2 is shown in Figure 6.

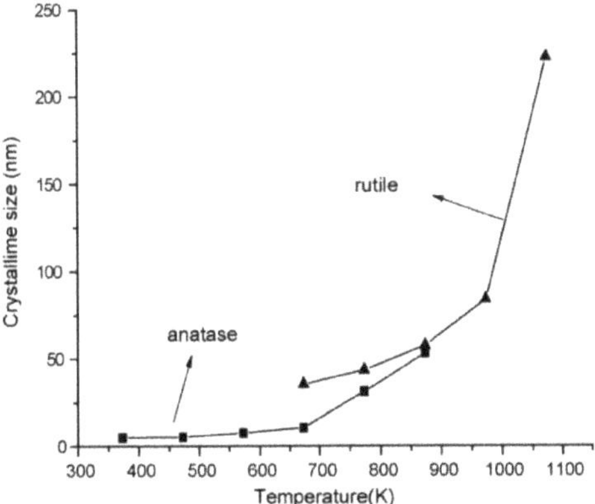

Figure 6. Effect of thermal treatment temperature on crystallite size of TiO_2. Reprinted with permission from Ref. [161]. Copyright 2007, Elsevier.

As evident, an increase in the temperature could result in an increase in the crystallite size of both anatase and rutile TiO_2 (though at different degrees). Although the transformation of anatase to rutile has been widely studied, the transformation of brookite to anatase or vice versa is a bit controversial. The transformation of anatase to rutile could be carried out either directly or indirectly. However, some reports suggest that the brookite phase transforms to the anatase phase at first. Then, anatase TiO_2 transforms to rutile TiO_2. On the other hand, there are some reports suggesting that the anatase phase transforms to the brookite and/or rutile phases prior to the transformation of brookite to rutile [156].

Accordingly, the anatase and brookite phases are not stable at large crystallite sizes. In another words, the most thermodynamically stable phases of TiO_2 are rutile, brookite, and anatase, respectively (by increasing the crystallite sizes). Thermodynamic studies have also confirmed the higher stability of the rutile phase than the anatase phase at all positive temperatures and pressures. The results are shown in Figure 7.

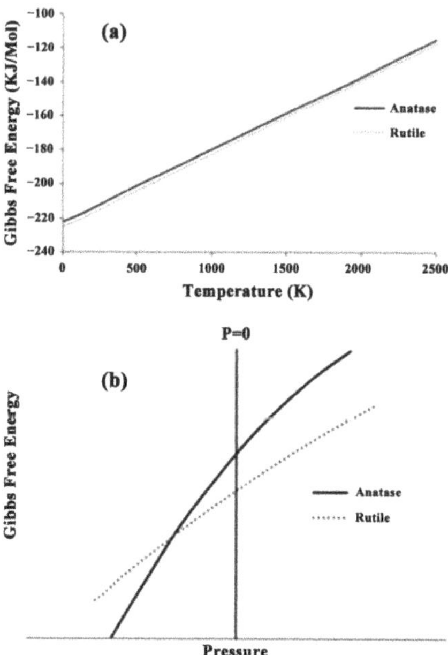

Figure 7. Plots of Gibbs free energy of anatase and rutile versus (**a**) temperature and (**b**) pressure (at room temperature). Reprinted with permission from Ref. [92]. Copyright 2011, Springer.

Bulk and/or surface doping of TiO_2 could affect the stability of anatase and rutile phases and change the temperature of the transition of anatase to rutile, which has been explained by a ceased crystal growth as the result of the formation of Ti-O-M bonds. For instance, doping with metal cations including Nd [162], La [163], Ni [164], Si, W, Cr [165], Al, Nb, Ga, and Ta [166] could retard the grain/particle growth of anatase and rutile phases and delays the transition of anatase to rutile. The addition of molybdenum and tungsten could completely eliminate the anatase phase at temperatures between 680 and 830 °C, respectively [102]. Doping with V, Ag, Mn [165], Zn, Fe and Cu [166] could generally promote this transformation. However, there are some dual behaviors which should be considered. For example, Wang et al. have reported the acceleration of anatase to rutile phase transition for Nb-doped TiO_2. It is while a depression effect has been observed on the grains growth of anatase for Nb-doped TiO_2 [165]. Using cationic/anionic dopants [167] and non-metal dopants, including carbon [168], nitrogen [169], fluorine [170,171] and chlorine [171] could affect the phase transformation of titania. Unlike interstitial atoms that could increase the strain energy that must be overcome and the bonds that must be broken, the creation of oxygen vacancies, originated from the incorporation of ions with relatively small ionic radius and lower valence than that of Ti^{4+}, could facilitate the phase transformation by reduction of the strain energy needed for the rearrangement of the Ti-O octahedra. The rate of the transition of anatase to rutile strongly depends on the incorporated dopants [172].

4.4. Photocatalytic Activity of Anatase Titania Compared with Its Other Polymorphs

It is generally believed that among the anatase, rutile, and brookite phases, the anatase phase is responsible for photocatalytic reactions. There are some statements that a pure rutile phase could not have any photocatalytic activity at all. However, some believe that it could possess photocatalytic activity, which depends on the nature of the organic reactant, preparation method of the photocatalyst and the nature of the precursor materials [150]. Considering pure phases, the anatase phase could provide higher photocatalytic activity than rutile TiO_2 [46,98]. The higher photocatalytic activity of anatase phase than other polymorphs could be attributed to its higher mobility of charge pairs and density of surface hydroxyl [46]. Notably, discrepancy with the recombination kinetics of e^-/h^+ pairs might also play an important role in the superior photocatalytic efficiency of anatase phase as photoluminescence (PL) studies for anatase TiO_2 have clarified its lower recombination rate of charge carriers than that of rutile phase under similar condition [173]. Brookite and rutile TiO_2 are direct band gap semiconductors, while anatase TiO_2 belongs to the category of indirect band gap semiconductors [174]. For indirect band gap semiconductors, the electronic transition from the VB to the CB is electrical dipole forbidden. Under these circumstances, the transition results in the change of both momentum and energy of the photogenerated e^-/h^+ pairs (transition is phonon assisted). On the other hand, the valence band to the conduction band electronic transition is electrical dipole allowed for direct band gap semiconductors. Due to the momentum change, the electronic absorption/emission of indirect band gap semiconductors is weaker than that of direct band gap semiconductors [175]. Thus, direct band gap semiconductors, such as rutile TiO_2, have more efficient absorption of solar energy than indirect ones (such as anatase TiO_2) [174,175]. However, a longer diffusion length as well as lifetime of photogenerated e^-/h^+ pairs could result in better photocatalytic performance of anatase TiO_2 than its rutile phase [174]. In addition, it has been reported that the anatase phase is more active than the rutile phase in absorption of hydroxyl groups and water [150] that could play an important role in photocatalysis. A schematic of recombination processes of photoexcited e^-/h^+ pairs within anatase TiO_2 (indirect gap) and rutile TiO_2 (direct gap) is shown in Figure 8 [174].

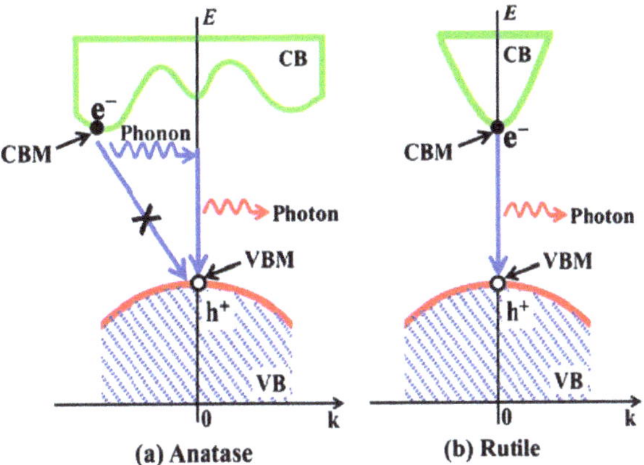

Figure 8. Schematic of recombination processes of photoexcited e^-/h^+ pairs within (**a**) anatase TiO_2 and (**b**) rutile TiO_2. Reprinted with permission from Ref. [174]. Copyright 2014, Royal Society of Chemistry.

In addition to the band gap nature, the diffusion of free charge carriers is of remarkable importance. To achieve appropriate photocatalytic efficiency, the photoexcited e^-/h^+ pairs should migrate to the active sites where they are used prior to the recombination process. In this context, the diffusion of photogenerated e^-/h^+ pairs is strongly linked to their mobility, which is in turn related to their effective mass. The ratio of the effective mass of electrons to effective mass of holes is used to predict the stability of photogenerated e^-/h^+ pairs with respect to their recombination. In this case, a larger effective mass difference could enhance the photocatalytic activity as the result of the reduction of the e^-/h^+ pair recombination [176]. In general, the transfer rate of photoexcited e^-/h^+ pairs that affects the quantum efficiency of photocatalysts could be assessed by the effective mass of electrons and holes [174].

As evident, there is an inverse relationship between the transfer rate of photoexcited e^-/h^+ pairs and the effective mass of charge carriers. To sum up, the smaller the effective mass of charge carriers, the faster the transfer rate of charge carriers. Anatase TiO_2 has a lighter average effective mass of photoexcited e^-/h^+ pairs than that of both the brookite and rutile phases, which leads to the faster migration of e^-/h^+ pairs from bulk titania to its surface. This results in its lower recombination rate of photoexcited e^-/h^+ pairs [174]. It should be noted that there is not yet consensus on the reasons for higher photocatalytic activity of the anatase phase than that of the rutile and brookite phases [177]. It is worth mentioning that there are various factors, including method of synthesis, concentration of defects, purity of phases, surface crystallographic orientation, particle size, dopants, specific surface area, and its crystal structure (type of polymorph), that could remarkably affect the photocatalytic activity of TiO_2 [46,98].

4.5. Colorful TiO_2 versus White TiO_2

Recently, colored titania materials such as green, red, yellow, blue, brown and grey of various shades have been developed. The colorful titania could be synthesized by electronic and structural changes such as the incorporation of Ti-OH and Ti-H species, formation of oxygen vacancies, formation of disordered surface layers, reduction of Ti^{4+} to Ti^{3+}, narrowed band gap energy, and the modified electron density [178]. These could result in the synthesis of green, red, black, and grey materials [179]. Black TiO_2, synthesized by Chen et al. using high pressure hydrogenation, has shown a largely narrowed band gap energy (for massive visible light absorption) and an enhanced photocatalytic activity. Since then, it has received remarkable attention in visible light utilization. Notably, the optical absorption of black titania could be extended to the IR region (approximately 1150 nm) [178]. The importance of black titania has resulted in its various applications in photocatalytic water splitting, photoelectron chemical water splitting, photocatalysis, fuel cells, Li-ion batteries, Na-ion batteries, Al-ion batteries, surface enhanced Raman active scattering substrate, supercapacitors, cancer photothermal therapy, solar desalination, microwave absorption, field emission, and dye sensitized solar cells (DSSC) [180,181]. Various methods, such as low pressure hydrogen treatment, high pressure hydrogen treatment, argon treatment, hydrogen–nitrogen treatment, hydrogen–argon treatment, hydrogen–plasma treatment, chemical oxidation, chemical reduction, electrochemical reduction, pulsed laser ablation, and hydroxylation, have been used to synthesize black titania nanomaterials [180].

Black titania, synthesized by hydrogenation, was firstly used for photocatalytic degradation of methylene blue and phenol [182]. Since then, photodegradation of methylene blue, as a model of organic pollutants, has been studied by many researchers using this polymorph of titania [182–184]. It is worth mentioning that black TiO_2 has shown a higher photocatalytic activity than that of white TiO_2 in several cases [184,185].

It has been stated that the formation of some surface defects, including oxygen vacancies and/or Ti^{3+} oxidation states, and disordered layers in the surface of highly crystalline TiO_2, could increase the interaction to the light irradiation and the lifetime of photogenerated electron/hole pairs [184]. However, Ti^{3+} ions have not been observed in some samples of black TiO_2. Thus, oxygen vacancies could be known as the main factor responsible for black coloration of titania [186]. The superior photocatalytic activity and/or photo electrochemical properties of black TiO_2 than pristine TiO_2 or P25 have also been reported by other researchers some of which are given in Table 3.

Table 3. Superior photocatalytic and photoelectrochemical properties of black TiO_2 than white TiO_2 in some case studies.

Synthesis Method	Light Source	Improvement of Photocatalytic Activity	Improvement of Photoelectrochemical Properties	References
Melted aluminum reduction of pristine anodized and air-annealed TiO_2 nanotube arrays	The simulated sunlight (intensity of 100 mW cm^{-2})	-	Approximately 5 times higher than pristine TiO_2 nanotube arrays	[187]
Electrospinning process	A 150 W xenon lamp	-	Approximately a 10-fold increase compared with pristine TiO_2 nanofibers	[188]
In situ plasma hydration of TiO_2 thin films	A 150 W xenon lamp (intensity of 100 mW cm^{-2})	-	Approximately 2.5 times higher than pristine TiO_2 thin films	[189]
Electrochemical reductive doping	• UV source: A UV lamp (intensity of 5.8 mW cm^{-2}) • Visible-light source: A xenon lamp (intensity of 100 mW cm^{-2})	-	Approximately 2.2 times higher than pristine anodic TiO_2 nanotubes (under both UV and simulated solar irradiation)	[190]
Using Ti_2O_3 as precursor for preparing Ti^{3+} self-doped TiO_2 nanowires	A 20 W UV lamp	Approximately 7.5 times higher than pure TiO_2 (P25) in photodegradation of methyl orange	-	[191]
hydrogen plasma assisted chemical vapour deposition	A 50 W simulated solar light source	Complete photodegradation of rhodamine B after approximately 30 min against partial photodegradation of rhodamine B even after 50 min for pure TiO_2	-	[192]
Annealing the TiO_2 nanobelts in hydrogen atmosphere	• UV source: A 350 W mercury lamp • Visible-light source: A 300 W xenon arc lamp	• UV illumination: An approximate increase of 24% compared with pristine TiO_2 nanobelts in decomposing methyl orange • Visible light irradiation: An approximate increase of 17% compared with pristine TiO_2 nanobelts in decomposing methyl orange	-	[193]
Annealing the TiO_2 nanobelts in hydrogen atmosphere	A 300 W xenon arc lamp	-	Approximately 9.2 times higher than pristine TiO_2 nanobelts	[193]

Oxygen vacancies are among the most important/common point defects originated by removal of some neutral oxygen atoms from the lattice structure of metal oxides including TiO_2 [194]. However, the major defect could become tetravalent titanium interstitials under extremely reduced conditions that are experimentally difficult to achieve [103]. In general, the driving force for the creation of vacancies is minimizing the Helmholtz free energy to establish thermodynamic equilibrium at a specific temperature. Cations and anions are two different types of vacancy defects in ionic crystals that lead to the formation of localized energy levels over the VB maximum and below the CB minimum, respectively [194]. The formation of oxygen vacancies could result in the development of Ti^{3+} species (by filling the empty states of titanium ions using excesses electrons originated from removal of oxygen atoms) [195,196] and the formation of shallow donor states under the CB of titania originating from Ti 3d orbits [195]. Owing to the higher active surface area of nanomaterials than that of bulk materials, a higher number of oxygen vacancies could be stimulated in the surface and subsurface regions of nanostructured titania than bulk titania. In addition, titania nanocrystals have less formation energy of oxygen vacancies than their bulk counterpart, which is why the creation of oxygen vacancies and evaluation of their effect in TiO_2 nanomaterials have received greater attention. Oxygen vacancies possess the minimum formation energy among the defects that could act as donors. Therefore, they have received great attention in defining chemical and physical properties of materials (including superconductivity, ferromagnetism, photocatalysis, resistive switching, phase transitions, redox activity, piezoelectric response, and photoelectrochemical performance) [194]. Hydrogen thermal treatment, high energy particle bombardment, doping of metal or non-metal ions, thermal treatment under oxygen depleted conditions, and some special reaction conditions are common methods used for development of TiO_2 with oxygen vacancies [195]. Thermal treatment under oxygen deficient conditions could be considered as an efficient method for creation of oxygen vacancies in TiO_2-based nanomaterials [194]. This process is performed at elevated temperatures, usually > 400 °C, in vacuum or pure Ar, He, and N_2 gas atmosphere [195].

The decrease of oxygen pressure results in an increase of the concentration of oxygen vacancies; hence, thermal treatment of titania under oxygen deficient conditions could facilitate the creation of oxygen vacancies. Therefore, the exposure of titania to air leads to the gradual removal of oxygen vacancies [195]. Doping with accept-type foreign ions, including Fe and Zn, could stabilize the as-formed oxygen vacancies [95,195]. Oxygen vacancies on anatase phase and rutile phase have been widely studied since they could be engineered/generated at mild conditions [196]. The calculation of the formation energy of oxygen vacancies has confirmed that creation of oxygen vacancies at the top surface of anatase (101) and (001) planes is more probable than other planes (because of their lower formation energies of vacancies). In comparison, rutile (110) planes have lower vacancy formation energy than that of other rutile crystalline planes [194]. Oxygen vacancies are capable of introducing localized states into the band gap structure (instead of the change of positions of valence band and conduction band). The driving force for development of these localized states is the Madelung potential of highly ionic crystal [195]. Transition of electrons from the oxygen vacancies to the valence band and/or from the valence band to the oxygen vacancies leads to infrared- and visible-light absorption [185,186]. Additionally, they act as potential shallow traps for significant improvement of the efficiency of electron–hole separation [184,186]. The energy level position of oxygen vacancies, which form a donor level under the conduction band of TiO_2, is from 0.75 to 1.18 eV [185,186,195]. It is worth mentioning that experimental and theoretical studies show that the excess electrons, originated from the formation of the oxygen vacancies, could affect the reactivity and surface absorption of important adsorbents including H_2O and O_2 on titania. These are some reasons why development of TiO_2 with the desired amount of oxygen vacancies is of great importance [195].

In addition to black TiO_2, other colored titania materials (including red, green, blue, yellow, and various shades of grey titania) have also found applications in photocatalysis and photoelectrochemical water splitting due to their specific features. For instance, Liu et al. have synthesized anatase titania microspheres with an interstitial $B^{\sigma+}$ ($\sigma \leq 3$) gradient shell, red TiO_2, that are capable of absorbing the full visible light spectrum. UV-visible absorption of the white TiO_2 and these red TiO_2 microspheres are compared in Figure 9 (black and red graphs, respectively) [197]. Noteworthy, the colorful region represents the visible light spectrum ranged from 400 to 700 nm.

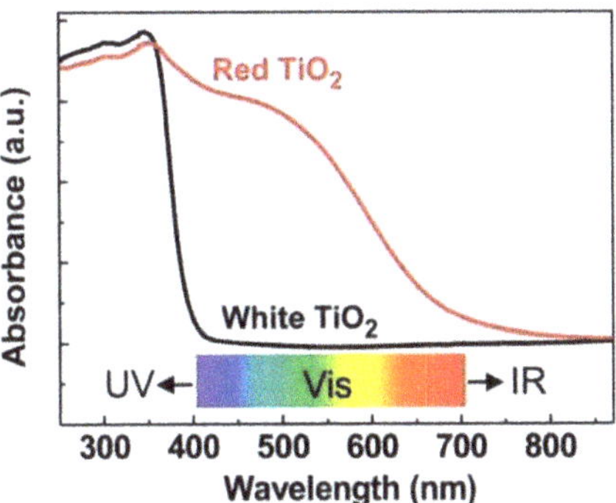

Figure 9. Comparison of UV-visible absorption of the white TiO_2 and red TiO_2 microspheres. Reprinted with permission from Ref. [197]. Copyright 2012, Royal Society of Chemistry.

As evident, red titania microspheres possess much higher photo absorption ability than white titania in the visible light range. These microspheres have provided a gradient of absorption band gap energy from 1.94 eV (on their surface) to 3.22 eV (in their core) [179,197]. It has also been reported that red anatase TiO_2-based materials could show an unusual visible-light absorption that is representative of new types of visible light absorption bands of titania [198]. Thus, red titania-based photoanodes could be used for photoelectrochemical water splitting under visible-light irradiation [197,198]. Some other colorful titania-based photoanodes, including blue TiO_2, could be inactive in photoelectrochemical water oxidation under visible-light irradiation [198].

5. Conclusions and Perspectives

The application of TiO_2 semiconductor nanoparticles for the photocatalytic decomposition of pollutants in wastewater has been considered in this study. Considering that TiO_2 has three different structures (anatase, rutile and brookite), their stability and photocatalytic efficiency were compared. It should be noted that the band gap energy of these three crystal structures is different. In addition to the effect of band gap energy, the photocatalytic activity is influenced by the degree of stability and separation of the produced electron–hole pairs which could be dependent on the mobility as well as the mass ratio of the electron to the hole. Among the existing phases, the rutile phase has relatively high stability at different temperatures and pressures. The photocatalytic activity of titanium depends on its structure and morphology. Investigations show that the photocatalytic activity of anatase phase is higher than that of rutile and brookite. Although some studies indicate that the rutile phase does not have any photocatalytic activity, studies show that the rutile phase can be used for photocatalytic processes, which depends on the structure of the pollutant, the preparation method, and the precursor material of the photocatalyst. In general, the photocatalytic activity depends on the stability of the produced electron–hole pairs. Increasing the stability and separation of the created charges leads to a decrease in their recombination and an increase in the photocatalytic efficiency. Studies show that the photocatalytic activity of titania is also dependent on its color. Electronic and structural changes such as the combination of Ti-OH and Ti-H species, the formation of oxygen vacancies, the formation of irregular surface layers, and the reduction of Ti^{4+} to Ti^{3+} are among the effective factors in the formation of colored titania. Black titania has a largely narrowed band gap energy which leads to the enhancement of photocatalytic activity.

Due to their stability and adverse effects on human health and environment, the application of colorful TiO_2 in the degradation of persistent organic pollutants is worth studying. It should be noted that photocatalysis cannot find practical applications without immobilization of catalysts, due to the several issues encountered by using nanoparticles including agglomeration and difficulty of recovery. Moreover, immobilization could facilitate the application of photoelectrocatalysis with its superior efficiency than custom photocatalysis. Hence, the application of surface engineering methods in the deposition of colorful TiO$_2$ films for photocatalytic degradation of persistent organic pollutants is of high importance in future studies.

Author Contributions: A.H.N.: conceptualization, writing—original draft, writing—review and editing; S.A.: writing—original draft, D.L.: writing—review and editing; A.M.: writing—review and editing; J.L.Z.: supervision, writing—review and editing. All authors have read and agreed to the published version of the manuscript.

Funding: This research received no external funding.

Data Availability Statement: Not applicable.

Acknowledgments: This work is supported by the University of Technology Sydney. The author Sedigheh Abbasi is grateful for the support by the Esfarayen University of Technology.

Conflicts of Interest: All authors undertake that they have no conflict of interest.

References

1. Liu, J.; Yang, H.; Gosling, S.N.; Kummu, M.; Flörke, M.; Pfister, S.; Hanasaki, N.; Wada, Y.; Zhang, X.; Zheng, C.; et al. Water scarcity assessments in the past, present, and future. *Earth's Future* **2017**, *5*, 545–559. [CrossRef] [PubMed]
2. Ong, C.B.; Ng, L.Y.; Mohammad, A.W. A review of ZnO nanoparticles as solar photocatalysts: Synthesis, mechanisms and applications. *Renew. Sustain. Energy Rev.* **2018**, *81*, 536–551. [CrossRef]
3. Wei, Z.; Spinney, R.; Ke, R.; Yang, Z.; Xiao, R. Effect of pH on the sonochemical degradation of organic pollutants. *Environ. Chem. Lett.* **2016**, *14*, 163–182. [CrossRef]
4. Debabrata, P.; Sivakumar, M. Sonochemical degradation of endocrine-disrupting organochlorine pesticide Dicofol: Investigations on the transformation pathways of dechlorination and the influencing operating parameters. *Chemosphere* **2018**, *204*, 101–108. [CrossRef] [PubMed]
5. Deng, Y.; Zhao, R. Advanced oxidation processes (AOPs) in wastewater treatment. *Curr. Pollut. Rep.* **2015**, *1*, 167–176. [CrossRef]
6. Navidpour, A.H.; Hosseinzadeh, A.; Huang, Z.; Li, D.; Zhou, J.L. Application of machine learning algorithms in predicting the photocatalytic degradation of perfluorooctanoic acid. *Catal. Rev. Sci. Eng.* **2022**, 1–26. [CrossRef]
7. Navidpour, A.H.; Fakhrzad, M. Photocatalytic activity of Zn$_2$SnO$_4$ coating deposited by air plasma spraying. *Appl. Surf. Sci. Adv.* **2021**, *6*, 100153. [CrossRef]
8. Navidpour, A.H.; Fakhrzad, M. Photocatalytic and magnetic properties of ZnFe$_2$O$_4$ nanoparticles synthesised by mechanical alloying. *Int. J. Environ. Anal. Chem.* **2022**, *102*, 690–706. [CrossRef]
9. Navidpour, A.H.; Salehi, M.; Amirnasr, M.; Salimijazi, H.R.; Azarpour Siahkali, M.; Kalantari, Y.; Mohammadnezhad, M. Photocatalytic iron oxide coatings produced by thermal spraying process. *J. Therm. Spray Technol.* **2015**, *24*, 1487–1497. [CrossRef]
10. Navidpour, A.H.; Salehi, M.; Salimijazi, H.R.; Kalantari, Y.; Azarpour Siahkali, M. Photocatalytic activity of flame-sprayed coating of zinc ferrite powder. *J. Therm. Spray Technol.* **2017**, *26*, 2030–2039. [CrossRef]
11. Dai, Y.; Zhang, N.; Xing, C.; Cui, Q.; Sun, Q. The adsorption, regeneration and engineering applications of biochar for removal organic pollutants: A review. *Chemosphere* **2019**, *223*, 12–27. [CrossRef] [PubMed]
12. Wang, J.; Zhuang, S. Removal of various pollutants from water and wastewater by modified chitosan adsorbents. *Crit. Rev. Environ. Sci. Technol.* **2017**, *47*, 2331–2386. [CrossRef]
13. Schwarze, M. Micellar-enhanced ultrafiltration (MEUF)—State of the art. *Environ. Sci. Water Res. Technol.* **2017**, *3*, 598–624. [CrossRef]
14. Hussain, K.I.; Usman, M.; Siddiq, M.; Rasool, N.; Nazar, M.F.; Ahmad, I.; Holder, A.A.; Altaf, A.A. Application of micellar enhanced ultrafiltration for the removal of sunset yellow dye from aqueous media. *J. Dispers. Sci. Technol.* **2017**, *38*, 139–144. [CrossRef]
15. Mojiri, A.; Ohashi, A.; Ozaki, N.; Shoiful, A.; Kindaichi, T. Pollutant removal from synthetic aqueous solutions with a combined electrochemical oxidation and adsorption method. *Int. J. Environ. Res. Public Health* **2018**, *15*, 1443. [CrossRef]
16. Paisa, J.D.; Chojaat, R. Removal of organic dye pollutants from wastewater by electrochemical oxidation. *Phys. Chem. Liq.* **2007**, *45*, 479–485. [CrossRef]
17. Kim, Y.; Osako, M.; Lee, D. Removal of hydrophobic organic pollutants by coagulation-precipitation process with dissolved humic matter. *Waste Manag. Res.* **2002**, *20*, 341–349. [CrossRef]
18. Ren, X.; Xu, X.; Xiao, Y.; Chen, W.; Song, K. Effective removal by coagulation of contaminants in concentrated leachate from municipal solid waste incineration power plants. *Sci. Total Environ.* **2019**, *685*, 392–400. [CrossRef]
19. Belhaj, D.; Baccar, R.; Jaabiri, I.; Bouzid, J.; Kallel, M.; Ayadi, H.; Zhou, J.L. Fate of selected estrogenic hormones in an urban sewage treatment plant in Tunisia (North Africa). *Sci. Total Environ.* **2015**, *505*, 154–160. [CrossRef]

20. Jorgensen, T.C.; Weatherley, L.R. Ammonia removal from wastewater by ion exchange in the presence of organic contaminants. *Water Res.* **2003**, *37*, 1723–1728. [CrossRef]

21. Feng, Y.; Yang, S.; Xia, L.; Wang, Z.; Suo, N.; Chen, H.; Long, Y.; Zhou, B.; Yu, Y. In-situ ion exchange electrocatalysis biological coupling (i-IEEBC) for simultaneously enhanced degradation of organic pollutants and heavy metals in electroplating wastewater. *J. Hazard. Mater.* **2019**, *364*, 562–570. [CrossRef] [PubMed]

22. Cui, Y.; Liu, X.-Y.; Chung, T.-S.; Weber, M.; Staudt, C.; Maletzko, C. Removal of organic micro-pollutants (phenol, aniline and nitrobenzene) via forward osmosis (FO) process: Evaluation of FO as an alternative method to reverse osmosis (RO). *Water Res.* **2016**, *91*, 104–114. [CrossRef]

23. Zhang, A.; Gu, Z.; Chen, W.; Li, Q.; Jiang, G. Removal of refractory organic pollutants in reverse-osmosis concentrated leachate by Microwave–Fenton process. *Environ. Sci. Pollut. Res.* **2018**, *25*, 28907–28916. [CrossRef] [PubMed]

24. Rasalingam, S.; Peng, R.; Koodali, R.T. Removal of hazardous pollutants from wastewaters: Applications of TiO$_2$-SiO$_2$ mixed oxide materials. *J. Nanomater.* **2014**, *2014*, 617405. [CrossRef]

25. Dewil, R.; Mantzavinos, D.; Poulios, I.; Rodrigo, M.A. New perspectives for advanced oxidation processes. *J. Environ. Manag.* **2017**, *195*, 93–99. [CrossRef]

26. Kurniawan, T.A.; Lo, W.-h. Removal of refractory compounds from stabilized landfill leachate using an integrated H$_2$O$_2$ oxidation and granular activated carbon (GAC) adsorption treatment. *Water Res.* **2009**, *43*, 4079–4091. [CrossRef] [PubMed]

27. Oller, I.; Malato, S.; Sánchez-Pérez, J.A. Combination of advanced oxidation processes and biological treatments for wastewater decontamination—A review. *Sci. Total Environ.* **2011**, *409*, 4141–4166. [CrossRef] [PubMed]

28. Bandara, J.; Pulgarin, C.; Peringer, P.; Kiwi, J. Chemical (photo-activated) coupled biological homogeneous degradation of p-nitro-o-toluene-sulfonic acid in a flow reactor. *J. Photochem. Photobiol. A Chem.* **1997**, *111*, 253–263. [CrossRef]

29. Paździor, K.; Bilińska, L.; Ledakowicz, S. A review of the existing and emerging technologies in the combination of AOPs and biological processes in industrial textile wastewater treatment. *Chem. Eng. J.* **2019**, *376*, 120597. [CrossRef]

30. Ahmed, S.N.; Haider, W. Heterogeneous photocatalysis and its potential applications in water and wastewater treatment: A review. *Nanotechnology* **2018**, *29*, 342001. [CrossRef]

31. Zhu, S.; Wang, D. Photocatalysis: Basic principles, diverse forms of implementations and emerging scientific opportunities. *Adv. Energy Mater.* **2017**, *7*, 1700841. [CrossRef]

32. Inoue, T.; Fujishima, A.; Konishi, S.; Honda, K. Photoelectrocatalytic reduction of carbon dioxide in aqueous suspensions of semiconductor powders. *Nature* **1979**, *277*, 637–638. [CrossRef]

33. Ghaderi, A.; Abbasi, S.; Farahbod, F. Synthesis, characterization and photocatalytic performance of modified ZnO nanoparticles with SnO$_2$ nanoparticles. *Mater. Res. Express* **2018**, *5*, 065908–065918. [CrossRef]

34. Abbasi, S.; Hasanpour, M. The effect of pH on the photocatalytic degradation of methyl orange using decorated ZnO nanoparticles with SnO$_2$ nanoparticles. *J. Mater. Sci. Mater. Electron.* **2017**, *28*, 1307–1314. [CrossRef]

35. Abbasi, S.; Ekrami-Kakhki, M.-S.; Tahari, M. Modeling and predicting the photodecomposition of methylene blue via ZnO–SnO$_2$ hybrids using design of experiments (DOE). *J. Mater. Sci. Mater. Electron.* **2017**, *28*, 15306–15312. [CrossRef]

36. Xu, B.; Liu, S.; Zhou, J.L.; Zheng, C.; Jin, W.; Chen, B.; Zhang, T.; Qiu, W. PFAS and their substitutes in groundwater: Occurrence, transformation and remediation. *J. Hazard. Mater.* **2021**, *412*, 125159. [CrossRef]

37. Nur, H.; Misnon, I.I.; Wei, L.K. Stannic oxide-titanium dioxide coupled semiconductor photocatalyst loaded with polyaniline for enhanced photocatalytic oxidation of 1-Octene. *Int. J. Photoenergy* **2007**, *2007*, 098548. [CrossRef]

38. Roozban, N.; Abbasi, S.; Ghazizadeh, M. The experimental and statistical investigation of the photo degradation of methyl orange using modified MWCNTs with different amount of ZnO nanoparticles. *J. Mater. Sci. Mater. Electron.* **2017**, *28*, 7343–7352. [CrossRef]

39. Abbasi, S.; Hasanpour, M.; Ekrami-Kakhki, M.-S. Removal efficiency optimization of organic pollutant (methylene blue) with modified multi-walled carbon nanotubes using design of experiments (DOE). *J. Mater. Sci. Mater. Electron.* **2017**, *28*, 9900–9910. [CrossRef]

40. Navidpour, A.H.; Hosseinzadeh, A.; Zhou, J.L.; Huang, Z. Progress in the application of surface engineering methods in immobilizing TiO$_2$ and ZnO coatings for environmental photocatalysis. *Catal. Rev. Sci. Eng.* **2021**, 1–52. [CrossRef]

41. Abbasi, S. Investigation of the enhancement and optimization of the photocatalytic activity of modified TiO$_2$ nanoparticles with SnO$_2$ nanoparticles using statistical method. *Mater. Res. Express* **2018**, *5*, 066302. [CrossRef]

42. Abbasi, S. The degradation rate study of methyl orange using MWCNTs@TiO$_2$ as photocatalyst, application of statistical analysis based on Fisher's F distribution. *J. Clust. Sci.* **2022**, *33*, 593–602. [CrossRef]

43. Mazinani, B.; Masrom, A.K.; Beitollahi, A.; Luque, R. Photocatalytic activity, surface area and phase modification of mesoporous SiO$_2$–TiO$_2$ prepared by a one-step hydrothermal procedure. *Ceram. Int.* **2014**, *40*, 11525–11532. [CrossRef]

44. Réti, B.; Major, Z.; Szarka, D.; Boldizsár, T.; Horváth, E.; Magrez, A.; Forró, L.; Dombi, A.; Hernádi, K. Influence of TiO$_2$ phase composition on the photocatalytic activity of TiO$_2$/MWCNT composites prepared by combined sol–gel/hydrothermal method. *J. Mol. Catal. A Chem.* **2016**, *414*, 140–147. [CrossRef]

45. Siah, W.R.; Lintang, H.O.; Shamsuddin, M.; Yuliati, L. High photocatalytic activity of mixed anatase-rutile phases on commercial TiO$_2$ nanoparticles. *IOP Conf. Ser. Mater. Sci. Eng.* **2016**, *107*, 012005. [CrossRef]

46. Etacheri, V.; Valentin, C.D.; Schneider, J.; Bahnemann, D.; C.Pilla, S. Visible-light activation of TiO$_2$ photocatalysts: Advances in theory and experiments. *J. Photochem. Photobiol. C Photochem. Rev.* **2015**, *25*, 1–29. [CrossRef]

47. Osterloh, F.E. Photocatalysis versus photosynthesis: A sensitivity analysis of devices for solar energy conversion and chemical transformations. *ACS Energy Lett.* **2017**, *2*, 445–453. [CrossRef]

48. Zhou, X.; Dong, H.; Ren, A.-M. Exploring the mechanism of water-splitting reaction in NiOx/β-Ga2O3 photocatalysts by first-principles calculations. *Phys. Chem. Chem. Phys.* **2016**, *18*, 11111–11119. [CrossRef]

49. Yu, F.; Zhou, H.; Huang, Y.; Sun, J.; Qin, F.; Bao, J.; Goddard, W.A.; Chen, S.; Ren, Z. High-performance bifunctional porous non-noble metal phosphide catalyst for overall water splitting. *Nat. Commun.* **2018**, *9*, 2551. [CrossRef]

50. Leung, D.Y.C.; Fu, X.; Wang, C.; Ni, M.; Leung, M.K.H.; Wang, X.; Fu, X. Hydrogen production over titania-based photocatalysts. *ChemSusChem* **2010**, *3*, 681–694. [CrossRef]

51. Yang, X.; Wang, D. Photocatalysis: From fundamental principles to materials and applications. *ACS Appl. Energy Mater.* **2018**, *1*, 6657–6693. [CrossRef]

52. Wang, R.; Hashimoto, K.; Fujishima, A.; Chikuni, M.; Kojima, E.; Kitamura, A.; Shimohigoshi, M.; Watanabe, T. Light-induced amphiphilic surfaces. *Nature* **1997**, *388*, 431–432. [CrossRef]

53. Otitoju, T.A.; Ahmad, A.L.; Ooi, B.S. Superhydrophilic (superwetting) surfaces: A review on fabrication and application. *J. Ind. Eng. Chem.* **2017**, *47*, 19–40. [CrossRef]

54. Kazemi, M.; Mohammadizadeh, M.R. Simultaneous improvement of photocatalytic and superhydrophilicity properties of nano TiO$_2$ thin films. *Chem. Eng. Res. Des.* **2012**, *90*, 1473–1479. [CrossRef]

55. Ohtani, B. Preparing articles on photocatalysis—Beyond the illusions, misconceptions, and speculation. *Chem. Lett.* **2008**, *37*, 216–229. [CrossRef]

56. Kong, D.; Zheng, Y.; Kobielusz, M.; Wang, Y.; Bai, Z.; Macyk, W.; Wang, X.; Tang, J. Recent advances in visible light-driven water oxidation and reduction in suspension systems. *Mater. Today* **2018**, *21*, 897–924. [CrossRef]

57. Abbasi, S.; Dastan, D.; Ţălu, Ş.; Tahir, M.B.; Elias, M.; Tao, L.; Li, Z. Evaluation of the dependence of methyl orange organic pollutant removal rate on the amount of titanium dioxide nanoparticles in MWCNTs-TiO$_2$ photocatalyst using statistical methods and Duncan's multiple range test. *Int. J. Environ. Anal. Chem.* **2022**, 1–15. [CrossRef]

58. Abbasi, S. Improvement of photocatalytic decomposition of methyl orange by modified MWCNTs, prediction of degradation rate using statistical models. *J. Mater. Sci. Mater. Electron.* **2021**, *32*, 14137–14148. [CrossRef]

59. Gui, M.M.; Chai, S.-P.; Xu, B.-Q.; Mohamed, A.R. Enhanced visible light responsive MWCNT/TiO$_2$ core–shell nanocomposites as the potential photocatalyst for reduction of CO$_2$ into methane. *Sol. Energy Mater. Sol. Cells* **2014**, *122*, 183–189. [CrossRef]

60. Wang, D.; Li, Y.; Li Puma, G.; Wang, C.; Wang, P.; Zhang, W.; Wang, Q. Mechanism and experimental study on the photocatalytic performance of Ag/AgCl @ chiral TiO$_2$ nanofibers photocatalyst: The impact of wastewater components. *J. Hazard. Mater.* **2015**, *285*, 277–284. [CrossRef]

61. Gentili, P.L.; Penconi, M.; Costantino, F.; Sassi, P.; Ortica, F.; Rossi, F.; Elisei, F. Structural and photophysical characterization of some La$_{2x}$Ga$_{2y}$In$_{2z}$O$_3$ solid solutions, to be used as photocatalysts for H$_2$ production from water/ethanol solutions. *Sol. Energy Mater. Sol. Cells* **2010**, *94*, 2265–2274. [CrossRef]

62. Babu, D.S.; Srivastava, V.; Nidheesh, P.V.; Kumar, M.S. Detoxification of water and wastewater by advanced oxidation processes. *Sci. Total Environ.* **2019**, *696*, 133961. [CrossRef]

63. Wang, W.; Huang, G.; Yu, J.C.; Wong, P.K. Advances in photocatalytic disinfection of bacteria: Development of photocatalysts and mechanisms. *J. Environ. Sci.* **2015**, *34*, 232–247. [CrossRef] [PubMed]

64. Linsebigler, A.L.; Lu, G.; Yates, J.T. Photocatalysis on TiO$_2$ surfaces: Principles, mechanisms, and selected results. *Chem. Rev.* **1995**, *95*, 735–758. [CrossRef]

65. Peng, L.; Xie, T.; Lu, Y.; Fan, H.; Wang, D. Synthesis, photoelectric properties and photocatalytic activity of the Fe$_2$O$_3$/TiO$_2$ heterogeneous photocatalysts. *Phys. Chem. Chem. Phys.* **2010**, *12*, 8033–8041. [CrossRef]

66. Guo, H.; Kemell, M.; Heikkilä, M.; Leskelä, M. Noble metal-modified TiO$_2$ thin film photocatalyst on porous steel fiber support. *Appl. Catal. B Environ.* **2010**, *95*, 358–364. [CrossRef]

67. Liu, L.; Gao, F.; Zhao, H.; Li, Y. Tailoring Cu valence and oxygen vacancy in Cu/TiO$_2$ catalysts for enhanced CO$_2$ photoreduction efficiency. *Appl. Catal. B Environ.* **2013**, *134–135*, 349–358. [CrossRef]

68. Akpan, U.G.; Hameed, B.H. The advancements in sol–gel method of doped-TiO$_2$ photocatalysts. *Appl. Catal. A Gen.* **2010**, *375*, 1–11. [CrossRef]

69. Wen, J.; Xie, J.; Chen, X.; Li, X. A review on g-C3N4-based photocatalysts. *Appl. Surf. Sci.* **2017**, *391*, 72–123. [CrossRef]

70. Diesen, V.; Jonsson, M. Formation of H$_2$O$_2$ in TiO$_2$ photocatalysis of oxygenated and deoxygenated aqueous systems: A probe for photocatalytically produced hydroxyl radicals. *J. Phys. Chem. C* **2014**, *118*, 10083–10087. [CrossRef]

71. Zhang, J.; Nosaka, Y. Mechanism of the OH radical generation in photocatalysis with TiO$_2$ of different crystalline types. *J. Phys. Chem. C* **2014**, *118*, 10824–10832. [CrossRef]

72. Navidpour, A.H.; Kalantari, Y.; Salehi, M.; Salimijazi, H.R.; Amirnasr, M.; Rismanchian, M.; Azarpour Siahkali, M. Plasma-sprayed photocatalytic zinc oxide coatings. *J. Therm. Spray Technol.* **2017**, *26*, 717–727. [CrossRef]

73. Montoya, J.F.; Peral, J.; Salvador, P. Comprehensive kinetic and mechanistic analysis of TiO$_2$ photocatalytic reactions according to the direct–indirect model: (I) Theoretical approach. *J. Phys. Chem. C* **2014**, *118*, 14266–14275. [CrossRef]

74. Monllor-Satoca, D.; Gómez, R.; González-Hidalgo, M.; Salvador, P. The "Direct–Indirect" model: An alternative kinetic approach in heterogeneous photocatalysis based on the degree of interaction of dissolved pollutant species with the semiconductor surface. *Catal. Today* **2007**, *129*, 247–255. [CrossRef]

75. Ollis, D.F. Kinetics of liquid phase photocatalyzed reactions: An illuminating approach. *J. Phys. Chem. B* **2005**, *109*, 2439–2444. [CrossRef] [PubMed]
76. Ollis, D.F. Kinetics of photocatalyzed reactions: Five lessons learned. *Front. Chem.* **2018**, *6*, 1–7. [CrossRef]
77. Nosaka, Y.; Nosaka, A.Y. Langmuir–Hinshelwood and light-intensity dependence analyses of photocatalytic oxidation rates by two-dimensional-ladder kinetic simulation. *J. Phys. Chem. C* **2018**, *122*, 28748–28756. [CrossRef]
78. Tseng, T.-K.; Lin, Y.; Chen, Y.; Chu, H. A review of photocatalysts prepared by sol-gel method for VOCs removal. *Int. J. Mol. Sci.* **2010**, *11*, 2336–2361. [CrossRef]
79. Emeline, A.V.; Ryabchuk, V.K.; Serpone, N. Dogmas and misconceptions in heterogeneous photocatalysis. Some enlightened reflections. *J. Phys. Chem. B* **2005**, *109*, 18515–18521. [CrossRef]
80. Mills, A.; Wang, J.; Ollis, D.F. Kinetics of liquid phase semiconductor photoassisted reactions: Supporting observations for a pseudo-steady-state model. *J. Phys. Chem. B* **2006**, *110*, 14386–14390. [CrossRef]
81. Boulamanti, A.K.; Philippopoulos, C.J. Photocatalytic degradation of C5–C7 alkanes in the gas–phase. *Atmos. Environ.* **2009**, *43*, 3168–3174. [CrossRef]
82. Jiang, Y.; Tian, B. Inorganic semiconductor biointerfaces. *Nat. Rev. Mater.* **2018**, *3*, 473–490. [CrossRef] [PubMed]
83. Dolgonos, A.; Mason, T.O.; Poeppelmeier, K.R. Direct optical band gap measurement in polycrystalline semiconductors: A critical look at the Tauc method. *J. Solid State Chem.* **2016**, *240*, 43–48. [CrossRef]
84. Viezbicke, B.D.; Patel, S.; Davis, B.E.; Birnie Iii, D.P. Evaluation of the Tauc method for optical absorption edge determination: ZnO thin films as a model system. *Phys. Status Solidi B* **2015**, *252*, 1700–1710. [CrossRef]
85. Bansal, A.; Kumar, A.; Kumar, P.; Bojja, S.; Chatterjee, A.K.; Ray, S.S.; Jain, S.L. Visible light-induced surface initiated atom transfer radical polymerization of methyl methacrylate on titania/reduced graphene oxide nanocomposite. *RSC Adv.* **2015**, *5*, 21189–21196. [CrossRef]
86. Stroyuk, O.; Kryukov, A.; Kuchmii, S.; Pokhodenko, V. Quantum size effects in semiconductor photocatalysis. *Theor. Exp. Chem.* **2005**, *41*, 207–228. [CrossRef]
87. Zhang, H.; Zhang, H.; Zhu, P.; Huang, F. Morphological effect in photocatalytic degradation of direct blue over mesoporous TiO_2 catalysts. *ChemistrySelect* **2017**, *2*, 3282–3288. [CrossRef]
88. Ola, O.; Maroto-Valer, M.M. Review of material design and reactor engineering on TiO_2 photocatalysis for CO_2 reduction. *J. Photochem. Photobiol. C Photochem. Rev.* **2015**, *24*, 16–42. [CrossRef]
89. Al-Kattan, A.; Wichser, A.; Vonbank, R.; Brunner, S.; Ulrich, A.; Zuin, S.; Nowack, B. Release of TiO2 from paints containing pigment-TiO_2 or nano-TiO_2 by weathering. *Environ. Sci. Process. Impacts* **2013**, *15*, 2186–2193. [CrossRef]
90. Jacobs, J.; Poel, I.; Osseweijer, P. Sunscreens with titanium dioxide (TiO_2) nano-particles: A societal experiment. *Nanoethics* **2010**, *4*, 103–113. [CrossRef]
91. Rompelberg, C.; Heringa, M.B.; van Donkersgoed, G.; Drijvers, J.; Roos, A.; Westenbrink, S.; Peters, R.; van Bemmel, G.; Brand, W.; Oomen, A.G. Oral intake of added titanium dioxide and its nanofraction from food products, food supplements and toothpaste by the Dutch population. *Nanotoxicology* **2016**, *10*, 1404–1414. [CrossRef] [PubMed]
92. Hanaor, D.A.H.; Sorrell, C.C. Review of the anatase to rutile phase transformation. *J. Mater. Sci.* **2011**, *46*, 855–874. [CrossRef]
93. Chen, X.; Mao, S.S. Titanium dioxide nanomaterials: Synthesis, properties, modifications, and applications. *Chem. Rev.* **2007**, *107*, 2891–2959. [CrossRef] [PubMed]
94. Abbasi, S. Photocatalytic activity study of coated anatase-rutile titania nanoparticles with nanocrystalline tin dioxide based on the statistical analysis. *Environ. Monit. Assess.* **2019**, *191*, 206–218. [CrossRef] [PubMed]
95. Pei, D.-N.; Gong, L.; Zhang, A.-Y.; Zhang, X.; Chen, J.-J.; Mu, Y.; Yu, H.-Q. Defective titanium dioxide single crystals exposed by high-energy {001} facets for efficient oxygen reduction. *Nat. Commun.* **2015**, *6*, 8696. [CrossRef] [PubMed]
96. Ye, F.X.; Ohmori, A.; Tsumura, T.; Nakata, K.; Li, C.J. Microstructural analysis and photocatalytic activity of plasma-sprayed titania-hydroxyapatite coatings. *J. Therm. Spray Technol.* **2007**, *16*, 776–782. [CrossRef]
97. Ren, K.; Liu, Y.; He, X.; Li, H. Suspension plasma spray fabrication of nanocrystalline titania hollow microspheres for photocatalytic applications. *J. Therm. Spray Technol.* **2015**, *24*, 1213–1220. [CrossRef]
98. Luttrell, T.; Halpegamage, S.; Tao, J.; Kramer, A.; Sutter, E.; Batzill, M. Why is anatase a better photocatalyst than rutile?—Model studies on epitaxial TiO_2 films. *Sci. Rep.* **2014**, *4*, 4043. [CrossRef]
99. Fresno, F.; Portela, R.; Suárez, S.; Coronado, J.M. Photocatalytic materials: Recent achievements and near future trends. *J. Mater. Chem. A* **2014**, *2*, 2863–2884. [CrossRef]
100. Setvin, M.; Aschauer, U.; Hulva, J.; Simschitz, T.; Daniel, B.; Schmid, M.; Selloni, A.; Diebold, U. Following the reduction of oxygen on TiO_2 anatase (101) step by step. *J. Am. Chem. Soc.* **2016**, *138*, 9565–9571. [CrossRef]
101. Hoffmann, M.R.; Martin, S.T.; Choi, W.; Bahnemann, D.W. Environmental applications of semiconductor photocatalysis. *Chem. Rev.* **1995**, *95*, 69–96. [CrossRef]
102. Ali, I.; Suhail, M.; Alothman, Z.A.; Alwarthan, A. Recent advances in syntheses, properties and applications of TiO_2 nanostructures. *RSC Adv.* **2018**, *8*, 30125–30147. [CrossRef] [PubMed]
103. Nowotny, M.K.; Bak, T.; Nowotny, J. Electrical properties and defect chemistry of TiO_2 single crystal. I. Electrical conductivity. *J. Phys. Chem. B* **2006**, *110*, 16270–16282. [CrossRef] [PubMed]
104. Fujitsu, S.; Hamada, T. Electrical properties of manganese-doped titanium dioxide. *J. Am. Ceram. Soc.* **1994**, *77*, 3281–3283. [CrossRef]

105. Lu, L.; Xia, X.; Luo, J.K.; Shao, G. Mn-doped TiO$_2$ thin films with significantly improved optical and electrical properties. *J. Phys. D Appl. Phys.* **2012**, *45*, 485102. [CrossRef]
106. Wang, Y.; Cheng, H.; Hao, Y.; Ma, J.; Li, W.; Cai, S. Preparation, characterization and photoelectrochemical behaviors of Fe(III)-doped TiO$_2$ nanoparticles. *J. Mater. Sci.* **1999**, *34*, 3721–3729. [CrossRef]
107. Bally, A.R.; Korobeinikova, E.; Schmid, P.; Lévy, F.; Bussy, F. Structural and electrical properties of Fe-doped TiO$_2$ thin films. *J. Phys. D Appl. Phys.* **1999**, *31*, 1149. [CrossRef]
108. Li, Z.; Ding, D.; Liu, Q.; Ning, C.; Wang, X. Ni-doped TiO$_2$ nanotubes for wide-range hydrogen sensing. *Nanoscale Res. Lett.* **2014**, *9*, 118. [CrossRef]
109. Li, Y.; Wlodarski, W.; Galatsis, K.; Moslih, S.; Cole, J.; Russo, S.; Rockelmann, N. Gas sensing properties of P-type semiconducting Cr-doped TiO$_2$ thin films. *Sens. Actuators B Chem.* **2002**, *83*, 160–163. [CrossRef]
110. Ruiz, A.; Cornet, A.; Sakai, G.; Shimanoe, K.; Morante, J.; Yamazoe, N. Preparation of Cr-doped TiO$_2$ thin film of p-type conduction for gas sensor application. *Chem. Lett.* **2002**, *31*, 892–893. [CrossRef]
111. Cao, J.; Zhang, Y.; Liu, L.; Ye, J. A p-type Cr-doped TiO$_2$ photo-electrode for photo-reduction. *Chem. Commun.* **2013**, *49*, 3440–3442. [CrossRef] [PubMed]
112. Shao, B.; Feng, M.; Zuo, X. Carrier-dependent magnetic anisotropy of cobalt doped titanium dioxide. *Sci. Rep.* **2014**, *4*, 7496. [CrossRef] [PubMed]
113. Lontio Fomekong, R.; Saruhan, B. Synthesis of Co^{3+} doped TiO$_2$ by co-precipitation route and its gas sensing properties. *Front. Mater.* **2019**, *6*, 1–6. [CrossRef]
114. Sathasivam, S.; Bhachu, D.S.; Lu, Y.; Chadwick, N.; Althabaiti, S.A.; Alyoubi, A.O.; Basahel, S.N.; Carmalt, C.J.; Parkin, I.P. Tungsten doped TiO$_2$ with enhanced photocatalytic and optoelectrical properties via aerosol assisted chemical vapor deposition. *Sci. Rep.* **2015**, *5*, 10952. [CrossRef]
115. He, S.; Meng, Y.; Cao, Y.; Huang, S.; Yang, J.; Tong, S.; Wu, M. Hierarchical Ta-doped TiO$_2$ nanorod arrays with improved charge separation for photoelectrochemical water oxidation under FTO side illumination. *Nanomaterials* **2018**, *8*, 983. [CrossRef] [PubMed]
116. Biedrzycki, J.; Livraghi, S.; Giamello, E.; Agnoli, S.; Granozzi, G. Fluorine- and niobium-doped TiO$_2$: Chemical and spectroscopic properties of polycrystalline n-type-doped anatase. *J. Phys. Chem. C* **2014**, *118*, 8462–8473. [CrossRef]
117. Eguchi, R.; Takekuma, Y.; Ochiai, T.; Nagata, M. Improving interfacial charge-transfer transitions in Nb-doped TiO$_2$ electrodes with 7,7,8,8-Tetracyanoquinodimethane. *Catalysts* **2018**, *8*, 367. [CrossRef]
118. Ansari, S.; Khan, M.M.; Ansari, M.; Cho, M.H. Nitrogen-doped titanium dioxide (N-doped TiO$_2$) for visible light photocatalysis. *New J. Chem.* **2016**, *40*, 3000–3009. [CrossRef]
119. Asahi, R.; Morikawa, T.; Irie, H.; Ohwaki, T. Nitrogen-doped titanium dioxide as visible-light-sensitive photocatalyst: Designs, developments, and prospects. *Chem. Rev.* **2014**, *114*, 9824–9852. [CrossRef]
120. Kim, T.H.; Go, G.-M.; Cho, H.-B.; Song, Y.; Lee, C.-G.; Choa, Y.-H. A novel synthetic method for N doped TiO$_2$ nanoparticles through plasma-assisted electrolysis and photocatalytic activity in the visible region. *Front. Chem.* **2018**, *6*, 458. [CrossRef]
121. Yang, G.; Jiang, Z.; Shi, H.; Xiao, T.; Yan, Z. Preparation of highly visible-light active N-doped TiO$_2$ photocatalyst. *J. Mater. Chem.* **2010**, *20*, 5301–5309. [CrossRef]
122. Zhang, X.; Zhou, J.; Gu, Y.; Fan, D. Visible-light photocatalytic activity of N-doped TiO$_2$ nanotube arrays on acephate degradation. *J. Nanomater.* **2015**, *2015*, 527070. [CrossRef]
123. Agyeman, D.A.; Song, K.; Kang, S.H.; Jo, M.R.; Cho, E.; Kang, Y.-M. An improved catalytic effect of nitrogen-doped TiO$_2$ nanofibers for rechargeable Li–O$_2$ batteries; the role of oxidation states and vacancies on the surface. *J. Mater. Chem. A* **2015**, *3*, 22557–22563. [CrossRef]
124. Asahi, R.; Morikawa, T. Nitrogen complex species and its chemical nature in TiO$_2$ for visible-light sensitized photocatalysis. *Chem. Phys.* **2007**, *339*, 57–63. [CrossRef]
125. Peng, F.; Cai, L.; Yu, H.; Wang, H.; Yang, J. Synthesis and characterization of substitutional and interstitial nitrogen-doped titanium dioxides with visible light photocatalytic activity. *J. Solid State Chem.* **2008**, *181*, 130–136. [CrossRef]
126. Zeng, L.; Song, W.; Li, M.; Jie, X.; Zeng, D.; Xie, C. Comparative study on the visible light driven photocatalytic activity between substitutional nitrogen doped and interstitial nitrogen doped TiO$_2$. *Appl. Catal. A Gen.* **2014**, *488*, 239–247. [CrossRef]
127. Lee, J.; Lee, S.J.; Han, W.B.; Jeon, H.; Park, J.; Kim, H.; Yoon, C.S.; Jeon, H. Effect of crystal structure and grain size on photocatalytic activities of remote-plasma atomic layer deposited titanium oxide thin film. *ECS J. Solid State Sci. Technol.* **2012**, *1*, Q63–Q69. [CrossRef]
128. Retamoso, C.; Escalona, N.; González, M.; Barrientos, L.; Allende-González, P.; Stancovich, S.; Serpell, R.; Fierro, J.L.G.; Lopez, M. Effect of particle size on the photocatalytic activity of modified rutile sand (TiO$_2$) for the discoloration of methylene blue in water. *J. Photochem. Photobiol. A Chem.* **2019**, *378*, 136–141. [CrossRef]
129. Kočí, K.; Obalová, L.; Matějová, L.; Plachá, D.; Lacný, Z.; Jirkovský, J.; Šolcová, O. Effect of TiO$_2$ particle size on the photocatalytic reduction of CO$_2$. *Appl. Catal. B Environ.* **2009**, *89*, 494–502. [CrossRef]
130. Amano, F.; Nogami, K.; Tanaka, M.; Ohtani, B. Correlation between surface area and photocatalytic activity for acetaldehyde decomposition over bismuth tungstate particles with a hierarchical structure. *Langmuir* **2010**, *26*, 7174–7180. [CrossRef]
131. Vorontsov, A.V.; Kabachkov, E.N.; Balikhin, I.L.; Kurkin, E.N.; Troitskii, V.N.; Smirniotis, P.G. Correlation of surface area with photocatalytic activity of TiO$_2$. *J. Adv. Oxid. Technol.* **2018**, *21*, 127–137. [CrossRef]

132. Cheng, H.; Wang, J.; Zhao, Y.; Han, X. Effect of phase composition, morphology, and specific surface area on the photocatalytic activity of TiO_2 nanomaterials. *RSC Adv.* **2014**, *4*, 47031–47038. [CrossRef]
133. Elgh, B.; Yuan, N.; Cho, H.S.; Magerl, D.; Philipp, M.; Roth, S.V.; Yoon, K.B.; Müller-Buschbaum, P.; Terasaki, O.; Palmqvist, A.E.C. Controlling morphology, mesoporosity, crystallinity, and photocatalytic activity of ordered mesoporous TiO_2 films prepared at low temperature. *APL Mater.* **2014**, *2*, 113313. [CrossRef]
134. Tanaka, K.; Capule, M.F.V.; Hisanaga, T. Effect of crystallinity of TiO_2 on its photocatalytic action. *Chem. Phys. Lett.* **1991**, *187*, 73–76. [CrossRef]
135. Chen, W.-T.; Chan, A.; Jovic, V.; Sun-Waterhouse, D.; Murai, K.-i.; Idriss, H.; Waterhouse, G.I.N. Effect of the TiO_2 crystallite size, TiO_2 polymorph and test conditions on the photo-oxidation rate of aqueous methylene blue. *Top. Catal.* **2015**, *58*, 85–102. [CrossRef]
136. Yuan, B.; Wang, Y.; Bian, H.; Shen, T.; Wu, Y.; Chen, Z. Nitrogen doped TiO_2 nanotube arrays with high photoelectrochemical activity for photocatalytic applications. *Appl. Surf. Sci.* **2013**, *280*, 523–529. [CrossRef]
137. Peng, F.; Cai, L.; Huang, L.; Yu, H.; Wang, H. Preparation of nitrogen-doped titanium dioxide with visible-light photocatalytic activity using a facile hydrothermal method. *J. Phys. Chem. Solids* **2008**, *69*, 1657–1664. [CrossRef]
138. Di Valentin, C.; Pacchioni, G.; Selloni, A.; Livraghi, S.; Giamello, E. Characterization of paramagnetic species in N-Doped TiO_2 powders by EPR spectroscopy and DFT calculations. *J. Phys. Chem. B* **2005**, *109*, 11414–11419. [CrossRef]
139. Dong, F.; Zhao, W.; Wu, Z.; Guo, S. Band structure and visible light photocatalytic activity of multi-type nitrogen doped TiO_2 nanoparticles prepared by thermal decomposition. *J. Hazard. Mater.* **2009**, *162*, 763–770. [CrossRef]
140. Di Valentin, C.; Finazzi, E.; Pacchioni, G.; Selloni, A.; Livraghi, S.; Paganini, M.C.; Giamello, E. N-doped TiO_2: Theory and experiment. *Chem. Phys.* **2007**, *339*, 44–56. [CrossRef]
141. Weibel, A.; Bouchet, R.; Knauth, P. Electrical properties and defect chemistry of anatase (TiO_2). *Solid State Ion.* **2006**, *177*, 229–236. [CrossRef]
142. Byl, O.; Yates, J.T. Anisotropy in the electrical conductivity of rutile TiO_2 in the (110) plane. *J. Phys. Chem. B* **2006**, *110*, 22966–22967. [CrossRef] [PubMed]
143. Demetry, C.; Shi, X. Grain size-dependent electrical properties of rutile (TiO_2). *Solid State Ion.* **1999**, *118*, 271–279. [CrossRef]
144. Johnson, G.; Weyl, W.A. Influence of minor additions on color and electrical properties of rutile*. *J. Am. Ceram. Soc.* **1949**, *32*, 398–401. [CrossRef]
145. Dang, Y.; West, A.R. Oxygen stoichiometry, chemical expansion or contraction, and electrical properties of rutile, $TiO_{2\pm\delta}$ ceramics. *J. Am. Ceram. Soc.* **2019**, *102*, 251–259. [CrossRef]
146. Hong, M.; Dai, L.; Li, H.; Hu, H.; Liu, K.; Linfei, Y.; Pu, C. Structural phase transition and metallization of nanocrystalline rutile investigated by high-pressure raman spectroscopy and electrical conductivity. *Minerals* **2019**, *9*, 441. [CrossRef]
147. Bak, T.; Nowotny, J.; Stranger, J. Electrical properties of TiO_2: Equilibrium vs dynamic electrical conductivity. *Ionics* **2010**, *16*, 673–679. [CrossRef]
148. Nowotny, J. *Oxide Semiconductors for Solar Energy Conversion*; CRC Press: Boca Raton, FL, USA, 2016.
149. Grundmann, M. *The Physics of Semiconductors: An Introduction Including Nanophysics and Applications*; Springer: Berlin/Heidelberg, Germany, 2010.
150. Ding, Z.; Lu, G.Q.; Greenfield, P.F. Role of the crystallite phase of TiO_2 in heterogeneous photocatalysis for phenol oxidation in water. *J. Phys. Chem. B* **2000**, *104*, 4815–4820. [CrossRef]
151. Li, Z.; Cong, S.; Xu, Y. Brookite vs anatase TiO_2 in the photocatalytic activity for organic degradation in water. *ACS Catal.* **2014**, *4*, 3273–3280. [CrossRef]
152. Tran, H.T.T.; Kosslick, H.; Ibad, M.F.; Fischer, C.; Bentrup, U.; Vuong, T.H.; Nguyen, L.Q.; Schulz, A. Photocatalytic performance of highly active brookite in the degradation of hazardous organic compounds compared to anatase and rutile. *Appl. Catal. B Environ.* **2017**, *200*, 647–658. [CrossRef]
153. Kaplan, R.; Erjavec, B.; Pintar, A. Enhanced photocatalytic activity of single-phase, nanocomposite and physically mixed TiO_2 polymorphs. *Appl. Catal. A Gen.* **2015**, *489*, 51–60. [CrossRef]
154. Vequizo, J.J.M.; Matsunaga, H.; Ishiku, T.; Kamimura, S.; Ohno, T.; Yamakata, A. Trapping-induced enhancement of photocatalytic activity on brookite TiO_2 powders: Comparison with anatase and rutile TiO_2 powders. *ACS Catal.* **2017**, *7*, 2644–2651. [CrossRef]
155. Tran, T.T.H.; Kosslick, H.; Schulz, A.; Nguyen, Q.L. Photocatalytic performance of crystalline titania polymorphs in the degradation of hazardous pharmaceuticals and dyes. *Adv. Nat. Sci. Nanosci. Nanotechnol.* **2017**, *8*, 015011. [CrossRef]
156. Kandiel, T.A.; Robben, L.; Alkaim, A.; Bahnemann, D. Brookite versus anatase TiO_2 photocatalysts: Phase transformations and photocatalytic activities. *Photochem. Photobiol. Sci.* **2013**, *12*, 602–609. [CrossRef]
157. Fischer, K.; Gawel, A.; Rosen, D.; Krause, M.; Latif, A.; Griebel, J.; Prager, A.; Schulze, A. Low-temperature synthesis of anatase/rutile/brookite TiO_2 nanoparticles on a polymer membrane for photocatalysis. *Catalysts* **2017**, *7*, 209. [CrossRef]
158. Farhadian Azizi, K.; Bagheri-Mohagheghi, M.M. Transition from anatase to rutile phase in titanium dioxide (TiO_2) nanoparticles synthesized by complexing sol–gel process: Effect of kind of complexing agent and calcinating temperature. *J. Sol-Gel Sci. Technol.* **2013**, *65*, 329–335. [CrossRef]
159. Zhou, Y.; Fichthorn, K.A. Microscopic view of nucleation in the anatase-to-rutile transformation. *J. Phys. Chem. C* **2012**, *116*, 8314–8321. [CrossRef]

160. Byrne, C.; Fagan, R.; Hinder, S.; McCormack, D.E.; Pillai, S.C. New approach of modifying the anatase to rutile transition temperature in TiO$_2$ photocatalysts. *RSC Adv.* **2016**, *6*, 95232–95238. [CrossRef]
161. Wu, Q.; Li, D.; Hou, Y.; Wu, L.; Fu, X.; Wang, X. Study of relationship between surface transient photoconductivity and liquid-phase photocatalytic activity of titanium dioxide. *Mater. Chem. Phys.* **2007**, *102*, 53–59. [CrossRef]
162. Bokare, A.; Pai, M.; Athawale, A.A. Surface modified Nd doped TiO$_2$ nanoparticles as photocatalysts in UV and solar light irradiation. *Sol. Energy* **2013**, *91*, 111–119. [CrossRef]
163. Zhang, W.; Li, X.; Jia, G.; Gao, Y.; Wang, H.; Cao, Z.; Li, C.; Liu, J. Preparation, characterization, and photocatalytic activity of boron and lanthanum co-doped TiO$_2$. *Catal. Commun.* **2014**, *45*, 144–147. [CrossRef]
164. Alijani, M.; Najibi ilkhechi, N. Effect of Ni doping on the structural and optical properties of TiO$_2$ nanoparticles at various concentration and temperature. *Silicon* **2018**, *10*, 2569–2575. [CrossRef]
165. Wang, G.Q.; Lan, W.; Han, G.J.; Wang, Y.; Su, Q.; Liu, X.Q. Effect of Nb doping on the phase transition and optical properties of sol–gel TiO$_2$ thin films. *J. Alloys Compd.* **2011**, *509*, 4150–4153. [CrossRef]
166. Fernández-García, M.; Martínez-Arias, A.; Hanson, J.C.; Rodriguez, J.A. Nanostructured oxides in chemistry: Characterization and properties. *Chem. Rev.* **2004**, *104*, 4063–4104. [CrossRef]
167. Akhtar, M.K.; Pratsinis, S.E.; Mastrangelo, S.V.R. Dopants in vapor-phase synthesis of titania powders. *J. Am. Ceram. Soc.* **1992**, *75*, 3408–3416. [CrossRef]
168. Nyamukamba, P.; Tichagwa, L.; Greyling, C. The influence of carbon doping on TiO$_2$ nanoparticle size, surface area, anatase to rutile phase transformation and photocatalytic activity. *Mater. Sci. Forum* **2012**, *712*, 49–63. [CrossRef]
169. Bu, X.; Zhang, G.; Zhang, C. Effect of nitrogen doping on anatase–rutile phase transformation of TiO$_2$. *Appl. Surf. Sci.* **2012**, *258*, 7997–8001. [CrossRef]
170. Yu, J.C.; Yu, J.; Ho, W.; Jiang, Z.; Zhang, L. Effects of F- doping on the photocatalytic activity and microstructures of nanocrystalline TiO$_2$ powders. *Chem. Mater.* **2002**, *14*, 3808–3816. [CrossRef]
171. Singh, V.; Rao, A.; Tiwari, A.; Yashwanth, P.; Lal, M.; Dubey, U.; Aich, S.; Roy, B. Study on the effects of Cl and F doping in TiO$_2$ powder synthesized by a sol-gel route for biomedical applications. *J. Phys. Chem. Solids* **2019**, *134*, 262–272. [CrossRef]
172. Shannon, R.D.; Pask, J.A. Kinetics of the anatase-rutile transformation. *J. Am. Ceram. Soc.* **1965**, *48*, 391–398. [CrossRef]
173. Qian, R.; Zong, H.; Schneider, J.; Zhou, G.; Zhao, T.; Li, Y.; Yang, J.; Bahnemann, D.W.; Pan, J.H. Charge carrier trapping, recombination and transfer during TiO$_2$ photocatalysis: An overview. *Catal. Today* **2019**, *335*, 78–90. [CrossRef]
174. Zhang, J.; Zhou, P.; Liu, J.; Yu, J. New understanding of the difference of photocatalytic activity among anatase, rutile and brookite TiO$_2$. *Phys. Chem. Chem. Phys.* **2014**, *16*, 20382–20386. [CrossRef] [PubMed]
175. Madhusudan Reddy, K.; Manorama, S.V.; Ramachandra Reddy, A. Bandgap studies on anatase titanium dioxide nanoparticles. *Mater. Chem. Phys.* **2003**, *78*, 239–245. [CrossRef]
176. Soares, G.B.; Ribeiro, R.A.P.; de Lazaro, S.R.; Ribeiro, C. Photoelectrochemical and theoretical investigation of the photocatalytic activity of TiO$_2$:N. *RSC Adv.* **2016**, *6*, 89687–89698. [CrossRef]
177. Odling, G.; Robertson, N. Why is anatase a better photocatalyst than rutile? The importance of free hydroxyl radicals. *ChemSusChem* **2015**, *8*, 1838–1840. [CrossRef]
178. Chatzitakis, A.; Sartori, S. Recent advances in the use of black TiO$_2$ for production of hydrogen and other solar fuels. *ChemPhysChem* **2019**, *20*, 1272–1281. [CrossRef]
179. Zhou, X.; Liu, N.; Schmuki, P. Photocatalysis with TiO$_2$ nanotubes: "Colorful" reactivity and designing site-specific photocatalytic centers into TiO$_2$ nanotubes. *ACS Catal.* **2017**, *7*, 3210–3235. [CrossRef]
180. Ullattil, S.G.; Narendranath, S.B.; Pillai, S.C.; Periyat, P. Black TiO$_2$ nanomaterials: A review of recent advances. *Chem. Eng. J.* **2018**, *343*, 708–736. [CrossRef]
181. Liu, Y.; Tian, L.; Tan, X.; Li, X.; Chen, X. Synthesis, properties, and applications of black titanium dioxide nanomaterials. *Sci. Bull.* **2017**, *62*, 431–441. [CrossRef]
182. Chen, X.; Liu, L.; Yu, P.Y.; Mao, S.S. Increasing solar absorption for photocatalysis with black hydrogenated titanium dioxide nanocrystals. *Science* **2011**, *331*, 746. [CrossRef]
183. Naldoni, A.; Altomare, M.; Zoppellaro, G.; Liu, N.; Kment, Š.; Zbořil, R.; Schmuki, P. Photocatalysis with reduced TiO$_2$: From black TiO$_2$ to cocatalyst-free hydrogen production. *ACS Catal.* **2019**, *9*, 345–364. [CrossRef] [PubMed]
184. Coto, M.; Divitini, G.; Dey, A.; Krishnamurthy, S.; Ullah, N.; Ducati, C.; Kumar, R.V. Tuning the properties of a black TiO$_2$-Ag visible light photocatalyst produced by a rapid one-pot chemical reduction. *Mater. Today Chem.* **2017**, *4*, 142–149. [CrossRef]
185. Song, H.; Li, C.; Lou, Z.; Ye, Z.; Zhu, L. Effective formation of oxygen vacancies in black TiO$_2$ nanostructures with efficient solar-driven water splitting. *ACS Sustain. Chem. Eng.* **2017**, *5*, 8982–8987. [CrossRef]
186. Chen, S.; Xiao, Y.; Wang, Y.; Hu, Z.; Zhao, H.; Xie, W. A Facile Approach to prepare black TiO$_2$ with oxygen vacancy for enhancing photocatalytic activity. *Nanomaterials* **2018**, *8*, 245. [CrossRef] [PubMed]
187. Cui, H.; Zhao, W.; Yang, C.; Yin, H.; Lin, T.; Shan, Y.; Xie, Y.; Gu, H.; Huang, F. Black TiO$_2$ nanotube arrays for high-efficiency photoelectrochemical water-splitting. *J. Mater. Chem. A* **2014**, *2*, 8612–8616. [CrossRef]
188. Lepcha, A.; Maccato, C.; Mettenbörger, A.; Andreu, T.; Mayrhofer, L.; Walter, M.; Olthof, S.; Ruoko, T.P.; Klein, A.; Moseler, M.; et al. Electrospun black titania nanofibers: Influence of hydrogen plasma-induced disorder on the electronic structure and photoelectrochemical performance. *J. Phys. Chem. C* **2015**, *119*, 18835–18842. [CrossRef]

189. Singh, A.P.; Kodan, N.; Mehta, B.R.; Dey, A.; Krishnamurthy, S. In-situ plasma hydrogenated TiO_2 thin films for enhanced photoelectrochemical properties. *Mater. Res. Bull.* **2016**, *76*, 284–291. [CrossRef]
190. Xu, C.; Song, Y.; Lu, L.; Cheng, C.; Liu, D.; Fang, X.; Chen, X.; Zhu, X.; Li, D. Electrochemically hydrogenated TiO_2 nanotubes with improved photoelectrochemical water splitting performance. *Nanoscale Res. Lett.* **2013**, *8*, 391. [CrossRef]
191. Wang, J.; Yang, P.; Huang, B. Self-doped TiO_2-x nanowires with enhanced photocatalytic activity: Facile synthesis and effects of the Ti^{3+}. *Appl. Surf. Sci.* **2015**, *356*, 391–398. [CrossRef]
192. Teng, F.; Li, M.; Gao, C.; Zhang, G.; Zhang, P.; Wang, Y.; Chen, L.; Xie, E. Preparation of black TiO_2 by hydrogen plasma assisted chemical vapor deposition and its photocatalytic activity. *Appl. Catal. B Environ.* **2014**, *148–149*, 339–343. [CrossRef]
193. Tian, J.; Leng, Y.; Cui, H.; Liu, H. Hydrogenated TiO_2 nanobelts as highly efficient photocatalytic organic dye degradation and hydrogen evolution photocatalyst. *J. Hazard. Mater.* **2015**, *299*, 165–173. [CrossRef] [PubMed]
194. Sarkar, A.; Khan, G.G. The formation and detection techniques of oxygen vacancies in titanium oxide-based nanostructures. *Nanoscale* **2019**, *11*, 3414–3444. [CrossRef] [PubMed]
195. Pan, X.; Yang, M.-Q.; Fu, X.; Zhang, N.; Xu, Y.-J. Defective TiO_2 with oxygen vacancies: Synthesis, properties and photocatalytic applications. *Nanoscale* **2013**, *5*, 3601–3614. [CrossRef]
196. Zhao, H.; Pan, F.; Li, Y. A review on the effects of TiO_2 surface point defects on CO_2 photoreduction with H_2O. *J. Materiomics* **2017**, *3*, 17–32. [CrossRef]
197. Liu, G.; Yin, L.-C.; Wang, J.; Niu, P.; Zhen, C.; Xie, Y.; Cheng, H.-M. A red anatase TiO_2 photocatalyst for solar energy conversion. *Energy Environ. Sci.* **2012**, *5*, 9603–9610. [CrossRef]
198. Yang, Y.; Yin, L.-C.; Gong, Y.; Niu, P.; Wang, J.-Q.; Gu, L.; Chen, X.; Liu, G.; Wang, L.; Cheng, H.-M. An unusual strong visible-light absorption band in red anatase TiO_2 photocatalyst induced by atomic hydrogen-occupied oxygen vacancies. *Adv. Mater.* **2018**, *30*, 1704479. [CrossRef] [PubMed]

Article

Oxidation Catalysis of Au Nano-Particles Immobilized on Titanium(IV)- and Alkylthiol-Functionalized SBA-15 Type Mesoporous Silicate Supports

Tomoki Haketa, Toshiaki Nozawa, Jun Nakazawa, Masaya Okamura and Shiro Hikichi *

Department of Material and Life Chemistry, Faculty of Engineering, Kanagawa University, 3-27-1 Rokkakubashi, Kanagawa-ku, Yokohama 221-8686, Japan
* Correspondence: hikichi@kanagawa-u.ac.jp; Tel.: +81-45-481-5661

Abstract: Novel Au nano-particle catalysts immobilized on both titanium(IV)- and alkylthiol-functionalized SBA-15 type ordered mesoporous silicate supports were developed. The bi-functionalized SBA-15 type support could be synthesized by a one-pot method. To the synthesized supports, Au was immobilized by the reaction of the alkylthiol groups on the supports with $AuCl_4^-$, following reduction with $NaBH_4$. The immobilized amount and the formed structures and the electronic property of the Au species depended on the loading of alkylthiol. The moderate size (2–3 nm) nano particulate Au sites formed on $Ti(0.5)$-$SBA^{SH}(0.5)$ were negatively charged. The aerobic alcohol oxidation activity of the catalysts depended on the loading of alkylthiol and the structure of the Au nano-particles. The non-thiol-functionalized catalyst ($Au/Ti(0.5)$-$SBA^{SH}(0)$) composed of the large (5–30 nm) and the higher thiol-loaded catalyst ($Au/Ti(0.5)$-$SBA^{SH}(8)$) composed of the small cationic Au species were almost inactive. The most active catalyst was $Au/Ti(0.5)$-$SBA^{SH}(0.5)$ composed of the electron-rich Au nano-particles formed by the electron donation from the highly dispersed thiol groups. Styrene oxidation activity in the presence of 1-phenylethanol with O_2 depended on the loadings of titanium(IV) on the $Au/Ti(x)$-$SBA^{SH}(0.5)$. The titanium(IV) sites trapped the H_2O_2 generated through the alcohol oxidation reaction, and also contributed to the alkene oxidation by activating the trapped H_2O_2.

Keywords: Au nano-particles; aerobic oxidation; functionalized mesoporous silicate

Citation: Haketa, T.; Nozawa, T.; Nakazawa, J.; Okamura, M.; Hikichi, S. Oxidation Catalysis of Au Nano-Particles Immobilized on Titanium(IV)- and Alkylthiol-Functionalized SBA-15 Type Mesoporous Silicate Supports. *Catalysts* **2023**, *13*, 35. https://doi.org/10.3390/catal13010035

Academic Editors: Gassan Hodaifa, Antonio Zuorro, Joaquín R. Dominguez, Juan García Rodríguez, José A. Peres, Zacharias Frontistis and Mha Albqmi

Received: 30 November 2022
Revised: 16 December 2022
Accepted: 21 December 2022
Published: 24 December 2022

1. Introduction

The catalysis of nano-scale Au particles has attracted much attention. In particular, the oxidation catalysis of Au nano-particles is interesting because of their applicability toward various oxidation reactions including dehydrogenative oxidation of alcohol and oxygenation of hydrocarbons [1–10]. Catalytic performances of Au particles are affected by the size of the particles. To control the particle sizes and prevent aggregating the particles, interactions between Au and soft donors have been utilized [11–16]. Particularly, organic thiol-modified silicates have been employed as precursors of solid supports of Au nano-particle catalysts [15,16]. During the immobilization process, Au(III) precursors are reduced to Au(I) species by thiols and stable Au–thiolate complexes are formed. Following reduction in the Au–thiolate compounds yield Au(0) nano-particles. In the Au catalysts prepared by these procedures, loadings of the thiols are relatively high (~ 1 mmol g^{-1}), and these thiol functionalities are removed by calcination [15,16]. The remaining thiols are believed to have negative effects on the catalysis due to changing the electronic properties of the Au particle and preventing the access of substrates to active sites. However, a correlation between the thiol loadings and properties of the supported Au nano-particle catalysts has not been investigated systematically.

To improve the activity toward hydrocarbon oxidation, Au nano-particle catalysts are combined with other active species, such as titanium(IV) sites of titanosilicates [17,18]. Several titanium(IV) compounds are known to catalyze the alkene epoxidation with hydrogen peroxide in homogeneous and heterogeneous system [19–25]. The heterogeneous titanosilicate compounds have been reported to catalyze not only the epoxidation of alkenes, but also the oxygenation of alkanes and aromatic rings using hydrogen peroxide as an oxidant [24–28]. During these catalytic reactions based on titanium(IV), peroxido complexes of titanium(IV) might form as the reaction intermediate. It is known that aerobic alcohol oxidation is catalyzed by Au nano-particle catalysts, and during the alcohol oxidation, H_2O_2 would be formed and the resulting generated H_2O_2 could be utilized as an oxidant in situ. Therefore, a combination of Au nano-particle catalysts and other species which activate H_2O_2 would become an aerobic oxidation catalyst applicable to various substrates including alkenes and alkanes. In this work, we have designed novel Au nano-particle catalysts immobilized on both titanium(IV)- and alkylthiol-functionalized SBA-15 type mesoporous silicates.

2. Results

2.1. Preparation and Characterization of the Catalysts

2.1.1. Preparation and Characterization of Bi-Functionalized SBA-15 Type Supports

The titanium(IV)- and alkylthiol-functionalized SBA-15 type supports Ti(x)-SBA^{SH}(y), of which x and y denote the molar ratio of the titanium and alkylthiol sources, respectively, were synthesized by the one-pot condensation of alkoxide precursors in the presence of polymer micelle templates. Molar ratio of precursors Si(OEt)$_4$: Ti(OiPr)$_4$: Si(C$_3$H$_6$SH)(OMe)$_3$ were controlled as $100 - (x + y)$: x: y (Scheme 1).

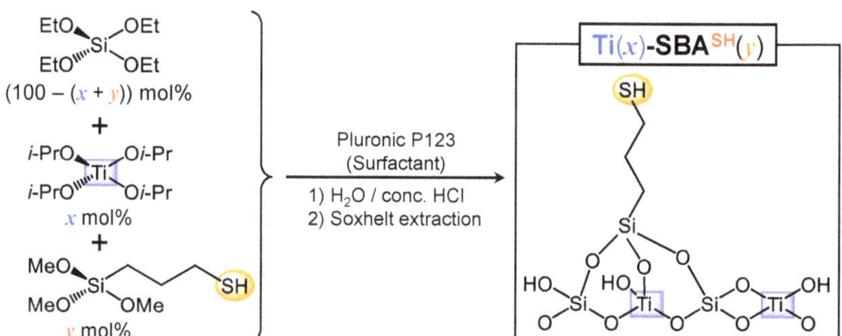

Scheme 1. Preparation of the bi-functionalized supports Ti(x)-SBA^{SH}(y).

The contents of titanium(IV) and alkylthiol, and the physicochemical properties of Ti(0.5)-SBA^{SH}(y), prepared by using 0.5 mol% of Ti(OiPr)$_4$ and 0–10 mol% of Si(C$_3$H$_6$SH)(OMe)$_3$ are shown in Table 1. The nitrogen adsorption/desorption isotherms of the obtained supports showed type IV hysteresis, confirming the presence of mesopores (Figure S1). The surface area, pore size distribution, and pore volume derived from the isotherms were also consistent with the characteristics of SBA-15 type mesoporous silica support except for Ti(0.5)-SBA^{SH}(10), of which the surface area and pore volume were decreased significantly (Figure 1a, see also XRD analysis data shown as Figure S2 in Supplementary Materials). The loading amounts of alkylthiol and titanium(IV) were quantified by analysis of the solution of the supports obtained by alkaline treatment with the method on ^{1}H NMR for alkylthiol and UV-vis spectroscopy for titanium(IV). The loadings of alkylthiol were proportional to the amount of Si(C$_3$H$_6$SH)(OMe)$_3$ applied on the preparation when y was less than 8. The loading amount of alkylthiol on Ti(0.5)-SBA^{SH}(10) was, however, decreased due to partial destruction of the SBA-15 type mesoporous channel structure. Furthermore, the anchored alkylthiol did not affect the loading of titanium(IV) as evidenced by the almost constant amounts of titanium involved in Ti(0.5)-SBA^{SH}(y).

Table 1. The contents of functionalities and the physicochemical properties on Ti(x)-SBA^{SH}(y).

Applied Ratio/mol%		Contents/mmol g^{-1}		Surface Area/m^2 g^{-1}	Average Pore Diameter/nm	Total Pore Volume /cm^3 g^{-1}
x	y	Ti	Alkylthiol			
0.5	0	0.064	—	847.3	8.57	0.82
	0.25	0.062	0.035	1013.1	10.09	0.95
	0.5 *	0.066 *	0.064 *	971.4 *	9.58 *	1.04 *
	0.75	0.063	0.101	1058.3	8.96	0.97
	1	0.068	0.121	948.1	9.54	1.15
	4	0.066	0.486	710.2	9.46	0.71
	8	0.056	0.900	726.3	8.14	0.99
	10	0.058	0.761	415.8	6.41	0.43
0	0.5	—	0.062	897.5	10.05	0.91
0.5 *		0.066 *	0.064 *	971.4 *	9.58 *	1.04 *
1		0.152	0.062	795.2	9.64	0.79
2		0.309	0.063	819.2	7.48	0.85
3		Not measured		732.0	Not calculated	0.56

* Same sample of Ti(0.5)-SBA^{SH}(0.5).

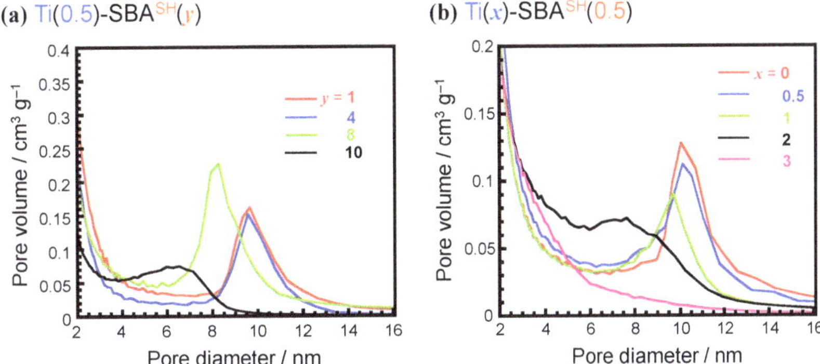

Figure 1. The distribution of the pore diameters of the bi-functionalized supports. (a) Ti(0.5)-SBA^{SH}(y); (b) Ti(x)-SBA^{SH}(0.5).

Titanium-loading controlled supports, Ti(x)-SBA^{SH}(0.5) were prepared with the applied amount of alkyl thiols fixed at 0.5 mol% and with varying applied amounts (x) of Ti(OiPr)$_4$ from 0 to 3 mol% on the preparation (Figures S2 and S4). The pore diameter distributions indicate that the loadings of titanium(IV) over 2.0 mol% lead to the destruction of the SBA-15 type structure as shown in Figure 1b because Ti(IV) favors a six-coordinate structure, and that leads to the formation of TiO$_2$ crystalline domains as reported previously [29,30]. The characteristics of Ti(x)-SBA^{SH}(0.5) are summarized in Table 1. The loading amounts of titanium(IV) were proportional to the applied ratio of Ti(OiPr)$_4$ on the preparation whereas the incorporated amounts of alkylthiol were almost constant.

2.1.2. Preparation and Characterization of Au-Immobilized Catalysts

The reaction of the prepared thiol-functionalized supports with yellow-colored EtOH solution of NaAuCl$_4$ (1.0 wt% based on the weight of the support was applied) yielded ionic Au-immobilized precursors. After filtration and washing, the color of the resulting ionic Au species-anchored supports was white. The observed color-changing behavior suggests that the thiols perform as a reducing agent (Au(III) + 2RSH → Au(I) + RSSR), and an anchor to form Au(I)-S(thiolate or disulfide) complexes.

Reduction in the Au(I)-anchored supports by NaBH$_4$ in EtOH yielded the catalysts Au/Ti(x)-SBA^{SH}(y). The color of the catalysts varied from reddish brown to pale brown on the thiol-functionalized supports. The non-thiol-functionalized support catalyst Au/Ti(0.5)-

SBA^{SH}(0)*, which was obtained by direct deposition of Au particles generated in situ through the reduction in the Au(III) ions by NaBH$_4$, colored purple. These colorations (shown in Figure 2c) are due to the surface plasmon resonance on the gold nano-particles, and the color difference is attributed to the difference in particle size.

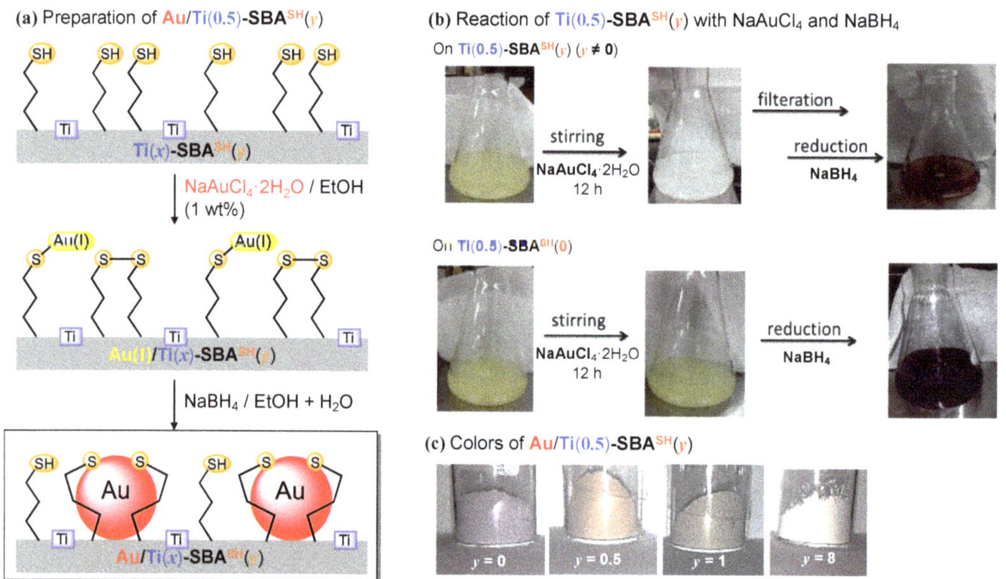

Figure 2. Preparation of Au-immobilized catalysts Au/Ti(x)-SBA^{SH}(y): (**a**) A synthetic procedure. (**b**) Reaction mixture of the supports with NaAuCl$_4$·2H$_2$O and reduction by NaBH$_4$. (**c**) Colors of the obtained Au-immobilized catalysts.

TEM and STEM images of Au/Ti(0.5)-SBA^{SH}(y) revealed the correlation between the loading amount of alkylthiol and the size of formed Au nano-particles (Figure 3). On the lower thiol loading support (y = 0.5), the particle sizes of Au were 2–5 nm (Figure S5 in Supplementary Materials). The Au nano-particle sizes were decreased with the increased loading amount of thiol on the supports. On the higher alkylthiol loading support (y = 8.0), no Au particles larger than 1 nm in diameter could be observed. On the non-thiol-functionalized support (y = 0), the formed Au particles were larger than 5 nm.

The structures of the Au sites of the catalysts and their precursors (non-reduced Au ions immobilized) were analyzed by EXAFS (Figure 4). On the precursors, Au–S bonds were observed. On the lower thiol functionalized support (y = 0.5), the reduction with NaBH$_4$ led to reducing the Au–S bonds, whereas an increase in the Au–Au bonds was due to the formation of the Au nano-particles. In contrast, many Au–S bonds remained on the higher thiol-functionalized support (y = 8) even after the treatment with NaBH$_4$. A small extent of Au–Au bonds appeared, and, therefore, strong Au–S interaction would prevent the forming of the Au clusters. Electronic states of the resulting Au species were also varied according to the amounts of the alkylthiol loadings y. On the support with y = 0.5, the binding energy of XPS of the formed Au particles was slightly shifted to lower compared to that of the metallic Au formed on the non-thiol-functionalized support (y = 0). In contrast, increasing the thiol functionalities (y = 8) yielded cationic Au with higher binding energy as found in Figure 5.

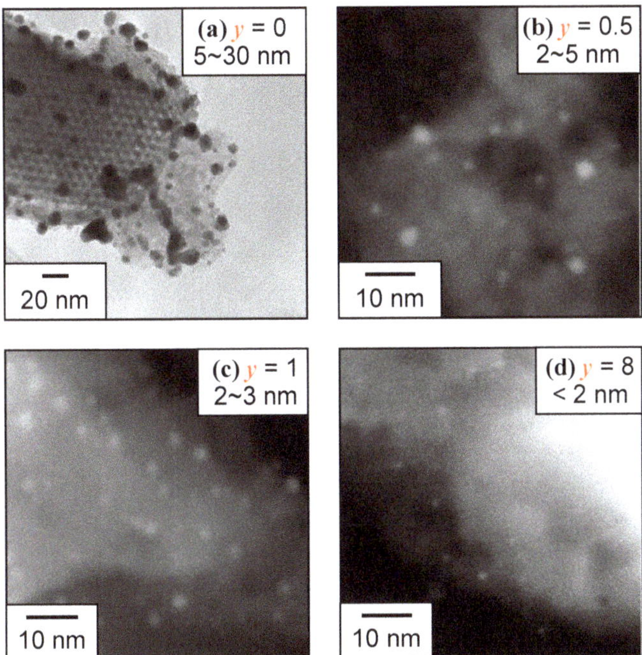

Figure 3. TEM (**a**) and STEM (**b–d**) images of Au/Ti(0.5)-SBA^{SH}(y) with y = 0, 0.5, 1, and 8.

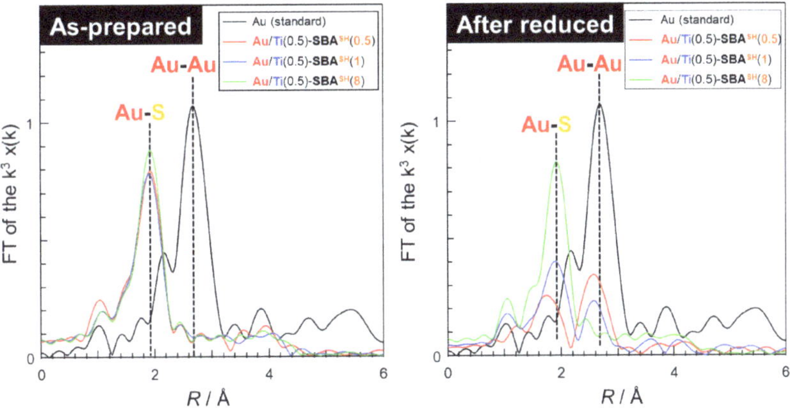

Figure 4. XAFS of Au/Ti(0.5)-SBA^{SH}(y) with y = 0, 0.5, 1, and 8.

The immobilized amount of Au also depended on the amount of alkylthiol on the supports. For the non-thiol-functionalized support (y = 0), a few Au species were immobilized on the solid filtered, and washed from the suspension of Ti(0.5)-SBA^{SH}(0) and NaAuCl$_4$ in EtOH. For Au/Ti(0.5)-SBA^{SH}(y) with $0 < y \leq 4$, the amount of the immobilized Au increased as increasing the alkylthiol content but was almost the same for the supports with y = 4 and y = 8 (Figure 6 and Table 2), respectively. On the titanium loadings varied catalysts Au/Ti(x)-SBA^{SH}(0.5) with $0 < x \leq 2$, the loading amount of Au was almost constant as shown in Table 2. The color of these catalysts was the same. Therefore, the loadings of titanium(IV) did not affect the immobilized amount and size of the Au nano-particles.

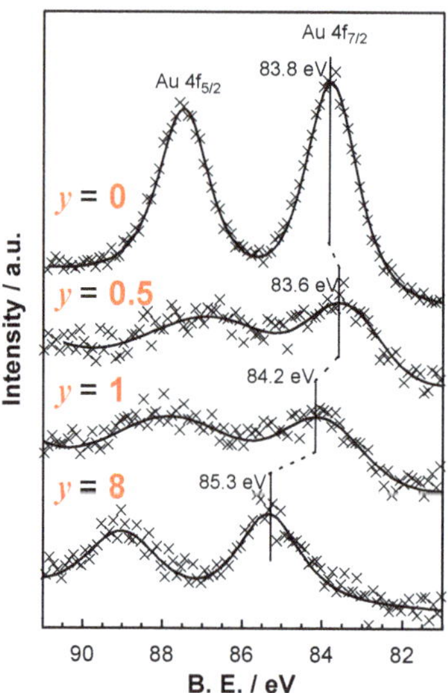

Figure 5. XPS of Au/Ti(0.5)-SBA^{SH}(y) with y = 0, 0.5, 1, and 8.

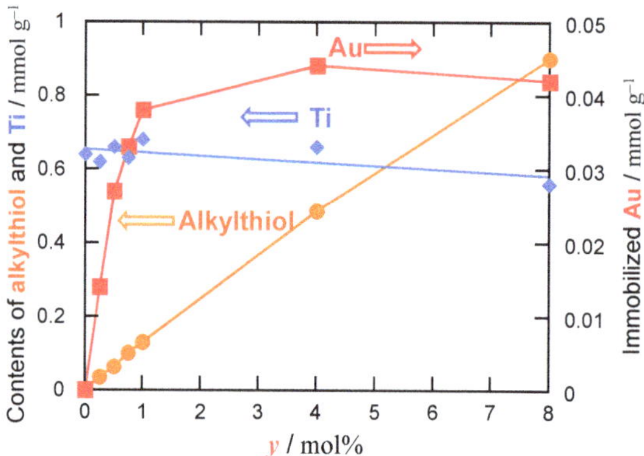

Figure 6. Correlation between the immobilized amount of Au and the contents of alkylthiol and titanium(IV) on Au/Ti(0.5)-SBA^{SH}(y).

Table 2. Immobilized amounts of Au on Au/Ti(x)-SBA^{SH}(y).

Catalyst	Alkylthiol/mmol g^{-1}	Immobilized Au/mmol g^{-1}
Au/Ti(0.5)-SBA^{SH}(0)	—	<0.001
Au/Ti(0.5)-SBA^{SH}(0) [1]	—	0.049
Au/Ti(0.5)-SBA^{SH}(0.25)	0.035	0.014
Au/Ti(0.5)-SBA^{SH}(0.5)	0.064	0.027
Au/Ti(0.5)-SBA^{SH}(0.75)	0.101	0.033
Au/Ti(0.5)-SBA^{SH}(1)	0.121	0.038
Au/Ti(0.5)-SBA^{SH}(4)	0.486	0.044
Au/Ti(0.5)-SBA^{SH}(8)	0.900	0.042
Au/Ti(0)-SBA^{SH}(0.5)	0.062	0.028
Au/Ti(1)-SBA^{SH}(0.5)	0.062	0.030
Au/Ti(2)-SBA^{SH}(0.5)	0.063	0.029

[1] Prepared by the deposition of in situ generated Au(0) particles.

2.1.3. Characterization of the Calcined Catalyst

TG analysis of the catalysts with $y = 1$ and 8, respectively, indicated that the weight loss of the catalyst occurred over 423 K due to the combustion of alkylthiol. As described above, the particle size and immobilized amount of Au depended on the loading amount of alkylthiol. In addition, the EXAFS of Au/Ti(0.5)-SBA^{SH}(8) showed the Au–S bond. These observations suggest that alkylthiol interacts with the Au particles formed on the support. To remove alkylthiol, the calcination of the catalyst was examined. The higher alkylthiol loaded Au/Ti(0.5)-SBA^{SH}(8) was calcined at 673 K for two hours under air. The resulting calcined Au/Ti(0.5)-SBA^{SH}(8) was characterized by TEM and XPS. Formation of the nano Au particles over 2 nm diameter was observed on the TEM image, as shown in Figure 7a. The XPS data also indicated that the cationic Au species converted to the metallic one by calcination (Figure 7b).

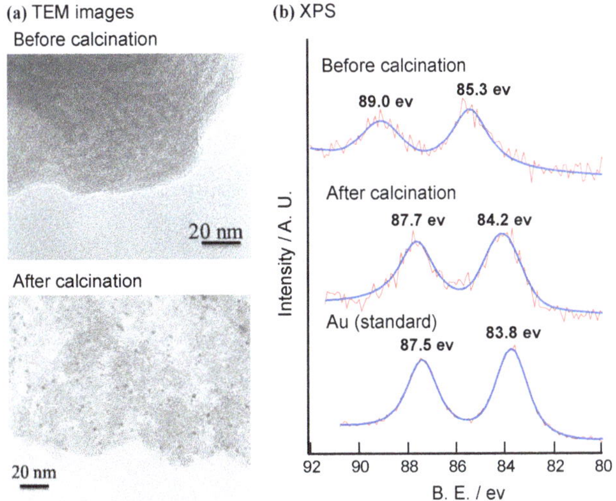

Figure 7. Comparison between the before and after calcined Au/Ti(0.5)-SBA^{SH}(8): (a) TEM images, (b) XPS data. The calcination was done at 673 K for 2 h.

2.2. Catalysis

2.2.1. Aerobic Alcohol Oxidation Activity

Catalysis of the Au/Ti(0.5)-SBA^{SH}(y) was assessed by oxidation of 1-phenylethanol with O_2 at 333 K (Figure 8). In our system, any base additives were not required. The non-thiol-functionalized ($y = 0$) and the higher thiol-loaded ($y = 8$) catalysts were almost inactive. The catalytic efficiency based on the immobilized amounts of Au atoms (=TON of Au) depended on the amounts of the thiol functionalities as $y = 0.25 \sim 0.5 > 0.75 > 1.0 >> 8.0 = 0$ (Figure 8c).

The calcination of the Au/Ti(0.5)-SBA^{SH}(y) ($y = 0.5$, 1, and 8) at 573 or 673 K, respectively, for 2 h under air resulted in changing the catalytic performance. The order of the alcohol oxidizing activity of the catalysts calcined at 573 K was $y = 1 > 0.5 > 8$. On the catalyst with $y = 0.5$, the calcination led to a decrease in the activity. In contrast, the calcination resulted in increasing the activity on the catalyst with $y = 1$, although the improved activity was still lower than the activity of the non-calcined catalyst with $y = 0.5$. Notably, the calcined catalyst with $y = 8$ showed acid catalysis rather than oxidation causing dehydrative condensation (giving compound **2**) and dehydration (giving compound **3**), as shown in Scheme 2 and Table 3. The origin of the acid catalysis might be the sulfate which was formed by the partial oxidation of thiol. Elongation of the calcination time resulted in increasing the alcohol oxidation activity with decreasing the acidic catalytic activity. Finally, washing the 2 h calcined catalyst with H_2O showed the highest activity in the calcined catalysts, although that was still lower than the non-calcinated catalyst with $y = 0.5$.

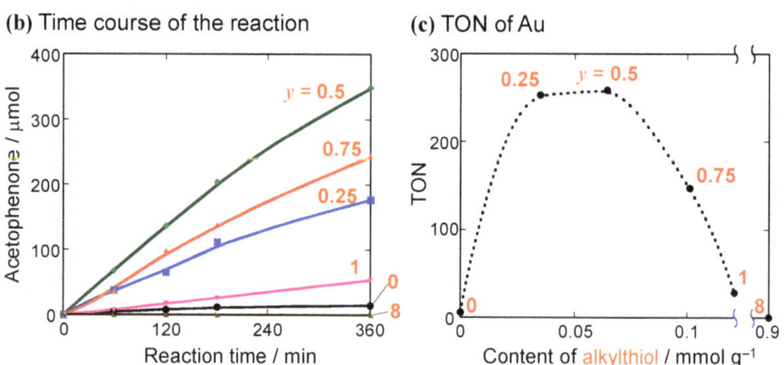

Figure 8. Aerobic oxidation of 1-phenylethanol catalyzed by Au/Ti(0.5)-SBA^{SH}(y): (**a**) Reaction condition; (**b**) time course of the reaction; (**c**) content of alkylthiol versus TON of Au.

Scheme 2. Reaction of 1-phenylethanol catalyzed by the calcined Au/Ti(0.5)-SBA^{SH}(y).

Table 3. Conversion of 1-phenylethanol mediated by the calcined catalysts under O_2.

Catalyst	Calcination Condition		Amount of Products/µmol		TON of Au for 1
	Temp./K	Time/h	1	2 + 3	
Au/Ti(0.5)-SBA^{SH}(0.5)	—	—	349.1	0	258.6
	573	2	99.7	trace	73.9
Au/Ti(0.5)-SBA^{SH}(1)	—	—	53.6	0	29.8
	573	2	135.1	trace	75.1
	673	2	114.7	trace	63.7
Au/Ti(0.5)-SBA^{SH}(8)			0.2	0	0.1<
	573	2	0.8	66.9	0.4
	673	2	4.7	166.1	2.3
	673	4	8.3	110.3	4.0
	673	8	26.7	trace	12.8
	673	12	0	0	0
	673 *	2 *	160.6	0	77.2

* This catalyst was washed with H_2O after the calcination.

2.2.2. Co-Oxidation of Alcohol and Alkene

Then we investigated the role of the titanium(IV) sites by the assay of the styrene oxidation activity in the presence of 1-phenylethanol with O_2 (Scheme 3 and Table 4). In this reaction system, 1-phenylethanol might be a sacrificial reducing reagent and a hydrogen source for H_2O_2 generation. The yields of the styrene-oxidized products were correlated with the loading amounts of titanium(IV) (i.e., value of x), although the major product derived from styrene was benzaldehyde, and the yields of styrene oxide were quite low on each catalyst. When the non-Ti(IV)-loaded catalyst was used, the oxidation of styrene occurred [31]. The styrene oxidation activity was improved on the titanium(IV)-containing catalysts: The most reactive one was $x = 2.0$ in a series of Au/Ti(x)-SBA^{SH}(0.5), although the major product was not epoxide **4**, but aldehyde **5**. In the absence of 1-phenylethanol, the yields of styrene oxidized products were low even on the catalyst with $x = 1.0$. Acetophenone was not formed in the absence of 1-phenylethanol, and that suggests Wacker-type alkene oxidation given ketone did not occur. Co-oxidation of 1-phenylethanol and styrene also proceeded on the mixture of the non-Au-immobilized support Ti(1)-SBA^{SH}(0.5) and the non-titanium (IV)-loaded catalyst Au/Ti(0)-SBA^{SH}(0.5), but the yields of the oxidized products were decreased compared to those on Au/Ti(1)-SBA^{SH}(0.5). This result suggests that synergistic catalysis appears efficiently on the close arrangement of Au and Ti(IV). A trend of the yields of acetophenone which was derived from 1-phenylethanol was similar to that of the yields of styrene oxidized products.

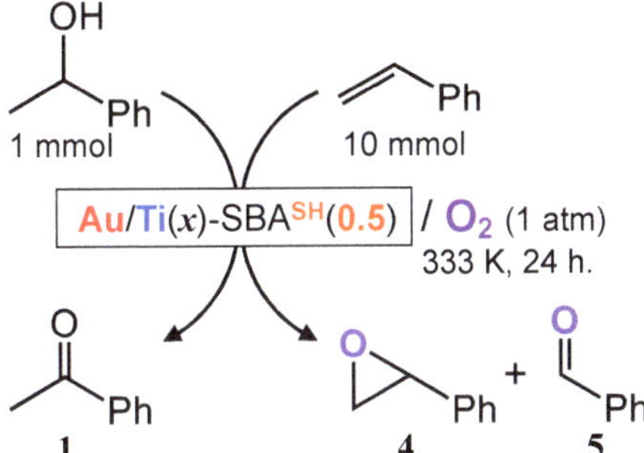

Scheme 3. Aerobic co-oxidation of 1-phenylethanol and styrene catalyzed by Au/Ti(x)-SBA^{SH}(0.5).

Table 4. Conversion of 1-phenylethanol mediated by the calcined catalysts under O_2.

Catalyst	Amount of Products/µmol			TON of Au for 1	TON for 4 + 5 (Styrene Oxidation)	
	1	4	5		Based on Ti	Based on Au
Au/Ti(0)-SBA^{SH}(0.5)	51.2	1.2	17.3	36.6	—	13.2
Au/Ti(0.5)-SBA^{SH}(0.5)	106.1	0.1	30.0	78.6	9.1	22.3
Au/Ti(1)-SBA^{SH}(0.5)	58.3	2.5	55.4	38.9	7.6	38.6
Au/Ti(2)-SBA^{SH}(0.5)	96.9	1.3	83.2	66.8	5.5	58.3
Au/Ti(1)-SBA^{SH}(0.5) [1]	0	4.8	9.1	—	1.8	9.3
Ti(1)-SBA^{SH}(0.5) + Au/Ti(0)-SBA^{SH}(0.5) [2]	39.2	0.4	30.2	28.0	4.0	21.9

[1] Reaction without 1-phenylethanol; [2] Mixture of 50 mg of non-Au-supported Ti(1)-SBA^{SH}(0.5) and 50 mg of Au-immobilized non-Ti-involving support was used.

We examined the detection of the generated H_2O_2 in the absence of styrene. In the suspension including the catalysts, the amounts of detected H_2O_2 were consistent with the styrene oxidizing activity correlated with the loading amount of Ti(IV) (Figure 9a). In the liquid phase, however, no or small amounts of H_2O_2 were detected. Therefore, the formed H_2O_2 was absorbed on the surface of the catalysts. The detection of H_2O_2 even on the non-Ti(IV)-loaded catalyst suggests that the absorption on the silicate surface occurs.

In addition, the detection of the higher amount of H_2O_2 on the Ti-loaded catalysts can be explained why the complexation of H_2O_2 with the surface titanium occurs. The H_2O_2 trapping on the titanium(IV) sites might affect the alcohol oxidation activity. The order of the TON of Au on 1-phenylethanol oxidation was inverse of the loadings of titanium(IV); i.e., the non-titanium(IV)-functionalized catalyst Au/Ti(0)-SBA^{SH}(0.5) was the most active (Figure 9b).

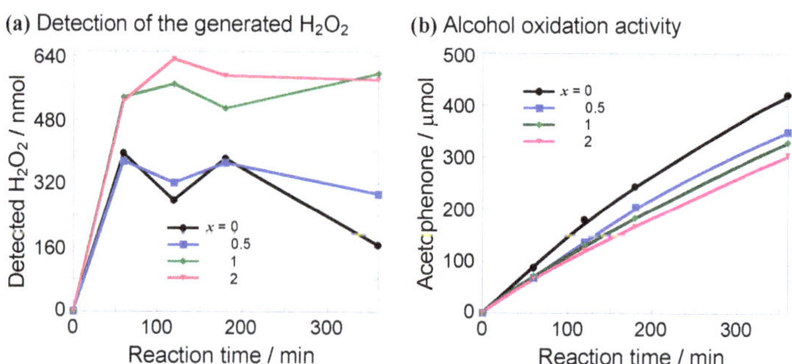

Figure 9. Effects of titanium sites on Au/Ti(x)-SBA^{SH}(0.5): (**a**) Detection of H_2O_2 generated during the oxidation of 1-phenylethanol, (**b**) time course of 1-phenylethanol oxidation with O_2.

3. Discussion

In this work, we developed bi-functionalized SBA-15 type mesoporous silica by using a one-pot synthetic method successfully. The merit of the one-pot synthesis is the highly dispersed arrangement of the introduced functional groups [32]. In our developed supports Ti(x)-SBA^{SH}(y), the alkylthiol groups would be located on the wall of mesopores because the hydrophobic part of the alkylthiol precursor interacted with the polymer micelle templates during the formation of the silicate wall. The titanium(IV) ions were incorporated into the SiO_2 framework of SBA-15 and that resulted in the lower limit of the loading amounts of titanium(IV) (i.e., $x \leq 2$) because titanium(VI) favors octahedral rather tetrahedral coordination geometry. The catalytically active titanium(IV) species might locate at the surface of the silicate wall, and such titanium(IV) species were part of the whole titanium(IV) on Ti(x)-SBA^{SH}(y).

The correlation between the loading amounts of the alkylthiol groups and the immobilized amounts and the size of the particles of Au demonstrates that the alkylthiol groups worked as the initial binding site of ionic Au species and the template of the formed Au nano-particles. Noteworthy, the negatively- and positively-charged Au species formed on Au/Ti(0.5)-SBA^{SH}(0.5) and Au/Ti(0.5)-SBA^{SH}(8), respectively, as shown in Figure 5. Therefore, alkylthiol tunes the electronic property of the immobilized Au species.

The catalytic activity of the non-calcined catalysts toward the aerobic 2-phenylethanol oxidation depended on the loadings of alkylthiol (i.e., y of Au/Ti(0.5)-SBA^{SH}(y)). The most active catalyst was Au/Ti(0.5)-SBA^{SH}(0.5), of which the Au particles had 2–5 nm diameters and were charged negatively. Tsukuda and coworkers reported that the electron donation from Poly(N-vinyl-2-pyrrolidone) to the Au cluster resulted in the negatively-charged Au species [33]. Ebitani and coworkers also reported the formation of negatively-charged Au species in the Au–Pd clusters [34]. The Au nano-particles on Au/Ti(0.5)-SBA^{SH}(0.5) were donated the electron from the thiolate donors. Those negatively-charged Au species activate the absorbed O_2 via the back-donation from Au 5d to π^* orbital of O_2 giving superoxide (O_2^-) or peroxide (O_2^{2-}) like species bound to Au [35]. Therefore, the negatively-charged Au nano-particles formed on the support with y = 0.5 was appropriate as the aerobic oxidation catalyst. On the higher alkylthiol-loaded supports with $0.5 < y \leq 8$, the formed Au species were rather positively charged and covered by alkylthiol. These electronic and structural properties might be a reason for the lower catalytic activity. The partial removal of alkylthiol by calcination improved the activity of the catalysts with $0.5 < y \leq 8$. However, the activity of the calcined catalysts could not exceed that of $Au/Ti(0.5)$-$SBA^{SH}(0.5)$.

The aerobic co-oxidation of 1-phenylethanol and styrene was efficiently promoted by the titanium(IV)-functionalized catalysts, whereas the activity of the non-titanium(IV) loaded catalyst was low. The styrene oxidation activity depended on the loading amounts of titanium(IV). Based on the assumption that the styrene oxidation proceeds only on the titanium site, the catalytic efficiencies of titanium(IV) estimated from the oxidized styrene products per the loaded titanium ion (i.e., TON of Ti(IV)) were inverse to the loading amounts x. As reported previously, higher loading of titanium led to the destruction of the ordered mesoporous structure of the titanosilicate due to the formation of titanium(IV) oxide (TiO_x) domain. In our catalysts, the higher loading of titanium(IV) might increase the inactive TiOx sites.

The results of the H_2O_2 detection experiments suggest the catalytic oxidation mechanism and the roles of the components of the present catalysts. Hydrogen peroxide is generated by the oxidation of alcohol by the gold catalyst. The resulting H_2O_2 is not only absorbed on the surface of the silica support but is also trapped by the titanium on the surface of the support. As a result of the trapping of H_2O_2 by titanium, the oxidation activity is reduced when alcohol is the only substrate. The highest 1-phenylethanol oxidizing activity on $Au/Ti(0)$-$SBA^{SH}(0.5)$ may imply that the formed H_2O_2 is also utilized as the oxidant for alcohol oxidation. In the simultaneous oxidation of alcohols and alkenes, however, alkene oxidation activity is enhanced in the presence of titanium. Therefore, we can conclude that the titanium moiety contributes to the capture and activation of H_2O_2. In our catalysts, titanium might exist both on the surface and in the framework of the silicate wall. The titanium-containing supports absorbed a higher amount of H_2O_2 and that is evidence of the presence of titanium on the surface of the silica wall. However, only some titanium(IV) might be present on the surface. In order to improve the catalytic performance, the amount of the surface titanium species must be increased.

Finally, we will comment on the stability of the present catalyst. Checking the reusability of $Au/Ti(0.5)$-$SBA^{SH}(0.5)$ on the aerobic oxidation of 1-phenylethanol revealed that the activity of the recovered catalyst was reduced (Figure S6 in Supplementary Materials). Measurement of the amount of immobilized gold onto the catalyst after the third use showed that approximately 8.2% of the initially loaded gold had been lost. The loss of gold may have been caused by a change in the Au-binding thiolate anion during the liquid phase oxidation reaction with a protic substrate, but the details are unknown. We are planning to investigate the catalytic activity of the gas-phase reaction in the future.

4. Materials and Methods

4.1. General

Nitrogen sorption/desorption studies were performed at liquid nitrogen temperature (77K) using a TriStar 3000 (Micromeritics Instrument Corporation, Norcross, GA, USA). Before the adsorption experiments, the samples were outgassed under reduced pressure for 3 h at 333 K. The X-ray diffraction data of the powder sample of the supports were collected a RINT-Ultima III (Rigaku, Tokyo, Japan). Inductively coupled plasma mass spectrometry (ICP-MS) was performed on a 7700 Series ICP-MS (Agilent Technologies, Santa Clara, CA, USA). Transmission electron microscopic (TEM) images were observed on a JEM-2010 (JEOL, Tokyo Japan) with an acceleration voltage of 200 kV and LaB_6 cathode. Scanned transmission electron microscopic (STEM) images were observed on a JEM-2100F (JEOL, Tokyo, Japan). Samples were prepared by suspending the catalyst powder ultrasonically in methanol and depositing a drop of the suspension on a standard copper grid covered with carbon monolayer film. EXAFS data were collected on BL01B1 at Spring-8 (Hyogo, Japan). X-ray photoelectron spectroscopy was measured on a JPS-9010 (JEOL, Tokyo, Japan) with a $MgK\alpha$ X-ray source (10 kV, 10 mA) for the analysis of the chemical states of the catalysts. The catalyst was pressed into a 20 mm diameter disk. Then, the disk was mounted on the sample holder of the XPS preparation chamber. The observed binding energy was calibrated by using an O_{1s} transition peak. Thermogravimetric analysis was performed on a Thermo plus EVO (Rigaku, Tokyo Japan). NMR spectra were recorded on an ECX-600 (JEOL, Tokyo, Japan). UV-vis spectra were measured on a V650 (Jasco, Tokyo, Japan). GC analysis was performed on a GC2010 (Shimazu, Kyoto, Japan) with an Rtx-1701 column (length = 30 m, i.d. = 0.25 mm, and thickness = 0.25 μm, Restek, Bellefonte, PA, USA).

All commercial reagents and solvents were used without further purification.

4.2. Preparation of the Bi-Functionalized SBA-15 Type Supports Ti(x)-SBA^{SH}(y)

The bi-functionalized mesoporous silica supports, Ti(x)-SBA^{SH}(y), were prepared similarly for the previously reported alkylthiol-functionalized SBA-15 by us [36]. To control the loading amounts of titanium(IV) and alkylthiol, the condensation ratio of tetraethoxysilane (Si(OEt)$_4$; $100 - (x + y)$ mol%), Titanium(IV) tetraisopropoxide (Ti(OEt)$_4$; x mol%), and mercaptopropyltrimethoxysilane (Si(C$_3$H$_6$SH)(OMe)$_3$; y mol%) were defined as x = 0, 0.5, 1, 2, or 3, respectively, and y = 0, 0.25, 0.5, 0.75, 1, 4, 8, or 10, respectively.

In a flask, 4.1 g of the surfactant Pluronic P123 (triblock copolymer EO$_{20}$PO$_{70}$EO$_{20}$ where EO = poly(ethylene oxide) and PO = poly(propylene oxide)) was placed, then dissolved in 150 mL of aqueous HCl solution (pH 3.0). To the resulting solution, appropriate amounts of the precursors (summarized in Table 1) were added. The mixture was stirred at 323 K for 20 h and subsequently heated and leave to stand at 373 K for 24 h. Once cooled, the solid product was filtered and washed with water and ethanol. The P123 surfactant was removed by Soxhlet extraction with a mixture of 100 mL of water and 200 mL of ethanol over a 24 h period. The resulting white solid was dried under a vacuum.

Loading amounts of titanium(IV) and alkylthiol were determined by UV-vis and 1H NMR spectrometry, respectively. The sample solution was prepared as follows. 10 mg of Ti(x)-SBA^{SH}(y) was dissolved in 2 mL of D$_2$O solution containing small amount of NaOH with heating. The quantitative analysis of titanium(IV) was done by using diantipyrylmethane as the indicator for colorimetric quantitative analysis. For the quantity of alkylthiol, the NMR sample was prepared as follows. To the alkaline D$_2$O solution, 10 mg (0.072 mmol) of p-nitrophenol was added as an internal standard. The loading amounts of the thiol groups were estimated by comparison of the integration values of 1H NMR signals of ethylene and phenyl groups.

4.3. Preparation of the Catalysts Au/Ti(x)-SBA^{SH}(y)

In a flask, 0.50 g of the support Ti(x)-SBA^{SH}(y) (except y = 0) were suspended in 30 mL of EtOH. In another flask, 0.01 g (0.025 mmol) of NaAuCl$_4$·2H$_2$O was dissolved in 20 mL of EtOH. The resulting EtOH solution of AuCl$_4^-$ was added dropwise to the suspension of the support, and the mixture was stirred for 12 h to adsorb gold ions on the support surface. The resulting solid was collected by filtration and washed with 400 mL of EtOH. The Au ions-absorbed support was suspended in 50 mL of EtOH. In a glass vessel, 20 mg (0.50 mmol) of NaBH$_4$ was dissolved in 5 mL of a mixture of H$_2$O and EtOH (1:1, v/v). The resulting NaBH$_4$ solution was added to the EtOH suspension of the Au ions-absorbed support. Then colored powder catalyst was collected, washed with 200 mL of EtOH, and dried under a vacuum.

The non-alkylthiol-functionalized support Ti(0.5)-SBA^{SH}(0) could not absorb Au ions. Therefore, Au/Ti(0.5)-SBA^{SH}(0) was prepared by the deposition of the in situ generated Au particles. To 0.5 g of Ti(0.5)-SBA^{SH}(0) suspended in EtOH, 0.01 g (0.025 mmol) of NaAuCl$_4$·2H$_2$O was added, then reduced by NaBH$_4$. The resulting purple-colored solid was collected, washed with EtOH, and dried under a vacuum.

The immobilized amount of Au was quantified by ICP-MS. The catalyst was dissolved in aqua regia, and the resulting solution was diluted with H$_2$O to apply ICP-MS.

4.4. Catalytic Aerobic Oxidation of 1-Phenylethanol

In the reaction vessel, 50 mg of the catalyst and 13 mg (100 mmol) of naphthalene (as an internal standard for GC analysis) were placed under an O$_2$ atmosphere. Then, 4 mL of toluene (solvent) was charged and warmed at 333 K. Finally, 122 mL of (1.0 mmol) of 1-phenylethanol (substrate) was injected.

The by-produced H_2O_2 was analyzed by the method reported by Takamura and co-workers [37].

4.5. Catalytic Co-Oxidation of 1-Phenylethanol and Styrene

The reaction procedures were essentially similar to those of the aerobic oxidation of 1-phenylethanol (see above), although this co-oxidation reaction was performed without solvent. In the reaction vessel, 50 mg of the catalyst and 13 mg (100 mmol) of naphthalene (as an internal standard for GC analysis) were placed under an O_2 atmosphere. Then, 1.1 mL (10 mmol) of styrene (substrate) was charged and warmed at 333 K. Finally, 122 mL of (1.0 mmol) of 1-phenylethanol (substrate) was injected.

5. Conclusions

The titanium(IV)- and alkylthiol-functionalized SBA-15 type mesoporous silicate supports, $Ti(x)$-$SBA^{SH}(y)$, were synthesized by the one-pot condensation of the alkoxide precursors in the presence of the polymer micelle templates. The molar ratio of the precursors $Si(OEt)_4$: $Ti(OiPr)_4$: $Si(C_3H_6SH)(OMe)_3$ were controlled as $100 - (x + y)$: x: y where $0 \leq x \leq 2$ and $0 \leq y \leq 8$, respectively. The supports synthesized with such composition showed the desired SBA-15 type characteristics, whereas the higher loadings of the functionalities (such as $x = 3$ or $y = 10$) destroyed the ordered mesoporous framework.

The catalysts $Au/Ti(x)$-$SBA^{SH}(y)$ were prepared by the reaction of the synthesized supports with $Au^{III}Cl_4{}^-$ and following reduction with $NaBH_4$. The structure and electronic properties of the immobilized Au species depended on the loadings of alkylthiol. The lower alkylthiol loaded support with $y = 0.5$ gave the negatively charged nano particulate Au species, whereas the higher loaded one with $y = 8$ yielded the positively charged Au clusters with the sulfur ligands. The aerobic alcohol oxidation activity also depended on the loadings of alkylthiol. The most active catalyst in a series of $Au/Ti(0.5)$-$SBA^{SH}(y)$ was that with $y = 0.5$, of which the negatively charged moderate-sized Au nano-particles might yield the active superoxide or peroxide adduct species. Both of the non-thiol-functionalized catalyst $Au/Ti(0.5)$-$SBA^{SH}(0)$ and higher alkylthiol-loaded catalyst $Au/Ti(0.5)$-$SBA^{SH}(8)$ were almost inactive due to the inactivity of the larger Au particles and the highly thiolate-covered small Au species.

The co-oxidation of styrene and 1-phenylethanol was promoted by $Au/Ti(x)$-$SBA^{SH}(0.5)$. The yields of the oxygenated compounds derived from styrene were increased on the increasing of the titanium loadings (i.e., x), although the major product was not styrene oxide but benzaldehyde. Hydrogen peroxide is generated by the oxidation of alcohol by the gold catalyst. The titanium moiety contributes to the capture and activation of H_2O_2.

Reuse of $Au/Ti(0.5)$-$SBA^{SH}(0.5)$ on the aerobic oxidation of alcohol resulted in reducing the catalytic activity due to the leaching of Au. The loss of Au might have been caused by a change in the Au-binding thiolate anion. We are planning to investigate the catalytic activity of the gas-phase reaction in the future.

Supplementary Materials: The following supporting information can be downloaded at: https://www.mdpi.com/article/10.3390/catal13010035/s1, Figure S1: N_2 adsorption isotherms of $Ti(0.5)$-$SBA^{SH}(y)$; Figure S2: N_2 adsorption isotherms of $Ti(x)$-$SBA^{SH}(0.5)$; Figure S3: XRD patterns of $Ti(0.5)$-$SBA^{SH}(y)$; Figure S4: XRD patterns of $Ti(x)$-$SBA^{SH}(0.5)$; Figure S5: Distributions of the particle size of the immobilized Au on $Ti(0.5)$-$SBA^{SH}(y)$ with $y = 0.5$ and 1; Figure S6: Reuse test of $Au/Ti(0.5)$-$SBA^{SH}(0.5)$ on the aerobic oxidation of 1-phenylethanol.

Author Contributions: Conceptualization, T.H., T.N., J.N. and S.H.; investigation, T.H. and T.N.; data curation, T.H., T.N., J.N. and M.O.; writing—original draft preparation, T.H.; writing—review and editing, M.O. and S.H.; supervision, S.H. All authors have read and agreed to the published version of the manuscript.

Funding: This research was funded by CREST, JST (JPMJCR16P1) and Kanagawa University (ordinary budget for 411).

Institutional Review Board Statement: Not applicable.

Data Availability Statement: The data presented in this study are contained within this article and are supported by the data in the Supplementary Materials.

Conflicts of Interest: The authors declare no conflict of interest.

References

1. German, D.; Pakrieva, E.; Kolobova, E.; Carabineiro, S.A.C.; Stucchi, M.; Villa, A.; Prati, L.; Bogdanchikova, N.; Cortés Corberán, V.; Pestryakov, A. Oxidation of 5-Hydroxymethylfurfural on Supported Ag, Au, Pd and Bimetallic Pd-Au Catalysts: Effect of the Support. *Catalysts* **2021**, *11*, 115. [CrossRef]
2. Cattaneo, S.; Stucchi, M.; Villa, A.; Prati, L. Gold Catalysts for the Selective Oxidation of Biomass-Derived Products. *ChemCatChem* **2019**, *11*, 309–323. [CrossRef]
3. Pakrieva, E.; Kolobova, E., German, D., Stucchi, M., Villa, A., Prati, L., Carabineiro, S.A.; Bogdanchikova, N.; Corberán, V.C.; Pestryakov, A. Glycerol Oxidation over Supported Gold Catalysts: The Combined Effect of Au Particle Size and Basicity of Support. *Processes* **2020**, *8*, 1016. [CrossRef]
4. Jouve, A.; Stucchi, M.; Barlocco, I.; Evangelisti, C.; Somodic, F.; Villa, A.; Prati, L. Carbon-Supported Au Nanoparticles: Catalytic Activity Ruled Out by Carbon Support. *Top. Catal.* **2018**, *61*, 1928–1938. [CrossRef]
5. Carabineiro, S.A.C. Supported Gold Nanoparticles as Catalysts for the Oxidation of Alcohols and Alkanes. *Front. Chem.* **2019**, *7*, 709. [CrossRef] [PubMed]
6. Ishida, T.; Murayama, T.; Taketoshi, A.; Haruta, M. Importance of Size and Contact Structure of Gold Nanoparticles for the Genesis of Unique Catalytic Processes. *Chem. Rev.* **2020**, *120*, 464–525. [CrossRef]
7. Okumura, M.; Fujitani, T.; Huang, J.; Ishida, T. A Career in Catalysis: Masatake Haruta. *ACS Catal.* **2015**, *5*, 4699–4707. [CrossRef]
8. Yamazoe, S.; Koyasu, K.; Tsukuda, T. Nonscalable Oxidation Catalysis of Gold Clusters. *Accounts Chem. Res.* **2014**, *47*, 816–824. [CrossRef]
9. Gutiérrez, L.-F.; Hamoudi, S.; Belkacemi, K. Synthesis of Gold Catalysts Supported on Mesoporous Silica Materials: Recent Developments. *Catalysts* **2011**, *1*, 97–154. [CrossRef]
10. Aprile, C.; Abad, A.; García, H.; Corma, A. Synthesis and catalytic activity of periodic mesoporous materials incorporating gold nanoparticles. *J. Mater. Chem.* **2005**, *15*, 4408–4413. [CrossRef]
11. Tsunoyama, H.; Sakurai, H.; Tsukuda, T. Size effect on the catalysis of gold clusters dispersed in water for aerobic oxidation of alcohol. *Chem. Phys. Lett.* **2006**, *429*, 528–532. [CrossRef]
12. Liu, Y.; Tsunoyama, H.; Akita, T.; Xie, S.; Tsukuda, T. Aerobic Oxidation of Cyclohexane Catalyzed by Size-Controlled Au Clusters on Hydroxyapatite: Size Effect in the Sub-2 nm Regime. *ACS Catal.* **2011**, *1*, 2–6. [CrossRef]
13. Miyamura, H.; Matsubara, R.; Miyazaki, Y.; Kobayashi, S. Aerobic Oxidation of Alcohols at Room Temperature and Atmospheric Conditions Catalyzed by Reusable Gold Nanoclusters Stabilized by the Benzene Rings of Polystyrene Derivatives. *Angew. Chem. Int. Ed.* **2007**, *46*, 4151–4154. [CrossRef]
14. Truttmann, V.; Drexler, H.; Stöger-Pollach, M.; Kawawaki, T.; Negishi, Y.; Barrabés, N.; Rupprechter, G. CeO$_2$ Supported Gold Nanocluster Catalysts for CO Oxidation: Surface Evolution Influenced by the Ligand Shell. *ChemCatChem* **2022**, *14*, e2022003. [CrossRef] [PubMed]
15. Wu, P.; Bai, P.; Lei, Z.; Loh, K.; Zhao, X.S. Gold nanoparticles supported on functionalized mesoporous silica for selective oxidation of cyclohexane. *Microporous Mesoporous Mater.* **2011**, *141*, 222–230. [CrossRef]
16. Wu, P.; Bai, P.; Yan, Z.; Zhao, G.X.S. Gold nanoparticles supported on mesoporous silica: Origin of high activity and role of Au NPs in selective oxidation of cyclohexane. *Sci. Rep.* **2016**, *6*, 18817. [CrossRef]
17. Bravo-Suárez, J.J.; Bando, K.K.; Akita, T.; Fujitani, T.; Fuhrer, T.J.; Oyama, S.T. Propane reacts with O$_2$ and H$_2$ on gold supported TS-1 to form oxygenates with high selectivity. *Chem. Commun.* **2008**, 3272–3274. [CrossRef]
18. Huang, J.; Akita, T.; Faye, J.; Fujitani, T.; Takei, T.; Haruta, M. Propene Epoxidation with Dioxygen Catalyzed by Gold Clusters. *Angew. Chem. Int. Ed.* **2009**, *48*, 7862–7866. [CrossRef] [PubMed]
19. Solé-Daura, A.; Zhang, T.; Fouilloux, H.; Robert, C.; Thomas, C.M.; Chamoreau, L.-M.; Carbó, J.J.; Proust, A.; Guillemot, G.; Poblet, J.M. Catalyst Design for Alkene Epoxidation by Molecular Analogues of Heterogeneous Titanium-Silicalite Catalysts. *ACS Catal.* **2020**, *10*, 4737–4750. [CrossRef]
20. Goto, Y.; Kamata, K.; Yamaguchi, K.; Uehara, K.; Hikichi, S.; Mizuno, N. Synthesis, Structural Characterization, and Catalytic Performance of Dititanium-Substituted g-Keggin Silicotungstate. *Inorg. Chem.* **2006**, *45*, 2347–2356. [CrossRef]
21. Bhadra, S.; Akakura, M.; Yamamoto, H. Design of a New Bimetallic Catalyst for Asymmetric Epoxidation and Sulfoxidation. *J. Am. Chem. Soc.* **2015**, *137*, 15612–15615. [CrossRef] [PubMed]
22. Talsi, E.P.; Samsonenko, D.G.; Bryliakov, K.P. Titanium Salan Catalysts for the Asymmetric Epoxidation of Alkenes: Steric and Electronic Factors Governing the Activity and Enantioselectivity. *Chem. Eur. J.* **2014**, *20*, 14329–14335. [CrossRef] [PubMed]

23. Berkessel, A.; Günther, T.; Wang, Q.; Neudörfl, J.-M. Titanium Salalen Catalysts Based on cis-1,2-Diaminocyclohexane: Enantioselective Epoxidation of Terminal Non-Conjugated Olefins with H_2O_2. *Angew. Chem. Int. Ed.* **2013**, *52*, 8467–8471. [CrossRef]
24. Yu, Y.; Fang, N.; Chen, Z.; Liu, D.; Liu, Y.; He, M. Catalyst Design for Alkene Epoxidation by Molecular Analogues of Heterogeneous Titanium-Silicalite Catalysts. *ACS Sustainable Chem. Eng.* **2022**, *10*, 11641–11654. [CrossRef]
25. Vinu, A.; Srinivasu, P.; Miyahara, M.; Ariga, K. Preparation and Catalytic Performances of Ultralarge-Pore TiSBA-15 Mesoporous Molecular Sieves with Very High Ti Content. *J. Phys. Chem. B* **2006**, *110*, 801–806. [CrossRef] [PubMed]
26. Wang, J.; Zhao, Y.; Yokoi, T.; Kondo, J.N.; Tatsumi, T. High-Performance Titanosilicate Catalyst Obtained through Combination of Liquid-Phase and Solid-Phase Transformation Mechanisms. *ChemCatChem* **2014**, *6*, 2719–2726. [CrossRef]
27. Huybrechts, D.R.C.; De Bruycker, L.; Jacobs, P.A. Oxyfunctionalization of alkanes with hydrogen peroxide on titanium silicalite. *Nature* **1990**, *345*, 240–242. [CrossRef]
28. Sasaki, M.; Sato, Y.; Tsuboi, Y.; Inagaki, S.; Kubota, Y. Ti-YNU-2: A Microporous Titanosilicate with Enhanced Catalytic Performance for Phenol Oxidation. *ACS Catal.* **2014**, *4*, 2653–2657. [CrossRef]
29. Li, G.; Zhao, X.S. Characterization and Photocatalytic Properties of Titanium-Containing Mesoporous SBA-15. *Ind. Eng. Chem. Res.* **2006**, *45*, 3569–3573. [CrossRef]
30. Zhang, W.-H.; Lu, J.; Han, B.; Li, M.; Xiu, J.; Ying, P.; Li, C. Direct Synthesis and Characterization of Titanium-Substituted Mesoporous Molecular Sieve SBA-15. *Chem. Mater.* **2002**, *14*, 3413–3421. [CrossRef]
31. Turner, M.; Golovko, V.B.; Vaughan, O.P.H.; Abdulkin, P.; Berenguer-Murcia, A.; Tikhov, M.S.; Johnson, B.F.G.; Lambert, R.M. Selective oxidation with dioxygen by gold nanoparticle catalysts derived from 55-atom clusters. *Nature* **2008**, *454*, 981–983. [CrossRef] [PubMed]
32. Nakazawa, J.; Stack, T.D.P. Controlled Loadings in a Mesoporous Material: Click-on Silica. *J. Am. Chem. Soc.* **2008**, *130*, 14360–14361. [CrossRef] [PubMed]
33. Tsunoyama, H.; Ichikuni, N.; Sakurai, H.; Tsukuda, T. Effect of Electronic Structures of Au Clusters Stabilized by Poly(*N*-vinyl-2-pyrrolidone) on Aerobic Oxidation Catalysis. *J. Am. Chem. Soc.* **2009**, *131*, 7086–7093. [CrossRef] [PubMed]
34. Nishimura, S.; Yakita, Y.; Katayama, M.; Higashimine, K.; Ebitani, K. The role of negatively charged Au states in aerobic oxidation of alcohols over hydrotalcite supported AuPd nanoclusters. *Catal. Sci. Technol.* **2013**, *3*, 351–359. [CrossRef]
35. Staykov, A.; Nishimi, T.; Yoshizawa, K.; Ishihara, T. Oxygen Activation on Nanometer-Size Gold Nanoparticles. *J. Phys. Chem. C* **2012**, *116*, 15992–16000. [CrossRef]
36. Nakamizu, A.; Kasai, T.; Nakazawa, J.; Hikichi, S. Immobilization of a Boron Center-Functionalized Scorpionate Ligand on Mesoporous Silica Supports for Heterogeneous Tp-Based Catalysts. *ACS Omega* **2017**, *2*, 1025–1030. [CrossRef]
37. Matsubara, C.; Iwamoto, T.; Nishikawa, Y.; Takamura, K.; Yano, S.; Yoshikawa, S. Coloured species formed from the titanium(IV)-4-(2′-pyridylazo)resorcinol reagent in the spectrophotometric determination of trace amounts of hydrogen peroxide. *J. Chem. Soc. Dalton Trans.* **1985**, 81–84. [CrossRef]

Article

The Electrochemical Reaction Kinetics during Synthetic Wastewater Treatment Using a Reactor with Boron-Doped Diamond Anode and Gas Diffusion Cathode

Mohammad Issa [1,*], Dennis Haupt [1], Thorben Muddemann [2], Ulrich Kunz [2] and Michael Sievers [1]

[1] CUTEC Research Center for Environmental Technologies, Clausthal University of Technology, 38678 Clausthal-Zellerfeld, Germany

[2] Institute of Chemical and Electrochemical Process Engineering, Clausthal University of Technology, 38678 Clausthal-Zellerfeld, Germany

* Correspondence: mohammad.issa@cutec.de

Citation: Issa, M.; Haupt, D.; Muddemann, T.; Kunz, U.; Sievers, M. The Electrochemical Reaction Kinetics during Synthetic Wastewater Treatment Using a Reactor with Boron-Doped Diamond Anode and Gas Diffusion Cathode. *Water* **2022**, *14*, 3592. https://doi.org/10.3390/w14223592

Academic Editors: Gassan Hodaifa, Antonio Zuorro, Joaquín R. Dominguez, Juan García Rodríguez, José A. Peres, Zacharias Frontistis and Mha Albqmi

Received: 22 September 2022
Accepted: 7 November 2022
Published: 8 November 2022

Publisher's Note: MDPI stays neutral with regard to jurisdictional claims in published maps and institutional affiliations.

Abstract: A system of boron-doped diamond (BDD) anode combined with a gas diffusion electrode (GDE) as a cathode is an attractive kind of electrolysis system to treat wastewater to remove organic pollutants. Depending on the operating parameters and water matrix, the kinetics of the electrochemical reaction must be defined to calculate the reaction rate constant, which enables designing the treatment reactor in a continuous process. In this work, synthetic wastewater simulating the vacuum toilet sewage on trains was treated via a BDD-GDE reactor, where the kinetics was presented as the abatement of chemical oxygen demand (COD) over time. By investigating three different initial COD concentrations ($C_{0,1} \approx 2 \times C_{0,2} \approx 4 \times C_{0,3}$), the kinetics was presented and the observed reaction rate constant $k_{obs.}$ was derived at different current densities (20, 50, 100 mA/cm^2). Accordingly, a mathematical model has derived $k_{obs.}$ as a function of the cell potential E_{cell}. Ranging from 1×10^{-5} to 7.4×10^{-5} s^{-1}, the $k_{obs.}$ is readily calculated when E_{cell} varies in a range of 2.5–21 V. Furthermore, it was experimentally stated that the highest economic removal of COD was achieved at 20 mA/cm^2 demanding the lowest specific charge (~7 Ah/g$_{COD}$) and acquiring the highest current efficiency (up to ~48%).

Keywords: kinetics; electrochemical reaction; rate constant; electrolysis system; boron-doped diamond electrode; gas diffusion electrode

1. Introduction

Studying the kinetics of the electrochemical mineralization of organic pollutants in wastewater is one of the most important research aspects helping to define the reaction rate constant (k), which is very essential in the matter of reactor design and optimization of the operating parameters. The residence time in the reactor can be readily defined if the value of k is known. Accordingly, the reactor volume can be obtained through the flowrate of the treated wastewater achieving a certain removal efficiency [1]. This arrangement is very beneficial for discharged wastewater on public transport services such as trains or cruise ships, where the wastewater properties change significantly according to the passenger numbers and travel time during the day.

Using boron-doped diamond electrodes (BDD) as an anode in the electrolysis cells to mineralize organic pollutants has drawn more attention in the last decades [2–8] due to the generation of hydroxyl radicals (•OH), which completely degrades water organics to CO$_2$ with zero-discharge and without the formation of toxic by-products. Those distinctive features are missing in other wastewater treatment methods, such as ozone, that forms undesired by-products [9] or biological treatment processes that produce sludge in considerable amounts. •OH generated at the BDD surface mineralizes organics (R) in the electrolyte, according to reaction Equations (1) and (2), while the energy-wasting undesired reaction (3) leads to oxygen evolution due to mass transport limitations [10].

$$BDD + H_2O \rightarrow BDD(\bullet OH) + H^+ + e^- \tag{1}$$

$$BDD(\bullet OH) + R \rightarrow BDD + \text{mineralization products} + H^+ + e^- \tag{2}$$

$$BDD(\bullet OH) \rightarrow BDD + O_2 + H^+ + e^- \tag{3}$$

A combination of BDD as anode with a gas diffusion electrode (GDE) in the same electrolysis cell showed higher effectivity to remove organic pollutants from wastewater, as more reactive species are released at low energy consumption [11,12]. These oxygen-reactive species contribute to mineralizing organics in the electrolyte solution, as well as $\bullet OH$ released at the BDD surface [13,14]. Produced by reduction of oxygen at GDE according to Equation (4), H_2O_2 decomposes by $\bullet OH$ generating more oxygen reactive species such as $\bullet O_2H$, $\bullet O_2$, $\bullet OH$, $^1O_2$ [15–17].

$$O_2 + 2H^+ + 2e^- \rightarrow H_2O_2 \tag{4}$$

The electrochemical mineralization of organics relates to the operating parameters in the electrolysis cell [18]. For example, the physico-chemical properties of the electrolyte [19] can increase or diminish the mineralization efficiency, which is affected by other effectors such as conductivity [20] and type of organics [21]. Current density and cell potential also play a very important role to determine the efficiency of the electrolysis system. Regarding the other operating parameters, increasing the current density can promote the mineralization rate and consequently the efficiency, as the $\bullet OH$ generation rate will be effectively accelerated [22]; however, it depends on the types of organics aimed to be degraded [23]. The application of very high current density leads to the formation of side products, such as oxygen, through the reaction of oxygen evolution (Equation (3)), decreasing the current efficiency, because reactive species are wasted without promoting the mineralization of the organic [24]. Treating the vacuum toilet sewage generated on trains or cruise ships is an environmental matter which has gained more attention from many suppliers during the last two decades [25]. This sewage is a kind of low-diluted blackwater with high organics concentration (COD = 3350–25,800 mg/L and BOD_5 = 3750–7424 mg/L), according to Wasielewski et al., 2016 [26]. Different treatment methods have been investigated to deal with such sewage, including a continuously stirred tank reactor (CSTR), an up-flow anaerobic sludge blanket (UASB), and a membrane bioreactor (MBR), but they need much more space than available on the mentioned transition means [27]. Place shortage and production of sludge has forced industry to look for other alternatives. Electrolysis systems can be an optimal choice due to their small occupation size and zero-discharge property [15]. Other methods to treat vacuum toilet sewage were investigated by Haupt et al., 2019 [11] to compare the removal effectivity of organics and the related energy demand. For this goal, Haupt et al. investigated ozonation, peroxone, and BDD-GDE reactor as treatment methods.

The purpose of this study is to investigate the treatment efficiency of organic pollutants, measured as COD and contained in synthetic wastewater, using an electrochemical BDD-GDE reactor. Different values of current density and initial COD concentration were tested. Experimental data were used to describe the kinetics of the electrochemical oxidation depending on COD degradation under the dominant operating conditions, aiming to calculate the related values of the rate constant. A mathematical model to determine the rate constant depending on the desired cell potential value was also proposed.

2. Materials and Methods

The investigated synthetic wastewater (SWW) used to simulate the real vacuum toilet wastewater was developed in cooperation with a train toilet operator [11]. The efficiency and the kinetics of the electrochemical treatment were presented through COD elimination over the experiment time. All conditions, including pH and temperature, were kept the same as that of the vacuum toilet wastewater on trains, to meet the real treatment conditions. The used SWW included different organic sources where glucose was the main organic source, contributing to 96% of the whole COD concentration. To test the effect of the initial COD concentration ($COD_{int.}$) of SWW on the electrochemical mineralization of organics, the original COD concentration of SWW ($C_{0,1}$ of ~4410 $\pm$ 17 mg/L) was diluted to $C_{0,2}$ = 2267 $\pm$ 99 mg/L and $C_{0,3}$ = 1175 $\pm$ 13 mg/L, whereby the electrolyte conductivity σ was reduced from $\sigma_{0,1}$ = 8.58 $\pm$ 0.24 to $\sigma_{0,2}$ = 4.62 $\pm$ 0.12, and $\sigma_{0,3}$ = 2.49 $\pm$ 0.06 mS/cm, respectively.

2.1. Experimental Set-Up

All experiments were conducted using the experimental setup shown in Figure 1a. The SWW as feed was stored in a tank of 3 L. A centrifugal pump (Schmitt Kreiselpumpen GmbH & Co. KG, Ettlingen, Germany) was used to supply the reactor at a flow velocity of 0.23 m/s, from the bottom to the top. The BDD-GDE reactor had flow channels to ensure homogeneous electrolyte flow in the reaction zone (10 cm $\times$ 10 cm). The flow rate profile between both electrodes at the highest flow velocity (0.23 m/s) in the current BDD-GDE reactor was simulated using the computational fluid dynamics (CFD) software and presented in Figure 1b. Air, containing oxygen, needed for the GDE to generate H_2O_2, was provided through a back compartment behind the GDE and its pressure was adjusted to approximately 35 mbar via the following water column. The current was supplied in three different densities 20, 50, and 100 mA/cm^2 by a TDK Lambda power supply (Gen 20–76; Tokyo, Japan), which also enabled controlling both voltage and current applied at the electrolysis reactor. Each current density was investigated at the three different concentrations ($C_{0,1}$, $C_{0,2}$, and $C_{0,3}$) for 4 h, while COD, pH, temperature, voltage, and current were measured over the time of the experiment. A pH-, temperature- and conductivity meter (Windaus-Labortechnik GmbH & Co. KG, Clausthal-Zellerfeld, Germany) was used for these measurements.

The electrolysis reactor had a combination of BDD and GDE without a separator (Figure 1c). In this reactor design, a BDD anode (type DIA-CHEM® by CONDIAS) was coupled with a carbon-based GDE (Printex L6 on Ag-plated Ni mesh; Covestro AG, Leverkusen, Germany). The printex catalyst was chosen for the GDE due to its high yield of H_2O_2 in acid and alkaline environments in comparison to other pure carbon catalysts. Detailed information on electrode characterization and manufacturing method was published by [12]. A PTFE-frame was located between both electrodes to seal the electrochemical reaction zone and keep a 3 mm distance.

2.2. Analysis and Calculations

NaHCO$_3$ (Th. Geyer GmbH & Co. KG, Renningen, Germany) was used to destroy H_2O_2 generated during the process and existed in wastewater samples to avoid overestimation in the COD test [28,29]. The samples were prepared at the ratio 22.5 mol-NaHCO$_3$/mol-H_2O_2 and were shaken at ambient temperature for 24 h by shaking the device (Edmund Bühler GmbH, Bodelshausen, Germany) before COD tests were carried out [30]. A rapid determination of H_2O_2 was conducted by test strips (Macherey-Nagel, Dürren, Germany) to ensure the total decomposition of H_2O_2 in wastewater samples before the COD test. COD was measured by cuvette tests (Macherey-Nagel, Dürren, Germany) and spectrophotometry (Nanoclor UV/VIS; Macherey-Nagel, Dürren, Germany) according to DIN 38 409-H41-1 [31].

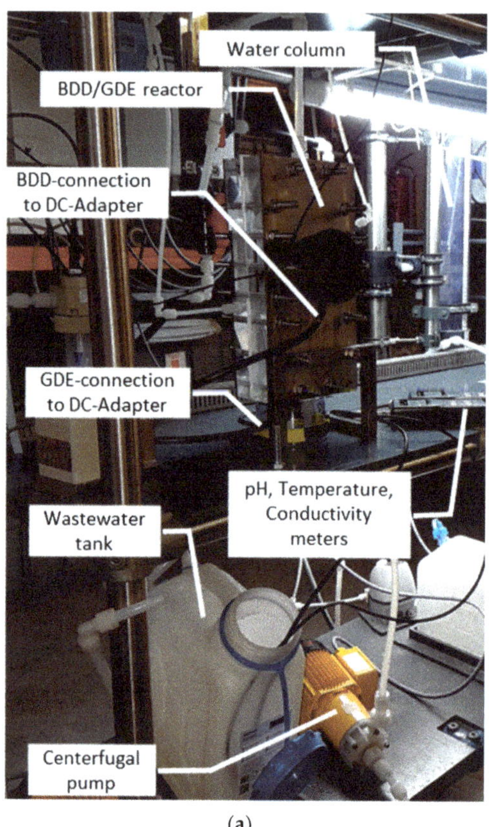

(a)

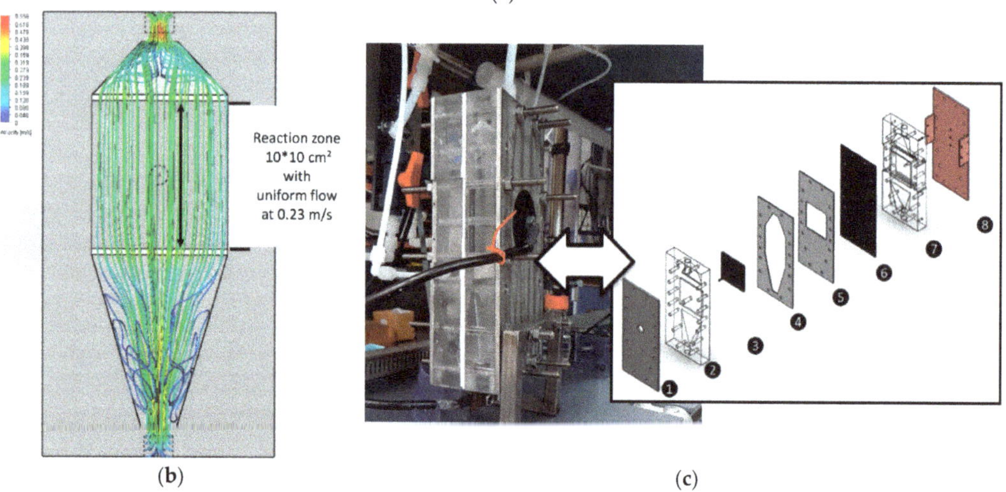

(b)   (c)

Figure 1. (a) Experimental set-up with BDD-GDE reactor to treat synthetic wastewater (SWW); (b) CFD-flow schema at the highest flow velocity (0.23 m/s); (c) BDD-GDE reactor: (1) front metal plate; (2) anode half-shell; (3) boron-doped diamond (BDD) electrode; (4) thin PTFE frame of anode side; (5) thick PTFE frame of cathode side; (6) gas diffusion electrode (GDE); (7) cathode half-shell; (8) front metal plate of cathode side.

The organic mineralization effectivity of such an electrolysis system can be evaluated through the organics removal and current efficiency, in addition to the charge demand needed to mineralize a specific organic quantity. The specific charge demand (SCD) of the electrochemical mineralization can be calculated via Equation (5).

$$SCD = \frac{j \cdot A \cdot t}{V \cdot (COD_0 - COD_t)}$$ (5)

where SCD is the specific charge demand (Ah/g_{COD}), j is current density (mA/cm^2), A is the active area of electrodes (cm^2), t is experiment time (h), V is wastewater volume (L). COD_0 and COD are the chemical oxygen demand at $t = 0$ and the time point t, respectively. Both concentrations are given in g/L.

Current efficiency (CE), specified for the amount of electrochemically removed COD, can be obtained from the equation used by Panizza et al., 2019 [32], expressed in Equation (6).

$$CE = F \cdot V \frac{(COD_0 - COD_t)}{8 \cdot j \cdot A \cdot t}$$ (6)

where F is Faraday's constant ($96{,}485\ C\ mol^{-1}$).

Conducting the oxidation process under certain operating parameters and presenting the experimental data as $\ln COD_0/COD_t$ proportioning linearly to the experiment time, the rate constant can be obtained as a slope of the linear plot of that direct proportionality. Accordingly, a mathematical model can be built on how the rate constant changes depending on the applied cell operation parameters [10]. Such a model can help to estimate the residence time in the reactor for a certain electrolyte at desired settings to achieve a certain removal efficiency, especially for wastewater with changed properties, such as that on public transport services (i.e., on trains). Accordingly, the reactor volume can be calculated beforehand to meet the maximal flow rate of the discharged wastewater.

On whole, each one of the operating parameters relatively contributes to controlling the electrochemical mineralization kinetics of the organics, which consequently determines the treatment efficiency. The kinetics can be experimentally investigated and graphically presented as COD degradation during the treatment time. Depending on what Schmidt, 2003 stated [33], most electrochemical reactions usually take place as a first-order reaction, which has the following mathematical form [34]:

$$k \cdot t = \ln \frac{COD_0}{COD_t}$$ (7)

where k is the rate constant (s^{-1}).

2.3. Results and Discussion

In the investigated BDD-GDE reactor, the effect of the different current densities (20, 50, and 100 mA/cm^2) at three different $COD_{int.}$ ($C_{0,1} > C_{0,2} > C_{0,3}$) on the organics mineralization rate and specific charge demand was investigated. At all current densities, and after 4 h treatment, lower initial concentration showed a lower total removed COD (TR-COD) and consequently higher SCD, as shown in Figure 2a,b, respectively. Although $C_{0,1}$ is approx. 4 times higher than $C_{0,3}$, the TR-COD at $C_{0,1}$ was only 21–34% higher than that at $C_{0,3}$ at all current densities. Taking the TR-COD value (380 mg/L) at 20 mA/cm^2 and $C_{0,1}$ as a set-point, this value of TR-COD increased by 84% and 163% by the rising of the current density to 50 mA/cm^2 and 100 mA/cm^2, respectively. For the other two concentrations, TR-COD also jumped 120% and 210% at $C_{0,2}$ and 74% and 186% at $C_{0,3}$ as the current density reached 50 and 100 mA/cm^2, respectively.

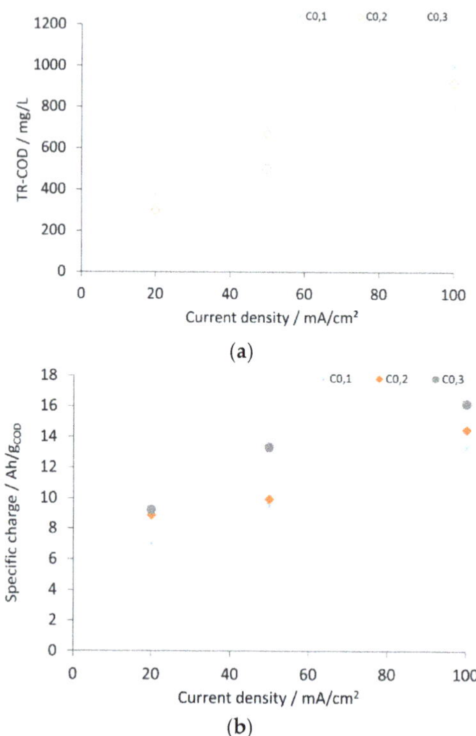

Figure 2. Influence of different $COD_{int.}$ ($C_{0,1}$ = 4410 ± 17 mg/L; $C_{0,2}$ = 2267 ± 99 mg/L; $C_{0,3}$ = 1175 ± 13 mg/L) on: (**a**) total removed COD (TR-COD) and (**b**) specific charge demand (SCD); at different current densities (20, 50, and 100 mA/cm^2) after 4 h treatment.

The interpretation of this result is that the application of a higher current density generated more reactive species contributing to the electrochemical mineralization of organics in the BDD-GDE reactor, which means a higher generation rate of •OH radicals at the BDD surface. One portion of the generated amount of •OH radicals degraded more organics in SWW [35,36], whereas the other portion accelerated also the decomposition rate of H_2O_2 generated by GDE, providing the electrochemical mineralization once again with more oxygen-reactive species, such as •O_2H, •O_2, •OH, $^1O_2$. This agrees with the results reported by Muddemann et al., 2021 for the electrochemical degradation of phenol [12]. It is also important to mention that the H_2O_2 amount generated by the GDE in the reaction medium increases as well by increasing the current density [37]. This contributes to enhancing the quantity of the reactive species •O_2H, •O_2, •OH, $^1O_2$ through the H_2O_2 decomposition by •OH radicals.

On the other side, the comparison between both effects of current density and $COD_{int.}$ increase on the TR-COD amount shows a dominant mass transfer limitation in the reaction medium which is hardly affected by changing the $COD_{int.}$, but is more affected by the applied current density. This result is close to that of [38], where mass transfer controlled the electrochemical mineralization rate at high current density and high reactants concentration.

The positively mentioned effect of increasing current density on the TR-COD led to a charge loss outside the mineralization reaction, indicating an increase in the SCD value. The increase of current density from 20 to 100 mA/cm^2 increased the SCD by 90%, 63%, and 75% at $C_{0,1}$, $C_{0,2}$, and $C_{0,3}$, respectively. Conversely, the SCD decreased by 18%, 29%, and 24% at 20, 50, and 100 mA/cm^2 when $COD_{int.}$ is quadrupled from $C_{0,3}$ to $C_{0,1}$. The reason is that energy consumption increases by higher current density, causing electrons (charge) transfer promoting acceleration of anodic polarization, and higher generation rate of •OH radicals [39]. The dominantly existing mass transfer obligates a portion of •OH to participate in some side (non-mineralization) reactions (i.e., oxygen evolution). The higher the current density is, the higher this participation in the non-mineralization reactions [40] which appears as an increase in SCD magnitude. Increasing the $COD_{int.}$ from $C_{0,3}$ to $C_{0,1}$ offers higher organic molecules that pick the generated •OH radicals via mineralization reaction and as a result lower the SCD slightly.

Charge loss due to side reactions can be also mirrored in Figure 3, by the diminishing current efficiency calculated depending on Equation (6), while the current density increases. Increasing current density (from 20 to 100 mA/cm^2) at the same conditions reduced the current efficiency by 47% (from 47.7% to 25.1%) at $C_{0,1}$ versus 40% and 43% for both $C_{0,2}$ and $C_{0,3}$, respectively. This reduction is attributed to the charge loss through the side reactions/products caused by mass transfer limitation. This result agrees with Zhang et al., 2013 who stated a diminishing current efficiency was due to the formation of side products [35].

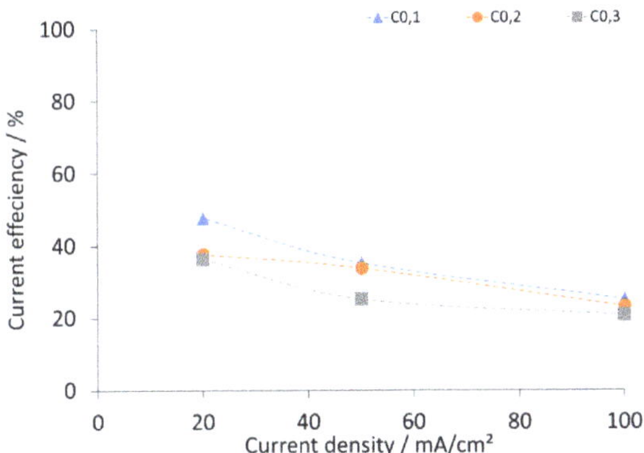

Figure 3. Current efficiency at different current density.

According to Equation (6), current efficiency is related to the TR-COD amount throughout the treatment process time (4 h), which is controlled by the kinetics conditions dominating in the reaction medium during the treatment. The kinetics presented as a timed COD concentration decreasing in the BDD-GDE reactor, as shown in Figure 4, whereby the pH decreased from 7 to 4 and the temperature increased from 10 to 34 °C during the 4 h treatment. It is assumed that the curves of the electrochemical reactions are fitted to the first-order pattern. Continued lines in Figure 4 present the measured COD concentration throughout the experiment time at 100 (a), 50 (b), and 20 mA/cm^2 (c).

At each current density, the COD concentration decreased over time in the approximately same pattern despite the different initial concentrations, whereas the stepwise decrease in current density from 100 to 50 and 20 mA/cm^2 at the same initial concentration softened the slope of the curves, indicating a lower degraded COD amount. The 4 h treatment agrees with the results in Figure 2a. Depending on Equation (7) for first-order reactions, $\ln(COD_0/COD)$ shows a linear relationship to the experiment time at all current densities and $COD_{int.}$ values as shown in Figure 5a–c. As glucose content in the treated wastewater contributes to 96% of the whole COD, glucose was considered as the fingerprint COD source to simplify the deriving of the reaction rate constant. The slope of each curve refers to the value of the rate constant at the related conditions and shows that the rising of current density increased the value of the rate. The effect of increasing current density on the value of the rate constant can be illustrated by the activation of ascending the number of locations on the catalyst (electrode) surface reducing the activation energy of reactants and consequently increasing the rate constants [41,42].

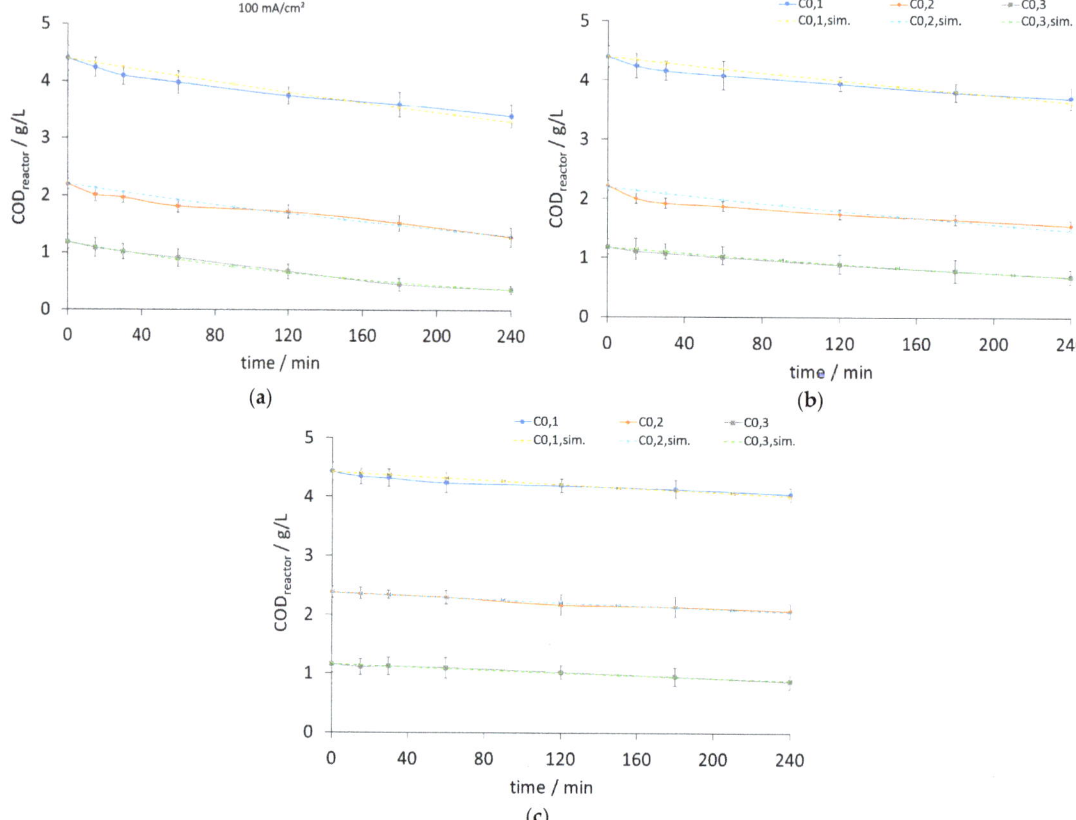

Figure 4. (a–c) COD concentration decrease at different current densities. The continued line is the measured value during the experiments. The dotted line is the theoretical value based on the rate constant calculated from measured values.

This proportionality of the rate constant is not only related to the current density but also the applied cell potential as stated by [33]. As mentioned, the dilution of the original solution ($C_{0,1}$) about 2-fold (generating $C_{0,2}$) and 4-fold (generating $C_{0,3}$) reduced the conductivity of both diluted solutions by an approximate extent from $\sigma_{0,1} = 8.58 \pm 0.24$ to $\sigma_{0,2} = 4.62 \pm 0.12$, and $\sigma_{0,3} = 2.49 \pm 0.06$ mS/cm, respectively, which increased the internal resistance in the treated wastewater. To overcome this difficulty and provide a certain charge input at different conductivities, an increasing initial potential ($V_{0,3} > V_{0,2} > V_{0,1}$) must be applied according to the decreasing conductivity ($\sigma_{0,3} < \sigma_{0,2} < \sigma_{0,1}$) as explained in Table 1.

This necessary increase in the cell potential was the reason for the value increase of the observed rate constant ($k_{obs.}$) which ranged in this work between 0.67×10^{-5} and 8.33×10^{-5} s^{-1}. In the literature, $k_{obs.}$ showed very different values according to the organics to be treated and the treatment conditions. For example, it showed the value 2.4×10^{-5} s^{-1} at 18.90 mA/cm^2 for degrading ethidium bromide [35], 8.8×10^{-5} s^{-1} at 15 mA/cm^2 for the treatment of bisphenol A [43], whereas it was 1.9×10^{-5} s^{-1} at 15 mA/cm^2 during reducing COD of diluted raw leachate [44]. Comparably, $k_{obs.}$ showed a lower value (1.83×10^{-5} s^{-1}) at a current density of 20 mA/cm^2 for glucose degradation close to the current work, indicating that the natural structure of organics considerably controls the rate constant besides the effect of the current density and other parameters [45].

Table 1. Reaction rate constant k and initial potential V_0 at different electrolyte conductivities and current densities.

Current Density (mA/cm^2)	k [1/s]			Initial Potential V_0 [volt]		
	Conductivity [mS/cm]			Conductivity [mS/cm]		
	8.58 ± 0.24	4.62 ± 0.12	2.49 ± 0.06	8.58 ± 0.24	4.62 ± 0.12	2.49 ± 0.06
100	2.00×10^{-5}	3.67×10^{-5}	8.33×10^{-5}	8.03	14.3	20.9
50	1.33×10^{-5}	2.83×10^{-5}	3.83×10^{-5}	6.59	9.51	12.3
20	0.67×10^{-5}	1.00×10^{-5}	1.83×10^{-5}	4.24	4.85	7.8

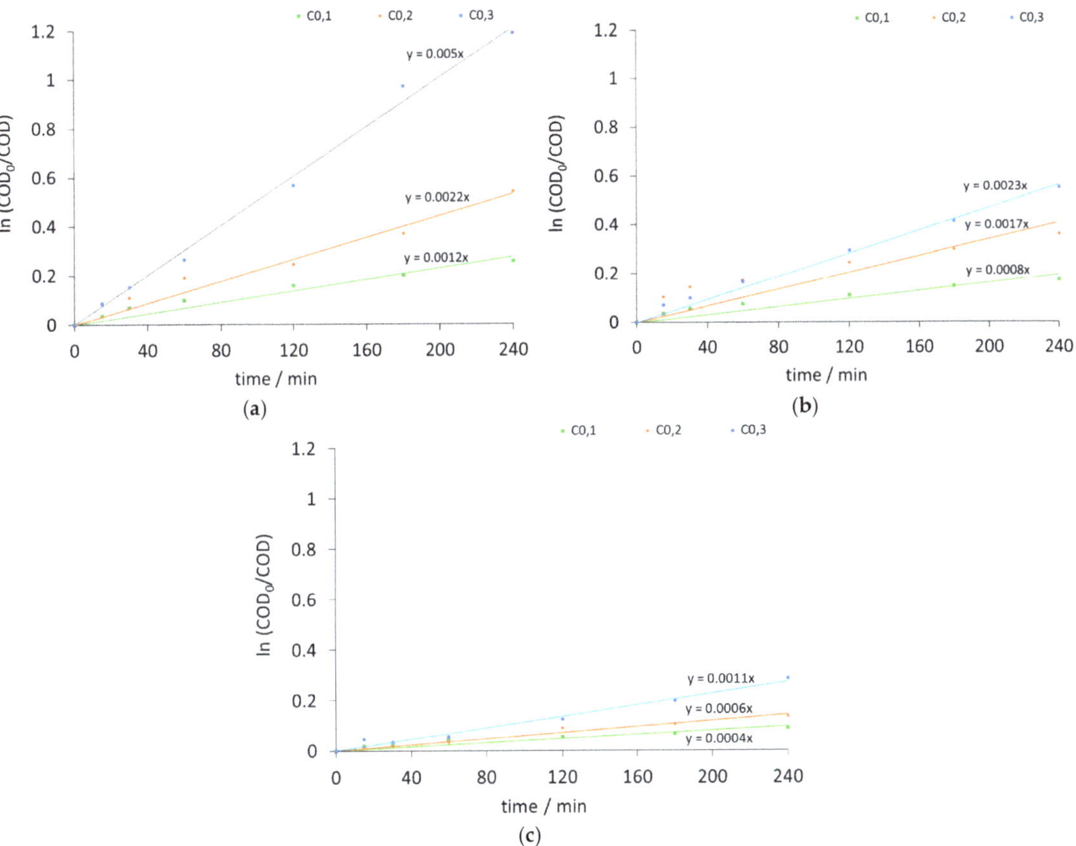

Figure 5. Kinetic curves of the electrochemical reaction as a first-order reaction at different current densities: (**a**) at 100 mA/cm^2; (**b**) at 50 mA/cm^2; (**c**) 20 mA/cm^2, and different concentrations: $C_{0,1}$ (square), $C_{0,2}$ (diamond), and $C_{0,3}$ (circle).

To validate the obtained values of the rate constant listed in Table 1, they were used in the kinetics equation of the first-order reaction (Equation (7)) to theoretically calculate COD concentration profiles in the reactor over time. The theoretical values are presented in Figure 4a–c as dotted lines. Regarding the standard deviation, the theoretical values are fitted to the experimental ones for glucose mineralization by BDD-GDE reactor very well. Depending on values listed in Table 1, three models at three different current densities are graphically represented in Figure 6a which provides a wide range of cell potential to calculate the rate constant.

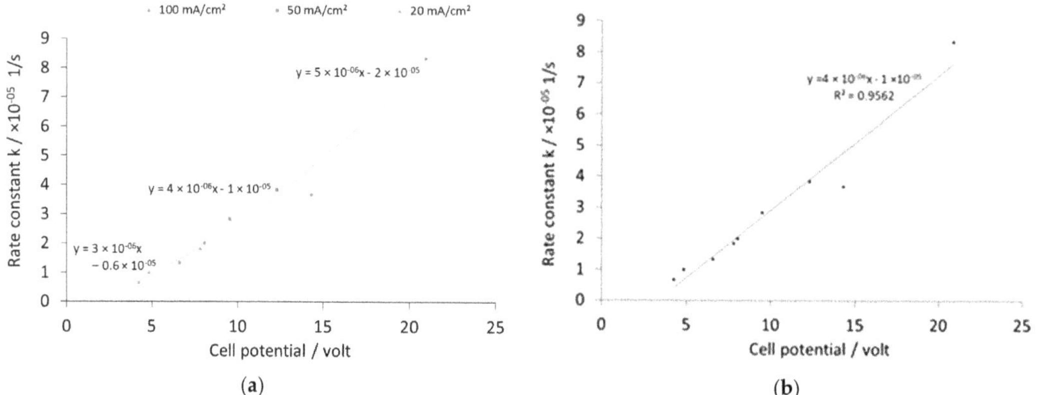

Figure 6. Three mathematical models (**a**) and its uniform model (**b**) to calculate the rate constant of glucose oxidation in BDD-GDE-cell on a wide range of current densities (20–100 mA/cm²) and cell potentials (5–20 V).

The three models at three different current densities (20, 50, and 100 mA/cm²) were assembled in one uniform model which is plotted in Figure 6b. This model allows calculating the rate constant of the electrochemical mineralization of organics in the SWW directly from the measured cell potential needed at a certain conductivity from the range of $2.49 \pm 0.06 - 8.58 \pm 0.24$ mS/cm to provide a certain applied current density ranging from 20 to 100 mA/cm². Indeed, changing the conductivity of the synthetic wastewater will just change the set-point value of the cell potential to meet the desired current density changing the rate constant. It means the rate constant can be directly and beforehand obtained from Equation (8) of the proposed model (Figure 6b) just by determining the cell potential.

$$k = 4 \times 10^{-6} \cdot E_{cell} - 10^{-5} \tag{8}$$

where E_{cell} is the applied cell potential at the BDD-GDE-cell (volt), whereby the model provides the value of $k_{obs.}$ on a range of 1×10^{-5}–7.4×10^{-5} s⁻¹ when the E_{cell} varies from 2.5 to 21 V to provide a current density ranging between 20 and 100 mA/cm².

The validity of the model was mathematically checked on a cell potential range of 6–21 V and showed a maximal standard deviation $\leq \pm 20\%$ of the value of the rate constant calculated via the three models with different current densities (Figure 6a). Additionally, the R-squared value (95.6%) shows the model highly adapting to the $k_{obs.}$ values drawn from the experimental measurements. Therefore, it can be stated that the proposed model (Figure 6b) is derived and checked to deliver trustable values for organics mineralization in the investigated synthetic wastewater at the mentioned temperature, pH, conductivity, and current density limitations.

For electrochemical COD mineralization in raw leachate and according to the model of [44], $k_{obs.}$ reached 7.19×10^{-5} s⁻¹, locating nearby the value 7.4×10^{-5} s⁻¹, obtained in this work at close settings of both cell potential and current density. Such differences in $k_{obs.}$ values are mainly to be interpreted by the different physico-chemical properties of types of organics in the wastewater, which emphasizes the fact that a certain mathematical model must be derived for each type of wastewater, under certain limitations of operating parameters, to be able to determine the rate constant as precisely as possible.

3. Conclusions

In this work, different settings concerning current density and initial COD concentrations were investigated to study the performance of the electrolysis BDD-GDE reactor used to mineralize organics in synthetic wastewater, prepared to simulate real toilet wastewater collected on board trains. In addition, this study enabled us to derive a mathematical equation to determine the rate constant of the electrochemical reaction in advance, which mainly helps design an electrochemical reactor with high accuracy for real situations.

Increasing current densities improves the removal efficiency of the electrolysis BDD-GDE reactor more than increasing initial COD concentrations. For example, increasing the current density from 20 to 50 and 100 mA/cm^2 raised the TR-COD 84% and 163% (at $C_{0,1}$), 120% and 210% (at $C_{0,2}$), as well as 74% and 186% (at $C_{0,3}$), respectively. Conversely, using a higher initial COD concentration such as $C_{0,1}$ (four times higher than $C_{0,3}$) helped increase the TR-COD by just 21–34%, depending on the adjusted current density at the BDD-GDE reactor.

Considering the energy side, the increase of current density from 20 to 100 mA/cm^2 increased the SCD by 90%, 63%, and 75% at $C_{0,1}$, $C_{0,2}$, and $C_{0,3}$, respectively. This means that the advancement in the quantity of mineralized organics was accompanied by a charge loss in the side reactions. Accordingly, the best current efficiency of the system is achieved at low current density (20 mA/cm^2) ranging between 36–48%, depending on the adjusted $COD_{int.}$ value, whereas it diminished to ~23% at all $COD_{int.}$ values when the density increases up to 100 mA/cm^2.

Depending on the degradation of COD in the wastewater during the time of the electrochemical treatment, the reaction kinetics presented by the experimental measurements adopts the pattern of the first-order reaction. The experimental data of the different treatment runs enabled the derivation of a simple mathematical model to calculate the rate constant as a function of the cell potential. According to the model suited for the investigated synthetic wastewater and applied operating parameters, the observed rate constant $k_{obs.}$ ranges between 1×10^{-5} and 7.4×10^{-5} s^{-1} depending on a cell potential of 2.5 to 21 V.

Author Contributions: M.I., conceptualization, methodology, investigation, validation, data curation, visualization, writing—original draft preparation, writing—review and editing. D.H. and T.M., investigation, writing—review and editing. U.K. and M.S., supervision, project administration, funding acquisition, writing—review and editing. All authors have read and agreed to the published version of the manuscript.

Funding: This research was funded by Federal Ministry of Education and Research (Bundesministerium für Bildung und Forschung), BMBF, contract numbers 03XP0107E.

Institutional Review Board Statement: Not applicable.

Informed Consent Statement: Not applicable.

Data Availability Statement: Not applicable.

Acknowledgments: This study was supported by the German Federal Ministry of Education and Research "Bundesministerium für Bildung und Forschung (BMBF)". "Niedersächsisches Ministerium für Wissenschaft und Kultur" (Lower Saxony Ministry for Science and Culture), Hannover, Germany is acknowledged for the financial support received by Mohammad Issa via the scholarship "Wissenschaft.Niedersachsen.Weltoffen".

Conflicts of Interest: The authors declare no conflict of interest. The funders had no role in the design of the study; in the collection, analyses, or interpretation of data; in the writing of the manuscript, or in the decision to publish the results.

References

1. Viessman, W.; Hammer, M. *Water Supply and Pollution Control*; Pearson Education Inc.: Upper Saddle River, NJ, USA, 2005.
2. Frontistis, Z.; Brebou, C.; Venieri, D.; Mantzavinos, D.; Katsaounis, A. BDD anodic oxidation as tertiary wastewater treatment for the removal of emerging micro-pollutants, pathogens and organic matter. *J. Chem. Technol. Biotechnol.* **2011**, *86*, 1233–1236. [CrossRef]
3. Pacheco Mj Santos, V.; Ciríaco, L.; Lopes, A. Electrochemical degradation of aromatic amines on BDD electrodes. *J. Hazard. Mater.* **2011**, *186*, 1033–1041. [CrossRef] [PubMed]
4. Rabaaoui, N.; Saad, M.E.K.; Moussaoui, Y.; Allagui, M.S.; Bedoui, A.; Elaloui, E. Anodic oxidation of o-nitrophenol on BDD electrode: Variable effects and mechanisms of degradation. *J. Hazard. Mater.* **2013**, *250–251*, 447–453. [CrossRef]
5. Chen, X.; Chen, G. Anodic oxidation of Orange II on Ti/BDD electrode: Variable effects. *Sep. Purif. Technol.* **2006**, *48*, 45–49. [CrossRef]
6. Urtiaga, A.; Fernandez-Castro, P.; Gómez, P.; Ortiz, I. Remediation of wastewaters containing tetrahydrofuran. Study of the electrochemical mineralization on BDD electrodes. *Chem. Eng. J.* **2014**, *239*, 341–350. [CrossRef]
7. Klidi, N.; Clematis, D.; Delucchi, M.; Gadri, A.; Ammar, S.; Panizza, M. Applicability of electrochemical methods to paper mill wastewater for reuse. Anodic oxidation with BDD and TiRuSnO$_2$ anodes. *J. Electroanal. Chem.* **2018**, *815*, 16–23. [CrossRef]
8. Ghazouani, M.; Akrout, H.; Bousselmi, L. Nitrate and carbon matter removals from real effluents using Si/BDD electrode. *Environ. Sci. Pollut. Res.* **2017**, *24*, 9895–9906. [CrossRef]
9. Wunderlin, P.; Grelot, J. *Abklärungen Verfahrenseignung Ozonung*; Empfehlung: Glattbrugg, Switerland, 2021.
10. He, Y.; Lin, H.; Guo, Z.; Zhang, W.; Li, H.; Huang, W. Recent developments and advances in boron-doped diamond electrodes for electrochemical oxidation of organic pollutants. *Sep. Purif. Technol.* **2019**, *212*, 802–821. [CrossRef]
11. Haupt, D.; Muddemann, T.; Kunz, U.; Sievers, M. Evaluation of a new electrochemical concept for vacuum toilet wastewater treatment—Comparison with ozonation and peroxone processes. *Electrochem. Commun.* **2019**, *101*, 115–119. [CrossRef]
12. Muddemann, T.; Neuber, R.; Haupt, D.; Graßl, T.; Issa, M.; Bienen, F.; Enstrup, M.; Möller, J.; Matthée, T.; Sievers, M.; et al. Improving the Treatment Efficiency and Lowering the Operating Costs of Electrochemical Advanced Oxidation Processes. *Processes* **2021**, *9*, 1482. [CrossRef]
13. Radjenovic, J.; Petrovic, M. Sulfate-mediated electrooxidation of X-ray contrast media on boron-doped diamond anode. *Water Res.* **2016**, *94*, 128–135. [CrossRef]
14. Busca, G.; Berardinelli, S.; Resini, C.; Arrighi, L. Technologies for the removal of phenol from fluid streams: A short review of recent developments. *J. Hazard. Mater.* **2008**, *160*, 265–288. [CrossRef]
15. Sievers, M. Advanced Oxidation Processes. *Treatise Water Sci.* **2011**, *4*, 377–408.
16. Bokare, A.D.; Choi, W. Singlet-Oxygen Generation in Alkaline Periodate Solution. *Environ. Sci. Technol.* **2015**, *49*, 14392–14400. [CrossRef]
17. Lim, J.; Hoffmann, M.R. Substrate oxidation enhances the electrochemical production of hydrogen peroxide. *Chem. Eng. J.* **2019**, *374*, 958–964. [CrossRef]
18. Upadhyay, S.K. *Chemical Kinetics and Reaction Dynamics*; Springer: New York, NY, USA, 2006.
19. Garcia-Segura, S.; Keller, J.; Brillas, E.; Radjenovic, J. Removal of organic contaminants from secondary effluent by anodic oxidation with a boron-doped diamond anode as tertiary treatment. *J. Hazard. Mater.* **2015**, *283*, 551–557. [CrossRef]
20. Lan, Y.; Coetsier, C.; Causserand, C.; Groenen Serrano, K. On the role of salts for the treatment of wastewaters containing pharmaceuticals by electrochemical oxidation using a boron doped diamond anode. *Electrochim. Acta* **2017**, *231*, 309–318. [CrossRef]
21. He, Y.P.; Chen, R.L.; Wang, C.N.; Li, H.D.; Huang, W.M.; Lin, H.B. Electrochemical oxidation of substituted phenols on a boron doped diamond electrode. *Wuli Huaxue Xuebao/Acta Phys. Chim. Sin.* **2017**, *33*, 2253–2260.
22. Comninellis, C.; Chen, G. *Electrochemistry for the Environment*; Springer: New York, NY, USA, 2010.
23. Cañizares, P.; Louhichi, B.; Gadri, A.; Nasr, B.; Paz, R.; Rodrigo, M.A.; Saez, C. Electrochemical treatment of the pollutants generated in an ink-manufacturing process. *J. Hazard. Mater.* **2007**, *146*, 552–557. [CrossRef]
24. Panizza, M.; Cerisola, G. Application of diamond electrodes to electrochemical processes. *Electrochim. Acta* **2005**, *51*, 191–199. [CrossRef]
25. EVAC. Wastewater Treatment. 2022. Available online: https://evac.com/solutions/wastewater-treatment/ (accessed on 21 September 2022).
26. Wasielewski, S.; Morandi, C. Impacts of blackwater co-digestion upon biogas production in pilot-scale UASB and CSTR reactor. In Proceedings of the 13th IWA Specialized Conference on Small Water and Wastewater Systems & 5th IWA Specialized Conference on Resources-Oriented Sanitation, Belo Horizonte, Brazil, 13–17 September 2020; IWA Publishing Allianceto: Athen, Greece, 2016.
27. Issa, M. *Optimierung Eines Aeroben Membran-Bioreaktor-Systems mit Querstrom-Membranabtrennung*; Papierflieger: Clausthal-Zellerfeld, Germany, 2006.
28. Wu, T.; Englehardt, J.D. A new method for removal of hydrogen peroxide interference in the analysis of chemical oxygen demand. *Environ. Sci. Technol.* **2012**, *46*, 2291–2298. [CrossRef] [PubMed]
29. Groele, J.; Foster, J. Hydrogen Peroxide Interference in Chemical Oxygen Demand Assessments of Plasma Treated Waters. *Plasma* **2019**, *2*, 294–302. [CrossRef]

30. Issa, M.; Muddemann, T.; Haupt, D.R.; Kunz, U.; Sievers, M. Simple Catalytic Approach for Removal of Analytical Interferences caused by Hydrogen Peroxide in a Standard Chemical Oxygen Demand Test. *J. Environ. Eng.* **2021**, *147*, 04021059. [CrossRef]
31. Kolb, M.; Bahadir, M.; Teichgräber, B. Determination of chemical oxygen demand (COD) using an alternative wet chemical method free of mercury and dichromate. *Water Res.* **2017**, *122*, 645–654. [CrossRef]
32. Panizza, M.; Cerisola, G. Electrochemical degradation of gallic acid on a BDD anode. *Chemosphere* **2009**, *77*, 1060–1064. [CrossRef]
33. Schmidt, V.M. *Elektrochemische Verfahrenstechnik*; Wiley-VCH Verlag GmbH & Co. KGaA: Weinheim, Germany, 2003.
34. Gottschalk, C.; Libra, J.A.; Saupe, A. *Ozonation of Water and Waste Water: A Practical Guide to Understanding Ozone and Its Applications*; John Wiley & Sons, Ltd.: Hoboken, NJ, USA, 2010.
35. Zhang, C.; Liu, L.; Wang, J.; Rong, F.; Fu, D. Electrochemical degradation of ethidium bromide using boron-doped diamond electrode. *Sep. Purif. Technol.* **2013**, *107*, 91–101. [CrossRef]
36. Saad Mek Rabaaoui, N.; Elaloui, E.; Moussaoui, Y. Mineralization of p-methylphenol in aqueous medium by anodic oxidation with a boron-doped diamond electrode. *Sep. Purif. Technol.* **2016**, *171*, 157–163. [CrossRef]
37. Muddemann, T.; Haupt, D.R.; Sievers, M.; Kunz, U. Improved Operating Parameters for Hydrogen Peroxide-Generating Gas Diffusion Electrodes. *Chem. Ing. Tech.* **2020**, *92*, 505–512. [CrossRef]
38. Zhang, Y.; Wei, K.; Han, W.; Sun, X.; Li, J.; Shen, J.; Wang, L. Improved electrochemical oxidation of tricyclazole from aqueous solution by enhancing mass transfer in a tubular porous electrode electrocatalytic reactor. *Electrochim. Acta* **2016**, *189*, 1–8. [CrossRef]
39. Zhou, B.; Yu, Z.; Wei, Q.; Long, H.; Xie, Y.; Wang, Y. Electrochemical oxidation of biological pretreated and membrane separated landfill leachate concentrates on boron doped diamond anode. *Appl. Surf. Sci.* **2016**, *377*, 406–415. [CrossRef]
40. Velegraki, T.; Balayiannis, G.; Diamadopoulos, E.; Katsaounis, A.; Mantzavinos, D. Electrochemical oxidation of benzoic acid in water over boron-doped diamond electrodes: Statistical analysis of key operating parameters, kinetic modeling, reaction by-products and ecotoxicity. *Chem. Eng. J.* **2010**, *160*, 538–548. [CrossRef]
41. Savéant, J.M. Molecular catalysis of electrochemical reactions. *Chem. Rev.* **2008**, *108*, 2348–2378. [CrossRef]
42. Jia, Y.; Zhang, L.; Du, A.; Gao, G.; Chen, J.; Yan, X.; Christopher, L.; Brown, X.Y. Defect Graphene as a Trifunctional Catalyst for Electrochemical Reactions. *Adv. Mater.* **2016**, *28*, 95432–99538. [CrossRef]
43. Li, H.; Long, Y.; Zhu, X.; Tian, Y.; Ye, J. Influencing factors and chlorinated byproducts in electrochemical oxidation of bisphenol A with boron-doped diamond anodes. *Electrochim. Acta* **2017**, *246*, 1121–1130. [CrossRef]
44. Urtiaga, A.; Ortiz, I.; Anglada, A. Kinetic modeling of the electrochemical removal of ammonium and COD from landfill leachates. *J. Appl. Electrochem.* **2012**, *42*, 779–786. [CrossRef]
45. Zhu, X.; Shi, S.; Wei, J.; Lv, F.; Zhao, H.; Kong, J.; He, Q.; Ni, J. Electrochemical oxidation characteristics of p-substituted phenols using a boron-doped diamond electrode. *Environ. Sci. Technol.* **2007**, *41*, 6541–6546. [CrossRef]

Article

Rapid AOP Method for Estrogens Removal via Persulfate Activated by Hydrodynamic Cavitation

Petra Přibilová [1,2], Klára Odehnalová [2], Pavel Rudolf [3], František Pochylý [3], Štěpán Zezulka [2], Eliška Maršálková [2], Radka Opatřilová [4] and Blahoslav Maršálek [2,*]

[1] Institute of Chemistry and Technology of Environmental Protection, Faculty of Chemistry, Brno University of Technology, Purkynova 118, 612 00 Brno, Czech Republic
[2] Institute of Botany, Czech Academy of Sciences, Lidická 25/27, 602 00 Brno, Czech Republic
[3] V. Kaplan Department of Fluid Engineering, Faculty of Mechanical Engineering, Brno University of Technology, Technická 2896/2, 616 69 Brno, Czech Republic
[4] Faculty of Pharmacy, Masaryk University, Žerotínovo nám. 1, 602 00 Brno, Czech Republic
* Correspondence: marsalek@sinice.cz

Abstract: The production and use of manufactured chemicals have risen significantly in the last few decades. With interest in preserving and improving the state of the environment, there is also growing interested in new technologies for water purification and wastewater treatment. One frequently discussed technological group is advanced oxidation processes (AOPs). AOPs using sulphur-based radicals appear to reduce the volume of organic contaminants in wastewater significantly. The use of persulfate has excellent potential to successfully eliminate the number of emerging contaminants released into the environment. The main disadvantage of sulphur-based AOPs is the need for activation. We investigated an economically and environmentally friendly solution based on hydrodynamic cavitation, which does not require heating or additional activation of chemical substances. The method was evaluated for emerging contaminant removal research, specifically for the group of steroid estrogens. The mixture of estrone (E1), 17β-estradiol (E2), estriol (E3), and 17α-ethinylestradiol (EE2) was effectively eliminated and completely removed during a treatment that lasted just a few seconds. This novel method can be used in a broad spectrum of water treatment processes or as the intensification of reactions in chemical engineering technologies.

Keywords: hydrodynamic cavitation; advanced oxidation processes; estrogens removal; water treatment; persulfate activation

Citation: Přibilová, P.; Odehnalová, K.; Rudolf, P.; Pochylý, F.; Zezulka, Š.; Maršálková, E.; Opatřilová, R.; Maršálek, B. Rapid AOP Method for Estrogens Removal via Persulfate Activated by Hydrodynamic Cavitation. *Water* **2022**, *14*, 3816. https://doi.org/10.3390/w14233816

Academic Editors: Gassan Hodaifa, Antonio Zuorro, Joaquín R. Dominguez, Juan García Rodríguez, José A. Peres, Zacharias Frontistis and Mha Albqmi

Received: 7 November 2022
Accepted: 21 November 2022
Published: 23 November 2022

Publisher's Note: MDPI stays neutral with regard to jurisdictional claims in published maps and institutional affiliations.

1. Introduction

Steroid estrogens (Figure 1) are representatives of a group of pollutants called endocrine disruptors. Their increased environmental presence poses a potential risk to wildlife and human health, even at low concentrations. Estrogens are suspected of causing the development of certain defects and diseases, such as reproduction dysfunction, metabolic diseases, cancer, and many others. The suggested link to the increased numbers of patients with breast cancer—the most diagnosed cancer in women—is alarming [1]. Besides harming humans and animals, steroid estrogens also affect plant growth [2].

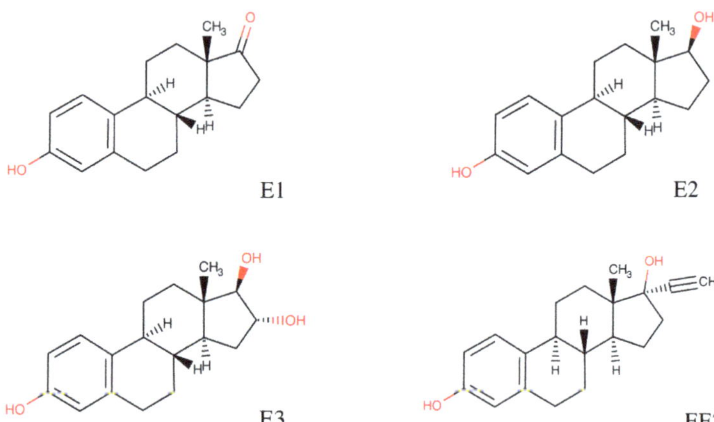

Figure 1. Chemical structure of selected estrogens.

The physical-chemical properties of these compounds play an essential role in predicting their fate in the environment. Estrogens are poorly soluble in water. The values of the octanol/water coefficient (K_{ow})—defined as the ratio of the concentration of a compound in n-octanol and water under equilibrium conditions at a specific temperature—indicate slightly hydrophobic behavior and, thereby, a tendency to sorb to the solid phase and thus be retained in the environment [3,4]. Most estrogens are excreted from the body in the urine in conjugated form. These polar conjugates are biologically inactive and more soluble in water. Nevertheless, at the influent of the wastewater treatment plant, primarily unconjugated estrogens are found, indicating hydrolysis of the conjugates prior to entering the treatment plant caused by bacteria like *E. coli* [4].

The crucial factor in polluting the environment with these substances is inadequate wastewater treatment. Although there are already methods that can satisfactorily reduce concentrations of various pollutants entering the environment, their application in practice are both technologically and economically demanding. On the other hand, wastewater treatment research has been actively producing new technologies based on various mechanisms with various efficiency levels. Traditional Fenton and Fenton-like processes, ozonization, UV-based methods, or other heterogenous photocatalyzed processes have been studied, modified, and intensified for over two decades [5]. Significant development has been achieved in advanced oxidation processes (AOPs) based on highly reactive radicals.

Sulphate radical ($SO_4\bullet^-$)-based AOPs play a significant role in advanced wastewater treatment development. Sulphate radicals are usually generated from persulfate (PS), supplied as $Na_2S_2O_8$, or less commonly from peroxymonosulphate (PMS) in the form of $2\,KHSO_5\cdot KHSO_4\cdot K_2SO_4$ [6]. It should be noted that the PS method is more cost-effective and, in practice, more user-friendly than PMS [7]. Sodium persulfate is a white crystalline compound highly soluble in water (73 g/100 g H_2O at 25 °C [6,8], providing easy manipulation. Although PS is a powerful oxidant, some form of activation needs to be used for pollutant degradation at a reasonable rate [7]. Many papers have already been published on PS activation; some introduce potentially environmentally responsible technologies, such as thermolysis or photolysis.

Heat- or UV-activated persulfate has a significant advantage as it does not require additional chemicals, thus it is a potentially environmentally responsible technology. However, it is necessary to consider the cost and impact of using the electric energy needed for UV lamps or other additional energy to heat the system. Moreover, it must be recognized that heating is not economical because thermal heating has been classified as pollution, nor is it ecological [6] in recent literature. Table 1 lists some examples of PS activation methods.

Table 1. Examples of PS activation methods.

Method	Pollutant	Efficiency	Commentary	Reference
UV and/or transition metals: Fe(II), Fe(III), Co(II), Ag(I)	2,4-dichlorophenol	99.9% within 4 h	The high scavenging effect, possible inhibition by dissolved oxygen, secondhand water contamination with high concentrations of metal ions, prolonged reaction time	[9]
Iron-based nanoparticles (bimetallic zero valent nanoparticles) Fe/PS process	trichloroethylene	>99% in 20 s reaction time	High cost and potential environmental risk caused by nanoparticles	[10]
PS and PMS activation by electrophilic transition metal cations (Ag^+ and Co^{2+}), UV ($300 < \lambda < 400$ nm) and/or heat ($T = 30\ ^\circ$C)	microcystin-LR	~77% in 10 min	Best results achieved at lower pH (pH = 3) and higher PS concentrations [PMS]/[MC-LR] molar ratio = 100	[11]
Magnetite nanoparticles/PS	norfloxacin	90% within 60 min	The concentration of PS 1 mM; dose of nanoparticles: 0.3 g L^{-1}; adjusted pH = 4.0	[12]
TiO_2/light/PS	acetaminophen	up to 100% in 9 h, pH 9	High costs and complicated in practice (high dose of PS, pH adjustment, prolonged reaction time)	[13]
Phenols/PS	nitrobenzene	over 60% in 8 h, pH 11.5	Addition of hazardous chemicals and the need for significant pH adjustment, prolonged reaction time	[14]
PS activated by quinones	PCBs	over 60% in 1 h, over 80% in 2 h	The mechanism of persulfate activation was primarily elucidated	[15]
PS activated by Fe^{2+}	diuron, ibuprofen and caffeine	>90% in 240 min	pH adjustment needed; kinetics model primarily evaluated	[16]
PS activated by nitrogen-modified carbon nanotubes	phenol	>90% under 30 min	Phenol concentration = 20 ppm; catalyst dose 0.2 g L	[17]
PS/activated carbon	Azo dye (orange 7)	80% degradation and 50% mineralization in 5 h	The activation effectiveness decreased by adsorption of the pollutant on the catalyst	[18]
Thermally activated PS	59 volatile organic compounds	>90% in 72 h	The best results were achieved in combination with 5 g 1^{-1} of $Na_2S_2O_8$ at 40 $^\circ$C for 72 h	[19]

Table 1. *Cont.*

Method	Pollutant	Efficiency	Commentary	Reference
Thermally activated PS	antipyrine	80% removal within 2 h	Anaerobic conditions favoured degradation (20%)	[20]
UV/PS	sulfamethazine	>95% in 45 min	Photolysis (22.0%), persulfate oxidation (15.10%), UV/H_2O_2 (87.5%) efficiencies were also investigated	[21]
UV/PS	cylindrospermopsin	>99% in 20 min	UV (less than 5%) and UV/H_2O_2 (~20%) efficiencies were also investigated	[22]
UV/PS	2,4,6-trichloroanisole	>80% in 30 min	Mechanism and kinetics were primarily investigated	[23]
PS/sonolysis	carbamazepine	89.4% in 120 min, pH 3.0	PS and ultrasound efficiencies were also investigated; PS alone with less than 50% and ultrasound with less than 5%	[24]
PS/sonolysis	bisphenol A	>90% under 60 min	High temperatures enhanced sulfate radical formation but impeded sonochemical activity.By-products were also investigated	[25]

Both environmentally and economically sustainable methods are still needed in optimal wastewater treatment technology. The methods mentioned above are only effective to a certain extent and are associated with high time requirements. Experiments are usually performed within tens of minutes, sometimes up to two or three hours, once for days [9–30]. Such a time delay is difficult to achieve in real-life water treatment. Therefore, we present results with a time allowance of a few seconds and an efficiency comparable to or higher than previously published alternatives.

We studied hydrodynamic cavitation (HC) as a persulfate activation process, which is presented as an essential step in persulfate-based AOPs. Hydrodynamic cavitation is based on lowering the pressure in the system, causing the formation of imploding bubbles and a local increase in temperature. The imploding process generates a shock wave with enough energy to produce radicals that are the basis of AOP [31,32]. The main advantage is that there is no need for other added substances nor pH adjustment, and it requires significantly shorter treatment time (seconds) and saves energy (the system does not need to be heated, and cavitation can be provided just with the gravitation-based flow).

2. Materials and Methods

2.1. Chemicals and Reagents

Estrone standards (E1; 99%): 17β-estradiol (E2; 98%), estriol (E3; 98%), and 17α-ethinylestradiol (EE2; 98%) were purchased from Sigma-Aldrich (St. Louis, MO, USA). As internal standards for the quantification of estrogens, deuterated 17β-estradiol (E2D) was used for the quantification of estrone, estradiol, and estriol, and deuterated 17α-ethinylestradiol (EE2D) for the quantification of ethinylestradiol cations, both purchased from C/D/N Isotopes Inc. (Pointe-Claire, QC, Canada). The solvents methanol and acetonitrile for LC-MS and acetone for HPLC were purchased from Sigma-Aldrich (St. Louis, MO, USA). Ultra-purified distilled water was produced directly in the laboratory using a Millipore system (Merck KGaA, Darmstadt, Germany).

Formic acid was used as the mobile phase (0.7 mM), hydrochloric acid to adjust the pH of the samples before analysis, dansyl chloride (1 mg mL in acetone) in the derivatization of estrogens to increase the sensitivity of the method, and sodium bicarbonate (100 mM; pH = 10.5) as a derivatization buffer.

For the experiments, $Na_2S_2O_8$ was purchased from Sigma-Aldrich (St. Louis, MO, USA), and KI and $NaHCO_3$, which was used in the spectrophotometric determination of PS, from Penta, s.r.o., (Czech Republic).

2.2. Experiment Design

The experiments were performed on two litres of spiked water with an estrogen concentration of 300 ng L^{-1} in the cavitation unit consisting of a tank, a pump, Venturi tube, and control valves (Figure 2). The unit operates in circulation mode at speed flow 0.45 L s^{-1}, inlet pressure 450 kPa, and pump power 0.75 kW. The experiment was performed in three variants: (i) PS activation by HC, (ii) with thermal activation of persulfate (60 °C) combined with HC, and (iii) with HC only.

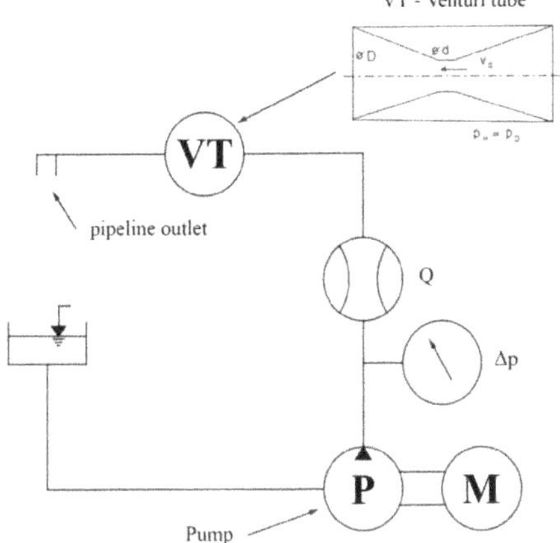

Figure 2. Design of laboratory cavitation equipment.

The PS dose was chosen to be 0.1 mM in accordance with the available literature dealing with a similar topic with a low time requirement and a task to keep the dose at a reasonable level [12,19–25,31].

Treated water was circulated through the system and sampled after 4, 8, 12, and 20 s of treatment (1, 2, 3, and 5 cycles through the system). The monitored parameters include pH, conductivity, persulfate, and estrogen concentrations. With regard to future research, neither the ionic strength nor the pH of the solutions was adjusted in any way to minimize the operational steps, hence the procedure was as economical as possible and potentially suitable for practical implementation. The pH, temperature, and conductivity were measured using a Combo pH/EC meter (Hanna, HI 98129). As the sulfate radical is more stable and, therefore, has a longer lifetime than the hydroxyl radical and a slower reaction rate [33], after collection, the samples were untouched for 3 h and 24 h, allowing the degradation of destabilized molecules sufficient time to take place.

2.3. Analytical Method

The analytical method of estrogen analysis has already been published and described [34]. The sample was analysed using the HPLC/MS (QQQ) system by Agilent

Technologies (Santa Clara, CA, USA). The column used for analysis was Poroshell 120 EC-C18 (2.1 × 100 mm, 2.7 μm); the mobile phase was a mixture of 7 mM HCOOH and acetonitrile with a flow of 0.35 mL min^{-1}. In short, the pH of 50 mL of the sample was adjusted to pH = 3 (±0.2), extracted with an SPE cartridge (Waters Oasis hydrophilic-lipophilic balance (HLB) cartridges) to 8 mL of methanol, dried, reconstituted in 20 μL of acetone, derivatized with dansyl chloride, dried again, and dissolved in 1 mL of 40% methanol.

Spectrophotometric analysis was used to determine the persulfate concentration. Exactly 1 mL of reagent (KI/NaHCO$_3$) was added to 200 μL of the sample [35], and the sample was mixed well and allowed to react for 20 min in the dark. The reaction product was analysed at 394 nm in a 96-well plate using a SparkTM multimode microplate reader (Tecan, Austria).

3. Results

All the experiments were performed using a persulfate concentration of 0.1 mM. In the first set, the system was activated by hydrodynamic cavitation only (without heating). As can be seen in Figure 3, after only one cycle through the cavitation unit (t = 4 s), the concentration of the estrogen mixture drops to approximately 60% of the initial concentration. After 24 h, the concentration lowered to a fraction of the initial amount. Simultaneously, a decrease in persulfate content in the mixture was observed, confirming its consumption in estrogens removal (see Figure 3B).

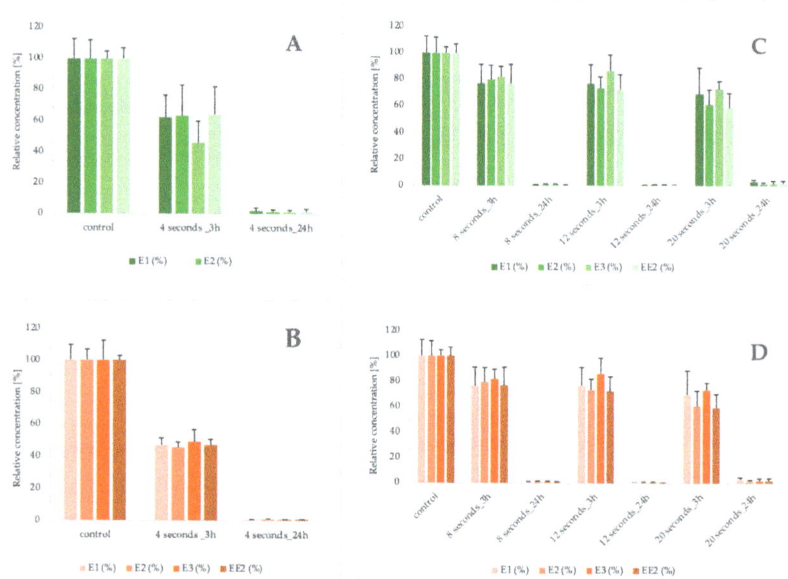

Figure 3. (**A**) Relative concentration of estrogens after treatment with persulfate (0.1 mM) activated by HC; (**B**) Relative concentration of estrogens after treatment with persulfate (0.1 mM) activated by HC combined with heat (60 °C); (**C**) Relative concentration of estrogens treated with persulfate (0.1 mM) activated by HC (8, 12, and 20 s treatment) analysed 3 and 24 h after reaction; (**D**) Relative concentration of estrogens treated with persulfate (0.1 mM) activated by the combination of HC (8, 12, and 20 s treatment) and heat (60 °C) analysed 3 and 24 h after the reaction.

The second set of experiments was performed by combining heat and HC activation. The combined activation showed slightly less pronounced results 3 h after treatment, but no significant difference was observed after 24 h (Figure 3B).

Although the most significant data obtained are related to a single flow through the cavitation device, degradation after 2, 3, and 5 cycles (8, 12, and 20 s) was also observed. Within 3 h, post-reaction processes occurred, and more intensive estrogen removal was observed. However, after 24 h, these differences disappeared, and PS activated by hydrodynamic cavitation destroyed 95–99% of the selected estrogenic compounds, similar to the PS activated by HC and 60 °C heating (compare Figure 3C,D).

The graphical results are supported by the calculation of the rate constants in Table 2. Based on the kinetic model of pseudo-first order, degradation constants (k) of estrogens were calculated according to Formula (1):

$$-\ln\left(\frac{c_t}{c_0}\right) = k \times t, \tag{1}$$

where c_0 and c_t represent the initial concertation and concentration at time t (min), respectively [32,36].

Table 2. The pseudo-first-rate constants of estrogens degradation; r > 0.97.

Conditions	k_{E1} (min^{-1})	k_{E2} (min^{-1})	k_{E3} (min^{-1})	k_{EE2} (min^{-1})
PS 0.1 mM + HC	1.24	1.51	0.94	1.2
PS 0.1 mM (heat activated 60 °C) + HC	1.15	1.40	1.68	1.54

PS concentration was also monitored during the experiments. The results show that only about half of the dosed PS was needed (Figure 4), and there is room for possible dose reduction. Moreover, no significant difference in PS concentrations was observed between the sets with and without thermal activation.

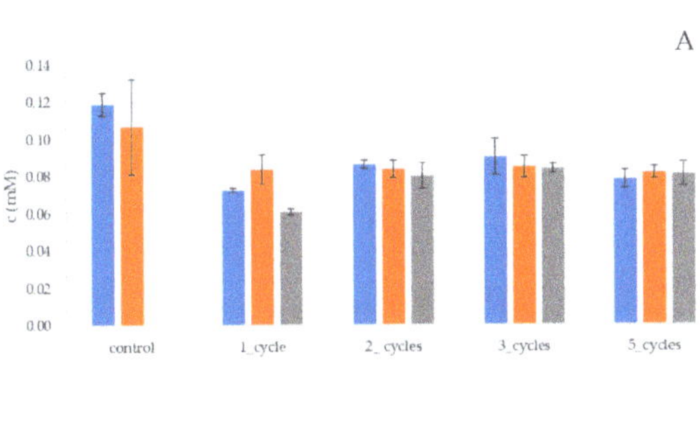

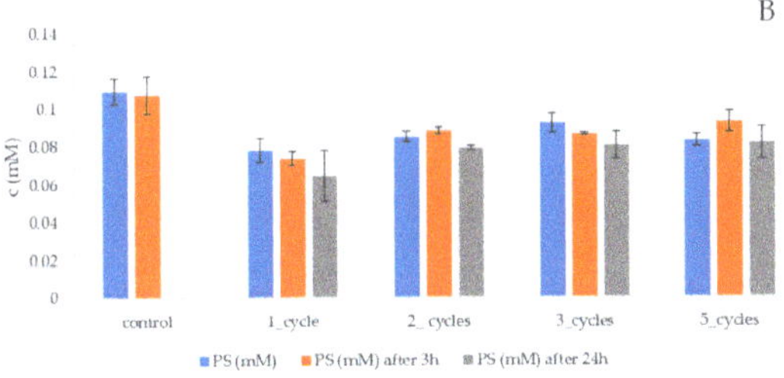

Figure 4. (**A**) Concentration of PS after treatment with PS (0.1 mM) activated by HC; 1, 2, 3, and 5 cycles through the unit corresponding with 4, 8, 12, and 20 s of treatment; (**B**) Concentration of PS after treatment with PS (0.1 mM) activated by HC combined with heat (60 °C); 1, 2, 3, and 5 cycles through the unit corresponding with 4, 8, 12, and 20 s of treatment.

To evaluate the effect of HC alone, a set of experiments was performed without added PS. Figure 5 shows that HC alone does not eliminate estrogens and only acts as a tool to activate PS.

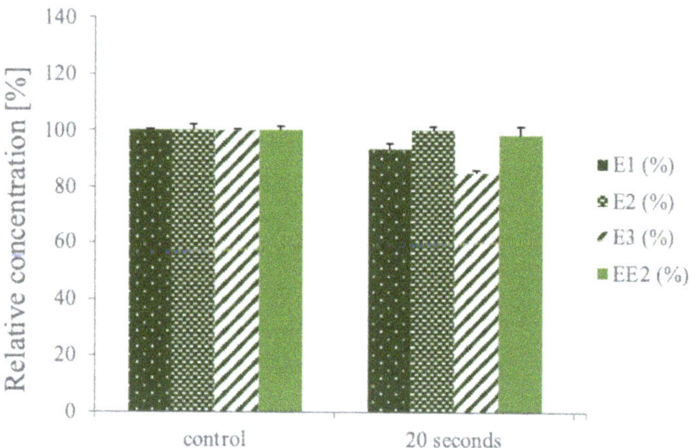

Figure 5. Relative concentration of estrogens after treatment with HC only.

Simultaneously, the pH value was monitored for all samples. It can be seen (Figure 6) that the pH value decreases slightly with an increasing number of cycles (longer reaction time). This phenomenon is possibly caused by the formation of sulphates in the aqueous solution [37]. This trend was observed for both the HC-only and HC-heat-activated sets. However, even in this case, no difference was observed between the two variants.

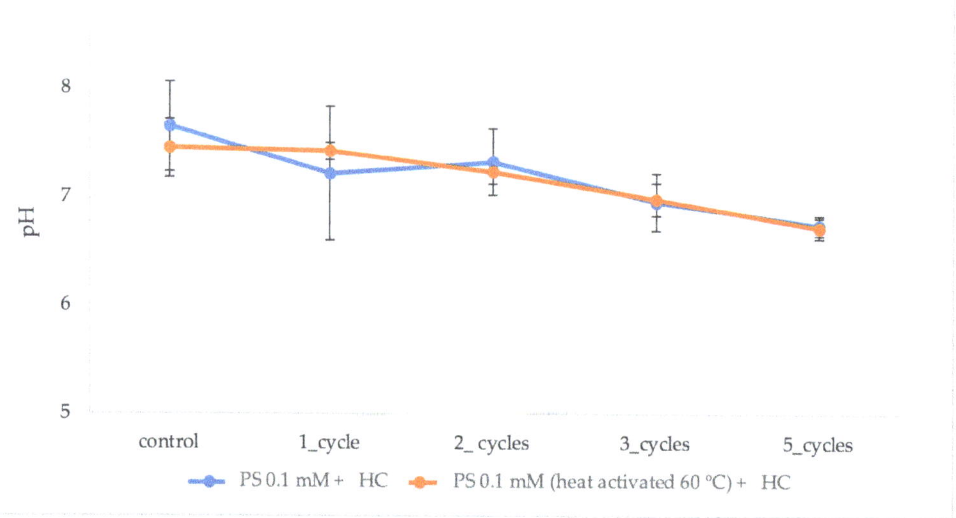

Figure 6. The pH levels of samples after treatment with PS (0.1 mM) activated by HC; 1, 2, 3, and 5 cycles through the unit corresponding with 4, 8, 12, and 20 s of treatment, and after treatment with PS (0.1 mM) activated by HC combined with heat (60 °C); 1, 2, 3, and 5 cycles through the unit corresponding with 4, 8, 12, and 20 s of treatment.

For the purpose of the cost comparison of some advanced oxidation processes, electricity consumption can be used [33]. Electric energy per mass (E_{EM}) and Electric energy per order (E_{EO}) were reported to be useful for calculating different types of treatment [38,39]. When contaminant concentrations are greater than 10 mg L^{-1}, E_{EM} should be applied, while E_{EO} should be applied when contaminant concentrations are less than 10 mg L^{-1}. To calculate treatment costs for the estrogen's concentrations, (300 µg L^{-1}) E_{EO} was selected. E_{EO} values (kWh m^{-3} order^{-1}) were calculated using the following Formula (2) [38].

$$E_{EO} = \frac{P_{HC} \times t \times 1000}{\log\left(\frac{c_i}{c_f}\right) \times V},\tag{2}$$

where P_{HC} is the rated power of the pump (kW) in the HC system, t is the treatment time (h), c_i and c_f are the initial and final concentrations (mol L^{-1}) of each estrogen, and V is the reaction volume (L).

Calculated values in kWh m^{-3} order^{-1} for PS 0.1 mM activated by HC, 4 s long treatment are E_{EO}(E1) = 2.02, E_{EO}(E2) = 2.10, E_{EO}(E3) = 1.12, and E_{EO}(EE2) = 2.15. In the case of PS 0.1 mM activated by the combination of HC and heat, the energy of external heating depending on the heating source must be considered, so the formula cannot be applied. Obviously, energy consumption is much higher in heat-combined activation as compared to the case of HC activation.

4. Discussion

Our experiment setup is unique due to the short time needed for the actual treatment and the lack of need for additional system heating. For comparison, the study performed on wastewater to eliminate frequently occurring pharmaceuticals using PS was accomplished at increased temperatures of 55, 64, and 75 °C. To achieve at least a 50% decrease in the concentration of the monitored drugs, the wastewater had to be heated up to 75 °C and allowed to react for 50 min (PS concentration $\leq$ 500 µM) [30].

Other studies focusing on eliminating estrogens by AOP produced results in reducing concentration, summarised in Table 3 [26].

Table 3. Estrogen removal based on AOP with focus on PS-based AOP.

Method	Estrogen	Efficiency	Reference
Fenton oxidation	EE2 (200 µg L^{-1})	100% in 10 min	[27]
Photo-Fenton	E2 (272 µg L^{-1})	86.4% in 8 h	[28]
Photo-Fenton	E2 (1 mg L^{-1})	98% in 60 min	[29]
PS/modified Fenton-like process	E2 (6 mg L^{-1})	100% in 90 min	[30]
PS/UV	E1, E2 and EE2 (5 µM)	over 95% in 5 min	[31]
UVC/PS/TiO$_2$ (on ceramic membrane)	E2 and EE2 (100 µg L^{-1})	uder 45% (radiation time 4.6 s)	[40]
PS activated on nanoscale zero-valent iron loaded porous graphitized biochar	E2 (3 mg L^{-1})	100% in 45 min	[41]
PS/visible light/Bi$_2$WO$_6$/Fe$_3$O$_4$	E2 (5 mg L^{-1})	~70% in 60 min	[42]
PS activated by reduced graphene oxide–elemental silver/magnetite nanohybrids	EE2 (10 µM)	~90% in 15 min	[43]
PS/ultrasound	E2 (5 mg L^{-1})	over 90% in 90 min	[44]
PS/ultrasound	E1, E2, E3 and EE2 (17–239 ng L^{-1}), real wastewater sample	over 95% in 10 min	[45]
PS/HC	E1, E2, E3 and EE2 (300 µg L^{-1})	99% in 4 s treatment	This study

Although these methods show promising efficiencies, some even in a relatively short time, Fenton-like oxidations are specific for the relatively high amount of waste and the demand for added chemicals and/or energy [46]. For example, to eliminate EE2 200 $\mu g\ L^{-1}$ within 10 min, 5 mg L^{-1} of Fe^{2+} and 8.6 mg L^{-1} of H_2O_2 is needed [26].

Promising results were observed in the literature: using UV-activated PS (c = 40 mg L^{-1}), 50% of the E2 concentration was removed in deionized water within 5 min. However, in natural wastewater, it was necessary to increase the concentration of PS to 200 mg L^{-1} to achieve similar results [47]. Comparable results were observed in a study degrading E1, E2, and EE2 (5 μM) in 5 min (PS dose 5 mM, pH = 6, UV-B) [31]. UV-based activation of PS has been shown to be fast and effective. The main disadvantage in comparison to a PS/HC system is the need for a UV source, which represents extra operation costs.

Furthermore, heating activation was performed to eliminate the common pharmaceutical drug ibuprofen. The temperature required to achieve the half-life of 3.6 min (initial concentration 20.36 μM) was 70 °C [48]. When using a PS concentration of 2 mM and 50 °C temperature conditions, more than 360 min were required to remove at least 50% of the sulfamethoxazole. For other sulfonamides, at least 6 min were required to halve the initial concentration [49].

Another drug, the antibiotic chloramphenicol, was degraded by combining PS/UV. The experiments were performed under natural conditions, and complete elimination was achieved within 1 h [50]. Similarly time-consuming is the successful degradation of the beta-blocker bisoprolol, which requires thermal activation of PS for at least 60 °C and a contact time of 1 h [51]. A study combining thermal and UV activation on municipal wastewater achieved E2 removal of over 90% within an hour [52].

Based on the available literature, it is assumed that $SO_4^-\bullet$ and $HO\bullet$ radicals are involved in removing estrogens by HC-activated PS [45,52]. HC-based treatment has also been reported to promote the generation of $HO_2\bullet$ and $O_2^{\bullet-}$ radicals [53,54]. Nevertheless $SO_4^-\bullet$ and $HO\bullet$ are significantly stronger oxidants than $HO_2\bullet$ and $O_2^-\bullet$ [53].

The positive synergy of PS and HC has already been proven on the degradation of polycyclic aromatic hydrocarbons in sediments removing PAH by 79% in 60 min [39]. Our set-up proves the ability of HC-activated PS to effectively eliminate estrogens in a short time, even in a flow-through-like system.

Since AOPs represent a large number of various processes, they are difficult to compare with each other from different points of view. Based on E_{EO} values, AOP can be classified into three groups:

- <1 kWh m^{-3} order^{-1} for representing a realistic range for full-scale application,
- 1–100 kWh m^{-3} order^{-1} for a group that is possibly too energy intensive for most practical applications, but that can still be recommended for further full-scale-application investigation,
- >100 kWh m^{-3} order^{-1}, which is considered as not (yet) energy efficient [55].

Our results show that the PS activated by HC should be classified in group 2. Nevertheless, the financial complexity of AOP processes is highly dependent on operating costs. Here, it is necessary to think about the equipment's lifespan. For comparison, this can be a limiting factor in PS activation in frequently used UV lamps (with a lifespan of around 12,000 h) and other UV-based AOPs. In addition, compared to similarly operating ultrasonic activation, HC has been reported to be 10 times more efficient in the means of electricity consumption [56].

The above examples show that heating, adding additional chemicals, and/or UV radiation are required to eliminate estrogens or other drugs using PS-based AOPs successfully. Compared to using hydrodynamic cavitation as PS activation, all these processes require higher initial costs and high operating costs, whether in the form of increasing energy prices or input chemicals. With the increasing demand for environmental responsibility, there is a growing need for functional "green" technologies, and cavitation activation has the potential to become an example of such technology.

5. Conclusions

We demonstrated that selected estrogens could be effectively eliminated from water during a short treatment time—within seconds. Venturi tube cavitation is an easy-to-install and easy-to-use economically and environmentally friendly technique compared to other known AOP (PS/AOP) alternatives. Based on the presented results, it can be assumed that cavitation acts as persulfate activation. Its main advantage is that it requires neither adding/dosing other substances into the treated water nor heating it, as opposed to methods described in earlier papers. This method can be used in a broad spectrum of water treatment processes or to intensify reactions in chemical engineering technologies. Calculated values of E_{EO} can be used for further comparison with other similar techniques and scale-up.

A lab-scale experiment, which proved the efficiency of PS activation, was conducted in this study. Pilot or other scale-up experiments are required to assess the different processes' efficacy on real wastewater fully. Nonetheless, the short treatment time (4 s), estrogens removal rate 99%, and flow rate of the lab-scale equipment 4.5 m^3 h^{-1} proved that this novel technology for removing estrogenic compounds is promising.

Author Contributions: Conceptualization, B.M. and P.P.; methodology, P.P., K.O., E.M. and B.M.; validation, K.O. and B.M.; formal analysis, P.P. and K.O.; investigation, P.P. and K.O.; resources, B.M. and R.O.; data curation, P.P., K.O. and Š.Z.; writing—original draft preparation, P.P.; writing—review and editing, P.P., K.O., E.M., Š.Z. and B.M.; visualization, P.P.; supervision, B.M. and R.O.; project administration, B.M., P.R. and F.P.; funding acquisition, B.M. and R.O. All authors have read and agreed to the published version of the manuscript.

Funding: This research was funded by the Czech Science Foundation through Project No. 19-10660S "Removal of estrogens from waste water using hydrodynamic cavitation in combination with advanced oxidation processes".

Data Availability Statement: Not applicable.

Acknowledgments: This work was supported by project FCH-S-22-8001 of the Ministry of Education, Youth and Sports of the Czech Republic.

Conflicts of Interest: The authors declare no conflict of interest.

References

1. Lecomte, S.; Habauzit, D.; Charlier, T.D.; Pakdel, F. Emerging Estrogenic Pollutants in the Aquatic Environment and Breast Cancer. *Genes* **2017**, *8*, 229. [CrossRef]
2. Adeel, M.; Yang, Y.; Wang, Y.; Song, X.; Ahmad, M.A.; Rogers, H.J. Uptake and transformation of steroid estrogens as emerging contaminants influence plant development. *Environ. Pollut.* **2018**, *243*, 1487–1497. [CrossRef]
3. Hamid, H.; Eskicioglu, C. Fate of estrogenic hormones in wastewater and sludge treatment. A review of properties and analytical detection techniques in sludge matrix. *Water Res.* **2012**, *46*, 5813–5833. [CrossRef] [PubMed]
4. Jones-Lepp, T.L.; Stevens, R. Pharmaceuticals and personal care products in biosolids/sewage sludge. the interface between analytical chemistry and regulation. *Anal. Bioanal. Chem.* **2007**, *387*, 1173–1183. [CrossRef]
5. Munter, R. Advanced oxidation processes—Current status and prospects, in: Proceedings of the Estonian Academy of Science. *Chemistry* **2001**, *50*, 59–80.
6. Ike, I.A.; Linden, K.G.; Orbell, J.D.; Duke, M. Critical review of the science and sustainability of persulphate advanced oxidation processes. *Chem. Eng. J.* **2018**, *338*, 651–669. [CrossRef]
7. Zhang, T.; Chen, Y.; Wang, Y.; Le Roux, J.; Yang, Y.; Croué, J.-P. Efficient Peroxydisulfate Activation Process Not Relying on Sulfate Radical Generation for Water Pollutant Degradation. *Environ. Sci. Technol.* **2014**, *48*, 5868–5875. [CrossRef]
8. FMC. Persulfates Technical Information, in, FMC Corporation. 2001. Available online: http://www.peroxychem.com/media/90826/aod_brochure_persulfate.pdf (accessed on 3 March 2022).
9. Anipsitakis, G.P.; Dionysiou, D.D. Transition metal/UV-based advanced oxidation technologies for water decontamination. *Appl. Catal. B Environ.* **2004**, *54*, 155–163. [CrossRef]
10. Al-Shamsi, M.A.; Thomson, N.R.; Forsey, S.P. Iron based bimetallic nanoparticles to activate peroxygens. *Chem. Eng. J.* **2013**, *232*, 555–563. [CrossRef]
11. Antoniou, M.G.; de la Cruz, A.A.; Dionysiou, D.D. Degradation of microcystin-LR using sulfate radicals generated through photolysis, thermolysis and e− transfer mechanisms. *Appl. Catal. B Environ.* **2010**, *96*, 290–298. [CrossRef]

12. Ding, D.; Liu, C.; Ji, Y.; Yang, Q.; Chen, L.; Jiang, C.; Cai, T. Mechanism insight of degradation of norfloxacin by magnetite nanoparticles activated persulfate: Identification of radicals and degradation pathway. *Chem. Eng. J.* **2017**, *308*, 330–339. [CrossRef]

13. Lin, J.C.-T.; de Luna, M.D.G.; Aranzamendez, G.L.; Lu, M.-C. Degradations of acetaminophen via a K2S2O8 -doped TiO2 photocatalyst under visible light irradiation. *Chemosphere* **2016**, *155*, 388–394. [CrossRef]

14. Ahmad, M.; Teel, A.L.; Watts, R.J. Mechanism of Persulfate Activation by Phenols. *Environ. Sci. Technol.* **2013**, *47*, 5864–5871. [CrossRef] [PubMed]

15. Fang, G.; Gao, J.; Dionysiou, D.D.; Liu, C.; Zhou, D. Activation of Persulfate by Quinones: Free Radical Reactions and Implication for the Degradation of PCBs. *Environ. Sci. Technol.* **2013**, *47*, 4605–4611. [CrossRef]

16. Rodriguez, S.; Santos, A.; Romero, A. Oxidation of priority and emerging pollutants with persulfate activated by iron: Effect of iron valence and particle size. *Chem. Eng. J.* **2017**, *318*, 197–205. [CrossRef]

17. Sun, H.; Kwan, C.; Suvorova, A.; Ang, H.M.; Tadé, M.O.; Wang, S. Catalytic oxidation of organic pollutants on pristine and surface nitrogen-modified carbon nanotubes with sulfate radicals. *Appl. Catal. B Environ.* **2014**, *154–155*, 134–141. [CrossRef]

18. Yang, S.; Yang, X.; Shao, X.; Niu, R.; Wang, L. Activated carbon catalyzed persulfate oxidation of Azo dye acid orange 7 at ambient temperature. *J. Hazard. Mater.* **2011**, *186*, 659–666. [CrossRef]

19. Huang, K.-C.; Zhao, Z.; Hoag, G.E.; Dahmani, A., Block, P.A. Degradation of volatile organic compounds with thermally activated persulfate oxidation. *Chemosphere* **2005**, *61*, 551–560. [CrossRef]

20. Tan, C.; Gao, N.; Deng, Y.; Rong, W.; Zhou, S.; Lu, N. Degradation of antipyrine by heat activated persulfate. *Sep. Purif. Technol.* **2013**, *109*, 122–128. [CrossRef]

21. Gao, Y.-Q.; Gao, N.-Y.; Deng, Y.; Yang, Y.-Q.; Ma, Y. Ultraviolet (UV) light-activated persulfate oxidation of sulfamethazine in water. *Chem. Eng. J.* **2012**, *195–196*, 248–253. [CrossRef]

22. He, X.; de la Cruz, A.A.; Dionysiou, D.D. Destruction of cyanobacterial toxin cylindrospermopsin by hydroxyl radicals and sulfate radicals using UV-254nm activation of hydrogen peroxide, persulfate and peroxymonosulfate. *J. Photochem. Photobiol. A Chem.* **2013**, *251*, 160–166. [CrossRef]

23. Luo, C.; Jiang, J.; Ma, J.; Pang, S.; Liu, Y.; Song, Y.; Guan, C.; Li, J.; Jin, Y.; Wu, D. Oxidation of the odorous compound 2,4,6-trichloroanisole by UV activated persulfate: Kinetics, products, and pathways. *Water Res.* **2016**, *96*, 12–21. [CrossRef] [PubMed]

24. Wang, S.; Zhou, N. Removal of carbamazepine from aqueous solution using sono-activated persulfate process. *Ultrason. Sonochem.* **2016**, *29*, 156–162. [CrossRef] [PubMed]

25. Darsinou, B.; Frontistis, Z.; Antonopoulou, M.; Konstantinou, I.; Mantzavinos, D. Sono-activated persulfate oxidation of bisphenol A: Kinetics, pathways and the controversial role of temperature. *Chem. Eng. J.* **2015**, *280*, 623–633. [CrossRef]

26. Frontistis, Z.; Xekoukoulotakis, N.P.; Hapeshi, E.; Venieri, D.; Fatta-Kassinos, D.; Mantzavinos, D. Fast degradation of estrogen hormones in environmental matrices by photo-Fenton oxidation under simulated solar radiation. *Chem. Eng. J.* **2011**, *178*, 175–182. [CrossRef]

27. Feng, X.; Ding, S.; Tu, J.; Wu, F.; Deng, N. Degradation of estrone in aqueous solution by photo-Fenton system. *Sci. Total Environ.* **2005**, *345*, 229–237. [CrossRef]

28. Mboula, V.M.; Héquet, V.; Andrès, Y.; Gru, Y.; Colin, R.; Doña-Rodríguez, J.; Pastrana-Martínez, L.; Silva, A.; Leleu, M.; Tindall, A.; et al. Photocatalytic degradation of estradiol under simulated solar light and assessment of estrogenic activity. *Appl. Catal. B Environ.* **2015**, *162*, 437–444. [CrossRef]

29. Yaping, Z.; Jiangyong, H. Photo-Fenton degradation of 17β-estradiol in presence of α-FeOOHR and H2O2. *Appl. Catal. B Environ.* **2008**, *78*, 250–258. [CrossRef]

30. Zhang, P.; Tan, X.; Liu, S.; Liu, Y.; Zeng, G.; Ye, S.; Yin, Z.; Hu, X.; Liu, N. Catalytic degradation of estrogen by persulfate activated with iron-doped graphitic biochar: Process variables effects and matrix effects. *Chem. Eng. J.* **2019**, *378*, 122141. [CrossRef]

31. Gabet, A.; Métivier, H.; de Brauer, C.; Mailhot, G.; Brigante, M. Hydrogen peroxide and persulfate activation using UVA-UVB radiation: Degradation of estrogenic compounds and application in sewage treatment plant waters. *J. Hazard. Mater.* **2021**, *405*, 124693. [CrossRef] [PubMed]

32. Gągol, M.; Przyjazny, A.; Boczkaj, G. Wastewater treatment by means of advanced oxidation processes based on cavitation—A review. *Chem. Eng. J.* **2018**, *338*, 599–627. [CrossRef]

33. Yi, L.; Li, B.; Sun, Y.; Li, S.; Qi, Q.; Qin, J.; Sun, H.; Wang, X.; Wang, J.; Fang, D. Degradation of norfloxacin in aqueous solution using hydrodynamic cavitation: Optimization of geometric and operation parameters and investigations on mechanism. *Sep. Purif. Technol.* **2021**, *259*, 118166. [CrossRef]

34. Sadílek, J.; Spálovská, P.; Vrana, B.; Vávrová, M.; Maršálek, B.; Šimek, Z. Comparison of extraction techniques for isolation of steroid oestrogens in environmentally relevant concentrations from sediment. *Int. J. Environ. Anal. Chem.* **2016**, *96*, 1022–1037. [CrossRef]

35. Wacławek, S.; Grübel, K.; Černík, M. Simple spectrophotometric determination of monopersulfate. *Spectrochim. Acta Part A Mol. Biomol. Spectrosc.* **2015**, *149*, 928–933. [CrossRef]

36. Perondi, T.; Michelon, W.; Junior, P.R.; Knoblauch, P.M.; Chiareloto, M.; Moreira, R.D.F.P.M.; Peralta, R.A.; Düsman, E.; Pokrywiecki, T.S. Advanced oxidative processes in the degradation of 17β-estradiol present on surface waters: Kinetics, byproducts and ecotoxicity. *Environ. Sci. Pollut. Res.* **2020**, *27*, 21032–21039. [CrossRef]

37. Xia, X.; Zhu, F.; Li, J.; Yang, H.; Wei, L.; Li, Q.; Jiang, J.; Zhang, G.; Zhao, Q. A Review Study on Sulfate-Radical-Based Advanced Oxidation Processes for Domestic/Industrial Wastewater Treatment: Degradation, Efficiency, and Mechanism. *Front. Chem.* **2020**, *8*, 592056. [CrossRef]

38. Bolton, J.R.; Bircher, K.G.; Tumas, W.; Tolman, C.A. Figures-of-merit for the technical development and application of advanced oxidation technologies for both electric- and solar-driven systems (IUPAC Technical Report). *Pure Appl. Chem.* **2001**, *73*, 627–637. [CrossRef]

39. Hung, C.-M.; Huang, C.-P.; Chen, C.-W.; Dong, C.-D. Hydrodynamic cavitation activation of persulfate for the degradation of polycyclic aromatic hydrocarbons in marine sediments. *Environ. Pollut.* **2021**, *286*, 117245. [CrossRef]

40. Castellanos, R.M.; Presumido, P.H.; Dezotti, M.; Vilar, V.J. Ultrafiltration ceramic membrane as oxidant-catalyst/water contactor to promote sulfate radical AOPs: A case study on 17β-estradiol and 17α-ethinylestradiol removal. *Environ. Sci. Pollut. Res.* **2022**, *29*, 42157–42167. [CrossRef]

41. Ding, J.; Xu, W.; Liu, S.; Liu, Y.; Tan, X.; Li, X.; Li, Z.; Zhang, P.; Du, L.; Li, M. Activation of persulfate by nanoscale zero-valent iron loaded porous graphitized biochar for the removal of 17β-estradiol: Synthesis, performance and mechanism. *J. Colloid Interface Sci.* **2021**, *588*, 776–786. [CrossRef]

42. Liu, Y.; Guo, H.; Zhang, Y.; Tang, W. Feasible oxidation of 17β-estradiol using persulfate activated by Bi 2 WO 6 /Fe 3 O 4 under visible light irradiation. *RSC Adv.* **2016**, *6*, 79910–79919. [CrossRef]

43. Park, C.M.; Heo, J.; Wang, D.; Su, C.; Yoon, Y. Heterogeneous activation of persulfate by reduced graphene oxide–elemental silver/magnetite nanohybrids for the oxidative degradation of pharmaceuticals and endocrine disrupting compounds in water. *Appl. Catal. B Environ.* **2018**, *225*, 91–99. [CrossRef]

44. Alvarez Corena, J.R.; Bergendahl, J.A. Effect of pH, temperature, and use of synergistic oxidative agents on the ultrasonic degradation of tris-2-chloroethyl phosphate, gemfibrozil, and 17β estradiol in water. *J. Environ. Chem. Eng.* **2021**, *9*, 105005. [CrossRef]

45. Choi, J.; Cui, M.; Lee, Y.; Ma, J.; Kim, J.; Son, Y.; Khim, J. Hybrid reactor based on hydrodynamic cavitation, ozonation, and persulfate oxidation for oxalic acid decomposition during rare-earth extraction processes. *Ultrason. Sonochem.* **2019**, *52*, 326–335. [CrossRef] [PubMed]

46. Domingues, E.; Silva, M.J.; Vaz, T.; Gomes, J.; Martins, R.C. Sulfate radical based advanced oxidation processes for agro-industrial effluents treatment: A comparative review with Fenton's peroxidation. *Sci. Total Environ.* **2022**, *832*, 155029. [CrossRef]

47. Angkaew, A.; Sakulthaew, C.; Satapanajaru, T.; Poapolathep, A.; Chokejaroenrat, C. UV-activated persulfate oxidation of 17β-estradiol: Implications for discharge water remediation. *J. Environ. Chem. Eng.* **2019**, *7*, 102858. [CrossRef]

48. Ghauch, A.; Tuqan, A.M.; Kibbi, N. Ibuprofen removal by heated persulfate in aqueous solution: A kinetics study. *Chem. Eng. J.* **2012**, *197*, 483–492. [CrossRef]

49. Ji, Y.; Fan, Y.; Liu, K.; Kong, D.; Lu, J. Thermo activated persulfate oxidation of antibiotic sulfamethoxazole and structurally related compounds. *Water Res.* **2016**, *87*, 1–9. [CrossRef]

50. Ghauch, A.; Baalbaki, A.; Amasha, M.; El Asmar, R.; Tantawi, O. Contribution of persulfate in UV-254 nm activated systems for complete degradation of chloramphenicol antibiotic in water. *Chem. Eng. J.* **2017**, *317*, 1012–1025. [CrossRef]

51. Ghauch, A.; Tuqan, A.M. Oxidation of bisoprolol in heated persulfate/H$_2$O systems: Kinetics and products. *Chem. Eng. J.* **2012**, *183*, 162–171. [CrossRef]

52. Sakulthaew, C.; Chokejaroenrat, C.; Satapanajaru, T.; Chirasatienpon, T.; Angkaew, A. Removal of 17β-Estradiol Using Persulfate Synergistically Activated Using Heat and Ultraviolet Light. *Water Air Soil Pollut.* **2020**, 231. [CrossRef]

53. Joshi, S.M.; Gogate, P.R. Intensification of industrial wastewater treatment using hydrodynamic cavitation combined with advanced oxidation at operating capacity of 70 L. *Ultrason. Sonochem.* **2019**, *52*, 375–381. [CrossRef]

54. Zhang, Q.; Zhao, H.; Dong, Y.; Zhu, X.; Liu, X.; Li, H. A novel ternary MQDs/NCDs/TiO$_2$ nanocomposite that collaborates with activated persulfate for efficient RhB degradation under visible light irradiation. *New J. Chem.* **2021**, *45*, 1327–1338. [CrossRef]

55. Miklos, D.B.; Remy, C.; Jekel, M.; Linden, K.G.; Drewes, J.E.; Hübner, U. Evaluation of advanced oxidation processes for water and wastewater treatment—A critical review. *Water Res.* **2018**, *139*, 118–131. [CrossRef] [PubMed]

56. Gogate, P.R. Cavitational reactors for process intensification of chemical processing applications: A critical review. *Chem. Eng. Process. Process Intensif.* **2008**, *47*, 515–527. [CrossRef]

International Journal of
Environmental Research and Public Health

MDPI

Article

Treatment of Winery Wastewater by Combined Almond Skin Coagulant and Sulfate Radicals: Assessment of HSO_5^- Activators

Nuno Jorge [1,2], Ana R. Teixeira [2], Lisete Fernandes [2], Sílvia Afonso [3], Ivo Oliveira [3], Berta Gonçalves [3], Marco S. Lucas [2] and José A. Peres [2,*]

1   Escuela Internacional de Doctorado (EIDO), Campus da Auga, Campus Universitário de Ourense, Universidade de Vigo, As Lagoas, 32004 Ourense, Spain
2   Centro de Química de Vila Real (CQVR), Departamento de Química, Universidade de Trás-os-Montes e Alto Douro (UTAD), Quinta de Prados, 5000-801 Vila Real, Portugal
3   Centre for the Research and Technology of Agro-Environmental and Biological Sciences (CITAB), Universidade de Trás-os-Montes e Alto Douro (UTAD), Quinta de Prados, 5000-801 Vila Real, Portugal
*   Correspondence: jperes@utad.pt

Abstract: The large production of wine and almonds leads to the generation of sub-products, such as winery wastewater (WW) and almond skin. WW is characterized by its high content of recalcitrant organic matter (biodegradability index < 0.30). Therefore, the aim of this work was to (1) apply the coagulation–flocculation–decantation (CFD) process with an organic coagulant based on almond skin extract (ASE), (2) treat the organic recalcitrant matter through sulfate radical advanced oxidation processes (SR-AOPs) and (3) evaluate the efficiency of combined CFD with UV-A, UV-C and ultrasound (US) reactors. The CFD process was applied with variation in the ASE concentration vs. pH, with results showing a chemical oxygen demand (COD) removal of 61.2% (0.5 g/L ASE, pH = 3.0). After CFD, the germination index (GI) of cucumber and corn seeds was $\geq$80%; thus, the sludge can be recycled as fertilizer. The SR-AOP initial conditions were achieved by the application of a Box–Behnken response surface methodology, which described the relationship between three independent variables (peroxymonosulfate (PMS) concentration, cobalt (Co^{2+}) concentration and UV-A radiation intensity). Afterwards, the SR-AOPs were optimized by varying the pH, temperature, catalyst type and reagent addition manner. With the application of CFD as a pre-treatment followed by SR-AOP under optimal conditions (pH = 6.0, [PMS] = 5.88 mM, [Co^{2+}] = 5 mM, T = 343 K, reaction time 240 min), the COD removal increased to 85.9, 82.6 and 80.2%, respectively, for UV-A, UV-C and US reactors. All treated wastewater met the Portuguese legislation for discharge in a municipal sewage network (COD $\leq$ 1000 mg O_2/L). As a final remark, the combination of CFD with SR-AOPs is a sustainable, safe and clean strategy for WW treatment and subproduct valorization.

Keywords: almond skin extract; Box–Behnken; coagulation–flocculation; response surface methodology; sludge valorization; SR-AOPs; wine production

Citation: Jorge, N.; Teixeira, A.R.; Fernandes, L.; Afonso, S.; Oliveira, I.; Gonçalves, B.; Lucas, M.S.; Peres, J.A. Treatment of Winery Wastewater by Combined Almond Skin Coagulant and Sulfate Radicals: Assessment of HSO_5^- Activators. *Int. J. Environ. Res. Public Health* **2023**, *20*, 2486. https://doi.org/10.3390/ijerph20032486

Academic Editor: Paul B. Tchounwou

Received: 30 December 2022
Revised: 26 January 2023
Accepted: 27 January 2023
Published: 30 January 2023

1. Introduction

The agro-industries account for 70% of all freshwater withdrawals worldwide, with the wine industry covering three sectors of the economy: agriculture, manufacturing and trade [1]. Due to the consumer demand for quality wines, washing and disinfection operations are necessary, and it is estimated that a winery produces about 1.3 to 1.5 kg of residue per liter of wine produced, 75% of which is winery wastewater (WW). Without a proper treatment process, the WW environmental impact is enormous due to the pollution of the water, degradation of the soil, damage to the vegetation, release of odors into the air, eutrophication of water resources, and consumption of oxygen from the rivers and lakes, leading to the suffocation of aquatic and amphibious life [2,3].

The almond (*Prunus dulcis* (Miller) D.A. Webb) is a member of the genus *Prunus*. The widespread distribution of the almond is related to its low chilling requirement, which makes almonds very suitable for regions with mild winters and dry, hot summers. World almond production has increased from 3.1 million to 4.1 million metric tons from 2011 to 2020 [4]. The almond industrial processes lead to the generation of large amounts of waste related to the removal of shells and skins [5]. Therefore, in this work, it is proposed to valorize the almond skin as a coagulant to treat WW via the CFD process.

The CFD process is a straightforward and economical process that has the capability to remove organic and inorganic substances and colloidal particles, depending on operational conditions, coagulant type and wastewater characteristics [6]. Although almond skin has never been used before in WW treatment, it has been observed that the extraction of active compounds from plants achieved significant results in wastewater treatment [7,8] and the sludge could be recycled for fertilization [9].

Considering the chemical composition of WW and the presence of recalcitrant organic matter in WW, one possible treatment strategy could be the use of advanced oxidation processes (AOPs). Among AOPs, hydroxyl-based AOPs (HR-AOPs) or sulfate-based AOPs (SR-AOPs) can be applied. In AOPs, hydroxyl radicals ($HO^{\bullet}$) are generated by a number of processes. These radicals have an oxidation potential of 2.8 V, they are non-selective, reacting with the pollutants and oxidizing them to CO_2, H_2O and partially oxidized species [10,11].

In the past years, persulfates, such as peroxymonosulfate (PMS, HSO_5^-) and peroxydisulfate (PDS, $S_2O_8^{2-}$), have attracted increasing attention because they are much more stable than hydrogen peroxide [12]. In addition, the sulfate radicals ($SO_4^{\bullet-}$) produced by the activation of PMS and PDS (1) are more selective than $HO^{\bullet}$ radicals for the oxidation of compounds with carbon-carbon double bonds and benzene rings, and (2) have a higher oxidation potential than $HO^{\bullet}$ radicals (E^0 = 2.5–3.1 V) [13,14]. PMS is the active ingredient of triple potassium salt ($KHSO_5 \bullet 0.5KHSO_4 \bullet 0.5K_2SO_4$): with an oxidation potential of 1.82 V, it is stable at ambient temperature, easy to handle and the bond energy is estimated to be in the range of 140–213.3 kJ/mol [15]. To generate the $SO_4^{\bullet-}$ radicals, the PMS can be activated by metal catalysts, heat, UV, visible light, ultrasound (US), alkali and photo-catalytic activation (h^+/e^-) (Equations (1)–(6)) [16].

$$M^{n+}+HSO_5^- \rightarrow M^{(n+1)+}+SO_4^{\bullet-}+HO^- \tag{1}$$

$$M^{(n+1)+}+HSO_5^- \rightarrow SO_5^{\bullet-}+M^{n+}+H^+ \tag{2}$$

$$HSO_5^- +UV, \text{ heat or US} \rightarrow SO_4^{\bullet-}+HO^{\bullet} \tag{3}$$

$$HSO_5^- +e^- \rightarrow SO_4^{\bullet-}+HO^- \tag{4}$$

$$HSO_5^- +e^- \rightarrow SO_4^{2-}+HO^{\bullet} \tag{5}$$

$$HSO_5^- +h^+ \rightarrow SO_5^{\bullet-}+2H^+ \tag{6}$$

One of the major problems detected in the treatment of WW is the presence of high values of turbidity, total suspended solids, total polyphenols and dissolved organic carbon (DOC) [17,18], which affect the efficiency of the AOPs due to radical scavenging (Equations (7) and (8)) [19]. In a search performed in the Web of Science and Scopus, results showed only 11 works published with the application of sulfate radicals for the treatment of WW. A background for the activation of PMS in real WW has not been established, particularly for WW with high contents of turbidity and TSS.

$$DOC + HO^{\bullet} \rightarrow DOC_{ox}^+ +HO^- \tag{7}$$

$$DOC + SO_4^{\bullet-} \rightarrow DOC_{ox}^+ +SO_4^{2-} \tag{8}$$

Considering this background, the main aim and novelty of this work is the production of a coagulant from almond skin to apply in a CFD process for WW treatment and to recycle the sludge as fertilizer. To decrease more recalcitrant organic matter, the aim is to use a response surface methodology (RSM) Box–Behnken design to define the best SR-AOP to adopt for WW treatment. The influence of pH, temperature, transition metals, single vs. multiple addition of reagents and different radiation sources (UV-A, UV-C and ultrasound) is studied in the removal of organic carbon.

2. Material and Methods

2.1. Reagents

Potassium peroxymonosulfate (PMS) was acquired from Alfa Aesar (Ward Hill, Massachusetts, US), aluminum sulfate 18-hydrate (10% w/w, $Al_2(SO_4)_3 \bullet 18H_2O$) was obtained from Scharlau (Barcelona, Spain), cobalt(II) sulfate heptahydrate, iron(II) sulfate heptahydrate ($FeSO_4 \bullet 7H_2O$) and manganese(II) sulfate monohydrate ($MnSO_4 \bullet H_2O$) were acquired from Panreac (Barcelona, Spain), Zinc(II) sulfate heptahydrate was acquired from Merck, (Darmstadt, Germany), copper(II) sulfate 2-hydrate ($CuSO_4 \bullet 2H_2O$) was acquired from Honeywell Riedel-de-Haën™ (Charlotte, North Caroline, USA), magnesium sulfate heptahydrate ($MgSO_4 \bullet 7H_2O$), 2,2-diphenyl-1-picrylhydrazyl (DPPH) radical, 6-hydroxy-2,5,7,8-tetramethylchroman-2-carboxylic acid (Trolox), catechin, gallic acid, caffeic acid and sodium chloride (NaCl) were purchased from Sigma-Aldrich (St. Louis, MO, USA). For pH adjustment, sodium hydroxide (NaOH) was used from Labkem (Barcelona, Spain) and sulfuric acid (H_2SO_4, 95%) from Scharlau (Barcelona, Spain). Deionized water was used to prepare the respective solutions.

2.2. Analytical Determinations

Different physical–chemical parameters were determined to characterize the WW, including turbidity, total suspended solids (TSS), volatile suspended solids (VSS), chemical oxygen demand (COD), biological oxygen demand (BOD_5), dissolved organic carbon (DOC) and total polyphenols (TPh). The main WW characteristics are shown in Table 1.

Table 1. Physicochemical characteristics of winery wastewater (mean $\pm$ SD).

Parameter	Values	
	WW1	**WW2**
pH (Sorensen scale)	3.95 $\pm$ 0.20	3.61 $\pm$ 0.24
Conductivity (μS cm^{-1})	45 $\pm$ 2.3	285 $\pm$ 14.3
Turbidity (NTU)	69 $\pm$ 4	649 $\pm$ 32
Total suspended solids—TSS (mg L^{-1})	200 $\pm$ 10	1405 $\pm$ 70
Dissolved organic carbon—DOC (mg C L^{-1})	138 $\pm$ 7	976 $\pm$ 49
Total nitrogen—TN (mg N L^{-1})	3.4 $\pm$ 0.2	10.7 $\pm$ 0.5
Chemical oxygen demand—COD (mg O$_2$ L^{-1})	616 $\pm$ 31	4925 $\pm$ 246
Biochemical oxygen demand—BOD$_5$ (mg O$_2$ L^{-1})	163 $\pm$ 8	1438 $\pm$ 72
Biodegradability—BOD$_5$/COD	0.26 $\pm$ 0.01	0.29 $\pm$ 0.02
Total polyphenols—TPh (mg gallic acid L^{-1})	1.90 $\pm$ 0.1	49.5 $\pm$ 2.5
Absorbance at 254 nm (diluted 1:25)	0.102 $\pm$ 0.005	0.198 $\pm$ 0.010
Absorbance at 254 nm (diluted 1:10)	0.124 $\pm$ 0.006	0.356 $\pm$ 0.018

The COD analysis was carried out in a COD reactor from Macherey-Nagel (Düren, Germany), and a HACH DR 2400 spectrophotometer (Loveland, CO, USA) was used for colorimetric measurements. The BOD_5 was determined according to the 5-day BOD test (Standard Method 5210B) using a respirometric OxiTop® IS 12 system (WTW, Yellow Springs, OH, USA). The pH was determined via a 3510 pH meter (Jenway, Cole-Parmer, UK) and conductivity was determined via a portable condutivimeter, VWR C030 (VWR, V. Nova de Gaia, Portugal), in accordance with the methodology of the Standard Methods [20]. The turbidity was determined via a 2100N IS Turbidimeter (Hach, Loveland, CO, USA), and the total suspended solids were measured through spectrophotometry according to Standard Method 2540D using a HACH DR/2400 portable spectrophotometer (Hach, Loveland, CO, USA). The total polyphenols were measured using the Folin–Ciocalteau method, adapted by Singleton and Rossi [21]. UV-vis measurements were performed using a Jasco V-530

UV/VIS spectrophotometer. The total nitrogen (TN) and DOC samples were analyzed via direct injection of the filtered samples into a Shimadzu TOC-L_{CSH} analyzer (Shimadzu, Kyoto, Japan) equipped with an ASI-L autosampler, provided with an NDIR detector, and calibrated with standard solutions of potassium phthalate.

2.3. Preparation of Almond Skin Extract (ASE) and Characterization Methodologies

Almond samples (*Prunus dulcis*) from cultivars commonly produced in Trás-os-Montes, northeastern Portugal, were obtained directly from producers located in this region, transported to the laboratory and the skin was removed from the almond. The skin was ground to a fine powder using a 150 W Princess grinder with 2 knife blades (Deluxe, Netherlands). All ground skins were sieved through a 0.4 mm sieve, and the resulting fraction with particle size less than 0.4 mm was used in the coagulation experiments. The extract was prepared by the addition of 12.5 g of prepared powder to 250 mL of a NaCl 1M solution, and the suspension was stirred for 5 h at ambient temperature (298 K) to extract the coagulation-active compounds. Finally, the suspension was allowed to rest for 5 min to sediment the solid parts, and the extract was collected through decantation [22]. The crude extracts were stored in the refrigerator (278 K) and used the following day to avoid aging phenomena and improve reproducibility. The ASE presented a pH of 5.0 ± 0.25.

The chemical spectrum of the almond powder (AP) was obtained via Fourier transform infrared spectroscopy (FTIR), and the KBr sample was analyzed using an IRAffinity-1S Fourier transform infrared spectrometer (Shimadzu, Kyoto, Japan) with the infrared spectrum in transmission mode recorded in the 4000–400 cm^{-1} frequency region. The microstructural characterization of the almond powder was carried out with scanning electron microscopy (FEI QUANTA 400 SEM/ESEM, Fei Quanta, Hillsboro, WA, USA) and the chemical composition was estimated using energy dispersive X-ray spectroscopy (EDS/EDAX, PAN'alytical X'Pert PRO, Davis, CA, USA). The pH of the point of zero charge (pH$_{PZC}$) was determined according to the method described by Oussalah et al. [23], in which 50 mL of 0.01 M NaCl solutions were adjusted to a pH range of 2–12. Then 200 mg of almond skin powder was added to each NaCl solution. The suspensions were stirred for 48 h at room temperature, and the final pH of the solutions (pH$_f$) was determined. The pH$_{PZC}$ was obtained from the plot of (pH$_f$-pH$_i$) versus pHi.

For the preparation of the methanolic extract, 40 mg of AP was weighed and mixed by vortexing with 1 mL of 70% methanol. The mixtures were heated at 70 °C for 30 min and centrifuged at 13,000 rpm at 1 °C for 15 min (Eppendorf Centrifuge 5804 R, Hamburg, Germany). The supernatants were collected and filtered with Spartan filters (0.2 mm) into HPLC amber vials. The methodology of Singleton and Rossi (1965) [21] was used for the quantification of total phenolics in a 96-well microplate. Total phenolics results were expressed as mg gallic acid equivalent (GAE)/g f.w. The total flavonoid content was determined in a 96-well microplate using the colorimetric method described in Dewanto et al. [24]. The total flavonoid content was expressed as mg catechin equivalent (CE)/g f.w. The total ortho-diphenol content was determined in a 96-well microplate, in accordance with the methodology of Soufi et al. [25], and the results were expressed as mg of caffeic acid equivalents/g of dry weight. The 2,2-diphenyl-1-picrylhydrazyl (DPPH) antioxidant activity assay was performed through spectrophotometry, as described by Siddhraju and Becker [26], in a 96-well microplate. The DPPH was expressed as μg trolox equivalent/g f.w.

2.4. Coagulation–Flocculation–Decantation Experiments

The coagulation–decantation–flocculation process was performed in a Jar test device (ISCO JF-4, Louisville, KY, USA), with four mechanical agitators powered by a regulated speed engine. The mixing of the ASE with WW samples was performed under a fast mix of 150 rpm/3 min and a slow mix of 20 rpm/20 min, at ambient temperature (298 K). Four different ASE concentrations (0.1, 0.5, 1.0 and 2.0 g/L) were tested against four different pH levels (3.0, 6.0, 9.0 and 11.0), and, after a sedimentation time of 4 h, samples were retrieved for analysis.

2.5. Box–Behnken Experimental Design

A Box–Behnken design was employed to assess the effect of different parameters on the UV-A LED SR-AOP treatment of WW1, such as the concentration of PMS (mM, X_1), the concentration of Co^{2+} (mM, X_2) and radiation intensity (W m^{-2}, X_3) under fixed conditions (COD = 616 mg O_2/L, temperature = 323 K, pH = 6.0, time = 120 min). For this study, 15 experiments were performed in triplicate. The levels considered for the Box–Behnken design are listed in Table 2, with three replicates at the center of the design (SR2, SR10 and SR11). Experiments were randomized to maximize the efforts of unexplained variability in the observed response due to external factors.

Table 2. Symbols and coded factor levels for the considered variables.

Independent Variables	Code	Levels		
		−1	0	1
[PMS] (mM)	X_1	0	5	10
[Co^{2+}] (mM)	X_2	0	2.5	5.0
Radiation	X_3	0	16.35	32.70

2.6. SR-AOP Set-Up

The experiments were performed on WW2 in a beaker with 500 mL capacity under constant agitation (350 rpm). The temperature of the reaction was controlled with a heating plate (Nahita blue model 692/1, Navarra, Spain) equipped with temperature probe. The following variables were studied: (1) pH (3.0–11.0), (2) temperature (298–363 K), (3) catalyst type (Zn^{2+}, Al^{3+}, Co^{2+}, Cu^{2+}, Fe^{2+}, Mg^{2+} and Mn^{2+}) and (4) reagent addition (single vs. multiple addition). Finally, three radiation sources were used:

(1) A UV-A LED system composed of 12 indium gallium nitride (InGaN) LED lamps (Roithner APG2C1-365E LEDs) with a λ_{max} = 365 nm. Each UV-A LED had a nominal consumption of 1.4 W when the current was 350 mA, with an optical power of 135 mW and an opening angle of 120°, making any shadow zone impossible. The radiation was emitted in continuous mode for all 12 UV-A LEDs and was controlled using a power MOSFET in six different current settings, resulting in irradiance levels from 5.2 to 32.7 W m^{-2} measured at a 5 cm distance with a UVA Light Meter (Linshang model LS126A);

(2) A Heraeus TNN 15/32 lamp (14.5 cm in length and 2.5 cm in diameter) mounted in the axial position inside the reactor, with 15 W power. The spectral output of the low-pressure mercury vapor lamp emitted mainly (85–90%) at 253.7 nm and about 7–10% at 184.9 nm;

(3) An ultrasonic system (Vibracell Ultrasonic processor VCX 500, Sonics & Materials Inc., Danbury, CT, USA) with 500 W power, equipped with a titanium alloy probe (136 mm diameter, 13 mm) and a temperature control probe. For temperature control, a water jacket was installed.

To determine the removal percentage of the parameters, Equation (9) was applied, as follows [27,28]:

$$X_i(\%) = \frac{C_i - C_F}{C_i} \times 100 \tag{9}$$

where C_i and C_f are the initial concentrations and 100 is the conversion factor.

2.7. Phytotoxicity Tests

Phytotoxicity tests were performed via the germination of *Zea mays* (corn) and *Cucumis sativa* (cucumber) seeds (standard species recommended by the US EPA, the US FDA and the OECD [29]). Seeds were immersed in a 10% sodium hypochloride solution for 10 min to ensure surface sterility, then they were soaked in pure water. One piece of filter paper (Whatman filter paper 9 cm, Maidstone, UK) was put into each 100 mm × 15 mm Petri dish, and 5 mL of test medium was added [30]. Seeds were then transferred onto the filter paper, with 10 seeds per dish and a 1 cm or larger distance between each seed. Petri dishes were covered and sealed with tape and placed in a controlled atmosphere with a constant temperature (25 °C), maintained during the course of the experiment with a WTM TS 608-G/2-i (Weilheim, Germany). After 7 days of darkness and 7 days of light, the germination index was determined by Equation (10), in accordance with Varnero et al. [31] and Tiquia and Tam [32], as follows:

$$\text{GI}(\%) = \frac{\overline{N}_{SG,T}}{\overline{N}_{SG,B}} \times \frac{\overline{L}_{R,T}}{\overline{L}_{R,B}} \times 100 \tag{10}$$

where GI is the germination index, $\overline{N}_{SG,T}$ is the arithmetic mean of the number of germinated seeds in each extract (wastewater), $\overline{N}_{SG,B}$ is the arithmetic mean of the number of germinated seeds in standard solution (distilled water), $\overline{L}_{R,T}$ is the mean root length of each extract (wastewater), and $\overline{L}_{R,B}$ is the mean root length in control (distilled water). If GI ≤ 50%, then there was a high concentration of phytotoxic substances; if 80% < GI > 50%, then there was a moderate presence of phytotoxic substances; and if GI ≥ 80%, then there were no phytotoxic substances (or they existed in very small dosages).

2.8. Statistical Analysis

The coefficients corresponding to the model equation were obtained using Minitab Statistical Software 2018 (State College, PA, USA). All experiments were performed in triplicate, and a one-way ANOVA was carried out to determine any significant differences ($p < 0.05$) using the Tukey test. Results are presented as mean ± standard deviation (SD).

3. Results and Discussion

3.1. Characterization of Almond Skin Powder

Before the production of the almond skin extract (ASE), it was necessary to characterize the powder obtained from the almond skin. Figure 1a shows the chemical spectrum of the AS by FTIR. The band at 3460 cm^{-1} can be attributed to O-H and N-H stretching vibrations of hydroxyl groups and amide A of polypeptides and amino acids, respectively. The O-H stretching is also associated with the water fraction and with the polyphenol content of the samples [33]. The band at 3190 cm^{-1} is related to the stretching vibration of C=CH cis olefinic groups of unsaturated fatty acids [34]. The bands at 2924 and 2895 cm^{-1} correspond to the symmetric and asymmetric stretching vibrations of aliphatic CH$_2$ functional groups, respectively, which are linked with the saturated fatty acid fraction [35]. The band at 1608 cm^{-1} corresponds with the stretching vibration of cis C=C of unsaturated acyl groups [36]. The band at 1508 cm^{-1} corresponds to the bending vibration of the N-H functional group mainly observed in Amide I and Amide II of protein compounds [33]. The bands at 1450 and 1255 cm^{-1} are associated with the presence of CH bending vibrations in CH$_2$ and CH$_3$, respectively [37]. Finally, a peak at 1016 cm^{-1} is related to the stretching vibration of C-O functional groups characteristic of the carbohydrate fraction [36].

The SEM images (Figure 1b) show that the almond skin powder presents dark spaces that correspond to empty spaces, similar to findings in other works involving plant-based materials [38]. These porous materials allow the adsorption of the NaCl solution and the desorption of material from the powder to the exterior, a necessary characteristic to produce the ASE.

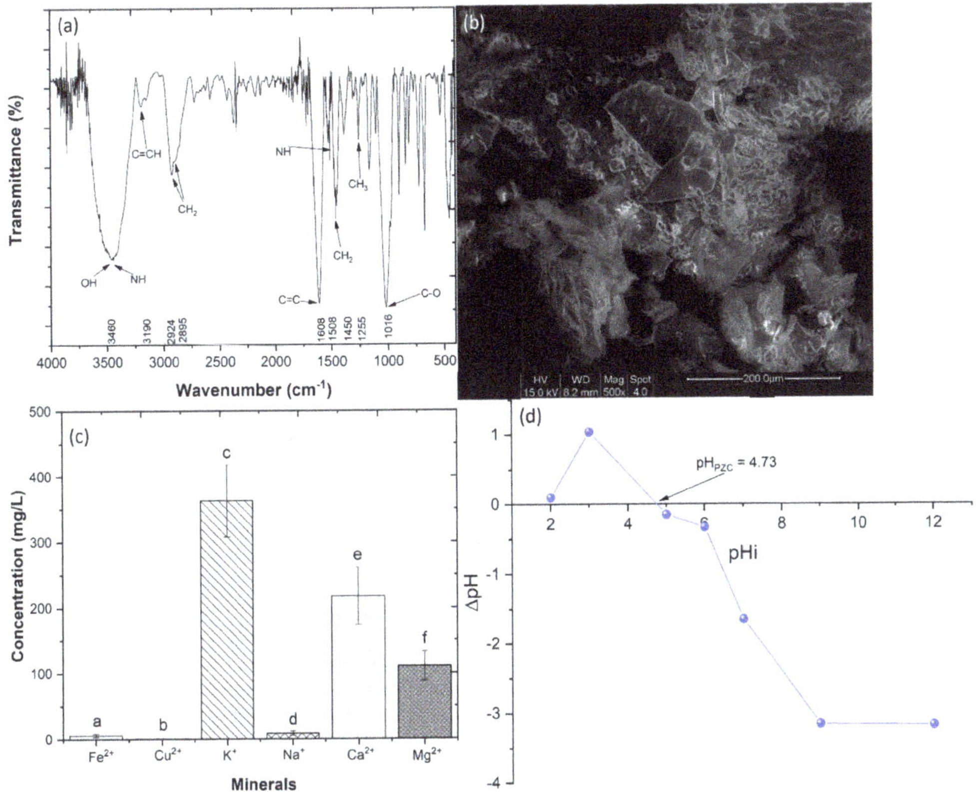

Figure 1. Almond skin powder characterization: (**a**) FTIR spectrum, (**b**) SEM image ($500\times$), (**c**) mineral analysis and (**d**) pH_{PZC} of almond powder. The pHi is the initial pH value of the solution, and ΔpH is the difference between final pH values (after contact with the AP) and pHi values. Means in bars with different letters represent significant differences ($p < 0.05$) between different minerals.

Figure 1c shows major concentrations of potassium (K^+), calcium (Ca^{2+}) and magnesium (Mg^{2+}), which are linked to the stress resistance of the plants [39]. The chemical analysis presented in Table S1 shows that AS revealed a total phenolic concentration of 6.51 ± 0.62 mg GAE/g, a flavonoid concentration of 4.30 ± 0.16 mg CE/g, an O-diphenol concentration of 0.65 ± 0.05 mg CAE/g and a DPPH concentration of 8.93 ± 0.34 µg Trolox/g. These results are in agreement with the work of Oliveira et al. [40], who observed similar concentrations in almonds. These results showed that the skin of the almond has antioxidant capacities.

Figure 1d shows the effect of pH on almond skin powder. The pH_{PZC} value determined for the ASP was 4.73. This means that at a pH < 4.73, the surface of the almond skin powder is positively charged, and the biosorption of anionic particles will be favored [41]. However, at a pH > 4.73, the surface of the ASP is negatively charged, increasing the attraction of cations [23].

3.2. Coagulation–Flocculation–Decantation Experiments

In this section, almond skin extract (ASE) was applied for the treatment of WW2 via the CFD process. Table 1 shows that WW2 has a complex composition with a high content of organic matter, polyphenols and turbidity. In order to optimize the CFD process and to understand the behavior of ASE, the pH of the wastewater and concentration of ASE were varied. The results showed that the ASE efficiency was affected by the pH of the wastewater. From Figure 2a, results show that a pH of 3.0 achieved a turbidity removal within the range of 94.8 to 96.8% and a TSS removal within the range of 96.9 to 98.1%. This turbidity and TSS removal decreased as the pH increased. On the other hand, the variation in ASE concentration showed little difference regarding turbidity and TSS removal. These results are not in agreement with the work of Hussain Haydar [42], who observed that the natural coagulant *Opuntia stricta* achieved high TSS removal at pH 10.3. Figure 2b shows the influence of ASE on TPh removal. The importance of TPh removal lies in the reduction in the color, which is caused mainly by the polyphenols [27]. The results showed that pH was the main factor that influenced TPh removal, with the highest removal observed at pH 3.0. These results agree with the pH_{PHZ} values, which showed that a pH < 4.73 increased the attraction of the proteins present in the ASE and the negatively charged polyphenols, such as tannins. As the pH increased above the pH_{PZC}, the electrostatic repulsion between the proteins in the ASE and the polyphenols increased, explaining the low removal [43]. The COD and DOC removals were studied (Figure 2c), with results showing high removals with 0.5 g/L ASE at pH 3.0 (61.2 and 56.8%, respectively). These results showed that ASE was able to reduce the organic carbon in suspension as well as the dissolved organic carbon. Considering the low levels of organic carbon removed in other works [44], the use of ASE proved to be competitive. Regarding the pH of the WW2 (3.61), the cost of pH changing is avoided, thus this becomes a cheaper process for wastewater treatment.

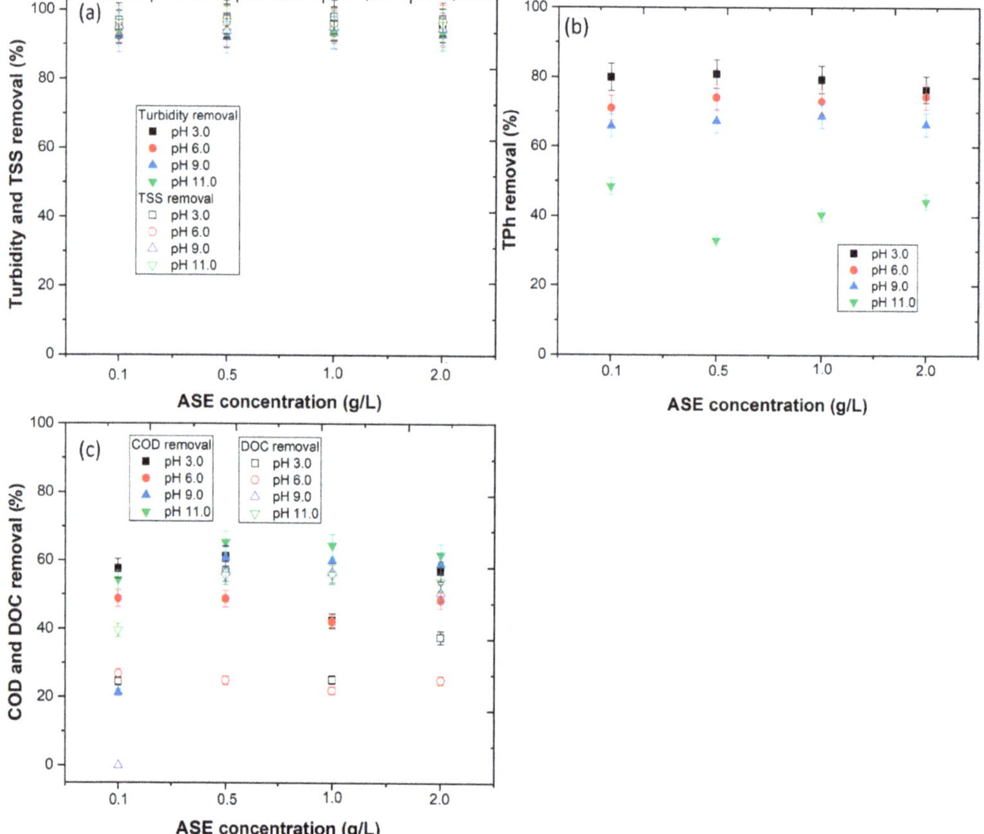

Figure 2. Effect of ASE concentration (g/L) vs. pH in (**a**) turbidity and TSS removal, (**b**) TPh removal, and (**c**) COD and DOC removal from winery wastewater. Operational conditions: fast mix (rpm/min) = 150/3, slow mix (rpm/min) = 20/20, sedimentation time of 4 h.

In order to decrease the environmental impact of sludge generation, this work tested the possibility of recycling the sludge as fertilizer. The recovered sludge was applied as a substrate for cucumber and corn seed germination. The results showed that the application of 1.0 and 2.0 g/L ASE at pH 3.0 and 6.0 could have a toxic effect on seed germination (Figure 3a,b). At the operational conditions selected (0.5 g/L ASE at pH 3.0), the results showed a GI $\geq$ 80% with good radicular growth; thus, the sludge can be recycled as fertilizer. These results are in agreement with the work of Jorge et al. [9], who observed that the sludge of WW could be recycled as fertilizer if organic coagulants were applied.

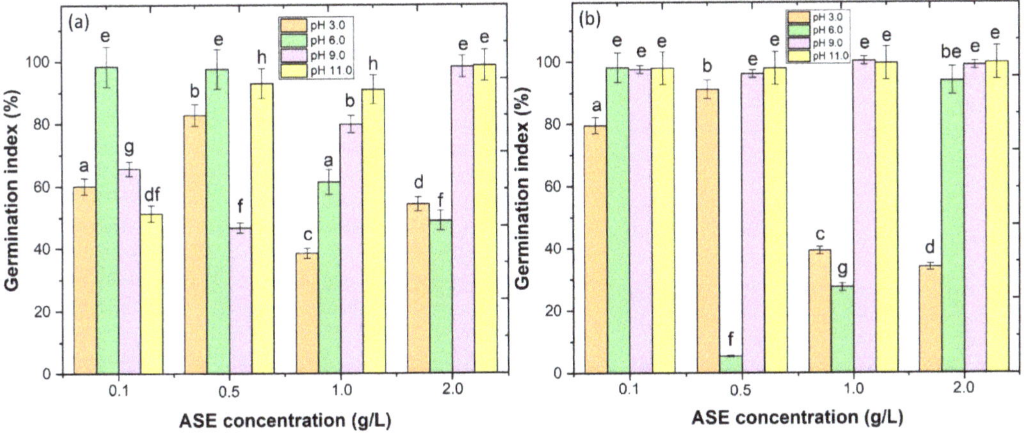

Figure 3. Effect of ASE concentration vs. pH in (**a**) germination index of cucumber and (**b**) corn seeds. Means in bars with different letters represent significant differences ($p < 0.05$) within each ASE concentration by comparing the pH.

3.3. SR-AOP Optimization through Response Surface Methodology

In the previous section, ASE showed high efficiency in the removal of turbidity and TSS; however, the coagulant showed limited capacity in the removal of COD. Therefore, in this section, the removal of COD by SR-AOPs was studied. Considering the limited information available regarding the oxidation of organic matter from WW by PMS and Co^{2+}, it was necessary to create a model that can help reach the concentration of PMS and Co^{2+} and, at the same time, allow the study of different SR-AOPs. In this section, a Box–Behnken design was applied to achieve the initial conditions of PMS and Co^{2+} concentrations. WW1 was used for this model due to its low COD, thus reducing the reagent requirement. The assessment of the COD removal was performed throughout a range of SR-AOP conditions (n = 15) based on distinct combinations of PMS concentration (X_1: 0–10 mM), Co^{2+} concentration (X_2: 0–5 mM) and radiation (X_3: 0–32.70 W m^{-2}). The results of the 15 runs are shown in Table 3. The range of PMS and Co^{2+} concentrations was consistent with those previously assayed in the treatment of agro-industrial wastewaters by other authors [45,46]. The LED lamps were selected as radiation sources due to (1) the LED efficiency, (2) the LED price and (3) the maximum emission wavelength [47,48].

Before optimizing the treatment process, it is necessary to understand which SR-AOP is best fitted to reduce the COD present in the WW. The RSM model allowed the study of different variables to understand how they affect the generation of sulfate radicals ($SO_4^{\bullet-}$). In Figure 4a, the removal efficiencies of different oxidation systems, which include (1) PMS, (2) UV-A, (3) Co^{2+}, (4) Co^{2+} + UV-A, (5) PMS + UV-A, (6) PMS + Co^{2+} and (7) PMS + Co^{2+} + UV-A, can be observed. The reactions involving Co^{2+}, PMS, Co^{2+} + UV-A and UV-A reached a COD removal of 7.7, 10.1, 18.3 and 23.1%, respectively. The low removal observed with the application of Co^{2+} and Co^{2+} + UV-A was expected since cobalt alone is not able to generate radicals [49]. The UV-A radiation alone was revealed to be inefficient, which is in agreement with the work of Jorge et al. [18]. The application of PMS alone achieved a reduced COD removal, a result that could be related to the low PMS oxidation potential (1.82 V) [16]. The PMS + UV-A reached higher COD removal than PMS and UV-A alone (33.7%), which is in agreement with the work of Huang et al. [50]. These results could be due to the conversion of the PMS by the UV-A radiation into $SO_4^{\bullet-}$ radicals. The highest COD removals were observed with the application of PMS + Co^{2+} and PMS + Co^{2+} + UV-A (44.3 and 62.9%, respectively). The results showed that the activation of PMS by cobalt was higher than the activation by UV-A radiation. It was also observed that a synergy effect occurs with the application of PMS + Co^{2+} + UV-A. This synergy could be linked to the regeneration of Co^{3+} to Co^{2+} by UV-A radiation, which further enhances the conversion of PMS and the generation of $SO_4^{\bullet-}$ radicals.

Table 3. Box–Behnken design: effect of operational variables on COD removal yield ([PMS] = 0–10 mM; [Co^{2+}] = 0–5 mM; radiation = 0–32.70 W m^{-2}). Operational conditions: COD = 616 mg O_2/L, temperature = 323 K, pH = 6.0, reaction time = 120 min.

Experiments	Coded Level			Response Values	
	[PMS] (mM)	[Co^{2+}] (mM)	Radiation	COD Removal (%)	
				Observed	Predicted
SR1	5	5.0	32.70	54	49.1
SR2	5	2.5	16.35	46.4	46.4
SR3	0	5.0	16.35	16.8	14.4
SR4	5	5.0	0.00	44.3	42.5
SR5	10	5.0	16.35	58.7	67.8
SR6	5	0.0	32.70	33.7	35.5
SR7	0	0.0	16.35	23.1	14.0
SR8	10	2.5	32.70	62.9	58.7
SR9	10	2.5	0.00	52.5	45.2
SR10	5	2.5	16.35	46.4	46.4
SR11	5	2.5	16.35	46.4	46.4
SR12	5	0.0	0.00	10.1	14.9
SR13	10	0.0	16.35	24.6	27.0
SR14	0	2.5	32.70	18.3	25.6
SR15	0	2.5	0.00	7.7	11.9

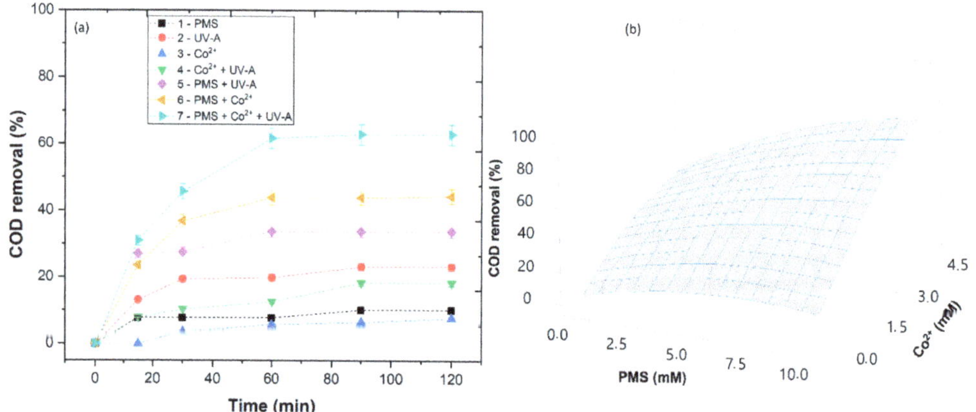

Figure 4. SR-AOP optimization: (**a**) experiments with different oxidation systems in the reduction in COD from WW1, (**b**) response surface methodology. Operational conditions: COD = 616 mg O_2/L, pH = 6.0, [PMS] = 5 mM, [Co^{2+}] = 2.5 mM, temperature = 323 K, radiation UV-A 32.7 W m^{-2}, reaction time = 120 min.

The RSM model also contributes to the optimization of PMS and Co^{2+} concentrations. The increase in PMS concentration up to 5 mM showed an increase in COD removal due to the higher generation of $SO_4^{\bullet-}$ radicals. As the PMS concentration increased to 10 mM, results showed a reduction in COD removal (Figure 4b) due to the excess of PMS in the solution, which competed with the organic matter for the $SO_4^{\bullet-}$ radicals (Equation (11)) [51]. These results are in agreement with the work of Govindan et al. [52], who observed that high concentrations of PMS led to the consumption of $SO_4^{\bullet-}$ radicals, decreasing the degradation of the contaminant pentachlorophenol.

$$HSO_5^- + SO_4^{\bullet-} \rightarrow HSO_4^- + SO_5^{\bullet-} \tag{11}$$

The Co^{2+} concentration was observed to have a significant effect on COD removal. Comparing the results obtained in SR1 and SR8, when the $PMS:Co^{2+}$ ratio decreased from 1:0.5 to 1:0.25, the COD removal increased. The excess of Co^{2+} present in SR1 was observed to compete for $SO_4^{\bullet-}$ radicals, decreasing the efficiency of the oxidation process (Equation (12)) [53]. These results are in agreement with the work of Rodríguez-Chueca et al. [45], who observed that an excess of cobalt led to the consumption of $SO_4^{\bullet-}$ radicals in the treatment of WW.

$$Co2^+ + SO_4^{\bullet-} \rightarrow Co3^+ + SO_4^{2-} \tag{12}$$

The quadratic model developed in this work permitted the adjustment of the theoretical values of COD removal to observed values with a low deviation (Table 3), suggesting a successful application of the RSM methodology. From the RSM model, the polynomial equation (Equation (13)) was obtained, and the regression coefficient (R^2) for this method was 0.931, which means that the model matches the COD removal adequately.

$$COD = 0.50 + 4.47X_1 + 7.64X_2 + 1.018X_3 - 0.316X_1X_1 - 1.234X_2X_2 - 0.0118X_3X_3 + 0.808X_1X_2 - 0.0006X_1X_3 - 0.085X_2X_3 \tag{13}$$

The regression coefficients of the intercept, linear, quadratic and interaction terms of the model were determined with the application of the least squares method. The effect of linear, quadratic or interaction coefficients on the response was studied via analysis of variance (ANOVA) (Table 4). The degree of significance of each factor is represented by its p-value, which indicates that the regression models for COD removal were statistically relevant with a level of significance of $p = 0.027 < 0.05$ (Table S2). The model did not display a significant lack of fit ($p > 0.05$), with $R^2 = 92.04\%$ (Table S3); thus, it can be considered a well-fitting model for the described variables. These statistical analyses revealed that the most important variables for the COD removal from the WW were the PMS (X_1) and Co^{2+} (X_2) (Table 4). Throughout this statistical model, the most relevant conditions were obtained: [PMS] = 5.88 mM, $[Co^{2+}]$ = 5 mM, radiation intensity = 32.7 W m^{-2}.

Table 4. F and p-values for selected responses for each obtained coefficient.

Variable	X_1	X_2	X_3	X_1X_1	X_1X_2	X_1X_3	X_2X_2	X_2X_3	X_3X_3
F-value	29.59	11.36	4.95	3.08	5.48	0.00	2.95	0.65	0.50
p-value	*	*	n.s.	n.s.	n.s.	n.s.	n.s.	n.s.	n.s.

X_1: PMS (mM); X_2: Co^{2+} (mM); X_3: radiation. n.s.: non-significant. Significant at * $p < 0.05$.

3.4. SR-AOPs Applied to a High Load WW

The operational conditions obtained with the RSM were applied in the treatment of WW2, with a COD of 4925 mg O_2/L. The advantage of creating a model with a WW with a lower COD content is the application of a lower concentration of reagents, decreasing both costs and scavenging reactions. The role of different parameters such as pH, temperature, transition metals, the dosing procedure of reagents and radiation sources were investigated to establish the optimal conditions for the treatment of WW2 with SR-AOPs.

Initially, the pH was varied from 3.0 to 11.0 over 240 min (Figure 5a). The results showed a COD removal of 47.3, 57.0, 59.4 and 56.4%, respectively, for pH levels 3.0, 6.0, 9.0

and 11.0. The low COD removal observed at pH 3.0 was consistent with the non-productive reactions under highly acidic pH levels (Equation (6)), resulting in reduced production of $SO_4^{\bullet-}$ radicals, which is in agreement with the work of Huling et al. [54], who observed no considerable differences between pH 3.0 and 6.0 in the degradation of methyl tert-butyl ether (MTBE) by persulfate oxidation. The results showed a favorable range from 6.0 to 9.0, which was in agreement with Yi et al. [55]. Above pH 9.0, a decrease in COD removal was observed. These results were due to the fact that the alkaline solution inhibited the dissolution of Co^{2+} while promoting the complexation and deposition of Co^{2+}. In this way, pH 11.0 inhibited the dissolution of Co^{2+}, slowing down the activation of PMS and reducing the catalytic activity [13,56]. Considering the initial pH of the WW (4.81) and considering that the difference in COD removal between pH 6.0 and 9.0 was not significant, pH 6.0 is a better choice, since a lower content of NaOH is required to be spent, which is in agreement with the work of Rodríguez-Chueca et al. [46], who observed that pH 6.5 could achieve significant COD removal from WW.

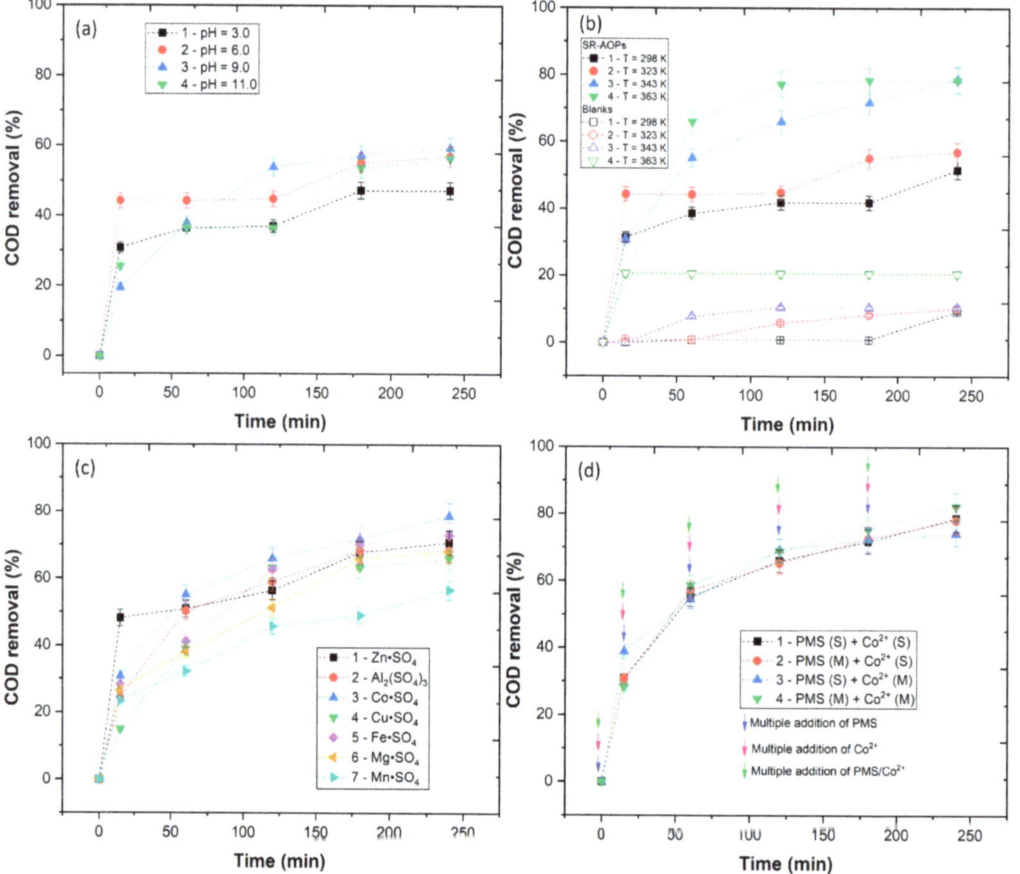

Figure 5. COD removal in the optimization of (**a**) pH ([PMS] = 5.88 mM, [Co^{2+}] = 5 mM, radiation UV-A, T = 323 K), (**b**) temperature (SR-AOPs: pH = 6.0, [PMS] = 5.88 mM, [Co^{2+}] = 5 mM, radiation UV-A; blanks: pH = 6.0, radiation UV-A), (**c**) type of metal catalyst (pH = 6.0, [PMS] = 5.88 mM, [M^{n+}] = 5 mM, radiation UV-A, T = 343 K) and (**d**) dosing of PMS and Co^{2+}. S—single addition, M—multiple addition (pH = 6.0, [PMS] = 5.88 mM, [Co^{2+}] = 5 mM, radiation UV-A, T = 343 K).

The temperature of the wastewater was observed by other authors to have an influence on the conversion of persulfate and the generation of $SO_4^{\bullet-}$ radicals [57,58]. Figure 5b shows the COD removal at different temperatures (298–363 K) using the optimal pH of 6.0. Before the application of the SR-AOPs, it was necessary to understand the effect of temperature without PMS and Co^{2+}, so blank experiments were performed. The results in Figure 5b show a maximum COD removal with the application of 363 K (20.6%). Thus, the organic matter appears to be stable at high temperatures. With the application of SR-AOPs, the results showed no significant differences between 298 and 323 K (51.6 and 57.0%, respectively); however, the COD increased significantly with 343 K (78.7%). In the work of Chen et al. [57], it was observed that persulfate was more easily converted in $SO_4^{\bullet-}$ radicals at 343 K, which agrees with the results obtained in this work. These results are also in agreement with the works of Rodríguez-Chueca et al. [59] and Jorge et al. [28], who observed that high temperatures are beneficial in activating PMS. Increasing the temperature to 363 K the COD removal is reduced (78.4%), which could suggest that temperatures above 343 K can inactivate the $SO_4^{\bullet-}$ radicals, decreasing the efficiency of the reaction. In previous studies, the reduction in organic matter via thermally activated PMS was observed to follow a pseudo first-order kinetic rate [46].

The effect of the nature of the metal catalyst was studied. In this section, seven different sulfate catalysts were tested ($ZnSO_4$, $Al_2(SO_4)_3$, $CoSO_4$, $CuSO_4$, $FeSO_4$, $MgSO_4$ and $MnSO_4$) to evaluate their effect on the activation of PMS (Figure 5c). The results showed the highest COD removal with catalysts $CoSO_4$, $FeSO_4$ and $ZnSO_4$, with 78.7, 73.0 and 70.8%, respectively. Cobalt has been reported as one of the most effective catalysts for the activation of PMS, promoting a high generation of $SO_4^{\bullet-}$ and $HO^{\bullet}$ radicals [55,60]. The regeneration of Co^{3+} to Co^{2+} in the PMS/Co (Equation (2)) is thermodynamically feasible (0.82 V), fast and the process proceeds cyclically many times until PMS is consumed [13]. The application of $FeSO_4$ was shown to be as efficient as $CoSO_4$. In the work of Latif et al. [61], iron was observed to be effective in the production of $SO_4^{\bullet-}$ radicals in the degradation of the organic contaminant bisphenol A (BPA). The $ZnSO_4$ catalyst was observed in this work to have a significant effect on the conversion of PMS in $SO_4^{\bullet-}$ and $HO^{\bullet}$ radicals, which is in agreement with the work of Fang et al. [62], who observed that zinc could efficiently react with PMS to generate $SO_4^{\bullet-}$ and $HO^{\bullet}$ radicals that degrade BPA. $Al_2(SO_4)_3$ was used as a catalyst in Al^{3+}/PMS reaction. The results showed a higher COD removal (68.4%) with the application of catalysts $MgSO_4$, $CuSO_4$ and $MnSO_4$ (68.1, 66.0 and 56.7%, respectively). Aluminum sulfate is highly used in coagulation–flocculation–decantation processes, posing a low risk to the environment with a higher discharge limit (10 mg Al/L) than cobalt (3 mg Co/L) and iron (2 mg Fe/L) [63]. Thus, $Al_2(SO_4)_3$ can be considered an alternative to cobalt.

To complement this optimization, the dosing procedure of the reagents was investigated, in which the reagents were added in a single step at the beginning of the reaction (S) or five times in a multiple addition (M). Four different ways were then selected to understand the effect of single vs. multiple addition in COD removal: (1) PMS (S) + Co^{2+} (S), (2) PMS (M) + Co^{2+} (S), (3) PMS (S) + Co^{2+} (M) and (4) PMS (M) + Co^{2+} (M) (Figure 5d). The results showed that the application of PMS (S) + Co^{2+} (M) achieved the lowest COD removal (74.2%). These results could be due to a higher concentration of PMS present in solution and insufficient Co^{2+} present to generate the $SO_4^{\bullet-}$. In addition, the excess of PMS could have a scavenger effect (Equation (11)), decreasing the efficiency of the reaction, which could explain the decrease in COD removal. The applications of PMS (S) + Co^{2+} (S) and PMS (M) + Co^{2+} (S) showed similar results, with 78.7 and 78.1% COD removal, respectively. These results showed that the catalyst addition had a significant effect on the conversion of PMS and the generation of $SO_4^{\bullet-}$ radicals. With the application of PMS (M) + Co^{2+} (M) the COD removal reached 82.3%, the highest of all addition methods. In contrast with PMS (S) + Co^{2+} (S), the multiple dosing of reagents kept the concentrations of PMS and Co^{2+} low in the reactor, suppressing the rate of scavenging reactions and, as a consequence, a more gradual supply of $SO_4^{\bullet-}$ radicals were generated, resulting in a more significant COD removal. The influence of multiple dosages of reagents was reported by other authors. For example, in the work of Sun et al. [64], it was observed that the addition of PMS in multiple dosages decreased the self-decomposition of the oxidant caused by the high concentrations of the addition in single mode.

In this work, the COD removal results were well-fitt into a pseudo first-order kinetic model (ln $[COD]_t = -kt + \ln [COD]_0$), and the corresponding rate constants were 2.16×10^{-3}, 2.49×10^{-3}, 5.82×10^{-3} and 6.40×10^{-3} min^{-1} for, respectively, 298, 323, 343 and 363 K (Figure 6a). To have a better understanding of the influence of temperature on PMS activation, the Arrhenius equation (ln$k = \ln A - E_a/RT$) was used, and a chart was plotted, fitting ln k vs. $1/T$ (Figure 6b), where A is the frequency factor, E_a is the activation energy, R is the universal gas constant and T is the absolute temperature [58]. The results showed a good fit of the data ($y = -1901.73X + 0.27647$, $R^2 = 0.967$). The average E_a (16.07 ± 0.58 kJ mol^{-1}) was observed to be lower than the activation of PMS reported by Rodriguez-Narvaez et al. [65] (29.9 kJ mol^{-1}) in the degradation of a contaminant by PMS/Co^{2+} at a temperature of 343 K. In further experiments, a temperature of 343 K was selected, considering that above this temperature no significant COD removal was observed.

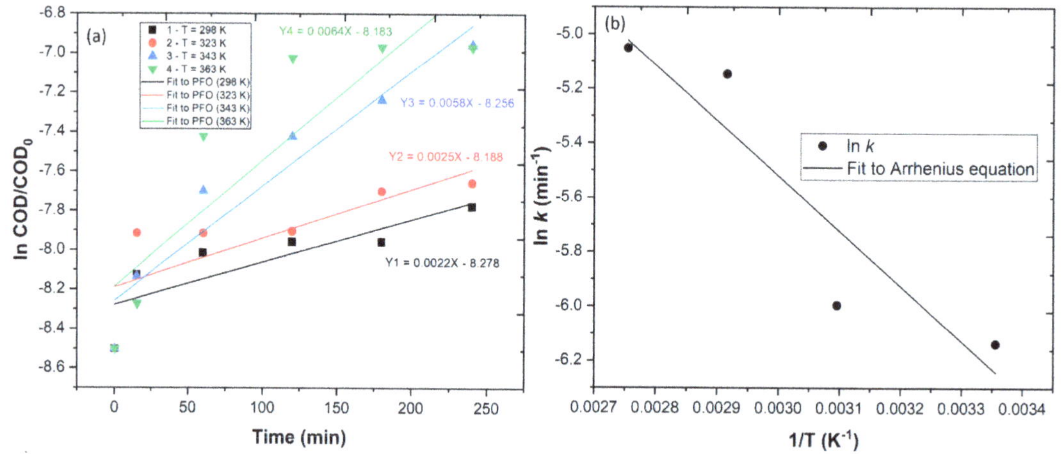

Figure 6. (a) Pseudo first-order of COD removal at different temperatures and (b) plot of ln k vs. $1/T$ for E_a estimation using the Arrhenius equation.

The PMS/Co^{2+}/UV system was shown to be effective in the removal of organic matter. In this section, the optimized conditions (pH = 6.0, [PMS] = 5.88 mM, [Co^{2+}] = 5 mM, T = 343 K, time = 240 min) were applied to the treatment of WW in three different reactors. A UV-A LED photosystem, a low-pressure UV-C mercury lamp and an ultrasound reactor were used. Figure 7a shows the COD and DOC removal after the application of the optimal PMS/Co^{2+}/UV system in the three reactors. Results showed a COD removal of 82.3, 76.0 and 52.2%, respectively, and a DOC removal of 75.8, 64.1 and 38.8%, respectively, for UV-A, UV-C and US. The high mineralization observed with the UV-C reactor can be explained by the photolysis of the PMS, which generates one mole of SO$_4^{\bullet-}$ radicals and one mole of HO$^{\bullet}$ radicals per mole of PMS (Equation (3)). In the work of Wang and Chu [66], it was observed that the system PMS/Fe^{2+}/UV (254 nm) achieved the highest degradation of 2,4,5-trichlorophenoxyacetic acid, which was attributed to the decomposition of PMS by the UV-C radiation. However, in this work, it was observed that the application of a UV-A reactor achieved higher mineralization than the UV-C reactor. The activation of PMS with UV-A radiation was shown to be effective in the removal of micropollutants from the wastewater [67] and in the removal of organic carbon from WW [46], which is in agreement with the results obtained in this work. Although the photolysis of PMS is negligible with UV-A radiation [68], the application of Co^{2+} appears to be the contributing factor that increased the efficiency of the reaction and higher mineralization of the organic carbon. The activation of PMS by ultrasound was also studied. The US can be used to generate $SO_4^{\bullet-}$ and $HO^{\bullet}$ radicals by the decomposition of PMS (Equation (3)) [69]. The use of ultrasound radiation showed a high removal of organic carbon, although the results were shown to be lower than those of UV-A and UV-C radiation. These results are in agreement with the work of Lu et al. [70], who observed a 71.4% atrazine degradation in the PMS/US system.

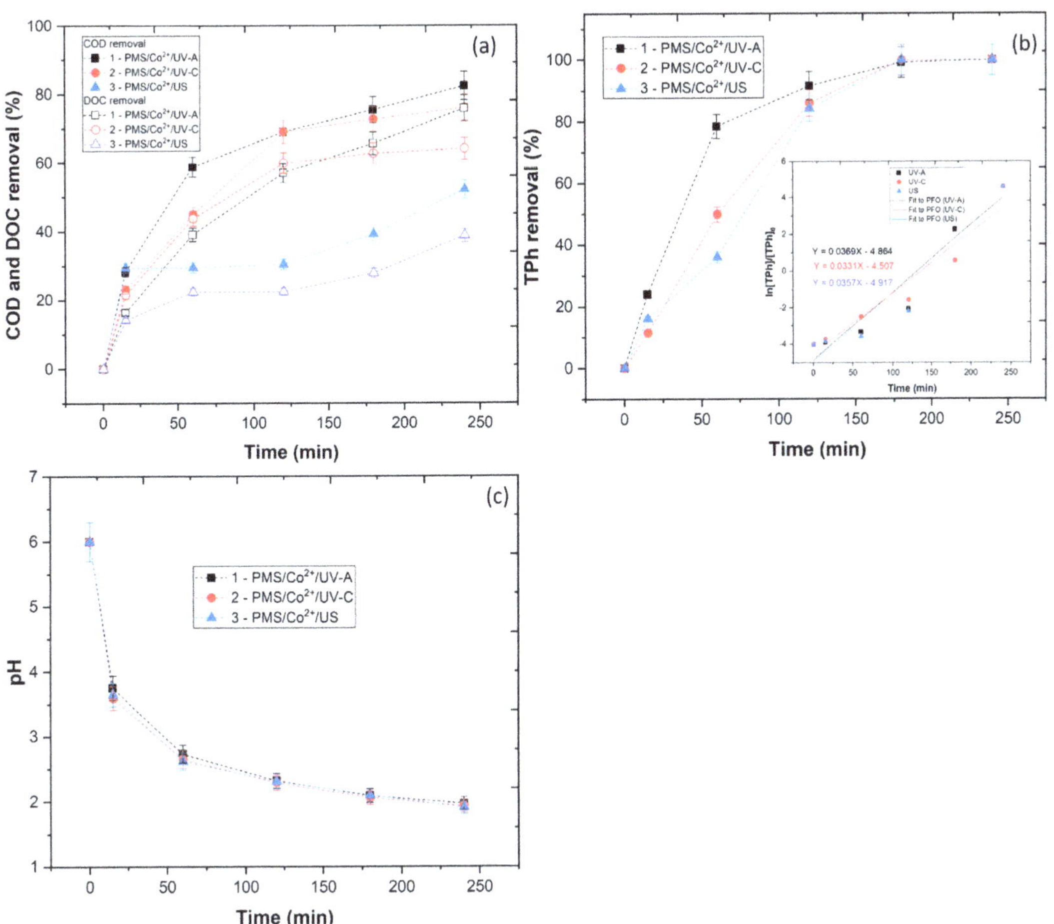

Figure 7. Evaluation of the PMS/Co^{2+}/UV system with three radiation reactors (UV-A, UV-C and US) in (**a**) COD and DOC removal, (**b**) TPh removal and (**c**) pH. Operational conditions: pH$_i$ = 6.0, [PMS] = 5.88 mM, [Co^{2+}] = 5 mM, T = 343 K, time = 240 min.

Among the organic compounds that are present in the WW, polyphenols are linked to its toxicity, since aromatic compounds are difficult to degrade by microorganisms. Polyphenols have a significant contribution to the dark color of the WW, which limits the penetration of UV radiation and the catalyst regeneration [71]. In this section, the TPh removal was evaluated as a function of the three reactors, with results showing a near complete removal after the 240 min of reaction (Figure 7b). These results are linked with the generation of $SO_4^{\bullet-}$ and $HO^{\bullet}$ radicals by the three systems, which are extremely powerful, non-selective and capable of oxidizing most organic compounds. The $SO_4^{\bullet-}$ and $HO^{\bullet}$ radicals are capable of attacking the phenol rings of phenolic compounds, yielding benzoic and cinnamic acids, flavonoids and anthocyanins, and then the rings of these compounds break up to give organic acids and finally CO_2 [28,72].

One of the main factors that affected the efficiency of organic carbon removal was the pH of the wastewater. Therefore, the pH of the reaction was monitored during the 240 min. The results in Figure 7c show that all three systems had a large drop in pH within the first 15 min, which could be linked to the production of $SO_4^{\bullet-}$ and $HO^\bullet$ radicals. This fact was previously reported by Esteves et al. [73], who observed a high decrease in the pH in the treatment of high-strength olive mill wastewater using a Fenton-like oxidation process. The follow-up of the pH is important because the type and rate of the produced radicals are among the important effects of pH changes. At pH 6.0, the $SO_4^{\bullet-}$ radicals were predominant [74,75]. As the reaction unfolded, a significant reduction in the pH took place, caused by the hydrolysis of the $SO_4^{\bullet-}$ and the cobalt ions. Therefore, it could be perceived that a higher concentration of $SO_4^{\bullet-}$ radicals leads to a more significant pH decline, which is in agreement with the work of Li et al. [76].

3.5. CFD + PMS/Co^{2+}/Radiation

In the previous section the legal limits for wastewater discharge in a sewage network were not achieved for the UV-C and US reactors (COD < 1000 mg O_2/L); therefore, in this section, the CFD + PMS/Co^{2+}/radiation combined treatment was studied. The CFD process was performed as a pre-treatment followed by oxidation with SR-AOPs. Figure 8a shows the evolution of COD and DOC removal with the combined system. Results showed a COD removal of 85.9, 82.6 and 80.2% and a DOC removal of 83.3, 79.1 and 74.5%, respectively, for CFD/PMS/Co^{2+}/UV-A, CFD/PMS/Co^{2+}/UV-C and CFD/PMS/Co^{2+}/US. Two factors were observed to increase the reactions' efficiency: (1) the removal of turbidity, TSS and TPh by the ASE, which clarified the wastewater, allowing better penetration of the radiation, and (2) the removal of the organic matter in suspension, which acted as a scavenger of $HO^\bullet$ and $SO_4^{\bullet-}$ radicals. These results are in agreement with the work of Jaafarzadeh et al. [77], who observed that the application of CFD as a pre-treatment increased the efficiency of electro-activated HSO_5^- to treat pulp and paper wastewater. In comparison to other works, such as Amor et al. [78], results showed that the application of the CFD/SR-AOP system achieved higher organic matter removal from WW with lower reagent consumption.

In order to evaluate the mineralization capacity of the different processes, an efficient parameter was applied, as a partial oxidation efficiency (μ_{partox}), which can be determined by Equations (14) and (15) [79], as follows:

$$COD_{partox} = \left(\frac{COD_0}{DOC_0} - \frac{COD}{DOC}\right) \times DOC \tag{14}$$

$$\mu_{partox} = \frac{COD_{partox}}{COD_0 - COD_t} \tag{15}$$

where μ_{partox} is the partial oxidation efficiency that is between 0 and 1. The μ_{partox} is 0 when only total oxidation occurs, and the value of 1 represents the ideal condition in which only partial oxidation occurs. In fact, total oxidation occurs in $\mu_{partox} < 0.5$, while partial oxidation is dominant in $\mu_{partox} > 0.5$. Figure 8b shows that with the application of UV-A and UV-C reactors, the reduction in organic carbon predominantly occurred via total oxidation reactions, becoming more pronounced at prolonged times, similar to the work of Papastefanakis et al. [80]. With application of the PMS/Co^{2+}/US and CFD/PMS/Co^{2+}/US systems, partial oxidation reactions occurred at 15 and 30 min, respectively, following total oxidation reactions until 240 min of reaction. These results can be explained by the high reduction in the organic matter by the $HO^\bullet$ and $SO_4^{\bullet-}$ radicals, which increased the mineralization.

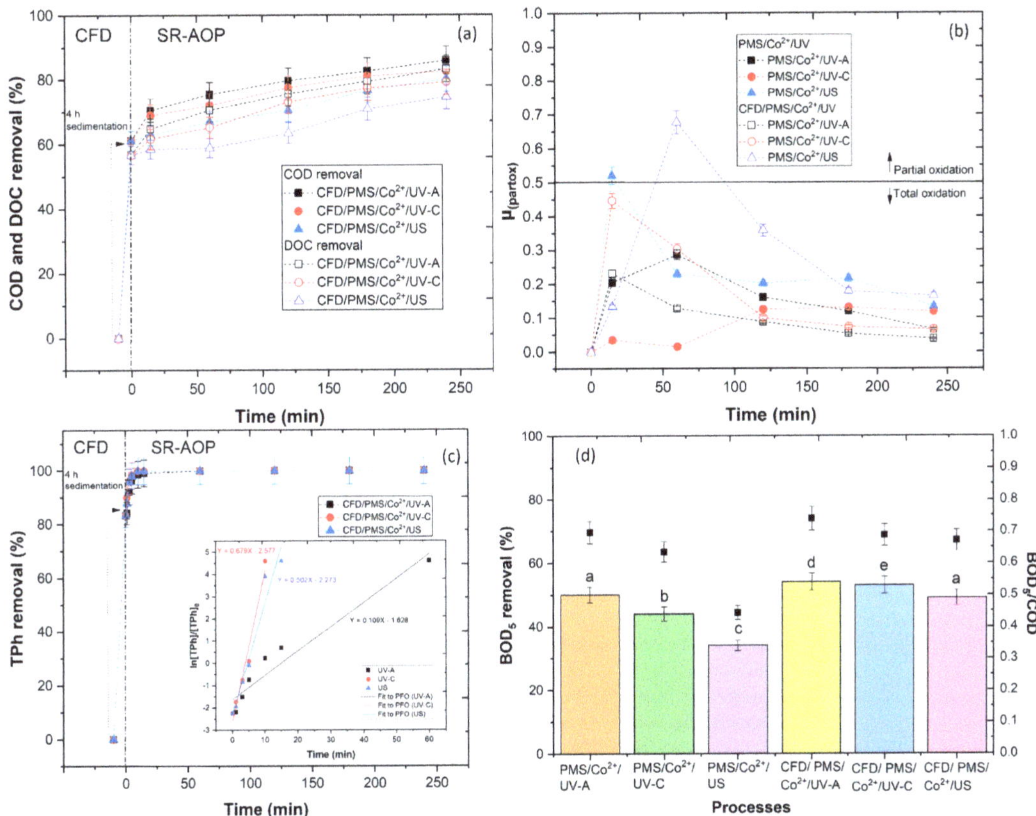

Figure 8. Effect of CFD/PMS/Co^{2+}/radiation in (**a**) COD and DOC removal, (**b**) $\mu_{(partox)}$, (**c**) TPh removal and (**d**) BOD$_5$ removal and BOD$_5$/COD. Means in bars with different letters represent significant differences ($p < 0.05$) within BOD$_5$/COD by comparing treatment processes. CFD operational conditions: pH = 3.0, [ASE] = 0.5 g/L, fast mix (rpm/min) = 150/3, slow mix (rpm/min) = 20/20, sedimentation time 4 h. SR-AOP operational conditions: pH = 6.0, [PMS] = 5.88 mM, [Co^{2+}] = 5 mM, radiation UV-A 32.7 W m^{-2}, UV-C 15 W, US 500 W, T = 343 K, reaction time = 240 min.

In Figure 8c, the effect of the combined treatment in the removal of TPh can be observed. The ASE was responsible for the removal of 83.2% of the TPh content from the WW, decreasing the dark color of the WW and allowing a better penetration of the radiation. With the application of the UV-A, UV-C and US reactors, the TPh removal rate increased significantly. These results were in agreement with the work of Jorge et al. [18], who observed that application of the CFD process boosted the TPh removal by the SR-AOPs.

Figure 8d shows the impact of the different treatment processes in the BOD$_5$ removal and, consequently, in the biodegradability. The applications of the PMS/Co^{2+}/UV-A, PMS/Co^{2+}/UV-C and PMS/Co^{2+}/US achieved a BOD$_5$ removal of 69.6, 63.5 and 44.4%, respectively. With the application of the CFD process, the BOD$_5$ removal was increased to 73.9, 68.7 and 67.0%, respectively. The biodegradability was also evaluated, with results showing an increase from 0.31 (CFD process) to 0.54, 0.53 and 0.49, respectively. The combination of processes, was able to remove part of the recalcitrant matter, increasing the biodegradable fraction, and allowing a subsequent biological treatment. These results were in agreement with the works of Amor et al. [81] and Rodríguez-Chueca et al. [82], who observed that the combined photo-Fenton and CFD processes increased the biodegradability of crystallized-fruit wastewater.

4. Conclusions

Wine and almond production are two major Portuguese agro-industries with enormous weight in Portugal's economy. The WW generated from wine production is of environmental concern due to the high content of organic matter and polyphenols. From almond production, the skin, often neglected by the food industry, is used in this work to produce an almond skin extract (ASE). The results show that ASE achieves the highest results regarding turbidity, TSS, TPh and COD removal with the application of 0.5 g/L ASE at pH 3.0. It is concluded that the pH has a considerable effect on the ASE efficiency, considering the isoelectric point (4.73). The sludge generated by the treatment with ASE can be recycled as fertilizer, allowing its valorization. The response surface methodology (RSM) associated with a Box–Behnken design was revealed to be one of the most appropriate methods for the optimization of the basic conditions (PMS and Co^{2+} concentration and radiation) for COD removal from WW. The efficiency of the SR-AOP is concluded to be dependent on factors, such as pH, temperature, type of transition metal and manner of addition. Under the optimal conditions, pH = 6.0, [PMS] = 5.88 mM, $[Co^{2+}]$ = 5.0 mM, T = 343 K, reaction time = 240 min, a COD removal was achieved of 82.3, 76.0 and 52.2%, respectively, for UV-A, UV-C and US reactors. The combination of CFD with SR-AOPs has a synergic effect, in which the ASE removes 61.2% of COD and the combined ASE/reactors remove 85.9, 82.6 and 80.2%, respectively, with UV-A, UV-C and US. Moreover, the combination of processes allows all reactors to achieve the Portuguese legal value of COD ($\leq$1000 mg O_2/L) for wastewater to be discharged as municipal wastewater in a wastewater treatment plant (WWTP).

Supplementary Materials: The following supporting information can be downloaded at: https://www.mdpi.com/article/10.3390/ijerph20032486/s1, Table S1: Concentration of total phenolic, flavonoids, O-diphenol and DPPH present in almond skin extract; Table S2: Analysis of variance; Table S3: Model summary.

Author Contributions: Conceptualization, N.J., A.R.T., L.F., S.A. and I.O.; methodology, N.J., S.A. and I.O.; software, N.J.; validation, N.J., B.G., M.S.L. and J.A.P.; formal analysis, N.J.; investigation, N.J.; resources, N.J. and L.F.; data curation, N.J.; writing—original draft preparation, N.J.; writing—review and editing, N.J., B.G., M.S.L. and J.A.P.; visualization, N.J., B.G., M.S.L. and J.A.P.; supervision, M.S.L. and J.A.P.; project administration, M.S.L. and J.A.P.; funding acquisition, M.S.L. and J.A.P. All authors have read and agreed to the published version of the manuscript.

Funding: The authors are grateful for the financial support of the Project AgriFood XXI NORTE-01-0145-FEDER-000041 and Fundação para a Ciência e a Tecnologia (FCT) to CQVR (UIDB/00616/2020). Ana R. Teixeira also thanks the FCT for the financial support provided through the doctoral scholarship UI/BD/150847/2020.

Institutional Review Board Statement: Not applicable.

Informed Consent Statement: Not applicable.

Data Availability Statement: Not applicable.

Acknowledgments: The authors are thankful to CITAB, supported by National Funds by FCT— Portuguese Foundation for Science and Technology, under the project UIDB/04033/2020. The authors are grateful for the support provided by the Unidade de Microscopia Eletrónica (UME) of UTAD.

Conflicts of Interest: The authors declare no conflict of interest.

References

1. Vlotman, D.E.; Key, D.; Bladergroen, B.J. Technological Advances in Winery Wastewater Treatment: A Comprehensive Review. *South Afr. J. Enol. Vitic.* **2022**, *43*, 58–80. [CrossRef]
2. Ioannou, L.A.; Puma, G.L.; Fatta-Kassinos, D. Treatment of Winery Wastewater by Physicochemical, Biological and Advanced Processes: A Review. *J. Hazard. Mater.* **2015**, *286*, 343–368. [CrossRef]

3. Jorge, N.; Santos, C.; Teixeira, A.R.; Marchão, L.; Tavares, P.B.; Lucas, M.S.; Peres, J.A. Treatment of Agro-Industrial Wastewaters by Coagulation-Flocculation-Decantation and Advanced Oxidation Processes—A Literature Review. *Eng. Proc.* **2022**, *19*, 33. [CrossRef]

4. FAOSTAT Almonds, with Shell. Available online: https://www.fao.org/faostat/en/#home (accessed on 26 October 2022).

5. Oliveira, I.; Meyer, A.S.; Afonso, S.; Sequeira, A.; Vilela, A.; Goufo, P.; Trindade, H.; Gonçalves, B. Effects of Different Processing Treatments on Almond (*Prunus dulcis*) Bioactive Compounds, Antioxidant Activities, Fatty Acids, and Sensorial Characteristics. *Plants* **2020**, *9*, 1627. [CrossRef]

6. Ahmad, A.; Kurniawan, S.B.; Abdullah, S.R.S.; Othman, A.R.; Hasan, H.A. Exploring the Extraction Methods for Plant-Based Coagulants and Their Future Approaches. *Sci. Total Environ.* **2022**, *818*, 151668. [CrossRef]

7. Madrona, G.S.; Serpelloni, G.B.; Vieira, A.M.S.; Nishi, L.; Cardoso, K.C.; Bergamasco, R. Study of the Effect of Saline Solution on the Extraction of the *Moringa oleifera* Seed's Active Component for Water Treatment. *Water Air Soil Pollut.* **2010**, *211*, 409–415. [CrossRef]

8. Teixeira, A.R.; Jorge, N.; Fernandes, J.R.; Lucas, M.S.; Peres, J.A. Textile Dye Removal by *Acacia dealbata Link.* Pollen Adsorption Combined with UV-A/NTA/Fenton Process. *Top. Catal.* **2022**, *65*, 1045–1061. [CrossRef]

9. Jorge, N.; Teixeira, A.R.; Lucas, M.S.; Peres, J.A. Agro-Industrial Wastewater Treatment with *Acacia dealbata* Coagulation/Flocculation and Photo-Fenton-Based Processes. *Recycling* **2022**, *7*, 54. [CrossRef]

10. Bello, M.M.; Raman, A.A.A.; Asghar, A. A Review on Approaches for Addressing the Limitations of Fenton Oxidation for Recalcitrant Wastewater Treatment. *Process. Saf. Environ. Prot.* **2019**, *126*, 119–140. [CrossRef]

11. Neyens, E.; Baeyens, J. A Review of Classic Fenton's Peroxidation as an Advanced Oxidation Technique. *J. Hazard. Mater.* **2003**, *98*, 33–50. [CrossRef] [PubMed]

12. Qi, C.; Liu, X.; Ma, J.; Lin, C.; Li, X.; Zhang, H. Activation of Peroxymonosulfate by Base: Implications for the Degradation of Organic Pollutants. *Chemosphere* **2016**, *151*, 280–288. [CrossRef] [PubMed]

13. Anipsitakis, G.P.; Dionysiou, D.D. Degradation of Organic Contaminants in Water with Sulfate Radicals Generated by the Conjunction of Peroxymonosulfate with Cobalt. *Environ. Sci. Technol.* **2003**, *37*, 4790–4797. [CrossRef] [PubMed]

14. Anipsitakis, G.P.; Dionysiou, D.D. Transition Metal/UV-Based Advanced Oxidation Technologies for Water Decontamination. *Appl. Catal. B Environ.* **2004**, *54*, 155–163. [CrossRef]

15. Wang, J.; Wang, S. Activation of Persulfate (PS) and Peroxymonosulfate (PMS) and Application for the Degradation of Emerging Contaminants. *Chem. Eng. J.* **2018**, *334*, 1502–1517. [CrossRef]

16. Giannakis, S.; Lin, K.A.; Ghanbari, F. A Review of the Recent Advances on the Treatment of Industrial Wastewaters by Sulfate Radical-Based Advanced Oxidation Processes (SR-AOPs). *Chem. Eng. J.* **2021**, *406*, 127083. [CrossRef]

17. Jorge, N.; Teixeira, A.R.; Marchão, L.; Lucas, M.S.; Peres, J.A. Application of Combined Coagulation–Flocculation–Decantation/Photo-Fenton/Adsorption Process for Winery Wastewater Treatment. *Eng. Proc.* **2022**, *19*, 22. [CrossRef]

18. Jorge, N.; Amor, C.; Teixeira, A.R.; Marchão, L.; Lucas, M.S.; Peres, J.A. Combination of Coagulation-Flocculation-Decantation with Sulfate Radicals for Agro-Industrial Wastewater Treatment. *Eng. Proc.* **2022**, *19*, 19. [CrossRef]

19. Vione, D.; Falletti, G.; Maurino, V.; Minero, C.; Pelizzetti, E.; Malandrino, M.; Ajassa, R.; Olariu, R.-I.; Arsene, C. Sources and Sinks of Hydroxyl Radicals upon Irradiation of Natural Water Samples. *Environ. Sci. Technol.* **2006**, *40*, 3775–3781. [CrossRef]

20. APHA; AWWA; WEF. *Standard Methods for the Examination of Water and Wastewater*, 20th ed.; American Public Health Association: Washington, DC, USA; American Water Works Association: Denver, CO, USA; Water Environment Federation: Alexandria, VA, USA, 1999.

21. Singleton, V.L.; Rossi, J.A. Colorimetry of Total Phenolics with Phosphomolybdic-Phosphotungstic Acid Reagents. *Am. J. Enol. Vitic.* **1965**, *16*, 144–158.

22. Šćiban, M.; Klašnja, M.; Antov, M.; Škrbić, B. Removal of Water Turbidity by Natural Coagulants Obtained from Chestnut and Acorn. *Bioresour. Technol.* **2009**, *100*, 6639–6643. [CrossRef]

23. Oussalah, A.; Boukerroui, A.; Aichour, A.; Djellouli, B. Cationic and Anionic Dyes Removal by Low-Cost Hybrid Alginate/Natural Bentonite Composite Beads: Adsorption and Reusability Studies. *Int. J. Biol. Macromol.* **2019**, *124*, 854–862. [CrossRef] [PubMed]

24. Dewanto, V.; Wu, X.; Adom, K.K.; Liu, R.H. Thermal Processing Enhances the Nutritional Value of Tomatoes by Increasing Total Antioxidant Activity. *J. Agric. Food Chem.* **2002**, *50*, 3010–3014. [CrossRef] [PubMed]

25. Soufi, O.; Romero, C.; Hayette, L. Ortho-Diphenol Profile and Antioxidant Activity of Algerian Black Olive Cultivars: Effect of Dry Salting Process. *Food Chem.* **2014**, *157*, 504–510. [CrossRef]

26. Siddhuraju, P.; Becker, K. Antioxidant Properties of Various Solvent Extracts of Total Phenolic Constituents from Three Different Agroclimatic Origins of Drumstick Tree (*Moringa Oleifera* Lam.) Leaves. *J. Agric. Food Chem.* **2003**, *51*, 2144–2155. [CrossRef] [PubMed]

27. Jorge, N.; Teixeira, A.R.; Matos, C.C.; Lucas, M.S.; Peres, J.A. Combination of Coagulation–Flocculation–Decantation and Ozonation Processes for Winery Wastewater Treatment. *Int. J. Environ. Res. Public Health* **2021**, *18*, 8882. [CrossRef]

28. Jorge, N.; Teixeira, A.R.; Guimarães, V.; Lucas, M.S.; Peres, J.A. Treatment of Winery Wastewater with a Combination of Adsorption and Thermocatalytic Processes. *Processes* **2022**, *10*, 75. [CrossRef]

29. OECD. *OECD Guidelines for the Testing of Chemicals: Terrestrial Plant Test*; OECD: Paris, France, 2004; Volume 208.

30. Lin, D.; Xing, B. Phytotoxicity of Nanoparticles: Inhibition of Seed Germination and Root Growth. *Environ. Pollut.* **2007**, *150*, 243–250. [CrossRef]

31. Varnero, M.T.; Rojas, C.; Orellana, R. Índices de Fitotoxicidad En Residuos Orgánicos Durante El Compostaje. *Rev. Cienc. Suelo Nutr. Veg.* **2007**, *7*, 28–37. [CrossRef]

32. Tiquia, S.M.; Tam, N.F.Y. Elimination of Phytotoxicity during Co-Composting of Spent Pig-Manure Sawdust Litter and Pig Sludge. *Bioresour. Technol.* **1998**, *65*, 43–49. [CrossRef]

33. Subramanian, A.; Harper, W.J.; Rodriguez-Saona, L.E. Rapid Prediction of Composition and Flavor Quality of Cheddar Cheese Using ATR–FTIR Spectroscopy. *J. Food Sci.* **2015**, *74*, C292–C297. [CrossRef]

34. Beltrán Sanahuja, A.; Moya, P.; Maestre Pérez, S.E.; Grané Teruel, N.; Martín Carratalá, M.L. Classification of Four Almond Cultivars Using Oil Degradation Parameters Based on FTIR and GC Data. *J. Am. Oil Chem. Soc.* **2009**, *86*, 51–58. [CrossRef]

35. Maqsood, S.; Benjakul, S. Synergistic Effect of Tannic Acid and Modified Atmospheric Packaging on the Prevention of Lipid Oxidation and Quality Losses of Refrigerated Striped Catfish Slices. *Food Chem.* **2010**, *121*, 29–38. [CrossRef]

36. Rohman, A.; Man, Y.C. Fourier Transform Infrared (FTIR) Spectroscopy for Analysis of Extra Virgin Olive Oil Adulterated with Palm Oil. *Food Res. Int.* **2010**, *43*, 886–892. [CrossRef]

37. García, A.V.; Beltran Sanahuja, A.; Garrigos Selva, M.D.C. Characterization and Classification of Almond Cultivars by Using Spectroscopic and Thermal Techniques. *J. Food Sci.* **2013**, *78*, C138–C144. [CrossRef]

38. Martins, R.B.; Jorge, N.; Lucas, M.S.; Raymundo, A.; Barros, A.I.; Peres, J.A. Food By-Product Valorization by Using Plant Based Coagulants Combined with AOPs for Agro-Industrial Wastewater Treatment. *Int. J. Environ. Res. Public Health* **2022**, *19*, 4134. [CrossRef]

39. Martin, N.; Bernard, D. Calcium Signaling and Cellular Senescence. *Cell Calcium* **2018**, *70*, 16–23. [CrossRef]

40. Oliveira, I.; Meyer, A.S.; Afonso, S.; Aires, A.; Goufo, P.; Trindade, H.; Gonçalves, B. Phenolic and Fatty Acid Profiles, A-tocopherol and Sucrose Contents, and Antioxidant Capacities of Understudied Portuguese Almond Cultivars. *J. Food Biochem.* **2019**, *43*, e12887. [CrossRef]

41. dos Santos Escobar, O.; de Azevedo, C.F.; Swarowsky, A.; Adebayo, M.A.; Netto, M.S.; Machado, F.M. Utilization of Different Parts of *Moringa oleifera* Lam. Seeds as Biosorbents to Remove Acid Blue 9 Synthetic Dye. *J. Environ. Chem. Eng.* **2021**, *9*, 105553. [CrossRef]

42. Hussain, G.; Haydar, S. Textile Effluent Treatment Using Natural Coagulant Opuntia Stricta in Comparison with Alum. *CLEAN–Soil Air Water* **2021**, *49*, 2000342. [CrossRef]

43. Wang, Y.; Li, Y.; Zhang, Y.; Wei, W. Enhanced Brilliant Blue FCF Adsorption Using Microwave-Hydrothermal Synthesized Hydroxyapatite Nanoparticles. *J. Dispers. Sci. Technol.* **2020**, *41*, 1346–1355. [CrossRef]

44. Beltrán-Heredia, J.; Sánchez-Martín, J.; Muñoz-Serrano, A.; Peres, J.A. Towards Overcoming TOC Increase in Wastewater Treated with *Moringa oleifera* Seed Extract. *Chem. Eng. J.* **2012**, *188*, 40–46. [CrossRef]

45. Rodríguez-Chueca, J.; Amor, C.; Mota, J.; Lucas, M.S.; Peres, J.A. Oxidation of Winery Wastewater by Sulphate Radicals: Catalytic and Solar Photocatalytic Activations. *Environ. Sci. Pollut. Res.* **2017**, *24*, 22414–22426. [CrossRef] [PubMed]

46. Rodríguez-Chueca, J.; Amor, C.; Silva, T.; Dionysiou, D.D.; Li Puma, G.; Lucas, M.S.; Peres, J.A. Treatment of Winery Wastewater by Sulphate Radicals: HSO_5^-/Transition Metal/UV-A LEDs. *Chem. Eng. J.* **2017**, *310*, 473–483. [CrossRef]

47. Ferreira, L.C.; Fernandes, J.R.; Rodriguez-Chueca, J.; Peres, J.A.; Lucas, M.S.; Tavares, P.B. Photocatalytic Degradation of an Agro-Industrial Wastewater Model Compound Using a UV LEDs System: Kinetic Study. *J. Environ. Manage.* **2020**, *269*, 110740. [CrossRef]

48. Jorge, N.; Teixeira, A.R.; Fernandes, J.R.; Oliveira, I.; Lucas, M.S.; Peres, J.A. Degradation of Agro-Industrial Wastewater Model Compound by UV-A-Fenton Process: Batch vs. Continuous Mode. *Int. J. Environ. Res. Public Health* **2023**, *20*, 1276. [CrossRef] [PubMed]

49. Li, J.; Lin, H.; Yang, L.; Zhang, H. Copper-Spent Activated Carbon as a Heterogeneous Peroxydisulfate Catalyst for the Degradation of Acid Orange 7 in an Electrochemical Reactor. *Water Sci. Technol.* **2016**, *73*, 1802–1808. [CrossRef] [PubMed]

50. Huang, Y.; Jia, Y.; Zuo, L.; Huo, Y.; Zhang, Y. Comparison of VUV/H_2O_2 and VUV/PMS (Peroxymonosulfate) for the Degradation of Unsymmetrical Dimethylhydrazine in Water. *J. Water Process Eng.* **2022**, *49*, 102970. [CrossRef]

51. He, B.; Yang, Y.; Liu, B.; Zhao, Z.; Shang, J.; Cheng, X. Degradation of Chlortetracycline Hydrochloride by Peroxymonosulfate Activation on Natural Manganese Sand Through Response Surface Methodology. *Environ. Sci. Pollut. Res.* **2022**, *29*, 82584–82599. [CrossRef]

52. Govindan, K.; Raja, M.; Noel, M.; James, E.J. Degradation of Pentachlorophenol by Hydroxyl Radicals and Sulfate Radicals Using Electrochemical Activation of Peroxomonosulfate, Peroxodisulfate and Hydrogen Peroxide. *J. Hazard. Mater.* **2014**, *272*, 42–51. [CrossRef]

53. Rastogi, A.; Al abed, S.R.; Dionysiou, D.D. Sulfate Radical-Based Ferrous–Peroxymonosulfate Oxidative System for PCBs Degradation in Aqueous and Sediment Systems. *Appl. Catal. B Environ.* **2009**, *85*, 171–179. [CrossRef]

54. Huling, S.G.; Ko, S.; Park, S.; Kan, E. Persulfate Oxidation of MTBE-and Chloroform-Spent Granular Activated Carbon. *J. Hazard. Mater.* **2011**, *192*, 1484–1490. [CrossRef] [PubMed]

55. Yi, Q.; Tan, J.; Liu, W.; Lu, H.; Xing, M.; Zhang, J. Peroxymonosulfate Activation by Three-Dimensional Cobalt Hydroxide/Graphene Oxide Hydrogel for Wastewater Treatment through an Automated Process. *Chem. Eng. J.* **2020**, *400*, 125965. [CrossRef]

56. Li, J.; Xu, M.; Yao, G.; Lai, B. Enhancement of the Degradation of Atrazine through $CoFe_2O_4$ Activated Peroxymonosulfate (PMS) Process: Kinetic, Degradation Intermediates, and Toxicity Evaluation. *Chem. Eng. J.* **2018**, *348*, 1012–1024. [CrossRef]

57. Chen, J.; Qian, Y.; Liu, H.; Huang, T. Oxidative Degradation of Diclofenac by Thermally Activated Persulfate: Implication for ISCO. *Environ. Sci. Pollut. Res.* **2016**, *23*, 3824–3833. [CrossRef]
58. Tan, C.; Gao, N.; Deng, Y.; An, N.; Deng, J. Heat-Activated Persulfate Oxidation of Diuron in Water. *Chem. Eng. J.* **2012**, *203*, 294–300. [CrossRef]
59. Rodríguez-Chueca, J.; Giannakisb, S.; Marjanovic, M.; Kohantorabi, M.; Gholami, M.R.; Grandjean, D.; de Alencastro, L.F.; Pulgarín, C. Solar-Assisted Bacterial Disinfection and Removal of Contaminants of Emerging Concern by Fe^{2+}-Activated HSO_5^- vs. $S_2O_8^{2-}$ in Drinking Water. *Appl. Catal. B Environ.* **2019**, *248*, 62–72. [CrossRef]
60. Ahmadi, M.; Ghanbari, F. Combination of UVC-LEDs and Ultrasound for Peroxymonosulfate Activation to Degrade Synthetic Dye: Influence of Promotional and Inhibitory Agents and Application for Real Wastewater. *Environ. Sci. Pollut. Res.* **2018**, *25*, 6003–6014. [CrossRef]
61. Latif, A.; Kai, S.; Si, Y. Catalytic Degradation of Organic Pollutants in Fe (III)/Peroxymonosulfate (PMS) System: Performance, Influencing Factors, and Pathway. *Environ. Sci. Pollut. Res.* **2019**, *26*, 36410–36422. [CrossRef]
62. Fang, X.; Gan, L.; Wang, L.; Gong, H.; Xu, L.; Wu, Y.; Lu, H.; Han, S.; Cui, J.; Xia, C. Enhanced Degradation of Bisphenol A by Mixed ZIF Derived CoZn Oxide Encapsulated N-Doped Carbon via Peroxymonosulfate Activation: The Importance of N Doping Amount. *J. Hazard. Mater.* **2021**, *419*, 126363. [CrossRef]
63. Grehs, B.W.N.; Lopes, A.R.; Moreira, N.F.F.; Fernandes, T.; Linton, M.A.O.; Silva, A.M.T.; Manaia, C.M.; Carissimi, E.; Nunes, O.C. Removal of Microorganisms and Antibiotic Resistance Genes from Treated Urban Wastewater: A Comparison between Aluminium Sulphate and Tannin Coagulants. *Water Res.* **2019**, *166*, 115056. [CrossRef]
64. Sun, J.; Li, X.; Feng, J.; Tian, X. Oxone/Co^{2+} Oxidation as an Advanced Oxidation Process: Comparison with Traditional Fenton Oxidation for Treatment of Landfill Leachate. *Water Res.* **2009**, *43*, 4363–4369. [CrossRef] [PubMed]
65. Rodriguez-Narvaez, O.M.; Pacheco-Alvarez, M.O.A.; Wróbel, K.; Páramo-Vargas, J.; Bandala, E.R.; Brillas, E.; Peralta-Hernandez, J.M. Development of a Co^{2+}/PMS Process Involving Target Contaminant Degradation and PMS Decomposition. *Int. J. Environ. Sci. Technol.* **2020**, *17*, 17–26. [CrossRef]
66. Wang, Y.R.; Chu, W. Photo-Assisted Degradation of 2,4,5-Trichlorophenoxyacetic Acid by Fe(II)-Catalyzed Activation of Oxone Process: The Role of UV Irradiation, Reaction Mechanism and Mineralization. *Appl. Catal. B Environ.* **2012**, *123–124*, 151–161. [CrossRef]
67. Guerra-Rodríguez, S.; Ribeiro, A.R.L.; Ribeiro, R.S.; Rodríguez, E.; Silva, A.M.; Rodríguez-Chueca, J. UV-A Activation of Peroxymonosulfate for the Removal of Micropollutants from Secondary Treated Wastewater. *Sci. Total Environ.* **2021**, *770*, 145299. [CrossRef]
68. Ao, X.; Li, Z.; Zhang, H. A Comprehensive Insight into a Rapid Degradation of Sulfamethoxazole by Peroxymonosulfate Enhanced UV-A LED/g-C_3N_4 Photocatalysis. *J. Clean. Prod.* **2022**, *356*, 131822. [CrossRef]
69. Yin, R.; Guo, W.; Wang, H.; Du, J.; Zhou, X.; Wu, Q.; Zheng, H.; Chang, J.; Ren, N. Enhanced Peroxymonosulfate Activation for Sulfamethazine Degradation by Ultrasound Irradiation: Performances and Mechanisms. *Chem. Eng. J.* **2018**, *335*, 145–153. [CrossRef]
70. Lu, Y.; Xu, W.; Nie, H.; Zhang, Y.; Deng, N.; Zhang, J. Mechanism and Kinetic Analysis of Degradation of Atrazine by US/PMS. *Int. J. Environ. Res. Public Health* **2019**, *16*, 1781. [CrossRef]
71. Lucas, M.S.; Peres, J.A.; Li, G. Treatment of Winery Wastewater by Ozone-Based Advanced Oxidation Processes (O_3, O_3/UV and O_3/UV/H_2O_2) in a Pilot-Scale Bubble Column Reactor and Process Economics. *Sep. Purif. Technol.* **2010**, *72*, 235–241. [CrossRef]
72. Marchão, L.; Fernandes, J.R.; Sampaio, A.; Peres, J.A.; Tavares, P.B.; Lucas, M.S. Microalgae and Immobilized TiO_2/UV-A LEDs as a Sustainable Alternative for Winery Wastewater Treatment. *Water Res.* **2021**, *203*, 117464. [CrossRef]
73. Esteves, B.M.; Rodrigues, C.S.D.; Maldonado-hódar, F.J.; Madeira, L.M. Treatment of High-Strength Olive Mill Wastewater by Combined Fenton-like Oxidation and Coagulation/Flocculation. *J. Environ. Chem. Eng.* **2019**, *7*, 103252. [CrossRef]
74. Norzaee, S.; Taghavi, M.; Djahed, B.; Kord Mostafapour, F. Degradation of Penicillin G by Heat Activated Persulfate in Aqueous Solution. *J. Environ. Manage.* **2018**, *215*, 316–323. [CrossRef] [PubMed]
75. Pouran, S.R.; Raman, A.A.A.; Daud, W.M.A.W. Review on the Application of Modified Iron Oxides as Heterogeneous Catalysts in Fenton Reactions. *J. Clean. Prod.* **2014**, *64*, 24–35. [CrossRef]
76. Li, M.; Yang, X.; Wang, D.; Yuan, J. Enhanced Oxidation of Erythromycin by Persulfate Activated Iron Powder–H_2O_2 System: Role of the Surface Fe Species and Synergistic Effect of Hydroxyl and Sulfate Radicals. *Chem. Eng. J.* **2017**, *317*, 103–111. [CrossRef]
77. Jaafarzadeh, N.; Ghanbari, F.; Alvandi, M. Integration of Coagulation and Electro-Activated HSO_5^- to Treat Pulp and Paper Wastewater. *Sustain. Environ. Res.* **2017**, *27*, 223–229. [CrossRef]
78. Amor, C.; Rodríguez-Chueca, J.; Fernandes, J.L.; Domínguez, J.R.; Lucas, M.S.; Peres, J.A. Winery Wastewater Treatment by Sulphate Radical Based-Advanced Oxidation Processes (SR-AOP): Thermally vs UV-Assisted Persulphate Activation. *Process. Saf. Environ. Prot.* **2019**, *122*, 94–101. [CrossRef]
79. Jaafarzadeh, N.; Omidinasab, M.; Ghanbari, F. Combined Electrocoagulation and UV-Based Sulfate Radical Oxidation Processes for Treatment of Pulp and Paper Wastewater. *Process. Saf. Environ. Prot.* **2016**, *102*, 462–472. [CrossRef]
80. Papastefanakis, N.; Mantzavinos, D.; Katsaounis, A. DSA Electrochemical Treatment of Olive Mill Wastewater on Ti/RuO_2 Anode. *J. Appl. Electrochem.* **2010**, *40*, 729–737. [CrossRef]

81. Amor, C.; Lucas, M.S.; Pirra, A.J.; Peres, J.A. Treatment of Concentrated Fruit Juice Wastewater by the Combination of Biological and Chemical Processes. *J. Env. Sci. Health Part A* **2012**, *47*, 1809–1817. [CrossRef]
82. Rodríguez-Chueca, J.; Amor, C.; Fernandes, J.R.; Tavares, P.B.; Lucas, M.S.; Peres, J.A. Treatment of Crystallized-Fruit Wastewater by UV-A LED Photo-Fenton and Coagulation-Flocculation. *Chemosphere* **2016**, *145*, 351–359. [CrossRef] [PubMed]

Article

The Variations and Influences of the Channel Centerline of the Zhenjiang-Yangzhou Reach of the Yangtze River Based on Archival and Contemporary Data Sets

Cunli Liu [1,2], Binglin Liu [3,*], Zhenke Zhang [1], Changfeng Li [4], Guoen Wei [1] and Shengnan Jiang [1]

[1] School of Geographic & Oceanographic Sciences, Nanjing University, Nanjing 210037, China
[2] School of Environmental Science, Nanjing Xiaozhuang University, Nanjing 211171, China
[3] College of Geography and Planning, Nanning Normal University, Nanning 530001, China
[4] School of Economics, Nanjing University of Finance and Economics, Nanjing 210003, China
* Correspondence: dg1827018@smail.nju.edu.cn; Tel.: +86-186-5185-2360

Abstract: The Zhenjiang-Yangzhou reach of the Yangtze River is located at the top of the Yangtze River Delta, which is one of the most dramatic changes in the lower reaches of the Yangtze River. The study on the migration characteristics of the channel centerline is crucial for a comprehensive understanding of the river channel changes in the Zhenjiang-Yangzhou reach. In this study, a detailed calculation method is proposed to extract the channel centerline of the Zhenjiang-Yangzhou reach by using old maps and remote sensing satellite map and decompose it into seven parts. The spatial and temporal changes of Net Shift Distance (*NSD*), Cumulative Moving Distance (*CMD*), Migration Rate of Channel Centerline (*MRCC*) and Linear Regression Change Rate of channel centerline (*LRCR*) from 1865 to 2019 on the cross-section scale are studied. The results show that: (1) from 1868 to 2019, the channel centerline of the Zhenjiang-Yangzhou reach kept shifting. The average net displacement distance of the section is 1103.47 m on the right bank, and the average cumulative displacement distance of the section is 2790.51 m. (2) According to the *NSD* and *CMD* data of each part, the long-term movement direction of the channel centerline is basically the same, and a small part of the channel centerline has periodic reverse swing. The probability of channel centerline moving right is about twice that of moving left. At the same time, some rivers have high erosion risk. (3) Through *MRCC* and *LRCR* data, the total number of channel centerline moving left and right is 156 and 329, respectively, and the erosion risk level of the near half of the shoreline is high. (4) The change of river boundary conditions and hydrodynamic force will affect the migration rate and direction of channel centerline. (5) This study proposes a method to extract channel centerline from a braided reach and study its changes, which can be applied to other similar reaches with a long history of human activities and high density. The results enrich people's understanding of the long-term changes of a braided reach in the lower reaches of the Yangtze River and have certain guiding significance for river regulation, navigation safety, and revetment construction.

Keywords: channel centerline; variations; the Zhenjiang-Yangzhou reach; the Yangtze River

Citation: Liu, C.; Liu, B.; Zhang, Z.; Li, C.; Wei, G.; Jiang, S. The Variations and Influences of the Channel Centerline of the Zhenjiang-Yangzhou Reach of the Yangtze River Based on Archival and Contemporary Data Sets. *Water* **2022**, *14*, 2478. https://doi.org/10.3390/w14162478

Academic Editor: Gassan Hodaifa

Received: 2 June 2022
Accepted: 6 August 2022
Published: 11 August 2022

Publisher's Note: MDPI stays neutral with regard to jurisdictional claims in published maps and institutional affiliations.

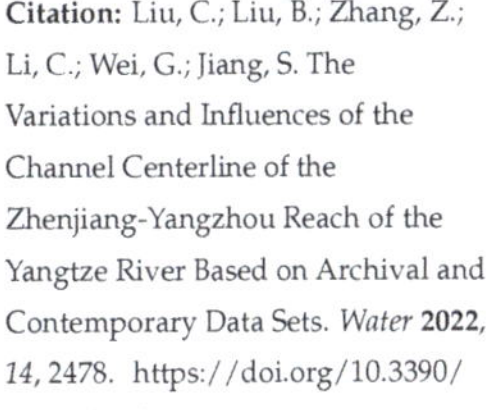

1. Introduction

River systems collect, transport, and distribute water, sediment, and biomass in basins, thereby tightly linking processes at multiple time scales, such as the hydrological cycle, landscape changes, and ecological evolution. Water bodies interact with the atmosphere, vegetation, environment, and landforms and play an important role in regional economic and environmental sustainability, drinking water safety, and ecological security [1–3]. River channel change can be very sensitive to environmental change and human activities, and it has been one of the main research topics in fluvial geomorphology [4].

There are many braided reaches in the middle and lower Yangtze River in China. These reaches have a number of central bars or islands, which lead to multiple channels at low flows, and such a planform geometry usually results in frequent lateral channel shifting [5]. These fluvial processes of braided reaches can cause unfavorable effects such as inducing the collapse of floodplain banks and threatening the safety of levees. Therefore, channel migration processes in braided rivers have aroused major concern in the study of river dynamics and geomorphology [6].

Multitemporal analyses of river course changes over long time are of crucial importance [7]. For example, the analysis of changes in water channel and floodplain of the lower Yuba River in California, USA, was performed over a period of 100 years. This study was mostly based on photogrammetric data and old maps, including a comparison of a digital elevation model (DEM) and planimetric change analysis [8]. A similar study was performed for the Basento River in Southern Italy. The researchers managed to carry out an analysis of the channel changes over 150 years [9]. Another research using different data sources concerns the Calore River in Northern Italy and change detection of its course since 1870. The data were processed using advanced GIS methods [10].

Remote sensing technology, as a multi-temporal, multi-spectral, low-cost, and high-resolution information source, has been widely used to study channel morphological changes in alluvial rivers, and specific methods or software have been developed to quantify channel characteristics [11,12]. It is theoretically possible to define the boundary of an extracted water surface area as the river bankline and then extract the channel centerline. However, the extracted water boundary is very sensitive to the water level changes in the same period, and the channel changes obtained from satellite images may be different when the water levels are in different value ranges. Low water levels may lead to substantial exposure of the riverbed, which will lead to errors in the extracted channels compared to those at high water levels [13]. Only satellite images with the lowest possible water level differences can be selected to reduce errors [14].

However, previous studies on channel evolution of braided rivers have generally focused on the study of inflow-sediment changes, channel deposition and degradation, and adjustments to cross-sectional geometry [15–17]. River channel migration plays a crucial role in the morphological changes of braided rivers, which usually include various lateral shifts of shoreline, water depth, main line, and channel centerline [18,19]. Several data sources from historical maps, cross-sectional topography, and remote sensing imagery can be used to determine these changes [20–22].

The Zhenjiang-Yangzhou reach is one of the most rapidly evolving braided reaches in the lower Yangtze River. Based on the measured hydrology and topographical data, extensive researches have been conducted on the evolution of the river, its influencing factors, and channel conditions [23–27]. Numerous studies have shown that the Zhenjiang-Yangzhou reach of the Yangtze River has undergone significant channel changes in the past century, including the continuous lateral movement of the channel centerline as well as the scouring and silting of the river banks. Therefore, to determine the temporal and spatial variation of the migration direction and rate of the channel centerline is helpful for a comprehensive understanding of the characteristics of the river process of the Zhenjiang-Yangzhou reach of the Yangtze River. Nowadays, with the popularization of aerial laser radar scanning technology and UAV aerial photography technology, the acquisition of high-quality river change data has become very simple, and the scanning resolution of the above method is also suitable for most software processing. Unfortunately, due to the slow speed of river change, short-term aerial data cannot provide enough information about river change. This means that in most cases, using one data source alone cannot obtain long-term river channel change data, and multi-source data are needed for research. However, remote sensing data and historical maps are not compatible in form, resolution, coordinate system, projection, and accuracy. Therefore, it is necessary to establish a more reliable semi-automatic program to standardize the two data sources. Based on the unified standard and process, the channel centerline in the historical maps and remote sensing satellite map was extracted, and the correlation analysis was completed.

In this study, we employed multi-source data fusion methods at different scales to study the spatiotemporal changes of the Net Shift Distance, Cumulative Moving Distance, Migration Rate of Channel Centerline, and Linear Regression Change Rate of the channel centerline from 1865 to 2019. The multi-source data include old maps and satellite imagery. The main objectives of the current study are to: (1) propose a detailed computational procedure to extract the channel centerline from old maps and satellite images and set the baseline and cross-section; (2) calculate the temporal and spatial changes of the channel centerline's shift distance, Cumulative Moving Distance, Migration Rate of Channel Centerline, and Linear Regression Change Rate on the cross-sectional scale to judge the risk of river bank erosion; and (3) analyze the key factors affecting the lateral movement of channel centerline and discuss the influence of river boundary conditions, hydrological conditions, human activities, and other factors on the lateral movement of channel centerline.

2. Materials and Methods

2.1. Study Area

The Zhenjiang-Yangzhou reach of the Yangtze River is located at the apex of the Yangtze River Delta, starting from Yizheng at the upstream to Sanjiangying at the downstream, of which the total length is 73.7 km [28]. This reach has a large range of changes and a complex evolution process (Figure 1) [29]. The north bank of Zhenjiang-Yangzhou reach is a vast delta alluvial plain, the south bank is close to the northern foot of the Nanjing-Zhenjiang mountain, and the Xiashu loess terrace was formed in the late Pleistocene, downstream of Zhenjiang Port; the river arm of bascule bridge bluff invades the river and controls the development of the river.

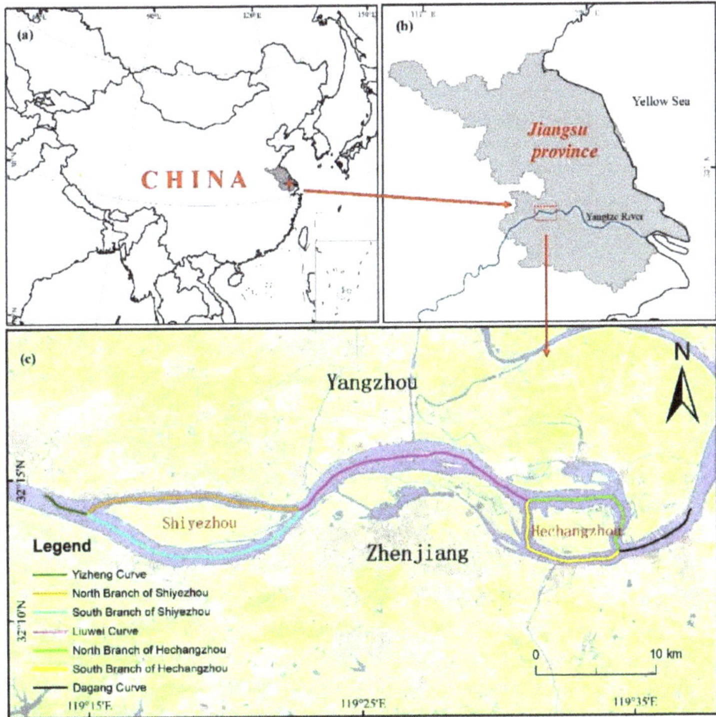

Figure 1. Location of the area of study (**a**) in Eastern China, (**b**) Southwestern Jiangsu Province, the apex of the Yangtze River Delta, and (**c**) the junction of Zhenjiang and Yangzhou.

Affected by Yellow River flood over Huaihe River from 1570, the lower reaches of Huaihe River gradually evolved from flowing eastward into the sea to flowing southward into the Yangtze River, becoming a tributary of the Yangtze River. Since then, the water inflow of the Huaihe River has gradually increased, reaching the maximum after the main stream of the lower reaches of the Huaihe River was diverted into the river in 1851. This change adds a new variable to the evolution of the river geomorphology of the Yangtze River and changes the original channel characteristics and flow structure of the Zhenjiang-Yangzhou reach of the Yangtze River. Beach scouring and silting and riverbank advancing and retreating also change greatly.

The total length of the Zhenjiang-Yangzhou reach is 73.3 km, consisting of Yizheng Curve, Shiyezhou Curve, Liuliu Curve, Hechangzhou Inlet, and Dagang Curve [30,31]. The reach plane form of Zhenjiang-Yangzhou is complex, with both branching sections and bending sections. The river course changes rapidly, and the regulation is difficult [32].

2.2. Data Source

2.2.1. Old Maps

Old maps presenting the Zhenjiang-Yangzhou reach of the Yangtze River valley were not a frequent subject of research. Old maps that were considered potentially useful include: (1) The sandbar merging north bank on the Jiangsu Province Atlas, compiled in 1868, was collected in the Library of Fudan University (No. 001126209), which recorded in detail the river course and sandbar of the Yangtze River in Zhenyang section. Combined with the historical data of Yangzhou City—"Jiangdu County Annals" and the data of The Harvard University Library in the United States does not publish the historical data of the old customs of China (1860–1949)", published by historian Wu Songdi in 2014. It can basically restore the change of the reach channel of the Zhenjiang-Yangzhou in the 1960s [33,34]. (2) In 1931, the Land Survey Bureau of the Ministry of Staff of the National Government completed the topographic map drawing of Jiangsu Province and sorted out the military topographic maps of various scales in Jiangsu Province. The relevant military topographic maps were collected and collated by the Department of Regional Affairs of the Ministry of Internal Affairs of the National Government. The relevant military topographic maps can be obtained online through the Institute of Modern History of the Central Academy in Taipei City, Taiwan Province, China [35]. The topographic map can be used as the basic data of the river morphology of the Zhenjiang-Yangzhou reach of the Yangtze River in the 1930s. (3) In 1953, the Cartographic Bureau of the United States Army Agency drew detailed topographic maps of the surrounding areas of Nanjing, Jiangsu Province, based on aerial imagery accumulated during the Anti-Japanese War and other mapping data from the 1950s. The military topographic map was kept by the University of Texas Library after decryption. The river morphology of the Zhenjiang-Yangzhou reach of the Yangtze River reflected in the military topographic map can be used as the basic data for the river change study of the Zhenjiang-Yangzhou reach in the 1950s [36].

2.2.2. Remote Sensing Satellite Images

Remote sensing images are real-time, accurate, and intuitive and are often used to identify water-land boundaries [37]. In order to ensure the representativeness of the extraction results, the selected remote sensing satellite images should be clear, cloudless, and located at similar water levels. We selected two remote sensing satellite images, including (1) TM image of Landsat 5 satellite, taken on 18 October 1984, with a spatial resolution of 30 m, and (2) OLI image of Landsat 8 satellite, taken on 16 November 2019, with the same as the TM image. The geometric correction was conducted by image-to-image registration, with the image in 2019 as the reference of all other images. The root mean square errors of correction are all less than 30 m.

2.3. Research Method

2.3.1. Research Process

The following flowchart (Figure 2) represents the dataset and processing methods used in this study. In some cases, it shows the relationship between data sets, the channel Centerline extraction method, and how to use them in the channel centerline of the Zhenjiang-Yangzhou reach of the Yangtze River. According to the flow chart, we used Arcgis10.5 and ENVI5.3 software to process the data step by step: (1) Each old map collected was registered and geometrically corrected several times according to the standard map published locally, and the quality inspection was carried out. The results with the highest accuracy were output, and the shoreline was vectorized. At the same time, we manually extracted coastlines from each satellite map by visual interpretation. (2) We extracted the five different years of vectorized shoreline map summary, with automatic extraction of their channel centerline. (3) We created a baseline and created appropriate interval sections. (4) Each section intersects with channel centerline to form an intersection. We calculated *NSD, CMD, MRCC,* and *LRCR* according to the distance between the intersection and other intersections of the same section and the distance between the intersection and baseline.

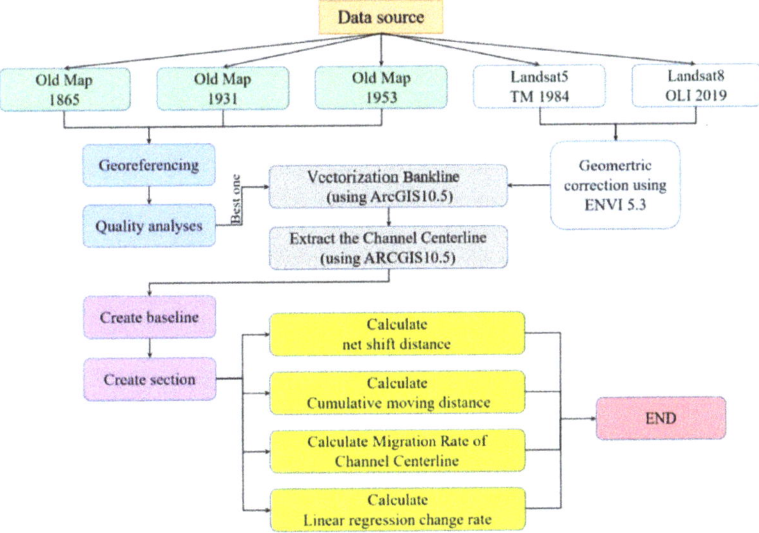

Figure 2. Flowchart of used data and methods.

2.3.2. Determination of Section

The channel centerline of the Zhenjiang-Yangzhou reach is divided into seven segments: Yizheng Curve, North Branch of Shiyezhou, South Branch of Shiyezhou, Six Curve, North Branch, of Hechangzhou, South Branch of Hechangzhou, and Dagang Curve (Figure 3). Then, we found the buffer that is closest to the channel centerline and created a baseline according to the buffer direction. Then, the cross-section was generated: the longest length of the cross section was set to 3500 m, and the sampling interval was set to 200 m. After smoothing, fitting, and adjustment, the cross-section was orthogonally projected from the baseline to the channel centerline in different years. A total of 510 equidistant vertical lines were generated, including 19 equidistant vertical lines in Yizheng Curve, 82 equidistant vertical lines in North Branch of Shiyezhou, 67 equidistant vertical lines in South Branch of Shiyezhou, and 103 equidistant vertical lines in Six Curve. In addition, the relevant parameters of this study are also shown in Figure 3.

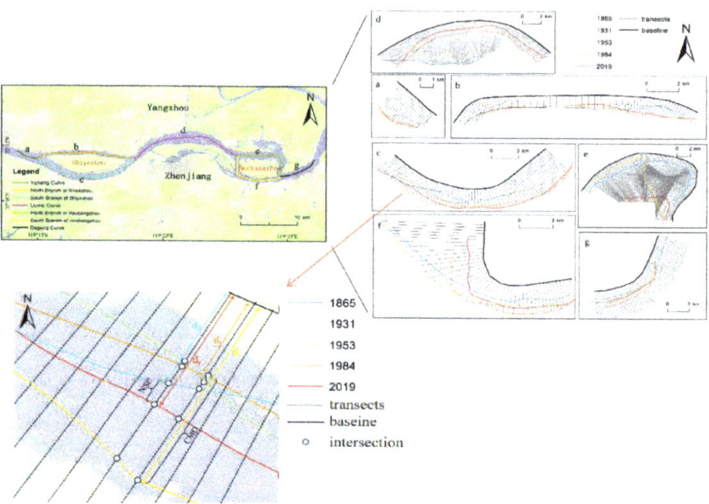

Figure 3. Map of the channel centerline and section position in Zhenyang Reach in different ages and related parameters.

2.3.3. Calculation of Net Shift Distance (*NSD*)

Net Shift Distance (*NSD*) is the distance between the channel centerline of the farthest and most recent years on each section. The *NSD* is expressed as follows:

$$NSD = d_r - d_0,$$

(1)

In this formula, d_r is the distance from the intersection point of the channel centerline and the section line to the baseline in the recent year, and d_0 is the distance from the intersection of the channel centerline and the section line to the baseline in the furthest year.

2.3.4. Calculation of Cumulative Moving Distance (*CMD*)

CMD is the distance between the channel centerline that is farthest from the baseline and the closest to the baseline on each section. *CMD* represents the sum of the motion distance of the channel centerline position in all phases, which is independent of time. The *CMD* is expressed as follows:

$$CMD = d_j - d_i,$$

(2)

In this formula, d_j is the distance from the nearest intersection of the channel centerline and the section line to the baseline; d_i is the distance from the intersection of the channel centerline and the section line to the baseline.

2.3.5. Calculation of Migration Rate of Channel Centerline (*MRCC*)

MRCC refers to the change rate of the distance between the channel centerline in the furthest year and the nearest year on each section. The *MRCC* is expressed as follows:

$$MRCC_{(i,j)} = \frac{d_j - d_i}{\Delta Y_{(j,i)}},$$

(3)

where i denotes the point where the channel centerline intersects the section line in the farthest year, and j denotes the point where the channel centerline intersects the section line in the recent year. $MRCC_{(i,j)}$ denotes the terminal rate of change from the channel centerline in the farthest year to the channel centerline in the latest year. d_i and d_j represent the distance between the channel centerline and the baseline in the furthest year and the nearest year, respectively; $\Delta Y_{(j,i)}$ represents the time interval between the nearest year and the farthest year.

2.3.6. Calculation of Linear Regression Change Rate (*LRCR*)

LRCR refers to the point where the cross-section line intersects the channel centerline by using the least square method, and finally, the change rate of the channel centerline is calculated. *LRCR* can fit the changes of the channel centerline in all years, which is simple in principle and easy to operate. The *LRCR* index is used in this study to calculate and analyze the multiple change rates the channel centerline over a long period of time. The *LRCR* is expressed as follows:

$$y = a + bx, \tag{4}$$

$$a = \sum_{i=1}^{n}(x_i - \overline{x})(y_i - \overline{y}), \tag{5}$$

$$LRCR = \overline{y} - a\overline{x}, \tag{6}$$

Among them, y is a dependent variable for the spatial location of the channel centerline; x represents the independent variable of the year; a is the fitted constant intercept; *LRCR* is the regression slope, which represents the y change corresponding to each unit x change.

3. Results and Discussion

In this study, the above methods were employed to quantify the section- and reach-scale morphometric parameters, covering the Net Shift Distance, Cumulative Moving Distance, Migration Rate of Channel Centerline, and Linear Regression Change Rate in the channel centerline of the Zhenjiang-Yangzhou reach in 1865–2019.

3.1. The Net Shift Distance and Cumulative Moving Distance of the Channel Centerline on the Section Scale

In the past 100 years, the channel centerline reached by Zhenjiang-Yangzhou has shown a trend of continuous swing on the whole (Figure 4). The average Net Shift Distance of the section is 1103.47 m for the right bank migration, and the average cumulative displacement distance of the section is 2790.51 m. The channel centerline has the largest net displacement distance of the channel curve, followed by the South Branch of Hechangzhou, North Branch of Shiyezhou, Yizheng Curve, and Curve, and South Branch of Hechangzhou, and the smallest is North Branch of Hechangzhou. The maximum Cumulative Moving Distance of the channel centerline is North Branch of Hechangzhou, followed by Six Curve, South Branch of Shiyezhou, South Branch of Hechangzhou, and North Branch of Shiyezhou.

Among them, for the channel centerline of Yizheng Curve during 1868–2019, the value of Net Shift Distance from upstream to downstream trend is high → low → medium, the average section of Net Shift Distance is right bank migration 302.01 m, the highest value appears in section 4, the right bank migration 650.63 m, and the lowest value appears in sections 1 and 2. The variation trend of Cumulative Moving Distance from upstream to downstream is low to high, with an average of 938.65 m. The highest value appeared in section 5, which was 1972.7 m, and the lowest value appeared in sections 1 and 2. Cumulative Moving Distance is much larger than Net Shift Distance, which indicates that Yizheng Curve has periodic reverse swing. The connection between Yizheng Curve and South Branch of Shiyezhou becomes the main channel in the upper reaches of the river, which receives the influence of the geostrophic bias force. The river continuously scours the right bank, resulting in the gradual movement of the channel center line to the right bank.

For the channel centerline of North Branch of Shiyezhou during 1868–2019, the value of Net Shift Distance from upstream to downstream changed from low to high. The average Net Shift Distance of the cross-section was 57.23 m, and the highest value appeared in section 71, which was right bank migration 683.79 m; the lowest value appeared in section 31, which was left bank migration, at 10.84 m. The change trend of Cumulative Moving Distance from upstream to downstream is high → low → high, with an average of 435.93 m. The highest value appears in section 4, which is 244.36 m, and the lowest value appears in section 1. The Cumulative Moving Distance of sections 1–14 is far greater than the Net Shift Distance, indicating that there is periodic reverse swing in this part of the river. The absolute value of Cumulative Moving Distance after section 41 is basically the same as that of Net Shift Distance, indicating that the long-term movement direction of this part of the river is basically the same. The right bank of North Branch of Shiyezhou was scoured, while the small sandbars in the area of Shiyezhou Sandbank continued to merge, resulting in an expanding area of Shiyezhou Sandbank and a dynamic balance on the right bank.

a：Yizheng Curve
b：North Branch of Shiyezhou
c：South Branch of Shiyezhou
d：Liuwei Curve
e：North Branch of Hechangzhou
f：South Branch of Hechangzhou
g：Dagang Curve

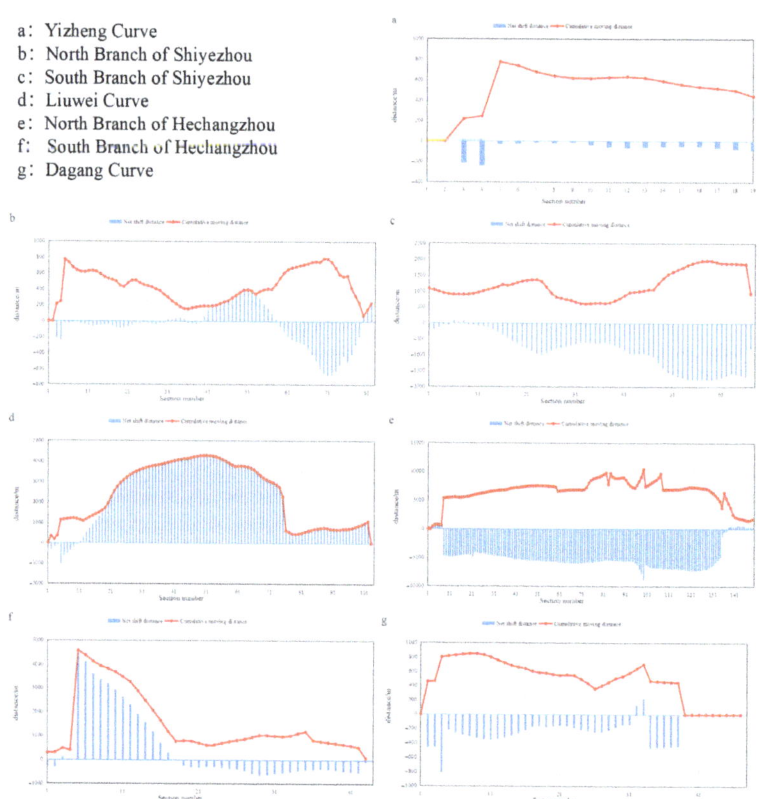

Figure 4. Changes of *NSD* and *CMD* of the channel centerline in Zhenjiang-Yangzhou reach. (**a**) Yizheng Curve; (**b**) North Branch of Shiyezhou; (**c**) South Branch of Shiyezhou; (**d**) Liuwei Curve; (**e**) North Branch of Hechangzhou; (**f**) South Branch of Hechangzhou; (**g**) Dagang Curve.

For the channel centerline of South Branch of Shiyezhou during 1868–2019, the value of Net Shift Distance from upstream to downstream changed from low to high. The average section Net Shift Distance was 869.04 m on the right bank. The highest value appeared in section 59, which was right bank migration, at 1777.36 m. The lowest value appeared in section 5, which was right bank migration, at 0.55 m. The change trend of Cumulative Moving Distance from upstream to downstream is high → low → high, with an average of 1186.31 m. The highest value appeared in section 59, which is 1958.07 m, and the lowest value appeared in section 33, at 606.65 m. Cumulative Moving Distance of sections 1–23 is greater than Net Shift Distance, indicating that there is periodic reverse swing in this part of the river. The absolute value of Cumulative Moving Distance after section 24 is basically the same as that of Net Shift Distance, indicating that the long-term movement direction of this part of the river is basically the same. At first, a large number of small sandbars were scattered in this area, which compressed the width of South Branch of Shiyezhou. Later, after the small sandbars were merged into Shiyezhou Sandbank, the river surface became wider.

For the channel centerline of Liuwei Curve during 1868–2019, the change trend of Net Shift Distance from upstream to downstream is low → high → low, and the average cross-section Net Shift Distance is left bank migration 2191.63 m. The highest value appears in section 51, which is 4273.94 m for left bank migration, and the lowest value appears in section 103. The variation trend of Cumulative Moving Distance from upstream to downstream is high → low → high, which is consistent with the variation trend of Net Shift Distance, with an average value of 2340.15 m. The maximum appears in section 59, which is 4273.94 m, and the minimum appears at section 103. The absolute values of Cumulative Moving Distance of most cross-sections are basically the same as those of Net Shift Distance, indicating that the long-term movement direction of boundary curve in each cross-section is basically the same. The reason for the dramatic change of the Six Curve is that the hydrodynamic force has changed greatly, resulting in the change of the top impact point, and the river changed from the initial concave bank to the convex bank.

For the channel centerline of North Branch of Hechangzhou during 1868–2019, the value of Net Shift Distance from upstream to downstream trend is low → high → low, the average section of Net Shift Distance is right bank migration at 4908.03 m, the highest value appears in section 59, and the right bank migration is 7851.58 m. The lowest value appears in section 2. The variation trend of Cumulative Moving Distance from upstream to downstream is low → high → low, which is consistent with the variation trend of Net Shift Distance, with an average value of 6465.19 m. The maximum appears at section 99, which is 9139.1 m, and the minimum appears at section 2. The absolute values of Cumulative Moving Distance of most sections are basically consistent with those of Net Shift Distance, indicating that the long-term movement direction of North Branch of Hechangzhou in each section is basically the same. Today's Hechangzhou Sandbank area is much smaller than in 1868. At first, there was a huge and scattered small sandbank, and the northern bank of the merged Hechangzhou Sandbank gradually shrank southward.

For the channel centerline of South Branch of Hechangzhou during 1868–2019, the value of Net Shift Distance from upstream to downstream trend is low → high → low, the average section of Net Shift Distance is left bank migration 498.38 m, the highest value appears in section 5, and left bank migration is 4480.37 m. The lowest value appears in section 4, with left bank migration of 20.15 m. The trend of Cumulative Moving Distance from upstream to downstream is low → high → low, which is basically consistent with the trend of Net Shift Distance, and the average value is 1474.11 m. The maximum value appears in section 5, which is 4575.81 m, and the minimum value appears in section 43, which is 80.86 m. The absolute values of Cumulative Moving Distance of most sections are basically consistent with those of Net Shift Distance, indicating that the long-term movement direction of South Branch of Hechangzhou in each section is basically the same. During the long-term evolution of Hechangzhou Sandbank, the North Bank and West Bank are shrinking, resulting in a part of the channel centerline of South Branch of Hechangzhou moving to the left bank.

For the channel centerline of Dagang Curve during 1868–2019, the change trend of Net Shift Distance from upstream to downstream is high → low → high → low, and the average cross-section Net Shift Distance is right bank migration 212.48 m. The highest value appears in section 4, which is 805.25 m of right bank migration, and the lowest value appears in section 1 and sections 39–46. The change trend of Cumulative Moving Distance from upstream to downstream is low → high → low, with an average of 475.66 m. The highest value appeared in section 4, which is 805.25 m, the lowest value appeared in sections 1 and 9–46. The Cumulative Moving Distance of sections 5–33 is greater than Net Shift Distance, indicating that this part of the river has periodic reverse swing. The Cumulative Moving Distance of sections 1–4 and sections 34–36 are basically the same as the absolute value of Net Shift Distance, indicating that the long-term movement direction of this part of the river is basically the same. Compared with other reaches, the hydrodynamic force of the Dagang Curve has little change, mainly because the left bank has built many ports and a large number of berths extend into the main channel.

3.2. Analysis and Risk Recognition Based on the Central Line of MRCC and LRCR

In this study, the Migration Rate of Channel Centerline (*MRCC*) and Linear Regression Change Rate (*LRCR*) were employed to quantify the section- and reach-scale morphometric parameters. It includes the moving distance and direction of channel centerline on all sections. Based on the above data, we identified the risk of riverbank erosion (Figures 5–11).

Studies have shown that the braided river channel has self-adjustment characteristics under basically unchanged climatic conditions and upstream water and sediment conditions. The probability of the channel centerline moving to the left bank should be roughly equal to that of moving to the right [38]. In this study, the total number of the channel centerline moving left and right in the Zhenjiang-Yangzhou reach of the Yangtze River is 156 and 329, respectively, indicating that the probability of the channel centerline moving right is about twice that of moving left. It shows that the upstream water and sediment conditions of the Zhenjiang-Yangzhou reach of the Yangtze River are unstable and vary greatly.

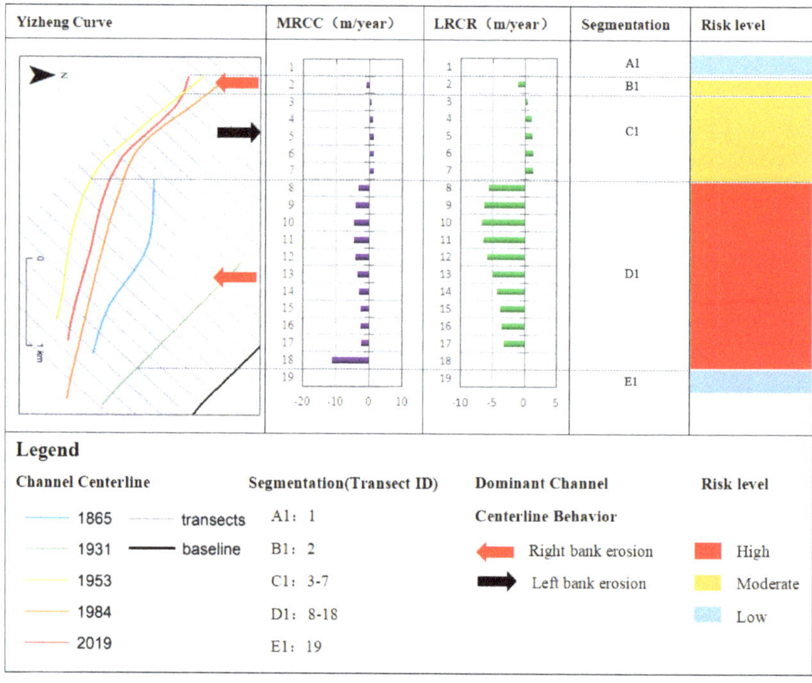

Figure 5. The Migration Rate of Channel Centerline, Linear Regression Change Rate, and erosion risk identification on Yizheng Curve.

As shown in Figure 5, the highest *MRCC* value of Yizheng Curve during 1868–2019 was 11.13 m/y, with an average of 2.19 m/y, and the highest *LRCR* value was 6.67 m/y, with an average of 2.45 m/y. The number of left shifts of the Yizheng Curve section is 5, and the number of right shifts is 12. Combining the direction and velocity of cross-section movement to identify the bank erosion risk, Yizheng Curve can be divided into five sections. Section A1 contains section 1, and the bank erosion risk level is low. Section B1 contains section 2, and the erosion risk level is moderate risk on the right bank. Section C1 contains sections 3–7, and the erosion risk level is moderate risk on the left bank. Section D1 contains sections 8–18, and the erosion risk level is high risk on the right bank. Section E1 contains section 19, and the risk level of riparian erosion is low.

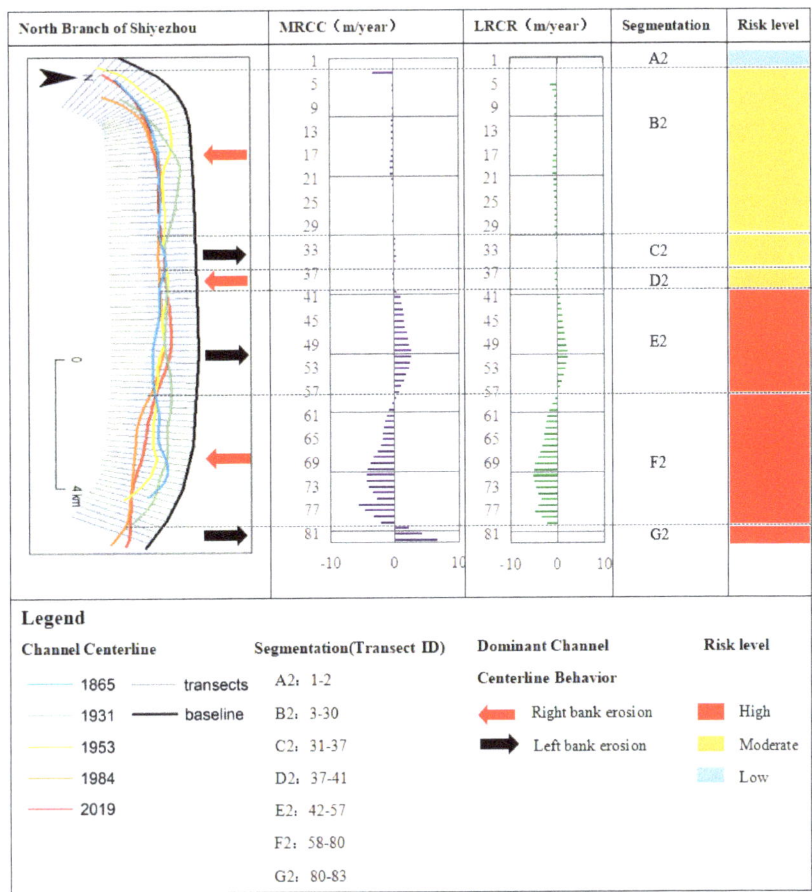

Figure 6. The Migration Rate of Channel Centerline, Linear Regression Change Rate, and erosion risk identification on North Branch of Shiyezhou.

As shown in Figure 6, the highest *MRCC* value of North Branch of Shiyezhou during 1868–2019 was 6.58 m/y, with an average of 0.38 m/y, and the highest *LRCR* value was 5.3 m/y, with an average of 0.94 m/y. The number of left shifts of North Branch of Shiyezhou section is 27, and the number of right shifts is 53. Combining the direction and velocity of cross-section movement to identify the bank erosion risk, Yizheng Curve can be divided into seven sections. Section A2 contains sections 1–2, and the bank erosion risk level is low. Section B2 contains section 3–30, and the erosion risk level is moderate on the right bank. Section C2 contains sections 31–37, and the erosion risk level is moderate on the left bank. Section D2 contains sections 37–41, and the erosion risk level is moderate on the right bank. Section E2 contains sections 42–57, and the erosion risk level is high on the left bank. Section F2 contains sections 58–80, and the erosion risk level is high on the right bank. Section G2 contains sections 80–83, with an erosion risk rating of high risk on the left bank.

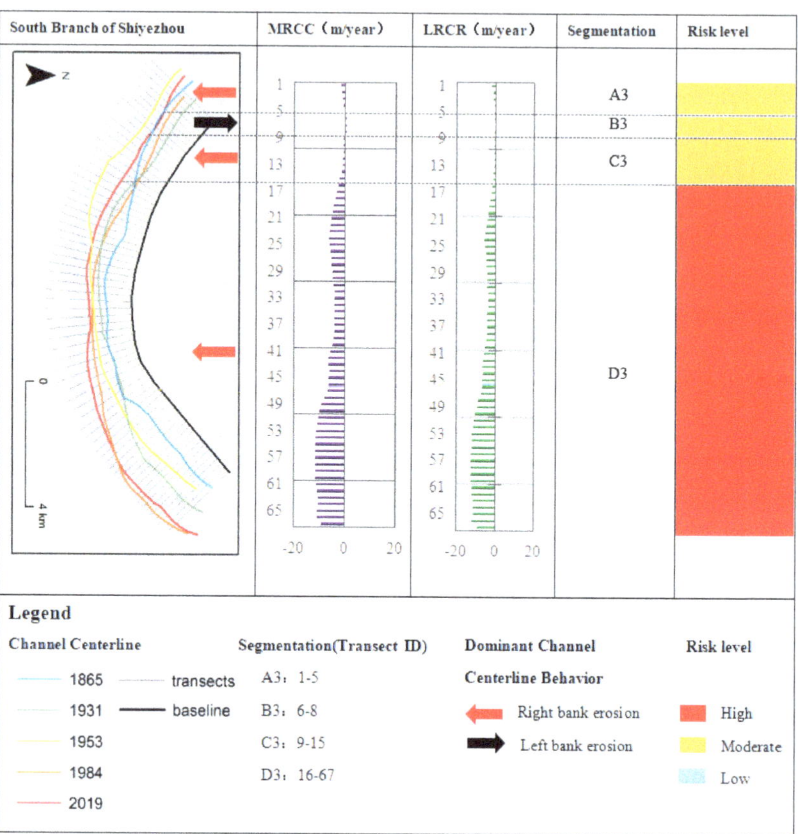

Figure 7. The Migration Rate of Channel Centerline, Linear Regression Change Rate, and erosion risk identification on south branch of Shiyezhou.

As shown in Figure 7, the maximum *MRCC* value of South Branch of Shiyezhou during 1868–2019 was 11.54 m/y, with an average of 5.7 m/y, and the maximum *LRCR* value was 12.63 m/y, with an average of 5.93 m/y. The number of left shifts of South Branch of Shiyezhou section is 3, and the number of right shifts is 63. Combining the direction and velocity of cross-section movement to identify the bank erosion risk, Yizheng Curve can be divided into four sections. Section A3 contains sections 1–5, and the erosion risk level is moderate on the right bank. Section B3 contains sections 6–8, and the erosion risk level is moderate on the left bank. Section C3 contains sections 9–15, and the erosion risk level is moderate on the right bank. Section D3 contains sections 16–67, and the erosion risk level is high on the right bank.

As shown in the Figure 8, the maximum value of *MRCC* of Liuwei Curve during 1868–2019 was 30.85 m/y, with an average of 18.32 m/y, and the maximum value of *LRCR* was 30.29 m/y, with an average of 13.76 m/y. The number of left shifts of Liuwei Curve section is 91, and the number of right shifts is 9. Combining the direction and velocity of cross-section movement to identify the bank erosion risk, Yizheng Curve can be divided into three sections. Section A4 contains sections 1–11, and the erosion risk level is moderate on the right bank. Section B4 contains sections 12–15, and the erosion risk level is moderate on the left bank. Section C4 contains sections 15–103, and the erosion risk level is high on the left bank.

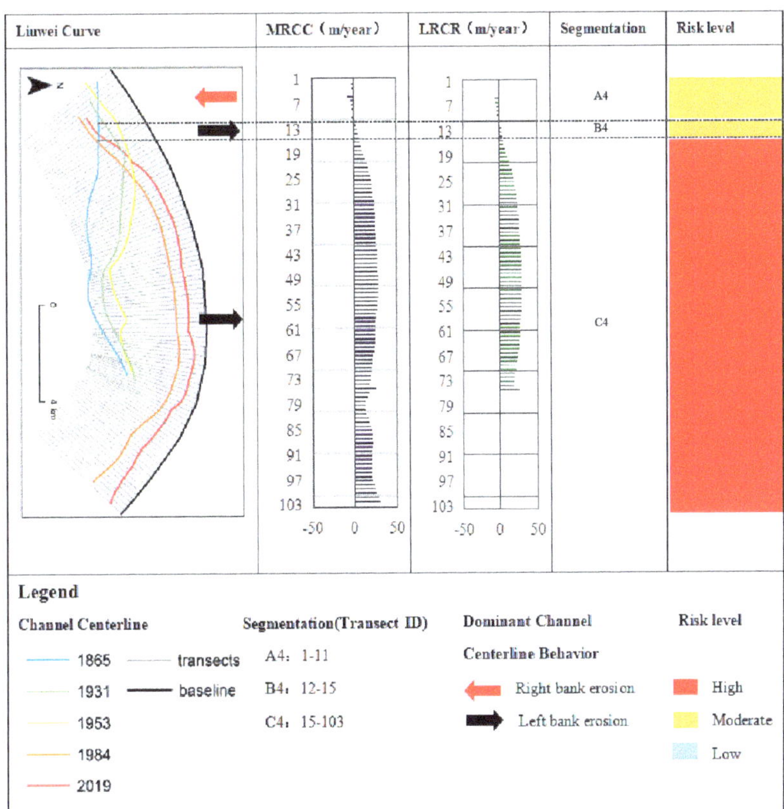

Figure 8. The Migration Rate of Channel Centerline, Linear Regression Change Rate, and erosion risk identification on Liuwei Curve.

As shown in Figure 9, the highest *MRCC* value of North Branch of Hechangzhou during 1868–2019 was 82.96 m/y, with an average of 40.41 m/y, and the highest *LRCR* value was 93.8 m/y, with an average of 26.83 m/y. The North Branch of Hechangzhou section moves left 12 times and right 130 times. Combining the direction and velocity of cross-section movement to identify the bank erosion risk, Yizheng Curve can be divided into three sections. Section A5 contains sections 1–7, and the erosion risk level is moderate on the left bank. Section B5 contains sections 8–137 with high risk of the right bank. Section C5 contains sections 138–149, and the erosion risk level is moderate on the left bank.

As shown in Figure 10, the maximum *MRCC* value of South Branch of Hechangzhou during 1868–2019 was 29.09 m/y, with an average of 2.69 m/y, and the maximum *LRCR* value was 28.16 m/y, with an average of 2.81 m/y. The number of left shifts of South Branch of Hechangzhou section is 16, and the number of right shifts is 27. Combining the direction and velocity of cross-section movement to identify the bank erosion risk, Yizheng Curve can be divided into five sections. Section A6 contains sections 1–2, and the erosion risk level is high on the right bank. Section B6 contains sections 3–4, and the erosion risk level is moderate on the left bank. Section C6 contains sections 5–16, and the erosion risk level is high on the left bank. Section D6 contains sections 17–18, and the erosion risk level is moderate on the left bank. Section E6 contains sections 19–43, and the erosion risk level is moderate on the left bank.

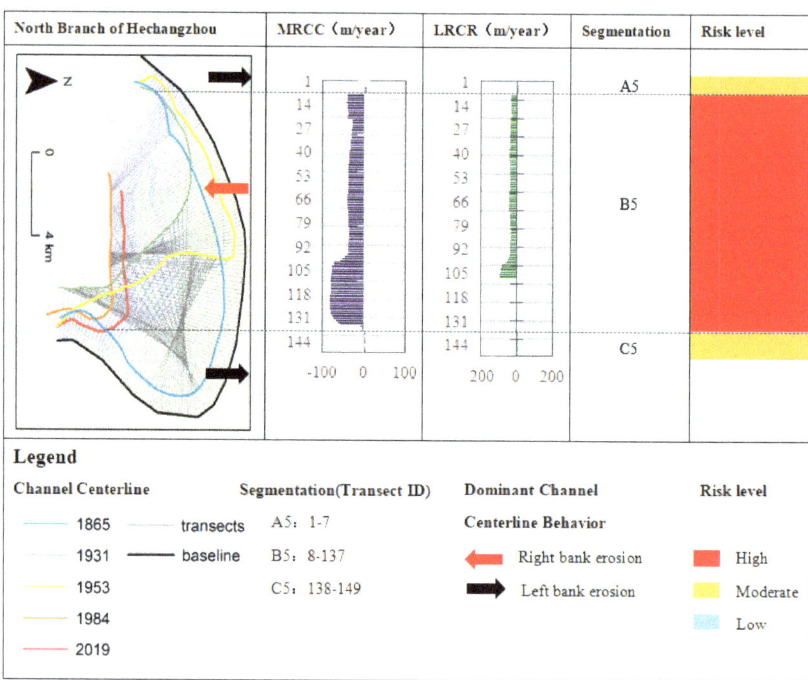

Figure 9. The Migration Rate of Channel Centerline, Linear Regression Change Rate, and erosion risk identification on North Branch of Hechangzhou.

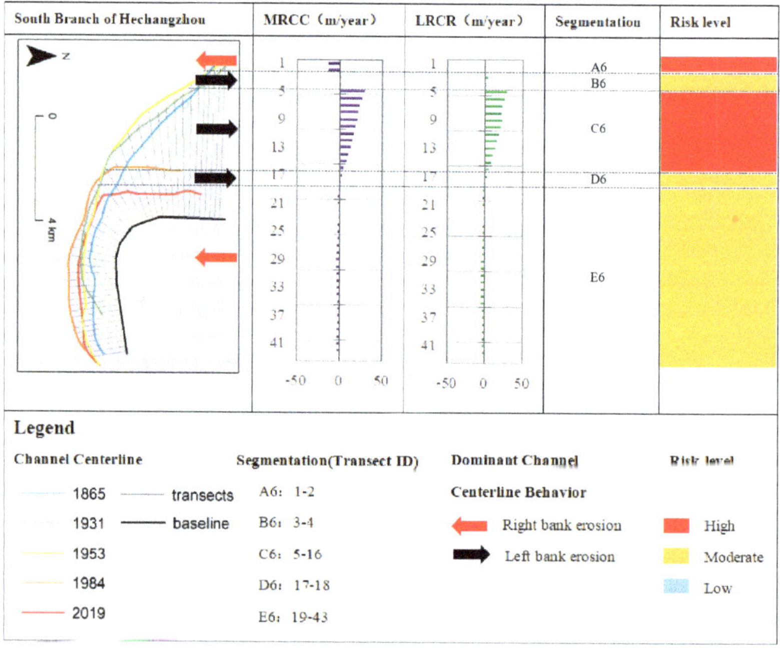

Figure 10. The Migration Rate of Channel Centerline, Linear Regression Change Rate, and erosion risk identification on South Branch of Hechangzhou.

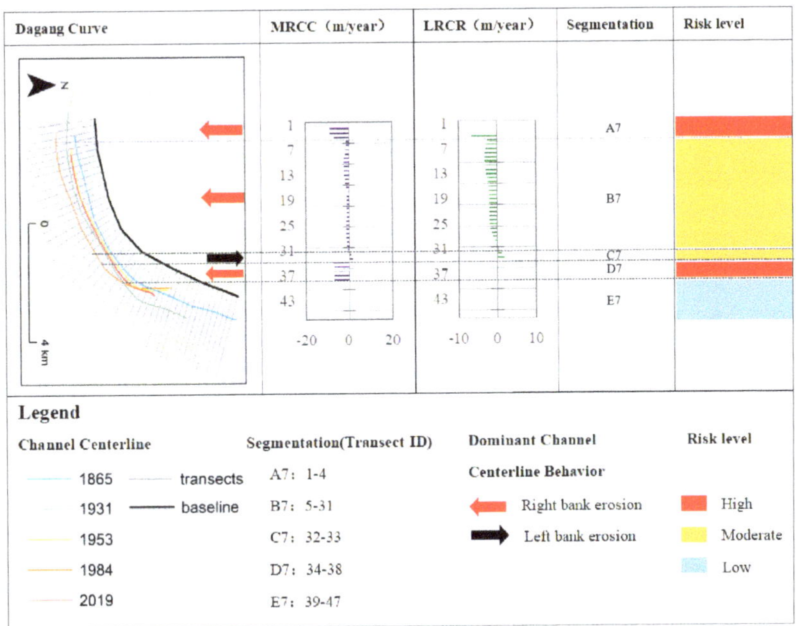

Figure 11. The Migration Rate of Channel Centerline, Linear Regression Change Rate, and erosion risk identification on Dagang Curve.

As shown in Figure 11, the maximum value of *MRCC* of Dagang Curve during 1868–2019 was 8.71 m/y, with an average of 2.05 m/y, and the maximum value of *LRCR* was 6.71 m/y, with an average of 1.26 m/y. The number of left shifts of Dagang Curve section is 2, and the number of right shifts is 35. Combining the direction and velocity of cross-section movement to identify the bank erosion risk, Yizheng Curve can be divided into five sections. Section A7 contains sections 1–4, and the erosion risk level is high on the right bank. Section B7 contains sections 5–31, and the erosion risk level is moderate on the right bank. Section C7 contains sections 32–33, and the erosion risk level is moderate on the left bank. Section D7 contains sections 34–38, and the erosion risk level is high on the right bank. Section E7 comprises sections 39–47, with a low risk rating for riparian erosion.

3.3. Influence Factors of Channel Centerline Changes

Channel evolution covers various deformation components such as adjustments in planform and cross-sectional geometries, and channel migration is a critical component of channel planform adjustments in alluvial rivers. Factors influencing channel migration mainly include changes in river boundary conditions, hydrological and sediment changes, and sandbanks merging. These influencing factors are presented herein to investigate the variation characteristics in the variation characteristics of the migration intensity of the channel centerline of the Zhenjiang-Yangzhou reach of the Yangtze River.

3.3.1. Changes of River Boundary Conditions

The Zhenjiang-Yangzhou reach is located at the top of the Yangtze River Delta, and the tidal current of the Yangtze River estuary can reach the reach to form a jacking effect. Therefore, the boundary conditions of the river will change with the dynamic changes of the material composition of the riverbed, topography, water level difference, and river slope.

The geological structure of the located Yangzhou reach belongs to the Yangzi quasi-platform, and the river trend is basically consistent with the geological structure trend, showing a west-east trend. The river valley reached by the Zhenjiang-Yangzhou reach is narrow in the upper part and wide in the lower, showing an asymmetric horn shape. Both sides are alluvial plains, and outward are loess terraces and low hills. The riverbed is mostly composed of fine sand or silt, and there are medium-coarse sand and gravel in the deep groove.

The southern bank of the reach is the Xiashu loess terrace and part of the impact plain at the northern foot of the Ningzhen Mountains. The Xiashu loess was formed about 100,000 years ago by river accumulation. All the riverbeds and bank walls made up of Xiashu soil are hard and have strong impact resistance. The riverbed on the south bank of the reach is mostly composed of Xiashu loess with good impact resistance except for Zhengzunzhou and a few bank walls. Therefore, all the rivers adjacent to the south bank (right bank) are not seriously eroded except the South Branch of Shiyezhou, where the south bank is continuously eroded.

The north bank of the reach can be divided into two parts from the terrain. Weiyang Loess Platform is located in the north of Yizheng to Yangzhou line. Loess is widely distributed and stretches for tens of kilometers. The southeast and east are the alluvial plains of the Yangtze River Delta. This area was once a shallow estuary area in history. The riverbed and bank wall there have poor erosion resistance and are easy to be washed by water. Therefore, after the crest point gradually moves to the northern bank of the adjacent curve opposite Jinshan Mountain, the central line of the channel moves northward rapidly.

The hard soil or bedrock outcrops in the river reach are not easy to scour. The bank wall has become a good diversion barrier, and even if there are some concave alluvial soil banks, due to the inertia of the flow, in general, in flood years, a large scour will not occur. However, the soil conditions of river floodplains are different. According to the investigation and study of the modern river floodplain sedimentary columnar samples collected in our field and the outcrop profile in the field, it was found that the mud deposits deposited at a high water level in the flood period of the river reach are prone to mud cracks in the dry and rainy environment and have obvious characteristics of mud cracks. Figure 12 shows the scene images of dense mud cracks found near Zhengrunzhou.

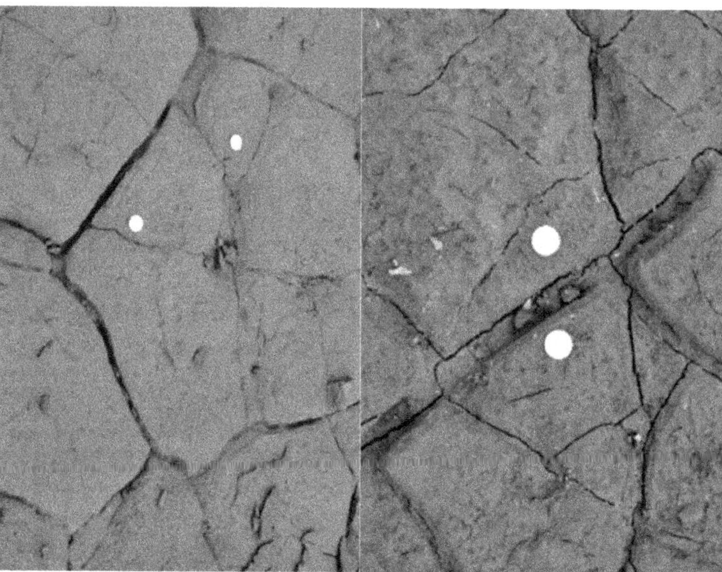

Figure 12. Mud fracture structure of floodplain near Zhengrunzhou.

The located Yangzhou reach is mainly composed of barrier sand, sandbars, and beaches, which originally had multiple rivers. Later, some Jiajiang Rivers were silted by sediment and became dark rivers. The floodplain sediments in front of the embankment have obvious horizontal bedding structures, most of which are horizontal laminas of millimeter thickness, as shown in Figure 13. Under the impact of river water and waves, the sand layer is very easy to be hollowed out, resulting in the loss of support and collapse of the upper floodplain sediments.

Figure 13. Pores in middle layer, sand layer, and plant root of columnar sample.

The swing of the river bank makes these dark rivers appear on the current river bank and even be pushed down by the river embankment. There will be a great deal of groundwater in and out of the dark river and will even take away silt, resulting in bank collapse and even embankment collapse into the river, forming an obvious gap on the bank. As shown in Figure 14, several examples of bank collapse show that large-scale bank collapse will lead to increased river width at the bank collapse, resulting in the channel centerline moving to the bank collapse.

Figure 14. Typical bank collapse cases in Zhenyang Reach.

At the same time, the two sides of the river are densely populated, and regular floods pose a huge threat to the safety of people's lives and property. The continuous deposition of Zhenjiang Port has seriously affected the local economic development. Therefore, various river regulation projects have been carried out in the river. These projects have gradually changed the boundary conditions of some river sections and have also had some impacts on the evolution of the river.

For example, the regulation of Zhenjiang port and the expansion of Jiaobei beach will lead to the evolution of the river regime in favor of the development of the South Branch of Hechangzhou. The problem of bank collapse on the north side of Hechangzhou is very active. Although Zhenjiang has carried out a dam project in the left bank of Hechangzhou to solve the problem of diversion ratio, which has a certain effect on improving the diversion ratio, the development of Jiaobei beach is continuing, and the change of river regime determines that it cannot fundamentally solve the problem of diversion ratio in Hechangzhou. Artificial river regulation leads to the development of Jiaobei beach and the narrow channel, which will aggravate the risk of bank collapse.

3.3.2. Changes in Hydrodynamic Conditions

The expansion of the Huaihe River channel and the increase of water quantity have a great influence on the change of the reach hydrodynamic conditions of the Zhenjiang-Yangzhou reach. The Zhenjiang-Yangzhou reach was narrow at both ends and broad in the middle hundreds of years ago. When the river channel changes from a narrow section to a wide section, the kinetic energy in the original mechanical energy of water and sediment flow will be significantly reduced, so the sediment will fall and silt accordingly. When the channel changes from a wide section to a narrow section, the backwater action of the narrow section causes the sediment to fall before the entrance to the narrow section. After the north branch of the Huaihe River enters the river, the water and sediment conditions change, and the main current line of the lower reach of the Zhenjiang-Yangzhou reach gradually moves to the southeast. The west entrance of the north branch and the Yangtze River are supported by each other, and the water flow becomes slow, and the sediment is deposited accordingly, resulting in the continuous shrinkage of the north branch, and the sandbar is parallel to the north, and finally, the north branch evolves into the Jiajiang River. The contraction of the north branch makes the east gate become the main channel of the Huaihe River and gradually evolved into the present channel of the Yangtze River, where the Yangtze River and the Huaihe River meet, making the sediment silt before entering the entrance of the narrow section, and the tail of the large sediment extends southward.

The expansion of the Huaihe River channel and the increase of water quantity have a great influence on the change of hydrodynamic conditions in the Zhenyang reach of the Yangtze River. In 1851, the Huaihe River flood destroyed the Xinli Dam, and now, the Huaihe River flows into the Yangtze River perennially. Zhengkezhou and Shiyezhou, two sandbars from Yizheng to Guazhou in the west reach of the Zhenjiang-Yangzhou reach, moved down rapidly. The upper end of Shiyezhou was concealed by Zhengkezhou and was not scoured. The sedimentation caused by backwater made the upper end rise. The lower end was controlled by Guazhou and Jinshan nodes and stopped the extension, which caused the upper and lower reaches to merge [39]. After crossing Guazhou, the mainstream of the Yangtze River turned to the right bank and then turned to the left bank through the mainstream of Beigu Mountain and Jiao Mountain. On the river surface, Xinzhou, Enyuzhou, and other sandbars were formed, and the formed Jiajiang River continued to bank northward. The river and sandbar information of the Zhenjiang-Yangzhou reach from 1868 to 2019 are digitized (Figure 15).

It can be seen from the figure that since the sandbank outside the estuary of Shiyezhou and Huaihe River merges with the left bank, the river channel changes from straight to curved. The Guazhou area becomes the concave bank, and the Zhenjiang area becomes the convex bank. Under the action of centrifugal force, the lateral circulation of the curved channel develops, and the lateral erosion occurs continuously, and at the same time, it moves downstream. Guazhou was originally the vertex of concave bank. With the downdraft of meandering stream, the east of Guazhou began to be eroded. Zhenrenzhou (the used name of Zhengrunzhou) was originally located in the west of Zhenjiang. Due to the influence of the downdraft of convex bank, Zhenjiang port is gradually covered. After 1868, the evolution of the Zhenjiang-Yangzhou reach was mainly characterized by a gradual change into an "S"-type shape.

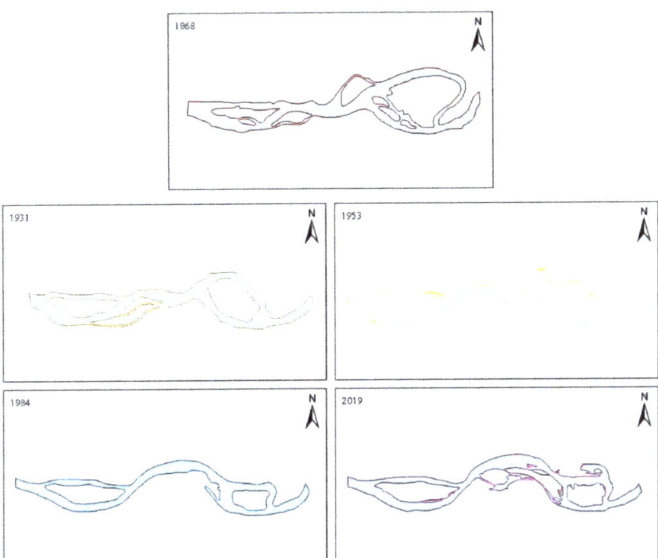

Figure 15. Changes of water areas and rivers in the upper reaches of the Yangtze River in Zhenjiang from 1868 to 2019.

3.3.3. Water and Sediment Changes after the Construction of the Three Gorges Dam

Water and sediment processes in the fluvial environment have a significant influence on the Zhenjiang-Yangzhou reach of the Yangtze River [40,41]. According to the results of the Changjiang Water Resources Commission and many other studies [42,43], water and sediment processes were drastically changed in the Zhenjiang-Yangzhou reach after the cutoff of Gezhouba Dam in 1981 and the impoundment of the Three Gorges Reservoir in 2003. A severe decrease in sediment load notably occurred in the Zhenjiang-Yangzhou reach. In general, the reduction of sediment would aggravate the lateral erosion and cause the channel centerline to move towards the eroded side.

Datong Hydrological Station in Chizhou City, Anhui Province, is the control station in the lower reaches of the Yangtze River. According to statistics, there are mainly small tributaries such as the Huaihe River, Chuhe River, Qingyi River, Shuiyang River, and Qinhuai River below Datong Station. The flow into the river in the mainstream section accounts for about 2–3% of the flow of Datong Station. Therefore, the flow and sediment characteristics of Datong Hydrological Station basically represent the characteristics of water and sediment in Zhenyang Reach. As shown in Table 1, since the impoundment of the Three Gorges Reservoir, the distribution and composition of water and sediment in Datong Station have been adjusted and changed to some extent during the year. The average annual runoff in 2003~2020 (after impoundment) is 2.95% less than that in 1950~2002 (before impoundment), and the average annual sediment discharge is 68.6% less than that in 1950~2002 [44]. As shown in Table 1. From the perspective of annual distribution, the inflow and sediment of Datong Station are mainly concentrated in the flood season (May–October). Among them, the inflow and sediment of the Three Gorges Reservoir in the flood season after impoundment account for 67.4% and 78.8% of the whole year, respectively, which are slightly reduced compared with those before impoundment by 3.5% and 8.7%, respectively.

Table 1. Annual average value statistics of flow and sediment at Dotong Station.

Times	Flow	Sediment Load
Before water storage	28700 m^3/s	427 million tons
After water storage (2003–2020)	27800 m^3/s	134 million tons

The change of incoming water and sediment caused by the construction of Three Gorges Dam has an important influence on the scouring and silting of the Zhenjiang-Yangzhou reach. Chen Jingru et al. collected underwater topographic maps of Zhenyang reach over the years and used SURFER8 software to calculate scouring and silting. The distribution of scouring and silting changes in this reach is shown in Table 2 [45].

Table 2. Annual average value statistics of flow and sediment at Dotong Station.

Year	1991–1998	1998–2001	2001–2006	2006–2011	2011–2016
Yizheng Curve	1111.9	−1253.0	−776.7	−339.7	−578.2
North Branch of Shiyezhou	1037.2	−1158.1	−1053.3	−2735.4	−1376.1
South Branch of Shiyezhou	1936.0	210.3	−556.8	−212.6	−445.4
Confluence reach	−209.5	666.5	411.9	−803.5	−371.6
Liuwei Curve	−1084.3	103.1	701.8	−1471.2	−2104.7
Before shunt	−1240.5	325.6	266.5	−710.7	−34.2
North Branch of Hechangzhou	−2420.2	−1704.0	809.0	−1161.7	−1700.6
South Branch of Hechangzhou	2068.1	437.6	677.6	121.8	−1399.4
Dagang Curve	618.3	−687.7	−166.5	−482.9	−571.3
Total	1817.1	−3059.6	646.7	−7795.8	−8581.5

In addition to the siltation of the river channel from 1991 to 1998, the river channel was scoured each year. Over the years, the cumulative scour was 183.57 million m^3, with an average brush depth of 1.8 m. The scour area was mainly located in the nearshore area below Siyuangou. Before the completion of the Three Gorges Dam, some rivers in Zhenyang reach were silted. After the completion of the Three Gorges Dam, the erosion became more and more intense. By 2011–2016, all parts of the Zhenjiang-Yangzhou reach had become scoured. In order to avoid severe erosion of the concave bank, two river and shoreline remediation projects have been carried out in the past two decades in the Zhenjiang-yangzhou reach, focusing on the construction or reinforcement of the South Branch of Shiyezhou and Six Curve embankments, repairing the damaged bank sections affected by a large number of bank collapse events, which greatly improves the anti-erosion ability of the concave bank of the river section. The trend of the river in the erosion of the key bank sections in this century has been slowed to a certain extent, and the effect of the revetment project is remarkable.

4. Conclusions and Foresight

In this study, we use multi source data at different scales, including old maps and satellite images, to study the spatiotemporal changes of the Net Shift Distance, Cumulative Moving Distance, Migration Rate of Channel Centerline, and Linear Regression Change Rate of the channel centerline from 1865 to 2019. The results show that:

(1) From 1868 to 2019, the channel centerline of the Zhenjiang-Yangzhou reach kept shifting. The average net displacement distance of the section is 1103.47 m on the right bank, and the average cumulative displacement distance of the section is 2790.51 m. The river section with the largest net displacement distance of the channel centerline is the Liuwei Curve, and the smallest is the North Branch of Hechangzhou. The

    maximum Cumulative Moving Distance of the river center line is North Branch of Hechangzhou, and the minimum is Yizheng Curve.

(2)   According to the *NSD* and *CMD* data of each part, it can be found that the long-term movement direction of the channel centerline is basically the same, and a small part of the channel centerline has periodic reverse swing.

(3)   The upstream runoff and sediment conditions of the Zhenjiang-Yangzhou reach of the Yangtze River are unstable and vary greatly. The probability of the channel centerline moving right is about twice that of moving left. At the same time, some rivers have high erosion risk.

(4)   Environmental changes and changes in river boundary conditions caused by human activities, the expansion of the Huaihe River into the Yangtze River, as well as changes in hydrodynamic forces caused by a large number of sandbanks, and changes in river scouring and silting after the completion of the construction of the Three Gorges Dam will affect the movement of the channel centerline in the Zhenyang reach.

This study proposes a calculation procedure to extract the channel centerline from a braided reach and study its changes. This calculation procedure selected Arcgis 10.5 as the implementation environment. The advantages of this calculation procedure are simple, efficient, and versatile. Computers will not have too much memory load, computational efficiency is relatively high, and accuracy and accuracy can be guaranteed. This method can also be understood as a standardized process for obtaining multi-source river channel change data, and the goal of this method can also be achieved by using other software. The method of this study can also be applied to other similar rivers with a long history of human activities and high density. In addition, based on the theoretical knowledge or research methods of natural geography, geology, geographic information science, and historical geography, this study reveals the changing trend of the river course change of the Yangtze River-Yangzhou reach of the Yangtze River and the complexity and difference of the river course change in space from the macro and micro perspectives and discusses the potential impact of human activities on river course change. The results enrich people's understanding of the long-term changes of a braided reach in the lower reaches of the Yangtze River and have certain guiding significance for river regulation, navigation safety, and revetment construction.

Of course, this study also has many deficiencies to be improved. First of all, for the Landsat medium-resolution remote sensing image used in this study, the satellite source is relatively single, and the resolution is slightly low, resulting in some errors in the extraction of Bankline; finally, the channel centerline generated there contains some errors. Secondly, the tidal Yangzhou reach of the Yangtze River is a channel weakly affected by the tide. Due to the difficulty in obtaining tidal information, the tidal level correction was not carried out in this study. Future research can combine higher resolution remote sensing images to obtain clearer shoreline contour, eliminate the influence of tides, and improve the output accuracy of the bankline and channel centerline. In addition, more on-site observation and investigation interviews should be carried out on the bank of the Yangzhou reach of the Yangtze River to confirm the specific influence of the production activities of many docks and industrial and mining enterprises on the bank, the ship traveling wave of the passing ships and the river, and shoreline renovation project on the change of river boundary conditions, which are also the main research objectives of the next stage.

Author Contributions: Conceptualization, methodology, and writing—original draft preparation, C.L. (Cunli Liu); data curation, B.L.; software, C.L. (Changfeng Li); visualization, B.L., S.J. and G.W.; supervision and funding acquisition, B.L. and Z.Z. All authors have read and agreed to the published version of the manuscript.

Funding: This research was funded by National Natural Science Foundation of China (No. 41371024) and Nanjing University Doctoral Promotion Program (No. 202101B035) and Jiangsu Water Conservancy Science and Technology Project (No. 2015081).

Institutional Review Board Statement: Not applicable.

Informed Consent Statement: Not applicable.

Data Availability Statement: The data presented in this study are available on request from the author. The data are not publicly available due to privacy. Images employed for the study will be available online for readers.

Conflicts of Interest: The authors declare no conflict of interest.

References

1. Lake, P.S. Ecological effects of perturbation by drought in flowing waters. *Freshw. Biol.* **2003**, *48*, 1161–1172. [CrossRef]
2. Alderman, K.; Turner, L.R.; Tong, S.L. Floods and human health: A systematic review. *Remote Sens. Environ.* **2012**, *140*, 23–35. [CrossRef] [PubMed]
3. Song, G.F. Evaluation on water resources and water ecological security with 2-tuple linguistic information. *Int. J. Knowl. Based Intell. Eng. Syst.* **2019**, *23*, 1–8. [CrossRef]
4. Liu, X.; Shi, C.; Zhou, Y.; Gu, Z.; Li, H. Response of Erosion and Deposition of Channel Bed, Banks and Floodplains to Water and Sediment Changes in the Lower Yellow River, China. *Water* **2019**, *11*, 357. [CrossRef]
5. Xia, J.; Wang, Y.; Zhou, M.; Deng, S.; Li, Z.; Wang, Z. Variations in Channel Centerline Migration Rate and Intensity of a Braided Reach in the Lower Yellow River. *Remote Sens.* **2021**, *13*, 1680. [CrossRef]
6. Eixoto, J.M.A.; Nelson, B.W.; Wittmann, F. Spatial and temporal dynamics of river channel migration and vegetation in central Amazonian white-water floodplains by remote-sensing techniques. *Remote Sens. Environ.* **2009**, *113*, 2258–2266.
7. Szombara, S.; Lewińska, P.; Żądło, A.; Róg, M.; Maciuk, K. Analyses of the Prądnik riverbed Shape Based on Archival and Contemporary Data Sets—Old Maps, LiDAR, DTMs, Orthophotomaps and Cross-Sectional Profile Measurements. *Remote Sens.* **2020**, *12*, 2208. [CrossRef]
8. Ghoshal, S.; James, L.A.; Singer, M.B.; Aalto, R. Channel and floodplain change analysis over a 100-year period: Lower Yuba river, California. *Remote Sens.* **2010**, *2*, 1797–1825. [CrossRef]
9. de Musso, N.M.; Capolongo, D.; Caldara, M.; Surian, N.; Pennetta, L. Channel changes and controlling factors over the past 150 years in the Basento river (southern Italy). *Water* **2020**, *12*, 307. [CrossRef]
10. Magliulo, P.; Valente, A. GIS-Based geomorphological map of the Calore River floodplain near Benevento (Southern Italy) overflooded by the 15th October 2015 event. *Water* **2020**, *12*, 148. [CrossRef]
11. Priestnall, G.; Aplin, P. Cover: Spatial and temporal remote sensing requirements for river monitoring. *Int. J. Remote Sens.* **2006**, *27*, 2111–2120.
12. Arnesen, A.S.; Silva, T.S.; Hess, L.L.; Novo, E.M.; Rudorff, C.M.; Chapman, B.D.; McDonald, K.C. Monitoring flood extent in the lower Amazon River floodplain using ALOS/PALSAR ScanSAR images. *Remote Sens. Environ.* **2013**, *130*, 51–61. [CrossRef]
13. Rozo, M.G.; Nogueira, A.C.R.; Castro, C.S. Remote sensing-based analysis of the planform changes in the upper Amazon River over the period 1986–2006. *J. S. Am. Earth Sci.* **2014**, *51*, 28–44. [CrossRef]
14. Rozo, M.G.; Nogueira, A.C.; Truckenbrodt, W. The anastomosing pattern and the extensively distributed scroll bars in the middle Amazon River. *Earth Surf. Process. Landf.* **2012**, *37*, 1471–1488. [CrossRef]
15. Miao, C.; Ni, J.; Borthwick, A.G.L. Recent changes of water discharge and sediment load in the Yellow River basin, China. *Prog. Phys. Geogr. Earth Environ.* **2010**, *34*, 541–561. [CrossRef]
16. Ma, Y.; Huang, H.Q.; Nanson, G.C.; Li, Y.; Yao, W. Channel adjustments in response to the operation of large dams: The upper reach of the lower Yellow River. *Geomorphology* **2012**, *147–148*, 35–48. [CrossRef]
17. Xia, J.; Li, X.; Zhang, X.; Li, T. Recent variation in reach-scale bankfull discharge in the Lower Yellow River. *Earth Surf. Process. Landf.* **2013**, *39*, 723–734. [CrossRef]
18. Julien, P.; Tuzson, J. River Mechanics. *Appl. Mech. Rev.* **2003**, *56*, B30–B31. [CrossRef]
19. Chen, J.-G.; Zhou, W.-H.; Chen, Q. Reservoir Sedimentation and Transformation of Morpho-Logy in the Lower Yellow River during 10 Year's Initial Operation of the Xiaolangdi Reservoir. *J. Hydrodyn.* **2012**, *24*, 914–924. [CrossRef]
20. Richard, G.A.; Julien, P.Y.; Baird, D.C. Case Study: Modeling the Lateral Mobility of the Rio Grande below Cochiti Dam, New Mexico. *J. Hydraul. Eng.* **2005**, *131*, 931–941. [CrossRef]
21. Kong, D.; Latrubesse, E.M.; Miao, C.; Zhou, R. Morphological response of the Lower Yellow River to the operation of Xiaolangdi Dam, China. *Geomorphology* **2020**, *350*, 106931. [CrossRef]
22. Cheng, Y.; Xia, J.; Zhou, M.; Deng, S.; Li, D.; Li, Z.; Wan, Z. Recent variation in channel erosion efficiency of the Lower Yellow River with different channel patterns. *J. Hydrol.* **2022**, *610*, 127962. [CrossRef]
23. Yang, X. Evolution and Cause Analysis of Sandbanks in the Center of the Yangtze River from 1570 to 1971. *J. Geogr.* **2020**, *75*, 1512–1522. (In Chinese)
24. Liu, X.; Pine, T.; Li, Z. Research on river evolution and regulation of Zhenyang reach in the lower reaches of the Yangtze River. *J. Yangtze River Acad. Sci.* **2011**, *28*, 1–9. (In Chinese)
25. Lu, L.; Liao, X.; Huang, W. Discussion on river channel evolution and control measures of Zhenyang reach of the Yangtze River. *People's Yangtze River* **2009**, *40*, 9–11+58+110. (In Chinese)
26. Chen, F.; Fu, Z.; Yang, F. Influence of river regime change on channel conditions in Zhenyang reach of the Yangtze River. *Water Transp. Proj.* **2011**, *6*, 112–116. (In Chinese)

27. Wang, J.; Fan, H.; Zhu, L.; Ying, H. Analysis on hydrodynamic improvement measures of Hechangzhou branch deepwater channel regulation right branch. *Water Transp. Proj.* **2014**, *9*, 11–17. (In Chinese)

28. Zhenjiang Municipal Bureau of Water Resources. Yangtze River. Available online: http://http://slj.zhenjiang.gov.cn/slj/hczhhjj/201806/ef443afe8e4349d38d0a1d469353ab08.shtml (accessed on 15 June 2018).

29. Wang, Z.; Chen, Z.; Li, M.; Chen, J.; Zhao, Y. Variations in downstream grain-sizes to interpret sediment transport in the middle-lower Yangtze River, China: A pre-study of Three-Gorges dam. *Geomorphology* **2009**, *113*, 217–229. [CrossRef]

30. Luan, H.; Liu, T.; Huang, W. The morphological evolution and trend of typical channel in the lower reaches of the Yangtze River under water and sediment conditions. *J. Yangtze Acad. Sci.* **2018**, *3*, 7–12. (In Chinese)

31. Wang, X.; Tang, L.; Ding, X. The river bed evolution in Zhenyang reach of Yangtze River is analyzed by GIS. *Sediment Study* **2008**, *6*, 68–73. (In Chinese)

32. Zhang, X.; Xie, R.; Fan, D.; Yang, Z.; Wang, H.; Wu, C.; Yao, Y. Sustained growth of the largest uninhabited alluvial island in the Changjiang Estuary under the drastic reduction of river discharged sediment. *Sci. China (Earth Sci.)* **2021**, *64*, 1687–1697. (In Chinese) [CrossRef]

33. Zhu, H. *Jiangdu County Chronicles*; Jiangsu People's Publishing House: Nanjing, China, 1996; pp. 47–49.

34. Wu, S. *Harvard University Library Unpublished History of Old Customs in China*; Guangxi Normal University Press: Guilin, China, 2014; pp. 13–38.

35. Institute of Modern History, Central Research Institute. Available online: https://map.rchss.sinica.edu.tw/cgi-bin/gs32/gsweb.cgi/login?o=dwebmge&cache=1652927698637 (accessed on 3 September 2021).

36. University of Texas Library. Available online: https://maps.lib.utexas.edu/maps/china.html (accessed on 7 September 2021).

37. Yang, C.; Cai, X.; Wang, X.; Yan, R.; Zhang, T.; Lu, X. Remotely Sensed Trajectory Analysis of Channel Migration in Lower Jingjiang Reach during the Period of 1983–2013. *Remote Sens.* **2015**, *7*, 16241–16256. [CrossRef]

38. Zhang, R.; Xie, J. *Sedimentation Research in China*; China Water and Power Press: Beijing, China, 1993; pp. 75–98.

39. Chen, J.; Yun, C. Channel process of the Changjiang River lower reach (from Nanjing to Wusong). *Acta Geogr. Sin.* **1959**, *25*, 221–239.

40. Leopold, L.B.; Wolman, M.G. *River Channel Patterns: Braided, Meandering, and Straight*; US Government Printing Office: Washington, DC, USA, 1957.

41. Kumar, M.; Kumar, P.; Kumar, A.; Elbeltagi, A.; Kuriqi, A. Modeling stage–discharge–sediment using support vector machine and artificial neural network coupled with wavelet transform. *Appl. Water Sci.* **2022**, *12*, 87. [CrossRef]

42. Xu, K.; Milliman, J.D. Seasonal variations of sediment discharge from the Yangtze River before and after impoundment of the Three Gorges dam. *Geomorphology* **2009**, *104*, 276–283. [CrossRef]

43. Changjiang Water Resources Commission. *Changjiang Sediment Bulletin*; Changjiang Press: Wuhan, China, 2012.

44. Dong, W.; Zhang, Y.; Zhang, L.; Chen, N.; Zou, Y.; Du, Y.; Liu, J. Study of the Three Gorges Dam's Impact on the Discharge of Yangtze River during Flood Season after Its Full Operation in 2009. *Water* **2022**, *14*, 1052. [CrossRef]

45. Chen, J.; Tian, Z.; Zhang, M. Analysis on the causes of bank collapse in the happy estuary section of the left branch of Shiyezhou in Zhenyang reach of the Yangtze River. *Water Conserv. Constr. Manag.* **2021**, *41*, 12–15. (In Chinese)

Article

Effects of Low-Level Organic Mercury Exposure on Oxidative Stress Profile

Radu Ciprian Tincu [1,2], Cristian Cobilinschi [2,3,*], Iulia Alexandra Florea [2,*], Ana-Maria Cotae [2,3], Alexandru Emil Băetu [3,4], Sebastian Isac [5,6], Raluca Ungureanu [2,3], Gabriela Droc [3,6], Ioana Marina Grintescu [2,3] and Liliana Mirea [2,3]

1   Clinical Toxicology, Department of Orthopedics and Anesthesiology, Faculty of Medicine, Carol Davila University of Medicine and Pharmacy, 020021 Bucharest, Romania
2   Bucharest Clinical Emergency Hospital, 014461 Bucharest, Romania
3   Anesthesiology, Department of Orthopedics and Anesthesiology, Faculty of Medicine, Carol Davila University of Medicine and Pharmacy, 020021 Bucharest, Romania
4   "Grigore Alexandrescu" Clinical Emergency Hospital for Children, 011743 Bucharest, Romania
5   Physiology, Functional Science, Faculty of Medicine, Carol Davila University of Medicine and Pharmacy, 020021 Bucharest, Romania
6   Fundeni Clinical Institute, 022328 Bucharest, Romania
*   Correspondence: cob_rodion@yahoo.com (C.C.); iulia.alexandra.florea@gmail.com (I.A.F.)

Citation: Tincu, R.C.; Cobilinschi, C.; Florea, I.A.; Cotae, A.-M.; Băetu, A.E.; Isac, S.; Ungureanu, R.; Droc, G.; Grintescu, I.M.; Mirea, L. Effects of Low-Level Organic Mercury Exposure on Oxidative Stress Profile. *Processes* **2022**, *10*, 2388. https://doi.org/10.3390/pr10112388

Academic Editors: Gassan Hodaifa, Antonio Zuorro, Joaquín R. Dominguez, Juan García Rodríguez, José A. Peres, Zacharias Frontistis, Mha Albqmi and Alessandro Trentini

Received: 20 August 2022
Accepted: 9 November 2022
Published: 14 November 2022

Publisher's Note: MDPI stays neutral with regard to jurisdictional claims in published maps and institutional affiliations.

Abstract: Background: The fish-based diet is known for its potential health benefits, but it is less known for its association with mercury (Hg) exposure, which, in turn, can lead to neurological and cardiovascular diseases through the exacerbation of oxidative stress. The aim of this study was to evaluate the correlations between Hg blood concentration and specific biomarkers for oxidative stress. Methods: We present a cross-sectional, analytical, observational study, including primary quantitative data obtained from 67 patients who presented with unspecific complaints and had high levels of blood Hg. Oxidative stress markers, such as superoxide dismutase (SOD), glutathione peroxidase (GPX), malondialdehyde (MLD), lymphocyte glutathione (GSH-Ly), selenium (Se), and vitamin D were determined. Results: We found positive, strong correlations between Hg levels and SOD (r = 0.88, $p < 0.0001$), GPx (r = 0.92, $p < 0.0001$), and MLD (r = 0.94, $p < 0.0001$). We also found inverted correlations between GSH-Ly and vitamin D and Hg blood levels (r = −0.86, r = −0.91, respectively, both with $p < 0.0001$). Se had a weak correlation with Hg plasma levels, but this did not reach statistical significance (r = −0.2, $p > 0.05$). Conclusions: Thus, we can conclude that low-level Hg exposure can be an inductor of oxidative stress.

Keywords: mercury; low-level exposure; oxidative stress

1. Introduction

Mercury (Hg) is included in the top ten chemicals of major public health concern by the World Health Organization (WHO), mainly because of its deleterious effects on the nervous, digestive, and immune systems, as well as on lungs, kidneys, skin, and eyes [1]. There are three forms of Hg in the environment: elemental (metallic), inorganic, and organic, each with its own chemical properties. One can be exposed to all three of these forms. Elemental mercury is the only liquid metal at normal pressure and ambient temperature, and it is mainly used in industrial processes, lightbulbs, and mining [2]. Not so long ago, it was used for various dental amalgams and thermometers, but nowadays, given its high risk, its everyday use is limited. Exposure to elemental mercury occurs through exposure to air containing mercury vapors. Inorganic mercury is a combination of mercury and other elements, and it is also mainly used in industrial processes. Therefore, exposure to inorganic mercury is usually related to the working environment. Exposure to organic mercury is the most frequent type of exposure, and it is caused by dietary intake, usually of fish and other types of seafood [3]. At a cellular level, consequences of Hg exposure include changes in membranes' permeability and macromolecular structure, mitochondrial metabolism, energy production, and DNA alterations [4,5]. Remodeling the oxidative stress balance through alterations in the structural integrity of the mitochondrial membrane, impairment of oxidative phosphorylation, adenosine 5′-triphosphate (ATP) depletion, porphyrinogen oxidation, depletion of reduced glutathione, and alterations in mitochondrial calcium (Ca^{2+}) homeostasis represent some of the most important cellular alterations caused by Hg [6]. Superoxide dismutase (SOD), glutathione peroxidase (GPx), malondialdehyde (MLD), selenium (Se), and lymphocyte glutathione (GSH-ly) are some of the biomarkers used to assess oxidative stress [7–11]. Few studies have been published in our country regarding the consequences of low-level exposure to mercury, and even fewer have described the possible associations between heavy metals and oxidative stress biomarkers.

This study aimed to evaluate the impact of mercury blood levels (HgBL) through low-level exposure on the oxidative stress balance, including GSH-ly, SOD, MLD, GPx, vitamin D, and Se. Obtaining an overview of the relationship between Hg and oxidative stress balance is one of the first steps in establishing directions regarding preventive measures that can be included in the day-to-day lifestyle, but also in hospitals/toxicology units, aiming to reduce the short- and long-term negative effects associated with Hg exposure.

2. Materials and Methods

Study design and subjects. This was a cross-sectional, observational study, performed in an outpatient clinic. It was conducted between the 1st of June 2021 and the 31st of December 2021. The study population was recruited from the pool of patients who addressed the outpatient clinic with non-specific signs and symptoms such as headache, muscle pain, insomnia, and peripheral neuropathy. After the exclusion of organic pathology, high mercury blood levels were revealed. Subjects having high blood levels of other heavy metals were excluded from the study. None of the included subjects had any known professional long-term exposure to Hg or other types of high-level exposure to Hg. Additionally, information regarding risk factors was included in a face-to-face survey on admission, providing data about age, lifestyle, smoking, and alcohol consumption. We analyzed age, sex, educational levels, cigarette smoking status, and alcohol consumption status as factors that could affect lifestyle profiles. Age categories were defined as follows: 19–39 years, 40–59 years, 60–69 years, and ≥70 years old. We used three educational levels, namely less than high school graduate, high school diploma, and college graduate. We assessed smoking status into three categories: past smoker, never smoked, and current smoker. Alcohol intake was addressed by yes or no questions.

Toxicology studies. Blood samples were provided from all patients through venous puncture, using royal blue cap containers with an anticoagulant (ethylenediaminetetraacetic acid, BD Vacutainer, ref 368381), filled up to 10 mL. HgBL was analyzed subsequently or after storage at 20 °C. The plasma coupled with the mass spectrometry wavelength used was of 254.65 nm, with a conversion factor of μg/L × 0.005 = μmol/L and μmol/L × 200= μg/L.

Oxidative stress biomarkers. We created a database that included GSH-ly, SOD, GPx, MLD, vitamin D, and Se blood levels. Using flow cytometry, glutathione levels in T lymphocytes were determined with the help of a non-fluorescent compound which, in reaction with intracellular thiol, becomes highly fluorescent (reference values: >355 median fluorescence intensity). Superoxide dismutase was determined through enzymatic photometry (reference levels: 1200–1800 U/ghb) in refrigerated whole blood. GPx's enzymatic activity in the erythrocytes was assessed from a 1 mL venous blood sample using photometry (reference values: 4171–10,881 U/L). Using high performance liquid chromatography and fluorescence detection, MLD levels in plasma were determined (reference values for the lab <1 μmol/L). Through electrochemiluminescence, we determined whole vitamin D levels in venous blood after centrifugation. Using atomic absorption spectroscopy, we determined selenium levels in venous blood (reference values: 50−120 mcg/L).

Statistics. Statistical analysis was performed using Graph Pad Prism 9.3.0 (Dotmatics, Boston, MA, USA), MedCalc 14 (MedCalc Software Ltd., Ostend, Belgium), and Microsoft Excel. We performed tests such as the D'Agostino-Pearson analysis, Spearman correlations, one-way ANOVA, and Dunn analysis.

3. Results

Subjects and Baseline Characteristics

We have included 67 patients in our study, with a median age of 46 years (SD = 7.55). Sex distribution showed 43.38% ($n = 29$) females. Most of the patients came from an urban area (88.05% ($n = 59$)). Median Hg blood concentration was 12 µg/L (SD = 7.71), with a minimum of 1 µg/L and a maximum of 25 µg/L. Regarding the Hg blood concentration distribution, when we take into consideration the presentation of the included patients, there is no further bias. The variable distribution is most likely caused by the small number of patients included. Other population characteristics which were studied are shown in Table 1.

Table 1. Baseline characteristics.

Variables	Total (*n*, %)	Males (*n*, %)	Females (*n*, %)	*p*
Total (*n*, %)	67	38 (56.71)	29 (43.38)	NS
Area				
Urban	59 (88.05)	34 (89.47)	25 (82.20)	NS
Rural	8 (11.94)	4 (10.52)	4 (13.79)	NS
Age (years)				
19–39	11 (16.41)	5 (13.15)	6 (20.68)	NS
39–59	23 (34.32)	12 (31.57)	11 (37.93)	NS
60–69	16 (23.88)	9 (23.68)	7 (24.13)	NS
≥70	17 (25.37)	12 (17.91)	5 (17.24)	<0.005
Education				
less than high school	19 (28.35)	9 (23.68)	10 (34.48)	NS
high school diploma	21 (31.34)	16 (42.13)	5 (17.24)	NS
college graduate	27 (40.29)	13 (34.21)	14 (48.27)	NS
Smoking				
past smoker	22 (32.83)	15 (39.47)	7 (24.13)	NS
never smoker	11 (16.41)	2 (5.26)	9 (31.03)	NS
current smoker	34 (50.74)	21 (55.26)	13 (44.82)	NS
Alcohol				
Yes	38 (56.71)	32 (84.21)	6 (20.68)	<0.005
No	29 (43.28)	6 (15.78)	23 (79.31)	<0.005

NS—not significant.

Further, we analyzed the distribution of the variables using the D'Agostino–Pearson test. None of the studied variables had a normal distribution: age ($K2 = 6.307$, $p = 0.04$), HgBL ($K2 = 24.5$, $p < 0.0001$), SOD ($K2 = 21.46$, $p < 0.001$), GPx ($K2 = 40.39$, $p < 0.0001$), MLD ($K2 = 15.3$, $p = 0.0005$), GSH-Ly ($K2 = 13.95$, $p = 0.0009$), Se ($K2 = 12.84$, $p = 0.0016$), or Vitamin D ($K2 = 7.168$, $p = 0.02$). Further description regarding the distribution of the studied variables is provided in Table 2.

Table 2. Distribution characteristics of the included variables.

	Age (years)	HgBL (mcg)	SOD (U/gHg)	GPx (U/I)	MLD (Micromol/L)	GSH-ly (mfi)	Se (mcg/L)	Vitamin D (ng/mL)
Median	46	12	2331	12,649	1.3	320	125	28
Minimum	37	1	1820	6347	0.3	202	43	19
Maximum	66	25	2590	16,899	2.2	389	146	35
95% CI of median lower limit	44	11	2267	11,270	1.2	300	122	27
95% CI of median upper limit	48	13	2360	13,502	1.4	334	132	29
Coefficient of variation	15.7%	40.17%	10.5%	30.55%	46.57%	20.79%	31.83%	16.35%

In order to assess the correlations between the HgBL and the oxidative stress biomarkers, we have used Spearman correlations and coefficients that were included in a comprehensive correlation matrix, which is graphically represented by the heat map shown in Figure 1.

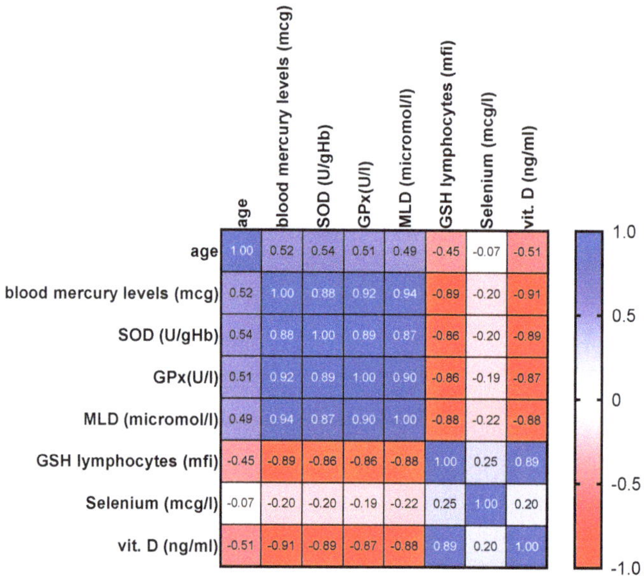

Figure 1. Heat map of the Spearman correlations between HgBL and oxidative stress biomarkers (vit. D—Vitamin D).

As can be seen in the provided picture, there are positive, statistically significant correlations between HgBL and SOD ($r = 0.88$, $p < 0.0001$), GPx ($r = 0.92$, $p < 0.0001$), and MLD ($r = 0.94$, $p < 0.0001$). We have found inverted correlations between HgBL and GSH-Ly ($r = -0.86$, $p < 0.000$), as well as Vitamin D ($r = -0.91$, $p < 0.0001$). There were no statistically significant correlations between HgBL and Se ($r = -0.2$, $p = 0.1$).

For further statistical studies, the study population was split into 3 groups, taking HgBL into consideration, as follows: group A (HgBL between 0 and 10 µg/L), group B (HgBL 11–20 µg/L) and group C (HgBL 21–30 µg/L). Group A included 23 patients, group B 28 patients, and Group C 16 patients.

We performed one-way ANOVA analysis for variables with an abnormal distribution (SOD, MLD, GPx, GSH-Ly, Se, and Vitamin D), and, using Dunn analysis, we compared the three groups. Mean SOD concentration in group A was 1942 µ/gHg (SD = 96.22); in group

B it was 2333 μ/gHg (SD = 48.7); and in group C it was 2499 μ/gHg (SD = 35.76), as it can be seen in Figure 2.

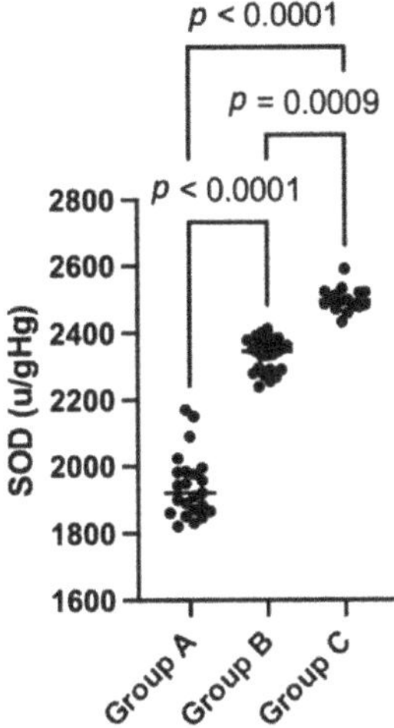

Figure 2. SOD statistical analysis.

Mean Gpx concentration was different between the three groups, as evidenced in Figure 3, was 7424 U/L (SD = 799.1) in group A, 12,934 U/L (SD = 1018) in group B, and 16,478 U/L (SD = 391.8) in group C.

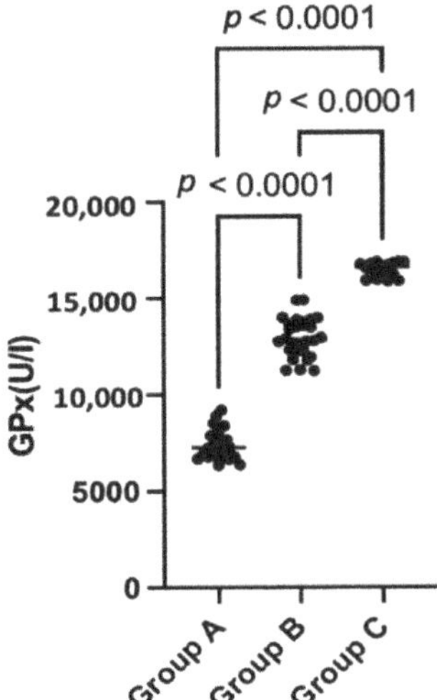

Figure 3. GPx statistical analysis.

Regarding MLD concentrations, there were statistically significant differences between the three groups, evidentiated in Figure 4: group A—mean concentration 0.5783 μmol/L (SD = 0.1476), group B—mean concentration 1.343 μmol/L (SD = 0.1260), and group C—mean concentration 2.073 μmol/L (SD = 0.1075).

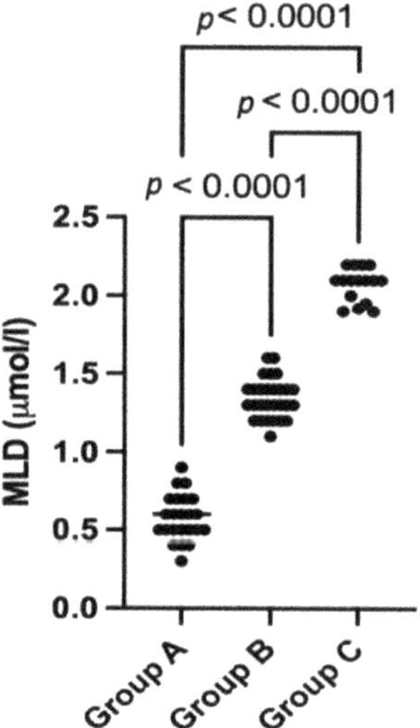

Figure 4. MLD statistical analysis.

Figure 5 ilustrates the differences between mean GSH-Ly concentration in group A, B, and C, which was 374.9 mfi (SD = 9.758), 311.8 mfi (SD = 17.27), and 208.6 mfi (SD = 4.761), respectively.

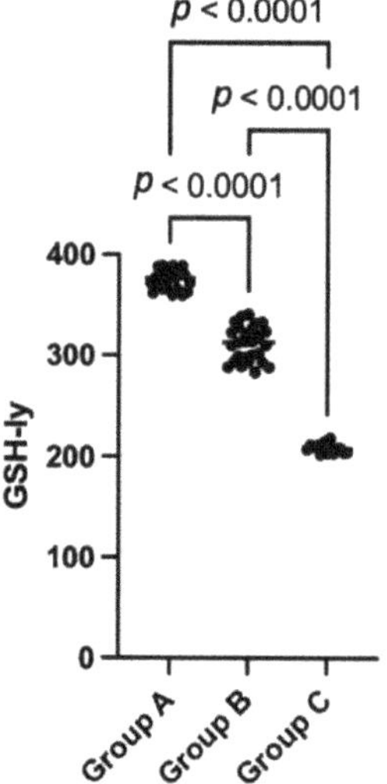

Figure 5. GSH-Ly statistical analysis.

Mean concentration of Se in group A was 123.6 μg/L (SD = 2.017), 137.1 μg/L (SD = 4.541) in group B, and 49.88 μg/L (SD = 3.704) in group C with statistically significant differences between the three groups as seen in Figure 6.

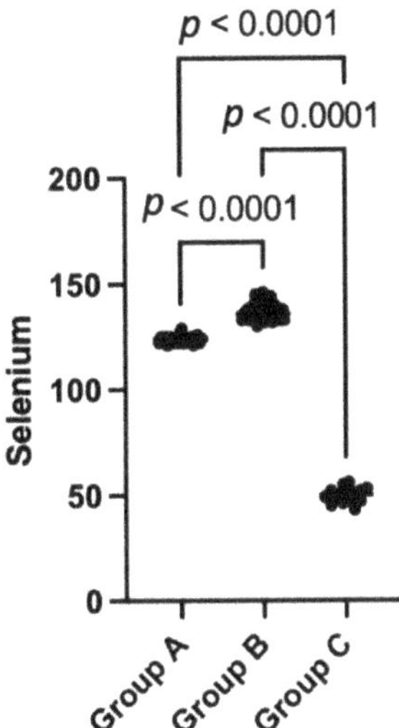

Figure 6. Se statistical analysis.

Regarding vitamin D concentrations, we have found statistically significant differences as it is shown in Figure 7. In group A, the mean Vitamin D concentration was 33.35 ng/mL (SD = 1.91), while in group B and C it was 27.64 ng/mL (SD = 0.9512) and 21.63 ng/mL (SD = 1.50), respectively.

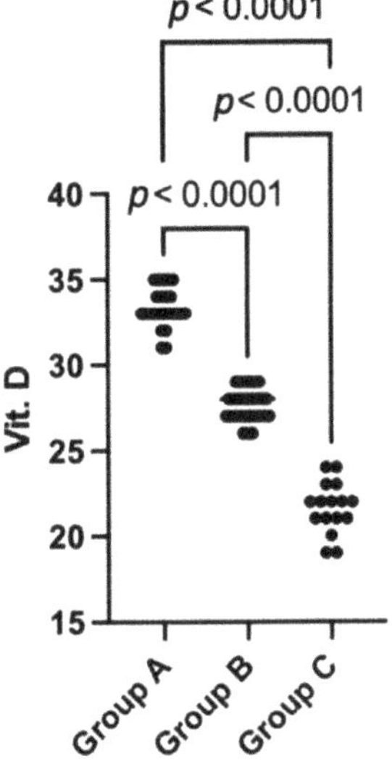

Figure 7. Vitamin D statistical analysis.

The mean for all analyzed oxidative stress biomarkers between the three groups was variable depending on the HgBL.

4. Discussion

Median Hg blood concentration in our study was higher than that in other such studies. For example, in an Austrian study which included 152 patients, the median Hg blood concentration was 2.38 µg/L [12]. The difference can be explained by the study population. In our study, we have assessed patients with previous exposures, while in the Austrian study, this was not the case. In a Brazilian study published in 2021, Hg blood concentration in the control group (non-exposed, non-fishermen) varied between 0.29–17.3 μg/L, which is, approximately, in the same range as our study. Evidently, in the exposed group, fishermen from the Mundau lagoon in Maceio had higher Hg blood concentrations (0.73–48.38 μg/L) [13]. In a Canadian study, which included 221 female immigrants of childbearing age, total Hg blood levels were between 0.4 and 26.05 μg/L, which is somewhat similar to our study [14]. These significant differences can be caused by the geographical area of residence, by the different types of diets, and by the Hg content in the water. In a study from the Czech Republic, which included 1069 patients aged over 61 years from 26 social care institutions, a negative correlation between age and blood Hg was found [15].

The impact on the antioxidant system varies among available studies. In the aforementioned study on Brazilian fishermen, SOD activity was decreased in the exposed group in comparison with the control group [16]. This somewhat contradicts our findings; the positive correlation between Hg blood level and SOD, is supported by another study, published in 2005, in which SOD activity was increased in the mercury exposed group [17].

A cross sectional study that included 211 patients from the Brazilian Amazon included a positive correlation between blood Se and Hg [18]. The main role of Se in Hg exposure appears to be related to neutralizing Hg toxicity, especially neurotoxicity. Its binding to Hg leads to a decrease in Se blood concentration [16]. Furthermore, an increase in urinary Se excretion can be attributed to mercury exposure [19]. Although there seems to be an important relationship between HgBL and Se, we did not find any statistically significant relationship between the two of them.

In a study with 889 subjects that aimed to evaluate the association between blood mercury, cadmium, and lead levels, as well as MLD and paraoxonase 1 activity, the conclusion reached regarding MLD stated that mercury levels were inversely correlated with MLD concentration [20], which contradicts the findings in our study. This contradiction can be partially explained by the difference in Hg blood concentrations between the studies, the cited study reaching a much higher mean Hg concentration.

As opposed to our study, glutathione peroxidase activity has been shown to be decreased in the setting of mercury exposure in several studies in humans, but also in animal models [13,21,22]. The conflicting results can be explained by the variability of Hg blood concentration, by the different methods used in assessing the GPx, and by the significant differences in the studied groups. In an experimental study, in which T lymphocytes were exposed to methylmercury, GSH was markedly decreased after said exposure, resulting in decreased activity of glutathione S-transferase [12].

Given the small patient sample and the type of the study, we cannot formulate special recommendations regarding the management of Hg-exposed patients. As was previously stated, available studies revolving around this subject have conflicting results, great limitations, and are not standardized. All of these are sufficient reasons to prevent the creation of strict guidelines. Nonetheless, mercury exposure is a potent inductor of oxidative stress, which, in turn, can be detrimental to the well-being of an individual. Neurological, cardiovascular, hematological, immunological, gastrointestinal, and renal systems depend on the balance between pro- and antioxidants. Inclining this balance in any direction leads to immense negative consequences. In the meantime, studies have shown some interest in modifications of the gut microbiota during low level exposure, generating clinical issues similar to functional abdominal disorders [23]. Thus, the potential role of modulating the intestinal population should be researched secondarily [24,25]. As we have described in the presented paper, mercury exposure can be secondary to fish consumption. Children's diets are also important due to high fructose corn syrup consumption, given that products containing the mercury cell chlor-alkali are largely used as food ingredients in the industry [26], raising additional metabolic issues in children suffering from intolerances [27]. There are known benefits associated with fish consumption, due to their nutritional value and richness in omega-3 fatty acids and vitamins. Fish intake has a pivotal role in influencing cardiovascular risk factors, and it appears to be associated with a lower risk of sudden cardiovascular-associated death. Neurocognitive development in children is related to the maternal fish intake; docosahexaenoic acid (DHA) from fish is beneficial for early neurodevelopment. Nonetheless, one must not forget that alongside omega-3 fatty acids, DHA, and vitamins, a fish diet means an increase in organic mercury exposure [28].

5. Conclusions

In conclusion, this study demonstrates that low-level Hg exposure may have an influence on the oxidative stress state, brought to light by its impact on a series of oxidative stress biomarkers. Long-term cardiovascular and neurological effects, secondary to a prolonged increased oxidative stress state, may outweigh the benefits of a fish diet.

Author Contributions: Conceptualization, R.C.T. and L.M.; methodology, R.C.T. and C.C.; software, S.I. and A.E.B.; validation, G.D., I.M.G. and L.M.; formal analysis, I.A.F., R.U. and A.-M.C.; investigation, R.C.T. and I.A.F.; resources, R.C.T.; data curation, A.E.B., A.-M.C. and L.M.; writing—original draft preparation, I.A.F. and C.C.; writing—review and editing, R.C.T. and R.U.; visualization, S.I. and C.C.; supervision, G.D. and I.M.G.; project administration, R.C.T. and C.C.; funding acquisition, I.A.F. and L.M. All authors have read and agreed to the published version of the manuscript.

Funding: This research received no external funding.

Institutional Review Board Statement: The study was conducted in accordance with the Declaration of Helsinki and it was approved by the Clinical Emergency Hospital of Bucharest Ethics Committee (protocol code 43428).

Data Availability Statement: The database will be available on demand.

Conflicts of Interest: The authors declare no conflict of interest.

References

1. Mercury and Health. Available online: https://www.who.int/news-room/fact-sheets/detail/mercury-and-health (accessed on 16 June 2022).
2. Sakamoto, M.; Nakamura, M.; Murata, K. Mercury as a Global Pollutant and Mercury Exposure Assessment and Health Effects. *Nippon. Eiseigaku Zasshi Jpn. J. Hyg.* **2018**, *73*, 258–264. [CrossRef] [PubMed]
3. WHO Regional Office for Europe. *Air Quality Guidelines*, 2nd ed.; Chapter 6.9; WHO Regional Office for Europe: Copenhagen, Denmark, 2000. Available online: https://www.euro.who.int/__data/assets/pdf_file/0004/123079/AQG2ndEd_6_9Mercury.PDF (accessed on 16 June 2022).
4. Boerleider, R.Z.; Roeleveld, N.; Scheepers, P.T. Human biological monitoring of mercury for exposure assessment. *AIMS Environ. Sci.* **2017**, *4*, 251–276. [CrossRef]
5. Mercury Factsheet | National Biomonitoring Program | CDC. Published 2 September 2021. Available online: https://www.cdc.gov/biomonitoring/Mercury_FactSheet.html (accessed on 16 June 2022).
6. Choi, H.; Park, S.-K.; Kim, M.-H. Risk Assessment of Mercury through Food Intake for Korean Population. *Korean J. Food Sci. Technol.* **2012**, *44*, 106–113. [CrossRef]
7. Ye, B.-J.; Kim, B.-G.; Jeon, M.-J.; Kim, S.-Y.; Kim, H.-C.; Jang, T.-W.; Chae, H.-J.; Choi, W.-J.; Ha, M.-N.; Hong, Y.-S. Evaluation of mercury exposure level, clinical diagnosis and treatment for mercury intoxication. *Ann. Occup. Environ. Med.* **2016**, *28*, 5. [CrossRef]
8. Posin, S.L.; Kong, E.L.; Sharma, S. *Mercury Toxicity*; StatPearls Publishing: Tampa, FL, USA, 2022. Available online: http://www.ncbi.nlm.nih.gov/books/NBK499935/ (accessed on 16 June 2022).
9. Rice, K.M.; Walker, E.M., Jr.; Wu, M.; Gillette, C.; Blough, E.R. Environmental Mercury and Its Toxic Effects. *J. Prev. Med. Public Health* **2014**, *47*, 74–83. [CrossRef]
10. Lund, B.-O.; Miller, D.M.; Woods, J.S. Studies on Hg(II)-induced H2O2 formation and oxidative stress in vivo and in vitro in rat kidney mitochondria. *Biochem. Pharmacol.* **1993**, *45*, 2017–2024. [CrossRef]
11. Younus, H. Therapeutic potentials of superoxide dismutase. *Int. J. Health Sci.* **2018**, *12*, 88–93.
12. Lubos, E.; Loscalzo, J.; Handy, D.E. Glutathione Peroxidase-1 in Health and Disease: From Molecular Mechanisms to Therapeutic Opportunities. *Antioxid. Redox Signal.* **2011**, *15*, 1957–1997. [CrossRef]
13. Linšak, Ž.; Linšak, D.T.; Špirić, Z.; Srebočan, E.; Glad, M.; Milin, Č. Effects of mercury on glutathione and glutathione-dependent enzymes in hares (*Lepus europaeus* Pallas). *J. Environ. Sci. Health Part A Tox. Hazard. Subst. Environ. Eng.* **2013**, *48*, 1325–1332. [CrossRef]

14. Del Rio, D.; Stewart, A.J.; Pellegrini, N. A review of recent studies on malondialdehyde as toxic molecule and biological marker of oxidative stress. *Nutr. Metab. Cardiovasc. Dis.* **2005**, *15*, 316–328. [CrossRef]
15. Agarwal, R.; Behari, J.R. Effect of Selenium Pretreatment in Chronic Mercury Intoxication in Rats. *Bull. Environ. Contam. Toxicol.* **2007**, *79*, 306–310. [CrossRef] [PubMed]
16. Silva-Filho, R.; Santos, N.; Santos, M.C.; Nunes, Á.; Pinto, R.; Marinho, C.; Lima, T.; Fernandes, M.P.; Santos, J.C.C.; Leite, A.C.R. Impact of environmental mercury exposure on the blood cells oxidative status of fishermen living around Mundaú lagoon in Maceió—Alagoas (AL), Brazil. *Ecotoxicol. Environ. Saf.* **2021**, *219*, 112337. [CrossRef] [PubMed]
17. Becker, A.; Soliman, K.F.A. The Role of Intracellular Glutathione in Inorganic Mercury-Induced Toxicity in Neuroblastoma Cells. *Neurochem. Res.* **2009**, *34*, 1677–1684. [CrossRef]
18. Schwalfenberg, G.K.; Genuis, S.J. Vitamin D, Essential Minerals, and Toxic Elements: Exploring Interactions between Nutrients and Toxicants in Clinical Medicine. *Sci. World J.* **2015**, *2015*, 318595. [CrossRef] [PubMed]
19. Wiseman, C.L.S.; Parnia, A.; Chakravartty, D.; Archbold, J.; Copes, R.; Cole, D. Total, methyl and inorganic mercury concentrations in blood and environmental exposure sources in newcomer women in Toronto, Canada. *Environ. Res.* **2018**, *169*, 261–271. [CrossRef]
20. Rambousková, J.; Krsková, A.; Slavíková, M.; Čejchanová, M.; Černá, M. Blood levels of lead, cadmium, and mercury in the elderly living in institutionalized care in the Czech Republic. *Exp. Gerontol.* **2014**, *58*, 8–13. [CrossRef]
21. Chen, C.; Qu, L.; Li, B.; Xing, L.; Jia, G.; Wang, T.; Gao, Y.; Zhang, P.; Li, M.; Chen, W.; et al. Increased Oxidative DNA Damage, as Assessed by Urinary 8-Hydroxy-2′-Deoxyguanosine Concentrations, and Serum Redox Status in Persons Exposed to Mercury. *Clin. Chem.* **2005**, *51*, 759–767. [CrossRef]
22. Lemire, M.; Fillion, M.; Frenette, B.; Mayer, A.; Philibert, A.; Passos, C.J.S.; Guimaraes, J.R.D.; Barbosa, F.; Mergler, N. Selenium and Mercury in the Brazilian Amazon: Opposing Influences on Age-Related Cataracts. *Environ. Health. Perspect.* **2010**, *118*, 1584–1589. [CrossRef]
23. Siblerud, R.; Mutter, J.; Moore, E.; Naumann, J.; Walach, H. A Hypothesis and Evidence That Mercury May be an Etiological Factor in Alzheimer's Disease. *Int. J. Environ. Res. Public Health* **2019**, *16*, 5152. [CrossRef]
24. Alexander, J.; Thomassen, Y.; Aaseth, J. Increased urinary excretion of selenium among workers exposed to elemental mercury vapor. *J. Appl. Toxicol.* **1983**, *3*, 143–145. [CrossRef]
25. Lopes, A.C.B.A.; Urbano, M.; de Souza-Nogueira, A.; Oliveira-Paula, G.H.; Michelin, A.P.; Carvalho, M.D.F.H.; Camargo, A.E.I.; Peixe, T.S.; Cabrera, M.A.S.; Paoliello, M.M.B. Association of lead, cadmium and mercury with paraoxonase 1 activity and malondialdehyde in a general population in Southern Brazil. *Environ. Res.* **2017**, *156*, 674–682. [CrossRef] [PubMed]
26. Branco, V.; Canário, J.; Lu, J.; Holmgren, A.; Carvalho, C. Mercury and selenium interaction in vivo: Effects on thioredoxin reductase and glutathione peroxidase. *Free Radic. Biol. Med.* **2011**, *52*, 781–793. [CrossRef] [PubMed]
27. Al-Azzawie, H.F.; Umran, A.; Hyader, N.H. Oxidative Stress, Antioxidant Status and DNA Damage in a Mercury Exposure Workers. *Br. J. Pharmacol. Toxicol.* **2013**, *4*, 80–88. [CrossRef]
28. Shenker, B.J.; Pankoski, L.; Zekavat, A.; Shapiro, I.M. Mercury-Induced Apoptosis in Human Lymphocytes: Caspase Activation Is Linked to Redox Status. *Antioxid. Redox Signal.* **2002**, *4*, 379–389. [CrossRef] [PubMed]

MDPI

St. Alban-Anlage 66

4052 Basel

Switzerland

www.mdpi.com

MDPI Books Editorial Office

E-mail: books@mdpi.com

www.mdpi.com/books